# Human Impact on Groundwater Environment

# Human Impact on Groundwater Environment

Topic Editors

**Zongjun Gao**
**Jiutan Liu**

MDPI

Basel • Beijing • Wuhan • Barcelona • Belgrade • Novi Sad • Cluj • Manchester

*Topic Editors*
Zongjun Gao
College of Earth Science and Engineering
Shandong University of Science and Technology
Qingdao
China

Jiutan Liu
College of Earth Science and Engineering
Shandong University of Science and Technology
Qingdao
China

*Editorial Office*
MDPI AG
Grosspeteranlage 5
4052 Basel, Switzerland

This is a reprint of the Topic, published open access by the journals *Energies* (ISSN 1996-1073), *Hydrology* (ISSN 2306-5338), *Sustainability* (ISSN 2071-1050), *Water* (ISSN 2073-4441) and *Earth* (ISSN 2673-4834), freely accessible at: https://www.mdpi.com/topics/32DU374GSR.

For citation purposes, cite each article independently as indicated on the article page online and as indicated below:

Lastname, A.A.; Lastname, B.B. Article Title. *Journal Name* **Year**, *Volume Number*, Page Range.

**ISBN 978-3-7258-3399-3 (Hbk)**
**ISBN 978-3-7258-3400-6 (PDF)**
**https://doi.org/10.3390/books978-3-7258-3400-6**

# Contents

# About the Editors

**Zongjun Gao**

Zongjun Gao is a professor and doctoral supervisor at the School of Earth Science and Engineering, Shandong University of Science and Technology. He is a member of the Hydrogeology Professional Committee of the Geological Society of China, a member of the Mine Water Prevention and Utilization Professional Committee of the Geological Society of China, an executive member of the Geothermal Professional Committee of the Shandong Energy Society, and a recipient of the Shandong Provincial Geological Science and Technology Award. He is engaged in teaching and research in hydrogeology, environmental geology, and ecological geochemistry. He has published 280 papers, 10 books and teaching materials, and has received more than 10 provincial and ministerial awards and 18 authorized national patents.

**Jiutan Liu**

Jiutan Liu graduated from Shandong University of Science and Technology, majoring in Hydrology and Water Resources Engineering and Geological Engineering. He holds a doctorate (postdoctoral) degree, is an academic professor and a master's supervisor. His main research directions are hydrogeology, environmental geology, hydrogeochemistry and ecological geochemistry. He was selected to participate in the Class A Talent Project of the "Elite Program" of Shandong University of Science and Technology. He has presided over three national and provincial projects and eight other projects. He won the second prize of Qingdao Science and Technology Progress Award. He has published 42 papers indexed by SCI and EI as the first author or corresponding author. He serves as a Topic Editor at MDPI and as an Editorial Board Member for the journal *Frontiers In Earth Science*.

*water*

MDPI

*Article*

# Characteristics and Controlling Factors of Groundwater Hydrochemistry in Dongzhi Tableland Area of the Loess Plateau of Eastern Gansu—A Case Study of Ning County Area, North China

**Mengnan Zhang [1,2], Shuangbao Han [2], Yushan Wang [1,2], Zhan Wang [1,3], Haixue Li [1,2], Xiaoyan Wang [1,2], Jiutan Liu [4,*], Changsuo Li [5,*] and Zongjun Gao [4]**

1 Hebei Center for Ecological and Environmental Geology Research, Hebei GEO University, Shijiazhuang 050031, China
2 Center for Hydrogeology and Environmental Geology Survey, China Geological Survey, Baoding 071051, China
3 School of Water Resources and Environment, Hebei GEO University, Shijiazhuang 050031, China
4 College of Earth Science and Engineering, Shandong University of Science and Technology, Qingdao 266590, China
5 801 Institute of Hydrogeology and Engineering Geology, Shandong Provincial Bureau of Geology & Mineral Resources, Jinan 250014, China
* Correspondence: ljtsdust@sdust.edu.cn (J.L.); lics120@163.com (C.L.)

**Citation:** Zhang, M.; Han, S.; Wang, Y.; Wang, Z.; Li, H.; Wang, X.; Liu, J.; Li, C.; Gao, Z. Characteristics and Controlling Factors of Groundwater Hydrochemistry in Dongzhi Tableland Area of the Loess Plateau of Eastern Gansu—A Case Study of Ning County Area, North China. *Water* **2022**, *14*, 3601. https://doi.org/10.3390/w14223601

Academic Editors: Alexander Yakirevich and Dedi Liu

Received: 13 October 2022
Accepted: 4 November 2022
Published: 8 November 2022

**Publisher's Note:** MDPI stays neutral with regard to jurisdictional claims in published maps and institutional affiliations.



**Abstract:** Groundwater plays an irreplaceable role in all aspects of the Loess Plateau. In this study, the loess phreatic water (LPW) and bedrock phreatic water (BPW) in the Ning County area (NCA) were sampled and analyzed, and the characteristics and controlling factors of groundwater were determined by using statistical analysis, hydrochemical methods, and hydrogeochemical simulation. The results indicated that the groundwater in the NCA was alkaline as a whole, and the average pH values of LPW and BPW were 8.1 and 7.8, respectively. The mean values of TDS concentrations of LPW and BPW were 314.9 mg/L and 675.3 mg/L, and the mean values of TH contents were 194.6 mg/L and 286.6 mg/L, respectively, which were mainly divided into hard fresh water. The Piper diagram illustrated that the hydrochemical type of groundwater in the NCA was mainly the $HCO_3 \cdot Ca$ type. The main recharge source of groundwater was atmospheric precipitation, and it was affected by evaporation to a certain extent. The linear relationships of $\delta^{18}O$ and $\delta^2H$ of LPW and BPW were $\delta^2H = 6.998\delta^{18}O - 3.802$ ($R^2 = 0.98$) and $\delta^2H = 6.283\delta^{18}O - 10.536$ ($R^2 = 0.96$), respectively. Hydrochemical analysis indicated that the groundwater in the NCA was mainly controlled by rock weathering and cation exchange. BPW was affected by the dissolution of gypsum. The possible mineral phases were identified on the basis of the main soluble minerals in the aquifer, and hydrogeochemical reverse simulations were performed. The dissolution of calcite, illite, and hornblende, and the precipitation of dolomite, plagioclase, and microcline occurred on both the LPW and BPW pathways.

**Keywords:** hydrochemical characteristics; major ions; formation mechanism; hydrogeochemical modeling; the Loess Plateau

## 1. Introduction

Water resources play a pivotal role in human production and life [1–3]. Industrial production, agricultural irrigation, residential life, and ecological maintenance all depend on stable and clean water sources [4–8]. As an important component of water resources, groundwater has the characteristics of stable water quantity and excellent water quality [9]. Groundwater is buried at a certain depth, easy to be exploited and not easily polluted, and has become the preferred source of water for residential and industrial use. Population growth and rapid socio-economic growth have led to increased consumption of groundwater resources and frequent occurrence of water pollution [10–14]. However, the contradiction between supply and demand of water resources is serious in many regions of the world. Problems such as shortage of groundwater resources, deterioration, and imbalance of groundwater resources limit the sustainable and healthy development of society and economy, having caused serious environmental problems in the past few decades, such as land subsidence, water environment pollution, and water ecosystem degradation [15–18]. To realize the sustainable utilization of regional groundwater resources and promote the coordinated development of the ecological environment and social economy, it is necessary to recognize the hydrochemical characteristics and causes of groundwater resources, as well as to develop and utilize groundwater resources scientifically and rationally [19–21].

The hydrochemical components of groundwater are affected by the long-term action of the environment and continuously interact with the surrounding medium during the migration process, resulting in a series of physical and chemical effects [22–24]. The chemical composition of water contains information such as climate change and the origin of water bodies. Therefore, the environmental quality of groundwater and its ecological functions can be understood by analyzing the chemical composition of water [25–27]. Groundwater under different hydrogeological conditions and human activities has different hydrochemical characteristics. Analyzing the hydrochemical characteristics of groundwater can effectively discriminate the formation process of hydrochemical components and reveal important information such as chemical behavior of elements and chemical weathering intensity in groundwater [28,29]. Therefore, the research on the formation mechanism of groundwater hydrochemistry has important practical significance.

The Chinese Loess Plateau (CLP) is located in the hinterland of China, in the middle reaches of the Yellow River. Affected by the continental monsoon climate, most of its areas belong to arid–semi-arid regions [30]. The CLP is also the region with the most serious soil erosion and the most fragile ecology in China, and the available surface water is very short [31]. The CLP has the widest distribution of loess in the world, and the thickest part of the soil layer is more than 350 m [32]. The deep soil layer forms a huge soil reservoir and groundwater reservoir, providing conditions for the exploitation of groundwater. For a long time, groundwater has become an important source of water supply for agricultural production, industrial development, and residents' domestic use in the CLP. Many researchers have carried out a large number of discussions on the study of groundwater in the Loess Plateau. For example, Li et al. [33] analyzed the hydrogeological conditions and groundwater characteristics, clarified the types and distribution characteristics of freshwater, and revealed the formation mechanism and influencing factors of freshwater on the CLP. Ling et al. [34] studied the isotopic significance of groundwater recharge, retention time, and hydrogeochemical evolution in the CLP of Eastern Gansu. Wang et al. [35] determined the spatial and seasonal variation characteristics and controlling factors of fluorine in natural water on the CLP and assessed its health risks. Xiao et al. [36] analyzed the characteristics and sources of trace elements in river water and well water in the CLP and evaluated water quality and health risks. Zhang et al. [37] studied the hydrochemical processes between surface water and groundwater in the northeastern CLP: implications for water chemistry and environmental evolution in a semi-arid region.

However, the research on the formation mechanism and evolution law of groundwater in the Ning County area (NCA) is relatively limited. Therefore, the purpose of this study was (1) to analyze the hydrochemical characteristics of groundwater in the area, and (2) to clarify the formation mechanism of groundwater hydrochemistry. Statistical analysis, hydrochemical methods, and hydrogeochemical modeling were comprehensively applied in this study, and the research results can provide some reference for the rational utilization of regional groundwater resources and ecological environment protection in the CLP or other similar regions in the world.

## 2. Study Area

The study area (Figure 1) is located in the middle reaches of the Malian River on the CLP of Eastern Gansu. Geographical coordinates: 35°30′–35°45′ east longitude, 107°45′–108°00′ north latitude. The region is located in the mid-latitude warm temperate zone, with a typical continental monsoon climate. The annual average precipitation is 525.8 mm, the annual maximum precipitation is 827.7 mm (1975), and the annual minimum precipitation is 357.7 mm (1995). The rainfall is mainly concentrated in July–September (Figure 2), and the spatial distribution of rainfall is extremely uneven after the flood season.

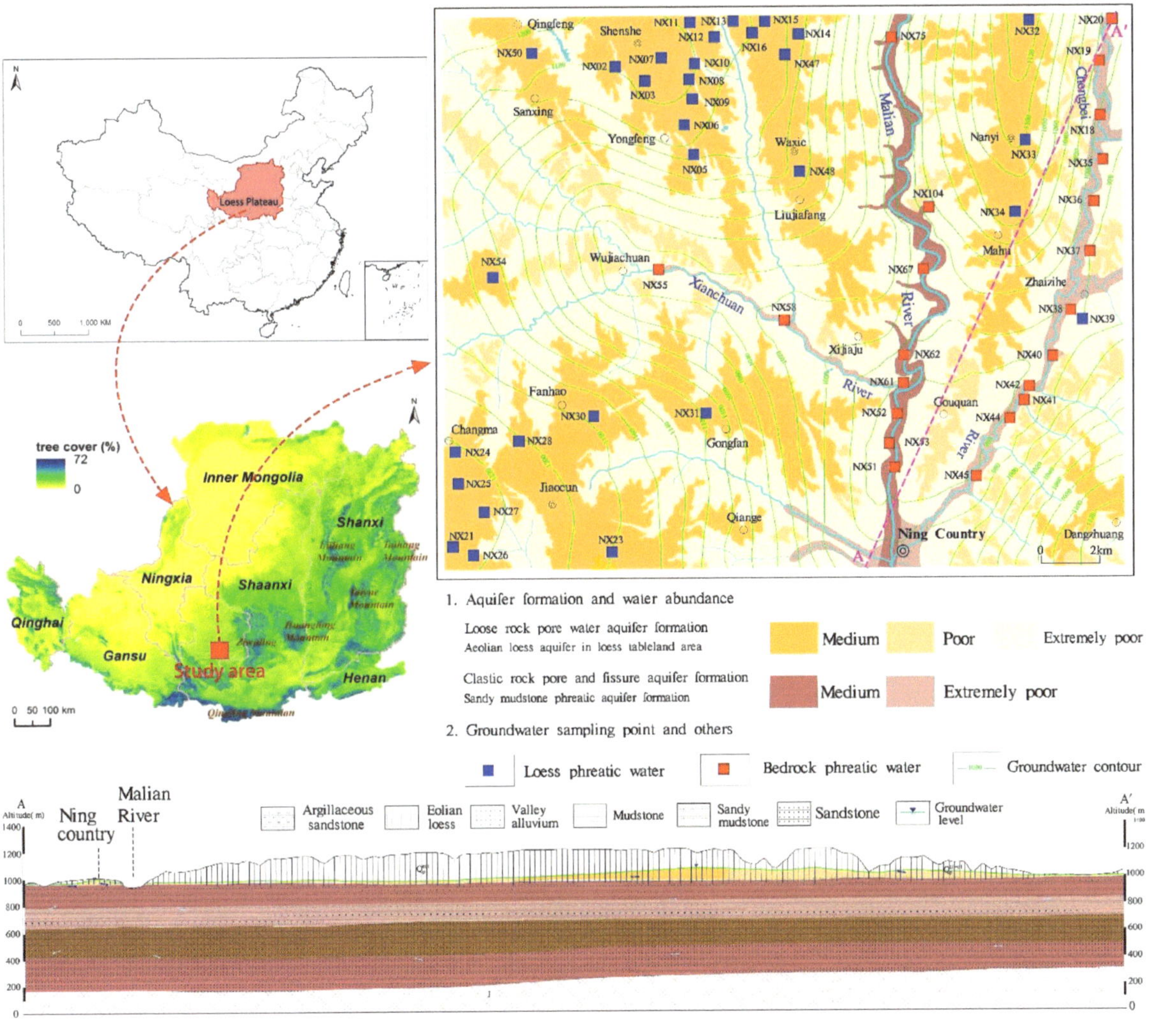


**Figure 1.** Location map of the study area and groundwater sampling points.

The study area belongs to Longdong Basin, located in the southwest of Ordos Basin, which is a Mesozoic asymmetric syncline basin with a south–north strike, gentle in the east and steep in the west (Figure 3). Its western boundary is the Zhuozi mountain-Pingliang fault zone, the southern part is the Weibei uplift, the eastern boundary is the Ziwu Ridge, and the northern boundary is the Baiyu mountain. Affected by the sedimentation of the CLP and the convoluted structure in Longxi, most of the area is covered by the Quaternary.

In addition, the drilling data in the study area indicate that the strata mainly include the Quaternary Pleistocene loose layer, Cretaceous Huanhe Formation sand mudstone interbed, Cretaceous Luohe Formation medium and coarse sandstone, and Jurassic Anding Formation sand mudstone interbed. Loess and valley sand–gravel layer of loose rock pore aquifer and Huanhe formation bedrock weathering crust pore aquifer formation are the main groundwater exploitation layers in the study area.

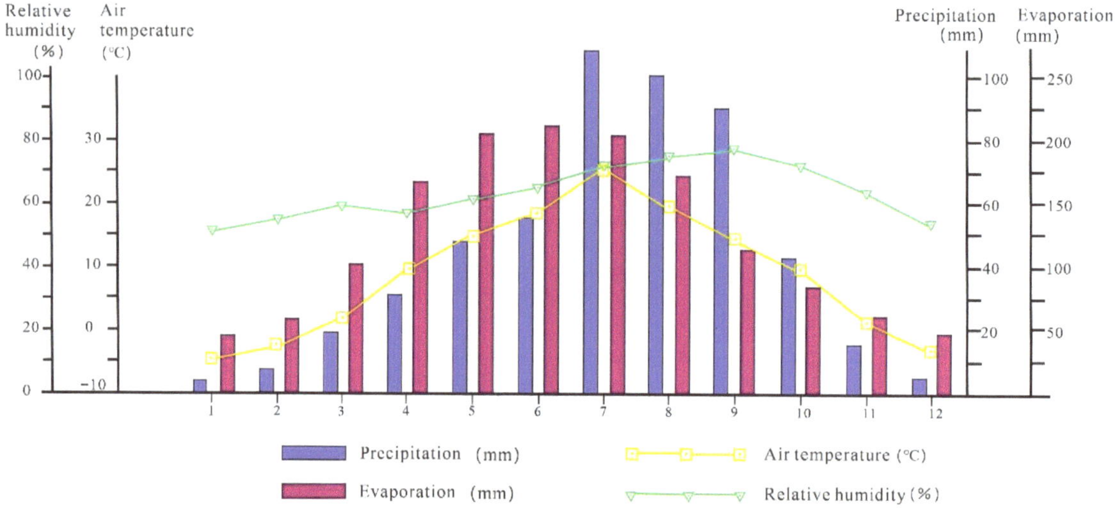


**Figure 2.** Regional meteorologic map.

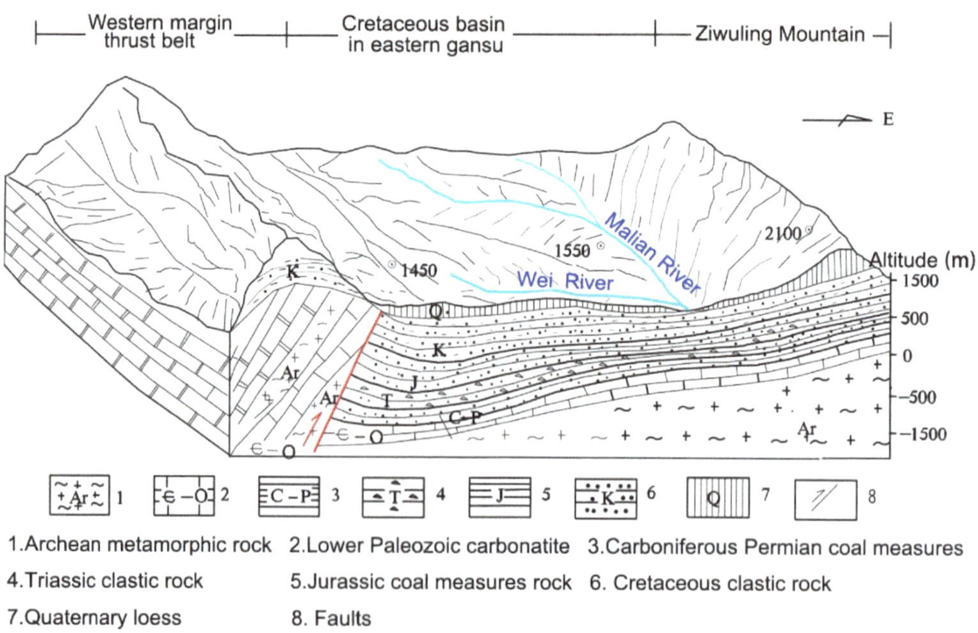


**Figure 3.** Stereoscopic sketch map of east–west geomorphology and geology in the Longdong basin of eastern Gansu.

The study area (Figure 1) is located in the loess gulled-hilly area in the east of Dongzhi tableland. The overall topography shows that the elevation gradually increases to both sides with the Malian River as the boundary. Quaternary loess is widely distributed in the area, with a thickness of 50–180 m. The region has ravines and gullies criss-cross, undulating terrain, and relatively simple landform, which is mainly divided into three geomorphic types: loess remnant plateau, hilly and gully, and river valley.

The flow direction of groundwater is consistent with the flow direction of surface water and generally flows from the tableland area to the nearby valley area. The recharge area is mainly distributed in the loess tableland, loess hills, and loess ridge area; the discharge area is gullies; and the Malian River valley is the final destination of diving discharge. The dynamic change of phreatic water in the loess tableland has obvious regional and seasonal changes, and the variation amplitude of phreatic water lags behind the precipitation time. The changes of phreatic water level and water volume in the river valley are mainly affected by meteorological and hydrological factors, with obvious seasonal changes. The change of water level lags behind the atmospheric precipitation for 1–2 months. The flow and water level dynamics of bedrock fissure pore phreatic water are mainly controlled by seasonality, and the annual variation range is small, generally 0.5–1.5 m, which is relatively stable.

## 3. Materials and Methods

### 3.1. Sampling and Testing

A total of 53 groundwater samples were obtained in this study, including 31 loess phreatic water (LPW) samples and 22 bedrock phreatic water (BPW) samples. The sampling time was June 2018, and the sampling position is shown in Figure 1. LPW is distributed in a large area of loess tableland area in the NCA, and the water-bearing medium is mainly "Lishi loess". BPW is mainly distributed in the river valley area, with few surface outcrops, and the aquifer is the weathering fissure and interlayer fissure in the surface of the Huanhe Formation. The sampling depth of LPW is mainly between 50 and 100 m, and that of BPW is between 5 and 20 m. All groundwater samples were collected in 2.5 L polyethylene plastic bottles. Bottles were rinsed with the water sample to be taken for more than 3 times before actual sampling. After sampling collection, the bottles were then sealed with sealing film. Groundwater samples were kept refrigerated during transport, and samples were sent to the Gansu geological engineering laboratory for water quality testing analysis within one week of completion of groundwater sampling.

The pH value of groundwater samples was determined by the pH potentiometric method, total dissolved solids (TDS) were determined by drying at 180 °C, and total hardness (TH) was determined by disodium EDTA titration method. The content of cations including $K^+$, $Na^+$, $Ca^{2+}$, and $Mg^{2+}$ was determined by atomic absorption spectrometry; the content of anions ($Cl^-$, $SO_4^{2-}$, $NO_3^-$, $F^-$) was determined by high-performance liquid chromatography; and the concentration of $HCO_3^-$ was determined by titration. Hydrogen ($\delta^2H$) and oxygen ($\delta^{18}O$) stable isotopes were measured using a liquid-water isotope analyzer (CSPC-I26), and Vienna Standard Seawater was used for the determination, using the $\delta^2H$ and $\delta^{18}O$ stable isotope content of Vienna standard mean ocean water as the standard. After the water quality test was completed, the ionic charge balance error (%CBE) calculation was carried out. The calculation results showed that the CBE% of all water samples was within the range of ±5%, which indicated the reliability of the water quality data.

$$\%CBE = \frac{\sum cation - \sum anion}{\sum cation + \sum anion} \times 100\% \quad (1)$$

### 3.2. Geochemical Data Analysis

In this study, descriptive statistics of groundwater hydrochemical data in the Ning County area were firstly carried out by means of statistical analysis, the general characteristics of hydrochemistry were determined, and the hydrochemical types of groundwater were divided by the Piper diagram. Furthermore, using the Gibbs model and the ion ratio relationship, the genetic mechanism of groundwater was revealed. Finally, the possible hydrogeochemical processes on the groundwater flow path and the amount of mineral transfer were determined on the basis of the hydrogeochemical inverse simulation.

### 3.3. Cation Exchange

The linear relationship between $Mg^{2+} + Ca^{2+} - HCO_3^- - SO_4^{2-}$ and $K^+ + Na^+ - Cl^-$ was used to identify cation exchange processes (Equations (2) and (3)) in groundwater. If the slope of its linear relationship is −1, it indicates that cation exchange is an important reaction affecting the hydrochemical characteristics of groundwater [38]. Further, the cation exchange direction can be identified using the chlor-alkali index (CAI) [38,39]. The calculation methods of CAI-1 and CAI-2 are shown in the Equations (4) and (5).

$$Na_2X + Ca^{2+} \rightarrow 2Na^+ + CaX \tag{2}$$

$$2Na^+ + CaX \rightarrow Na_2X + Ca^{2+} \tag{3}$$

$$CAI - 1 = \frac{Cl^- - Na^+ - K^+}{Cl^-} \tag{4}$$

$$CAI - 2 = \frac{Cl^- - Na^+ - K^+}{HCO_3^- + SO_4^{2-} + CO_3^{2-} + NO_3^-} \tag{5}$$

### 3.4. Hydrogeochemical Modeling

Hydrogeochemical simulation is an important method to analyze various geochemical interactions in the water–rock interaction system on the basis of equilibrium thermodynamics and chemical kinetics [40,41]. To further analyze the evolution law of groundwater in the study area, two representative groundwater runoff paths were selected, and PHREEQC software (USGS, Virginia State, USA) was used to perform a reverse hydrogeochemical simulation to quantitatively explain the changes of groundwater chemical composition and the main water–rock interactions.

Inverse hydrogeochemical simulations were performed on the basis of PHREEQC software, and the uncertainty of data analysis was set to 0.05. After entering the water quality parameters of the water samples at the initial and terminal points into the software, the SI of minerals and the amount of mineral transfer can be calculated. In the process of operation, considering various factors comprehensively, we adjusted the parameters to obtain the least number of solutions, and finally, we selected the optimal solution.

#### 3.4.1. Saturation Index (SI)

The saturation index [10,26] is used to describe the equilibrium state of minerals and aqueous solutions, referring to the logarithm of the ratio of the ionic activity product (IAP) to the solubility product constant ($K_s$).

$$SI = \lg\frac{IAP}{K_S} \tag{6}$$

SI = 0, SI > 0, and SI < 0 indicate that the minerals are in the equilibrium state of dissolution–precipitation, supersaturated state, and unsaturated state, respectively.

#### 3.4.2. Mass Balance

The theoretical basis for reverse hydrogeochemical modeling is mass balance. On the basis of the difference of the water chemical composition between the initial point and the terminal point on the groundwater flow path, the possible hydrogeochemical reactions are deduced backward [42,43]. This relationship can be expressed by the mass balance and electron conservation equations, which are as follows:

$$\sum_{n=1}^{p} a_n b_{n,k} = m_{T,k(terminal\ point)} - m_{T,k(initial\ point)}\ (k = 1,2,3)\cdots,j) \tag{7}$$

where $a_n$ represents the moles of phase transfer (dissolution or precipitation) of the $n$th mineral phase, $b_{n,k}$ represents the stoichiometric number of the $k$th element of the nth mineral phase, and $m_{T,k}$ represents the total molar concentration of the $k$th element in the solution (initial or terminal point).

$$\sum_{n=1}^{p} unan = \sum_{n=1}^{I} u_i m_{i(terminal\ point)} - \sum_{n=1}^{I} u_i m_{i(initial\ point)} \tag{8}$$

where $u_n$ represents the effective valence state in the $n$th mineral phase, $u_i$ represents the effective valence state of the $i$th component, and $m_i$ represents the molar concentration of the $i$th component in the solution (initial or terminal point).

## 4. Results

### 4.1. Statistical Characteristics of Hydrochemistry

Groundwater is a complex mixed solution, and statistical analysis of its chemical composition can help identify its source and cause [10]. The statistical results of the main chemical components of groundwater are shown in Figure 4. The average pH values of LPW and BPW were 8.1 and 7.8, respectively, indicating that the groundwater in NCA was weakly alkaline as a whole. On average, the dominant cations in LPW and BWP were $Na^+$ and $Ca^{2+}$, and the predominant anion was $HCO_3^-$. In terms of mean value, cations in LPW and BPW were characterized by $Na^+ > Ca^{2+} > Mg^{2+} > K^+$, while anions were characterized by $HCO_3^- > SO_4^{2-} > NO_3^- > Cl^- > F^-$ and $HCO_3^- > SO_4^{2-} > Cl^- > NO_3^- > F^-$ (Figure 4), respectively.

TH and TDS in groundwater are important parameters that reflect the quality of water [39,44]. The TH concentrations of LPW and BPW ranged from 77.6 to 906.7 mg/L and 152.1 to 567.5 mg/L, with mean values of 194.6 mg/L and 286.6 mg/L, respectively. The TDS contents of LPW and BPW were in the ranges of 238.2~1142 mg/L and 349.4~1826 mg/L, with averages of 314.9 mg/L and 675.3 mg/L, respectively.

### 4.2. Hydrochemical Type of Groundwater

The Piper diagram [45] is often used as an effective graphical representation of the chemical composition of water samples in hydrogeological research and is the most widely used diagram in groundwater hydrochemistry research [24,44]. On the basis of the Piper diagram, water chemical types can be divided into four categories (Figure 5), namely, ① $SO_4$·Cl-Ca·Mg type, ② $HCO_3$-Na type, ③ $HCO_3$-Ca type, and ④ $SO_4$·Cl-Na type. The 53 groundwater samples in the NCA were drawn into the Piper diagram, as shown in Figure 4. It can be seen that the groundwater samples were mainly distributed in zone ③, followed by zone ②, indicating that the hydrochemical type of groundwater in the study area was mainly $HCO_3$-Ca type.

### 4.3. $\delta^2H$ and $\delta^{18}O$ Composition Characteristics

Analysis of the $\delta^2H$ and $\delta^{18}O$ composition characteristics of groundwater and comparison with the atmospheric precipitation line in the region can be used to determine the source of groundwater recharge [46,47]. The statistical results indicate that the values of $\delta^{18}O$ and $\delta^2H$ of LPW were between −10.85~−9.46 and −78.40~−70.20, and the mean values were −9.94 and −73.00, respectively. The values of $\delta^{18}O$ and $\delta^2H$ of BPW were in the range of −12.70~−8.70 and −92.89~−63.34, and the mean values were −9.88 and −72.95, respectively. In addition, the linear relationships of the hydrogen and oxygen isotopes of LPW and BPW were $\delta^2H = 6.998\delta^{18}O - 3.802$ ($R^2 = 0.98$) and $\delta^2H = 6.283\delta^{18}O - 10.536$ ($R^2 = 0.96$), respectively.

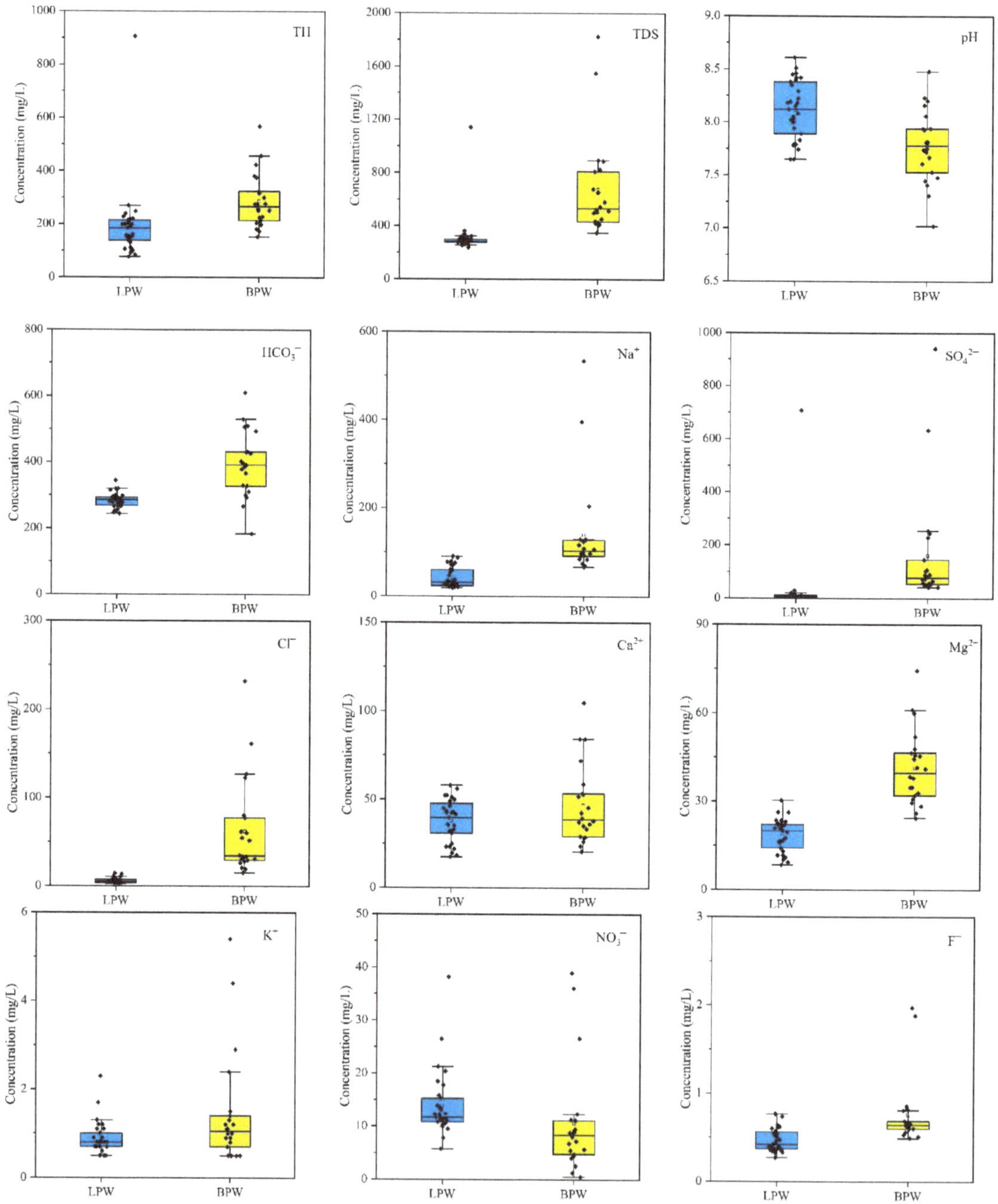


**Figure 4.** Box diagram of the main chemical components of groundwater.

### *4.4. Hydrogeochemical Reverse Simulation*

#### 4.4.1. Selection of Simulation Path

The precondition for reverse hydrogeochemical simulation is that the initial point and the terminal point are in the upstream and downstream relationship, namely, they are on the same groundwater flow path [42]. It can be seen from Figure 1 that the groundwater flow direction in the north of the study area was NX32 to NX34, and in the middle valley was along the river (NZ55→NX58). Therefore, according to the location and type of groundwater sampling points, combined with the characteristics of groundwater flow field in the study area (Figure 1), LPW and BPW each selected one profile (LPW path: NX32→NX34, BPW path: NX55→NX58) for simulation. Table 1 illustrates the main chemical components of groundwater samples at the initial and terminal points of different simulation paths.

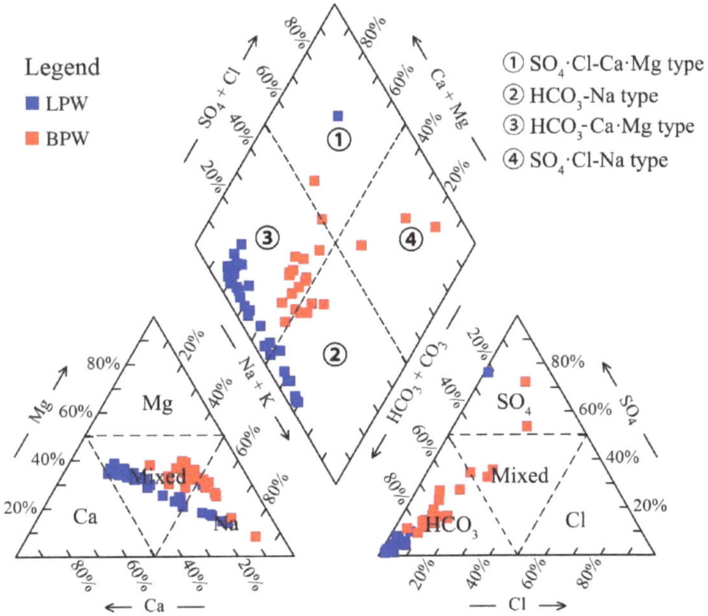


**Figure 5.** Piper diagram of groundwater hydrochemistry.

**Table 1.** Main chemical components at the initial point and terminal point of different paths.

| | LPW Path (NX32→NX34) | | BPW Path (NX55→NX58) | |
|---|---|---|---|---|
| | Initial Point | Terminal Point | Initial Point | Terminal Point |
| pH | 7.9 | 8.0 | 7.0 | 7.7 |
| TDS (mg/L) | 323.5 | 318.3 | 581.7 | 545.8 |
| $K^+$ (mg/L) | 0.8 | 1.1 | 2.9 | 0.5 |
| $Na^+$ (mg/L) | 21.1 | 25.3 | 97.5 | 129.2 |
| $Ca^{2+}$ (mg/L) | 55.9 | 52.3 | 36.2 | 28.7 |
| $Mg^{2+}$ (mg/L) | 26.2 | 26.3 | 45.4 | 30.8 |
| $HCO_3^-$ (mg/L) | 278 | 295 | 427 | 393 |
| $SO_4^{2-}$ (mg/L) | 11.0 | 19.7 | 64.8 | 103.7 |
| $Cl^-$ (mg/L) | 13.5 | 6.4 | 51.8 | 33.3 |
| $F^-$ (mg/L) | 0.33 | 0.34 | 0.6 | 0.63 |
| $NO_3^-$ (mg/L) | 38.2 | 21.3 | 26.6 | 8.0 |
| $H_4SiO_4$ (mg/L) | 28.4 | 29.2 | 36.7 | 25.0 |

4.4.2. Selection of Possible Mineral Phases

The selection of possible mineral phases is the key to determining the hydrogeochemical reactions that may occur on the groundwater flow path, and is also the core of the mass balance model of water–rock interaction [43]. Many studies have shown that there are two main factors for selecting possible minerals: one is the analysis of the mineral composition in the aquifer, and the main mineral composition in the aquifer should be given priority; the other is the chemical composition of groundwater and the occurrence conditions of groundwater.

The results of previous studies [48] have indicated that there are differences in the mineral composition and average content of loess soil samples and bedrock rocks in the region. Whole rock mineral X-ray diffraction analysis results showed that the main water-soluble minerals in loess were plagioclase (16.7%), microline (5.8%), illite (10%), chlorite (13.7%), calcite (8.1%) %), dolomite (4.9%), and hornblende (3.44%), while the main water-soluble minerals in bedrock rocks were plagioclase (18.6%), microline (8.8%), illite (10%), chlorite (4.0%), calcite (4.9%), dolomite (6.6%), and gypsum (10%). The possible mineral phases and main reaction equations are shown in Table 2.

**Table 2.** Possible mineral phases and their reaction equations.

| Mineral Name | Reaction Equation |
|---|---|
| Calcite | $CaCO_3 = Ca^{2+} + CO_3{}^{2-}$ |
| Dolomite | $CaMg(CO_3)_2 = Ca^{2+} + Mg^{2+} + 2CO_3{}^{2-}$ |
| Gypsum | $CaSO_4{\cdot}2H_2O = Ca^{2+} + SO_4{}^{2-} + 2H_2O$ |
| Cation exchange | $2NaX + Ca^{2+} = 2Na^+ + CaX_2$ |
| Chlorite | $Mg_5Al_2Si_3O_{10}(OH)_8 + 16H^+ = 2Al^{3+} + 6H_2O + 3H_4SiO_4 + 5Mg^{2+}$ |
| Illite | $K_{0.6}Mg_{0.25}Al_{2.3}Si_{3.5}O_{10}(OH)_2 + 11.2H_2O = 0.6K^+ + 0.25Mg^{2+} + 2.3Al(OH)_4{}^- + 3.5H_4SiO_4 + 1.2H^+$<br>$2.3Al(OH)_4{}^- + 3.5H_4SiO_4 + 1.2H^+$ |
| Plagioclase | $Na_{0.62}Ca_{0.38}Al_{1.38}Si_{2.62}O_8 + 5.52H^+ + 2.48H_2O = 1.38Al^{3+} + 0.38Ca^{2+} + 2.62H_4SiO_4 + 0.62Na^+$<br>$+0.38Ca^{2+} + 2.62H_4SiO_4 + 0.62Na^+$ |
| Hornblende | $Ca_2Mg_5Si_8O_{22}(OH)_2 + 14H^+ + 8H_2O = 2Ca^{2+} + 8H_4SiO_4 + 5Mg^{2+}$ |
| Microline | $KAlSi_3O_8 + H_2O = K^+ + Al(OH)_4{}^- + 3H_4SiO_4$ |

## 5. Discussion

### 5.1. Hydrochemical Characteristics of Groundwater

The order of $Na^+ > Ca^{2+} > Mg^{2+} > K^+$ existed in the cations in the two types of groundwater, while the anions were slightly different. The relationship of anions in LPW was $HCO_3{}^- > SO_4{}^{2-} > NO_3{}^- > Cl^- > F^-$, and the order of anions in BWP $HCO_3{}^- > SO_4{}^{2-} > Cl^- > NO_3{}^- > F^-$. When we compared the content of main chemical components in LPW and BPW (Figure 4), except for the fact that the pH value in LPW was higher than that in BPW, we found that other components were lower than BPW.

According to the classification [39,44] of TH and TDS (Figure 6a), the groundwater in Ning County was mainly divided into two categories: soft fresh water (18.9%) and hard fresh water (75.5%). In addition, one LPW sample and two BPW samples belonged to hard brackish water. All BWP samples were classified as hard water, while the proportion of LPW samples classified as hard water was 67.7%. The total ionic salinity (TIS) of LPW and BPW were in the range of 8.58–38.86 meq/L and 12.44–54.29 meq/L (Figure 6b), respectively. In addition, the $HCO_3$-Ca type was the predominant type of groundwater in the study area, indicating that the hydrochemical characteristics of groundwater may be influenced by the dissolution of carbonate rock minerals.

### 5.2. Hydrochemical Genetic Mechanism

#### 5.2.1. Rock Weathering

The Gibbs diagram [49] divides the main formation mechanisms that control water chemistry into three categories: precipitation, evaporation, and rock weathering. At present, the Gibbs graphical model is widely used to identify the controlling factors of groundwater hydrochemical characteristics [12,29]. The 53 groundwater samples in the NCA were projected onto the Gibbs map, as shown in Figure 7. It can be seen that the water sample points were mainly located in the rock weathering control area, that is, rock weathering was found to be the main formation mechanism of groundwater hydrochemistry in the NCA.

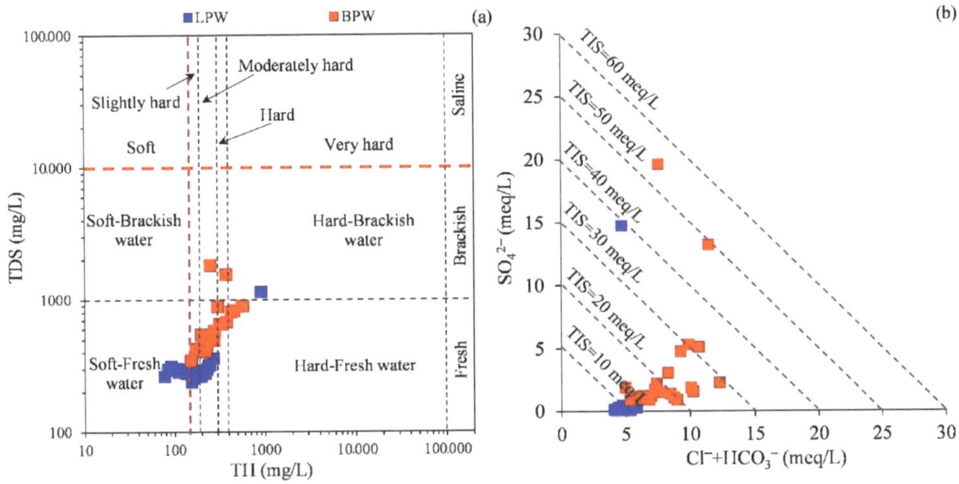


**Figure 6.** Scatter diagram of TH vs. TDS (**a**) and $SO_4^{2-}$ vs. $Cl^-$ + $HCO_3^-$ (**b**) of groundwater samples in the study area.

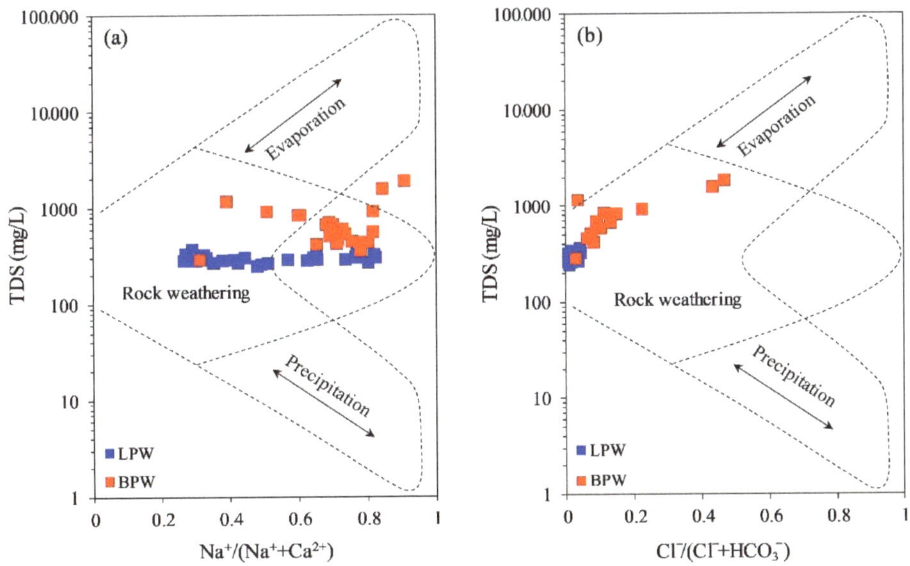


**Figure 7.** Gibbs diagram of groundwater hydrochemistry TDS vs. $Na^+/(Na^++Ca^{2+})$ (**a**) TDS vs. $Cl^-/(Cl^-+HCO_3^-)$ (**b**).

5.2.2. Mineral Dissolution

The main ion ratio relationship in groundwater can be used to further determine the rock weathering source type and mineral dissolution [22,24]. The $Na^+$ to $Cl^-$ ratio relationship can be used to determine the effect of halite dissolution on groundwater chemistry, since the dissolution of halite produces equal amounts of $Na^+$ and $Cl^-$ (Equation (9)). Figure 8a illustrates that all groundwater samples in the Ning County area were distributed on the right side of the line $Na^+/Cl^- = 1$, namely, $Na^+/Cl^- > 1$ in groundwater, indicating that the dissolution of halite was not the main source of $Na^+$ and $Cl^-$ in groundwater in the area. In addition, excess $Na^+$ may originate from other hydrogeochemical processes such as dissolution of silicate minerals or cation exchange.

$$NaCl \rightarrow Na^+ + Cl^- \quad (9)$$

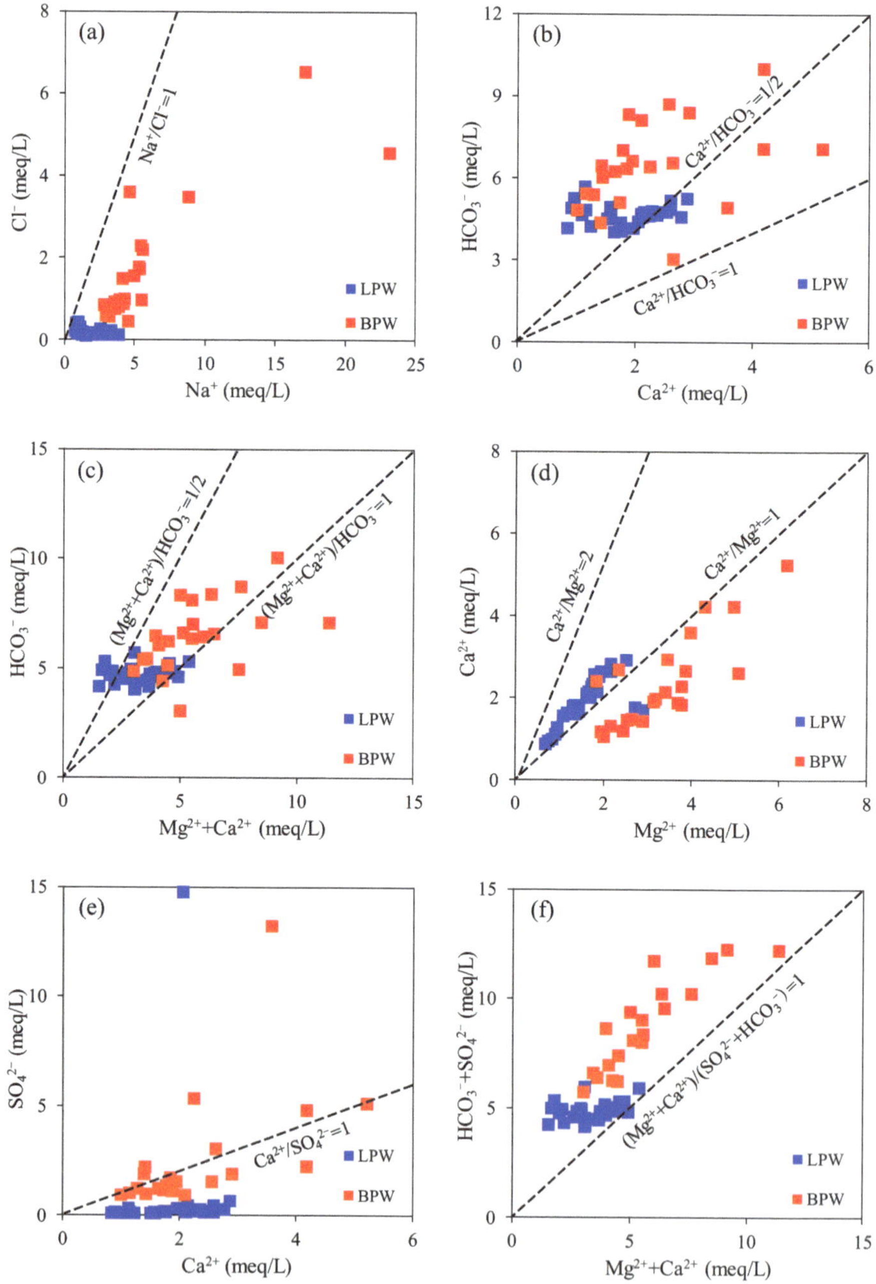


**Figure 8.** The relationship between the ion ratios of the main chemical components in the groundwater. $Na^+$ vs. $Cl^-$ (**a**), $Ca^{2+}$ vs. $HCO_3^-$ (**b**), $Mg^{2+}$+$Ca^{2+}$ vs. $HCO_3^-$ (**c**), $Mg^{2+}$ vs. $Ca^{2+}$ (**d**), $SO_4^{2-}$ vs. $Ca^{2+}$ (**e**) and $Mg^{2+}$+$Ca^{2+}$ vs. $SO_4^{2-}$+$HCO_3^-$ (**f**).

The ratio relationship between $Ca^{2+}$, $Mg^{2+}$, and $HCO_3^-$ is widely used to determine the influence of the dissolution of carbonate minerals (calcite and dolomite) in groundwater on hydrochemical characteristics [6,39]. As shown in the Figure 8b, some water samples were distributed near the $Ca^{2+}/HCO_3^-$ = 1/2 line or between the $Ca^{2+}/HCO_3^-$ = 1/2 line and $Ca^{2+}/HCO_3^-$ = 1 line, indicating that the chemical characteristics of groundwater in the study area are affected by the dissolution of calcite (Equations (10) and (11)). Furthermore, most of the water samples were distributed near the line $Ca^{2+}$ + $Mg^{2+}/HCO_3^-$ = 1 or between the line $Ca^{2+}$ + $Mg^{2+}/HCO_3^-$ = 1 and the line $Ca^{2+}$ + $Mg^{2+}/HCO_3^-$ = 1/2 (Figure 8c), illustrating that the dissolution of dolomite (Equations (12) and (13)) is an important process affecting the chemical characteristics of groundwater in the study area.

$$CaCO_3 + H^+ \rightarrow Ca^{2+} + HCO_3^- \tag{10}$$

$$CaCO_3 + CO_2 + H_2O \rightarrow Ca^{2+} + 2HCO_3^- \tag{11}$$

$$CaMg(CO_3)_2 + 2H^+ \rightarrow Ca^{2+} + Mg^{2+} + 2HCO_3^- \tag{12}$$

$$1CaMg(CO_3)_2 + 2CO_2 + 2H_2O \rightarrow Ca^{2+} + Mg^{2+} + 4HCO_3^- \tag{13}$$

The ratio of $Ca^{2+}/Mg^{2+}$ is often used to reveal the source of $Ca^{2+}$ and $Mg^{2+}$ in groundwater [50]. $Ca^{2+}/Mg^{2+}$ = 1 indicates that it comes from the dissolution of dolomite, while if 1 < $Ca^{2+}/Mg^{2+}$ < 2, it means that there is more dissolution of calcite. $Ca^{2+}/Mg^{2+}$ > 2 indicates that more $Ca^{2+}$ is provided by the dissolution of silicate minerals or gypsum [50]. As shown in Figure 8d, the LPW samples were mainly distributed between $Ca^{2+}/Mg^{2+}$ = 1 and $Ca^{2+}/Mg^{2+}$ = 2, indicating that the $Ca^{2+}$ in LPW mainly originated from the dissolution of calcite. In addition, the BPW samples were mainly concentrated below the line of $Ca^{2+}/Mg^{2+}$ = 1, indicating that the dissolution of dolomite or Mg-rich carbonate minerals had an important influence on the chemical composition of the BPW.

The dissolution of gypsum may be one of the important sources of $SO_4^{2-}$ in groundwater in a region. If gypsum is the only source of $Ca^{2+}$ and $SO_4^{2-}$ in groundwater, the ratio of $Ca^{2+}$ to $SO_4^{2-}$ should be 1 (Equation (13)). As shown in Figure 8e, most of the BPW samples were distributed close to the $Ca^{2+}/SO_4^{2-}$ = 1 line, while the LPW samples deviated from the $Ca^{2+}/SO_4^{2-}$ = 1 line, indicating that the dissolution of gypsum had a more significant contribution to the chemical characteristics of the BPW. In addition, the oxidation of pyrite or other sulfur-bearing minerals may also be a source of $SO_4^{2-}$ in groundwater [8]. However, the Cretaceous strata in the study area were generally a set of complex sediments under arid and semi-arid climate conditions, in which the upper part of sandstone and mudstone of the Huanhe formation was mostly sandwiched with a salt marsh facies gypsum layer, with high salt content and being rich in gypsum, sodium sulfate, halite, and other evaporite minerals [51]. Thus, the main source of $SO_4^{2-}$ in the groundwater of the NCA is the dissolution of gypsum and other sulfate minerals. The relationship between $Mg^{2+}$ + $Ca^{2+}$ and $HCO_3^-$ + $SO_4^{2-}$ is also often used to explain the contribution of the weathering of carbonate and silicate and the dissolution of gypsum to groundwater hydrochemical characteristics [50]. As shown in Figure 8f, the groundwater sample points were mainly distributed in the upper part of the line ($Mg^{2+}$ + $Ca^{2+}$)/($HCO_3^-$ + $SO_4^{2-}$) = 1, indicating that the weathering of carbonate and the dissolution of gypsum were important processes that affect the chemical composition of groundwater in the NCA.

$$CaSO_4{\cdot}2H_2O \rightarrow Ca^{2+} + SO_4^{2-} + 2H_2O \tag{14}$$

5.2.3. Cation Exchange

In general, cation exchange may also be an important hydrogeochemical reaction in a regional groundwater system that affects the chemical composition of groundwater [16,38]. As shown in Figure 9a, groundwater sample points in Ning County were all distributed along the $x/y = -1$ line, and only one LPW sample deviated from the $x/y = -1$ line, suggesting that cation exchange is an important hydrochemical process in the groundwater system in the region. If the CAI value is greater than 0, it indicates that reverse cation exchange has occurred, resulting in the exchange of $Na^+$ in groundwater with $Ca^{2+}$ in the aquifer (Equation (2)). If the $Ca^{2+}$ in groundwater is exchanged with $Na^+$ in the aquifer, direct cation exchange occurs (Equation (3)), and the value of CAI is less than 0. Figure 9b illustrates that the calculated CAI values of the water samples were all less than 0, indicating that direct cation exchange occurred in the groundwater system in the NCA.

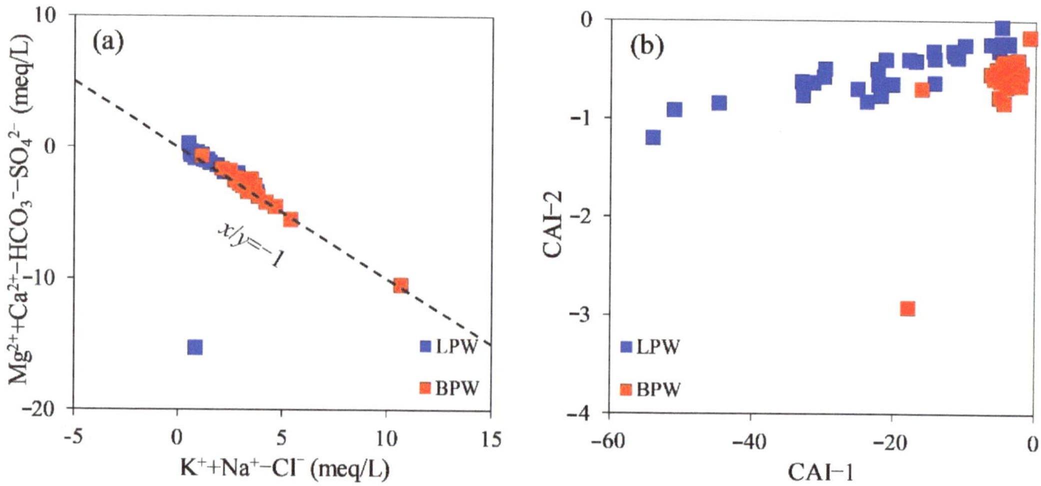


**Figure 9.** Diagrams of (**a**) $K^+ + Na^+ - Cl^-$ vs. $Mg^{2+} + Ca^{2+} - HCO_3^- - SO_4^{2-}$ and (**b**) CAI-1 vs. CAI-2.

*5.3. Analysis of Groundwater Recharge Sources*

The $\delta^2H$ and $\delta^{18}O$ data of groundwater samples in the study area were compared and analyzed with the Global Meteoric Water Line (GMWL) and the Local Meteoric Water Line (LMWL), as shown in Figure 10. Generally, the water sample points were distributed close to the atmospheric precipitation line, indicating that the main source of the water body was atmospheric precipitation. The GMWL ($\delta D = 8\delta^{18}O + 10$) in Figure 10 was obtained by Craig [52] in 1961. In addition, the LMWL in this study adopted the atmospheric precipitation line ($\delta D = 8.10\delta^{18}O + 6.97$) in the Pingliang area [53], which was close to the study area. It can be seen from Figure 10 that the groundwater sample points in the study area were mainly distributed near the lower part of the GMWL and along the LMWL as a whole, indicating that the main recharge source of groundwater in the region was atmospheric precipitation, which is affected by evaporation to a certain extent.

*5.4. Hydrogeochemical Simulation Results*

5.4.1. Saturation Index (SI)

The saturation indices and charge balance coefficients of minerals in groundwater at starting points on different simulation paths are shown in Table 3. Charge balance is a fundamental condition for performing reverse hydrogeochemical simulations. Table 4 illustrates that the charge balance coefficients of the water samples on the groundwater flow path were all close to 0, which indicates that the charges were essentially balanced. Table 3 indicates that there was a difference in the SI between the initial point and terminal point. The SI was less than 0, indicating that these minerals in the groundwater did not reach a saturated state, and that these minerals have a tendency to dissolve. The SI was greater than 0, indicating that it is in a supersaturated state and there is a tendency for precipitation to occur.

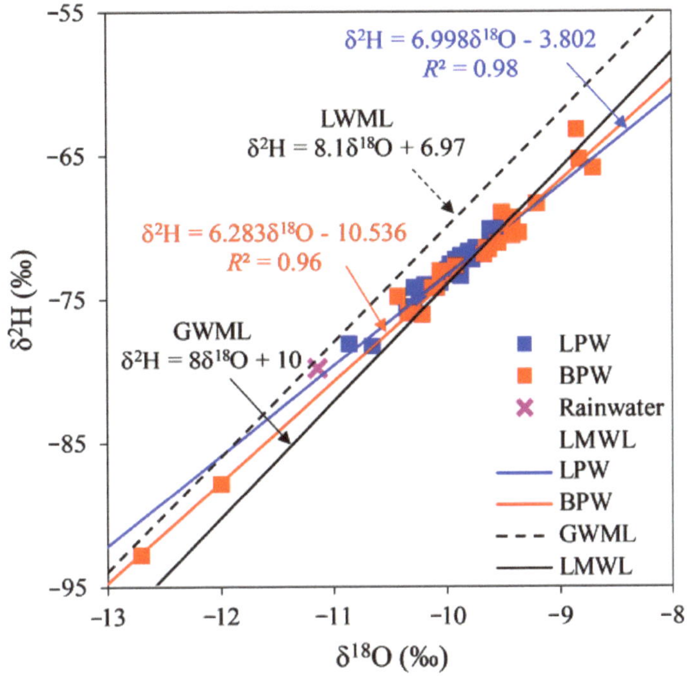


**Figure 10.** $\delta^2H$-$\delta^{18}O$ relationship diagram of groundwater.

**Table 3.** Calculated mineral saturation index and charge balance coefficient.

| | LPW Path (NX32→NX34) | | BPW Path (NX55→NX58) | |
|---|---|---|---|---|
| | **Initial Point** | **Terminal Point** | **Initial Point** | **Terminal Point** |
| Calcite | 0.68 | 0.80 | −0.24 | 0.26 |
| Dolomite | 1.39 | 1.65 | −0.04 | 0.90 |
| Gypsum | −2.66 | −2.44 | −2.16 | −2.04 |
| Chlorite | −0.22 | 0.81 | −6.00 | −2.15 |
| Illite | −1.23 | −1.28 | 0.50 | −1.31 |
| Plagioclase | 9.16 | 9.21 | 9.72 | 9.35 |
| Hornblende | 7.22 | 9.49 | −3.11 | 3.52 |
| Microline | −2.21 | −2.06 | −1.36 | −2.61 |
| Charge balance coefficient | $-2.5 \times 10^{-4}$ | $-5.88 \times 10^{-4}$ | $-1.48 \times 10^{-3}$ | $-1.35 \times 10^{-3}$ |

**Table 4.** Mineral transfer mass on the LPW path and BPW path.

| | Mineral Transfer Mass (mol/L) | |
|---|---|---|
| | **LPW Path** | **BPW Path** |
| Calcite | $8.197 \times 10^{-4}$ | $3.667 \times 10^{-3}$ |
| Dolomite | $-2.658 \times 10^{-4}$ | $-2.826 \times 10^{-3}$ |
| Gypsum | / | $3.297 \times 10^{-4}$ |
| Chlorite | $-2.170 \times 10^{-4}$ | $2.850 \times 10^{-4}$ |
| Illite | $1.231 \times 10^{-4}$ | $4.899 \times 10^{-4}$ |
| Plagioclase | $-1.208 \times 10^{-4}$ | $-9.721 \times 10^{-4}$ |
| Microline | $-7.311 \times 10^{-4}$ | $-3.554 \times 10^{-4}$ |
| Hornblende | $2.127 \times 10^{-4}$ | $1.233 \times 10^{-4}$ |
| NaX | $1.150 \times 10^{-3}$ | $2.537 \times 10^{-3}$ |
| $CaX_2$ | $-5.750 \times 10^{-4}$ | $-1.269 \times 10^{-3}$ |

#### 5.4.2. Mineral Transfer Mass

By using the PHREEQC software, the amount of mineral transfer mass on the simulated path was calculated. Mineral transfer masses greater than 0 or less than 0 represent mineral dissolution and precipitation, respectively. It can be seen from Table 4 that the dissolution of calcite, illite, and hornblende, as well as the precipitation of dolomite, plagioclase, and microcline occurred on both the LPW and BPW simulation paths. In addition, cation exchange occurred on both paths, and the $Na^+$ in the aquifer was replaced with $Ca^{2+}$ in the groundwater. The dissolution of gypsum and chlorite also occurred on the BWP path, while the precipitation of chlorite occurred on the LPW path. On different groundwater flow paths, the hydrogeochemical reactions of mineral transfer mass were different, which was the result of the combined effects of the lithology, geology, and hydrogeological conditions of the aquifer and its surrounding physical, chemical, and biological environments.

### *5.5. Suggestions for Regional Groundwater Research*

This study analyzed the chemical characteristics and controlling factors of groundwater in the NCA, and the research results can provide reference for the development and utilization of regional groundwater resources. However, the data from this study are only from a sampling period, and future studies should carry out sampling for different periods, depths, and types of groundwater in order to obtain more water quality data. In addition, other chemical components in groundwater such as trace elements, heavy metals, and organic matter should also be analyzed in order to enrich the results of regional groundwater-related studies and provide more comprehensive support for the development and utilization of groundwater resources.

## 6. Conclusions

The study of groundwater hydrochemistry is helpful for the rational development, utilization, and management of groundwater resources in a region, and can also provide a reference for similar regions in the world. To understand the hydrochemical characteristics and formation mechanism of groundwater in the NCA of the Loess Plateau, the LPW and BPW were sampled and analyzed, respectively, and the following main conclusions were obtained:

(1) The groundwater of NCA was weakly alkaline as a whole, mainly classified as hard fresh water, and $HCO_3{\cdot}Ca$ was the dominant hydrochemical type. Except for pH, the concentrations of major chemical components in BPW were higher than those in LPW.

(2) $\delta^2H$ and $\delta^{18}O$ analysis illustrated that the main recharge source of groundwater in NCA was atmospheric precipitation and was affected by evaporation. The linear relationships of the $\delta^2H$ and $\delta^{18}O$ of LPW and BPW were $\delta^2H = 6.998\delta^{18}O{-}3.802$ ($R^2 = 0.98$) and $\delta^2H = 6.283\delta^{18}O - 10.536$ ($R^2 = 0.96$), respectively.

(3) The main controlling factors of LPW and BPW in the study area were rock weathering and cation exchange. $Ca^{2+}$, $Mg^{2+}$, and $HCO_3^-$ in groundwater mainly originated from the dissolution of calcite and dolomite. In addition, the dissolution of gypsum also had a significant effect on the water chemical composition of BPW.

(4) Hydrogeochemical inverse simulations indicated that cation exchange; the dissolution of calcite, illite, and hornblende; and the precipitation of dolomite, plagioclase, and microcline occurred on both the LPW and BPW pathways. In addition, the dissolution of gypsum and chlorite also occurred on the BWP path, and the precipitation of chlorite occurred on the LPW path.

**Author Contributions:** Conceptualization, M.Z., J.L., C.L. and Z.G.; Methodology, M.Z., J.L., C.L., and Z.G.; Software, S.H., Y.W., Z.W., H.L. and X.W.; Validation, M.Z., J.L., C.L. and Z.G.; Formal Analysis, M.Z., J.L., C.L. and Z.G; Investigation, S.H., Y.W., Z.W., H.L. and X.W.; Resources, S.H., Y.W., Z.W., H.L. and X.W.; Data Management, S.H., Y.W., Z.W., H.L. and X.W.; Writing—Manuscript Preparation, M.Z., J.L., C.L. and Z.G.; Writing—Review and Editing, M.Z., J.L., C.L. and Z.G.; Visualization, M.Z., J.L., C.L. and Z.G.; Supervision, S.H., Y.W., Z.W., H.L. and X.W; Project Management, M.Z.; Funding Acquisition, M.Z. All authors have read and agreed to the published version of the manuscript.

**Funding:** The study involved in this paper was supported by the Open Project Program of the Hebei Center for Ecological and Environmental Geology Research (no. JSYF-202204); 1:50,000 Hydrogeological Survey in the Old Revolutionary Areas of Shanxi, Gansu, and Ningxia (no. DD20160288); and Investigation and Monitoring of Hydrogeology and Water Resources in Key Areas of Yellow River Basin (no. DD20221754). We thank the editors and anonymous reviewers for their valuable suggestions on the manuscript.

**Institutional Review Board Statement:** Not applicable.

**Informed Consent Statement:** Not applicable.

**Data Availability Statement:** Not applicable.

**Conflicts of Interest:** The authors declare no conflict of interest.

## References

1. Jaydhar, A.K.; Pal, S.C.; Saha, A.; Islam, A.R.M.T.; Ruidas, D. Hydrogeochemical evaluation and corresponding health risk from elevated arsenic and fluoride contamination in recurrent coastal multi-aquifers of eastern India. *J. Clean. Prod.* **2022**, *369*, 133150. [CrossRef]
2. Pal, S.C.; Ruidas, D.; Saha, A.; Islam, A.R.M.T.; Chowdhuri, I. Application of novel data-mining technique based nitrate concentration susceptibility prediction approach for coastal aquifers in India. *J. Clean. Prod.* **2022**, *346*, 131205. [CrossRef]
3. Ruidas, D.; Pal, S.C.; Saha, A.; Chowdhuri, I.; Shit, M. Hydrogeochemical characterization based water resources vulnerability assessment in India's first Ramsar site of Chilka lake. *Mar. Pollut. Bull.* **2022**, *184*, 114107. [CrossRef] [PubMed]
4. Ruidas, D.; Pal, S.C.; Islam, A.R.M.T.; Saha, A. Characterization of groundwater potential zones in water-scarce hardrock regions using data driven model. *Environ. Earth Sci.* **2021**, *80*, 809. [CrossRef]
5. Vaiphei, S.P.; Kurakalva, R.M. Hydrochemical characteristics and nitrate health risk assessment of groundwater through seasonal variations from an intensive agricultural region of upper Krishna River basin, Telangana, India. *Ecotoxicol. Environ. Saf.* **2021**, *213*, 112073. [CrossRef]
6. Wang, X.; Zheng, W.; Tian, W.; Gao, Y.; Wang, X.; Tian, Y.; Li, J.; Zhang, X. Groundwater hydrogeochemical characterization and quality assessment based on integrated weight matter-element extension analysis in Ningxia, upper Yellow River, northwest China. *Ecol. Indic.* **2022**, *135*, 108525. [CrossRef]
7. Keum, G.; Kim, Y.; Lee, K.-S.; Jeong, J. The geochemistry and isotopic compositions of the Nakdong River, Korea: Weathering and anthropogenic effects. *Environ. Monit. Assess.* **2022**, *194*, 487. [CrossRef]
8. Fuoco, I.; De Rosa, R.; Barca, D.; Figoli, A.; Gabriele, B.; Apollaro, C. Arsenic polluted waters: Application of geochemical modelling as a tool to understand the release and fate of the pollutant in crystalline aquifers. *J. Environ. Manag.* **2022**, *301*, 113796. [CrossRef]
9. Li, C.; Gao, Z.; Chen, H.; Wang, J.; Liu, J.; Li, C.; Teng, Y.; Liu, C.; Xu, C. Hydrochemical analysis and quality assessment of groundwater in southeast North China Plain using hydrochemical, entropy-weight water quality index, and GIS techniques. *Environ. Earth Sci.* **2021**, *80*, 523. [CrossRef]
10. Liu, J.; Peng, Y.; Li, C.; Gao, Z.; Chen, S. Characterization of the hydrochemistry of water resources of the Weibei Plain, Northern China, as well as an assessment of the risk of high groundwater nitrate levels to human health. *Environ. Pollut.* **2021**, *268*, 115947. [CrossRef]
11. Peng, H.; Yang, W.; Ferrer, A.S.N.; Xiong, S.; Li, X.; Niu, G.; Lu, T. Hydrochemical characteristics and health risk assessment of groundwater in karst areas of southwest China: A case study of Bama, Guangxi. *J. Clean. Prod.* **2022**, *341*, 130872. [CrossRef]
12. Xiao, Y.; Hao, Q.; Zhang, Y.; Zhu, Y.; Yin, S.; Qin, L.; Li, X. Investigating sources, driving forces and potential health risks of nitrate and fluoride in groundwater of a typical alluvial fan plain. *Sci. Total Environ.* **2022**, *802*, 149909. [CrossRef] [PubMed]
13. Saha, A.; Pal, S.C.; Chowdhuri, I.; Roy, P.; Chakrabortty, R. Effect of hydrogeochemical behavior on groundwater resources in Holocene aquifers of moribund Ganges Delta, India: Infusing data-driven algorithms. *Environ. Pollut.* **2022**, *314*, 120203. [CrossRef] [PubMed]
14. Schwarzenbach, R.P.; Egli, T.; Hofstetter, T.B.; Von Gunten, U.V.; Wehrli, B. Global Water Pollution and Human Health. *Annu. Rev. Environ. Resour.* **2010**, *35*, 109–136. [CrossRef]
15. Misstear, B.; Vargas, C.R.; Lapworth, D.; Ouedraogo, I.; Podgorski, J. A global perspective on assessing groundwater quality. *Hydrogeol. J.* **2022**. [CrossRef]
16. Mohammed, A.M.; Refaee, A.; El-Din, G.K.; Harb, S. Hydrochemical characteristics and quality assessment of shallow groundwater under intensive agriculture practices in arid region, Qena, Egypt. *Appl. Water Sci.* **2022**, *12*, 92. [CrossRef]
17. Lapworth, D.; Boving, T.; Kreamer, D.; Kebede, S.; Smedley, P. Groundwater quality: Global threats, opportunities and realising the potential of groundwater. *Sci. Total Environ.* **2022**, *811*, 152471. [CrossRef]
18. Fuoco, I.; Marini, L.; De Rosa, R.; Figoli, A.; Gabriele, B.; Apollaro, C. Use of reaction path modelling to investigate the evolution of water chemistry in shallow to deep crystalline aquifers with a special focus on fluoride. *Sci. Total Environ.* **2022**, *830*, 154566. [CrossRef]

19. Liu, Y.; Engel, B.A.; Flanagan, D.C.; Gitau, M.W.; McMillan, S.K.; Chaubey, I. A review on effectiveness of best management practices in improving hydrology and water quality: Needs and opportunities. *Sci. Total Environ.* **2017**, *601–602*, 580–593. [CrossRef]
20. Sarkar, M.; Pal, S.C.; Islam, A.R.M.T. Groundwater quality assessment for safe drinking water and irrigation purposes in Malda district, Eastern India. *Environ. Earth Sci.* **2022**, *81*, 52. [CrossRef]
21. Rivera-Rodríguez, D.A.; Beltrán-Hernández, R.I.; Lucho-Constantino, C.A.; Coronel-Olivares, C.; Hernández-González, S.; Villanueva-Ibáñez, M.; Nolasco-Arizmendi, V.; Vázquez-Rodríguez, G.A. Water quality indices for groundwater impacted by geogenic background and anthropogenic pollution: Case study in Hidalgo, Mexico. *Int. J. Environ. Sci. Technol.* **2019**, *16*, 2201–2214. [CrossRef]
22. Karkhaneh, T.; Sarikhani, R.; Dehnavi, A.G.; Baker, L.L.; Moradpour, A. Controlling factors of groundwater chemistry and statistical analysis of the samples related to ophiolite of Nourabad-Harsin, West of Iran. *Environ. Monit. Assess.* **2022**, *194*, 123. [CrossRef] [PubMed]
23. Chen, X.; Jiang, C.; Zheng, L.; Zhang, L.; Fu, X.; Chen, S.; Chen, Y.; Hu, J. Evaluating the genesis and dominant processes of groundwater salinization by using hydrochemistry and multiple isotopes in a mining city. *Environ. Pollut.* **2021**, *283*, 117381. [CrossRef] [PubMed]
24. Liu, J.; Peng, Y.; Li, C.; Gao, Z.; Chen, S. An investigation into the hydrochemistry, quality and risk to human health of groundwater in the central region of Shandong Province, North China. *J. Clean. Prod.* **2021**, *282*, 125416. [CrossRef]
25. Abdulsalam, A.; Ramli, M.F.; Jamil, N.R.; Ashaari, Z.H.; Umar, D.U.A. Hydrochemical characteristics and identification of groundwater pollution sources in tropical savanna. *Environ. Sci. Pollut. Res. Int.* **2022**, *29*, 37384–37398. [CrossRef]
26. Liu, J.; Gao, Z.; Wang, Z.; Xu, X.; Su, Q.; Wang, S.; Qu, W.; Xing, T. Hydrogeochemical processes and suitability assessment of groundwater in the Jiaodong Peninsula, China. *Environ. Monit. Assess.* **2020**, *192*, 384. [CrossRef]
27. Apollaro, C.; Di Curzio, D.; Fuoco, I.; Buccianti, A.; Dinelli, E.; Vespasiano, G.; Castrignanò, A.; Rusi, S.; Barca, D.; Figoli, A.; et al. A multivariate non-parametric approach for estimating probability of exceeding the local natural background level of arsenic in the aquifers of Calabria region (Southern Italy). *Sci. Total Environ.* **2022**, *806*, 150345. [CrossRef]
28. Tyagi, S.; Sarma, K. Expounding major ions chemistry of groundwater with significant controlling factors in a suburban district of Uttar Pradesh, India. *J. Earth Syst. Sci.* **2021**, *130*, 169. [CrossRef]
29. Chitsazan, M.; Aghazadeh, N.; Mirzaee, Y.; Golestan, Y. Hydrochemical characteristics and the impact of anthropogenic activity on groundwater quality in suburban area of Urmia city, Iran. *Environ. Dev. Sustain.* **2019**, *21*, 331–351. [CrossRef]
30. Liu, R.-P.; Zhu, H.; Liu, F.; Dong, Y.; El-Wardany, R.M. Current situation and human health risk assessment of fluoride enrichment in groundwater in the Loess Plateau: A case study of Dali County, Shaanxi Province, China. *China Geol.* **2021**, *4*, 487–497. [CrossRef]
31. Cheng, L.; Si, B.; Wang, Y.; Liu, W. Groundwater recharge mechanisms on the Loess Plateau of China: New evidence for the significance of village ponds. *Agric. Water Manag.* **2021**, *257*, 107148. [CrossRef]
32. Li, Z.; Coles, A.E.; Xiao, J. Groundwater and streamflow sources in China's Loess Plateau on catchment scale. *Catena* **2019**, *181*, 104075. [CrossRef]
33. Li, H.X.; Han, S.B.; Wu, X.; Wang, S.; Liu, W.P.; Ma, T.; Zhang, M.N.; Wei, Y.T.; Yuan, F.Q.; Yuan, L. Distribution, characteristics and influencing factors of fresh groundwater resources in the Loess Plateau, China. *China Geol.* **2021**, *4*, 509–526. [CrossRef]
34. Ling, X.; Ma, J.; Chen, P.; Liu, C.; Horita, J. Isotope implications of groundwater recharge, residence time and hydrogeochemical evolution of the Longdong Loess Basin, Northwest China. *J. Arid Land* **2022**, *14*, 34–55. [CrossRef]
35. Wang, W.; Li, Z.; Su, H.; Xiao, J.; Han, F.; Li, Z. Spatial and seasonal variability, control factors and health risk of fluoride in natural water in the Loess Plateau of China. *J. Hazard. Mater.* **2022**, *434*, 128897. [CrossRef]
36. Xiao, J.; Wang, L.; Deng, L.; Jin, Z. Characteristics, sources, water quality and health risk assessment of trace elements in river water and well water in the Chinese Loess Plateau. *Sci. Total Environ.* **2019**, *650*, 2004–2012. [CrossRef]
37. Zhang, F.; Jin, Z.; Yu, J.; Zhou, Y.; Zhou, L. Hydrogeochemical processes between surface and groundwaters on the northeastern Chinese Loess Plateau: Implications for water chemistry and environmental evolutions in semi-arid regions. *J. Geochem. Explor.* **2015**, *159*, 115–128. [CrossRef]
38. Gao, Z.; Han, C.; Xu, Y.; Zhao, Z.; Luo, Z.; Liu, J. Assessment of the water quality of groundwater in Bohai Rim and the controlling factors—A case study of northern Shandong Peninsula, north China. *Environ. Pollut.* **2021**, *285*, 117482. [CrossRef]
39. Li, Y.-Z.; Gao, Z.-J.; Liu, J.-T.; Wang, M.; Han, C. Hydrogeochemical and isotopic characteristics of spring water in the Yarlung Zangbo River Basin, Qinghai-Tibet Plateau, Southwest China. *J. Mt. Sci.* **2021**, *18*, 2061–2078. [CrossRef]
40. Rajendiran, T.; Sabarathinam, C.; Chandrasekar, T.; Keesari, T.; Senapathi, V.; Sivaraman, P.; Viswanathan, P.M.; Nagappan, G. Influence of variations in rainfall pattern on the hydrogeochemistry of coastal groundwater—An outcome of periodic observation. *Environ. Sci. Pollut. Res.* **2019**, *26*, 29173–29190. [CrossRef]
41. Manoj, S.; Thirumurugan, M.; Elango, L. Hydrogeochemical modelling to understand the surface water–groundwater interaction around a proposed uranium mining site. *J. Earth Syst. Sci.* **2019**, *128*, 49. [CrossRef]
42. Wang, Y.; Tong, J.; Hu, B.X.; Dai, H. Combining Isotope and Hydrogeochemistry Methods to Study the Seawater Intrusion: A Case Study in Longkou City, Shandong Province, China. *Water* **2022**, *14*, 789. [CrossRef]
43. Yu, F.; Zhou, D.; Li, Z.; Li, X. Hydrochemical Characteristics and Hydrogeochemical Simulation Research of Groundwater in the Guohe River Basin (Henan Section). *Water* **2022**, *14*, 1461. [CrossRef]

44. Liu, J.; Wang, M.; Gao, Z.; Chen, Q.; Wu, G.; Li, F. Hydrochemical characteristics and water quality assessment of groundwater in the Yishu River basin. *Acta Geophys.* **2020**, *68*, 877–889. [CrossRef]
45. Piper, A.M. A graphic procedure in the geochemical interpretation of water-analyses. *Eos Trans. Am. Geophys. Union* **1944**, *25*, 914–928. [CrossRef]
46. Qian, H.; Li, P.; Wu, J.; Zhou, Y. Isotopic characteristics of precipitation, surface and ground waters in the Yinchuan plain, Northwest China. *Environ. Earth Sci.* **2012**, *70*, 57–70. [CrossRef]
47. Liu, J.; Gao, Z.; Wang, M.; Li, Y.; Yu, C.; Shi, M.; Zhang, H.; Ma, Y. Stable isotope characteristics of different water bodies in the Lhasa River Basin. *Environ. Earth Sci.* **2019**, *78*, 71. [CrossRef]
48. Li, H.; Cheng, X.; Ma, Y.; Liu, W.; Zhou, B. Characteristics and Formation Mechanism of Strontium-rich Groundwater in Malian River Drainage Basin. *Geoscience* **2021**, *35*, 682–692.
49. Gibbs, R.J. Mechanisms Controlling World Water Chemistry. *Science* **1970**, *170*, 1088–1090. [CrossRef]
50. Xiao, J.; Jin, Z.; Wang, J.; Zhang, F. Major ion chemistry, weathering process and water quality of natural waters in the Bosten Lake catchment in an extreme arid region, NW China. *Environ. Earth Sci.* **2015**, *73*, 3697–3708. [CrossRef]
51. Wang, Y.; Cheng, X.; Zhang, M. Hydrochemistry and chemical weathering processes of Malian River Basin. *Earth Environ.* **2018**, *46*, 15–22.
52. Craig, H. Isotopic Variations in Meteoric Waters. *Science* **1961**, *133*, 1702–1703. [CrossRef] [PubMed]
53. Huang, J.Z.; Tan, H.B.; Wang, R.A.; Wen, X.W. Hydrogen and oxygen isotopic analysis of perennial meteoric water in northwest China. *J. China Hydrol.* **2015**, *35*, 33–39.

water

MDPI

Article

# Study of Single Fracture Seepage Characteristics of Fault-Filled Materials Based on CT Technology

Wenbin Sun [1,2,3,*], Shaoyu Wang [1], Faxu Dong [1] and Yandong Xue [1]

1 School of Energy and Mining Engineering, Shandong University of Science and Technology, Qingdao 266590, China
2 State Key Laboratory of Coal Mining and Clean Utilization, Beijing 100013, China
3 College of Earth and Mineral Sciences, The Pennsylvania State University, State College, PA 16801, USA
* Correspondence: swb@sdust.edu.cn

**Abstract:** In order to study the matrix loss process and skeleton seepage law in the fracture of the fault rock, the three-dimensional model of the skeletal rock sample of the fault rock was obtained by CT scan, and the porous media seepage model was established with different structural types of natural fractures, and the flow rate and pressure distribution law of the seepage in the fracture was obtained by FLUENT software simulation. The results show that: the seepage under different pressure conditions is approximately the same, and the velocity increases continuously with the increase in pressure; The water seepage in different directions of the fracture channels under the same pressure conditions is not exactly the same, which is caused by the different microstructures of the pores. For the pressure distribution, it gradually decreases along the direction of water seepage, and for the speed distribution, it shows the law of changing from large to small and then increasing.

**Keywords:** fissure; seepage; three-dimensional reconstruction; porous media; numerical simulation

**Citation:** Sun, W.; Wang, S.; Dong, F.; Xue, Y. Study of Single Fracture Seepage Characteristics of Fault-Filled Materials Based on CT Technology. *Water* **2022**, *14*, 3679. https://doi.org/10.3390/w14223679

Academic Editor: Jiutan Liu

Received: 26 October 2022
Accepted: 8 November 2022
Published: 14 November 2022



## 1. Introduction

With the gradual increase in the depth of coal mining in China, the safety risks faced are also increasing, and coal mine permeability accidents are still an important part of the accidents affecting coal mine production safety [1–5] and the economic losses caused by coal mine permeability have been the top of coal mine disasters. The fault is a weak geological structure that affects the safety of coal mining, and it is easy to form a water inrush channel in which the pressurized water of the aquifer layer conducts to the working face. Fault structure fill often consists of fault mud as a matrix and fault rock as the skeleton. The substrate in fault rock is very vulnerable to latent intrusion to form seepage channels [6,7], and under the support of the skeleton, the seepage channels expand to form a pipe surge through the fault, and the fault waterproof coal column is improperly retained or structural damage, and other protective measures fail, which will form water damage. Fault sudden water in coal mines [8–11] has gradually become a key scientific and technical problem that restricts the economic development of coal in China. From the case of sudden water of coal mine fault in past years, the sudden water of fault is influenced by the nature of its filling medium, especially the structure and permeability of the filling medium, so the study of the seepage characteristics of the fault filler is of great significance for the prediction of sudden water and its prevention and control.

Along with the rapid development of science and technology, computed tomography (CT) technology, three-dimensional microscopic imaging technology, focused ion beam-scanning electron microscopy (FIB-SEM), and magnetic resonance imaging (MRI) technology have become popular and widely used [12–17]. With the help of 3D digital core technology, porous media can be scanned from the nanometer to the micrometer level, enabling the reproduction of the real porous media pore structure with high accuracy, facilitating the observation of the degree of inhomogeneous changes in the internal structure of pores, visualizing the internal structure of pores and achieving fine characterization, which will help to promote the progress of 3D structure measurement and characterization of porous media [18–23]. Wang Gang [24,25] established a 3D coal skeleton structure model containing pore fissures by using CT scanning and 3D reconstruction techniques, and the simulations demonstrated that a velocity peak and a stable flow rate value exist at each point in the pore fissures during seepage. Yao Banghua [26] established a multi-field coupled deformation-seepage erosion mechanics model for trapped column sudden water deformation based on dual pore medium seepage theory. Li Liping [27] used a computational procedure that can consider the dual damage criterion of rock tension shear and compression shear under the action of groundwater to reproduce the catastrophic evolution process of mining fracture sprouting, expansion, and penetration until rupture channel formation in the water-isolated rock body between the pressurized aquifer and the roadway.

Therefore, this paper explores the influence of the fracture structure of the original rock on the seepage law through the numerical simulation of the actual rock single fracture model. Combined with CT scanning and 3D reconstruction technology, a three-dimensional rock skeleton structure model containing pore fractures was established. On this basis, a porous medium model is constructed, and the flow velocity and pressure distribution of different seepage paths are obtained by simulating the seepage of a single fracture structure to explore the effects of different seepage paths and pressure and other factors on the seepage characteristics.

## 2. CT Scan and 3D Model Reconstruction

### 2.1. Pre-Processing of CT Scan Images

In this paper, the skeleton rocks of small fault filling collected during the roadway development in the Jining mining area of Shandong Province (the location is shown in Figure 1) were selected and made as dynamic metamorphic schist samples with a size of $7 \times 7 \times 30$ mm. The terrain in the mining area is undulating. The terrain is high in the south and low in the north, high in the west and low in the east, and high in the southwest and low in the northeast. The ground elevation is between 506.2~229.3 m, the relative height difference is 276.9m, and the average slope is about 3°46′. The CT scanning device nanoVoxel-3502E cross-scale core scanning imaging system is used to import the raw data file obtained by the CT scanning device into the AVIZO software. The CT scan reconstructs a model size of $600 \times 600 \times 1400$ voxels with 0.02381770 mm opaque voxel units and scans to obtain a 3D image of the rock sample. A 3D digital core model is constructed by region selection, sequence adjustment, angle correction, and black spot elimination of image slices. In order to extract the pore structure information from the scanned images, it is necessary to first separate the pore structure from the matrix skeleton, mainly using image filtering and threshold segmentation binarization techniques. After the core has been scanned, isolated pixel points with different grey scale values will appear due to interference from noise, etc. Noise is generally excluded by means of an islanding algorithm. For skeletons and pores, even the same rock components may show different greyscale values due to subtle differences in the nature of the mineral grains and their degree of metamorphism; Therefore, it is necessary to adopt appropriate filtering algorithms to remove noise interference and at the same time make the gray value distribution of the same component more concentrated, and the gray value distinction of different components more obvious, highlighting the pore fracture structure surface characteristics of rock samples, which is convenient for later threshold segmentation.

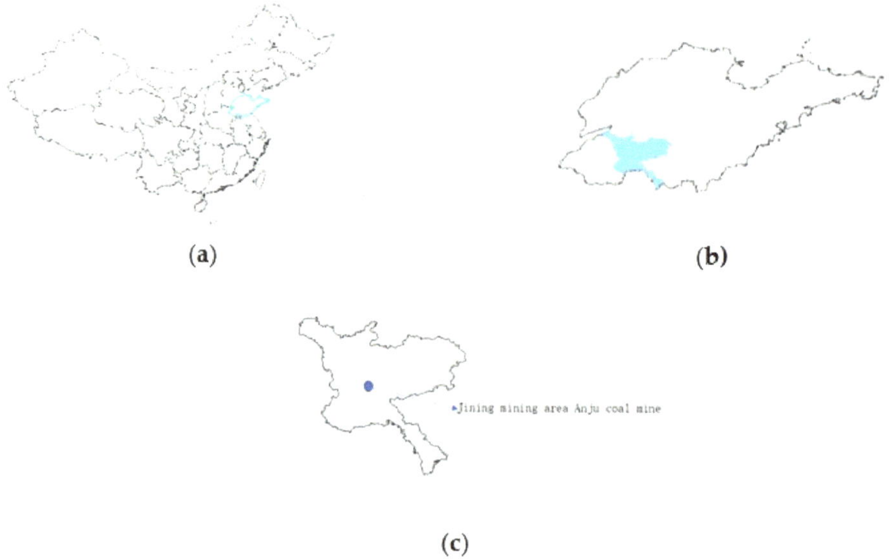


**Figure 1.** Mine Location Map: (**a**) Map of China. (**b**) Map of Shandong. (**c**) Diagram of Anju Coal Mine in Jining Mining District.

### *2.2. Fracture Extraction*

The CT images are image processed, and 3D reconstructed to obtain a 3D pore model. After processing the model with Median Filtering and the threshold splitting binarization method, the appropriate fissure extraction was selected in the cropped model by AVIZO software (as shown in Figure 2). The cropped section is $150 \times 150 \times 150$ voxels in size, and the sample contains a full fissure. After extracting the fissures, the surface of the fissures is smoothed, the entrance and exit of the fissures and the boundary conditions are set, the intersections are repaired, the number of faces is simplified so as to improve the quality of the mesh, and finally, a triangular mesh is generated on the surface of the fissures. After the meshing is completed, the quality of the mesh is checked, and low-quality, unconnected meshes are removed to ensure that a high-quality mesh is generated and finally saved as a cas format file, which is then imported into FLUENT software for the next numerical simulation.

## **3. Seepage Simulation**

### *3.1. Condition Setting*

Convert the network structure of the pores into a file in CAS format and import it into FLUENT. As the units in AVIZO and FLUENT software are different, the grid units need to be unified. Check the quality of the grid to ensure that there are no negative volumes. Set boundary conditions for walls and entrances. This paper mainly considers the effect of different paths and pressure gradient changes on the distribution of pore fissure seepage and flow rate changes, so the simplified model ignores the compressibility of water and the role of gas, does not consider the erosion of water on the rock boundary during seepage and the fracturing effect, measured after the seepage is stable to ignore the error of surface tension. In setting the boundary conditions, the model is divided into three parts: pressurized water, porous media structure, and atmosphere, where the atmospheric pressure is not considered to be influenced by the stratigraphic structure, the pore structure is represented as an airfield, and the pressurized water is set to be pressure driven, essentially a water-driven gas flow simulation. A gravity field of 9.8 $m/s^2$ was also set in the y-negative direction, and simulations of seepage under different pressure conditions were carried out in the z-positive and negative directions of the model, respectively, to study the seepage characteristics of the fissure model.

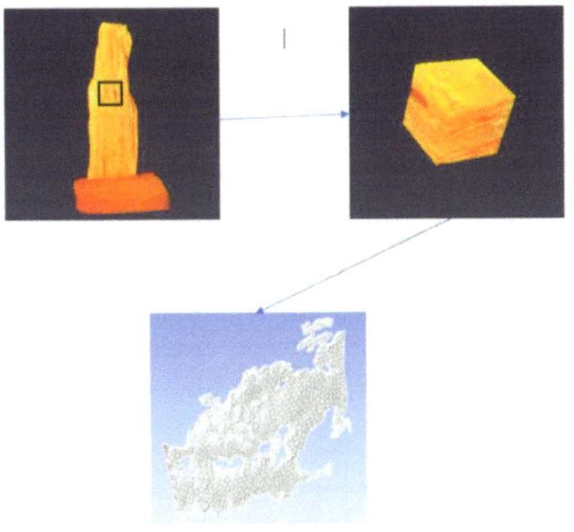

**Figure 2.** CT scan preprocessing.

### 3.2. Solver Model Selection

The rough structure of the fracture surface leads to non-Darcy flow, and the exact solution must be obtained by solving the N-S equation [28–30]. When the pore structure of the seepage medium is not considered, the seepage model is commonly used, and the method of studying fluid dynamics is introduced to the entire seepage area. The controlling equations for general incompressible flow have the continuity equation and the N-S component, namely:

$$\frac{\partial u}{\partial t} + u \cdot \nabla u = -\frac{1}{\rho}\nabla p + v\nabla^2 u + f \tag{1}$$

$$\nabla \cdot \mathrm{u} = 0 \tag{2}$$

The first of these formulas is the N-S equation, and the second is an incidental formula for incompressible fluids. Where: $u$ is the average velocity of the micro-element section u = nu'; $t$ is the time; ρ is the fluid density; $p$ is the permeation pressure; $f$ is the force per unit mass, and $v$ is the kinematic viscosity coefficient of the fluid.

$$\mathrm{S} = \mathrm{gJ} \tag{3}$$

Substitution of Darcy's theorem:

$$\mathrm{v} = -\mathrm{kJ} \tag{4}$$

$$\mathrm{S} = -\mathrm{g}\frac{v}{k} \tag{5}$$

where: S is the resistance to seepage per unit mass, J is the infiltration ratio, g is the acceleration of gravity.

Because the flow of fluid media within the pores of porous media is predominantly turbulent, simulations are carried out using the standard k-ε model. The advantages of the standard k-ε model are its wide applicability, its low cost, and the high accuracy it can achieve.

### 3.3. Experimental Design of Seepage Simulation

The control of variables in this experiment mainly lies in the condition setting of the entrance boundary and the selection of different paths. Since the sample fissures are laminar in distribution and there are no obvious through-hole fissures between the fissures, the entrance is set to a certain cross-section of the fissures, and it is also possible to compare the difference of seepage development in different scale fissures by selecting different fissures under the premise of ensuring seepage development. Other factors affecting the effect of seepage include fracture density, connectivity, and tortuosity, which are considered quantitative when the model is determined, and only two variables, pressure gradient, and different fractures, are considered in the experiments of this paper for their effect on the seepage characteristics. The initial pressure at the entrance boundary was divided into several stages, each through a different fissure channel, to see the effect of the two variables in the simulation results on the seepage of the specimen. The DPM model in FLUENT is applied to inject particles and, after post-processing, to observe the trajectory of the particles to see the changes in the flow field and to study the particle flow pattern.

### 3.4. Seepage Simulation Results and Analysis

The single-fracture structure model of the fault filling was established by means of region selection, the seepage simulation was carried out, and the pressure cloud map and velocity vector diagram at the initial pressure of 0.1 MPa, 0.5 MPa, 1 MPa, 3 MPa, 5 MPa, and 10 MPa at different entrances in the z forward direction and z negative direction were obtained respectively (as shown in Figures 3–6), and the seepage characteristics of the pores of the fault filling and the loss law of the matrix in the skeleton were analyzed.

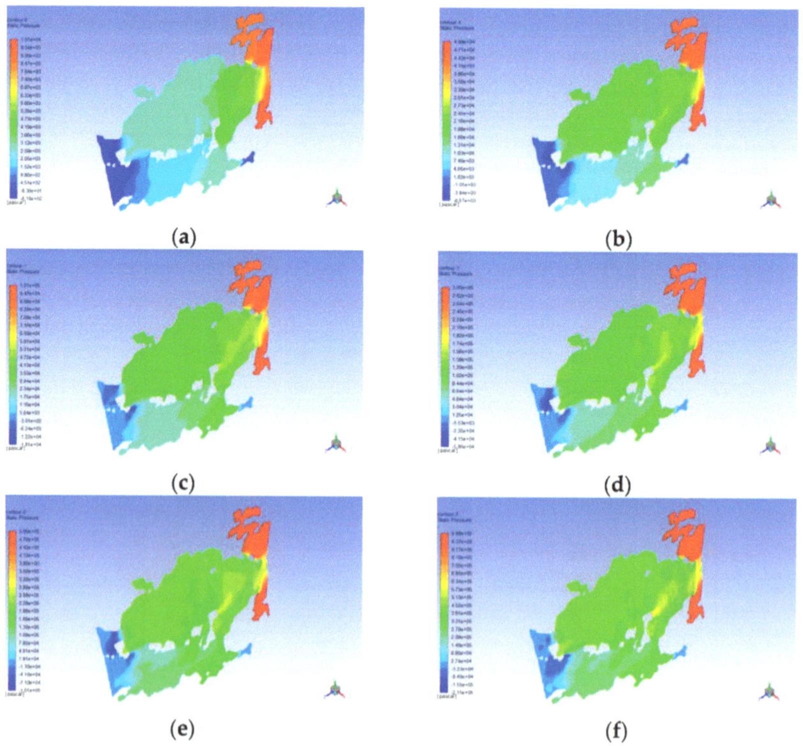

**Figure 3.** z forward inlet pressure cloud map: (**a**) 0.1 Mpa pressure cloud map. (**b**) 0.5 Mpa pressure cloud map. (**c**) 1 Mpa pressure cloud map. (**d**) 3 Mpa pressure cloud map. (**e**) 5 Mpa pressure cloud map. (**f**) 10 Mpa pressure cloud map.

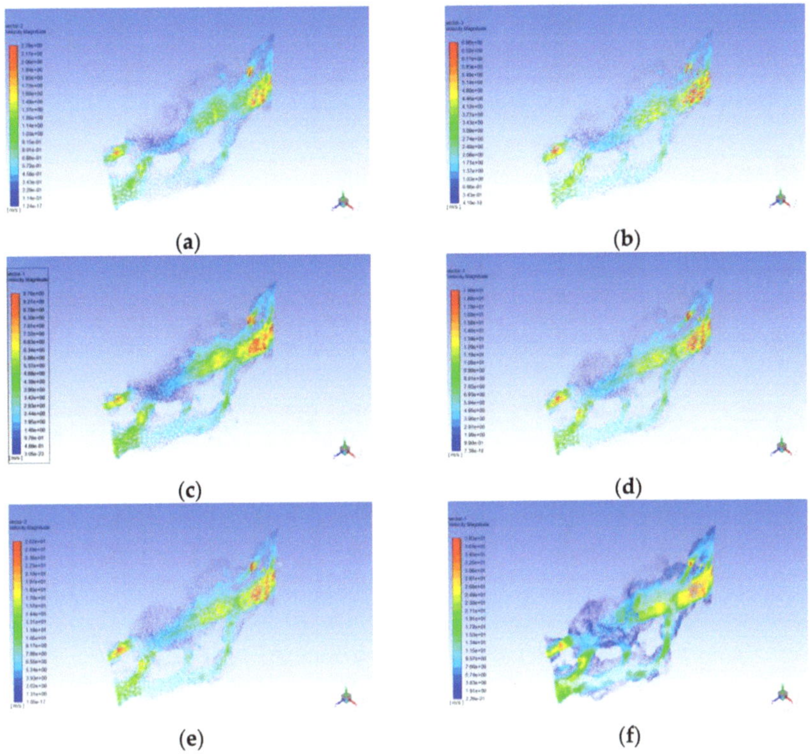

**Figure 4.** z forward inlet velocity vector diagram:(**a**) 0.1 Mpa speed vector. (**b**) 0.5 Mpa speed vector. (**c**) 1 Mpa speed vector. (**d**) 3 Mpa speed vector. (**e**) 5 Mpa speed vector. (**f**) 10 Mpa speed vector illustration.

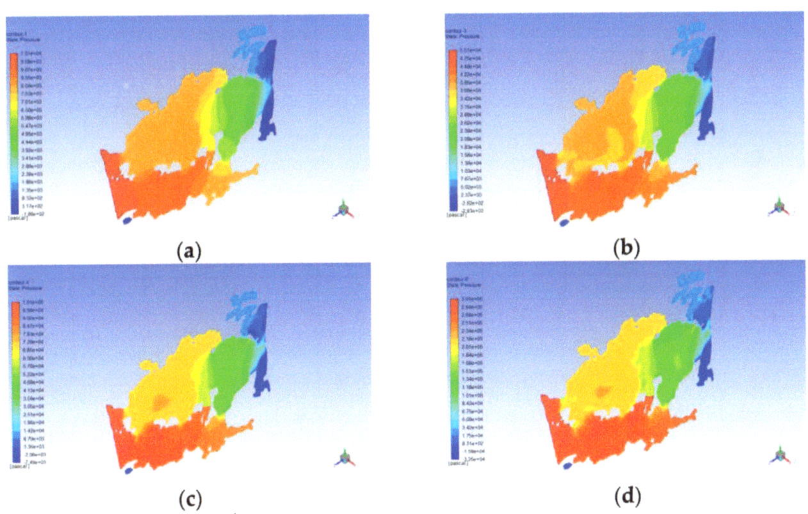

**Figure 5.** *Cont.*

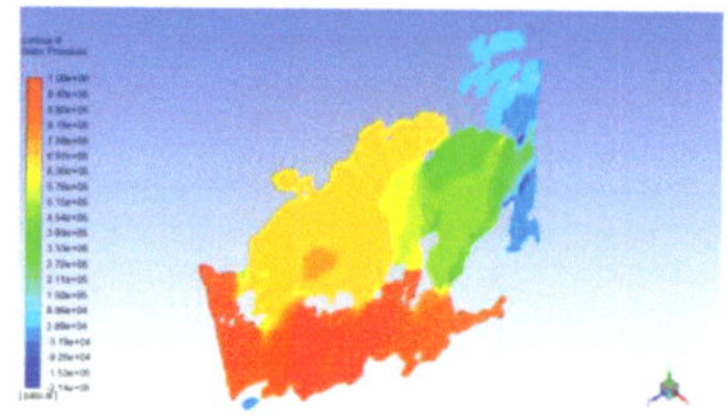

(e) (f)

**Figure 5.** z negative inlet pressure cloud map: (**a**) 0.1 Mpa pressure cloud map. (**b**) 0.5 Mpa pressure cloud map. (**c**) 1 Mpa pressure cloud map. (**d**) 3 Mpa pressure cloud map. (**e**) 5 Mpa pressure cloud map. (**f**) 10 Mpa pressure cloud map.

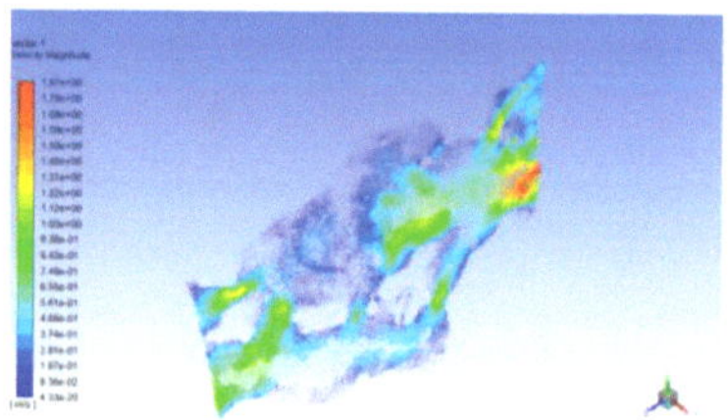

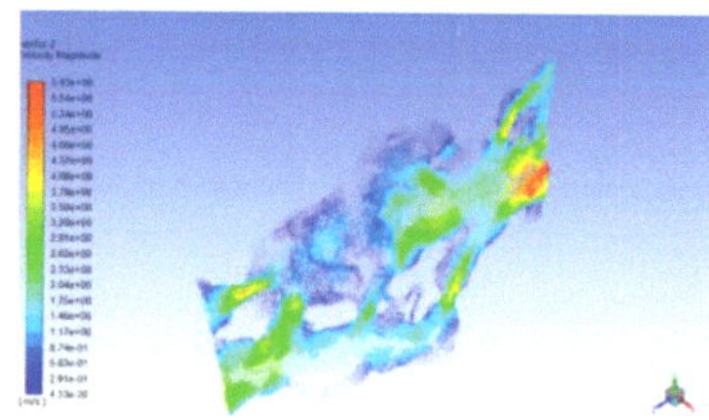

(**a**) (**b**)

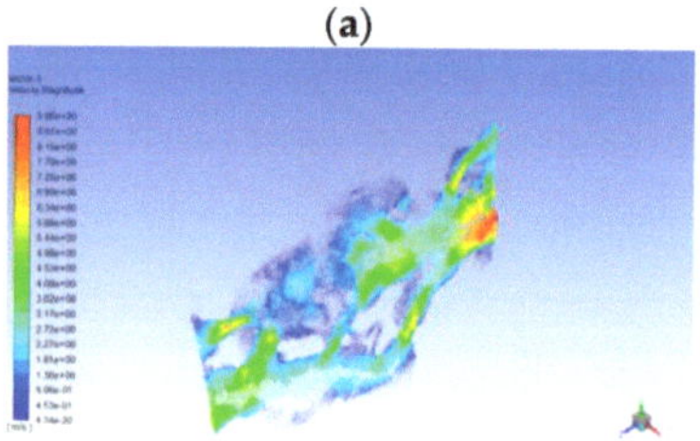

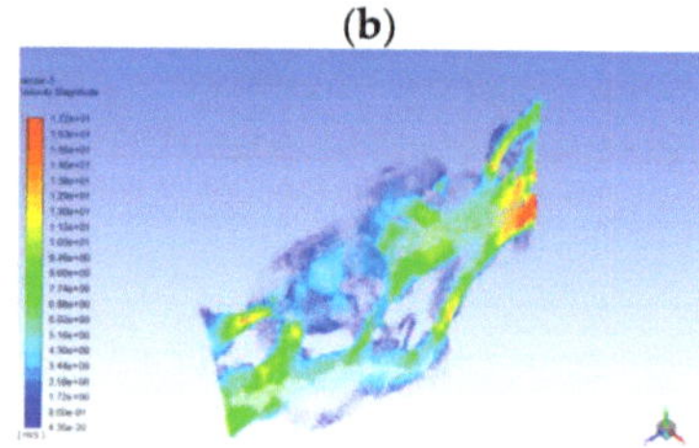

(**c**) (**d**)

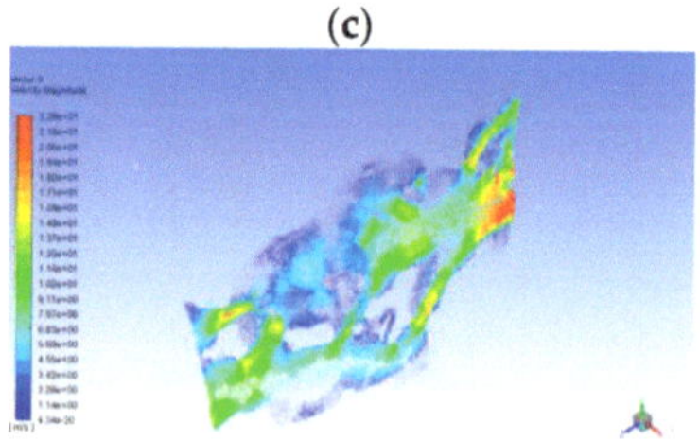

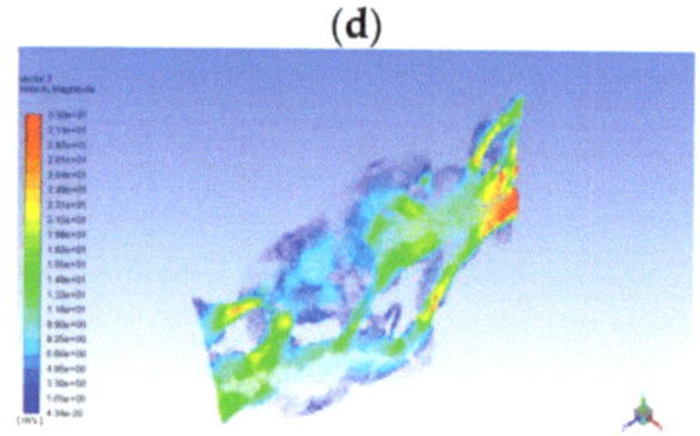

(e) (f)

**Figure 6.** z negative inlet velocity vector diagram:(**a**) 0.1 Mpa speed vector. (**b**) 0.5 Mpa speed vector. (**c**) 1 Mpa speed vector. (**d**) 3 Mpa speed vector. (**e**) 5 Mpa speed vector. (**f**) 10 Mpa speed vector.

The simulation results show that under the same pressure conditions, the seepage of water in different directions of the channel is not exactly the same, which is due to the different pore structure, permeability coefficient, and tortuosity inside the fill. For the pressure distribution, there is a gradual decrease in the direction of water seepage, with pressure changes decreasing before increasing when the pore channels become abruptly large and relatively small pressure changes in areas where the channels change relatively gently. There is no significant change in water seepage in the same direction under different pressure conditions, but as the pressure continues to increase, the rate of water seepage increases.

From Figures 3 and 5, it can be seen that the overall pressure distribution gradually decreases along the direction of water seepage while the velocity increases, which is due to the conversion of fluid pressure potential energy into kinetic energy in the back section of the model and the larger total energy loss due to wall friction and backflow factors.

From Figures 4 and 6, it can be seen that the seepage is divided into three regions: laminar, transitional, and turbulent flow. The flow velocity in the laminar region is low, the viscous force is dominant, and Darcy's law applies. The flow velocity in the transition zone increases, the viscous force becomes smaller, the inertial force increases, and the flow becomes nonlinear laminar flow, and Darcy's law does not apply. Turbulent zone flow velocity is higher, inertial forces dominate, and Darcy's law does not apply. Nonlinear seepage occurs when the seepage velocity is too large or too small. When the seepage velocity is too large, it is the inertial force that dominates and cannot be ignored, while when it is too small, the reason is that the adsorption and hydration membranes affect the linear relationship of seepage, which have different mechanisms and different descriptions, but both cause nonlinear seepage. In multi-phase seepage, the fluid interaction in the porous medium complicates the seepage conditions.

It can be seen from Figures 4–7 that under different pressure conditions, the seepage of water in the same direction does not change significantly, and with the gradual increase in pressure, the seepage speed of water also increases; When the initial pressure increases from 1 MPa to 3 MPa, the maximum velocity increases from 9.7 m/s to 19.8 m/s along the z-positive direction and from 9.1 m/s to 17.2 m/s along the z-negative direction; when the pressure increases from 3 MPa to 5 MPa, the maximum velocity increases from 19.8 m/s to 26.2 m/s in the z-positive direction and from 17.2 m/s to 22.7 m/s. It can be seen that as the initial pressure increases, the water seepage rate eventually plateaus. From Figures 4 and 6, it is obtained that the water seepage in different directions is different under the same pressure conditions, but the overall velocity tends to increase, then decrease, and then increase. This is due to the different tortuosity and permeability coefficients of the microscopic pore structure itself.

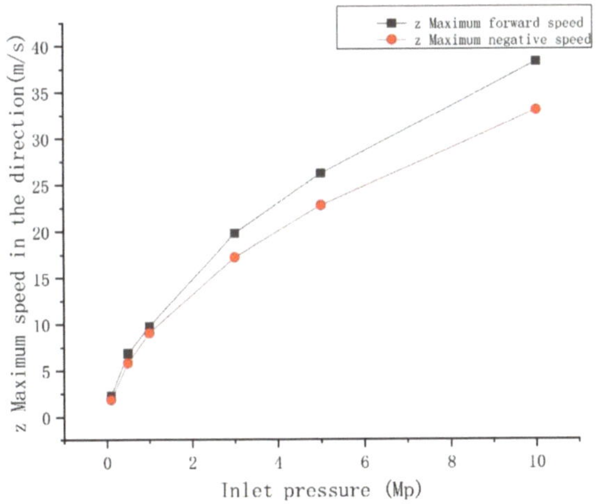


**Figure 7.** Velocity change diagram under different inlet pressure conditions.

In order to further study the characteristics of water seepage under different fissure model channels and analyze the seepage law, another fissure was selected for simulation, and the pressure cloud and velocity vector diagrams of the model at an initial pressure of 3 MPa were obtained, as shown in Figure 8.

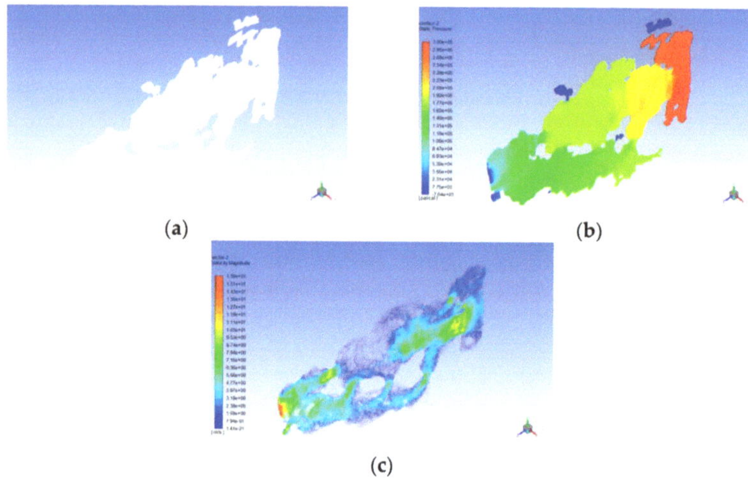

Figure 8. The 3 Mp pressure cloud and speed vector illustration: (a) fracture mesh model. (b) pressure cloud map. (c) speed vector illustration.

From Figures 3, 4 and 8, it can be seen that under the same pressure conditions, the seepage of different fracture models is approximately the same, and for the pressure distribution, different fracture models are gradually decreasing along the direction of water seepage. For the velocity distribution, the seepage flow as a whole tends to increase, then decrease, then increase again. When the pore space suddenly increases, the seepage velocity will decrease; when the pore channel suddenly decreases, the seepage velocity will increase.

In order to visually observe the matrix loss and skeletal seepage routes within the pore space, the DPM modeling method in FLUENT software was used to set the particles to be injected from the entrance surface and set the boundary to reflective particle type, and then the particles were tracked to show the seepage trajectory of the particles within the pore space in a more intuitive way, and the changes in the flow field could be seen through the particle trajectory in the post-processing. This is shown in Figure 9.

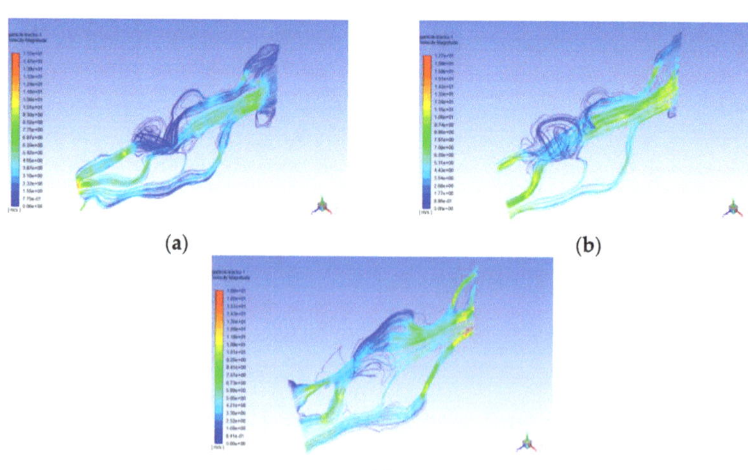

Figure 9. The 3 Mp pressure cloud and speed vector illustration: (a) fracture grid motion trajectory. (b) z forward motion trajectory diagram. (c) z negative motion trajectory diagram.

From Figure 9, it can be found that the particle motion trajectory is in the same direction as the water seepage. From Figure 9a, it can be seen that when the pore channel suddenly increases, the particle motion will generate vortices, and when the pore channel suddenly decreases, the motion velocity will increase. Although there are slight differences in the trajectories of the particles, the overall seepage is more or less the same, with the seepage velocity first increasing, then decreasing, and then increasing again.

## 4. Discussion

In the change of fault geological structure, the size, shape, and distribution of the fault-fill skeletal pore fractures differ greatly from those of sedimentary and magmatic rocks due to stress and metamorphism and are related to their mineral composition and degree of metamorphism, which inevitably has an impact on the general evaluation of fault-fill skeletal pore fractures. Compared with mudstone, kinetic metamorphic schist has more stable structural characteristics under water-rock action and contains more fractures along the grain direction, making it easy to obtain fracture models. Therefore, in the simulation experiments, CT scans and 3D model reconstructions were carried out using a sample of dynamically metamorphosed schist in the skeleton of the fill. The numerical simulation of seepage flow is used to analyze the evolutionary patterns affecting the seepage characteristics of the fractures from different directional channels and under different initial pressure conditions.

The seepage characteristics of the fracture skeleton can be controlled or influenced by a variety of factors. The tortuosity of the fractures can lead to differences in flow rates in the seepage field, which inevitably affect the evolution of the seepage characteristics. Differences in the pore structure within the fill are also inherent conditions affecting the evolution of the seepage characteristics. Typically, seepage characterization studies rarely involve aspects of differently oriented channels, so this simulated experimental study presents the seepage characteristics of channels in opposite directions in the same fissure.

As shown in Figures 3–7, the pressure and velocity distributions for the z-positive and z-negative inlet models show that the channel water seepage characteristics are similar in different directions. The overall pressure distribution pattern is similar in both positive and negative directions, with the pressure gradually decreasing along the direction of water seepage and the pressure change showing a trend of first decreasing and then increasing when the fissure channel suddenly becomes larger. The overall trend of velocity change in the positive and negative directions is to decrease and then increase. There is no significant change in the seepage of water in the same direction under different pressure conditions. As the pressure continues to increase, the seepage rate of water continues to increase but eventually tends to a steady state. As shown in Figures 3, 4 and 8, the seepage is approximately the same for the different fracture models under the same pressure conditions.

## 5. Conclusions

(1) A 3D skeleton model containing the fracture structure was constructed based on CT images of the coal mine fault fill. On this basis, the fracture structure data of the fault fill were obtained, and the seepage simulation of the fracture structure with control variables was combined with FLUENT. The seepage characteristics of the fill were analyzed, and data on the pressure and flow velocity distribution during seepage were obtained for the fracture structure model of the fill.

(2) In the microscopic case of filling fractures, seepage under different pressure conditions is approximately the same, and the rate increases continuously with increasing pressure. Under the same pressure conditions, the water seepage of fracture channels in different directions is not exactly the same, which is caused by the different microstructure of pores. For the pressure distribution, gradually decreases along the direction of water seepage, and for the speed distribution, it shows the law of changing from large to small and then increasing.

(3) By tracing the particle trajectory, it is found that vortices occur when the fissure channel suddenly increases, and the velocity of motion increases when the pore channel suddenly becomes smaller. Although there are subtle differences in particle trajectories, the overall seepage pattern is roughly the same, with the seepage velocity first increasing, then decreasing, and then increasing again.

(4) The seepage of the rock fracture structure in the fault fill is consistent with the characteristics of groundwater seepage, which is characterized by a decrease in pressure and an increase in flow rate when the fracture opening decreases due to the conversion of fluid pressure potential energy into kinetic energy in the back section of the model, and a larger total energy loss due to wall friction and backflow.

**Author Contributions:** S.W.: software, investigation, data curation, writing review. W.S.: Conceptualization, data curation, methodology, writing—original draft. F.D.: supervision, software, funding acquisition, validation. Y.X.: writing-review and editing, investigation. All authors have read and agreed to the published version of the manuscript.

**Funding:** This study was financially supported by the National Natural Science Foundation of China (No.51974172, 52274131), Shandong Higher Education Youth Innovation and Technology Program (2020KJH001), State Key Laboratory of Coal Mining and Clean Utilization (2021-CMCU-KF017).

**Institutional Review Board Statement:** Not applicable.

**Informed Consent Statement:** Not applicable.

**Data Availability Statement:** Not applicable.

**Acknowledgments:** Special thanks to the Key Laboratory of Intelligent Control and Green Mining of Mining Formations, Shandong University of Science and Technology, for providing the support of cross-scale core scanning experimental equipment.

**Conflicts of Interest:** The authors declare no conflict of interest.

## References

1. Zhao, J.; Bo, L.; Chen, J.; Ning, J. Mechanism of Seepage-Stress Fault Water Inrush and Grouting Seal. *Arab. J. Geosci.* **2020**, *13*, 4259–4279. [CrossRef]
2. Qian, Z.; Huang, Z.; Song, J. A case study of water inrush incident through fault zone in China and the corresponding treatment measures. *Arab. J. Geosci.* **2018**, *11*, 381. [CrossRef]
3. Zheng, Z.; Yu, F.; Hao, W.; Jia, R.; Yu, L. The Non-Darcy Characteristics of Fault Water Inrush in Karst Tunnel Based on Flow State Conversion Theory. *Therm. Sci.* **2021**, *25*, 4415–4421.
4. Wang, P.; Xu, J.; Li, C. Similar Simulation Test Study on Permeability Evolution Mechanism of Fault Sliding Fracture Zone. *Arab. J. Geosci.* **2022**, *15*, 548. [CrossRef]
5. Shao, J.; Zhou, F.; Sun, W. Evolution Model of Seepage Characteristics in the Process of Water Inrush in Faults. *Geofluids* **2019**, *2019*, 4926768. [CrossRef]
6. Zhu, Z.; Niu, Z.; Que, X.; Liu, C.; He, Y.; Xie, X. Study on Permeability Characteristics of Rocks with Filling Fractures Under Coupled Stress and Seepage Fields. *Water* **2020**, *12*, 2782. [CrossRef]
7. Shao, J.; Zhang, Q.; Wu, X.; Lei, Y.; Wu, X.; Wang, Z. Investigation on the Water Flow Evolution in a Filled Fracture under Seepage-Induced Erosion. *Water* **2020**, *12*, 3188. [CrossRef]
8. Xue, S.; Yuan, L.; Wang, Y.; Xie, J. Numerical Analyses of the Major Parameters Affecting the Initiation of Outbursts of Coal and Gas. *Rock Mech. Rock Eng.* **2014**, *47*, 1505–1510. [CrossRef]
9. Hao, Z.; Sun, G.; Zhang, G. Mechanism and Inducing Factors of Rockburst Events of Roadways Under Ultrathick Strata. *Front. Earth. Sc-Switz.* **2022**, *10*, 860929. [CrossRef]
10. Zhang, Q.; Jiang, Q.; Zhang, X.; Wang, D. Model test on development characteristics and displacement variation of water and mud inrush on tunnel in fault fracture zone. *Nat. Hazards.* **2019**, *99*, 467–492. [CrossRef]
11. Li, S.; Bu, L.; Shi, S.; Li, L.; Zhou, Z. Prediction for Water Inrush Disaster Source and CFD-Based Design of Evacuation Routes in Karst Tunnel. *Int. J. Geomech.* **2022**, *22*, 05022001. [CrossRef]
12. Salzer, M.; Prill, T.; Spettl, A. Quantitative comparison of segmentation algorithms for FIB-SEM images of porous media. *J. Microsc. Oxford* **2014**, *257*, 23–30. [CrossRef] [PubMed]
13. Wu, Y.; Li, Y.Z.; Qiao, W.G.; Fan, Z.W.; Zhang, S.; Chen, K.; Zhang, L. Water Seepage in Rocks at Micro-Scale. *Water* **2022**, *14*, 2827. [CrossRef]

14. Ju, Y.; Xi, C.; Zheng, J.; Gong, W.; Wu, J.; Wang, S.; Mao, L. Study on three-dimensional immiscible water–Oil two-phase displacement and trapping in deformed pore structures subjected to varying geostress via in situ computed tomography scanning and additively printed models. *Int. J. Eng. Sci.* **2022**, *171*, 103615. [CrossRef]
15. Wang, M.; Yang, S.; Li, J.; Zheng, Z.; Wen, J.; Ma, Q.; Wang, Q.; Chen, H. Cold water-flooding in a heterogeneous high-pour-point oil reservoir using computerized tomography scanning: Characteristics of flow channel and trapped oil distribution. *J. Pet. Sci. Eng.* **2021**, *202*, 108594. [CrossRef]
16. Dong, C.; Nemkumar, B.; Wang, Q.; Sun, W. Investigation on porosity of partly carbonated paste specimens blended with fly ash through dual CT scans. *Constr. Build. Mater.* **2019**, *196*, 692–702.
17. Yang, Y.; Yang, H.; Tao, L.; Yao, J.; Wang, W.; Zhang, K. Microscopic Determination of Remaining Oil Distribution in Sandstones with Different Permeability Scales Using Computed Tomography Scanning. *J. Energy Resour. Technol.* **2019**, *141*, 092903. [CrossRef]
18. Zhou, G.; Zhang, Q.; Bai, R.; Ni, G. Characterization of Coal Micro-Pore Structure and Simulation on the Seepage Rules of Low-Pressure Water Based on CT Scanning Data. *Minerals* **2016**, *6*, 78. [CrossRef]
19. Choi, C.; Lee, Y.; Song, J. Equivalent Pore Channel Model for Fluid Flow in Rock Based on Microscale X-ray CT Imaging. *Materials* **2020**, *13*, 2619. [CrossRef]
20. Qiu, L.; Zhou, G.; Zhang, W.; Han, W. Simulations on the micro-seepage rules of gas and water based on micro-CT/CFD and the related contrastive analysis. *Arab. J. Geosci.* **2019**, *12*, 549. [CrossRef]
21. Liu, W.; Wang, G.; Han, D.; Xu, H.; Chu, X. Accurate characterization of coal pore and fissure structure based on CT 3D reconstruction and NMR. *J. Nat. Gas. Sci. Eng.* **2021**, *96*, 104242. [CrossRef]
22. Zhao, L.; Ni, G.; Wang, Y.; Jiang, H.; Wen, Y.; Dou, H.; Jing, M. Semi-homogeneous model of coal based on 3D reconstruction of CT images and its seepage-deformation characteristics. *Energy* **2022**, *259*, 125044. [CrossRef]
23. Liu, J.; Wang, Y.; Song, R. A Pore Scale Flow Simulation of Reconstructed Model Based on the Micro Seepage Experiment. *Geofluids* **2017**, *2017*, 7459346. [CrossRef]
24. Wang, G.; Jiang, C.; Shen, J.; Han, D.; Qin, X. Deformation and water transport behaviors study of heterogenous coal using CT-based 3D simulation. *Int. J. Coal. Geol.* **2019**, *211*, 103204. [CrossRef]
25. Wang, G.; Shen, J.; Liu, S.; Jiang, C.; Qin, X. Three-dimensional modeling and analysis of macro-pore structure of coal using combined X-ray CT imaging and fractal theory. *Int. J. Rock. Mech. Min.* **2019**, *123*, 104082. [CrossRef]
26. Yao, B.; Chen, Z.; Wei, J.; Bai, T.; Liu, S. Predicting Erosion-Induced Water Inrush of Karst Collapse Pillars Using Inverse Velocity Theory. *Geofluids* **2018**, *2018*, 2090584. [CrossRef]
27. Li, L.; Hu, J.; Li, S.; Qin, C.; Liu, H.; Chen, D.; Wang, J. Development of a Novel Triaxial Rock Testing Method Based on Biaxial Test Apparatus and Its Application. *Rock. Mech. Rock. Eng.* **2021**, *54*, 1597–1607. [CrossRef]
28. Wang, Y.; Sun, S.; Yu, B. On Full-Tensor Permeabilities of Porous Media from Numerical Solutions of the Navier-Stokes Equation. *Adv. Mech. Eng.* **2013**, *5*, 137086. [CrossRef]
29. Shi, R.; Lin, J.; Yang, H. Distribution of Nanoparticles in a Turbulent Taylor–Couette Flow Considering Particle Coagulation and Breakage. *Processes* **2021**, *9*, 1789. [CrossRef]
30. Zou, L.; Jing, L.; Cvetkovic, V. Modeling of flow and mixing in 3D rough-walled rock fracture intersections. *Adv. Water Resour.* **2017**, *107*, 1–9. [CrossRef]

*water*

MDPI

*Article*

# Numerical Simulation of the Wormhole Propagation in Fractured Carbonate Rocks during Acidization Using a Thermal-Hydrologic-Mechanics-Chemical Coupled Model

**Piyang Liu [1], Chaoping Huang [2], Lijing Jia [1], Weijing Ji [1], Zhao Zhang [3,*] and Kai Zhang [1,*]**

1 School of Civil Engineering, Qingdao University of Technology, Qingdao 266520, China; piyang.liu@qut.edu.cn (P.L.); lijing.jia@hotmail.com (L.J.); jiweijing2022@163.com (W.J.)
2 School of Science, Qingdao University of Technology, Qingdao 266520, China; huang90512021@163.com
3 Research Centre for Mathematics and Interdisciplinary Sciences, Shandong University, Qingdao 266237, China
* Correspondences: zhaozhang@sdu.edu.cn (Z.Z.); zhangkai@qut.edu.cn (K.Z.)

**Abstract:** Acidizing is a widely adopted approach for stimulating carbonate reservoirs. The two-scale continuum (TSC) model is the most widely used model for simulating the reactive process in a carbonate reservoir during acidizing. In realistic cases, there are overburden pressure and pore pressure at present. When the injected acid reacts with the rock, the dissolution of the rock and the consumption of the acid in the pore will break the mechanical balance of the rock. Many experimental studies show that cores after acidizing have lower strength. However, it is still not clear how the deformation of rocks by the change of ground stress influences the acidizing dynamics. For fractured carbonate reservoirs, fractures play a leading role in the flow of injected acid, which preferentially flows into the fractures and dissolves the fracture walls. The effect of the combined action of rock mechanical balance broken and fracture wall dissolution on the formation of wormholes in fractured carbonate reservoirs remains to be studied. To address the above-mentioned issues, a thermal-hydrologic-mechanical-chemical coupled model is presented based on the TSC model for studying the wormhole propagation in fractured carbonate reservoirs under practical conditions. Linear and radial flow cases are simulated to investigate the influences of fracture distribution, reaction temperature, and effective stress on acidizing dynamics. The simulation results show that more wormhole branches are formed by acidizing if the fractures are perpendicular to the flow direction of acid. Temperature is a key parameter affecting the acidification dissolution patterns, so the influence of temperature cannot be ignored during the acidification design. As the effective stress of the formation increases, the diameter of the wormhole gradually decreases, and the branching decreases. More acid is needed for the same stimulation result under higher effective stress.

**Keywords:** acidization; effective stress; dissolution patterns; fracture; carbonate

**Citation:** Liu, P.; Huang, C.; Jia, L.; Ji, W.; Zhang, Z.; Zhang, K. Numerical Simulation of the Wormhole Propagation in Fractured Carbonate Rocks during Acidization Using a Thermal-Hydrologic-Mechanics-Chemical Coupled Model. *Water* **2022**, *14*, 4117. https://doi.org/10.3390/w14244117

Academic Editors: Jiutan Liu and Zongjun Gao

Received: 27 October 2022
Accepted: 13 December 2022
Published: 16 December 2022



## 1. Introduction

Acidizing is one of the most common enhanced oil recovery (EOR) methods for oil and gas reservoirs. In matrix acidizing [1], the wormhole [2] penetrates the contaminated area near the wellbore, connecting the formation and the wellbore to increase production. There are typically five types of dissolution patterns for acidizing carbonate cores according to laboratory studies [3–7]. They are named face dissolution, conical dissolution, wormhole, ramified wormhole, and uniform dissolution. The patterns of dissolution depend on injection rates. Face dissolution takes place at a low injection rate, but uniform dissolution is formed when acid is injected at a high rate. Ramified wormholes and conical wormholes can be observed at other injection rates.

To understand the mechanism of wormhole propagation in carbonate reservoirs, extensive research has been conducted [3,4,8–21]. The two-scale continuum (TSC) model developed by Panga et al. (2005) [22] is the most commonly used one due to its computational efficiency. In the past decade, the TSC model has been continuously extended to better simulate wormhole propagation. Up to now, the TSC model can examine the influence of the reservoir conditions (temperature [23–25], pressure [25], domain geometry [26], rock properties (permeability [18], heterogeneity [18,27–29], anisotropy [27], minerals proportions [30], fracture geometry [31] and acid system characteristics (reaction kinetics and mass transfer [22], reaction rate and viscosity [32] on the wormhole formation. In addition, carbonate reservoirs usually have natural fractures. The injected acid penetrates the formation guided by the fractures, increasing the permeability of the formation. The productivity enhancement effect of acidization on fractured carbonate reservoirs is usually greater than that on nonfractured reservoirs [33–35]. Natural fractures have considerable effects on the acid distribution and wormhole propagation. Hence, it is important to study the acidizing process of naturally fractured carbonate reservoirs.

The models used to simulate the reactive flow process in fractured porous media can be divided into four categories. (1) the single fracture model [36–43]; (2) the fracture network model [44,45]; (3) the pseudo-fracture model [26,46,47] and (4) the discrete fracture model (DFM) [33,34,48–52]. The single fracture model only studies the reaction and flow in a single fracture. The fracture network model assumes that the acid flows only along the fractures while ignoring the fractures that are not directly connected to the wellbore. The DFM method treats the fractures explicitly and can characterize the effect of each fracture on fluid flow and solute transport. Due to the complexity of fracture geometry, an unstructured mesh is needed. However, when the distance or the angle between the adjacent fractures is very small, tiny mesh elements are required to discretize the area between fractures, leading to computational difficulties. The pseudo-fracture model treats fractures as high-porosity areas, which is simple in concept and can effectively model fractures in the dissolution process. Therefore, the pseudo-fracture model is adopted in the current study.

In realistic formation, the rock bears the internal fluid pressure in the pore and the overburdened rock pressure. When the injected acid reacts with the rock, the rock is constantly dissolved and the consumption of the acid in the pore will break the mechanical balance of the rock, resulting in the change of effective stress. When the effective stress changes, the pore structure [53] of the reservoir is deformed, which leads to the change of geophysical properties, hence the reservoir is stress sensitivity [54,55]. The stress sensitivity of fractured carbonate rocks is greater than that of non-fractured carbonate rocks and results mainly from that of fractures in the reservoir. Therefore, stress plays an important role in the acidification process of fractured carbonate rocks. It is necessary to study the formation mechanism of wormholes with formation pressure.

This paper establishes a Thermal-Hydrologic-mechanics-Chemical (THmC) coupled model to study the dissolution dynamics of acidizing fractured carbonate rocks. We use the capital letters T, H, and C to represent the full coupling of the thermal-hydrologic-chemical process, and lowercase (m) to indicate simplified mechanics. The model is focused on the Thermal-Hydrologic-Chemical processes and their coupling, with additional mechanical processes considered within a simplified geomechanically model. The main aim of this paper is to establish a comprehensive and practical simulation method for simulating the acidizing process in fractured carbonate reservoirs. The model takes into account the effect of reaction kinetics, rock heterogeneity, stress, and temperature effects on wormhole formation during acidization. Using realistic reservoir conditions, the dissolution process can be investigated by our model accurately.

This paper is organized as follows. In Section 2, the mathematical model is presented to describe the reaction involving the chemical system, fluid flow, mass and energy transport, rock dissolution, etc. In Section 3, the dimensionless mathematical model is derived. In Sections 4 and 5, the numerical method is presented. In Section 6, a sensitivity analysis of fracture parameters, reaction exotherm, and effective stress is conducted. In the last section, the conclusions of the paper are summarized.

## 2. Mathematical Model

The assumptions regarding the thermal-hydrologic-mechanics-chemical coupled model in carbonate rocks are as follows. First, both fluid and solid materials are assumed to be incompressible, while the rock itself can still deform under stress due to a modification of the pore space. Second, the thermal equilibrium between fluid and rock can be achieved instantaneously. Third, $CO_2$ generated from the acid reaction is dissolved in the liquid phase.

### 2.1. Chemical System

The main mineral component of the carbonate rock is $CaCO_3$ and $CaMg(CO_3)_2$. Therefore, hydrochloric acid (HCl) [56] is appropriate for stimulating the formation. Hydrochloric acid can be completely ionized into hydrogen ions $H^+$ and chloride ions ($Cl^-$) in aqueous solution, and the reaction between the hydrochloric acid and calcite, occurring at the fluid-solid surface can be expressed as:

$$2H^{+}+CO_3^{2-} \rightarrow H_2O+CO_2 \tag{1}$$

### 2.2. Darcy Scale Model

The governing equations of Darcy flow in porous media are as follows:

$$\mathrm{v} = -\frac{\mathrm{K}}{\mu} \cdot \nabla P \tag{2}$$

$$\frac{\partial \phi}{\partial t} + \nabla \cdot \mathrm{v} = 0 \tag{3}$$

where v denotes the velocity, $t$ is the time, K is the permeability tensor of the formation, $P$ is the pressure of formation, $\phi$ is the formation porosity and $\mu$ is the fluid viscosity. The material balance equation for the solute transport in the liquid phase is

$$\frac{\partial(\phi C_f)}{\partial t} + \nabla \cdot (\mathrm{v}C_f) = \nabla \cdot (\phi \mathrm{D}_e \cdot \nabla C_f) - k_c a_{\mathrm{v}}(C_f - C_s) \tag{4}$$

where $C_f$ denotes the concentration of acid in the liquid phase, $C_s$ is the concentration of acid at the fluid–solid surface, $\mathrm{D}_e$ is the dispersion tensor, $k_c$ is the mass transfer coefficient and $a_v$ is the specific surface area. In general, a chemical reaction occurs on the liquid-solid surface and the amount of acid liquid per unit area transmitted to the liquid-solid surface is the same as that consumed by the surface reaction, expressed as:

$$k_c(C_f - C_s) = R(C_s) \tag{5}$$

where $R(C_s)$ denotes the reaction power, $R(C_s)$ is equal to $k_s \cdot C_s$ when the chemical reaction rate is proportional to the concentration of the reactants. $k_s$ is the surface reaction rate constant that is dependent on temperature, calculated by the Arrhenius equation as:

$$k_s = A \cdot \exp(-\frac{E_a}{RT}) \tag{6}$$

where $A$ denotes the pre-exponential factor, $E_a$ is the activation energy and the value of both of them can be determined experimentally [57]; $R$ is the universal gas constant; $T$ is the temperature (in Kelvins). Let $k_s^0$ indicate the rate at temperature $T_0$. The rate at other temperatures can be calculated by the following formula

$$k_s = k_s^0 \cdot \exp\left(-\frac{E_a}{R}\left(\frac{1}{T} - \frac{1}{T^0}\right)\right) \tag{7}$$

Porosity is changed by chemical reactions over time as:

$$\frac{\partial(\phi)}{\partial t} = \frac{R(C_s)a_v\alpha}{\rho} \tag{8}$$

where $\rho_s$ denotes the density of solid and $\alpha$ is the power of the acid dissolving.

### 2.3. Pore Scale Model

Coefficient $k_c$ indicates the mass transfer rate from the liquid phase to the liquid-solid interface. Its value is related to the velocity and physical properties of the fluid along with the pore structure. The dimensionless form of the mass transfer coefficient is named the Sherwood number $Sh$ and is written as:

$$Sh = \frac{2k_c r_p}{D_m} = Sh_\infty + 0.7\,\mathrm{Re}_p^{1/2}\,Sc^{1/3} \tag{9}$$

where $r_p$ is the pore radius; $Sh_\infty$ is the asymptotic Sherwood number; The value of $Sh_\infty$ is taken as 3 in this article. $\mathrm{Re}_p$ is the pore Reynolds number expressed as $\mathrm{Re}_p = 2r_p u/\nu$, where $\nu$ denotes the kinematic viscosity; $S_c$ is the Schmidt number defined as $S_c = \nu/D_m$ and $D_m$ is the molecular diffusivity.

The fractal dimension for pore spaces $D_f$ and tortuosity $D_T$ can be expressed respectively as:

$$D_f = d_E - \frac{\ln\phi}{\ln\frac{\lambda_{\min}}{\lambda_{\max}}} \tag{10}$$

$$D_T = \frac{\ln\left(1 - \frac{\phi}{2} + \frac{\sqrt{1-\phi}}{4} + \frac{(\phi+1+\sqrt{1-\phi}).\sqrt{9-5\phi-8\sqrt{1-\phi}}}{8\phi}\right)}{\ln\left(\frac{D_f-1}{\sqrt{D_f}}\frac{\lambda_{\max}}{\lambda_{\min}}\sqrt{\frac{1-\phi}{\phi}\frac{\pi}{8-4D_f}}\right)} \tag{11}$$

where $d_E$ is the Euclidean dimension. $\lambda_{\min}$ and $\lambda_{\max}$ are the smallest and largest pore diameter, respectively. Sensitivity analysis shows that the value of $\lambda_{\min}$ and $\lambda_{\max}$ has little influence on the fractal dimension [28]. In this paper, the ratio of $\lambda_{\max}/\lambda_{\min}$ is set to be a typical value 0.01 following in Liu et al., 2017 [28]. Permeability is expressed by:

$$K = \frac{\left(\pi D_f\right)^{\frac{(1-D_T)}{2}}\left(8-4D_f\right)^{\frac{(1+D_T)}{2}} d^2}{128\left(3+D_T-D_f\right)}\left(\frac{\phi}{1-\phi}\right)^{\frac{3+D_T}{2}} \tag{12}$$

During the chemical reaction, the variation of permeability is correlated with the change in pore radius. The permeability, pore radius, and rock-acid interfacial surface area are expressed as follows:

$$\frac{K}{K_0} = \frac{\left(\pi D_f\right)^{\frac{(1-D_T)}{2}}\left(8-4D_f\right)^{\frac{(1+D_T)}{2}}\left(3+D_{T0}-D_{f0}\right)\left(\frac{\phi}{1-\phi}\right)^{\frac{3+D_T}{2}}}{\left(\pi D_{f0}\right)^{\frac{(1-D_{T0})}{2}}\left(8-4D_{f0}\right)^{\frac{(1+D_{T0})}{2}}\left(3+D_T-D_f\right)\left(\frac{\phi_0}{1-\phi_0}\right)^{\frac{3+D_{T0}}{2}}} \tag{13}$$

$$\frac{r_p}{r_0} = \sqrt{\frac{K}{K_0}\frac{\phi_0}{\phi}} \tag{14}$$

$$\frac{a_v}{a_0} = \frac{\phi}{\phi_0}\frac{r_0}{r_p} \tag{15}$$

where $K_0$ indicates the initial permeability of the matrix, the subscript 0 is the initial value of the corresponding parameter, $\beta$ is the constant that relates with the structure of the medium, $r_0$ and $a_0$ are the initial values of average pore radius and interfacial area, respectively.

The diffusion coefficient following the exponential Arrhenius relation is given by:

$$D_m = D_m^0 \exp\left(-\frac{E_D}{R}\left(\frac{1}{T} - \frac{1}{T^0}\right)\right) \tag{16}$$

where $E_D$ indicates the activation energy of diffusion, and its value can be determined experimentally [58], varying from 12.6 to 28.1 kJ/mol. $R$ is the universal gas constant.

The dispersion tensors depend on the pore structure, the pore-scale flow, and the properties of the fluid in the fluid flow. The relative ratio of convection to diffusion is expressed by the Peclet number, defined as:

$$Pe_p = \frac{|\mathrm{v}| d_h}{\phi D_m} \tag{17}$$

where $|\mathrm{v}|$ denotes the magnitude of the Darcy velocity, $d_h$ is the pore diameter defined as $d_h = 2r_p$. When pore connectivity is good, Panga [22] gave the calculation formula of diffusion coefficient as:

$$D_{eX} = \alpha_{os} D_m + D_m \lambda_X Pe_p \tag{18}$$

$$D_{eT} = \alpha_{os} D_m + D_m \lambda_T Pe_p \tag{19}$$

where the subscripts $X$ and $T$ indicate acid injection direction and vertical direction respectively. The $X$ denotes the $x$ direction, and $T$ is the $y$ and $z$ direction in the 3D orthogonal coordinate system. However, in 3D radial coordinate system, the $X$ is the $r$ direction and $T$ is the $\theta$ and $z$ direction; $\alpha_{os}$, $\lambda_X$ and $\lambda_T$ are the constants related to the pore structure. In this paper, we set typical values for them, which are 0.5, 0.5, and 0.1 respectively.

### 2.4. Energy Transport

First, the chemical reaction between HCl and carbonate rocks releases heat, resulting in a change in formation temperature. Second, the chemical reaction and acid transport rates are influenced by temperature. Third, there is energy transfer between the injected acid and the formation. The energy transport equation is written as:

$$\begin{aligned} &\frac{\partial}{\partial t}\left(\phi \rho_f C_f^P T + (1-\phi)\rho_s C_s^P T\right) + \rho_f C_f^P \nabla \cdot (\mathrm{v}T) \\ &= \left(\phi \Gamma_f + (1-\phi)\Gamma_s\right)\nabla \cdot (\nabla T) + 2(-\Delta H_r(T)) a_v k_e C_f \end{aligned} \tag{20}$$

where $\rho_f$ indicates the density of the liquid phase; $\rho_s$ is the density of the solid phase, $C_f^p$ is the specific heat capacity of the liquid phase; $C_s^p$ is the specific heat capacity of the solid phase. $T$ is the temperature; $\Gamma_f$ is the coefficient of heat conduction in the liquid phase, $\Gamma_s$ is the coefficient of heat conduction in the solid phase; $\Delta H_r(T)$ is the heat released by reaction at temperature $T$, defined as the heat generated per mole of $H^+$ consumed and expressed as:

$$\Delta H_r(T) = \sum |\Delta H_{resultants}(T)| - \sum |\Delta H_{reactants}(T)| \tag{21}$$

Equation (21) indicates the enthalpy difference between resultants and reactants. Kirchhoff's law determines the value of enthalpy at different temperatures. At temperature $T$, the exothermic heat of reaction Equation (1) derived as follows:

$$\Delta H_r(T) = -6846 + 8.038T - 3.22 \times 10^{-3} T^2 - 870.3 T^{-1} \tag{22}$$

## 3. Dimensionless Model

The governing equations are rewritten into their dimensionless forms using the following parameters:

$$x_D = \frac{x}{L}, y_D = \frac{y}{L}, z_D = \frac{z}{L}, r_D = \frac{r}{L}, r_{wD} = \frac{r_w}{L}, \mathrm{U} = \frac{\mathrm{v}}{v_0}, t_D = \frac{t}{L/v_0}, r_{pD} = \frac{r_p}{r_0}, a_{vD} = \frac{a_v}{a_0},$$
$$T_D = \frac{T}{T^o}, \kappa = \frac{\mathrm{K}}{k_0}, D_{mD} = \frac{D_m}{D_m^0}, C_D = \frac{\mathrm{C}}{C_0}, C_{fD}^P = \frac{C_f^P}{C_s^P}, P_D = \frac{P - P_e}{(\mu v_0 L)/K_0}, h_T^2 = \frac{2K_s(T^0) r_0}{D_m^0},$$
$$D_a = \frac{K_s(T^0) a_{v0} L}{u_0}, N_{ac} = \frac{\alpha C_0}{\rho_s}, \rho_{fD} = \frac{\rho_f}{\rho_s}, A_E = \frac{E_a}{RT^0}, D_{mD} = \frac{D_m}{D_m^0}, a_{vD} = \frac{a_v}{a_{v0}},$$
$$Pe_L = \frac{u_0 L}{D_m^0}, \eta = \frac{2r_0}{L}, H_T^2 = \frac{K_s(T^0) a_{v0} L^2}{D_m^0}, Pe_H = \frac{L u_0 \rho_s C_s^p}{f\Gamma_{ef} + (1-f)\Gamma_{es}}, H_D = \frac{-\Delta H_r c_0}{T_0 \rho_s C_s^p}.$$

where subscript $D$ indicates that the variable is dimensionless; $L$ is the characteristic length set to be the core length for linear flow and the core radius for radial flow; U is the dimensionless velocity; $\kappa$ is the dimensionless permeability. $\eta$ is the ratio of the pore-to-domain scale. $A_E$ is the dimensionless activation energy; $H_D$ is the dimensionless exothermic reaction; $H_T^2$ and $h_T^2$ are Thiele modulus, which is the reaction velocity divided by the diffusion velocity at the initial pore volume. $Da$ is the Damkohler number defined as the reaction rate divided by the convection velocity; $P_{eL}$ and $P_{eH}$ are the Peclet number defined on a core scale and heat transport scale, respectively. However,$P_{eL}$ is the convective transport rate divided by the diffusive transport rate. $P_{eH}$ is the ratio of heat transfer by convection to that by thermal conduction. The acid capacity constant $N_{ac}$ is defined as the volume of dissolved solids per unit volume of consumed acid. Therefore, the governing equations can be rewritten as:

$$\mathrm{U} = -\kappa \cdot \nabla P_D \tag{23}$$

$$\frac{\partial \phi}{\partial t_D} + \nabla \cdot (-\kappa \cdot \nabla P_D) = 0 \tag{24}$$

$$\frac{\partial(\phi C_D)}{\partial t_D} + \nabla \cdot (\mathrm{U} C_D) = \nabla \cdot (\mathrm{D}_{eD} \nabla C_D) - \frac{Sh \cdot D_{mD} \cdot Da \cdot a_{vD} \cdot C_D \cdot \exp(A_E(1 - 1/T_D))}{Sh \cdot D_{mD} + h_T^2 r_{pD} \cdot \exp(A_E(1 - 1/T_D))} \tag{25}$$

$$\frac{\partial \phi}{\partial t_D} = \frac{Sh \cdot Da \cdot a_{vD} \cdot C_D \cdot D_{mD} \cdot N_{ac} \cdot \exp(A_E(1 - 1/T_D))}{Sh \cdot D_{mD} + h_T^2 r_{pD} \cdot \exp(A_E(1 - 1/T_D))} \tag{26}$$

$$\begin{aligned} &\frac{\partial T_D}{\partial t_D}\left(\phi \rho_{fD} C_{fD}^P + (1 - \phi)\right) + \rho_{fD} C_{fD}^P \nabla \cdot (\mathrm{U} T_D) \\ &= \frac{1}{Pe_H} \nabla \cdot (\nabla T_D) + \frac{H_D \cdot Sh \cdot D_{mD} \cdot Da \cdot a_{vD} \cdot C_D \cdot \exp\left(A_E\left(1 - \frac{1}{T_D}\right)\right)}{Sh \cdot D_{mD} + h_T^2 r_{pD} \cdot \exp\left(A_E\left(1 - \frac{1}{T_D}\right)\right)} \end{aligned} \tag{27}$$

$$D_{eD}^X = \frac{\phi \alpha_{os} Da D_{mD}}{H_T^2} + \lambda_X |\mathrm{U}| r_{pD} \cdot \eta \tag{28}$$

$$D_{eD}^Y = \frac{\phi \alpha_{os} Da D_{mD}}{H_T^2} + \lambda_Y |\mathrm{U}| r_{pD} \cdot \eta \tag{29}$$

The dimensionless form of the initial and boundary conditions can be written as:

$$-\kappa \cdot \nabla P_D \cdot \mathrm{n} = 1, \text{ on } \partial\Omega_{in} \tag{30}$$

$$C_D - \left(\frac{\phi \alpha_{os} Da D_{mD}}{H_T^2} + \lambda_X |U| r_{pD} \cdot \eta\right) \nabla C_D \cdot \mathrm{n} = 1, \text{ on } \partial\Omega_{in} \tag{31}$$

$$T_D = T_{fD}, \text{ on } \partial\Omega_{in} \tag{32}$$

$$P_D = 0, \text{ on } \partial\Omega_{out} \tag{33}$$

$$\nabla c_D \cdot \text{n} = 0, \text{ on } \partial\Omega_{out}, \text{ and } \partial\Omega_{tran} \tag{34}$$

$$\nabla T_D \cdot \text{n} = 0, \text{ on } \partial\Omega_{out}, \text{ and } \partial\Omega_{tran} \tag{35}$$

$$\nabla P_D \cdot \text{n} = 0, \text{ on } \partial\Omega_{tran} \tag{36}$$

$$c_D = 0, T_D = T_{sD}, \text{ at } ,t = 0 \tag{37}$$

## 4. Numerical Method

The governing equations are solved sequentially and described as follows. First, the permeability, mass transfer coefficient, and diffusion coefficient are calculated based on initial conditions. Then, they are brought into Equation (24) to solve the pressure distribution at $t^{n+1}$. Next, the pressure distribution is brought in Equation (23) to obtain the Darcy velocity. The reactive transport procedure is divided into two parts, i.e., a physical transport process and a process involving chemical reactions. In the physical transport process, the distribution of temperature and acid concentration is calculated by physical transport, followed by their values modified by reactions, as shown below:

$$\begin{cases} \dfrac{\partial T_D}{\partial t_D}\left(\phi\rho_{fD}C_{fD}^P + (1-\phi)\right) + \rho_{fD}C_{fD}^P\nabla\cdot(\text{U}T_D) = \dfrac{1}{Pe_H}\nabla\cdot(\nabla T_D) \\ \dfrac{\partial(\phi C_D)}{\partial t_D} + \nabla\cdot(\text{U}C_D) = \nabla\cdot(\text{D}_{eD}\cdot\nabla C_D) \\ \dfrac{\partial\phi}{\partial t_D} = 0 \end{cases} \tag{38}$$

$$\begin{cases} \dfrac{\partial}{\partial t_D}\left[\left(\phi\rho_{fD}C_{fD}^P + (1-\phi)\right)T_D - \dfrac{H_D\phi}{N_{ac}}\right] = 0 \\ \dfrac{\partial}{\partial t_D}\left(\phi C_D + \dfrac{\phi}{N_{ac}}\right) = 0 \\ \dfrac{\partial\phi}{\partial t_D} = N_{ac}\cdot\dfrac{Sh\cdot Da\cdot a_{vD}\cdot C_D\cdot D_{mD}\cdot\exp(A_E(1-1/T_D))}{Sh\cdot D_{mD} + h_T^2 r_{pD}\cdot\exp(A_E(1-1/T_D))} \end{cases} \tag{39}$$

The finite volume method is employed to discretize these governing equations. Orthogonal and radial grids are used for the linear and radial cases, respectively. The diffusion terms are discretized by the central difference scheme and the TVD scheme is used to discretize the convection terms. The backward Euler scheme is used for time integration to help the convergence of solutions at larger time steps. For the linear case, the simulation is performed on a 5 × 2 cm rectangle core. This rectangular domain is discretized with a 500 × 200 orthogonal grid. For the radial case, the simulation is performed on an annular domain, which is discretized with a 160 × 640 radial grid system (160 cells in $r$ direction and 640 cells in $\theta$ direction). The fractures are introduced by the following procedure:

(1) Generate randomly distributed discrete fractures.
(2) Calculate the starting point and end point coordinates of each fracture.
(3) Loop all faces of the grid cell to check whether the fracture segment intersects the face segment.
(4) If the fracture intersects the face segment of the grid cell, set the two cells sharing the face as the fracture cell, and change its porosity to 0.99.

The porosity value 0.99 used to characterize fractures is obtained by comparing the Darcy formula with the laminar flow formula, through which the cubic law can be derived as:

$$K = \frac{b^2}{12} \tag{40}$$

In the current work, we set the fracture aperture $b$ = 1 mm, which means that the permeability of fracture is $8 \times 10^7$ md. For a porous medium with an average porosity of

0.35 and an average permeability of 100 md, it can be calculated with Equation (13) that the porosity of fracture is 0.99.

## 5. Effect of Stress

Prior to the development of oil and gas reservoirs, the pressure of overburden formation is balanced by the pressure of pore fluid and the support of the rock skeleton. With the dissolution of rocks, the porosity increases consequentially. If the influence of stress is ignored, the variation of porosity can be obtained through the procedure described in Section 4. However, stress exists in the actual formation. The influence of stress should be considered on the basis of the original porosity increment induced by rock dissolution. The influence of stress on porosity increment can be expressed as follows:

$$\Delta\phi = \Delta\phi_0 e^{-a_0\sigma} \tag{41}$$

where $\Delta\phi$ indicates the increment of porosity under the current effective formation stress; $\Delta\phi_0$ is the porosity increment induced by rock dissolution; $a_0$ is the pore compression coefficient of rock at the original pressure and its value is 0.08; $\sigma$ is the effective stress of rock and its value can be calculated by

$$\sigma = P_0 - P \tag{42}$$

where $P_0$ is the pressure of the primordial formation; $P$ is the current pore pressure of the rock.

## 6. Results and Discussions

This section presents simulation results for the sensitivity analysis of fracture parameters, temperature, and stress. The values of the parameters are as in Table 1 unless otherwise stated.

**Table 1.** List of values of parameters used in the simulation.

| Parameters | Values |
|---|---|
| $L$ | 5 cm |
| $H$ | 2 cm |
| $r_e$ | 10 cm |
| $r_w$ | 0.5 cm |
| $\phi_0$ | 0.35 |
| $K_0$ | 100 md |
| $r_0$ | $5 \times 10^{-6}$ |
| $k_s$ | 0.2 |
| $v$ | 0.0286 |
| $Sh_\infty$ | 3 |
| $Sc$ | 1000 |
| $N_{ac}$ | 0.1 |
| $\eta$ | $1 \times 10^{-6}$ |
| $v_0$ | 0.004 cm/s |
| $\alpha_{os}$ | 0.5 |
| $\lambda_T$ | 0.1 |
| $\lambda_X$ | 0.5 |
| $m$ | 1 |
| $\beta$ | 1 |
| $h_T^2$ | 0.07 |

### 6.1. *Effect of Fracture Parameters*

#### 6.1.1. Effect of Fracture Orientation

The effects of fracture orientation on the dissolution structure are studied from both radial and linear aspects (Figure 1), while keeping the other parameters fixed. It is evident in Figure 1 that the wormhole is formed along the fractures which become part of the wormhole. However, the direction of the wormhole is not along fractures if the fractures are perpendicular to the injection direction (Figure 1 row(c), column (A)). The calculated $PV_{BT}$ is plotted against fracture inclinations (Figure 2). It can be seen from Figure 2 that as the fracture inclination increases, the breakthrough pore volume increases, but then decreases. When the dip angle between the fracture and the flow direction is small, the fracture becomes part of the wormhole, and the breakthrough can be achieved by dissolving a small amount of rock. When fractures are perpendicular to the flow direction, wormholes spread along both the fracture direction and fluid flow direction. Furthermore, the presence of the fractures makes the wormholes produce more branches, which means more acid is needed to dissolve the rock. So, the breakthrough pore volume reaches its maximum when the fracture inclination is 90°. However, fracture orientation has little effect on wormhole structures in radial acidizing simulation as shown in Figure 1 column (B).

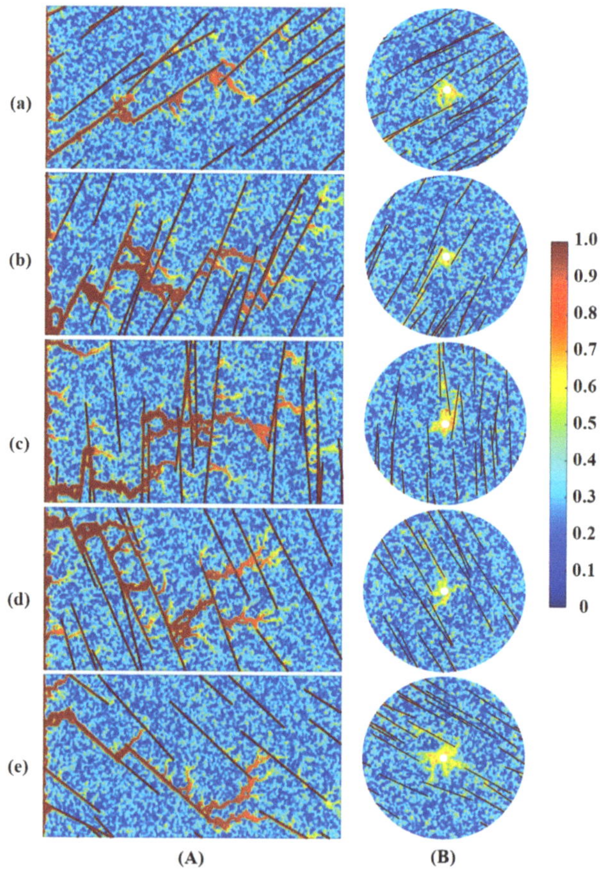


**Figure 1.** Porosity fields show the impact of fracture orientation on wormhole structures (**A**) Linear case, (**B**) Radial case and at different fracture dips: (**a**) 30°, (**b**) 60°, (**c**) 90°, (**d**) 120° (−60°), (**e**) 150° (−30°).

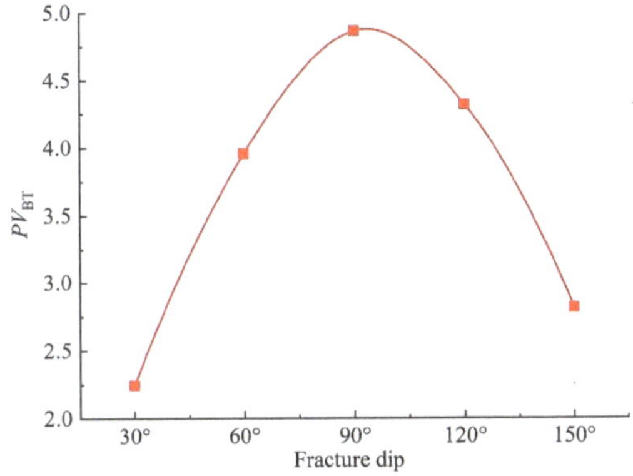


**Figure 2.** The effect of fracture dip on dimensionless breakthrough volumes corresponds to the dissolution structures in Figure 1 column (A).

### 6.1.2. Effect of Fracture Length

The distribution of natural fractures is an important factor for acidization Figures 3 and 4 show the influence of fracture length on dissolution structure. $L_L$ and $L_R$ are the average fracture lengths of linear and radial acidizing simulations, respectively. The values are placed under the picture. Related parameters are shown in Table 1. In Figures 3A and 4A, the wormholes are more irregular, wider, and more branched. Obviously, as the fracture length is increased, wormholes become thinner with fewer branches (Figures 3 and 4). When the fractures are longer, they can be interconnected to allow acid advances faster through fractures and prevent the formation of ramified wormholes. The relationship is shown between $PV_{BT}$ against fracture length in Figure 5. The $PV_{BT}$ for the longest fractures is the smallest, demonstrating that longer fractures can significantly improve the efficiency of wormhole propagation.

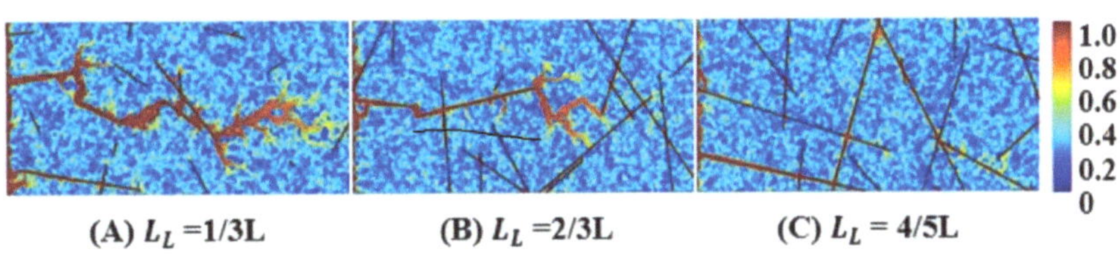


**Figure 3.** Effect of fracture length on wormhole structures in linear acidizing simulation.

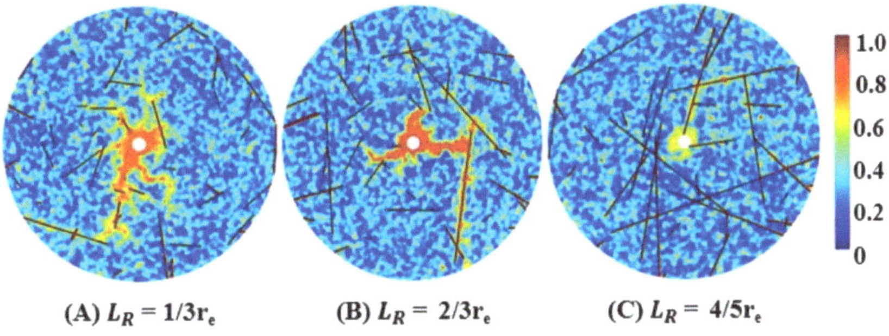


**Figure 4.** Effect of fracture length on wormhole structures in radial acidizing stimulation.

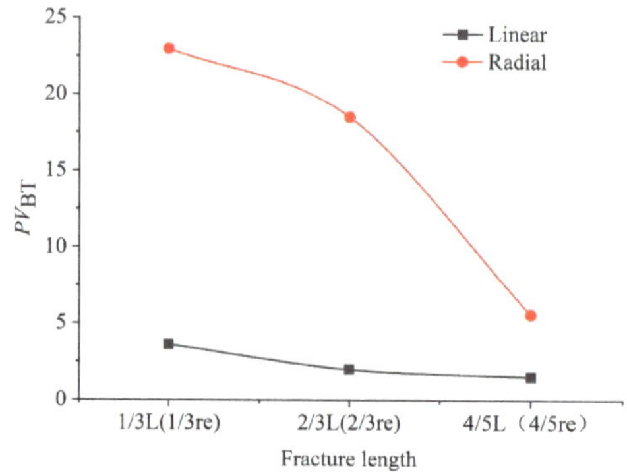


**Figure 5.** The dimensionless breakthrough volumes correspond to the dissolution structures in Figures 3 and 4.

### 6.1.3. Effect of Fracture Density

The density of natural fractures has a significant influence on wormhole propagation. The reactive flow simulation of cores with a different number of fractures was carried out for both the linear and radial cases (Figures 6 and 7). It is obvious that when the fracture density is small, wormholes are created with more branches. As the increase of the number of fractures, competition appears between fractures. When the number of fractures is large, the injected acid mainly flows along the fractures. This is attributable to the fact that the permeability of the fracture is much larger than that of the matrix. The injected acid preferentially flows along the path with the least resistance during wormhole propagation, resulting in a smaller breakthrough volume of injected acid and much higher efficiency in wormhole formation.

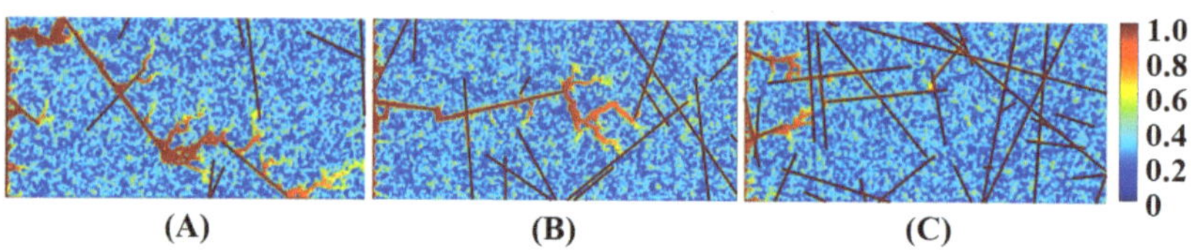


**Figure 6.** Influence of fracture density on wormhole structures, the number of fractures: (**A**) 12; (**B**) 17; (**C**) 30.

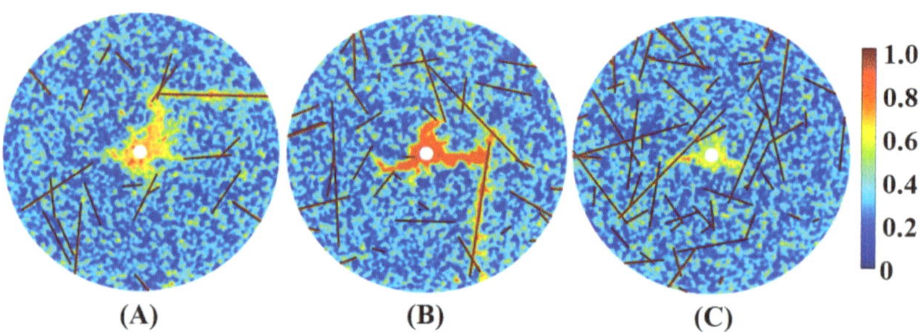


**Figure 7.** Influence of fracture density on wormhole structures, the number of fractures: (**A**) 20; (**B**) 35; (**C**) 50.

## *6.2. Effect of Reaction Temperature*

In this section, the acid is injected into the rock radially or linearly at the same formation temperature (375 K). The effect of temperature on the dissolution pattern is investigated.

The acid reaction with carbonate minerals releases heat and increases the temperature of the formation, which in its turn accelerates the reaction. To investigate the effect of this heating on the dissolution pattern, in this section, the formation temperature is set to a constant value of 375 K. The dissolution structures are shown in Figures 8 and 9. It is obvious that different wormhole patterns are generated at the breakthrough stage under the influence of temperature. This is because acid reaction becomes faster at a higher temperature. The diffusivity and reactivity also increase with higher temperatures. Hence, different dissolution modes are produced at the same injection rate.

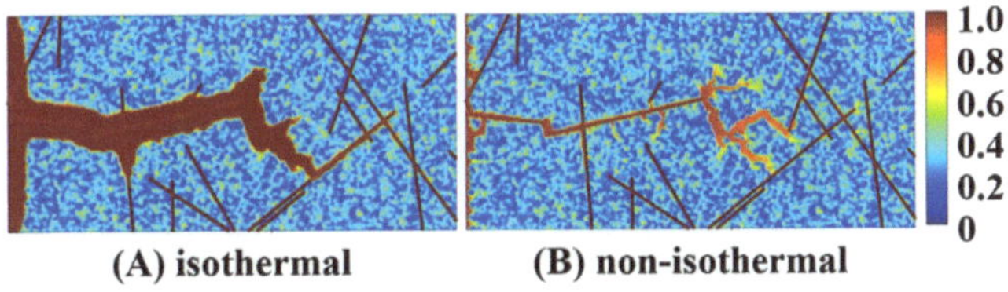


**Figure 8.** Comparison of dissolution structures for non-isothermal and isothermal cases in linear acidizing simulation.

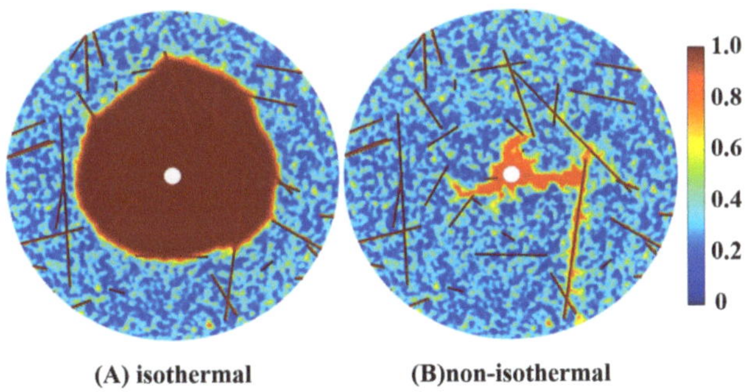


**Figure 9.** Comparison of dissolution structures for non-isothermal and isothermal cases in radial acidizing stimulation.

### 6.3. *Effect of Stress Sensitivity*

Stress sensitivity is a common characteristic in oil and gas reservoirs. Many theoretical, numerical, and experimental studies have been conducted to study the effect of stress sensitivity [59–62]. The distribution, variation, and stress sensitivity of fractures along with the hydraulic properties of the matrix are regarded as important parameters for thermal–hydraulic–mechanical modeling [63–65].

Although many studies have been carried out on the stress sensitivity of fractured reservoirs, they mainly focus on the relationship between reservoir permeability and stress sensitivity but neglect the effect on the dissolution patterns. The stress sensibility of the reservoir results mainly from the presence of fractures. The influence of the effective stress on wormhole patterns is analyzed in fractured carbonate rocks from the radial and linear aspects. This is evident in Figures 10 and 11 that the range of the wormhole propagation is reduced at low injection rates. While at high injection rates, the range of the wormhole propagation is extended (Figure 10d). It is worth noting that at a low injection rate, the change of effective stress has little effect on wormhole structures in Figure 11d. The rapid injection and incomplete reaction allow some reactants to propagate further downstream, resulting in a more uniform dissolution. The effective stress has little effect on the dissolution patterns of wormhole in linear simulation (Figure 10). However, the dissolution pattern is changed by the conical dissolution of the wormhole in radial simulation (Figure 11 row (b)). The reason is that the value of the formation porosity decreases as the effective stress increases. The value of $PV_{BT}$ is plotted against $1/Da$ in

Figure 12. It could be observed in Figure 12 that $PV_{BT}$ increases with higher effective stress regardless of the injection rate.

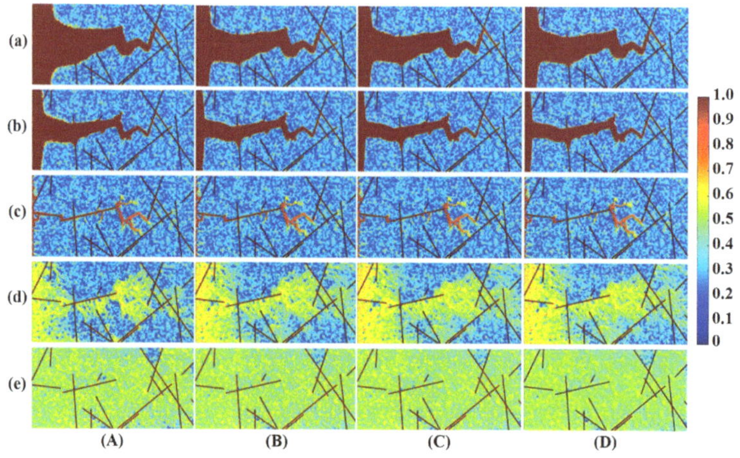

**Figure 10.** Comparison of dissolution structures at different effective stress: (**A**) $\sigma = 0$ MPa, (**B**) $\sigma = 15$ MPa, (**C**) $\sigma = 30$ MPa, (**D**) $\sigma = 45$ MPa and at different injection rates: $1/Da$ = (**a**) $10^{-6}$ (**b**) $2 \times 10^{-6}$, (**c**) $10^{-4}$, (**d**) 0.005, (**e**) 0.05.

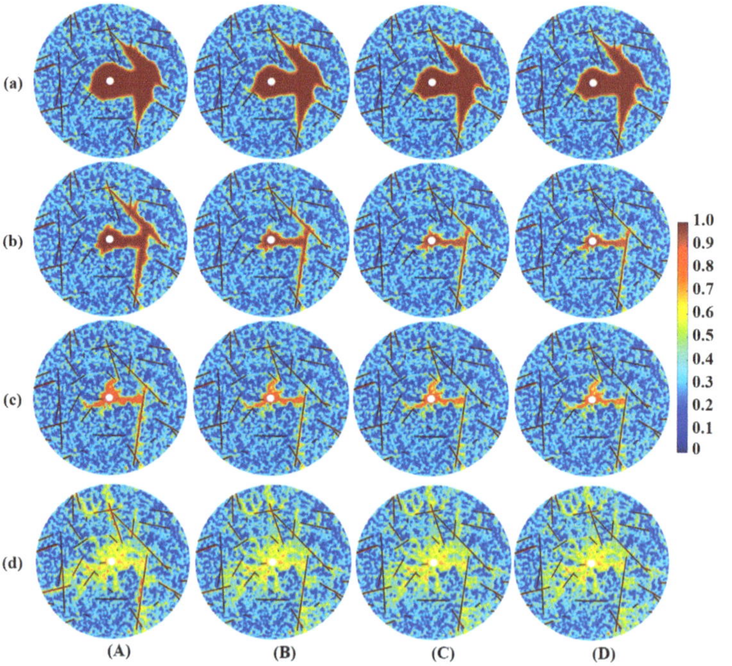

**Figure 11.** Comparison of dissolution structures at different effective stress: (**A**) $\sigma = 0$ MPa, (**B**) $\sigma = 15$ MPa, (**C**) $\sigma = 30$ MPa, (**D**) $\sigma = 45$ MPa and at different injection rates: $1/Da$ = (**a**) 0.00075, (**b**) 0.0225, (**c**) 0.075, (**d**) 0.75.

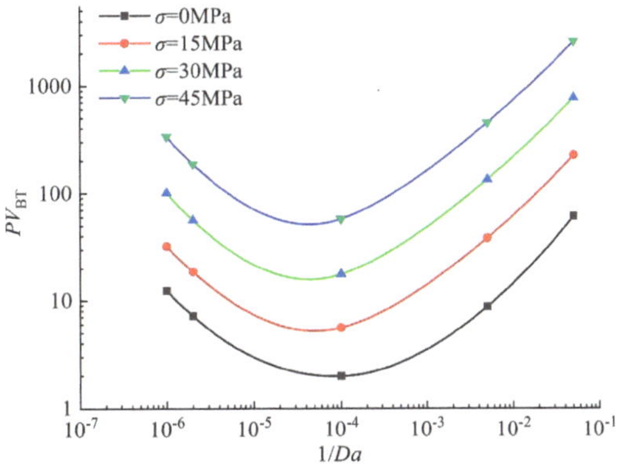


**Figure 12.** Comparison of the $PV_{BT}$ calculated under different effective stress conditions.

## 7. Conclusions

In this study, we developed a Thermal-Hydrologic-mechanics-Chemical coupled model to simulate the dissolution process during acidizing in carbonate reservoirs. The following conclusions are made after analyzing the influence of fracture parameters, temperature, and stress on the dissolution process.

1. Wormhole propagation is affected by the orientation of fractures. The wormhole grows along the fracture direction in both linear and radial cases. When acid is injected into the formation radially, no matter which direction the fracture is, the injected acid can always flow along the fracture direction. Therefore, the direction of wormhole propagation is determined by the fracture distribution of the formation.

2. With the increase in fracture length, thinner wormholes with fewer branches are obtained. If there is a long fracture in the formation, the wormhole will expand along this fracture. Therefore, to make all zones near the wellbore be stimulated sufficiently, long fractures need to be plugged before the acid is injected into the formation.

3. The presence of natural fractures enhanced the efficiency of acidizing. As increasing natural-fracture density, wormholes are formed with fewer branches. The injected acid preferentially selectively flows into fractures with the least resistance, resulting in much higher efficiency in wormhole propagation.

4. The influence of temperature on the dissolution pattern cannot be ignored, and hence should be considered in the acidizing design. With the increase in temperature, not only the acid reaction becomes faster, but also the diffusivity and reactivity also increase. Hence, different dissolution patterns are produced at the same injection rate by Comparing the non-isothermal and isothermal dissolution structures.

5. As the effective stress increases, the increment of porosity decreases, and the permeability enhancement is inhibited, resulting in the increase of $PV_{BT}$. Therefore, the mechanical compaction effects should be considered when designing the acidizing stimulation treatment.

It should be noted that in this work we are mainly concerned with coupling in Thermal-Hydrologic-Chemical(THC) processes and only local effects are considered for the mechanics, so only limited simulation result is presented. In practice, the mechanics should be coupled to THC processes fully. A fully coupled Thermal-Hydrologic-Mechanics-Chemical model for simulating the wormhole formation during acidizing will be developed in our future work.

**Author Contributions:** Z.Z. and K.Z. designed the research. P.L. performed the numerical calculations and wrote the main manuscript. C.H., L.J. and W.J. contributed to the interpretation of results and writing. All authors reviewed the manuscript. All authors have read and agreed to the published version of the manuscript.

**Funding:** The authors gratefully acknowledge the support from the National Natural Science Foundation of China (No. 51804325, No. 52274057), the Major Scientific and Technological Projects of CNPC under Grant ZD2019-183-008, the Major Scientific and Technological Projects of CNOOC under Grant CCL2022RCPS0397RSN, the Science and Technology Support Plan for Youth Innovation of University in Shandong Province under Grant 2019KJH002 , 111 Project under Grant B08028, and the China Postdoctoral Science Foundation (No. 2019M652508).

**Data Availability Statement:** The datasets generated during the current study are available from the corresponding author on reasonable request.

**Conflicts of Interest:** The authors declare no conflict of interest.

## Abbreviations

The following abbreviations are used in this manuscript:

| | |
|---|---|
| THmC | Thermal-Hydrologic-mechanics-Chemical |
| TSC | Two-Scale Continuum |
| TVD | Total Variation Diminishing |
| EOR | Enhanced Oil Recovery |

## References

1. Wei, W.; Varavei, A.; Sanaei, A.; Sepehrnoori, K. Geochemical modeling of wormhole propagation in carbonate acidizing considering mineralogy heterogeneity. *SPE J.* **2019**, *24*, 2163–2181. [CrossRef]
2. Qiu, X.; Aidagulov, G.; Ghommem, M.; Edelman, E.; Brady, D.; Abbad, M. Towards a better understanding of wormhole propagation in carbonate rocks: Linear vs. radial acid injection. *J. Pet. Sci. Eng.* **2018**, *171*, 570–583. [CrossRef]
3. Fredd, C.; Fogler, H.S. Alternative stimulation fluids and their impact on carbonate acidizing. *SPE J.* **1998**, *3*, 34–41. [CrossRef]
4. Fredd, C.N.; Fogler, H.S. Influence of transport and reaction on wormhole formation in porous media. *AIChE J.* **1998**, *44*, 1933–1949. [CrossRef]
5. Fredd, C.N.; Fogler, H. Optimum conditions for wormhole formation in carbonate porous media: Influence of transport and reaction. *Spe J.* **1999**, *4*, 196–205. [CrossRef]
6. McDuff, D.R.; Shuchart, C.E.; Jackson, S.K.; Postl, D.; Brown, J.S. Understanding wormholes in carbonates: Unprecedented experimental scale and 3-D visualization. In *Proceedings of the SPE Annual Technical Conference and Exhibition*; OnePetro: Dubai, United Arab Emirates, 2010.
7. Aljawad, M.S.; Aboluhom, H.; Schwalbert, M.P.; Al-Mubarak, A.; Alafnan, S.; Mahmoud, M. Temperature impact on linear and radial wormhole propagation in limestone, dolomite, and mixed mineralogy. *J. Nat. Gas Sci. Eng.* **2021**, *93*, 104031. [CrossRef]
8. Jeffry, S.J.M.; Trjangganung, K.; Chandrakant, A.A.; Madon, B.; Katende, A.; Ismail, I. Selection of suitable acid chemicals for matrix stimulation: A Malaysian Brown field scenario. *J. Pet. Sci. Eng.* **2020**, *186*, 106689. [CrossRef]
9. Aldhayee, K.; Ali, M.T.; Nasr-El-Din, H.A. Modeling wormhole propagation during closed-fracture-acidizing stimulation in tight-carbonate formations. *SPE J.* **2020**, *25*, 2373–2400. [CrossRef]
10. Al-Arji, H.; Al-Azman, A.; Le-Hussain, F.; Regenauer-Lieb, K. Acid stimulation in carbonates: A laboratory test of a wormhole model based on Damköhler and Péclet numbers. *J. Pet. Sci. Eng.* **2021**, *203*, 108593. [CrossRef]
11. Cheng, H.; Zhu, D.; Hill, A. The effect of evolved $CO_2$ on wormhole propagation in carbonate acidizing. *SPE Prod. Oper.* **2017**, *32*, 325–332. [CrossRef]
12. Daccord, G. Chemical dissolution of a porous medium by a reactive fluid. *Phys. Rev. Lett.* **1987**, *58*, 479. [CrossRef]
13. Daccord, G.; Lenormand, R.; Lietard, O. Chemical dissolution of a porous medium by a reactive fluid—I. Model for the "wormholing" phenomenon. *Chem. Eng. Sci.* **1993**, *48*, 169–178. [CrossRef]
14. Kelemen, P.B.; Whitehead, J.; Aharonov, E.; Jordahl, K.A. Experiments on flow focusing in soluble porous media, with applications to melt extraction from the mantle. *J. Geophys. Res. Solid Earth* **1995**, *100*, 475–496. [CrossRef]
15. Golfier, F.; Zarcone, C.; Bazin, B.; Lenormand, R.; Lasseux, D.; Quintard, M. On the ability of a Darcy-scale model to capture wormhole formation during the dissolution of a porous medium. *J. Fluid Mech.* **2002**, *457*, 213–254. [CrossRef]
16. Siddiqui, S.; Nasr-El-Din, H.A.; Khamees, A.A. Wormhole initiation and propagation of emulsified acid in carbonate cores using computerized tomography. *J. Pet. Sci. Eng.* **2006**, *54*, 93–111. [CrossRef]
17. Machado, A.; Oliveira, T.; Cruz, F.; Lopes, R.; Lima, I. X-ray microtomography of hydrochloric acid propagation in carbonate rocks. *Appl. Radiat. Isot.* **2015**, *96*, 129–134. [CrossRef]
18. Maheshwari, P.; Ratnakar, R.; Kalia, N.; Balakotaiah, V. 3-D simulation and analysis of reactive dissolution and wormhole formation in carbonate rocks. *Chem. Eng. Sci.* **2013**, *90*, 258–274. [CrossRef]
19. Frick, T.; Mostofizadeh, B.; Economides, M. Analysis of radial core experiments for hydrochloric acid interaction with limestones. In *Proceedings of the SPE Formation Damage Control Symposium*; OnePetro: Dubai, United Arab Emirates, 1994.

20. Dong, K.; Jin, X.; Zhu, D.; Hill, A. The effect of core dimensions on the optimum acid flux in carbonate acidizing. In *Proceedings of the SPE International Symposium and Exhibition on Formation Damage Control*; OnePetro: Dubai, United Arab Emirates, 2014.
21. Karale, C.; Beuterbaugh, A.; Pinto, M.; Hipparge, G.; Prakash, A. HP/HT Carbonate Acidizing—Recent Discoveries and Contradictions in Wormhole Phenomenon. In *Proceedings of the Offshore Technology Conference Asia*; OnePetro: Dubai, United Arab Emirates, 2016.
22. Panga, M.K.; Ziauddin, M.; Balakotaiah, V. Two-scale continuum model for simulation of wormholes in carbonate acidization. *AIChE J.* **2005**, *51*, 3231–3248. [CrossRef]
23. Li, Y.; Liao, Y.; Zhao, J.; Peng, Y.; Pu, X. Simulation and analysis of wormhole formation in carbonate rocks considering heat transmission process. *J. Nat. Gas Sci. Eng.* **2017**, *42*, 120–132. [CrossRef]
24. Liu, P.; Li, J.; Sun, S.; Yao, J.; Zhang, K. Numerical investigation of carbonate acidizing with gelled acid using a coupled thermal–hydrologic–chemical model. *Int. J. Therm. Sci.* **2021**, *160*, 106700. [CrossRef]
25. Xue, H.; Huang, Z.; Zhao, L.; Wang, H.; Kang, B.; Liu, P.; Liu, F.; Cheng, Y.; Xin, J. Influence of acid-rock reaction heat and heat transmission on wormholing in carbonate rock. *J. Nat. Gas Sci. Eng.* **2018**, *50*, 189–204. [CrossRef]
26. Liu, P.; Couples, G.D.; Yao, J.; Huang, Z.; Song, W.; Ma, J. A general method for simulating reactive dissolution in carbonate rocks with arbitrary geometry. *Comput. Geosci.* **2018**, *22*, 1187–1201. [CrossRef]
27. Liu, P.; Yan, X.; Yao, J.; Sun, S. Modeling and analysis of the acidizing process in carbonate rocks using a two-phase thermal-hydrologic-chemical coupled model. *Chem. Eng. Sci.* **2019**, *207*, 215–234. [CrossRef]
28. Liu, P.; Yao, J.; Couples, G.D.; Ma, J.; Iliev, O. 3-D modelling and experimental comparison of reactive flow in carbonates under radial flow conditions. *Sci. Rep.* **2017**, *7*, 17711. [CrossRef] [PubMed]
29. Jia, C.; Sepehrnoori, K.; Huang, Z.; Zhang, H.; Yao, J. Numerical studies and analysis on reactive flow in carbonate matrix acidizing. *J. Pet. Sci. Eng.* **2021**, *201*, 108487. [CrossRef]
30. Liu, P.; Xue, H.; Zhao, L.; Zhao, X.; Cui, M. Simulation of 3D multi-scale wormhole propagation in carbonates considering correlation spatial distribution of petrophysical properties. *J. Nat. Gas Sci. Eng.* **2016**, *32*, 81–94. [CrossRef]
31. Qi, N.; Chen, G.; Liang, C.; Guo, T.; Liu, G.; Zhang, K. Numerical simulation and analysis of the influence of fracture geometry on wormhole propagation in carbonate reservoirs. *Chem. Eng. Sci.* **2019**, *198*, 124–143. [CrossRef]
32. Jia, C.; Huang, Z.; Sepehrnoori, K.; Yao, J. Modification of two-scale continuum model and numerical studies for carbonate matrix acidizing. *J. Pet. Sci. Eng.* **2020**, *197*, 107972. [CrossRef]
33. Liu, P.; Yao, J.; Couples, G.D.; Huang, Z.; Sun, H.; Ma, J. Numerical modelling and analysis of reactive flow and wormhole formation in fractured carbonate rocks. *Chem. Eng. Sci.* **2017**, *172*, 143–157. [CrossRef]
34. Liu, P.; Yao, J.; Couples, G.D.; Ma, J.; Huang, Z.; Sun, H. Modeling and simulation of wormhole formation during acidization of fractured carbonate rocks. *J. Pet. Sci. Eng.* **2017**, *154*, 284–301. [CrossRef]
35. Mou, J.; Yu, X.; Wang, L.; Zhang, S.; Ma, X.; Lyu, X. Effect of Natural Fractures on Wormhole-Propagation Behavior. *SPE Prod. Oper.* **2019**, *34*, 145–158. [CrossRef]
36. Deng, H.; Molins, S.; Steefel, C.; DePaolo, D.; Voltolini, M.; Yang, L.; Ajo-Franklin, J. A 2.5 D reactive transport model for fracture alteration simulation. *Environ. Sci. Technol.* **2016**, *50*, 7564–7571. [CrossRef]
37. Hill, A.; Zhu, D.; Dong, C.; Luna-Garcia, A. Computer model predicts matrix acidizing of naturally fractured carbonate. *J. Pet. Technol.* **2001**, *53*, 20–25. [CrossRef]
38. Hanna, R.B.; Rajaram, H. Influence of aperture variability on dissolutional growth of fissures in karst formations. *Water Resour. Res.* **1998**, *34*, 2843–2853. [CrossRef]
39. Upadhyay, V.K.; Szymczak, P.; Ladd, A.J. Initial conditions or emergence: What determines dissolution patterns in rough fractures? *J. Geophys. Res. Solid Earth* **2015**, *120*, 6102–6121. [CrossRef]
40. Dong, C.; Zhu, D.; Hill, A. Modeling of the acidizing process in naturally fractured carbonates. *SPE J.* **2002**, *7*, 400–408. [CrossRef]
41. Detwiler, R.L.; Rajaram, H. Predicting dissolution patterns in variable aperture fractures: Evaluation of an enhanced depth-averaged computational model. *Water Resour. Res.* **2007**, *43*, 1–14. [CrossRef]
42. Szymczak, P.; Ladd, A. Wormhole formation in dissolving fractures. *J. Geophys. Res. Solid Earth* **2009**, *114* 1–22. [CrossRef]
43. O'Brien, G.; Bean, C.; McDermott, F. Numerical investigations of passive and reactive flow through generic single fractures with heterogeneous permeability. *Earth Planet. Sci. Lett.* **2003**, *213*, 271–284. [CrossRef]
44. Dong, C.; Zhu, D.; Hill, A. Acid penetration in natural fracture networks. *SPE Prod. Facil.* **2002**, *17*, 160–170. [CrossRef]
45. Dong, C.; Zhu, D.; Hill, A. Acidizing in Naturally Fractured Carbonate Reservoirs. In *Proceedings of the SPE/DOE Improved Oil Recovery Symposium*; OnePetro: Dubai, United Arab Emirates, 2002.
46. Li, Y.; Liao, Y.; Zhao, J. Wormhole dissolution pattern study in complicated carbonate rock based on two-scale continuum model and equivalent seepage theory. *Nat. Gas Geosci.* **2016**, *27*, 121–127.
47. Kalia, N.; Balakotaiah, V. Effect of medium heterogeneities on reactive dissolution of carbonates. *Chem. Eng. Sci.* **2009**, *64*, 376–390. [CrossRef]
48. Garipov, T.; Karimi-Fard, M.; Tchelepi, H. Discrete fracture model for coupled flow and geomechanics. *Comput. Geosci.* **2016**, *20*, 149–160. [CrossRef]
49. Sandve, T.H.; Berre, I.; Nordbotten, J.M. An efficient multi-point flux approximation method for discrete fracture–matrix simulations. *J. Comput. Phys.* **2012**, *231*, 3784–3800. [CrossRef]

50. Therrien, R.; Sudicky, E. Three-dimensional analysis of variably-saturated flow and solute transport in discretely-fractured porous media. *J. Contam. Hydrol.* **1996**, *23*, 1–44. [CrossRef]
51. Alotaibi, M.; Chen, H.; Sun, S. Generalized multiscale finite element methods for the reduced model of Darcy flow in fractured porous media. *J. Comput. Appl. Math.* **2022**, *413*, 114305. [CrossRef]
52. Milliotte, C.; Jonoud, S.; Wennberg, O.P.; Matthäi, S.K.; Jurkiw, A.; Mosser, L. Well-data-based discrete fracture and matrix modelling and flow-based upscaling of multilayer carbonate reservoir horizons. *Geol. Soc. Lond. Spec. Publ.* **2018**, *459*, 191–210. [CrossRef]
53. Yang, Y.; Tao, L.; Yang, H.; Iglauer, S.; Wang, X.; Askari, R.; Yao, J.; Zhang, K.; Zhang, L.; Sun, H. Stress sensitivity of fractured and vuggy carbonate: An X-Ray computed tomography analysis. *J. Geophys. Res. Solid Earth* **2020**, *125*, e2019JB018759. [CrossRef]
54. Bohnsack, D.; Potten, M.; Freitag, S.; Einsiedl, F.; Zosseder, K. Stress sensitivity of porosity and permeability under varying hydrostatic stress conditions for different carbonate rock types of the geothermal Malm reservoir in Southern Germany. *Geotherm. Energy* **2021**, *9*, 1–59. [CrossRef]
55. Wang, L.; Yang, S.; Meng, Z.; Chen, Y.; Qian, K.; Han, W.; Wang, D. Time-dependent shape factors for fractured reservoir simulation: Effect of stress sensitivity in matrix system. *J. Pet. Sci. Eng.* **2018**, *163*, 556–569. [CrossRef]
56. Mustafa, A.; Alzaki, T.; Aljawad, M.S.; Solling, T.; Dvorkin, J. Impact of acid wormhole on the mechanical properties of chalk, limestone, and dolomite: Experimental and modeling studies. *Energy Rep.* **2022**, *8*, 605–616. [CrossRef]
57. Kalia, N.; Glasbergen, G. Wormhole formation in carbonates under varying temperature conditions. In *Proceedings of the 8th European Formation Damage Conference*; OnePetro: Dubai, United Arab Emirates, 2009.
58. Wilke, C.; Chang, P. Correlation of diffusion coefficients in dilute solutions. *AIChE J.* **1955**, *1*, 264–270. [CrossRef]
59. Vairogs, J.; Hearn, C.; Dareing, D.W.; Rhoades, V. Effect of rock stress on gas production from low-permeability reservoirs. *J. Pet. Technol.* **1971**, *23*, 1161–1167. [CrossRef]
60. Tian, Y. Experimental study on stress sensitivity of naturally fractured reservoirs. In *Proceedings of the SPE Annual Technical Conference and Exhibition*; OnePetro: Dubai, United Arab Emirates, 2014.
61. Abass, H.; Ortiz, I.; Khan, M.; Beresky, J.; Sierra, L. Understanding stress dependant permeability of matrix, natural fractures, and hydraulic fractures in carbonate formations. In *Proceedings of the SPE Saudi Arabia Section Technical Symposium*; OnePetro: Dubai, United Arab Emirates, 2007.
62. Tan, Q.; Kang, Y.; You, L.; Xu, C.; Zhang, X.; Xie, Z. Stress-sensitivity mechanisms and its controlling factors of saline-lacustrine fractured tight carbonate reservoir. *J. Nat. Gas Sci. Eng.* **2021**, *88*, 103864. [CrossRef]
63. Brehme, M.; Blöcher, G.; Cacace, M.; Kamah, Y.; Sauter, M.; Zimmermann, G. Permeability distribution in the Lahendong geothermal field: A blind fault captured by thermal–hydraulic simulation. *Environ. Earth Sci.* **2016**, *75*, 1–11. [CrossRef]
64. Cacace, M.; Jacquey, A.B. Flexible parallel implicit modelling of coupled thermal–hydraulic–mechanical processes in fractured rocks. *Solid Earth* **2017**, *8*, 921–941. [CrossRef]
65. Konrad, F.; Savvatis, A.; Wellmann, F.; Zosseder, K. Hydraulic behavior of fault zones in pump tests of geothermal wells: A parametric analysis using numerical simulations for the Upper Jurassic aquifer of the North Alpine Foreland Basin. *Geotherm. Energy* **2019**, *7*, 1–28. [CrossRef]

*water*

MDPI

*Article*

# Hydrogeochemical Characteristics, Water Quality, and Human Health Risks of Groundwater in Wulian, North China

**Min Wang [1], Wenxiu Zhang [1], Peng Yang [2], Jianguo Feng [1], Ruilin Zhang [1], Zongjun Gao [1,*], Hongjie Jin [1], Xiaoyu Song [1] and Xiaobing Gao [1]**

[1] College of Energy and Mining Engineering, Shandong University of Science and Technology, Qingdao 266590, China

[2] Shandong Eighth Geological and Mineral Exploration Institute, Rizhao 276800, China

* Correspondence: zongjungao1964@163.com; Tel.: +86-13792900575

**Abstract:** Groundwater shortage and pollution are critical issues of global concern. In Wulian County, a typical hilly area, groundwater is the main source of water supply. This study investigates the current situation of groundwater pollution in Wulian City through the analysis of groundwater water chemistry characteristics, water quality evaluation, and health risk evaluation. After the analysis of the controlling factors of chemical components in groundwater and the analysis of ion sources, the main ion sources in groundwater were determined. The results showed that the major cations in groundwater were $Ca^{2+}$ and $Na^+$ and the major anions were $HCO_3^-$ and $SO_4^{2-}$. Nevertheless, $NO_3^-$ exceeded the standard to different degrees in pore water (PW), fissure pore water (FPW), and fissure water (FW). The minimum $NO_3^-$ concentration exceeded the standard in FW. Under the influence of rock weathering and salt rock dissolution, the main hydrochemical types of groundwater were the $HCO_3$-Ca, $HCO_3$-Ca·Mg, and $SO_4$·Cl-Ca·Mg types. According to the water quality evaluation and health risk assessment, the FW area in the south had the highest water quality, where Class I water appeared and potable water was more widely distributed. The PW and FPW areas in the north had lower water quality, with higher health risks. Category V water gradually appeared in the FPW area, which is not suitable as a water supply source. Factor analysis and ion ratio analysis showed that the study area is strongly affected by anthropogenic factors. These research methods have important reference value to the research of groundwater pollution status.

**Keywords:** groundwater hydrochemistry; water quality evaluation; health risk assessment; nitrate ion source identification; Wulian

**Citation:** Wang, M.; Zhang, W.; Yang, P.; Feng, J.; Zhang, R.; Gao, Z.; Jin, H.; Song, X.; Gao, X. Hydrogeochemical Characteristics, Water Quality, and Human Health Risks of Groundwater in Wulian, North China. *Water* **2023**, *15*, 359. https://doi.org/10.3390/w15020359

Academic Editor: Maurizio Barbieri

Received: 3 December 2022
Revised: 30 December 2022
Accepted: 31 December 2022
Published: 15 January 2023



## 1. Introduction

Suitable groundwater characteristics, such as wide distribution and good quality, are of great significance to water supply, which is crucial for economic, industrial, and agricultural development [1–3]. However, in many cities, economic development is achieved at the cost of the over-exploitation of groundwater and excessive use of chemical fertilizers and pesticides [4–6], resulting in groundwater shortage and pollution and even affecting human health [3,7–10]. In recent years, the issue of groundwater pollution has become increasingly serious with the progress of social development [11–16]. Under the influence of anthropogenic factors and climate change, many cities around the world are facing serious groundwater pollution [17,18]. Therefore, water quality and water security have become issues of global concern.

The chemical composition of groundwater is attributable to the long-term interaction between groundwater and its surrounding environment, and the type and concentration of substances in groundwater are significantly affected by geological, climatic, and human factors [19,20]. Groundwater chemical characteristics and their controlling factors can be analyzed using many widely used and effective methods, including mathematical analyses, Piper diagrams [21], correlation analyses, Gibbs diagrams [22], and ion ratios [23]. Among various common methods for water quality evaluation, the entropy-weighted water quality index (EWQI) is an objective method that converts a large number of measured water quality data into a simple water quality index and determines the index weight according to the variation degree of each index value. This method can avoid deviations caused by human factors [24,25]. The risk of groundwater to human health can be quantitatively described, and the risk level of nitrate ions to human health can be classified through health risk assessments [26,27]. Factor analysis (FA) is a suitable tool for determining key control factors through multivariate dimension reduction, analyzing the relationship between principal components, and preliminarily determining the source of nitrate ions in groundwater.

Located in the south of the Shandong Peninsula, China, Wulian is a hilly area with groundwater as the primary source of water. With the rapid agricultural development, the use of chemical fertilizers and pesticides has also increased, which has led to serious pollution of the water environment. In order to understand the current situation of groundwater pollution and the sustainable utilization of groundwater resources, it is necessary to study the chemical characteristics of the groundwater environment in the area [28,29]. Previous studies have investigated water pollution in the south of the Shandong Peninsula. For example, Song et al. [30] studied the pollution status, distribution, and source of heavy metals in groundwater in the south of the Shandong Peninsula. He et al. [31] analyzed the hydrochemical characteristics of groundwater in the southern Shandong Peninsula. However, these studies paid little attention to groundwater pollution in the Wulian area.

In this study, the quality of different types of groundwater and their risks to human health are investigated using hydrochemical analysis, the entropy weight method, health risk assessments, and other methods. FA is also conducted to identify the source of nitrate ions in groundwater. The main objectives of this study are to (1) analyze the chemical characteristics and controlling factors of different types of groundwater in the study area; (2) analyze the water quality of different types of groundwater; (3) evaluate the health risks of different types of groundwater; and (4) identify the source of nitrate ions in groundwater. This study provides a theoretical basis for the prevention and control of groundwater pollution and the establishment of a high-quality groundwater source.

## 2. Materials and Methods

### 2.1. Study Area

The study area (Figure 1) is located in Hongning Town and Gaoze Town of Wulian County in the south of the Shandong Peninsula (119°12′–119°20′ E, 35°39′–35°51′ N). The region has a warm and semi-humid continental monsoon climate with four distinct seasons. The temperature of this area has a wide annual range, with a monthly average temperature from −4.6 to 26.8 °C over many years. The annual average precipitation is 784.1 mm, decreasing from 825 mm/a in the south to 725 mm/a in the north. The annual precipitation shows obvious seasonal changes, with an average of one rainstorm a year, mostly concentrated in the period from June to September. During this period, precipitation can reach 210 mm, often causing flood events. However, the precipitation from November to April is only 10–40 mm.

Rivers in the study area are vertically and horizontally distributed, most of which belong to seasonal, intermittent rivers and are mainly fed by rainfall–runoff. The county has a large basin area comprising the Shu River system, the Wei River system, the Futuna River system, and the Chaobai River system. The rivers originate from the mountains in the central part of the county, and the tributaries radiate outward. According to the regional geological survey report of Wulian at the 1:50,000 scale, Wulian is located in the low hilly area of southeast Shandong Province on the coast of the Yellow Sea. The overall terrain is high in the middle and low in the north and south. The hilly area with low mountains covers 86% of Wulian, and the plain area is smaller, covering only 14% of the area. The topography of the area fluctuates widely, with developed gullies and foundation stones at the bottom of exposed gullies. Wulian County has a complex geological structure with folds and faults.

Metamorphic rocks are multi-stage intrusive rocks with large distribution, complex strata, large changes in lithology and lithofacies, and uneven quaternary sedimentary distribution.

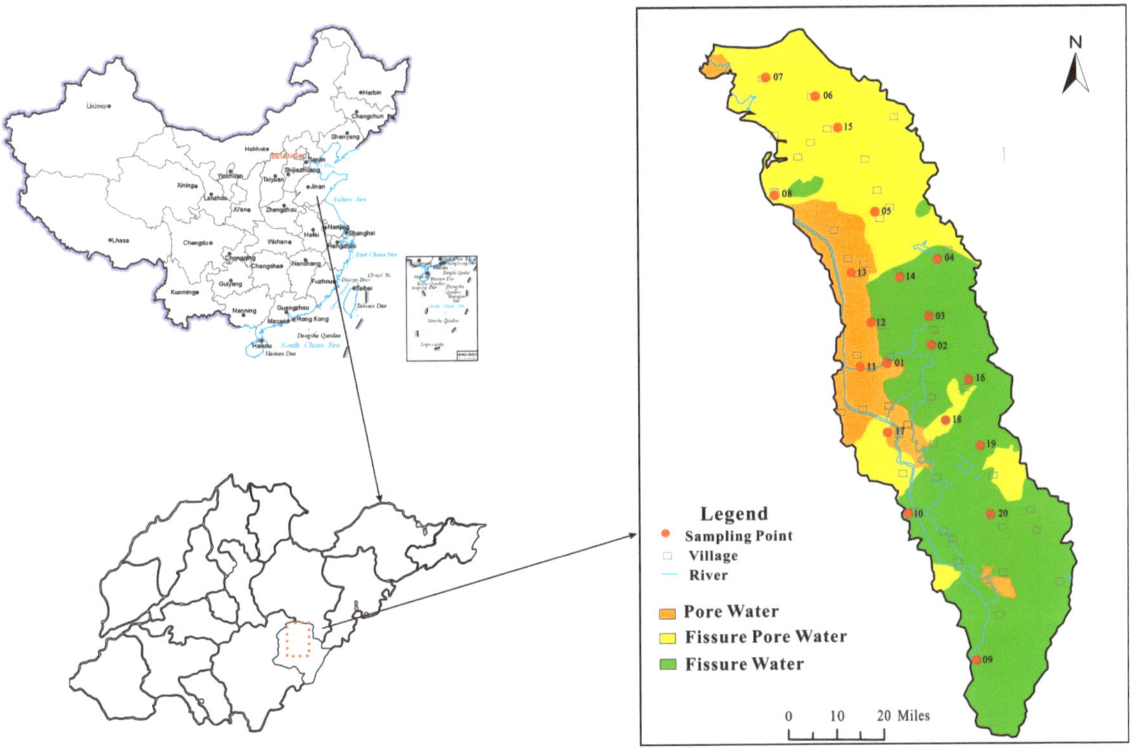


**Figure 1.** Geographical locations of the study area and sampling sites for three different types of groundwater.

Based on the type of void ratio, stratigraphic combinations, groundwater conditions, and hydrodynamic characteristics, the groundwater in Wulian County is divided into three types: pore water (PW), fissure pore water (FPW), and fissure water (FW). PW aquifers have a limited distribution area, with a maximum burial depth of 3 m, annual variation of 0.2–1.2 m, and thickness of more than 3 m. The water inflow in a single well is generally less than 1000 $m^3/d$. FPW aquifers are characterized by developed fractures and pores, and most of them are distributed in high terrain, with a thickness of more than 100 m, maximum burial depth of 3 m, and poor water yield. The water inflow in a single well is generally less than 100 $m^3/d$ or even zero. FW aquifers comprise pure marble, metamorphic rocks, and intrusive rocks, with a maximum burial depth of 4 m. The marble aquifers have relatively developed karst fissures, high water abundance, and a developed water-blocking water-storage structure, which is suitable for water supply. The metamorphic and intrusive water-bearing formations, including various stages of magmatic rocks such as diorite and granite, are dense, hard, and resistant to weathering, with undeveloped fractures. These rocks have poor water abundance, and the water inflow in a single well is less than 100 $m^3/d$. In this area, groundwater is mainly replenished by precipitation recharge, near-shore recharge, and lateral recharge from downstream to upstream.

### 2.2. Sample Collection and Measurements

In this study, 40 groundwater samples were collected from representative wells. Employing uniform sampling, groundwater samples were collected according to the

DZ/T0133-94 standard of the "Procedures for Dynamic Monitoring of Groundwater". PW is mainly distributed in the western part of the study area, and samples were collected at depths ranging from 5 to 10 m. FPW is mainly distributed in the northern part of densely populated residential areas, and the sampling points were mostly civil wells, with sampling depths ranging from 5 to 20 m. FW is distributed in the southern area, where farmlands are concentrated, and the sampling points were mostly mechanical wells, with sampling depths ranging from 50 to 100 m. According to the principle of uniform sampling, 9 PW, 9 FPW, and 22 FW samples were collected. Bottles were rinsed with the sample water 2–3 times before sampling. In this manner, the purity of the samples was ensured. The sampling bottle was then filled with flowing water to avoid the interference of external conditions such as air. For unstable components of water, a stabilizer such as $HgCl_2$ was added to the samples to maintain their relative stability during transportation. After extraction, the samples were sealed and labeled, and their details were recorded. Anti-shock and anti-freezing measures were implemented while avoiding direct sunlight. The samples were immediately sent to a specialized laboratory for composition analysis. pH and total dissolved solids (TDSs) were measured directly on-site using a portable instrument (HQ40D, HACH). Groundwater samples were analyzed in the laboratory for cations by inductively coupled plasma-optical emission spectrometry (ICP-OSE) (Optima 7000 Valencia, DV, USA) and for anions through ion chromatography (Thermo Fisher ICS600, Waltham, MA, USA) and chemical oxygen demand (COD); total hardness (TH) was determined by titration [32]. All the water samples were divided into four parts, one of which was temporarily used as a backup water sample and the other three for triplicate measurements. The mean value of the three measurements was taken as the final result. The error calculated using the charge balance method was within ±5%, and the measurement method showed a small deviation and high precision [20]. The formula for calculating the error is shown in Equation (1).

$$\mathrm{CBE} = \frac{\sum \mathrm{cation} - \sum \mathrm{anions}}{\sum \mathrm{cation} + \sum \mathrm{anions}} \times 100 \quad (1)$$

### 2.3. *Analysis Methods*

#### 2.3.1. Analysis of Hydrochemical Composition

Firstly, the basic properties of the chemical components of groundwater were analyzed using mathematical statistics [33]. Secondly, a Piper diagram [21] was used to classify the hydrochemical types of groundwater. Thirdly, water quality assessment [18,24] and health risk assessment [27] were conducted to understand the status of groundwater quality. Fourthly, correlation maps were used to analyze the correlation among chemical components, and a Gibbs diagram [22] and the ion ratio [34] were combined to analyze the controlling factors of the ion components. Finally, FA [35] was performed to identify the sources of nitrate ions in groundwater.

#### 2.3.2. Water Quality Analysis

In this study, the EWQI method was used to evaluate the general status of groundwater quality. This method can be used to generally classify groundwater quality and assess the suitability of groundwater for drinking water and irrigation water [36]. The basic principle is to integrate a large amount of water quality data and convert them into an easily understandable indicator of groundwater quality that can be qualitatively expressed [37]. To apply this method, relevant formulas were adopted [25,34], and the specific steps are as follows:

(1) Establish an initial water quality matrix [25,34].

In the initial data matrix X, n represents the number of groundwater samples, m is the number of evaluation indexes, and $x_{ij}$ represents the initial value of the $j$th evaluation index of the $i$th water sample point.

$$X = \begin{bmatrix} x_{11} & x_{12} & \cdots & x_{1m} \\ x_{21} & x_{22} & \cdots & x_{2m} \\ \vdots & \vdots & \vdots & \vdots \\ x_{n1} & x_{n2} & \cdots & x_{nm} \end{bmatrix} \quad (2)$$

(2) Data standardization

In the original water quality data, the matching of different evaluation indicators and water sample points is quite different, and large errors will be generated during the calculation. Therefore, it is necessary to standardize the original data [25,34].

$$y_{ij} = \frac{x_{ij} - (x_{ij})_{min}}{(x_{ij})_{max} - (x_{ij})_{min}} \quad (3)$$

where $(x_{ij})_{min}$ and $(x_{ij})_{max}$ represent the minimum and maximum values of the evaluation index in the initial matrix, respectively [25,34]. The normalized matrix is shown in Equation (4):

$$Y = \begin{bmatrix} y_{11} & y_{12} & \cdots & y_{1m} \\ y_{21} & y_{22} & \cdots & y_{2m} \\ \vdots & \vdots & \vdots & \vdots \\ y_{n1} & y_{n2} & \cdots & y_{nm} \end{bmatrix} \quad (4)$$

(3) Calculation and determination of entropy weight [25,34]

In this study, the entropy weight method is used to determine the weight of different indicators.

$$P_{ij} = \frac{y_{ij}}{\sum_{i=1}^{n} y_{ij}} \quad (5)$$

$$e_j = -\frac{1}{\ln n} \sum_{i=1}^{n} P_{ij} \ln P_{ij} \quad (6)$$

where in $P_{ij}$ represents the entropy value in the evaluation index; $e_j$ represents the information entropy in the evaluation index. Then, the entropy weight ($w_i$) can be calculated, as shown in Equation (7):

$$w_i = \frac{1 - e_j}{\sum_{j=1}^{m} (1 - e_j)} \quad (7)$$

(4) Calculation of quality ($q_i$) [25,34]

$$q_i = \frac{C_i}{S_j} \times 100 \quad (8)$$

$$q_{pH} = \begin{cases} \frac{C_{pH} - 7}{8.5 - 7} \times 100 \; C_{pH} > 7 \\ \frac{7 - C_{pH}}{8.5 - 7} \times 100 \; C_{pH} < 7 \end{cases} \quad (9)$$

where $C_i$ and $C_{pH}$ are the measured concentration and the pH of the $i$th water sample point, respectively; $S_j$ is the groundwater quality standard.

(5) Solution of water quality index [25,34]

$$\mathrm{WQI} = \sum_{j=1}^{n} \mathrm{w_j q_i} \quad (10)$$

(6) Water quality classification [25,38]

According to the calculation results, water quality was divided into five categories: Class I to Class V, representing groundwater of excellent (WQI < 25), good (WQI = 25–50), medium (WQI = 50–100), poor (WQI = 100–150), and very poor (WQI > 150) quality.

### 2.3.3. Health Risk Assessment

In this study, the health risk assessment model recommended by the United States Environmental Protection Agency (USEPA) was used to evaluate and analyze the hazard of nitrate ions in groundwater. The health risk assessment method is a typical "four-step method" [39], including hazard identification, dose–response relationship, exposure assessment, and risk characterization [40].

$$HQ = \frac{I_{Oral}}{D_{Rf}} \tag{11}$$

In Equation (11), $HQ$ is the non-carcinogenic risk index, $I_{Oral}$ is the dose of non-carcinogens in drinking water mg/(kg·days), $D_{Rf}$ is the reference dose of non-carcinogens in water mg/(kg·days), and the reference dose of nitrate in drinking water is 1.6 mg/(kg·days) [41]. According to the USEPA [42,43], when $HQ < 1$, the non-carcinogenic pollution in groundwater is in the acceptable range of health risk, while $HQ > 1$ is in the unacceptable range.

The dose for non-carcinogens (nitrates) entering the body via drinking water is calculated as follows [44]:

$$I_{Oral} = \frac{C \times IR \times EF \times ED}{BW \times AT} \tag{12}$$

In Equation (12), $C$ is the laboratory-measured concentration of nitrate in groundwater (mg/L). $IR$ is the daily water intake (L/d), which is taken as 0.78 and 2.5 L/d for children and adults, respectively [11,45]. $EF$ is exposure frequency (d/a), for which the standard is 365 d/a for both adults and children, and $ED$ is exposure duration (a), for which the standard is 45 and 12 a for adults and children, respectively. $BW$ is the mean body weight, for which the standard is 65 and 18.7 kg for adults and children, respectively. $AT$ is the average time, calculated as $ED \times 365$ days, for which the standard is 16,425 and 4380 d for adults and children, respectively [46].

### 2.3.4. Source Identification of Nitrate

FA is an effective method for the dimensionality reduction of variables. This method can be used to restore and analyze the multivariate data of groundwater, reflect most of the original information of many indicators, and reduce the loss of key information. FA is helpful for the extraction and classification of potential factors in the original data, such that a more intuitive understanding of the groundwater system can be achieved. In this study, 11 indicators of the three groundwater types were standardized and their dimensions were reduced. Principal component analysis was used to analyze the variance of common factors and the interpretation of total variance. Variance with a characteristic value greater than 1 was extracted. The absolute value of factor loadings of 0.30–0.50, 0.50–0.75, and $\geq$0.75, respectively, represent weak, medium, and strong grades. Factor scores were calculated by regression methods, which are convenient and easy to apply [33] and can also reflect the degree of influence. A positive factor score indicates that the area is most affected by the process represented by the factor, a negative factor score indicates that it is not affected by the representative factor, and a score close to zero indicates that the area is affected on an average scale. FA can be used to extract and analyze pollution factors in groundwater. Combined with the chemical analysis of groundwater, this method can be used to preliminarily assess the nitrate source to provide a theoretical basis for establishing a high-quality water supply area and to carry out pollution prevention and water quality treatment in later development activities and the utilization of groundwater.

## 3. Results and Discussion

### 3.1. Groundwater Chemical Characteristics

The mathematical and statistical analysis of groundwater chemical compositions of different groundwater samples in the study area was carried out, and boxplots of different chemical compositions are shown in Figure 2.

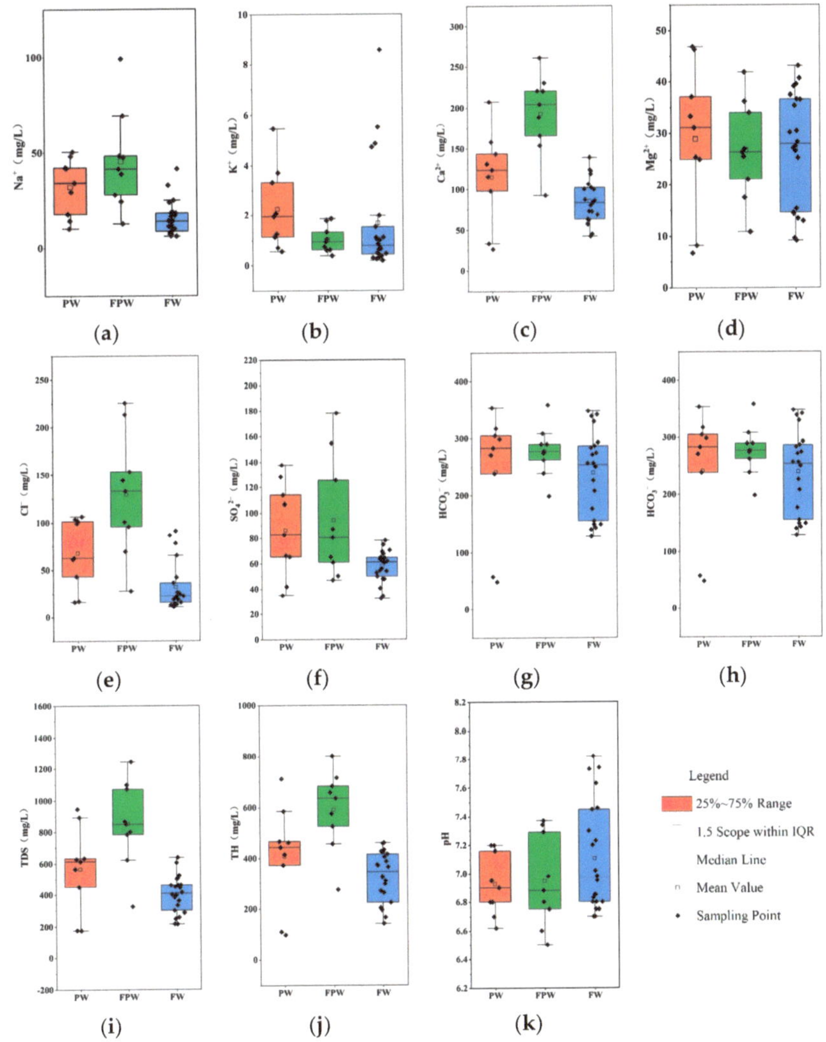


**Figure 2.** Box plots of the chemical contents of different groundwater types. (**a**) $Na^+$ box plot; (**b**) $K^+$ box plot; (**c**) $Ca^{2+}$ box plot; (**d**) $Mg^{2+}$ box plot; (**e**) $Cl^-$ box plot; (**f**) $SO_4^{2-}$ box plot; (**g**) $HCO_3^-$ box plot; (**h**) $NO_3^-$ box plot; (**i**) TDS box plot; (**j**) TH box plot; (**k**) pH box plot.

According to the box plots, the minimum pH values of PW and FPW were 6.62 and 6.5, respectively, with mean values of 6.93 and 6.95, showing an overall weak acidity. The maximum pH of FW was 7.82, with a mean value corresponding to weakly alkaline water. Total dissolved solids (TDSs) and total hardness (TH) are not only the main indexes of groundwater chemical analysis but also the main indexes of groundwater quality. TDS and TH showed wide variation among different types of groundwater in the study area. TDS showed the strongest dispersion in PW, with a maximum value of 946.88 mg/L, a minimum value of 175.47 mg/L, and a mean value of 565.27 mg/L. TDS generally ranged from 25% to 75%, but there was one concentration outlier. FPW showed the highest TDS, with a peak value of 1243.51 mg/L, an outlier value of only 325.29 mg/L, and a mean value of 851.37 mg/L. The overall TDS was at a high level, and the level of water pollution was relatively serious. TDS was relatively concentrated in FW, ranging from 215.42 to 638.55 mg/L, with a mean value of 403.08 mg/L. TH showed a similar degree of dispersion to TDS, with PW and FPW generally corresponding to brackish hard water, which is closely related to the dissolution of soluble salts and minerals. Nevertheless, FW showed better water quality, generally corresponding to soft freshwater. This situation is speculated to be more related to human activities [47].

The main cations in different types of groundwater were found to be $Ca^{2+}$ and $Na^{+}$. Groundwater with high contents of $Ca^{2+}$ and $Mg^{2+}$ is known to affect human health [48]. According to the box chart, the $Ca^{2+}$ + $Mg^{2+}$ concentrations in the three types of groundwater were 254.45, 302.95, and 182.36 mg/L, respectively. This may be related to water–rock interactions within the permissible concentration range for drinking water [38,48]. The concentrations of $HCO_3^-$ were 48.82–353.92, 198.32–358.32, and 128.14–347.81 mg/L in PW, FPW, and FW, respectively, with the highest concentration among the anions. The concentration of $SO_4^{2-}$ showed little variation among the three types of groundwater, and the maximum concentration was 178.11 mg/L in FPW, which meets the drinking water standard. The concentration of $Cl^-$ widely varied, with mean values of 67.98, 129.20, and 32 mg/L in PW, FPW, and FE, respectively, and the overall concentration was relatively low. $NO_3^-$ can be used as a reflection of the impact of human activities on groundwater, and it was detected in all water samples, with concentration ranges of 19.96–262.04, 12.5–344.79, and 0.15–188.62 mg/L in the three water samples. $NO_3^-$ was the most dispersed in FW, with three nitrate ion outliers, including two extreme outliers, indicating the poor stability and large spatial variability of $NO_3^-$ in groundwater. The maximum concentrations of $NO_3^-$ in FPW and PW were higher than that in FW. In the statistical analysis, the standard exceedance rates of FPW, PW, and FW were 88.9%, 44.4%, and 18.2%, respectively, which are all at a high level and indicate serious nitrate pollution. According to Figure 1, settlements near FW are relatively scattered, with small distribution density and many farmlands, while areas near FPW have more concentrated settlements, with shallow groundwater depth and dense population. Therefore, the $NO_3^-$ variability in groundwater was initially attributed to human production and living activities.

### 3.2. Hydrochemical Types

The Piper diagram [21] is a generally used and effective method for the classification of groundwater hydrochemical types [7]. In this study, AqQA software was used to arrange and project the water sample points to the corresponding positions of the Piper diagram according to the groundwater types. As shown in Figure 3, the three groundwater samples were all located in Zones I and II of the rhombus area and were distributed in strips, representing the $HCO_3$-Ca·Mg and $SO_4$·Cl-Ca·Mg types, respectively. The A, B, and E zones in the triangle represent the calcium type, no dominant type, and the bicarbonate type, respectively. Accordingly, the groundwater of the study area mainly corresponds to the $HCO_3$-Ca type, the $HCO_3$-Ca·Mg type, and the $SO_4$·Cl-Ca·Mg type. $Ca^{2+}$ and $Mg^{2+}$ in the water samples account for the vast majority of total ions in the groundwater. However, the percentage distribution of the anion content is relatively dispersed, and $HCO_3^-$ + $CO_3^{2-}$ range from 75% to 25%. According to the distribution areas of different types of groundwater, FW is distributed in the lower left corner of the strip area, and the distribution is concentrated, indicating that this type of groundwater occurs in an environment with strong hydrodynamic conditions. Gradually to the upper right of the strip area, FPW and PW appear in turn. In addition, the distribution of FPW is scattered, and the projection points appear in the middle and upper right of the strip, indicating that the hydrodynamic conditions of this type of groundwater are widely varied across different regions.

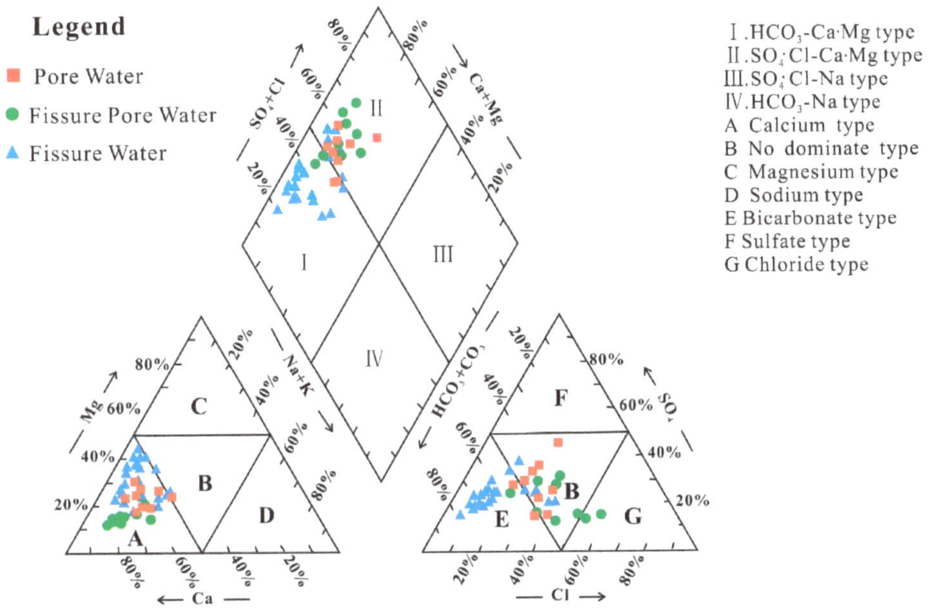


**Figure 3.** Piper diagram showing the hydrochemical characteristics of groundwater in the Wulian area.

*3.3. Groundwater Quality Assessment*

The water quality of the 40 samples of the three groundwater types in the study area was calculated and evaluated through the entropy weight formula, and their water quality conditions were found to vary widely. Among the nine samples of PW, two, three, and four samples corresponded to Classes II, III, and IV, accounting for 22%, 33%, and 45%, respectively, as shown in Figure 4a. Overall, the quality of PW was poor. According to the water quality standard (GB 5749-2006), only Class II water in the water sample can be used as central drinking water, and the rest is generally used as industrial and agricultural water (GB/T 14848-2017). Regarding the nine samples of FPW, 11%, 22%, 45%, and 22% of samples were of Classes II, III, IV, and V, respectively, as shown in Figure 4b. The Class V standard was identified only in FPW, and the concentration of nitrate ion was the highest in this class, reaching 345 mg/L, which is more than six times the WHO limit [48]. Most FPW samples were collected near villages, and the impact of domestic sewage on underground water quality was thus fully considered. Regarding the 22 samples of FW, 5%, 68%, and 27% corresponded to Classes I, II, and III, respectively, as shown in Figure 4c. Accordingly, FW exhibited the highest water quality among the three types. Class I water was identified only in FW, and the nitrate ion concentration was generally low. Considering the quality assessment results of FPW, groundwater quality is closely related to nitrate ion concentration. According to the GB 5749-2006 standard, Class I and II water can be used as drinking water. Therefore, the FW distribution area features a large amount of drinkable Class I and II water. On the whole, water quality deteriorates from the FW distribution area in the south and northeast of the study area to the PW and FPW distribution areas in the northwest. The FPW distribution area has poor water quality, mainly of Class IV.

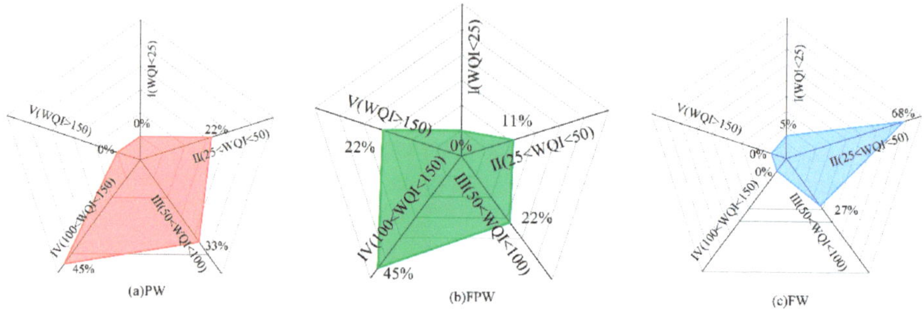


**Figure 4.** Radar maps of water quality of different groundwater types. (**a**) PW, (**b**) FPW, (**c**) FW.

### 3.4. Health Risk Assessment

Many countries pay close attention to the risks of $NO_3^-$ to the human body [48]. After $NO_3^-$ enters the human body through drinking water, nitrate ions will decompose into nitrite, resulting in methemoglobinemia [49], which poses a great risk to the human body. In this study, the health risk of nitrate ions in groundwater to adults and children was evaluated using the model recommended by USEPA. The magnitude of risk in different types of groundwater varies for adults and children.

The calculated results (Table 1) showed that the range of HQ in PW was 0.54–6.86 for adults and 0.52–6.83 for children, with mean values of 2.37 and 2.27, respectively. The proportion of water samples exceeding the risk limit of HQ (HQ = 1) was relatively high, being 44%. Regarding FPW, the range of HQ for adults and children was 0.34–9.37 and 0.33–8.99, with mean values of 5.41 and 5.19, respectively. The proportion of samples exceeding the risk limit of HQ (HQ = 1) reached 89%. This type of groundwater poses the biggest threat to human health. In contrast, FW posed the smallest risk, ranging from 0.004 to 5.31 for adults and 0.004 to 4.92 for children, with mean values of 1.19 and 1.14, respectively. The HQ value was mostly located near the risk limit value, and the proportion of water samples exceeding the HQ limit (HQ = 1) was 45% and 41%. The low risk of nitrate ions in FW is consistent with the water quality evaluation results, which are closely related to the distribution of villages and farmland near different types of groundwater.

**Table 1.** Health risk indicators of different groundwater types.

| | | HQ (Min) | HQ (Max) | HQ (Mean) | Proportion of HQ > 1 |
|---|---|---|---|---|---|
| Adult | PW | 0.54 | 6.86 | 2.37 | 44% |
| | FPW | 0.34 | 9.37 | 5.41 | 89% |
| | FW | 0.004 | 5.13 | 1.19 | 45% |
| Children | PW | 0.52 | 6.83 | 2.27 | 44% |
| | FPW | 0.33 | 8.99 | 5.19 | 89% |
| | FW | 0.004 | 4.92 | 1.14 | 41% |

According to the spatial distribution map (Figure 5) of the health risk index, high-risk areas were mainly located in the south and northeast, with higher risk for both adults and children, reaching up to 8.988 at the extreme risk level. The health risk index in the central area was relatively small, with a minimum of 0.04 for adults and children, representing a risk-free area. However, the health risk of groundwater is relatively higher in the areas with a higher density of village distribution. FPW presented the largest risk, without any risk-free area in its area of distribution, and most of the risk indices were at HQ = 3 and above. On the whole, groundwater in the study area presented a high risk. Therefore, groundwater pollution prevention and control are needed in the study area.

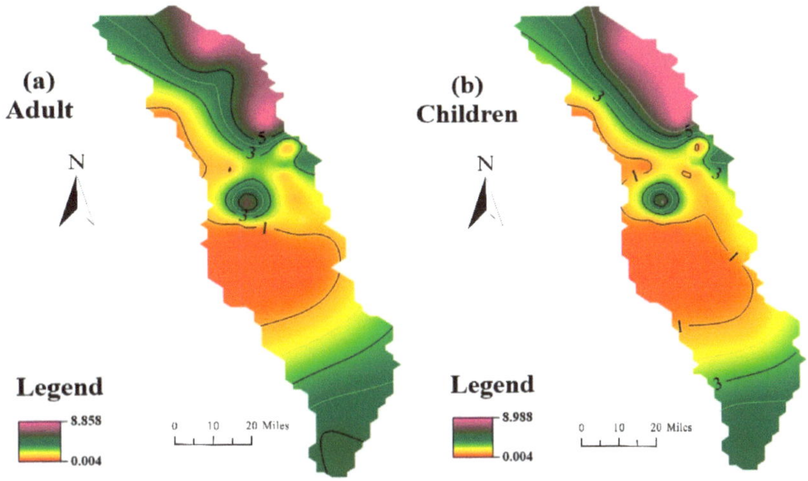


**Figure 5.** Health risk index distribution maps for adults and children.

### 3.5. Correlation Analysis

The composition and hydrochemical types of ions in groundwater are subject to many factors. Changes in solutes are caused by complex chemical reactions in groundwater, which are eventually expressed in the form of ion content and parameters. Therefore, the change of ion content in groundwater is closely related to the chemical reaction of groundwater and the mutual conversion of ions. In the correlation analysis of groundwater parameters, strong and weak correlations between parameters are compared after the standardized treatment of groundwater parameter data, and a quantitative representation is finally obtained. Through correlation analysis, the internal relationship among parameters can be analyzed to further study the causes of hydrochemistry. Correlation plots (Figure 6) can directly and clearly show the degree of correlation among various ions in groundwater. Origin software was used to analyze the hydrochemical parameters and ions in the water samples, through which the ion sources in the study area could be inferred. The correlation factors are distributed diagonally, with red and blue ellipses indicating positive and negative correlations, respectively, and slimmer ellipses and darker colors indicating stronger correlations (Figure 6).

All ions in PW were strongly correlated. Among the three basic indicators, pH was negatively correlated with $K^+$ and positively correlated with other indicators, but the correlation was weak and was speculated to be related to the dissolution of carbonate rocks. TDS was significantly correlated with other parameters except for $K^+$, which is consistent with the distribution of evaporation in the study area. TH showed a positive correlation with $Ca^{2+}$ and $Mg^{2+}$ (correlation coefficient of up to 0.9), with $Cl^-$ and $HCO_3^-$ (correlation coefficient above 0.8), and with $SO_4^{2-}$ and $NO_3^-$ (correlation coefficient around 0.7), indicating that the total hardness of the groundwater is affected by both carbonate and non-carbonate solutes. The ionic correlation was more specific for $K^+$, which showed a low level of negative correlation with all parameters, which may be attributable to the low content of $K^+$ and its susceptibility to perturbation by the groundwater dynamic field.

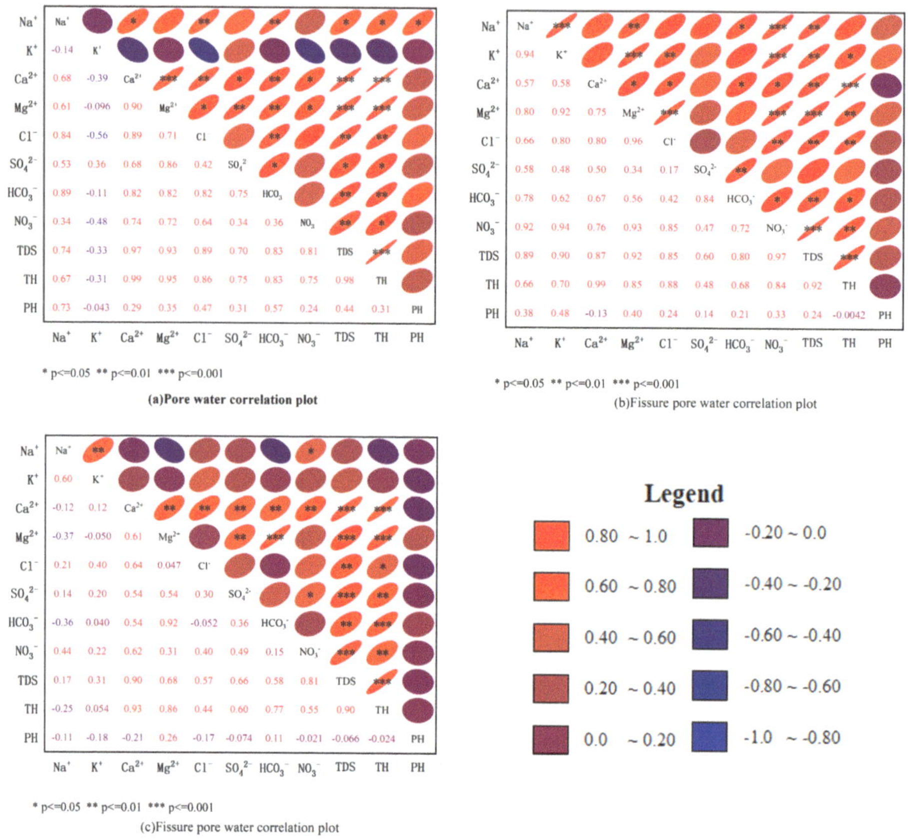

**Figure 6.** Correlation plots between ionic components of three groundwater types.

Ions in FPW also showed a strong correlation with other parameters. pH was negatively correlated with $Ca^{2+}$ but positively correlated with other indicators, with a correlation coefficient of less than 0.5. TDS was significantly correlated with other parameters, except $SO_4^{2-}$, which is consistent with the distribution of evaporation. The positive correlation between TH and $Ca^{2+}$, $Mg^{2+}$, and $Cl^-$, with correlation coefficients around 0.9, indicates that the variation of total hardness in this groundwater type is influenced by non-carbonate solutes.

Ions in FW showed a weak correlation with other parameters. Except for a positive correlation with $Mg^{2+}$ and $HCO_3^-$, pH showed a negative correlation with all other ions. This type of groundwater contradicts the dissolution effect of carbonate rocks, and its origin requires further investigation. TDS showed a strong positive correlation with $Ca^{2+}$ and $NO_3^-$, which was presumed to be related to the distribution of evaporation. The positive correlations of TH with $Ca^{2+}$ and $Mg^{2+}$ (correlation coefficient around 0.9) and with $HCO_3^-$ (correlation coefficient around 0.7) indicate that the variation of total hardness in this groundwater type is influenced by carbonate solutes.

### *3.6. Gibbs Analysis of Controlling Factors of Hydrochemical Characteristics of Groundwater*

The main natural controlling factors of groundwater hydrochemical characteristics are rock weathering, precipitation, and evaporative crystallization [37]. Gibbs [22] established a semi-logarithmic coordinate model for analyzing the influence of the three controlling factors, in which TDS (unit: mg/L) is taken as the ordinate and $Na^+/(Na^+ + Ca^{2+})$ or $Cl^-/(Cl^- + HCO_3^-)$ is taken as the abscissa. As shown in Figure 7, the ratios of $Na^+/(Na^+ + Ca^{2+})$ and $Cl^-/(Cl^- + HCO_3^-)$ of all the water samples were lower than 0.4, and all the water samples fell into the rock weathering zone, indicating that the influencing factor of groundwater chemical characteristics in the study area was rock weathering. Moreover, the main groundwater hydrochemical types in this area were $HCO_3$-Ca and $HCO_3$-Ca·Mg, indicating the relatively strong weathering of carbonate rocks. As the Gibbs diagram cannot determine the influence of human activities on the chemical characteristics of groundwater, the controlling factors should be further analyzed in combination with ion ratios and other methods.

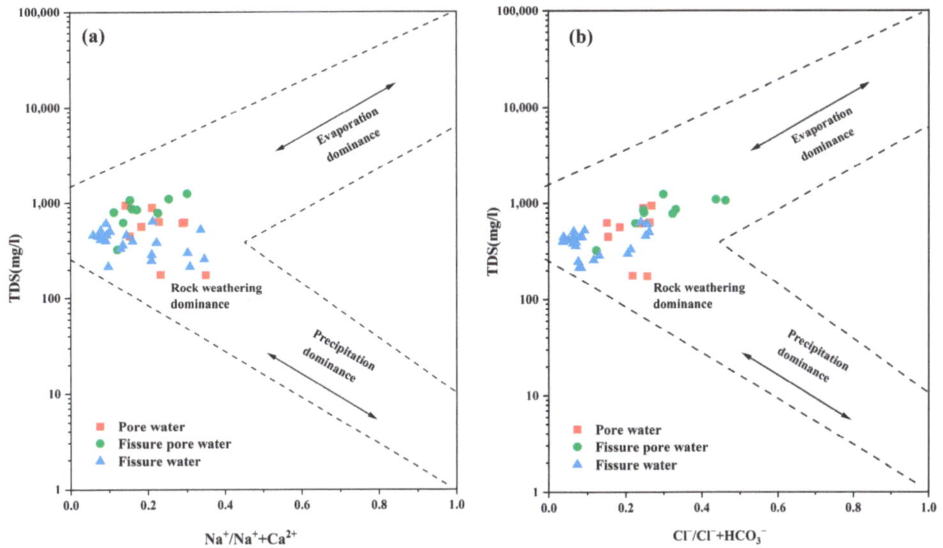


**Figure 7.** Gibbs plots of factors influencing the chemical characterization of groundwater in the Wulian. (TDS vs. $Na^+/(Na^+ + Ca^{2+})$ (**a**) TDS vs. $Cl^-/(Cl^- + HCO_3^-)$ (**b**)).

## *3.7. Ion Ratio Analysis*

The dissolution of soluble minerals is an important part of the ionic composition of groundwater. The composition and amount of dissolved minerals can be inferred from the law of charge conservation and ion concentration in groundwater. Salt rock (NaCl) is the most common dissolved mineral, and it can be used to assess the source of ions in groundwater according to the $Cl^-/Na^+$ ratio. If the ratio is equal to 1, all $Na^+$ and $Cl^-$ in groundwater are from the dissolution of salt rocks [37,50]. As shown in Figure 8a, the ratio of $Cl^-/Na^+$ widely varied among the different groundwater types. Most FW sample points were located below the y = x line, indicating that $Na^+$ in this type of groundwater had other important sources. As the study area is located far from the ocean, the source is speculated to be related to the dissolution or chemical reaction of other minerals. The PW water samples were located above the ratio line, and the concentration of $Cl^-$ was the highest among the three groundwater types, indicating a strong anthropogenic influence. FPW water samples were close to the ratio line, and the main source of this groundwater type was the dissolution of salt rock. As shown in Figure 8b, almost all water samples were in the sulfate weathering region, which indicated that the $Ca^{2+}$ and $Mg^{2+}$ sources were closely related to weathering. The FW samples were mostly distributed in that dissolution range of carbonate rocks, and the sources of $Ca^{2+}$ and $Mg^{2+}$ in this groundwater type were inferred to be closely related to ion exchange [51].

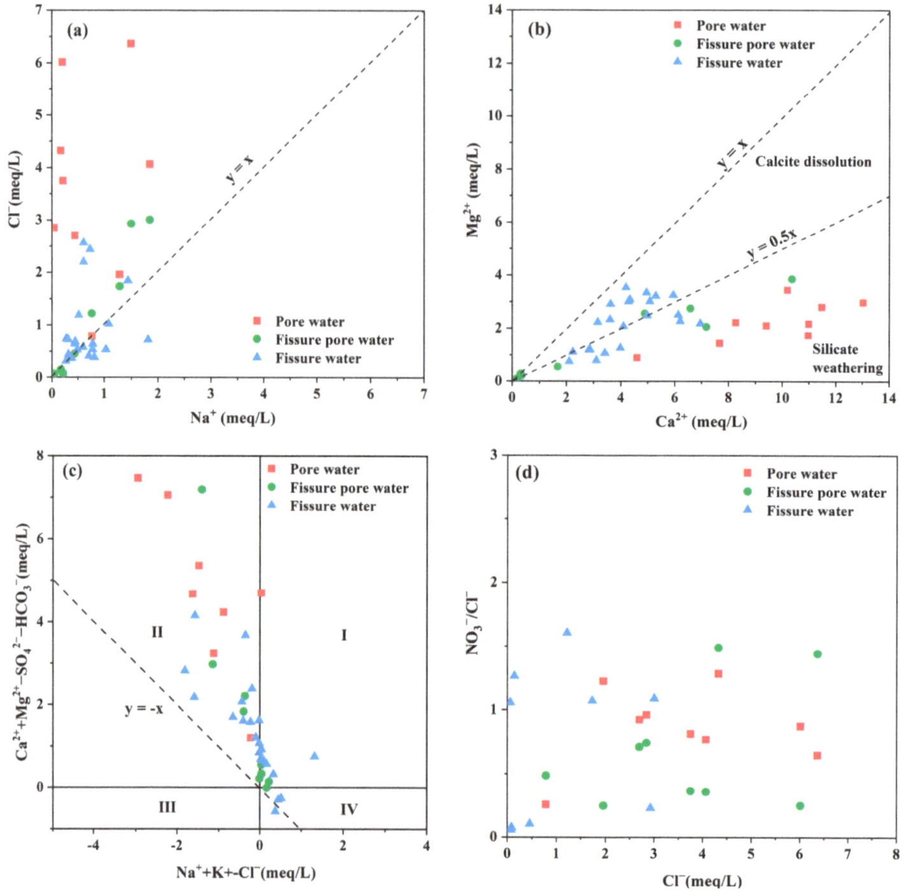


**Figure 8.** Analysis of the ion ratio in groundwater. (**a**) $Cl^-$ vs. $Na^+$; (**b**) $Mg^{2+}$ vs. $Ca^{2+}$; (**c**) $Ca^{2+}$ + $Mg^{2+}$ − $SO4^{2-}$ − $HCO_3^-$ vs. $Na^+$ + K + -$Cl^-$; (**d**) $NO_3^-/Cl^-$ vs. $Cl^-$.

The ion ratio in Figure 8c was used to determine the direction and intensity of cation exchange in groundwater, with Zone II and Zone IV representing the direction of cation exchange and the distance from the intersection of the zones representing the intensity of cation exchange. The PW samples were distributed in Zone II and far from the intersection, which indicates that cation exchange occurred in the positive direction with low intensity. In contrast, the FW and FPW samples were mostly distributed uniformly in Zone II, and they were close to the intersection, which indicates that cation exchange in groundwater mostly occurred in the positive direction with high intensity. The difference in $Cl^-$ concentration is generally related to the influence of domestic sewage and septic tanks or sea sources. Considering the distance of the study area from the ocean, $Cl^-$ can be used as an indicator of the impact of domestic sewage as well as septic tanks [52]. The ratio of $NO_3^-/Cl^-$ to $Cl^-$ can indicate the effect of human activities on nitrate pollution [53]. The high $NO_3^-/Cl^-$ and low $Cl^-$ concentrations in FW, shown in Figure 8d, indicate that nitrate ions were mainly affected by atmospheric precipitation and fertilizers. In contrast, the low $NO_3^-/Cl^-$ and high $Cl^-$ concentrations in FW indicate that the nitrate ion content in groundwater was mainly affected by domestic sewage and human and animal feces, which is consistent with the distribution of villages and farmlands in the study area.

### 3.8. Source Identification of Nitrate Ion

FA can be used to analyze potential sources of nitrate ions in groundwater and was used to extract principal factors based on eigenvalues larger than 1. The results showed that the three principal factors in PW explain 93.3% of the total variance (Table 2a). The $NO_3^-$ parameter was 0.72 in this analysis model, which indicated that the absolute loading of this model was in the middle level and can be used to indicate the source of nitrate ions in groundwater, with less information loss and more reliable significance. Therefore, the original component matrix can be used to analyze the three extracted principal factors. Specifically, Factor 1 explained 67.25% of the total variance and had significant contributions to the variables except for $K^+$ and pH. Combined with the ion ratio, Factor 1 can be regarded as a dual factor corresponding to rock weathering and human influence. Factor 2 accounted for 15.01% of the total variance and contributed significantly to $K^+$, followed by $SO_4^{2-}$. Therefore, Factor 2 can be regarded as the effect of cation exchange in groundwater and domestic sewage discharge. Factor 3 accounted for 11.04% of the total variance and had a significant contribution to pH. Therefore, Factor 3 can be regarded as an acid-base balance factor. It can be inferred that most of the ions in PW groundwater originate from natural weathering and anthropogenic factors.

The three principal factors in FPW explained 94.9% of the total variance (Table 2b). In this analytical model, the $NO_3^-$ parameter was 0.97, which indicated that the absolute loading of the model was at a strong level and had high significance. Factor 1 explained 70.96% of that total variance and had a significant contribution to all variables except $SO_4^{2-}$ and pH. Therefore, Factor 1 can be considered a dual factor from atmospheric precipitation and anthropogenic influence. Factor 2 accounted for 12.54% of the total variance, which contributed significantly to the pH value. Therefore, Factor 2 could be regarded as a factor of acid-base balance. Factor 3 accounted for 11.44% of the total variance and had a significant contribution to $SO_4^{2-}$. Therefore, Factor 3 can be regarded as the influence of atmospheric precipitation. FPW and PW showed similar results, with most nitrate ions in groundwater originating from anthropogenic influences, including industrial sewage and domestic sewage discharge, septic tanks, and other factors.

The three principal factors in FW explained 79.0% of the total variance (Table 2c). In this analysis model, the $NO_3^-$ parameter was 0.69, which indicated that the absolute loading of the model was at the medium level; it could be used to indicate the source of nitrate ions in groundwater, and the significance was reliable. Factor 1 explained 46.78% of the total variance and had a significant influence on TDS, $Mg^{2+}$, TH, and $Ca^{2+}$, followed by $HCO_3^-$ and $SO_4^{2-}$. The contribution of this factor to most variables was significant, and it could be regarded as a combination of natural and anthropogenic factors, including natural weathering and domestic sewage. Factor 2 accounted for 22.41% of the total variance and contributed significantly to $Na^+$, followed by $K^+$. Therefore, Factor 2 can be regarded as the effect of cation exchange in groundwater. Factor 3 accounted for 11.04% of the total variance and contributed significantly to the pH value. Therefore, Factor 3 can be considered an acid-base balance factor. In the three types of groundwater, nitrate originates from anthropogenic factors, but nitrate pollution was low in FW because of the weak influence of human activities in this area. The development of water supply sources should therefore be carried out in the southern part of the study area.

**Table 2.** Factorial load (absolute strong load (≥0.75) in bold) and the corresponding communality (top half of the table). Cumulative percentage of factor eigenvalues, percentages, and variances (bottom half of the table).

| | Factor 1 | Factor 2 | Factor 3 | Communalities |
|---|---|---|---|---|
| TDS | **0.99** | | 0.1 | 0.99 |
| $NO_3^-$ | 0.72 | −0.44 | 0.27 | 0.78 |
| $Mg^{2+}$ | **0.93** | 0.14 | 0.3 | 0.97 |
| $K^+$ | −0.3 | **0.9** | 0.22 | 0.95 |
| TH | **0.97** | | 0.22 | 0.99 |
| $Na^+$ | **0.82** | 0.21 | −0.49 | 0.95 |
| $Cl^-$ | **0.91** | −0.26 | −0.23 | 0.95 |
| $Ca^{2+}$ | **0.96** | −0.14 | 0.17 | 0.97 |
| $HCO_3^-$ | **0.9** | 0.27 | −0.18 | 0.92 |
| pH | 0.51 | 0.27 | −0.69 | 0.81 |
| $SO_4^{2-}$ | 0.73 | 0.58 | 0.34 | 0.98 |
| Eigenvalues | 7.4 | 1.65 | 1.22 | (a) PW |
| Percentage of variance (%) | 67.25 | 15.01 | 11.04 | |
| Cumulative percentage of variance (%) | 67.25 | 82.26 | 93.3 | |
| | **Factor 1** | **Factor 2** | **Factor 3** | **Communalities** |
| TDS | **1** | | | 1 |
| $NO_3^-$ | **0.97** | | | 0.96 |
| $Mg^{2+}$ | **0.94** | 0.15 | −0.29 | 0.99 |
| $K^+$ | **0.92** | 0.32 | | 0.94 |
| TH | **0.91** | −0.36 | −0.17 | 0.98 |
| $Na^+$ | **0.9** | 0.22 | 0.18 | 0.89 |
| $Cl^-$ | **0.86** | | −0.49 | 0.98 |
| $Ca^{2+}$ | **0.84** | −0.5 | −0.12 | 0.97 |
| $HCO_3^-$ | **0.79** | −0.12 | 0.54 | 0.94 |
| pH | 0.3 | **0.88** | | 0.88 |
| $SO_4^{2-}$ | 0.59 | −0.16 | 0.66 | 0.93 |
| Eigenvalues | 7.4 | 1.65 | 1.22 | (b) FPW |
| Percentage of variance (%) | 67.25 | 15.01 | 11.04 | |
| Cumulative percentage of variance (%) | 67.25 | 82.26 | 93.3 | |
| | **Factor 1** | **Factor 2** | **Factor 3** | **Communalities** |
| TDS | **0.97** | 0.18 | | 0.98 |
| $NO_3^-$ | 0.69 | 0.42 | 0.22 | 0.69 |
| $Mg^{2+}$ | **0.79** | −0.53 | 0.21 | 0.96 |
| $K^+$ | 0.23 | 0.67 | 0.21 | 0.55 |
| TH | **0.97** | −0.22 | −0.11 | 0.99 |
| $Na^+$ | | **0.85** | 0.41 | 0.9 |
| $Cl^-$ | 0.52 | 0.52 | −0.32 | 0.64 |
| $Ca^{2+}$ | **0.92** | | −0.31 | 0.95 |
| $HCO_3^-$ | 0.7 | −0.54 | 0.12 | 0.79 |
| pH | | −0.37 | **0.75** | 0.71 |
| $SO_4^{2-}$ | 0.71 | 0.12 | 0.14 | 0.54 |
| Eigenvalues | 5.15 | 2.44 | 1.11 | (c) FW |
| Percentage of variance (%) | 46.78 | 22.14 | 10.12 | |
| Cumulative percentage of variance (%) | 46.78 | 68.92 | 79.04 | |

The FA results (Figure 9) showed that Factor 1 in PW was strongly affected by nitrate ions, suggesting that $NO_3^-$ was strongly affected by anthropogenic factors. Moreover, Factor 1 and Factor 3 in FPW were also affected by nitrate ions. As residential areas are densely distributed and the groundwater is shallow in the FPW distribution area, FPW is highly susceptible to anthropogenic factors. According to the FA results, $NO_3^-$ in this area was most seriously affected by anthropogenic factors. Factor 1 and Factor 2 in FW were affected by nitrate ions, which indicated that $NO_3^-$ in this type of groundwater is also controlled by anthropogenic factors. The FW distribution area primarily comprises farmlands, and the pollution is mainly attributable to the use of nitrogenous fertilizers. Therefore, to establish a high-quality water source, relevant systems should be modified to reduce pollution caused by anthropogenic factors and prevent and control pollution from the root. According to the multivariate analysis, the groundwater environment deteriorates from the FW distribution area in the south to the PW and FPW distribution area in the north. As a high-quality groundwater source area, the FW distribution area in the south, with sparse settlements, should be selected.

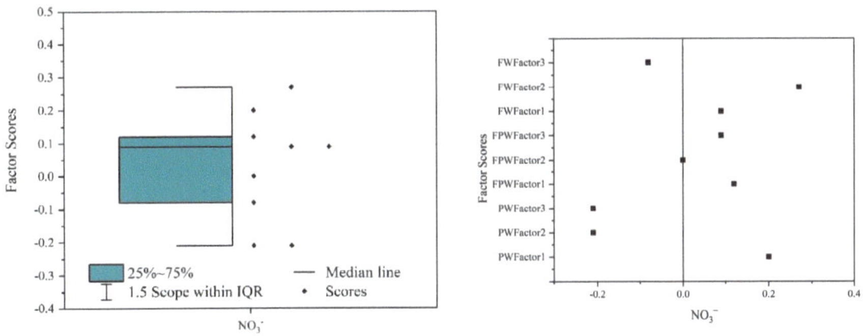


**Figure 9.** Factor scores of the three groundwater types. (The square indicates that the factor score value is between 25% and 75%).

### 3.9. Suggestions on Groundwater Development and Pollution Control in the Study Area

Through the analysis of the hydrochemical characteristics and controlling factors of groundwater, the physical and chemical properties of groundwater were preliminarily investigated in this study. Water quality assessments and health risk assessments of nitrate were carried out to further determine the status of groundwater pollution. The source control of groundwater pollution can be realized by analyzing the source of nitrate ions in groundwater. This study provides data support and a theoretical basis for the pollution control, development, and utilization of groundwater resources in the Wulian area. However, the scope of this study is relatively small. In order to establish a large water supply source, a large number of samples should be collected and the status of heavy metal pollution should be analyzed. In this manner, our understanding of groundwater pollution and pollution sources can be improved and a more comprehensive theoretical basis for groundwater pollution prevention, development, and utilization can be obtained.

## 4. Conclusions

This study has investigated the status and sources of pollution in three different types of groundwater in Wulian through the comprehensive analysis of groundwater chemical characteristics, controlling factors, and health risks. The main cations in groundwater are $Ca^{2+}$ and $Na^+$, and the anions are $HCO_3^-$ and $SO_4^{2-}$. However, $NO_3^-$ was detected in all three types of groundwater, with standard exceedance rates following the order FPW > PW > FW. The main groundwater hydrochemical types are $HCO_3$-Ca, $HCO_3$-Ca·Mg, and $SO_4$·Cl-Ca·Mg. The results show that rock weathering, salt rock dissolution, and human factors are the main factors controlling the chemical differences of groundwater in this area.

The EWQI and health risk assessment results show that the groundwater quality in the study area is closely related to the distribution of nitrate, and the water quality followed the order FW > PW > FPW. In addition, nitrate pollution was more serious in densely populated areas. Factor analysis results show that the main source of nitrate is anthropogenic input. Therefore, to establish a high-quality water source, relevant systems should be modified to reduce the pollution caused by anthropogenic factors and prevent and control pollution from the root. This study can provide data support and a theoretical basis for the pollution control, development, and utilization of groundwater resources in the Wulian area.

**Author Contributions:** Conceptualization, M.W., W.Z., J.F. and Z.G.; Methodology, M.W., W.Z. and J.F.; Software, P.Y., R.Z., H.J., X.S. and X.G.; Validation, M.W., W.Z., J.F. and Z.G.; Formal Analysis, M.W., W.Z., J.F. and Z.G.; Investigation, P.Y., R.Z., H.J., X.S. and X.G.; Resources, P.Y., R.Z., H.J., X.S. and X.G.; Data Management, P.Y., R.Z., H.J., X.S. and X.G.; Writing—Manuscript Preparation, M.W., W.Z. and J.F.; Writing—Review and Editing, M.W., W.Z., J.F. and Z.G.; Visualization, M.W., W.Z. and J.F.; Supervision, P.Y., R.Z., H.J., X.S. and X.G.; Project Management, J.F.; Funding Acquisition, Z.G. All authors have read and agreed to the published version of the manuscript.

**Funding:** This study was supported by the National Natural Science Foundation of China (No. 41772257; 41472216).

**Data Availability Statement:** Not applicable.

**Acknowledgments:** This study was financially supported by the National Natural Science Foundation of China (No. 41772257; 41472216). We thank the reviewers and editors for their pertinent comments and Jutan Liu and Zongjun Gao for their guidance.

**Conflicts of Interest:** The authors declare no conflict of interest.

## References

1. Wang, D.; Wu, J.H.; Wang, Y.H.; Ji, Y.J. Finding High-Quality Groundwater Resources to Reduce the Hydatidosis Incidence in the Shiqu County of Sichuan Province, China: Analysis, Assessment, and Management. *Expo. Health* **2020**, *12*, 307–322. [CrossRef]
2. Elsayed, S.; Gad, M.; Masoud, M.; Osta, M.; Alqarawy, A. Groundwater Suitability for Drinking and Irrigation Using Water Quality Indices and Multivariate Modeling in Makkah Al-Mukarramah Province, Saudi Arabia. *Water* **2022**, *14*, 483. [CrossRef]
3. Zhou, Y.H.; Li, P.Y.; Chen, M.J.; Dong, Z.H.; Lu, C.Y. Groundwater quality for potable and irrigation uses and associated health risk in southern part of Gu'an County, North China Plain. *Environ. Geochem. Health* **2021**, *43*, 813–835. [CrossRef]
4. Gad, M.; Saleh, A.; Hussein, H.; Farouk, M.; Elsayed, S. Appraisal of Surface Water Quality of Nile River Using Water Quality Indices, Spectral Signature and Multivariate Modeling. *Water* **2022**, *14*, 1131. [CrossRef]
5. Yu, F.; Zhou, D.; Li, Z.; Li, X. Hydrochemical Characteristics and Hydrogeochemical Simulation Research of Groundwater in the Guohe River Basin (Henan Section). *Water* **2022**, *14*, 1461. [CrossRef]
6. Li, P.Y.; Wu, J.H. Sustainable living with risks: Meeting the challenges. *Hum. Ecol. Risk Assess.* **2019**, *25*, 1–10. [CrossRef]
7. Liu, J.T.; Gao, Z.J.; Wang, M.; Li, Y.Z.; Yu, C.; Shi, M.J.; Zhang, H.Y.; Ma, Y.Y. Hydrochemical and isotopic characteristics of surface water in the Lhasa River basin. *Arab. J. Geosci.* **2019**, *12*, 520. [CrossRef]
8. Hua, K.; Jun, X.; Li, S.J.; Li, Z. Analysis of hydrochemical characteristics and their controlling factors in the Fen River of China. *Sustain. Cities Soc.* **2019**, *52*, 101827. [CrossRef]
9. Yousefi, M.; Ghoochani, M.; Mahvi, A. Health risk assessment to fluoride in drinking water of rural residents living in the Poldasht city, Northwest of Iran. *Ecotoxicol. Environ. Saf.* **2017**, *148*, 426–430. [CrossRef]
10. He, S.; Wu, J.H. Hydrogeochemical Characteristics, Groundwater Quality, and Health Risks from Hexavalent Chromium and Nitrate in Groundwater of Huanhe Formation in Wuqi County, Northwest China. *Expo. Health* **2019**, *11*, 125–137. [CrossRef]
11. Narsimha, A.; Qian, H. Groundwater quality evaluation using water quality index (WQI) for drinking purposes and human health risk (HHR) assessment in an agricultural region of Nanganur, south India. *Ecotoxicol. Environ. Saf.* **2019**, *176*, 153–161. [CrossRef]
12. Ahmed, I.; Tariq, N.; Muhery, A. Hydrochemical characterization of groundwater to align with sustainable development goals in the Emirate of Dubai, UAE. *Environ. Earth Sci.* **2019**, *78*, 44. [CrossRef]
13. Eslami, F.; Yaghmaeian, K.; Mohammadi, A.; Salari, M.; Faraji, M. An integrated evaluation of groundwater quality using drinking water quality indices and hydrochemical characteristics: A case study in Jiroft, Iran. *Environ. Earth Sci.* **2019**, *78*, 314. [CrossRef]
14. Li, P.Y.; Tian, R.; Xue, C.Y.; Wu, J.H. Progress, opportunities, and key fields for groundwater quality research under the impacts of human activities in China with a special focus on western China. *Environ. Sci. Pollut. Res.* **2017**, *24*, 13224–13234. [CrossRef]
15. Liu, J.T.; Gao, Z.J.; Wang, M.; Li, Y.Z.; Ma, Y.Y.; Shi, M.J.; Zhang, H.Y. Study on the dynamic characteristics of groundwater in the valley plain of Lhasa City. *Environ. Earth Sci.* **2018**, *77*, 646. [CrossRef]
16. Yuan, Y.Y.; Liu, Y.L.; Luo, K.L.; Shahid, M.Z. Hydrochemical characteristics and a health risk assessment of the use of river water and groundwater as drinking sources in a rural area in Jiangjin District, China. *Environ. Earth Sci.* **2020**, *79*, 160. [CrossRef]
17. Qian, H.; Chen, J.; Howard, K. Assessing groundwater pollution and potential remediation processes in a multi-layer aquifer system. *Environ. Pollut.* **2020**, *263*, 114669. [CrossRef] [PubMed]

18. Gharbi, A.; Ibn Ali, Z.; Zairi, M. Groundwater suitability for drinking and agriculture purposes using irrigation water quality index and multivariate analysis: Case of Sidi Bouzid aquifer, central Tunisia. *Environ. Earth Sci.* **2019**, *78*, 692. [CrossRef]
19. Everest, T.; Özcan, H. Applying multivariate statistics for identification of groundwater resources and qualities in NW Turkey. *Environ. Monit. Assess.* **2019**, *191*, 47. [CrossRef]
20. Sajil Kumar, P.J.; James, E.J. Geostatistical and geochemical model-assisted hydrogeochemical pattern recognition along the groundwater flow paths in Coimbatore district, South India. *Environ. Dev. Sustain.* **2019**, *21*, 369–384. [CrossRef]
21. Piper, A.M. A graphic procedure in the geochemical interpretation of water-analyses. *EOS Trans. Am. Geophys. Union* **1944**, *25*, 914–928. [CrossRef]
22. Gibbs, R. Mechanisms Controlling World Water Chemistry. *Science* **1971**, *170*, 1088–1090. [CrossRef]
23. Zhai, Y.Z.; Zheng, F.X.; Zhao, X.B.; Teng, Y.G. Identification of hydrochemical genesis and screening of typical groundwater pollutants impacting human health: A case study in Northeast China. *Environ. Pollut.* **2019**, *252*, 1202–1215. [CrossRef] [PubMed]
24. Abbasnia, A.; Yousefi, N.; Mahvi, A.; Nabizadeh, R.; Radfarad, M.; Yousefi, M.; Alimohammadi, M. Evaluation of groundwater quality using water quality index and its suitability for assessing water for drinking and irrigation purposes: Case study of Sistan and Baluchistan province (Iran). *Hum. Ecol. Risk Assess.* **2018**, *25*, 73–82. [CrossRef]
25. Liu, J.T.; Peng, Y.M.; Li, C.S.; Gao, Z.J.; Chen, S.J. An investigation into the hydrochemistry, quality and risk to human health of groundwater in the central region of Shandong Province, North China. *J. Clean. Prod.* **2021**, *282*, 125416. [CrossRef]
26. Su, H.; Kang, W.D.; Xu, Y.J.; Wang, J.D. Assessing Groundwater Quality and Health Risks of Nitrogen Pollution in the Shenfu Mining Area of Shaanxi Province, Northwest China. *Expo. Health* **2018**, *10*, 77–97. [CrossRef]
27. Zhang, Q.Y.; Xu, P.P.; Qian, H. Assessment of Groundwater Quality and Human Health Risk (HHR) Evaluation of Nitrate in the Central-Western Guanzhong Basin, China. *Int. J. Environ. Res. Public Health* **2019**, *16*, 4246. [CrossRef]
28. Zhang, M.N.; Han, S.B.; Wang, Y.S.; Wang, Z.; Li, H.X.; Wang, X.Y.; Liu, J.T.; Li, C.S.; Gao, Z.J. Characteristics and Controlling Factors of Groundwater Hydrochemistry in Dongzhi Tableland Area of the Loess Plateau of Eastern Gansu—A Case Study of Ning County Area, North China. *Water* **2022**, *14*, 3601. [CrossRef]
29. Duraisamy, S.; Govindhaswamy, V.; Duraisamy, K.; Krishnaraj, S.; Balasubramanian, A.; Subramani, T. Hydrogeochemical characterization and evaluation of groundwater quality in Kangayam taluk, Tirupur district, Tamil Nadu, India, using GIS techniques. *Environ. Geochem. Health* **2019**, *41*, 851–873. [CrossRef]
30. Song, H.Y.; Liu, J.Q.; Yin, P.; Zhang, Y. Distribution, enrichment and source of heavy metals in Rizhao offshore area, southeast Shandong Province. *Mar. Pollut. Bull.* **2017**, *119*, 175–180. [CrossRef]
31. He, Z.K.; Ma, C.M.; Zhou, A.G.; Qi, H.H.; Liu, C.F.; Cai, H.S.; Zhu, H.C. Using hydrochemical and stable isotopic ($\delta^2$H, $\delta^{18}$O, $\delta^{11}$B, and $\delta^{37}$Cl) data to understand groundwater evolution in an unconsolidated aquifer system in the southern coastal area of Laizhou Bay, China. *Appl. Geochem.* **2018**, *90*, 129–141. [CrossRef]
32. WHO. *Guidelines for Drinking-Water Quality*, 4th ed.; World Health Organization: Geneva, Switzerland, 2017.
33. Hardle, W.; Simar, L. *Applied Multivariate Statistical Analysis*; Springer: Berlin/Heidelberg, Germany, 2007.
34. Liu, J.T.; Peng, Y.M.; Li, C.S.; Gao, Z.J.; Chen, S.J. Characterization of the hydrochemistry of water resources of the Weibei Plain, Northern China, as well as an assessment of the risk of high groundwater nitrate levels to human health. *Environ. Pollut.* **2020**, *268*, 115947. [CrossRef] [PubMed]
35. Cao, Y.J.; Tang, C.Y.; Song, X.F.; Liu, C.M.; Zhang, Y.H. Identifying the hydrochemical characteristics of rivers and groundwater by multivariate statistical analysis in the Sanjiang Plain, China. *Appl. Water Sci.* **2016**, *6*, 169–178. [CrossRef]
36. Cude, C.G. Oregon water quality index a tool for evaluating water quality management effectiveness. *J. Am. Water Resour. Assoc.* **2001**, *37*, 125–137. [CrossRef]
37. Wu, C.; Wu, X.; Qian, C.; Zhu, G. Hydrogeochemistry and groundwater quality assessment of high fluoride levels in the Yanchi endorheic region, northwest China. *Appl. Geochem.* **2018**, *98*, 404–417. [CrossRef]
38. WHO (World Health Organization). *Guidelines for Drinking Water Quality*, 4th ed.; World Health Organization: Geneva, Switzerland, 2011; Volume 38, pp. 104–108.
39. Yousefi, M.; Ghalehaskar, S.; Asghari, F.; Ghaderpoury, A.; Dehghani, M.H.; Ghaderpoori, M.; Mohammadi, A. Distribution of fluoride contamination in drinking water resources and health risk assessment using geographic information system, northwest Iran. *Regul. Toxicol. Pharmacol.* **2019**, *107*, 104408. [CrossRef]
40. Chen, J.; Wu, H.; Qian, H. Groundwater Nitrate Contamination and Associated Health Risk for the Rural Communities in an Agricultural Area of Ningxia, Northwest China. *Expo. Health* **2016**, *8*, 349–359. [CrossRef]
41. Li, P.Y.; Qian, H.; Wu, J.H. Conjunctive use of groundwater and surface water to reduce soil salinization in the Yinchuan Plain, North-West China. *Int. J. Water Resour. D* **2018**, *34*, 337–353. [CrossRef]
42. USEPA (US Environmental Protection Agency). *Baseline Human Health Risk Assessment Vasquez Boulevard and I-70 Superfund Site*; USEPA (US Environmental Protection Agency): Denver, CO, USA, 2001. Available online: https://nepis.epa.gov/Exe/ZyNET.exe/P1006STM.TXT?ZyActionD=ZyDocument&Client=EPA&Index=2000+Thru+2005&Docs=&Query=&Time=&EndTime=&SearchMethod=1&TocRestrict=n&Toc=&TocEntry=&QField=&QFieldYear=&QFieldMonth=&QFieldDay=&IntQFieldOp=0&ExtQFieldOp=0&XmlQuery=&File=D%3A%5Czyfiles%5CIndex%20Data%5C00thru05%5CTxt%5C00000023%5CP1006STM.txt&User=ANONYMOUS&Password=anonymous&SortMethod=h%7C-&MaximumDocuments=1&FuzzyDegree=0&ImageQuality=r75g8/r75g8/x150y150g16/i425&Display=hpfr&DefSeekPage=x&SearchBack=ZyActionL&Back=ZyActionS&BackDesc=Results%20page&MaximumPages=1&ZyEntry=1&SeekPage=x&ZyPURL (accessed on 20 January 2011).

43. USEPA. 2004. Available online: http://www.epa.gov/oswer/riskassessment/ragse/pdf/introduction (accessed on 1 July 2004).
44. Bempah, C.; Ewusi, A. Heavy metals contamination and human health risk assessment around Obuasi gold mine in Ghana. *Environ. Monit. Assess.* **2016**, *188*, 261. [CrossRef]
45. Narsimha, A.; Rajitha, S. Spatial distribution and seasonal variation in fluoride enrichment in groundwater and its associated human health risk assessment in Telangana State, South India. *Hum. Ecol. Risk Assess.* **2018**, *24*, 2119–2132. [CrossRef]
46. Zhang, Y.T.; Wu, J.H.; Xu, B. Human health risk assessment of groundwater nitrogen pollution in Jinghui canal irrigation area of the loess region, northwest China. *Environ. Earth Sci.* **2018**, *77*, 273. [CrossRef]
47. Bouzourra, H.; Bouhlila, R.; Lakshmanan, E.; Slama, F.; Ouslati, N. Characterizaton of mechanisms and processes of groundwater salinization in irrigated coastal area using statistics, GIS, and hydrogeochemical investigations. *Environ. Sci. Pollut. Res.* **2015**, *22*, 2643–2660. [CrossRef] [PubMed]
48. Li, P.Y.; Wu, J.H.; Tian, R.; He, S.; He, X.D.; Xue, C.Y.; Zhang, K. Geochemistry, Hydraulic Connectivity and Quality Appraisal of Multilayered Groundwater in the Hongdunzi Coal Mine, Northwest China. *Mine Water Environ.* **2018**, *37*, 222–237. [CrossRef]
49. Narsimha, A. Controlling factors and mechanism of groundwater quality variation in semiarid region of South India: An approach of water quality index (WQI) and health risk assessment (HRA). *Environ. Geochem. Health* **2020**, *42*, 1725–1752. [CrossRef]
50. Narsimha, A. Spatial distribution, exposure, and potential health risk assessment from nitrate in drinking water from semi-arid region of South India. *Hum. Ecol. Risk Assess.* **2018**, *26*, 310–334. [CrossRef]
51. Keesari, T.; Saha, D. Groundwater salinization processes: Pitfalls of inferences from $Na^+/Cl^-$ versus $Cl^-$ correlation plots. *Environ. Geochem. Health* **2021**, *43*, 949–969. [CrossRef]
52. Liu, X.; Wang, X.L.; Zhang, L.; Fan, W.Y.; Yang, C.; Li, E.H.; Wang, Z. Impact of land use on shallow groundwater quality characteristics associated with human health risks in a typical agricultural area in Central China. *Environ. Sci. Pollut. Res.* **2021**, *28*, 1712–1724. [CrossRef]
53. Jin, Z.F.; Zheng, Q.; Zhu, C.Y.; Wang, Y.; Cen, J.R.; Li, F.L. Contribution of nitrate sources in surface water in multiple land use areas by combining isotopes and a Bayesian isotope mixing model. *Appl. Geochem.* **2018**, *93*, 10–19. [CrossRef]

hydrology

Article

# Estimation of Nitrate Background Value in Groundwater under the Long-Term Human Impact

**Patricia Buškulić, Jelena Parlov *, Zoran Kovač and Zoran Nakić**

Faculty of Mining, Geology and Petroleum Engineering, University of Zagreb, 10000 Zagreb, Croatia
* Correspondence: jelena.parlov@rgn.unizg.hr

**Abstract:** This study demonstrates an approach to estimate the background value of nitrate as a basis for better groundwater management and protection in areas under long-term human impact. The aim was to determine the ambient background value (ABV) of nitrate in the catchment area of the Velika Gorica well field, a hydrogeologically homogeneous area within the Zagreb aquifer. ABVs are determined using four well-known model-based objective methods (the iterative 2-σ technique, IT; the calculated distribution function, CDF; the cumulative frequency curve, CFC; and the probability plot, PP), while simultaneously testing the reliability of the results of each method. If the results are not statistically significant, data selection is performed. The results show that using data without selection can lead to statistically non-significant ABVs, but with the additional selection of data, a statistically non-significant result became a statistically significant one. In summary, all final ABVs must be statistically significant and determined using as large a data set as possible. Reducing the size of the data set is acceptable only in the case of a statistically non-significant result.

**Keywords:** ambient background value; nitrate; iterative 2-σ technique; calculated distribution function; cumulative frequency curve; probability plot; Zagreb aquifer

**Citation:** Buškulić, P.; Parlov, J.; Kovač, Z.; Nakić, Z. Estimation of Nitrate Background Value in Groundwater under the Long-Term Human Impact. *Hydrology* **2023**, *10*, 63. https://doi.org/10.3390/hydrology10030063

Academic Editors: Zongjun Gao and Jiutan Liu

Received: 3 February 2023
Revised: 1 March 2023
Accepted: 3 March 2023
Published: 4 March 2023



## 1. Introduction

Groundwater presents the most common and valuable source of drinking and irrigation water in the world [1]. Sustainable management of water and sanitation for the world population is one of the main goals established by the United Nations under the Sustainable Development Goals [2]. A large amount of research has been focused on the evaluation of groundwater quality, which has been diminished by different pollutants [1,3–6], while others emphasize the importance of implementing measures that will slow down the deterioration of water bodies [7–9]. Nitrate pollution of groundwater is one of the main environmental issues worldwide [10–18] and nitrate's increasing trend in natural waters has made it a contaminant of concern [4,19,20]. Burkart and Kolpin [21] have shown that a nitrate concentration greater than 13 mg/L as $NO_3^-$ in groundwater is a sign of natural groundwater quality degradation due to anthropogenic activities. In order to be sure that groundwater pollution is present, it is important to distinguish between geogenic and anthropogenic influences on groundwater quality by the estimation of geochemical background values that reflect natural processes [22]. Estimating the natural background value in groundwater enables us to recognize the concentration of the geochemical compound that is free from human influence [23]. The Groundwater Directive (2006/118/EC) introduced the term "background value" as *"the concentration of a substance or the value of an indicator in a body of groundwater corresponding to no, or only very minor, anthropogenic alterations to undisturbed conditions"*. Due to the ubiquitous human influence, which is also reflected in the chemical composition of groundwater, the natural composition of groundwater, especially in shallow aquifers, almost does not exist today. Since, in many parts of the world, human influence on the environment has been present for such a long time that it is futile to try to determine the natural background concentrations of substances in groundwater, Reiman and Garrett [24] devised the concept of the ambient background value (ABV), which describes an immeasurably disturbed and no longer intact background concentration of substances in groundwater. Because background values are location- and scale-dependent, i.e., they change regionally as well as locally [24], it is necessary to divide data sets into homogeneous groups to increase the probability of detecting an environmental problem. Sometimes, it is very difficult to clearly identify the populations related to the background processes as opposed to those reflecting contamination [25–28]. It has been shown that areas that are unaffected by human activities typically have natural background values of nitrate that do not exceed 5–10 mg/L as $NO_3^-$ [29].

Many researchers are using approaches based on the analysis of the distribution of many measured data using different statistical methods [30–32]. It has been shown that probability density functions can be different in background and anomalous samples, but the main issue is how to recognize these differences with confidence [33]. Although model-based subjective methods of background determination that include the mean plus two standard deviations are the best for data that follow a normal distribution and were frequently used, they are far from ideal when used for regional geochemical data [33]. The problem arises from the fact that the samples represent more than one population or process, resulting in multimodal distribution, where the use of the mean can give overestimated values [34]. On the contrary, model-based objective methods of background determination differ from subjective methods in that the background values and the upper limit of background range, i.e., threshold values, are defined by the data themselves, rather than by an arbitrary decision of the researcher. Recently, model-based objective methods have been gaining more and more importance. They are based on the division of a complete set of measured data into subgroups with characteristic probability density functions that reflect relevant natural geochemical processes and/or pollution events in an area. Background populations can be approximated by normal or lognormal density functions, while different processes have produced an anomalous population that differs from the background [33]. In recent decades, various model-based objective methods (the iterative 2-σ technique, the calculated distribution function, the cumulative frequency curve, and the probability plot) and their combinations have been used worldwide to determine background concentrations and threshold values [29,35–41]. Furthermore, there are also other statistical approaches. For example, the lithology of aquifers was used to define the natural background values of different chemical parameters in six EU countries [42]. In some cases, natural background values of selected species were estimated by means of the pre-selection and the component separation method [23]. Other studies combined pre-selection methods and statistical approaches [26,43–45].

The groundwater of the Zagreb aquifer is threatened due to increasing concentrations of pollutants and decreasing water quantity. Nitrates have been recognized as one of the most common groups of contaminants in the Zagreb aquifer [22], which is designated as a vulnerable area to nitrates, due to agricultural activities, especially on the right bank of the Sava River. Thus far, the elevated nitrate concentrations have been confirmed by various studies [22,46,47], while recent research was focused on the determination of areas with the highest risk for nitrate contamination [48], the definition of nitrate origin [49], the estimation of nitrate trends [50], and the in situ monitoring of multiple soil horizons in Eutric Cambisols at the Velika Gorica site [51]. Furthermore, it has been shown that nitrate concentrations present one of the most heterogeneous pieces of data, which is due to the pronounced anthropogenic influence in different parts of the Zagreb aquifer [52]. The concept of ABVs was implemented by Nakić et al. [25] and Kovač [52] to calculate the ABVs of nitrate in the groundwater of an alluvial aquifer in the Zagreb area. The upper limit of the ABV for the Zagreb aquifer ranged from 4.5 to 10.7 mg/L. Due to the obvious human influence and heterogeneity of the Zagreb aquifer, four different methods (the iterative 2-σ technique, IT; the calculated distribution function, CDF; the cumulative frequency curve, CFC; and the probability plot, PP), which belong to the model-based objective methods, were used to estimate the ABV of nitrate in the catchment area of the Velika

Gorica well field, which is located in the southern part of the Zagreb aquifer and complies with the principle of hydrogeological homogeneity. We are referring here to the results of research by a group of authors [53] who revealed that the background concentrations of substances, determined at the level of the regional aquifer, can significantly differ from the background concentrations of the same substances, determined in homogeneous areas within the same aquifer. The importance of applying the principle of homogeneity when determining background concentrations for substances that are sensitive to changes in oxidation-reduction conditions in the subsoil is emphasized by Molinari et al. [43], while Matschullat et al. [40] emphasize that the geochemical background concentration of a chemical substance needs to be determined in spatial conditions that are characterized by homogeneous, mainly climatological, lithological, and pedological features. The objectives of this study are (a) to modify the graphical interpretation of used methods to reduce subjectivity in determining a background value, (b) to show the differences in results when estimating ABVs using all data without selection compared to selected non-normally distributed (NON-ND) data, and (c) to evaluate the reliability of used methods in the anthropogenically influenced area.

## 2. Study Area

The Zagreb aquifer is situated in the north-western part of the Republic of Croatia along the Sava River, between the Medvednica Mountain in the north and the hills of Vukomeričke Gorice in the south. It is an unconfined aquifer composed of unconsolidated Quaternary sediments consisting mainly of sand, gravel and silt or silty clays and covers an area of approximately 350 km$^2$. The research area is located in the southern part of the Zagreb aquifer (Figure 1). The average annual precipitation for the period 2001 to 2020 is 967 mm. Precipitation is evenly distributed throughout the year, with a monthly average of around 80 mm. The annual average air temperature ranges from 10.5 °C to 12.7 °C, with a minimum in January and a maximum in July (Figure 2).

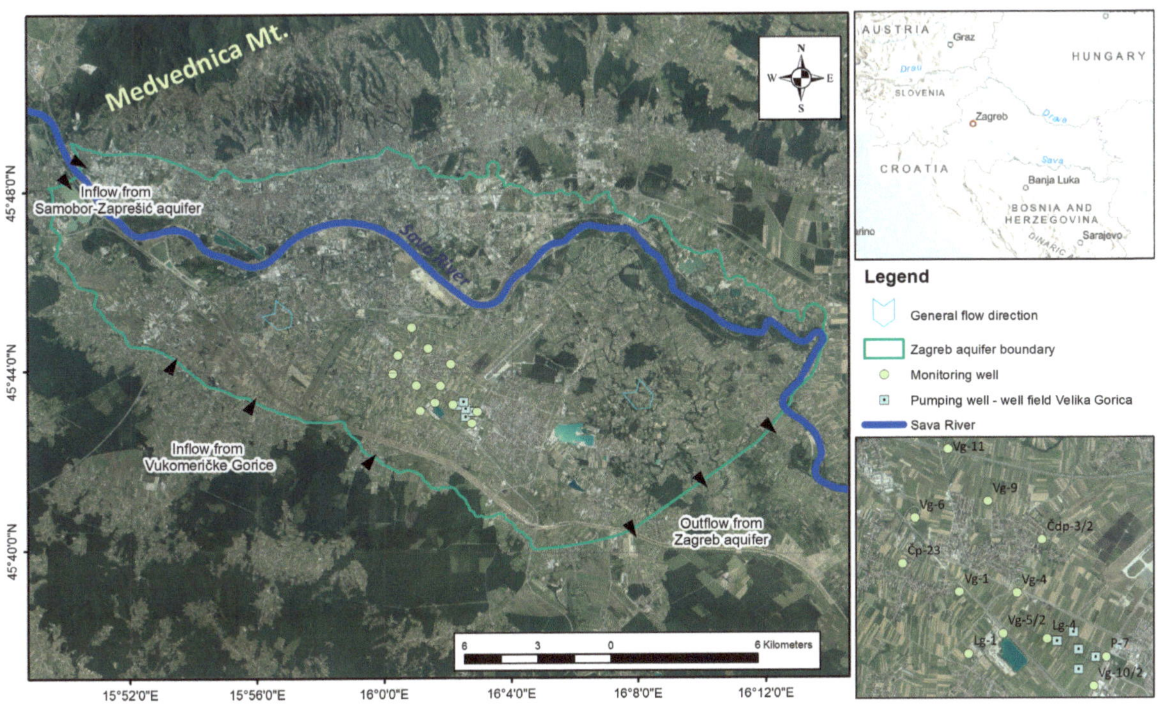


**Figure 1.** Research area.

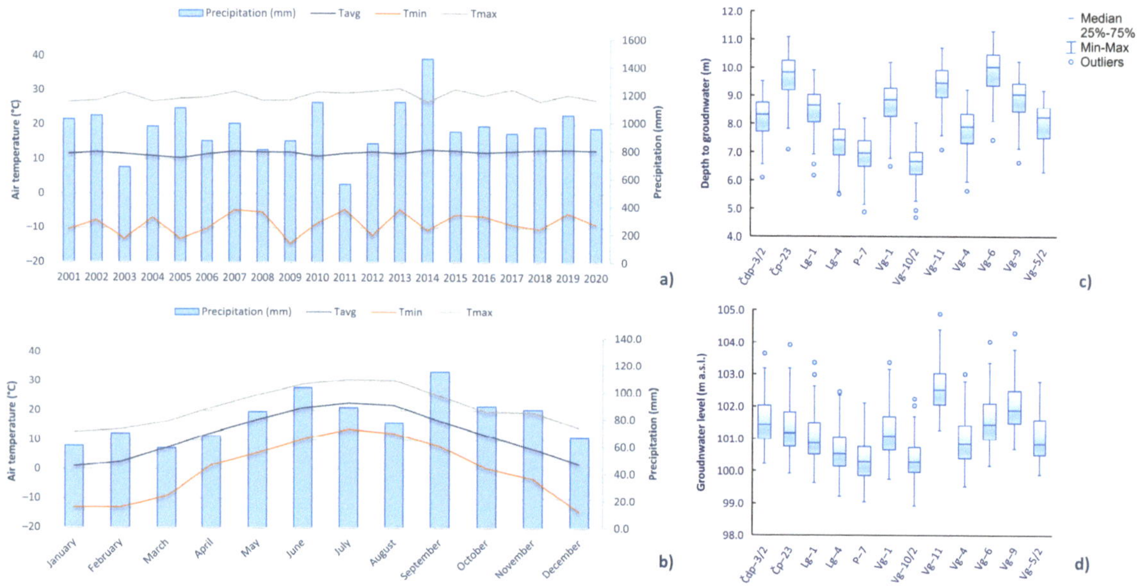


**Figure 2.** The annual precipitation (**a**) and the average monthly precipitation (**b**) with average, minimum and maximum air temperature; depth to groundwater (**c**) and groundwater levels (**d**) for the period 2001 to 2020.

Quaternary sediments were deposited during the Middle and Upper Pleistocene and the Holocene [54]. The shallow layer contains Holocene alluvial deposits of the Sava River, while the deep layer contains Pleistocene lacustrine–marshy deposits [55,56]. The unsaturated zone consists of gravels in the lower part, while the upper part consists mainly of silty to sandy material, interbedded with clay layers in places. The thickness of the unsaturated zone varies from two to eleven meters [57].

The shallow aquifer is in direct hydraulic connection with the Sava River, which is the main source of groundwater recharge [58–61]. The groundwater level changes mainly due to water level fluctuations of the Sava River, which recharges the aquifer during high water levels, while it discharges the aquifer during low and medium water levels. Other recharge mechanisms include infiltration from precipitation, infiltration from the porous water supply system and the sewage network, but also the water inflow from the Samobor-Zaprešić aquifer in the west and the water inflow across the southern boundary in the Vukomeričke Gorice area [62]. The general groundwater flow direction concurs with the flow direction of the Sava River from west to east (south-east) (Figure 1). The Zagreb aquifer area consists of three major pedological units: Fluvisols, Stagnosols and Eutric Cambisols on the Holocene deposits [63,64].

The Velika Gorica well field is one of the most important well fields in Zagreb County, supplying drinking water to the City of Zagreb and the City of Velika Gorica. The well field consists of five wells drilled to a depth of 39 m to 50 m, drawing water from the shallow aquifer. Near the well field, there are 12 monitoring wells in an area of approximately 9 km$^2$, generally used to monitor groundwater quality in the shallow aquifer. In this small area, the maximum elevation difference is 7 m. The maximum thickness of the unsaturated zone is 9 to 11 m (Figure 2), and most of the land is used for agriculture, mixed with urban areas. Figure 2c,d show that the depths to groundwater and groundwater level fluctuations are similar for all 12 monitoring wells. According to the results of previous research [65–67], the groundwater in the influence area of the Velika Gorica well field belongs to the calcium–magnesium–hydrogen carbonate ($CaMgHCO_3$) water type. All this makes the catchment area of the Velika Gorica well field a hydrogeologically homogeneous area relevant for the determination of background values.

## 3. Materials and Methods

### 3.1. Available Data Set

The ABV of nitrate was estimated using groundwater chemical data from the national groundwater quality monitoring program of Croatian Waters. Nitrate concentration data from 12 monitoring wells, located in the catchment area of the Velika Gorica well field, have been used for this purpose. Nitrate data sets of the past twenty years (2001–2020) have been analyzed. The data set included three analyses with limit of quantification (LOQ) or limit of detection (LOD) values, one analysis that was not measured and written as "0", and 135 analyses with unavailable data. Since only four analyses were recorded as "0" or as below the LOQ/LOD values, it was decided that these data would not be used. After data processing, approximately 944 analyses, out of 1083 chemical analyses, were used for background estimation. For the evaluation of the nitrate concentrations in the Sava River and precipitation, 15-year time series data (2001–2015) were used. Data included 627 analyses from the samples of river water (provided by Croatian Waters) and 1225 analyses from the samples of precipitation (provided by Croatian Meteorological and Hydrological Service).

### 3.2. Description of Methods

In this work, four well-known model-based objective methods were used to estimate the ABV of nitrate.

The iterative 2-$\sigma$ technique, IT, and the calculated distribution function method, CDF, have been implemented by the Visual Basic (VB) macro entitled BACKGROUND [25], which calculates the background values of chemical parameters and evaluates threshold values that separate background data from anomalous values. The macro integrates these two methods into a widely accessible platform (i.e., Microsoft Excel) and calculates the normal range of background values with 95% confidence. Both methods aim to define the background and threshold by approaching a normal range. The methods process the measured data (i.e., remove values) until a normal distribution is achieved. Thus, what is left in the normally distributed data are the background values and what has been removed are the non-background values [25,40]. The background values are estimated based on data derived from a population, which amounts to a large proportion of the total data set. Statistical methods include the Lilliefors test [68] for testing the normality of data sets and significance of results. If the test statistic D is smaller than the critical value $D_{crit}$, the result is statistically significant, i.e., the determined distribution of the background values corresponds to the normal distribution, which means that the results are acceptable. If the statistic D is greater than the critical value $D_{crit}$, the result is not statistically significant and the obtained background values are overestimated [25]. The advantage of using these two methods is that their application does not require normally or lognormally distributed data, they can be applied to relatively small data sets ($n > 30$), and they are applicable to unimodal and skewed distributions. Both methods work well only when the generally prevailing background data are used to define the background populations [25]. The IT is suited for the calculation of the threshold value as the outer limit of background variation and considers both the low and high values to define anomalies. The disadvantage of this method is that it cannot be applied to a polymodal distribution. The CDF method is applicable to the polymodal distributions if the data unaffected by human activities are dominant. The method is convenient for use if anthropogenic influences tend to lead to enrichments in natural systems. The lower values should thus be free from anthropogenic influences [40]. The IT and CDF methods differ in that the IT method can also be applied if it is necessary to determine the lower limit of the background value of some pollution indicator. This is the case for dissolved oxygen, which can be an indicator of extreme oxygen consumption due to groundwater pollution [25]. However, research related to the definition of the natural background values of chemical substances is usually focused on the determination of the upper limit of the background value [38,39,41,69,70].

The cumulative frequency curve, CFC, or Lepeltier's method [71], is a graphical method that analyzes the cumulative sum on a graph with a logarithmic scale. On the graph constructed in this way, a significant deviation from the lognormal distribution is visible as an abrupt change in the graph, i.e., as a point x where the appearance of the elbow is observed, and all values that are less than or equal to x are the result of the cumulative effect of natural processes that cause significant natural variability [40]. In his original work, Lepeltier suggested cumulating high to low values. The advantage of this method is that it enables the identification of meaningful ABVs for relatively small data sets with a minimum of 50 data [71]. A limitation of using this approach is that care must be taken to avoid the temptation to interpret the visual deviations between natural and anthropogenic populations in the lower part of the curve as significant [30]. This method contains an element of subjectivity due to visual detection and is highly dependent on the experience of the researcher [70].

The probability plot, PP, assumes that different processes generate data with different probability distributions and aims to identify several distributions into distinct components and points that separate them, i.e., the threshold values. If there is a partial overlap of background and non-background distributions, then a change in distribution on the PP can be considered as an inflection point [70]. Inflection points can be identified visually as changes in slope [29], or as an intersection of different trend lines [37]. The concentration at the inflection point is defined as the threshold value, below which all measured values of the substance belong to the background concentrations [33,72]. The advantage of this approach is that it enables the identification of multiple populations on the graph. The limitations of using this method are that a minimum of 100 values is needed and the determination of the threshold represents an estimation procedure, subject to random and systematic error [29,33].

In this paper, the inflection points were identified as the intersection of two "linear" segments, as observed in Rahman et al. [37]. The first inflection point separating the first two subsets is assumed to be ABV, while the first population is considered a natural event. Multiple inflection points on the plots may indicate multiple populations of natural or anthropogenic origin contributing to nitrate concentrations in groundwater [37]. To reduce the subjectivity of visual interpretation, subsets were selected to describe the population with the highest possible coefficient of determination. The minimum number of points in a subset was three and was increased by one point each time until the highest possible coefficient of determination was reached.

The first step was to estimate the ABV for data without prior selection to check the reliability of different methods. If one of the selected methods leads to a statistically non-significant result, data selection is required in the second step and re-estimation of the ABV for selected data in the third step. This is based on the fact that the values of a single normally or lognormally distributed population form a straight line on the cumulative probability plots, without the possibility of determining a background value, whereas the values of more mixed populations lead to a curved line with a pronounced inflection point [29]. Data selection was performed using the Lilliefors test [68] by testing normal distribution. TIBCO StatisticaTM (version 13.5.0.17) was utilized to perform the normality tests, while an alpha value of 0.05 was used as the statistical significance level in all data analyses. The Lilliefors test is an adaptation of the Kolmogorov–Smirnov test and uses the same calculations, but the critical value table in the Lilliefors test table is used instead of the Kolmogorov–Smirnov table, which gives the critical value $D_{crit}$. If the test statistic D is less than the critical value $D_{crit}$, the normal distribution cannot be rejected. Under the condition that the test statistic D is greater than the critical value $D_{crit}$, the normal distribution can be rejected.

All final ABVs must be statistically significant and determined using as large as possible a data set. A statistically significant result obtained in the first step is considered acceptable, while, in case of a statistically non-significant result, data selection is required in order to obtain a statistically significant result.

## 4. Results and Discussion

The descriptive statistics of nitrate concentrations (presented as mg $NO_3$/L) in groundwater in the catchment area of the Velika Gorica well field and Sava River and the precipitation are reported in Table 1. The nitrate concentration of groundwater ranges from 0.2 to 35.7 mg/L, with a mean value of 19.6 mg/L and a median of 19.5 mg/L. It is worth noting that the average value for each monitoring well does not exceed the maximum permissible level for drinking water (50 mg $NO_3$/L), stipulated by EU and Croatian regulations. Nevertheless, these higher concentrations indicate possible natural groundwater quality degradation, considering that, in areas not affected by human activities, the nitrate concentration in groundwater should be much lower [21]. The mean nitrate concentration in the Sava River, as the main source of recharge in the Zagreb aquifer, recorded in the period from 2001 to 2015, is 6.8 mg/L, with the maximum value of 14.6 mg/L. The mean nitrate concentration in precipitation from 2001 to 2015 is 0.8 mg/L, while the maximum measured concentration is 5.2 mg/L. Due to the low nitrate concentrations in precipitation, it can be assumed that precipitation has a very small effect on the nitrate content in the Zagreb aquifer [52]. Furthermore, it has been shown that the local meteoric water line of the Ljubljana area (Slovenia) better describes the stable isotopic composition of groundwater from the Zagreb aquifer than the local meteoric water line of the Zagreb area [22,49,59,73]. This confirms that the Sava River is the main source of recharge of the Zagreb aquifer, since the source of the river water is the precipitation that falls in Slovenia.

**Table 1.** Main statistics for nitrate in mg/L = milligram per liter as $NO_3$. SD is standard deviation, CV is coefficient of variation in %, N is number of samples.

| | Min | Max | Mean | SD | Median | CV | N | Interval |
|---|---|---|---|---|---|---|---|---|
| **All wells** | 0.2 | 35.7 | 19.6 | 5.7 | 19.5 | 29.0 | 944 | 2001–2020 |
| **Čdp-3/2** | 0.2 | 19.8 | 12.4 | 4.3 | 13.7 | 34.6 | 63 | 2001–2020 |
| **Čp-23** | 14.0 | 28.3 | 19.8 | 3.5 | 19.0 | 17.9 | 73 | 2001–2020 |
| **Lg-1** | 4.0 | 35.4 | 25.5 | 5.5 | 25.2 | 21.4 | 100 | 2001–2020 |
| **Lg-4** | 0.2 | 31.0 | 21.1 | 7.0 | 22.6 | 33.2 | 92 | 2001–2002; 2004; 2006–2020 |
| **P-7** | 6.2 | 33.1 | 17.8 | 6.0 | 18.6 | 33.9 | 99 | 2001–2020 |
| **Vg-1** | 15.1 | 32.0 | 20.3 | 3.0 | 19.5 | 14.6 | 112 | 2001–2020 |
| **Vg-10/2** | 14.6 | 34.7 | 20.5 | 2.9 | 20.5 | 14.4 | 66 | 2009–2020 |
| **Vg-11** | 10.2 | 24.9 | 15.1 | 2.5 | 15.1 | 16.8 | 63 | 2001–2020 |
| **Vg-4** | 15.5 | 27.4 | 21.3 | 2.5 | 21.2 | 11.7 | 110 | 2001–2020 |
| **Vg-6** | 12.4 | 27.5 | 17.3 | 3.0 | 17.1 | 17.3 | 73 | 2001–2020 |
| **Vg-9** | 11.5 | 20.2 | 15.1 | 1.8 | 15.1 | 11.9 | 58 | 2003–2020 |
| **Vg-5/2** | 17.7 | 35.7 | 28.9 | 3.3 | 29.2 | 11.5 | 35 | 2014–2020 |
| **Sava River** | 1.3 | 14.6 | 6.8 | 1.9 | 6.6 | 28.0 | 627 | 2001–2015 |
| **Precipitation** | 0.03 | 5.2 | 0.8 | 0.7 | 0.5 | 94.8 | 1225 | 2001–2015 |

First, the ABV of nitrate was estimated in groundwater for data without selection referring to all monitoring wells, using four different methods (Figure 3). The inflection points were determined according to the principle of the highest possible coefficient of determination using the CFC and PP methods. Estimated values vary from 7.1 to 11.8 mg/L. Usage of the IT method resulted in the highest estimate of the background value, while the CDF method gave a non-statistically significant result. As observed in Nakić et al. [25], if D > $D_{crit}$, the result is not statistically significant and the obtained background values are overestimated. Considering that the application of the CDF method gave a noticeable difference in terms of non-statistically significant results, the second step was data selection, performed using the Lilliefors test and re-estimation of the ABV.

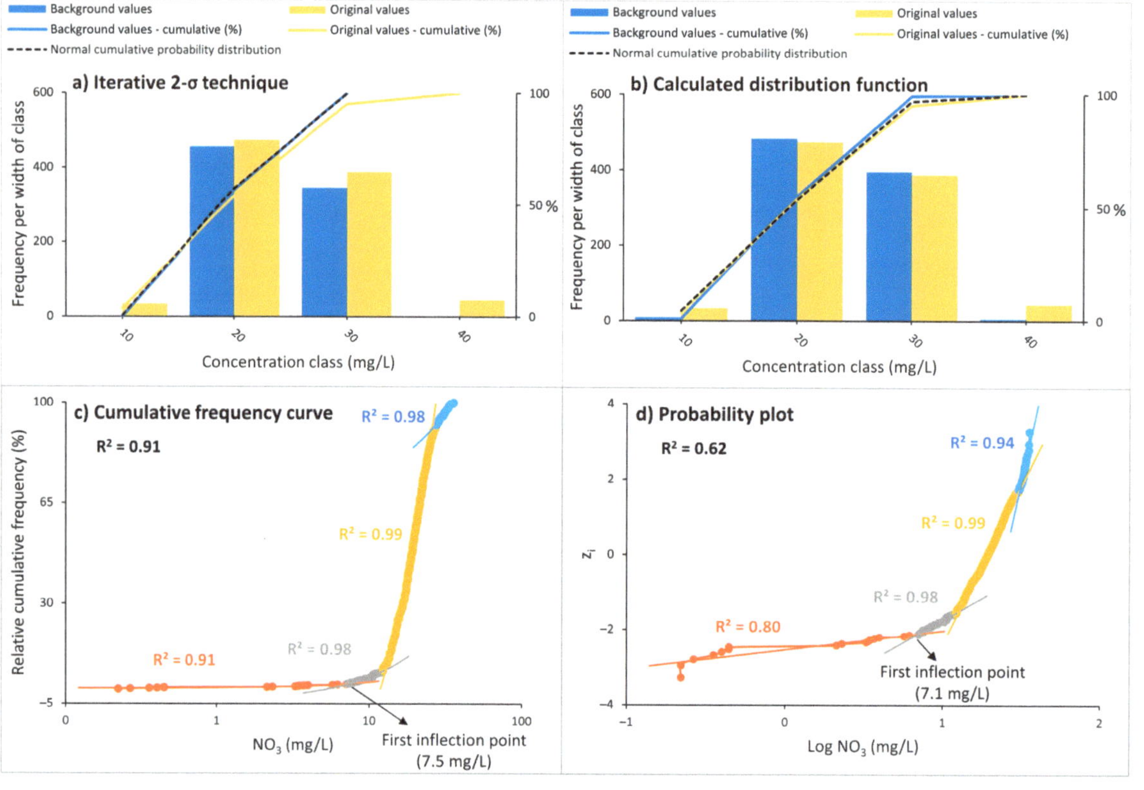


**Figure 3.** Four different methods to estimate ABV of nitrate for data without selection: (**a**) IT: lower background value: 11.8; D < $D_{crit}$, (**b**) CDF: lower background value: 8.3; D > $D_{crit}$, (**c**) CFC: background value: 7.5 and (**d**) PP: background value: 7.1. Arrow indicates first inflection point that separates background and non-background population.

The Lilliefors test was used to test the normality for each well separately (Table 2). This test showed normally distributed data for monitoring wells Vg-11, Vg-4, Vg-6, Vg-9, and Vg-5/2, which are not further considered for the background value estimation procedure in the third step, because, as observed in Panno et al. [29], a single normally or lognormally population forms a straight line on the cumulative probability plots, without the possibility of determining a background value. The data that were not normally distributed (wells Čdp-3/2, Čp-23, Lg-1, Lg-4, P-7, Vg-1 and Vg-10/2) were grouped for further processing as NON-ND data, because it is expected that the cumulative probability plots contain a curved line with at least one inflection point that could indicate the existence of a minimum of two different populations, natural and anthropogenically conditioned. Histograms and normal distribution curves for each well separately are shown in Figure 4.

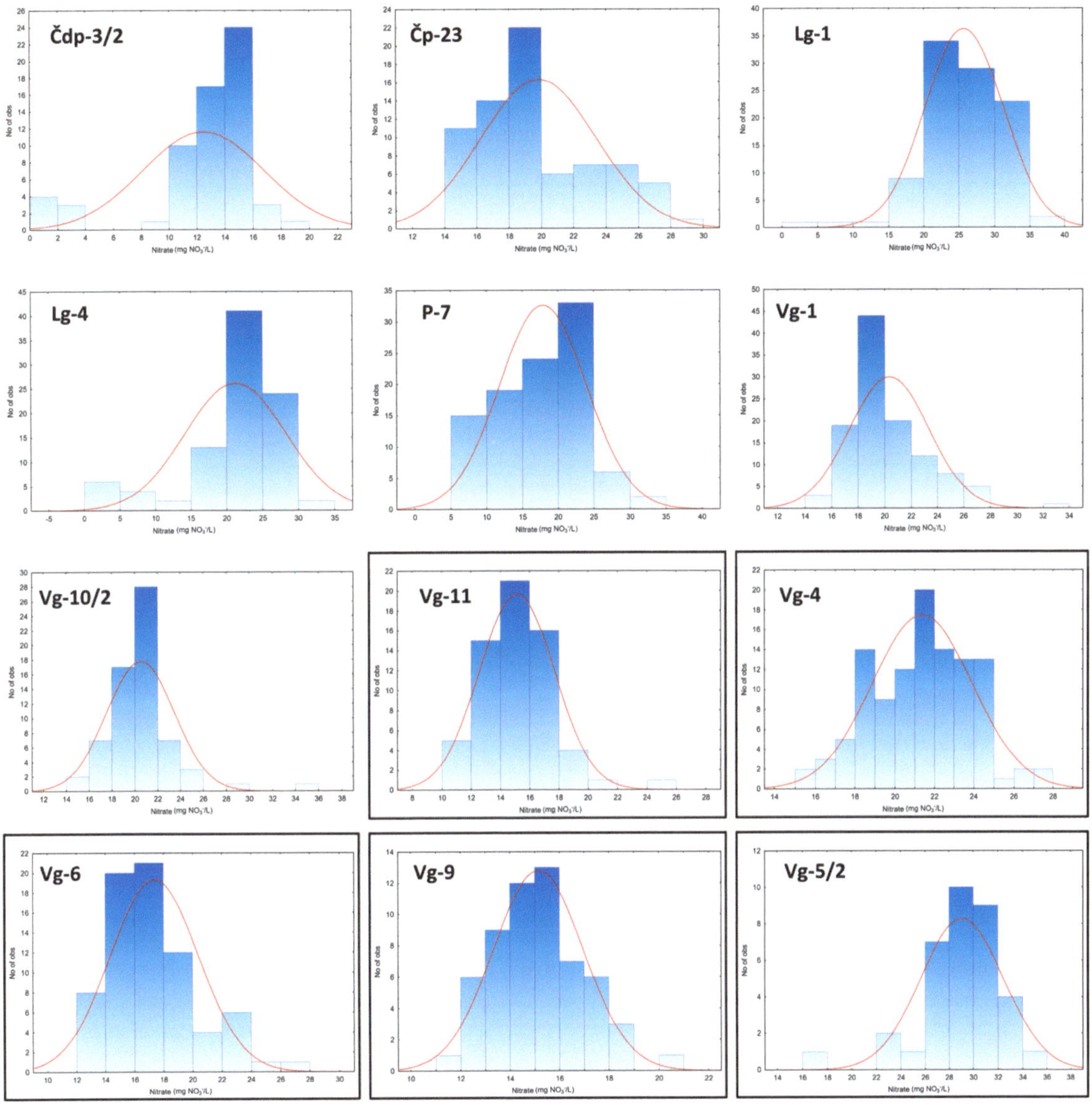


**Figure 4.** Histograms and normal distribution curves. Normally distributed data are bolded.

The third step was to re-estimate the ABV of nitrate in groundwater for selected data referring to NON-ND data, using the same methods. The use of selected data resulted in statistically significant results for all methods (Figure 5). It is important to emphasize that the result obtained with data without selection that was not statistically significant became significant.

**Table 2.** Lilliefors test of normality. Normally distributed data are bolded ($D < D_{crit}$).

| Monitoring Well | Lilliefors Test | |
|---|---|---|
| | D | $D_{crit}$ |
| Čdp-3/2 | 0.232 | 0.112 |
| Čp-23 | 0.135 | 0.104 |
| Lg-1 | 0.098 | 0.089 |
| Lg-4 | 0.212 | 0.092 |
| P-7 | 0.106 | 0.089 |
| Vg-1 | 0.140 | 0.084 |
| Vg-10/2 | 0.148 | 0.109 |
| **Vg-11** | **0.086** | **0.112** |
| **Vg-4** | **0.070** | **0.084** |
| **Vg-6** | **0.099** | **0.104** |
| **Vg-9** | **0.087** | **0.116** |
| **Vg-5/2** | **0.149** | **0.150** |

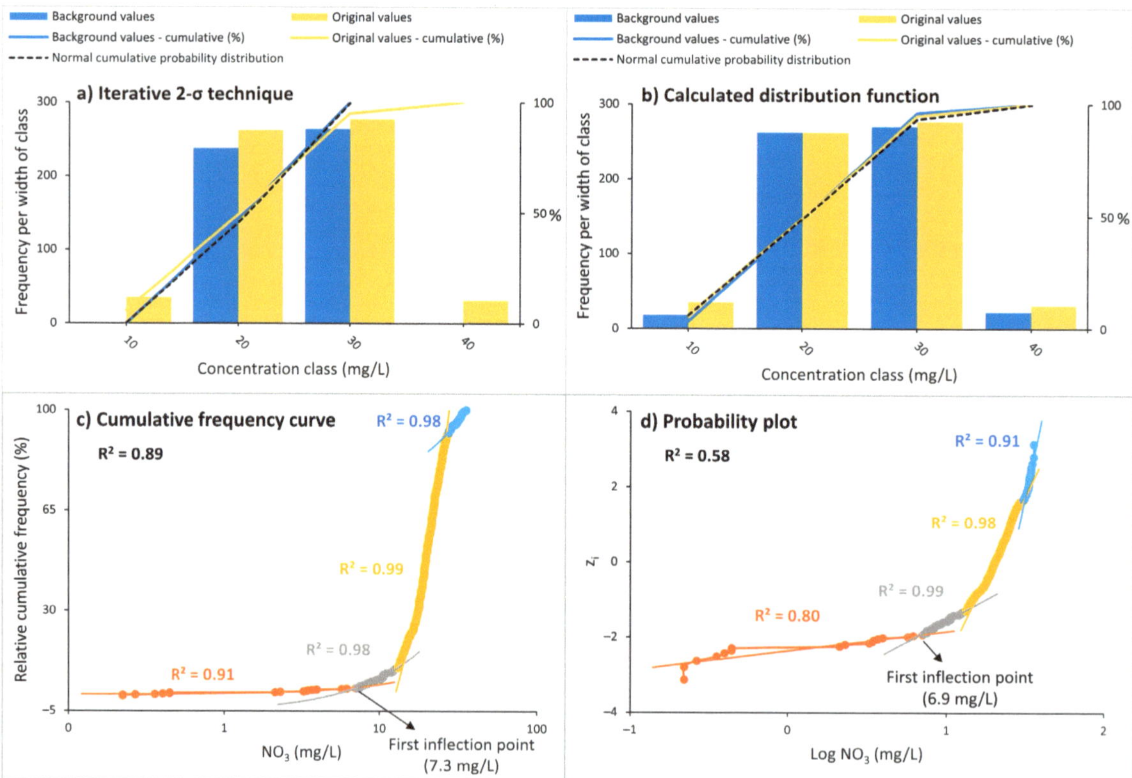


**Figure 5.** Four different methods to estimate ABV of nitrate for selected NON-ND data: (**a**) IT: lower background value: 13.2; $D < D_{crit}$, (**b**) CDF: lower background value: 7.2; $D < D_{crit}$, (**c**) CFC: background value: 7.3 and (**d**) PP: background value: 6.9. Arrow indicates first inflection point that separates background and non-background population.

Using the IT method again resulted in the highest background value and both results were statistically significant, while the estimated value increased from 11.8 to 13.2. Compared to other methods, IT gave a higher estimation of the nitrate background value because it considers both the low and high values to define anomalies and cannot be applied to a polymodal distribution. The CDF method gave a noticeable difference in terms of non-statistically significant results for data without selection (8.3 mg/L) and statistically significant results for a set of selected data (7.2 mg/L). Although this suggests that the CDF method depends more on which data are used for the calculation with respect to the IT, our results correspond to the results of previous research, where it was shown that it is applicable to polymodal distributions and convenient for use in conditions of anthropogenic activity [40], which are present in the catchment area of the Velika Gorica well field. Results suggest that the usage of CDF is the most applicable when using only NON-ND data, which can be confirmed in the estimation of the statistically significant background value of 7.2 mg/L, unlike the estimate calculated with the all data without selection. Furthermore, Urresti-Estalla et al. [39] suggested that the usage of both IT and CDF methods depends on the nature of the parameter to be estimated as well as the type of frequency distribution curve presented by the data series. From this perspective, the most important criterion to define is whether the low concentration of the parameter that is observed represents a possible indicator of contamination. If not, then CDF should be used instead of the IT only when aquifers are undergoing little pressure.

The usage of lower limits with the first two methods is more convenient in areas under long-term human impacts and corresponds to the usage of the first inflection points from CFCs and PPs [74], which, in the end, makes the results estimated by all methods comparable. The first inflection points observed at plots were directly considered as ABVs, because the first population of the data set was considered to be the consequence of the natural conditions that prevail in the investigated part of the aquifer. The ABV determined by the CFC method decreased from 7.5 to 7.3 mg/L, while the background value determined by the PP method decreased from 7.1 to 6.9 mg/L, for all data and for selected data, respectively. The nitrate concentration showed more than one inflection point on plots, indicating multiple local events of nitrate contamination in groundwater. This is consistent with previous research where it was defined that elevated nitrate concentrations are the result of different sources of contamination, i.e., wastewater and manure [49]. In general, both CFCs and PPs resulted in the extraction of four sets, i.e., populations. Additionally, the aim was also to increase the coefficient of determination of the chosen subset compared to the initial value that was calculated for the entire population. In all cases, the significant improvement in the coefficient of determination was present when dividing entire population into more sub-populations due to different inflection points. However, it must be emphasized that, in our case, CFCs with the entire population had very high values for the coefficient of determination of 0.91 and 0.89 (Figures 3 and 5). This was not the case with PPs, which had starting values for the coefficient of determination of 0.62 and 0.58 (Figures 3 and 5). As observed in Panno et al. [29], the CFC and PP methods can be used in the estimation of nitrate ABVs only when observation wells with non-normally distributed data are included. If these data are not present within the observed data, it is possible that the inflection point will not be recognized.

The study has shown that the CFC and PP methods for data without selection and the CDF method for selected NON-ND data gave consistent and realistic estimations of ABVs that are comparable to nitrate concentrations in the main source of recharge of the Zagreb aquifer, i.e., Sava River [58–61]. When observing all acceptable (statistically significant) results together, it is evident that all methods, except the IT, gave very similar estimations of nitrate ABVs ranging from 7.1 mg/L to 7.5 mg/L. Preziosi et al. [53] and this research confirmed that the background value of nitrate determined at the level of the Zagreb aquifer [25,52] differs from the background value of nitrate determined in the catchment area of the Velika Gorica well field, a homogeneous area within the same aquifer.

## 5. Conclusions

The ABVs and threshold values are crucial for identifying groundwater quality and pollution migration. The nitrate concentrations, which are a potential pressure and concern in the Zagreb aquifer, were analyzed. The new approach to estimate the background value of nitrate as a basis for better groundwater management and protection in areas under long-term human impact was applied. The study is based on statistical methodologies developed to separate the natural and anthropogenic populations in monitored concentrations from the national groundwater quality monitoring program of Croatian Waters over a 20-year period. The results show that using data without selection can lead to statistically non-significant results, but, with further data selection, a statistically non-significant result became a statistically significant one. In conclusion, all final results must be statistically significant and determined using as wide as possible a data set. Reducing the size of the data set is acceptable only in the case of statistically non-significant results.

**Author Contributions:** Conceptualization and methodology, P.B., J.P. and Z.K.; formal analysis and writing—original draft preparation, P.B.; writing—review and editing, J.P., Z.K. and Z.N.; visualization, P.B. and J.P. All authors have read and agreed to the published version of the manuscript.

**Funding:** This research received no external funding.

**Data Availability Statement:** The data presented in this study are available on request from Croatian Waters and the Croatian Meteorological and Hydrological Service.

**Acknowledgments:** We would like to thank Croatian Waters and the Croatian Meteorological and Hydrological Service for providing the data. The present research was supported by the Croatian Science Foundation (HRZZ) under the Young Researches' Career Development Project—Training New Doctoral Students; DOK-2020-01: The use of nitrogen and oxygen stable isotopes in the determination of nitrate origin in the unsaturated and saturated zone of the Velika Gorica well field.

**Conflicts of Interest:** The authors declare no conflict of interest.

## References

1. Sunitha, V.; Sudharshan Reddy, Y.; Suvarna, B.; Reddy, B.M. Human Health Risk Assessment (HHRA) of Fluoride and Nitrate Using Pollution Index of Groundwater (PIG) in and around Hard Rock Terrain of Cuddapah, A.P. South India. *Environ. Chem. Ecotoxicol.* **2022**, *4*, 113–123. [CrossRef]
2. United Nations. *The Sustainable Development Goals Report 2021*; United Nations Publications: New York, NY, USA, 2021.
3. Adimalla, N.; Qian, H.; Nandan, M.J. Groundwater Chemistry Integrating the Pollution Index of Groundwater and Evaluation of Potential Human Health Risk: A Case Study from Hard Rock Terrain of South India. *Ecotoxicol. Env. Saf.* **2020**, *206*, 111217. [CrossRef] [PubMed]
4. Abascal, E.; Gómez-Coma, L.; Ortiz, I.; Ortiz, A. Global Diagnosis of Nitrate Pollution in Groundwater and Review of Removal Technologies. *Sci. Total Environ.* **2022**, *810*, 152233. [CrossRef]
5. Pérez-Indoval, R.; Rodrigo-Ilarri, J.; Cassiraga, E.; Rodrigo-Clavero, M.E. PWC-Based Evaluation of Groundwater Pesticide Pollution in the Júcar River Basin. *Sci. Total Environ.* **2022**, *847*, 157386. [CrossRef]
6. Rezaei, A.; Hassani, H.; Tziritis, E.; Fard Mousavi, S.B.; Jabbari, N. Hydrochemical Characterization and Evaluation of Groundwater Quality in Dalgan Basin, SE Iran. *Groundw. Sustain. Dev.* **2020**, *10*, 100353. [CrossRef]
7. Mekonnen, M.M.; Hoekstra, A.Y. Four Billion People Facing Severe Water Scarcity. *Sci. Adv.* **2016**, *2*, e1500323. [CrossRef] [PubMed]
8. Díaz-Alcaide, S.; Martínez-Santos, P. Review: Advances in Groundwater Potential Mapping. *Hydrogeol. J.* **2019**, *27*, 2307–2324. [CrossRef]
9. Majkić-Dursun, B.; Oros, I.; Oparušić, I.; Petković, A. Assessing Groundwater Quality for Multiuse and Geochemical Evolution in the South Banat Area of Serbia, Pannonian Basin. *Pol. J. Env. Stud.* **2019**, *28*, 3269–3278. [CrossRef]
10. Erisman, J.W.; Galloway, J.N.; Seitzinger, S.; Bleeker, A.; Dise, N.B.; Roxana Petrescu, A.M.; Leach, A.M.; de Vries, W. Consequences of Human Modification of the Global Nitrogen Cycle. *Philos. Trans. R. Soc. B Biol. Sci.* **2013**, *368*, 20130116. [CrossRef] [PubMed]
11. Hosono, T.; Tokunaga, T.; Kagabu, M.; Nakata, H.; Orishikida, T.; Lin, I.T.; Shimada, J. The Use of Δ15N and Δ18O Tracers with an Understanding of Groundwater Flow Dynamics for Evaluating the Origins and Attenuation Mechanisms of Nitrate Pollution. *Water Res.* **2013**, *47*, 2661–2675. [CrossRef]
12. Liu, A.; Ming, J.; Ankumah, R.O. Nitrate Contamination in Private Wells in Rural Alabama, United States. *Sci. Total Environ.* **2005**, *346*, 112–120. [CrossRef] [PubMed]

13. Ma, Z.; Yang, Y.; Lian, X.; Jiang, Y.; Xi, B.; Peng, X.; Yan, K. Identification of Nitrate Sources in Groundwater Using a Stable Isotope and 3DEEM in a Landfill in Northeast China. *Sci. Total Environ.* **2016**, *563–564*, 593–599. [CrossRef]
14. Rockström, J.; Steffen, W.; Noone, K.; Persson, Å.; Chapin, F.S., III; Lambin, E.F.; Lenton, T.M.; Scheffer, M.; Folke, C.; Schellnhuber, H.J.; et al. A Safe Operation Space for Humanity. *Nature* **2009**, *461*, 472–475. [CrossRef] [PubMed]
15. Romanelli, A.; Soto, D.X.; Matiatos, I.; Martínez, D.E.; Esquius, S. A Biological and Nitrate Isotopic Assessment Framework to Understand Eutrophication in Aquatic Ecosystems. *Sci. Total Environ.* **2020**, *715*, 136909. [CrossRef]
16. Xue, D.; de Baets, B.; van Cleemput, O.; Hennessy, C.; Berglund, M.; Boeckx, P. Use of a Bayesian Isotope Mixing Model to Estimate Proportional Contributions of Multiple Nitrate Sources in Surface Water. *Environ. Pollut.* **2012**, *161*, 43–49. [CrossRef]
17. Xue, D.; Botte, J.; de Baets, B.; Accoe, F.; Nestler, A.; Taylor, P.; van Cleemput, O.; Berglund, M.; Boeckx, P. Present Limitations and Future Prospects of Stable Isotope Methods for Nitrate Source Identification in Surface- and Groundwater. *Water Res.* **2009**, *43*, 1159–1170. [CrossRef] [PubMed]
18. Xu, S.; Kang, P.; Sun, Y. A Stable Isotope Approach and Its Application for Identifying Nitrate Source and Transformation Process in Water. *Environ. Sci. Pollut. Res.* **2016**, *23*, 1133–1148. [CrossRef]
19. Fenech, C.; Rock, L.; Nolan, K.; Tobin, J.; Morrissey, A. The Potential for a Suite of Isotope and Chemical Markers to Differentiate Sources of Nitrate Contamination: A Review. *Water Res.* **2012**, *46*, 2023–2041. [CrossRef]
20. Ahada, C.P.S.; Suthar, S. Groundwater Nitrate Contamination and Associated Human Health Risk Assessment in Southern Districts of Punjab, India. *Environ. Sci. Pollut. Res.* **2018**, *25*, 25336–25347. [CrossRef]
21. Burkart, M.R.; Kolpin, D.W. Hydrologic and Land-Use Factors Associated with Herbicides and Nitrate in Near-Surface Aquifers. *J. Environ. Qual.* **1993**, *22*, 646–656. [CrossRef]
22. Nakić, Z.; Ružičić, S.; Posavec, K.; Mileusnić, M.; Parlov, J.; Bačani, A.; Durn, G. Conceptual Model for Groundwater Status and Risk Assessment—Case Study of the Zagreb Aquifer System. *Geol. Croat.* **2013**, *66*, 55–77. [CrossRef]
23. de Caro, M.; Crosta, G.B.; Frattini, P. Hydrogeochemical Characterization and Natural Background Levels in Urbanized Areas: Milan Metropolitan Area (Northern Italy). *J. Hydrol.* **2017**, *547*, 455–473. [CrossRef]
24. Reimann, C.; Garrett, R.G. Geochemical Background—Concept and Reality. *Sci. Total Environ.* **2005**, *350*, 12–27. [CrossRef] [PubMed]
25. Nakić, Z.; Posavec, K.; Bačani, A. A Visual Basic Spreadsheet Macro for Geochemical Background Analysis. *Ground Water* **2007**, *45*, 642–647. [CrossRef] [PubMed]
26. Sellerino, M.; Forte, G.; Ducci, D. Identification of the Natural Background Levels in the Phlaegrean Fields Groundwater Body (Southern Italy). *J. Geochem. Explor.* **2019**, *200*, 181–192. [CrossRef]
27. Cruz, J.; Andrade, C. Natural Background Groundwater Composition in the Azores Archipelago (Portugal): A Hydrogeochemical Study and Threshold Value Determination. *Sci. Total Environ.* **2015**, *520*, 127–135. [CrossRef] [PubMed]
28. Biddau, R.; Cidu, R.; Lorrai, M.; Mulas, M.G. Assessing Background Values of Chloride, Sulfate and Fluoride in Groundwater: A Geochemical-Statistical Approach at a Regional Scale. *J. Geochem. Explor.* **2017**, *181*, 243–255. [CrossRef]
29. Panno, S.; Kelly, W.R.; Martinsek, A.T.; Hackley, K.C. Estimating Background and Threshold Nitrate Concentrations Using Probability Graphs. *Ground Water* **2006**, *44*, 697–709. [CrossRef]
30. Parslow, G.R. Determination of Background and Threshold in Exploration Geochemistry. *J. Geochem. Explor.* **1974**, *3*, 319–336. [CrossRef]
31. Kelly, W.R.; Panno, S.V. Some Considerations in Applying Background Concentrations to Ground Water Studies. *Ground Water* **2008**, *46*, 790–792. [CrossRef]
32. Filzmoser, P.; Garrett, R.G.; Reimann, C. Multivariate Outlier Detection in Exploration Geochemistry. *Comput. Geosci.* **2005**, *31*, 579–587. [CrossRef]
33. Sinclair, A.J. A Fundamental Approach to Threshold Estimation in Exploration Geochemistry: Probability Plots Revisited. *J. Geochem. Explor.* **1991**, *41*, 1–22. [CrossRef]
34. Reimann, C.; Filzmoser, P. Normal and Lognormal Data Distribution in Geochemistry: Death of a Myth. Consequences for the Statistical Treatment of Geochemical and Environmental Data. *Environ. Geol.* **2000**, *39*, 1001–1014. [CrossRef]
35. Geranian, H.; Mokhtari, A.R.; Cohen, D.R. Bivariate Probability Plots: A Method for Delineating Different Populations in Soil Geochemical Data. *Sci. Total Environ.* **2019**, *671*, 1047–1055. [CrossRef]
36. Kim, K.-H.; Yun, S.-T.; Kim, H.-K.; Kim, J.-W. Determination of Natural Backgrounds and Thresholds of Nitrate in South Korean Groundwater Using Model-Based Statistical Approaches. *J. Geochem. Explor.* **2015**, *148*, 196–205. [CrossRef]
37. Rahman, A.; Tiwari, K.K.; Mondal, N.C. Assessment of Hydrochemical Backgrounds and Threshold Values of Groundwater in a Part of Desert Area, Rajasthan, India. *Environ. Pollut.* **2020**, *266*, 115150. [CrossRef] [PubMed]
38. Bulut, O.F.; Duru, B.; Çakmak, Ö.; Günhan, Ö.; Dilek, F.B.; Yetis, U. Determination of Groundwater Threshold Values: A Methodological Approach. *J. Clean. Prod.* **2020**, *253*. [CrossRef]
39. Urresti-Estala, B.; Carrasco-Cantos, F.; Vadillo-Pérez, I.; Jiménez-Gavilán, P. Determination of Background Levels on Water Quality of Groundwater Bodies: A Methodological Proposal Applied to a Mediterranean River Basin (Guadalhorce River, Málaga, Southern Spain). *J. Environ. Manag.* **2013**, *117*, 121–130. [CrossRef] [PubMed]
40. Matschullat, J.; Ottenstein, R.; Reimann, C. Geochemical Background—Can We Calculate It? *Environ. Geol.* **2000**, *39*, 990–1000. [CrossRef]

41. Amiri, V.; Nakhaei, M.; Lak, R.; Li, P. An Integrated Statistical-Graphical Approach for the Appraisal of the Natural Background Levels of Some Major Ions and Potentially Toxic Elements in the Groundwater of Urmia Aquifer, Iran. *Environ. Earth Sci.* **2021**, *80*, 432. [CrossRef]
42. Lions, J.; Devau, N.; Elster, D.; Voutchkova, D.D.; Hansen, B.; Schullehner, J.; Petrović Pantić, T.; Samolov, K.A.; Camps, V.; Arnó, G.; et al. A Broad-Scale Method for Estimating Natural Background Levels of Dissolved Components in Groundwater Based on Lithology and Anthropogenic Pressure. *Water* **2021**, *13*, 1531. [CrossRef]
43. Molinari, A.; Guadagnini, L.; Marcaccio, M.; Guadagnini, A. Natural Background Levels and Threshold Values of Chemical Species in Three Large-Scale Groundwater Bodies in Northern Italy. *Sci. Total Environ.* **2012**, *425*, 9–19. [CrossRef] [PubMed]
44. Parrone, D.; Ghergo, S.; Preziosi, E. A Multi-Method Approach for the Assessment of Natural Background Levels in Groundwater. *Sci. Total Environ.* **2019**, *659*, 884–894. [CrossRef] [PubMed]
45. Preziosi, E.; Parrone, D.; del Bon, A.; Ghergo, S. Natural Background Level Assessment in Groundwaters: Probability Plot versus Pre-Selection Method. *J. Geochem. Explor.* **2014**, *143*, 43–53. [CrossRef]
46. Kovač, Z.; Nakić, Z.; Pavlić, K. Influence of Groundwater Quality Indicators on Nitrate Concentrations in the Zagreb Aquifer System. *Geol. Croat.* **2017**, *70*, 93–103. [CrossRef]
47. Kovač, Z.; Cvetković, M.; Parlov, J. Gaussian Simulation of Nitrate Concentration Distribution in the Zagreb Aquifer. *J. Maps* **2017**, *13*, 727–732. [CrossRef]
48. Huljek, L.; Perković, D.; Kovač, Z. Nitrate Contamination Risk of the Zagreb Aquifer. *J. Maps* **2019**, *15*, 570–577. [CrossRef]
49. Kovač, Z.; Nakić, Z.; Barešić, J.; Parlov, J. Nitrate Origin in the Zagreb Aquifer System. *Geofluids* **2018**, *2018*, 2789691. [CrossRef]
50. Kovač, Z.; Nakić, Z.; Špoljarić, D.; Stanek, D.; Bačani, A. Estimation of Nitrate Trends in the Groundwater of the Zagreb Aquifer. *Geosciences* **2018**, *8*, 159. [CrossRef]
51. Kovač, Z.; Bačani, L.; Ružičić, S.; Parlov, J.; Posavec, K.; Buškulić, P. Using Water Stable Isotopes and Cross-Correlation Analysis to Characterize Infiltration of Precipitation through Unsaturated Zone at the Velika Gorica Site of Zagreb Aquifer. *J. Hydrol. Eng.* **2023**, *28*, 04023002. [CrossRef]
52. Kovač, Z. *Podrijetlo nitrata u podzemnoj vodi zagrebačkoga aluvijalnoga vodonosnika (Nitrate Origin in Groundwater of the Zagreb Alluvial Aquifer)*; University of Zagreb, Faculty of Mining, Geology and Petroleum Engineering: Zagreb, Croatia, 2017.
53. Preziosi, E.; Giuliano, G.; Vivona, R. Natural Background Levels and Threshold Values Derivation for Naturally As, V and F Rich Groundwater Bodies: A Methodological Case Study in Central Italy. *Environ. Earth Sci.* **2010**, *61*, 885–897. [CrossRef]
54. Velic, J.; Saftic, B. Subsurface Spreading and Facies Characteristics of Middle Pleistocene Deposits between Zaprešić and Samobor. *Geološki Vjesn* **1991**, *44*, 69–82.
55. Velić, J.; Saftić, B.; Malvić, T. Lithologic Composition and Stratigraphy of Quaternary Sediments in the Area of the "Jakuševec" Waste Depository (Zagreb, Northern Croatia). *Geol. Croat.* **1999**, *52*, 119–130.
56. Velić, J.; Durn, G. Alternating Lacustrine-Marsh Sedimentation and Subaerial Exposure Phases during Quaternary: Prečko, Zagreb, Croatia. *Geol. Croat.* **1993**, *46*, 71–90.
57. Ružičić, S.; Mileusnić, M.; Posavec, K. Building Conceptual and Mathematical Model for Water Flow and Solute Transport in the Unsaturated Zone at Kosnica Site. *Rud. -Geološko-Naft. Zb.* **2012**, *25*, 21–31.
58. Posavec, K.; Vukojević, P.; Ratkaj, M.; Bedeniković, T. Cross-Correlation Modelling of Surface Water—Groundwater Interaction Using the Excel Spreadsheet Application. *Rud. -Geološko-Naft. Zb.* **2017**, *32*, 25–32. [CrossRef]
59. Parlov, J.; Kovač, Z.; Nakić, Z.; Barešić, J. Using Water Stable Isotopes for Identifying Groundwater Recharge Sources of the Unconfined Alluvial Zagreb Aquifer (Croatia). *Water* **2019**, *11*, 2177. [CrossRef]
60. Barešić, J.; Parlov, J.; Kovač, Z.; Sironić, A. Use of Nuclear Power Plant Released Tritium as a Groundwater Tracer. *Rud. -Geološko-Naft. Zb.* **2020**, *35*, 25–35. [CrossRef]
61. Kovač, Z.; Barešić, J.; Parlov, J.; Sironić, A. Impact of Hydrological Conditions on the Isotopic Composition of the Sava River in the Area of the Zagreb Aquifer. *Water* **2022**, *14*, 2263. [CrossRef]
62. Posavec, K. *Identifikacija i prognoza minimalnih razina podzemne vode zagrebackoga aluvijalnog vodonosnika modelima recesijskih krivulja (Identification and Prediction of Minimum Ground Water Levels of Zagreb Alluvial Aquifer Using Recession Curve Models)*; University of Zagreb, Faculty of Mining, Geology and Petroleum Engineering: Zagreb, Croatia, 2006.
63. Bogunović, M.; Vidaček, Ž.; Husnjak, S.; Sraka, M. Inventory of Soils in Croatia. *Agric. Conspec. Sci.* **1998**, *63*, 105–112.
64. Ružičić, S. *Model Transporta Potencijalno Toksičnih Elemenata Kroz Nesaturiranu Zonu na Području Regionalnoga Vodocrpilišta Kosnica (Transport Model of Potentially Toxic Elements through Unsaturated Zone at Regional Wellfield Kosnica)*; Faculty of Mining, Geology and Petroleum Engineering, University of Zagreb: Zagreb, Croatia, 2013.
65. Nakić, Z.; Horvat, S.; Bačani, A. Statistical Indicators of Groundwater Geochemical Characteristics in a Quaternary Aquifer from the Mala Mlaka Well Field Catchment Area (Zagreb, Croatia). *Geol. Croat.* **2005**, *58*, 87–99. [CrossRef]
66. Vlahović, T.; Bačani, A.; Posavec, K. Hydrogeochemical Stratification of the Unconfined Samobor Aquifer (Zagreb, Croatia). *Environ. Geol.* **2009**, *57*, 1707–1722. [CrossRef]
67. Buškulić, P.; Parlov, J. Hydrogeochemical Facies of Groundwater in the Influence Area of the Velika Gorica Well Field. In *Proceedings of the Mathematical Methods and Terminology in Geology 2022, 4th Croatian Scientific Congress about Geomathematics and Terminology in Geology*; Malvić, T., Ivšinović, J., Eds.; University of Zagreb, Faculty of Mining, Geology and Petroleum Engineering: Zagreb, Croatia, 2022; pp. 1–196.

68. Lilliefors, H.W. On the Kolmogorov-Smirnov Test for Normality with Mean and Variance Unknown. *J. Am. Stat. Assoc.* **1967**, *62*, 399–402. [CrossRef]
69. de Lima Rodrigues, A.S.; Malafaia, G.; Costa, A.T.; Nalini Junior, H.A. Background Values for Chemical Elements in Sediments of the Gualaxo Do Norte River Basin, MG, Brazil. *Rev. De Ciências Ambient. —RCA* **2013**, *7*, 15–32. [CrossRef]
70. Nakić, Z.; Kovač, Z.; Parlov, J.; Perković, D. Ambient Background Values of Selected Chemical Substances in Four Groundwater Bodies in the Pannonian Region of Croatia. *Water* **2020**, *12*, 2671. [CrossRef]
71. Lepeltier, C. A Simplified Statistical Treatment of Geochemical Data by Graphical Representation. *Econ. Geol.* **1969**, *64*, 538–550. [CrossRef]
72. Nakić, Z.; Posavec, K.; Parlov, J. Model-Based Objective Methods for the Estimation of Groundwater Geo- Chemical Background. *Aqua Mundi* **2010**, *1*, 65–72. [CrossRef]
73. Marković, T.; Brkić, Ž.; Larva, O. Using Hydrochemical Data and Modelling to Enhance the Knowledge of Groundwater Flow and Quality in an Alluvial Aquifer of Zagreb, Croatia. *Sci. Total Environ.* **2013**, *458–460*, 508–516. [CrossRef]
74. Rahman, A.; Mondal, N.C.; Tiwari, K.K. Anthropogenic Nitrate in Groundwater and Its Health Risks in the View of Background Concentration in a Semi Arid Area of Rajasthan, India. *Sci. Rep.* **2021**, *11*, 9279. [CrossRef] [PubMed]

*sustainability*

MDPI

*Article*

# Investigating the Potential Impact on Shallow Groundwater Quality of Oily Wastewater Injection in Deep Petroleum Reservoirs: A Multidisciplinary Evaluation at the Val d'Agri Oilfield (Southern Italy)

**Pietro Rizzo [1,*], Antonio Bucci [2], Pamela Monaco [2], Anna Maria Sanangelantoni [1], Gino Naclerio [2], Mattia Rossi [1], Paola Iacumin [1], Federica Bianchi [1], Claudio Mucchino [1], Nicolò Riboni [1], Dario Avagliano [3], Francesco Coraggio [3], Antonella Caputi [4] and Fulvio Celico [1]**

[1] Department of Chemistry, Life Sciences and Environmental Sustainability, University of Parma, Parco Area delle Scienze 157/A, 43124 Parma, Italy; fulvio.celico@unipr.it (F.C.)

[2] Department of Biosciences and Territory, University of Molise, C.da Fonte Lappone, 86090 Pesche, Italy; antonio.bucci@unimol.it (A.B.)

[3] ENI S.p.A., 20097 Milan, Italy

[4] Italconsult, 85059 Viggiano, Italy

* Correspondence: pietro.rizzo@unipr.it; Tel.: +39-0521-905330

**Abstract:** The increase in oil production from petroleum reservoirs has led to studies examining the effects of these activities on groundwater quality. Oily wastewater associated with oil production is often reinjected through abandoned wells into the unproductive portions of the reservoir to avoid its discharge on the surface. The reinjection process is designed to be environmentally friendly and to exclude direct interactions between injected fluids and the surrounding groundwater; nevertheless, the evaluation of the compatibility between this process and the protection of the surrounding environment is of utmost importance when oilfields are located within sensitive and protected areas. The present work aimed to evaluate the impact of the oily wastewater reinjection into a long-term and high-rate disposal well in the Val d'Agri oilfield (Southern Italy). Previous preliminary investigations carried out at the study site led researchers to hypothesize the possible hydrocarbon contamination of the shallower aquifer caused by reinjection well integrity issues. Our strategy is based on an integrated and multidisciplinary approach involving isotopic (stable isotopes $^{2}$H and $^{18}$O), chemical, and microbiological (characterization of bacterial and archaeal communities) analyses. After a comprehensive and meticulous examination of the research data, it has been ascertained that significant discrepancies exist between the shallow and reinjection water systems. This allowed us to clarify the area's complex flow dynamics and exclude hydrocarbon contamination of spring waters caused by the reinjection process.

**Keywords:** oily wastewater reinjection; petroleum reservoir; groundwater quality; microbial community characterization; stable isotopes; Southern Italy

**Citation:** Rizzo, P.; Bucci, A.; Monaco, P.; Sanangelantoni, A.M.; Naclerio, G.; Rossi, M.; Iacumin, P.; Bianchi, F.; Mucchino, C.; Riboni, N.; et al. Investigating the Potential Impact on Shallow Groundwater Quality of Oily Wastewater Injection in Deep Petroleum Reservoirs: A Multidisciplinary Evaluation at the Val d'Agri Oilfield (Southern Italy). *Sustainability* **2023**, *15*, 9161. https://doi.org/10.3390/su15129161

Academic Editors: Zongjun Gao and Jiutan Liu

Received: 5 April 2023
Revised: 29 May 2023
Accepted: 31 May 2023
Published: 6 June 2023



## 1. Introduction

Oily wastewaters associated with oil production are often reinjected into unproductive portions of petroleum reservoirs through abandoned wells. The reinjection process is recognized worldwide and is environmentally friendly, without any impact on groundwater flowing within the shallow geological sequence through which the reinjection wells have been drilled. Obviously, the screened interval of the reinjection wells involves only the deep petroleum reservoir to avoid direct interactions between the injected fluids and the surrounding groundwater.

The compatibility between the wastewater reinjection process and the protection of the surrounding environment is of utmost importance, primarily when oilfields are located within a sensitive and protected area, such as the Agri Valley in Southern Italy. At this site, wastewater has been reinjected since 2006 into the Costa Molina 2 (CM2) injection well, with a well-head pressure ranging from 13 to 14 MPa and an injection rate of approximately 2000 $m^3$/day. Since then, no significant interruptions have been recorded; therefore, the CM2 injection well can be classified, on the whole, as a long-term and high-rate disposal well [1].

The impact of the oily wastewater reinjection process has been studied at the Val d'Agri oilfield [2–5] and worldwide [6] in terms of induced seismicity. However, few studies have been conducted to assess the possible interactions between oily wastewater and the local shallow environment [7,8]. Abou-Sayed et al. [7] performed a three-day injection test, by measuring quantitative parameters, such as surface and downhole pressures, ground surface deflections, and wellbore hydraulic impedance. Differently, in Kholy et al. [8] analyzed the feasibility of wastewater injection by assessing the subsurface geology and the petrophysical features of rocks to create a geomechanical earth model and a three-dimensional fracture simulator. The simulations allowed forecasting fracture growth within a candidate injection site at different injection scenarios, propaedeutic to forecast possible interactions between deep and shallow environments. In relation to the Val d'Agri oilfield, one published contribution [9] hypothesized that reinjected wastewaters in the CM2 well could contaminate shallow aquifers, probably due to well integrity issues. Nevertheless, this conference paper synthesized very preliminary results, in contrast to those obtained in more recent research carried out by Avagliano et al. [10] at the same study site.

In light of the differences that emerged between the two previous papers that focused on the Val d'Agri oilfield [9,10], a more detailed and exhaustive study was needed to experimentally verify whether the reinjection process can influence shallow groundwater and/or affect its quality. With the aim of obtaining conclusive evidence, a purpose-designed multidisciplinary approach based on chemical, isotopic, and biomolecular analyses has been developed.

In a wider context, the present study also provides a significant contribution from a methodological perspective, introducing new arguments into the discussion related to the experimental (and not only numerical) approaches that can be effective in analyzing the possible environmental impacts of oily wastewaters injection.

In the following paragraphs, after the description of the main geological and hydrogeological features of the study area, the main results of all investigations (chemical, isotopic, and biomolecular) are presented and then discussed, in order to achieve a final conclusion for the considered study site and to verify the effectiveness of the proposed multidisciplinary approach from a methodological point of view.

*Study Area*

The Val d'Agri is a Quaternary NW–SE trending intramountain basin located within the southern Apennines thrust belt (Southern Italy) (Figure 1), whose formation and evolution are controlled by brittle tectonics. The intense and recent deformation is testified by seismic activity in the last 40 ka, such as the M7, 1857 Basilicata earthquake [10–13].

The presence of petroleum in Val d'Agri is well known. Some explorations have been carried out in the area of Tramutola's municipality, in which there are still natural hydrocarbon springs, recently studied in terms of bioremediation and migration of fluids [14–16], but the main Val d'Agri oilfield is hosted in a reservoir made of fractured, low-porosity carbonates belonging to the buried inner Apulian Platform belt [17–19]. Light-to-medium crude oil and gas are stored in limestone and dolomite (Miocene to Cretaceous ages) [20]. The carbonate reservoir lies below the Pliocene siliciclastic foredeep deposits and a thick mélange layer (Figure 1). Hydrocarbons have been extracted in large quantities since the mid-1990 through several wells at depths ranging from 1.8 to 3.5 km below sea level. The actual production rates are up to 65,000 barrels (bbl)/d with $2.6 \times 10^5$ $m^3$/d of gas.

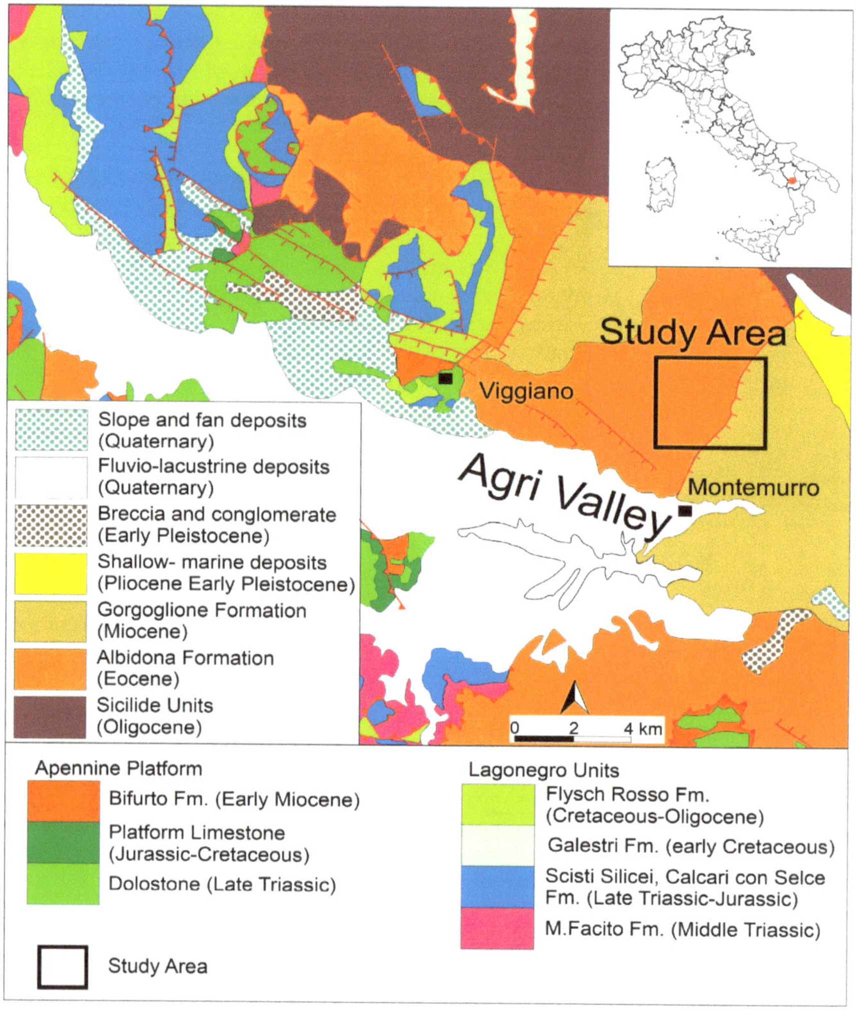


**Figure 1.** Schematic map of the Agri Valley showing the location of the study area (modified by Rizzo et al. [21]).

The Apulian Platform is about 7000 m thick with a bottom part made up of evaporites, sandstones, and conglomerates (Triassic age), and lies above the crystalline basement. Well data were used to establish the progressive movement of the front of the chain towards NE during the Pliocene and early Pleistocene.

The study area is located on the left orographic side in the southeastern edge of the valley, near the municipality of Montemurro, and includes reinjection well CM2 (Costa Molina 2) (screened into the petroleum reservoir, between 2890 and 3096 m below ground). On top of the carbonate reservoir, formations belonging to the Lagonegro basin (lower Triassic–lower Cretaceous), the Liguridi–Sicilidi basin (upper Jurassic–lower Miocene), and middle-late Miocene turbiditic siliciclastic successions can be found [19,22–27] (Figure 2).

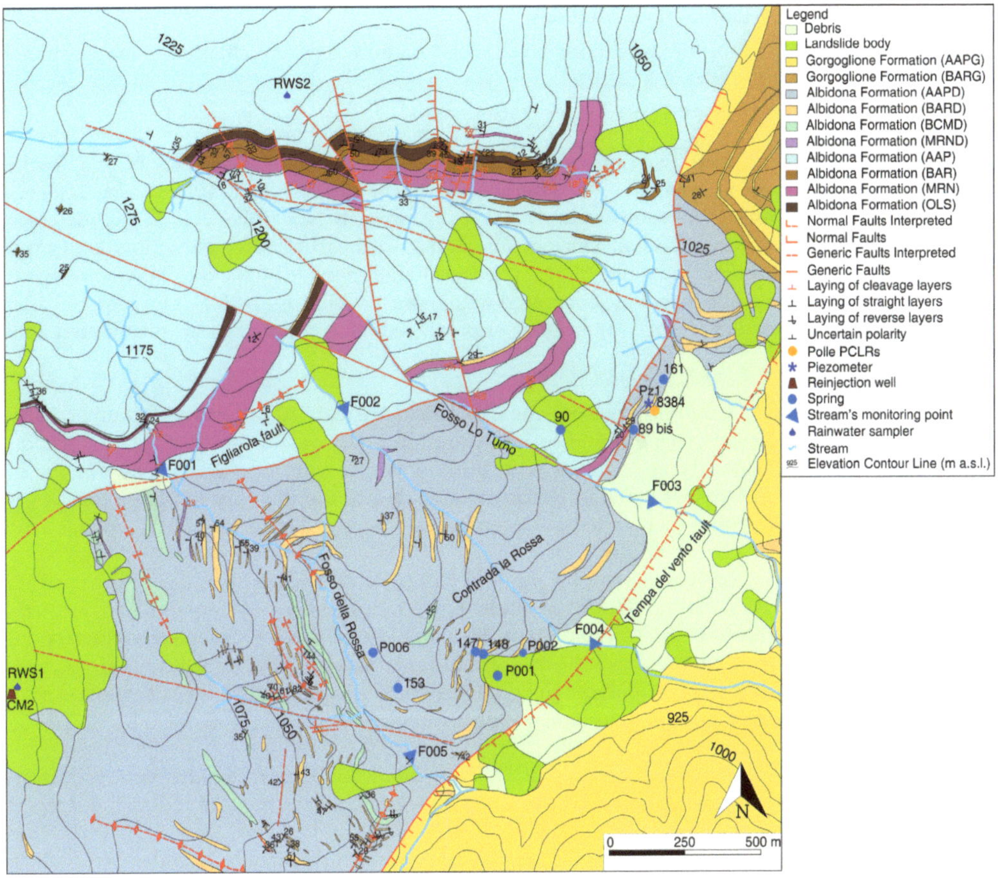


**Figure 2.** Geological map and monitored springs/streams (see location in Figure 1) (modified by Rizzo et al. [21]).

In the study area, the sedimentary successions belong to the so-called turbiditic siliciclastic hydrogeological complex (*sensu* [28]). A detailed hydrogeological study of the outcropping formations [21] showed the absence of vertical heterogeneity due to rock alteration and/or stress-release fracturing, thereby excluding the existence of a shallower and widespread horizon characterized by higher rock permeability. On the whole, the presence of perched groundwater due to the permeability contrast with depth was excluded, and all the analyzed springs were related to unique groundwater flowing within a low-permeability continuum at the basin scale. The hydraulic continuity at this scale is enhanced by the fracture network associated with folds and faults, which can interrupt the spatial continuity of the low-permeability layers within the heterogeneous sedimentary succession. As for the location of the springs, an overall adaptation of the potentiometric surface to the ground morphology was observed. As a matter of fact, the spatial variations in the topographic gradient and the thin unsaturated aquifer cause the shallower groundwater to flow out where the phreatic surface intersects the topographic gradient. In greater detail, Rizzo et al. [21] hypothesized that the relatively high hydraulic gradient observed within the heterogeneous system is due to the back-dipping low-permeability layers (opposite to the main groundwater flow lines) that act as hydraulic barriers. According to findings related to low-permeability fault zones [29], the relatively high head loss observed within the turbidite succession would be distributed across several aquitards characterized by relatively low integrity.

## 2. Materials and Methods

The investigations were based on an integrated chemical-isotopic-microbiological approach.

The research was developed from December 2017 to March 2019 and involved (i) oily wastewaters before their injection in well CM2 and (ii) groundwater flowing towards the former two outflows (the so-called "Polle" of C.da La Rossa; PCLRs) associated with methane emission [9]. A purpose-designed piezometer (Pz1; 14 m deep, screened between 8 and 13 m below ground) was drilled in early November 2017, a few meters upgradient from the PCLRs. During this study, no measurable or collectible water outflow was observed at the two PCLRs after an alteration made by the area's owner. Nevertheless, the data obtained were compared with those available for both PCLRs acquired after previous surveys (carried out in 2016; unpublished data) before they were tampered with.

### 2.1. Isotopic Analyses

A monthly monitoring campaign was carried out from December 2017 to March 2019 by collecting Pz1 groundwater and oily wastewaters before reinjection. The data were analyzed together with those obtained from previous surveys of the PCLRs and other spring waters. Unfortunately, due to the inability to access the study area, the campaign that started in December 2017 stopped until May 2018, when it resumed with monthly continuity.

Rainwater samples for $\delta(^{18/16}O)$ (indicated as $\delta^{18}O$) and $\delta(^{2/1}H)$ (indicated as $\delta^{2}H$) analyses were collected monthly in two rain samplers, RWS1 and RWS2, located within the Val d'Agri oilfield (Figure 2) at 1047 and 1290 m above sea level (m a.s.l.), respectively from May 2016 to April 2017 during a different study on the same area. Rainfall was collected using ten-liter polyethylene bottles containing about 300 mL of vaseline oil to prevent evaporation. In addition, oil contamination was carefully avoided by syringing the water samples out of the bottle.

For stable isotope ($\delta^{18}O$, $\delta^{2}H$) analysis, oily wastewater and groundwater from Pz1 were collected to compare the isotopic signature detected in the deeper and shallower aquifer.

After collection, the samples were transported in a refrigerated box to the laboratory.

Electrical conductivity, temperature, and pH were measured in situ using portable equipment (Hanna Instruments 9829, Roccafranca Padovana, Italy).

Stable isotope ($\delta^{18}O$ and $\delta^{2}H$) analyses were carried out at the Isotope Geochemistry Laboratory of the University of Parma, Italy, using a water equilibrator at 18 °C online with a Finnigan Delta XP spectrometer. For the oxygen isotope determination, 5 $cm^3$ of water was equilibrated with pure $CO_2$, while for the hydrogen isotope, 5 $cm^3$ of water was equilibrated with pure $H_2$ (platinum wire was used as a catalyzer for gas–liquid water equilibration). The isotope ratios were expressed as

$$^{A/B}\delta_{s/RF} = \frac{^{A/B}R_s}{^{A/B}R_{RF}} - 1 = \left[\left(\frac{^{A/B}R_s}{^{A/B}R_{RF}} - 1\right) 10^3\right] ‰ \quad (1)$$

where A is $^{18}O$ or $^{2}H$, B is $^{16}O$ or $^{1}H$, R is the ratio of the isotopic abundances, s is the sample of interest, ‰ = $10^{-3}$, and RF indicates that the data refer to the VSMOW-SLAP scale.

### 2.2. Chemical Analyses: Organic Compounds

Oily wastewaters before their injection into well CM2 and groundwater in Pz1 were collected for chemical analyses, together with samples for isotopic determinations. The analyses were carried out at the Analytical Chemistry Laboratory of the University of Parma, Italy, using the following methods: EPA 5021A + 8260C, EPA 3511 + 8270D, Internal method + ISO 17943, and APAT CNR IRSA 5160B Man 29 2003 [30–35].

### 2.3. Preliminary Characterization of Microbial Communities through Biomolecular Analyses Carried Out in 2016: 16S rDNA Polymerase Chain Reaction-Denaturing Gradient Gel Electrophoresis (PCR-DGGE) and Next-Generation Sequencing (NGS)

Water samples for microbiological investigations were first collected in April 2016 to preliminarily analyze the bacterial and archaeal communities of PCLRs (sample codes 83 and 84), before being tampered with, other springs (sample codes P001, P002, P006, 89BIS, 90, 147, 148, 153, and 161) and stream monitoring points (sample codes F001, F002, F003, F004, and F005).

#### 2.3.1. DNA Extraction

One L water samples were filtered through mixed esters of cellulose filters (S-Pak™ Membrane Filters, 47 mm diameter, 0.22 µm pore size, Millipore Corporation, Billerica, MA, USA) within eight hours after collection. Subsequently, the filters were stored at −80 °C until DNA extraction, which was performed using a PowerWater® DNA Isolation Kit (MO BIO Laboratories, Inc., Carlsbad, CA, USA).

#### 2.3.2. 16S rDNA Polymerase Chain Reaction-Denaturing Gradient Gel Electrophoresis (PCR-DGGE)

Nested PCR was carried out to amplify bacterial 16S rDNA fragments. The primer pairs 27F/1492R and 341F-GC (with a 40 bp GC-clamp)/907R [36] were used for the primary and secondary amplifications, respectively. PCRs were performed according to the protocols reported by Hernàndez-Diaz et al. [37] using 5 µL of the template.

Nested PCR was carried out to amplify also archaeal 16S rDNA fragments, by using the primer pairs 21F/958R [38] and Parch519F/ARC915R-GC (with 40 bp GC-clamp) [39]. Reagent concentrations in the PCR mixtures were the same as those reported by Hernàndez-Diaz et al. [37], but 5 µL and 6 µL of the template were used in the primary and secondary amplifications, respectively.

Primary amplification of archaeal 16S rDNA fragments was performed using the following process: 94 °C for 2 min, 30 cycles of denaturation at 94 °C for 45 s, annealing at 55 °C for 1 min, and extension at 72 °C for 1 min and 30 s, and a single final extension at 72 °C for 4 min. The secondary amplification consisted of an initial denaturation step at 94 °C for 2 min, 30 cycles of denaturation at 94 °C for 30 s, annealing at 57 °C for 40 s, extension at 72 °C for 40 s, and a final elongation step at 72 °C for 4 min.

DGGE and cluster analysis of microbial community patterns were performed as reported by Hernàndez-Diaz et al. [37].

#### 2.3.3. Next-Generation Sequencing (NGS) Carried Out in 2016 for Samples Collected before the PCLRs Were Tampered with

A preliminary NGS sequencing was performed at the BMR Genomics srl (Padova, Italy) for water samples collected at PCLRs, other springs (sample codes 89BIS, 90, 148, 153, 161), and one stream monitoring point (sample code F003) by amplifying the V3-V4 regions of the 16S rRNA genes of *Bacteria* and *Archaea*, using the primers Pro341F and Pro805R [40], modified with universal tails.

PCR products were purified with Agencourt XP 0.8× magnetic beads and amplified using the Nextera XT Index (Illumina, 5200 Illumina Way, San Diego, CA 92122, USA) in the second PCR step. Amplicons were then normalized using SequalPrep (Thermo Fisher Scientific, 168 Third Avenue, Waltham, MA 02451, USA) and multiplexed. The pool was purified using Agencourt XP 1× magnetic beads. The library was run on an Illumina MiSeq and sequenced with V3 chemistry—300PE strategy. The forward and reverse sequences (R1 and R2 reads) were merged using the software FLASH v.1.2.11 [41] and filtered for quality ($Q > 30$). The clustering of the Operational Taxonomic Units (OTUs) (cluster of similar sequences with 97% identity) was based on the method *pick_closed_reference_otus.py* of Qiime 1.9.1 [42] and on the database Greengenes v.13–8. OTUs representing fewer than 0.005% of all sequences were filtered as recommended by Bokulich et al. [43].

### 2.4. Characterization of Bacterial Communities through Next-Generation Sequencing (NGS) after the PCLRs Were Tampered with

For microbial community analyses carried out after the PCLRs were tampered with, one L water samples were filtered through sterile mixed esters of cellulose filters (S-Pak™ Membrane Filters, 47 mm diameter, 0.22 μm pore size, Millipore Corporation, Billerica, MA, USA) within 24 h of collection. Subsequently, DNA was extracted using the commercial kit FastDNA SPIN Kit for soil (MP Biomedicals, LLC, Solon, OH, USA) and the FastPrep®Instrument (MP Biomedicals, LLC, Solon, OH, USA). After the extraction, DNA integrity and quantity were evaluated by electrophoresis on 0.8% agarose gel containing 1 μg/mL of Gel-Red™ (Biotium, Inc., Fremont, CA, USA). The microbial community profiles in the samples were generated using NGS (Next-Generation Sequencing) technologies at Genprobio S.r.l. Laboratory, following the protocol reported by Rizzo et al. [14,21].

## 3. Results

### 3.1. Isotopic Analyses

The isotopic content of local precipitation gives the following regression $\delta^2H$ on $\delta^{18}O$:

$$\begin{gathered}\delta^2H = [6.48\ (\pm 0.66)\ \delta^{18}O + 4.58\ (\pm 5.19)]\ ‰ \\ s(yx) = 4.56\ ‰,\ n = 21,\ p\ (A = 0) = 0.39\end{gathered} \quad (2)$$

where s(yx) is the standard error of the regression, n is the number of couples of data, and A is the regression line's intercept.

The regression calculated at the study site is very close to that calculated by Longinelli and Selmo [44] using precipitation collected in Southern Italy, as well as those calculated at other study sites in the same macro-area [45–47].

All the isotopic data of groundwaters and PCLRs fall close to regression (1), demonstrating a meteoric origin of the analyzed waters (Figure 3). On the contrary, the oily wastewaters are far from the same regression line, being enriched more in $^{18}O$ than in $^2H$ (Figure 3). The data for CM2 water are consistent with a fossil source being modified by lithology-dependent diagenetic processes. This lithology is represented by the carbonate formation of a deep petroleum reservoir at high temperatures [48].

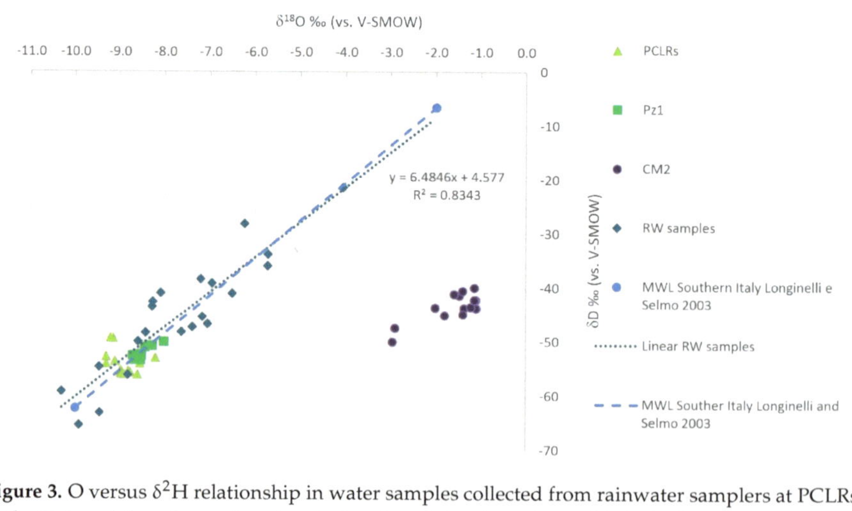


**Figure 3.** O versus $\delta^2H$ relationship in water samples collected from rainwater samplers at PCLRs, at the Pz1, and the oily wastewaters before the reinjection. Linear RW is the local meteoric water line obtained by local rainwater. MWL is the Meteoric Water Line of Southern Italy obtained from Longinelli and Selmo [44].

### 3.2. Chemical Analyses of Organic Compounds

Chemical analyses of organic compounds in the oily wastewaters before their injection in well CM2 and in waters collected at Pz1 revealed, as shown in Tables 1 and 2 (example of available data), that the content of benzene, toluene, ethylbenzene, xylenes, and styrene was as follows: (i) always lower than the detection limit in groundwater sampled at the piezometer drilled immediately upstream of PCLRs and, as expected, (ii) much higher in the reinjection waters.

**Table 1.** Results of the chemical analyses carried out on the CM2 well reinjection waters of the March 2019 sampling campaign.

| Analytes | Results (μg/L) | Methods | Limit by Italian Law (μg/L) |
|---|---|---|---|
| Benzene | 9610 ± 1100 | EPA 5021A + 8260C | 1 |
| Toluene | 8960 ± 540 | EPA 5021A + 8260C | 15 |
| Ethylbenzene | 720 ± 70 | EPA 5021A + 8260C | 50 |
| M- + p-xylene | 1510 ± 220 | EPA 5021A + 8260C | 10 |
| Styrene | 45 ± 6 | EPA 5021A + 8260C | 25 |
| Benzo (a) pyrene | <0.003 | EPA 3511 + 8270D | 0.01 |
| Benzo (b) fluoranthene | <0.01 | EPA 3511 + 8270D | 0.1 |
| Benzo (k) fluoranthene | <0.005 | EPA 3511 + 8270D | 0.05 |
| Benzo (a) anthracene | <0.01 | EPA 3511 + 8270D | 0.1 |
| Benzo (g, h, i) perylene | <0.003 | EPA 3511 + 8270D | 0.01 |
| Dibenzo (a, h) anthracene | <0.003 | EPA 3511 + 8270D | 0.01 |
| Indeno (1,2,3-c, d) pyrene | <0.01 | EPA 3511 + 8270D | 0.1 |
| Pyrene | <0.05 | EPA 3511 + 8270D | 50 |
| Chrysene | <0.05 | EPA 3511 + 8270D | 5 |
| Chlorobenzene | <0.1 | EPA 5021A + 8260C | 40 |
| 1,2-dichlorobenzene | <0.1 | EPA 5021A + 8260C | 270 |
| 1,4-dichlorobenzene | <0.1 | EPA 5021A + 8260C | 0.5 |
| 1,2,4-trichlorobenzene | <0.5 | EPA 5021A + 8260C | 190 |
| 1,2,4,5-tetrachlorobenzene | <0.1 | EPA 3511 + 8270D | 1.8 |
| Pentachlorobenzene | <0.1 | EPA 3511 + 8270D | 5 |
| 2- chlorophenol | <0.5 | Internal method + ISO 17943 | 180 |
| 2,4- dichlorophenol | <0.5 | Internal method + ISO 17943 | 110 |
| 2,4,6- trichlorophenol | <0.5 | Internal method + ISO 17943 | 5 |
| Pentachlorophenol | <0.05 | Internal method + ISO 17943 | 0.5 |
| Total hydrocarbons | 70.8 ± 4.3 (as n-hexane) | APAT CNR IRSA 5160B Man 29 2003 | 350 (as n-hexane) |

### 3.3. Preliminary Investigations on Microbial Communities through PCR-DGGE and Next-Generation Sequencing

A preliminary assessment of the bacterial and archaeal communities of PCLRs (sample codes 83 and 84), other springs (sample codes P001, P002, P006, 89BIS, 90, 147, 148, 153, and 161), and stream monitoring points (sample codes F001, F002, F003, F004, and F005) was conducted in April 2016 using a PCR-DGGE-based approach (Figure 4). In addition, NGS was also performed to obtain a more comprehensive identification of microorganisms constituting the microbial communities of PCLRs, some of the springs, and one stream monitoring point analyzed through PCR-DGGE (sample codes 89BIS, 90, 148, 153, 161, and F003). Unfortunately, the use of the primer pair Pro341F and Pro805R did not allow a detailed resolution of archaeal communities, unlike the bacterial ones. Nevertheless, taken together, the results of the preliminary biomolecular investigations provided an interesting contribution to further clarify if there could be hydrocarbon contamination of groundwater due to issues linked to the reinjection process. In general, the main bacterial phyla found after NGS data analysis were *Proteobacteria*, *Bacteroidota*, and *Verrucomicrobiota*. These taxa were present in almost all the samples, although with different relative abundance values. On the other hand, *Archaea* belonging to the genera *Methanosarcina*, *Methanosaeta*, and *Nitrosopumilus* were detected only in PCLRs and springs 90, 161, and 153. In detail, *Methanosarcina* and *Methanosaeta* are genera comprising species that are able to perform methanogenesis, an anaerobic process that generates methane as the final product of metabolism [49]. Their presence in PCLRs and in spring 161 suggests the following: (i) the methane emissions recorded in PCLRs have a biological origin (further supporting the suggestions made by Avagliano et al. [10], using the isotopic signature of methane sampled in PCLRs), and (ii) methanogenesis is not limited to PCLRs within the studied hydrogeological system. From a wider perspective, the DGGE results suggest that there is a hydraulic interconnection between the sub-basin hosting PCLRs and the larger hydrostructure to which this sub-basin belongs, in agreement with the whole hydrogeological model pointed out by Rizzo et al. [21] at the same site. In fact, the analysis of the DGGE profiles (Figure 5) shows, on the whole, a rather high similarity among the aquatic microbial communities.

**Table 2.** Results of the chemical analyses carried out on the Pz1 groundwater of the March 2019 sampling campaign.

| Analytes | Results (μg/L) | Methods | Limit by Italian Law (μg/L) |
|---|---|---|---|
| Benzene | <0.1 | EPA 5021A + 8260C | 1 |
| Toluene | <1 | EPA 5021A + 8260C | 15 |
| Ethylbenzene | <1 | EPA 5021A + 8260C | 50 |
| M- + p-xylene | <1 | EPA 5021A + 8260C | 10 |
| Styrene | <1 | EPA 5021A + 8260C | 25 |
| Benzo (a) pyrene | <0.003 | EPA 3511 + 8270D | 0.01 |
| Benzo (b) fluoranthene | <0.01 | EPA 3511 + 8270D | 0.1 |
| Benzo (k) fluoranthene | <0.005 | EPA 3511 + 8270D | 0.05 |
| Benzo (a) anthracene | <0.01 | EPA 3511 + 8270D | 0.1 |
| Benzo (g, h, i) perylene | <0.003 | EPA 3511 + 8270D | 0.01 |
| Dibenzo (a, h) anthracene | <0.003 | EPA 3511 + 8270D | 0.01 |
| Indeno (1,2,3-c, d) pyrene | <0.01 | EPA 3511 + 8270D | 0.1 |
| Pyrene | <0.05 | EPA 3511 + 8270D | 50 |
| Chrysene | <0.05 | EPA 3511 + 8270D | 5 |
| Chlorobenzene | <0.1 | EPA 5021A + 8260C | 40 |
| 1,2-dichlorobenzene | <0.1 | EPA 5021A + 8260C | 270 |
| 1,4-dichlorobenzene | <0.1 | EPA 5021A + 8260C | 0.5 |
| 1,2,4-trichlorobenzene | <0.5 | EPA 5021A + 8260C | 190 |
| 1,2,4,5-tetrachlorobenzene | <0.1 | EPA 3511 + 8270D | 1.8 |
| Pentachlorobenzene | <0.1 | EPA 3511 + 8270D | 5 |
| 2- chlorophenol | <0.5 | Internal method + ISO 17943 | 180 |
| 2,4- dichlorophenol | <0.5 | Internal method + ISO 17943 | 110 |
| 2,4,6- trichlorophenol | <0.5 | Internal method + ISO 17943 | 5 |
| Pentachlorophenol | <0.05 | Internal method + ISO 17943 | 0.5 |
| Total hydrocarbons | <35 | APAT CNR IRSA 5160B Man 29 2003 | 350 (as n-hexane) |

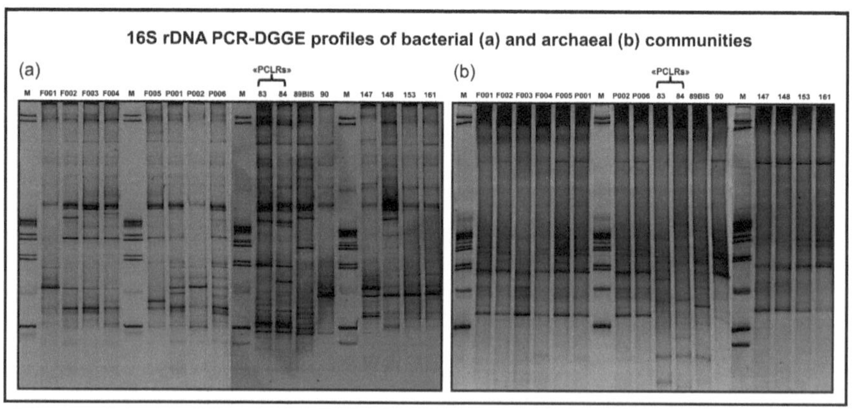


**Figure 4.** Bacterial (**a**) and archaeal (**b**) 16S rDNA PCR-DGGE community profiles of PCLRs (sample codes 83 and 84) before being tampered with (April 2016), other springs (sample codes P001, P002, P006, 89BIS, 90, 147, 148, 153, and 161), and stream monitoring points (sample codes F001, F002, F003, F004, and F005).

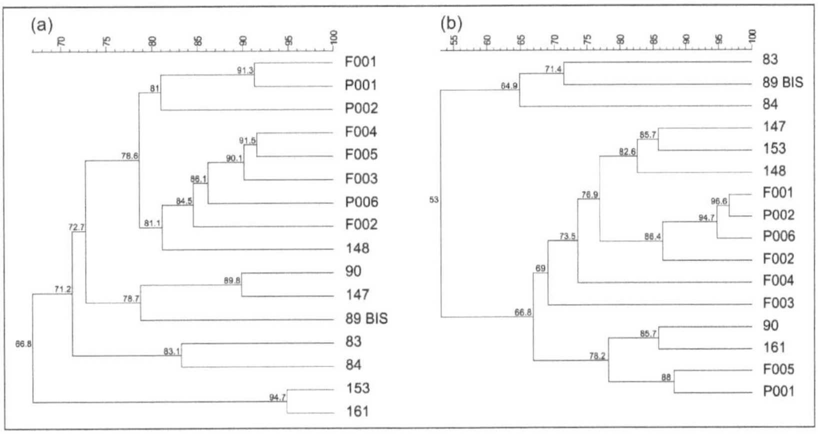

**Figure 5.** Cluster analysis of bacterial (**a**) and archaeal (**b**) 16S rDNA PCR-DGGE profiles of PCLRs (sample codes 83 and 84) before being tampered with (April 2016), other springs (sample codes P001, P002, P006, 89BIS, 90, 147, 148, 153, and 161), and stream monitoring points (sample codes F001, F002, F003, F004, and F005) (Dice coefficient with the UPGMA clustering algorithm). Similarity values are shown.

### *3.4. Bacterial Community Analysis through Next-Generation Sequencing (NGS) after the PCLRs Were Tampered with*

The results of the characterization of microbial communities of the springs through Next-Generation Sequencing have already been reported by Rizzo et al. [21]; accordingly, only the data relating to CM2 and PZ1, used as a representative of the groundwater formerly feeding the PCLRs, are discussed here. The final read numbers of the sequences obtained for each sample are shown in Table 3. Unfortunately, the PZ1 sample taken in January 2019 was damaged during transport and could not be analyzed in the laboratory.

**Table 3.** The number of 16S rDNA bacterial sequences obtained after NGS analysis for the four sampling campaigns.

| Sample | Final Read Number | | | |
|---|---|---|---|---|
| | Sampling Campaigns | | | |
| | **June 2018** | **September 2018** | **January 2019** | **March 2019** |
| **CM2** | 84,618 | 34,947 | 63,407 | 57,527 |
| **PZ1** | 75,283 | 57,530 | - | 65,482 |

The 16S rRNA gene sequences have been deposited in the NCBI Sequence Read Archive (SRA: accession numbers PRJNA629324 and PRJNA877189).

The rarefaction analysis, a measure used to estimate the alpha diversity in samples and gauge whether or not sequencing efforts captured the microbial diversity, revealed Shannon index values ranging from 3.6 and 4.0 for the CM2 samples. For the Pz1 samples, lower biodiversity was recorded in September 2018 (Shannon index value of 4.0) compared to June 2018 and March 2019 (Shannon index values of 7.3 and 7.0, respectively) (Figure 6).

Bacterial community analyses at the genus level (Figure 7) revealed that the waters of the CM2 well were characterized by a microbial signature very different from that found in the PZ1 waters; the most abundant genera were *Desulfomicrobium* (relative abundance values ranging from 38.40% to 46.75%), *Thermoanaerobacterium* (relative abundance values ranging from 10.88% to 28.92%), *Acetomicrobium* (relative abundance values ranging from 6.39% to 16.71%), and *Thermovirga* (relative abundance values ranging from 3.23% to 14.61%). These results testify to the presence of thermophilic bacteria in the reinjection waters, with anaerobic metabolism (fermentation or anaerobic respiration with sulfate, sulfite, and thiosulfate as electron acceptors) compatible with the characteristics of the waters in which they were found [50–52]. These four genera alone represented more than 67% of all the sequences obtained using NGS analyses in the different monitoring campaigns.

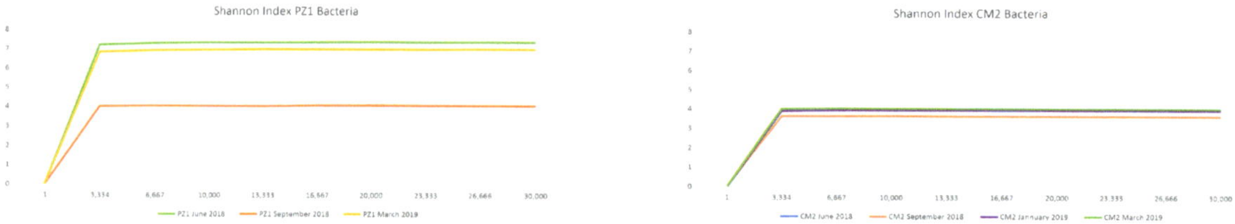


**Figure 6.** Rarefaction curves with Shannon index for *Bacteria* domain.

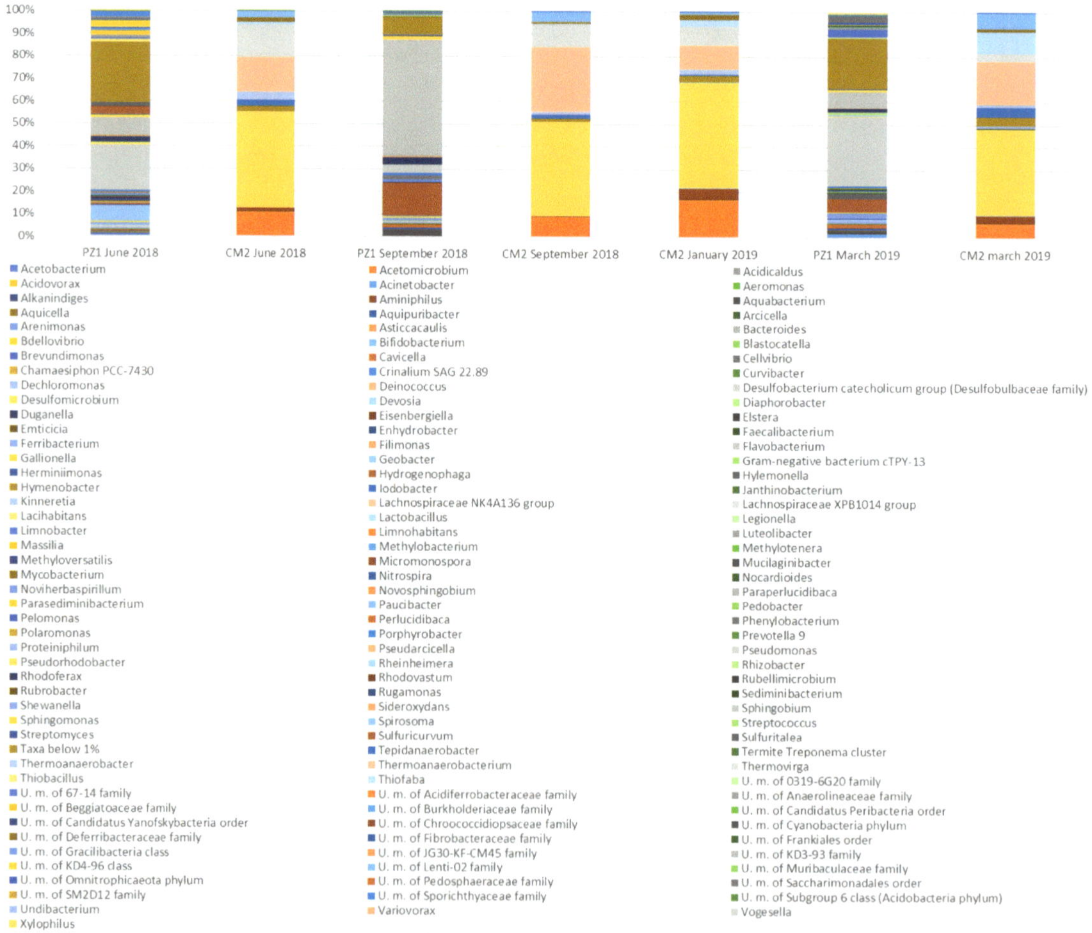


**Figure 7.** Genus level bacterial community composition.

On the other hand, a predominance of entirely different genera, comprising species living at lower temperatures and under aerobic conditions, such as *Sphingobium* (relative abundance values ranging from 7.25% to 51.56%) or *Pseudomonas* (relative abundance values ranging from 3.70% to 30.64%), was detected in PZ1.

## 4. Discussion and Conclusions

The isotopic content ($\delta^2H$ on $\delta^{18}O$) of the ground- (PZ1) and springwaters (including the PCLRs) is exclusively compatible with the isotopic signature of local rainwater, suggesting the absence of mixing between the shallow groundwater and possible deep fluids typical of the petroleum reservoir (in the case study, the reinjection waters). This hydrogeological scenario is in agreement with the conceptual model pointed out by Rizzo et al. [21]. As a matter of fact, the PCLRs (as well as the other springs) were located along an abrupt variation in the topographic gradient, where the hydraulic head intercepts the ground surface, thereby causing shallow groundwater to flow out. Their location was not anomalous within the local hydrogeological system, so much so that both PCLRs flowed out very close to springs 89 bis and 161. Moreover, the reinjection waters had an isotopic signature incompatible with that of the local precipitations, whereas they were consistent with a fossil source being modified by lithology-dependent diagenetic processes.

Biomolecular and chemical analyses further support this scenario. In fact, the biomolecular investigations showed very different microbial communities between the shallow and deep fluids. CM2 reinjection waters were characterized by thermophilic genera with anaerobic metabolism, whereas shallow ground- and springwaters were characterized by psychrophilic or mesophilic aerobic communities. Moreover, methanogens were found not only in the former PCLRs, but also in another spring (161) flowing out close to the PCLRs, further demonstrating that methane at the study site is biogenic (and not thermogenic), and is typical of some of the sub-basins coexisting within the whole hydrogeological system analyzed in the present study.

As for the chemical analyses, hydrocarbons have never been detected in the PZ1-waters in any monitoring campaign. Conversely, the reinjection fluids in CM2 were always characterized by high concentrations of hydrocarbons. In oil contaminant plumes, the analysis of only total petroleum hydrocarbons (TPH) could be ineffective due to the potential biological effects of degradation [53]; in the case study, a set of 25 different analytes were analyzed as possible tracers.

On the whole, the synergic analysis of the chemical, isotopic, and biomolecular data obtained through the developed multidisciplinary approach made it possible to demonstrate the total absence of interaction between the reinjection water in CM2 and the shallow groundwater system, including the former PCLRs.

From a methodological point of view, this research (together with that carried out by Rizzo et al. [21]) further demonstrates the importance of applying interdisciplinary approaches when studying the possible hydraulic interconnections between deep and shallow fluids in heterogeneous geological systems, with an emphasis on those systems where reinjection processes of oily wastewaters can potentially cause the contamination of groundwater and aquatic ecosystems. The synergy between different disciplines and the utilization of experimental data coming from independent investigations is the only way to obtain an exhaustive and effective response.

**Author Contributions:** Conceptualization, P.R., A.B., A.M.S., G.N., P.I., F.B., C.M. and F.C. (Fulvio Celico); methodology, P.R., A.B., P.I., F.B. and F.C. (Fulvio Celico); software, P.R. and A.B.; validation, P.R., A.B., P.M. and M.R.; formal analysis, P.R., A.B., P.I., F.B. and N.R.; investigation, P.R., A.B., P.M., N.R. and A.C.; resources, A.M.S., P.I., F.B., C.M., G.N., D.A., F.C. (Francesco Coraggio), A.C. and F.C. (Fulvio Celico); data curation, P.R., A.B. and M.R.; writing—original draft preparation, P.R. and A.B.; writing—review and editing, P.R., A.B., D.A., F.C. (Francesco Coraggio) and F.C. (Fulvio Celico); visualization, P.R. and A.B.; supervision, A.M.S., P.I., F.B., C.M. and F.C. (Fulvio Celico); project administration, F.C. (Fulvio Celico); funding acquisition, F.C. (Fulvio Celico). All authors have read and agreed to the published version of the manuscript.

**Funding:** This work has benefited from the equipment and framework of the COMP-HUB and COMP-R Initiatives, funded by the 'Departments of Excellence' program of the Italian Ministry for University and Research (MIUR, 2018–2022, and MUR, 2023–2027).

**Data Availability Statement:** The data presented in this study are all the available data.

**Acknowledgments:** We are thankful to three anonymous reviewers for their interesting comments/questions.

**Conflicts of Interest:** The authors declare no conflict of interest.

## References

1. Frohlich, C. Two-year survey comparing earthquake activity and injection-well locations in the Barnett Shale, Texas. *Proc. Natl. Acad. Sci. USA* **2012**, *109*, 13934–13938. [CrossRef] [PubMed]
2. Buttinelli, M.; Improta, L.; Bagh, S.; Chiarabba, C. Inversion of inherited thrusts by wastewater injection induced seismicity at the Val d'Agri oilfield (Italy). *Sci. Rep.* **2016**, *6*, 37165. [CrossRef] [PubMed]
3. Improta, L.; Bagh, S.; De Gori, P.; Valoroso, L.; Pastori, M.; Piccinini, D.; Chiarabba, C.; Anselmi, M.; Buttinelli, M. Reservoir structure and wastewater-induced seismicity at the Val d'Agri Oilfield (Italy) shown by three-dimensional Vp and Vp/Vs local earthquake tomography. *J. Geophys. Res. Solid Earth* **2017**, *122*, 9050–9082. [CrossRef]
4. Hager, B.H.; Dieterich, J.; Frohlich, C.; Juanes, R.; Mantica, S.; Shaw, J.H.; Bottazzi, F.; Caresani, F.; Castineira, D.; Cominelli, A.; et al. A process-based approach to understanding and managing triggered seismicity. *Nature* **2021**, *595*, 684–689. [CrossRef] [PubMed]
5. Stabile, T.A.; Vlček, J.; Wcisło, M.; Serlenga, V. Analysis of the 2016–2018 fluid-injection induced seismicity in the High Agri Valley (Southern Italy) from improved detections using template matching. *Sci. Rep.* **2021**, *11*, 20630. [CrossRef] [PubMed]
6. Rubinstein, J.L.; Ellsworth, W.L.; McGarr, A.; Benz, H.M. The 2001–present induced earthquake sequence in the Raton Basin of northern New Mexico and southern Colorado. *Bull. Seismol. Soc. Am.* **2014**, *104*, 2162–2181. [CrossRef]
7. Abou-Sayed, A.S.; Andrews, D.E.; Buhidma, I.M. Evaluation of oily waste injection below the permafrost in Prudhoe Bay field. In Proceedings of the SPE California Regional Meeting, Bakersfield, CA, USA, 5–7 April 1989. [CrossRef]
8. Kholy, S.M.; Sameh, O.; Mounir, N.; Shams, M.; Mohamed, I.M.; Abou-Sayed, A.; Abou-Sayed, O. Evaluating the Feasibility of Waste Slurry Injection in an Oil Prospect in the Western Desert, Egypt. In Proceedings of the SPE Middle East Oil and Gas Show and Conference, Manama, Bahrain, 18–21 March 2019. [CrossRef]
9. Colella, A. Anomalous deep waters gurgling to the surface and impacting soils in the Val d'Agri oil field, Southern Italy. *Int. J. Ecosyst. Ecol. Sci.* **2014**, *4*, 533–542.
10. Avagliano, D.; Coraggio, F.; Martinenghi, C.P.; Barbieri, C.; Previde Massara, E.; Musto, M.A.; Caputi, A.; Palladino, G.; Prosser, G.; Celico, F. L'importanza del modello geologico-concettuale negli studi ambientali: Il caso delle manifestazioni sorgive nell'area di Contrada La Rossa (Montemurro—Potenza). In *Geologia dell'Ambiente*; SIGEA Italian Society of Environmental Geology: Rome, Italy, 2022; ISSN 1591-5352. (In Italian)
11. Giano, S.I.; Maschio, L.; Alessio, M.; Ferranti, L.; Improta, S.; Schiattarella, M. Radiocarbon dating of active faulting in the Agri high valley, Southern Italy. *J. Geodyn.* **2000**, *29*, 371–386. [CrossRef]
12. Daddezio, G.; Karner, D.B.; Burrato, P.; Insinga, D.; Maschio, L.; Ferranti, L.; Renne, P.R. Tephrochronology in faulted Middle Pleistocene tephra layer in the Val d'Agri area (Southern Italy). *Ann. Geophys.* **2009**, *49*, 1029–1040. Available online: http://hdl.handle.net/2122/2122 (accessed on 31 March 2023).
13. Giocoli, A.; Stabile, T.A.; Adurno, I.; Perrone, A.; Gallipoli, M.R.; Gueguen, E.; Norelli, E.; Piscitelli, S. Geological and geophysical characterization of the south-eastern side of the High Agri Valley (southern Apennines, Italy). *Nat. Hazards Earth Syst. Sci. Discuss* **2014**, *2*, 6271–6294. [CrossRef]
14. Rizzo, P.; Bucci, A.; Sanangelantoni, A.M.; Iacumin, P.; Celico, F. Coupled Microbiological–Isotopic Approach for Studying Hydrodynamics in Deep Reservoirs: The Case of the Val d'Agri Oilfield (Southern Italy). *Water* **2020**, *12*, 1483. [CrossRef]
15. Rizzo, P.; Malerba, M.; Bucci, A.; Sanangelantoni, A.M.; Remelli, S.; Celico, F. Potential enhancement of the in-situ bioremediation of contaminated sites through the isolation and screening of bacterial strains in natural hydrocarbon springs. *Water* **2020**, *12*, 2090. [CrossRef]
16. Remelli, S.; Rizzo, P.; Celico, F.; Menta, C. Natural surface hydrocarbons and soil faunal biodiversity: A bioremediation perspective. *Water* **2020**, *12*, 2358. [CrossRef]
17. Merlini, S.; Mostardini, F. Appennino centro-meridionale: Sesioni geologiche e proposta di modello strutturale. In *Geologia Dell'Italia*; Centrale Congresso Nazionale: Rome, Italy, 1986; Volume 73, Available online: http://pascal-francis.inist.fr/vibad/index.php?action=getRecordDetail&idt=8170990 (accessed on 31 March 2023).
18. Butler, R.W.H.; Corrado, S.; Mazzoli, S.; De Donatis, M.; Di Bucci, D.; Naso, G.; Scrocca, D.; Nicolai, C.; Zucconi, V. Time and space variability of «thin-skinned» and «thick-skinned» thrust tectonics in the Apennines (Italy). *Rend. Lincei* **2000**, *11*, 5–39. [CrossRef]
19. Menardi Noguera, A.; Rea, G. Deep structure of the Campanian–Lucanian arc (southern Apennine, Italy). *Tectonophysics* **2000**, *324*, 239–265. [CrossRef]

20. Wavrek, D.A.; Mosca, F. Compositional grading in the oil column: Advances from a mass balance and quantitative molecular analysis. *Geol. Soc. Lond. Spec. Publ.* **2004**, *237*, 207–220. [CrossRef]
21. Rizzo, P.; Severini, E.; Bucci, A.; Bocchia, F.; Palladino, G.; Riboni, N.; Sanangelantoni, A.M.; Francese, R.; Giorgi, M.; Iacumin, P.; et al. How do turbidite systems behave from the hydrogeological point of view? New insights and open questions coming from an interdisciplinary work in Southern Italy. *PLoS ONE* **2022**, *17*, e0268252. [CrossRef]
22. Boenzi, F.; Ciaranfi, N. Stratigrafia di dettaglio del Flysch di Gorgoglione (Lucania). *Mem. Della Soc. Geol. Ital.* **1970**, *9*, 65–79.
23. Lentini, F.; Carbone, S.; Di Stefano, A.; Guarnieri, P. Stratigraphical and structural constraints in the Lucanian Apennines (Southern Italy): Tools for reconstructing the geological evolution. *J. Geodyn.* **2002**, *34*, 141–158. [CrossRef]
24. Tavarnelli, E.; Prosser, G. The complete Apennines orogenic cycle preserved in a transient single outcrop near San Fele, Lucania, Southern Italy. *J. Geol. Soc.* **2003**, *160*, 429–434. [CrossRef]
25. Giannandrea, P.; Loiacono, F.; Maiorano, P.; Liler, F. *Carta Geologica del Settore Orientale del Bacino di Gorgoglione (Eastern Sector)*; Litografia Artistica Cartografica Srl: Florence, Italy, 2009.
26. Maffione, M.; Speranza, F.; Cascella, A.; Longhitano, S.G.; Chiarella, D. A~125 post-early Serravallian counterclockwise rotation of the Gorgoglione Formation (Southern Apennines, Italy): New constraints for the formation of the Calabrian Arc. *Tectonophysics* **2013**, *590*, 24–37. [CrossRef]
27. Romano, G.; Balasco, M.; Siniscalchi, A.; Gueguen, E.; Petrillo, Z.; Tripaldi, S. Geological and geo-structural characterization of the Montemurro area (Southern Italy) inferred from audiomagnetotelluric survey. *Geomat. Nat. Hazards Risk* **2018**, *9*, 1156–1171. [CrossRef]
28. Pantaleone, D.V.; Vincenzo, A.; Fulvio, C.; Silvia, F.; Cesaria, M.; Giuseppina, M.; Ilaria, M.; Vincenzo, P.; Rosa, S.A.; Gianpietro, S.; et al. Hydrogeology of continental Southern Italy. *J. Maps* **2018**, *14*, 230–241. [CrossRef]
29. Petrella, E.; Aquino, D.; Fiorillo, F.; Celico, F. The effect of low-permeability fault zones on groundwater flow in a compartmentalized system. Experimental evidence from a carbonate aquifer (Southern Italy). *Hydrol. Process.* **2014**, *29*, 1577–1587. [CrossRef]
30. U.S. EPA. *Method 5021A (SW-846 Update V): Volatile Organic Compounds in Various Sample Matrices Using Equilibrium Headspace Analysis*; U.S. Environmental Protection Agency: Washington, DC, USA, 2014.
31. U.S. EPA. *Method 8260C: Volatile Organic Compounds by Gas Chromatography/Mass Spectrometry (GC/MS)*; U.S. Environmental Protection Agency: Washington, DC, USA, 2006.
32. U.S. EPA. *Method 3511 (SW-846 Update V): Organic Compounds in Water by Microextraction*; U.S. Environmental Protection Agency: Washington, DC, USA, 2014.
33. U.S. EPA. *Method 8270D (SW-846 Update V): Semivolatile Organic Compounds by Gas Chromatography/Mass Spectrometry*; U.S. Environmental Protection Agency: Washington, DC, USA, 2014.
34. *ISO 17943*; Water quality—Determination of Volatile Organic Compounds in Water—Method Using Headspace Solid-Phase Micro-Extraction (HS-SPME) followed by Gas Chromatography-Mass Spectrometry (GC-MS). International Organization for Standardization: Geneva, Switzerland, 2016.
35. *APAT CNR IRSA 5160B Man 29/2003*; Metodi Analitici per le Acque, Vol. 2—Sez. 5000: Costituenti Organici. The Agency for the Protection of the Environment and for Technical Services: Roma, Italy, 2004; ISBN 88-448-0083-7. Available online: https://www.irsa.cnr.it/wp/wp-content/uploads/2022/04/Vol2_Sez_5000_Organici.pdf (accessed on 31 March 2023). (In Italian)
36. Chong, C.W.; Dunn, M.J.; Convey, P.; Tan, G.Y.A.; Wong, R.C.S.; Tan, I.K.P. Environmental influences on bacterial diversity of soils on Signy Island, maritime Antarctic. *Polar Biol.* **2009**, *32*, 1571–1582. [CrossRef]
37. Hernàndez-Diaz, R.; Petrella, E.; Bucci, A.; Naclerio, G.; Feo, A.; Sferra, G.; Chelli, A.; Zanini, A.; Gonzalez-Hernandez, P.; Celico, F. Integrating hydrogeological and microbiological data and modelling to characterize the hydraulic features and behaviour of coastal carbonate aquifers: A case in western Cuba. *Water* **2019**, *11*, 1989. [CrossRef]
38. DeLong, E.F. Archaea in coastal marine environments. *Proc. Natl. Acad. Sci. USA* **1992**, *89*, 5685–5689. [CrossRef]
39. Coolen, M.J.; Hopmans, E.C.; Rijpstra, W.I.C.; Muyzer, G.; Schouten, S.; Volkman, J.K.; Damsté, J.S.S. Evolution of the methane cycle in Ace Lake (Antarctica) during the Holocene: Response of methanogens and methanotrophs to environmental change. *Org. Geochem.* **2004**, *35*, 1151–1167. [CrossRef]
40. Takahashi, S.; Tomita, J.; Nishioka, K.; Hisada, T.; Nishijima, M. Development of a prokaryotic universal primer for simultaneous analysis of Bacteria and Archaea using next-generation sequencing. *PLoS ONE* **2014**, *9*, e105592. [CrossRef]
41. Magoč, T.; Salzberg, S.L. FLASH: Fast length adjustment of short reads to improve genome assemblies. *Bioinformatics* **2011**, *27*, 2957–2963. [CrossRef]
42. Caporaso, J.G.; Kuczynski, J.; Stombaugh, J.; Bittinger, K.; Bushman, F.D.; Costello, E.K.; Fierer, N.; Gonzalez Peña, A.; Goodrich, J.K.; Gordon, J.I.; et al. QIIME allows analysis of high-throughput community sequencing data. *Nat. Methods* **2010**, *7*, 335–336. [CrossRef] [PubMed]
43. Bokulich, N.A.; Subramanian, S.; Faith, J.J.; Gevers, D.; Gordon, J.I.; Knight, R.; Mills, D.A.; Caporaso, J.G. Quality-filtering vastly improves diversity estimates from Illumina amplicon sequencing. *Nat. Methods* **2013**, *10*, 57–59. [CrossRef] [PubMed]
44. Longinelli, A.; Selmo, E. Isotopic composition of precipitation in Italy: A first overall map. *J. Hydrol.* **2003**, *270*, 75–88. [CrossRef]
45. Paternoster, M.; Liotta, M.; Favara, R. Stable isotope ratios in meteoric recharge and groundwater at Mt. Vulture volcano, Southern Italy. *J. Hydrol.* **2008**, *348*, 87–97. [CrossRef]

46. Petrella, E.; Capuano, P.; Carcione, M.; Celico, F. A high-altitude temporary spring in a compartmentalized carbonate aquifer: The role of low-permeability faults and karst conduits. *Hydrol. Process.* **2009**, *23*, 3354–3364. [CrossRef]
47. Petrella, E.; Celico, F. Mixing of water in a carbonate aquifer, Southern Italy, analysed through stable isotope investigations. *Int. J. Speleol.* **2013**, *42*, 25–33. [CrossRef]
48. Candela, S.; Mazzoli, S.; Megna, A.; Santini, S. Finite element modelling of stress field perturbations and interseismic crustal deformation in the Val d'Agri region, southern Apennines, Italy. *Tectonophysics* **2015**, *657*, 245–259. [CrossRef]
49. Lyu, Z.; Shao, N.; Akinyemi, T.; Whitman, W.B. Methanogenesis. *Curr. Biol.* **2018**, *28*, R727–R732. [CrossRef]
50. Guan, J.; Xia, L.-P.; Wang, L.-Y.; Liu, J.-F.; Gu, J.-D.; Mu, B.-Z. Diversity and distribution of sulfate-reducing bacteria in four petroleum reservoirs detected by using 16S rRNA and dsrAB genes. *Int. Biodeterior. Biodegrad.* **2012**, *76*, 58–66. [CrossRef]
51. Grassia, G.S.; McLean, K.M.; Glénat, P.; Bauld, J.; Sheehy, A.J. A systematic survey for thermophilic fermentative bacteria and archaea in high temperature petroleum reservoirs. *FEMS Microbiol. Ecol.* **1996**, *21*, 47–58. [CrossRef]
52. Dahle, H.; Birkeland, N.K. *Thermovirga lienii* gen. nov., sp. nov., a novel moderately thermophilic, anaerobic, amino-acid-degrading bacterium isolated from a North Sea oil well. *Int. J. Syst. Evol. Microbiol.* **2006**, *56*, 1539–1545. [CrossRef] [PubMed]
53. Bekins, B.A.; Brennan, J.C.; Tillitt, D.E.; Cozzarelli, I.M.; Illig, J.M.; Martinović-Weigelt, D. Biological effects of hydrocarbon degradation intermediates: Is the total petroleum hydrocarbon analytical method adequate for risk assessment? *Environ. Sci. Technol.* **2020**, *54*, 11396–11404. [CrossRef] [PubMed]

water

MDPI

*Article*

# Response Characteristics and Water Inflow Prediction of Complex Groundwater Systems under High-Intensity Coal Seam Mining Conditions

**Zhaolai Hua [1,2], Yao Zhang [1,*], Shihao Meng [1], Lu Wang [1,2], Xuejun Wang [3], Yang Lv [1,2], Jinming Li [2], Shaofeng Ren [2], Han Bao [1], Zhihao Zhang [1], Linger Zhao [1] and Yifan Zeng [1,*]**

1 Nation Engineering Research Center of Coal Mine Water Hazard Controlling, China University of Mining and Technology, Beijing 100083, China; huazhaolai110@163.com (Z.H.); meng_shihao@outlook.com (S.M.)
2 Shaanxi Shanmei Caojiatan Mining Co., Ltd., Yulin 719000, China; sqt2100203063@student.cumtb.edu.cn (J.L.)
3 Tangshan Kailuan Construction (Group) Co., Ltd., Tangshan 063000, China
* Correspondence: zqt2100203105@student.cumtb.edu.cn (Y.Z.); zengyf@cumtb.edu.cn (Y.Z.)

**Abstract:** With the gradual improvement in coal mining efficiency, the disturbance of groundwater systems caused by high-intensity mining also increases, leading to challenges in maintaining mine safety and protecting water resources in mining areas. How to accurately describe the dynamic changes in the groundwater system under mining and quantitatively predict mine water inflow are currently major problems to be addressed. Based on a full analysis of the response characteristics of a groundwater system to the extraction disturbance, this paper presents a new method to establish a mine hydrogeological conceptual model that can accurately represent the water inrush process. The unstructured-grid package of MODFLOW is used to accurately characterize the formation structure and finally make accurate water inflow predictions. Taking the Caojiatan coal mine in Shaanxi Province, China, as an example, a numerical model of unstructured water inflow is established, and the changes in the water inflow source and intensity are quantitatively evaluated. Compared with the traditional water inflow prediction method, the prediction accuracy of the new model is improved by 12–17%, which is achieved by detailing the response of the complex groundwater system under high-intensity mining conditions. The method presented in this paper has great significance and applicatory value for obtaining a comprehensive understanding of the disturbance characteristics of human underground engineering activities (e.g., coal mining) on groundwater systems, as well as accurately predicting water inflow.

**Keywords:** coal mine water hazard; MODFLOW; non-structural mesh; mine hydrogeology

**Citation:** Hua, Z.; Zhang, Y.; Meng, S.; Wang, L.; Wang, X.; Lv, Y.; Li, J.; Ren, S.; Bao, H.; Zhang, Z.; et al. Response Characteristics and Water Inflow Prediction of Complex Groundwater Systems under High-Intensity Coal Seam Mining Conditions. *Water* **2023**, *15*, 3376. https://doi.org/10.3390/w15193376

Academic Editor: Yeshuang Xu

Received: 31 August 2023
Revised: 18 September 2023
Accepted: 21 September 2023
Published: 26 September 2023



## 1. Introduction

In recent years, engineering activities have no longer been limited to the Earth's surface and have gradually extended underground. However, the development of underground engineering has caused serious damage to geological structures and groundwater systems. With the increase in human activity, it has become crucial to evaluate the impact of underground engineering on the environment and their interaction. Mining activities are common, and long-term, high-intensity coal mining has become one of the most typical underground engineering activities [1]. Water inrush in mines is a frequent occurrence during the mining process, posing serious threats to the safety of miners and causing significant economic losses [2]. Moreover, it also seriously damages groundwater systems [3,4]. Weather and drainage wells are the main driving factors for the impact of general underground engineering activities on groundwater systems, whereas those of coal mining are mainly water inflow and overlying strata damage caused by mining activities. Therefore, establishing a conceptual model under high-strength coal mining is of great importance to analyze how mining disturbs groundwater systems [5].

Currently, there are many studies on how mining disturbs groundwater systems, which mainly focus on the circulation, chemical characteristics, and evolution of the groundwater system [6,7]. Regarding the water inrush problems, Zeng et al. [8–10] achieved a series of research results on the water-richness in the roof, the identification of water sources for water inrushes, and mine water drainage. Ta et al. developed a numerical groundwater model, based on a conceptual hydrogeologic model of the Thach Khe iron ore mine and its surrounding area in Vietnam, and evaluated various scenarios to estimate groundwater ingress at different stages of the mine life [11]. Duan et al. [12] studied the tracing of groundwater by environmental isotopes and artificial radioisotopes, providing a basis for the application of isotopes in water cycling, and evaluated the impact of different factors. Under the dual carbon background, Wu et al. [13–15] proposed a theory of positive synergistic and coordinated coal-water-heat co-mining, providing an approach for the green and coordinated utilization of groundwater, coal resources, and geothermal resources. In recent years, many scholars have also made significant efforts toward the precise prediction of mine water inflow [16,17]. The main methods for predicting water inflow include the virtual large-diameter well method [18,19], analytical methods [20,21], the hydrogeological analogy method [22], and numerical simulation methods [23,24]. Unfortunately, these traditional methods often cannot accurately describe the response characteristics of groundwater systems under the conditions of high-intensity coal mining [25–27]. Therefore, establishing a hydrogeological conceptual mine model is critical to studying groundwater systems.

Based on comprehensive investigations of the Caojiatan coal mine in Shaanxi, China, this study summarizes a set of methods for constructing a model that accurately reflects the process of water inflow. Combined with numerical simulation, it accurately characterizes the process and intensity of water inflow at the working face. Our results have reference value for studying the disturbance of underground engineering activities on groundwater systems.

## 2. Study Area

The Caojiatan coal mine is located in the northern part of Yulin City, Shaanxi Province, China (Figure 1). It lies on the northeastern edge of the Ordos Plateau, adjacent to the southeastern edge of the Mu Us Desert, and features sand dunes, wind-blown sand plains, and loess landforms. The climate is characterized by fluctuating temperatures throughout the year, low rainfall, and high evaporation, and precipitation is extremely unevenly distributed throughout the year. The mining area belongs to the Tuwei River Basin and the Yuxi River Basin. There is a north–south watershed in the middle of the study area, which divides the confined aquifer into two relatively independent hydrogeological units, east and west.

The stratigraphic layers in the study area, from top to bottom, are as follows: the Quaternary Upper Member Salarwusu Formation ($Q_3s$), the Quaternary Middle Member Lishi Formation ($Q_2l$), the Neogene Upper Member Baode Formation ($N_2b$), the Jurassic Middle Member Anding Formation ($J_2a$), the Jurassic Middle Member Zhiluo Formation ($J_2z$), the Jurassic Middle Member Yan'an Formation ($J_2y$), and the coal seam (Figure 2).

The Salawusu Formation ($Q_3s$) is widely distributed across the study area, with only a partial absence observed at the watershed. The sand layer is characterized by significant thickness, high permeability, and a high capacity to be recharged from rainfall. Groundwater is moderately abundant in this formation, and the overall direction of groundwater flow is from northeast to southwest.

The Lishi Formation ($Q_2l$) is distributed throughout the western flank of the study area. At the watershed, the layer partially emerges at the surface. The lithology of this layer is heterogeneous, mainly consisting of sandy loam and silt, and it exhibits good vertical permeability.

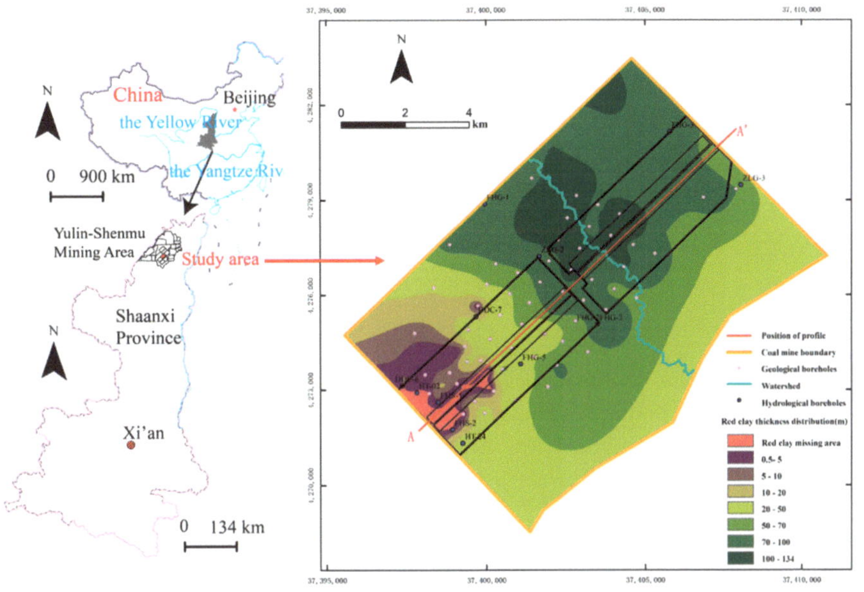


**Figure 1.** Location of Caojiatan well field and distribution map of laterite aquifer.

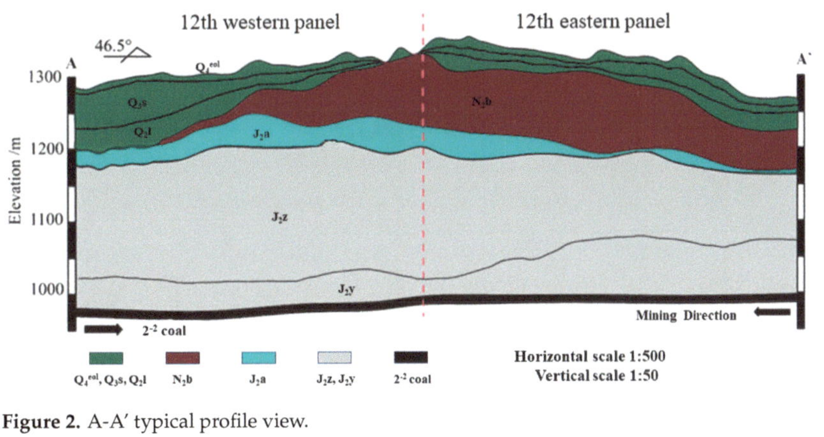


**Figure 2.** A-A' typical profile view.

The Baode Formation ($N_2b$) becomes thinner from the watershed toward the west and southwest directions. The lithology mainly consists of reddish-brown clay and sub-clay. The soil gradually decreases in the western wing of the mining area, and there is some red soil erosion loss at the western boundary, with a cumulative area of 1.31 km$^2$. The distribution of the red clay layer and the hydrological parameters of the strata were mainly obtained by collecting geological and hydrological borehole data from the study area. The boreholes are shown in Figure 1.

The Anding Formation ($J_2a$) is distributed throughout the western part of the oilfield, with a relatively small abundance of groundwater.

The Zhiluo Formation ($J_2z$) shows no weathering at the top and mainly consists of blue-gray mudstone and siltstone. Its fissures are poorly developed, resulting in a low overall abundance of groundwater.

The Yanan Formation ($J_2y$) mainly consists of medium- to fine-grained sandstone with mud or calcareous cementation. Its fissures have poor connectivity, leading to a low abundance of groundwater. The thickness of $2^{-2}$ coal in the study area is 11.72~12.7 m.

## 3. Study Methods

We used the mine hydrogeological conceptual modeling method, the traditional hydrogeological conceptual modeling method, the hydrogeological analog method, and the Grey Model (1,1) in this study.

### 3.1. Mine Hydrogeological Conceptual Model

The mine conceptual hydrogeological model has different driving forces compared to traditional hydrogeological models: those of the latter are mainly weather and drainage wells, whereas that of the former is overlying strata damage caused by mining activities and water inflow. At the same time, because mining activities have more intense characteristics, the mine hydrogeological conceptual model has features of instantaneous change in addition to static simulations, which are typical of traditional hydrogeological models.

Specifically, to establish the mine hydrogeological conceptual model, the following three aspects need to be addressed:

1. The geological characteristics of the mining area must be analyzed and studied, including geological structure, rock type, and stratigraphy. Through the investigation of geological characteristics, the lithological characteristics, spatial structure, water-richness, permeability, and stratigraphic structure of each aquifer or aquitard in the research area can be determined. This is the geological basis for establishing a hydrogeological conceptual model for mining areas.
2. The development height of the water-bearing fractured zone, as well as the water inflow intensity and whether there is leaking recharge, must be analyzed and studied in order to understand where the water comes from and where it goes. After coal mining, the underground area will suffer mining damage; at this time, the thickness of the bedrock of the roof of the mining working face and the development height of the water-bearing fractured zone determine the range of water inflow channels. An empirical formula [28] for determining the height of the water-bearing fractured zone is given as follows:

$$H_f = 20M + 10 \tag{1}$$

   where $H_f$ is the height of the water-conducting fracture zone (L) and $M$ is the thickness of the coal seam (L).
3. The chemical characteristics of the mine water must be analyzed and studied, including the pH value, electrical conductivity, hardness, and the contents of anions and cations, in order to identify the source of water inflows. By considering the characteristics of mine water inflow, we can determine the recharge impact of the overlying aquifers on the mine water inflow.

In summary, establishing the mine conceptual hydrogeological model involves comprehensive analyses of the hydrogeological characteristics, water chemistry, water inflow sources, height of water-conductive fracture zones, and water inflow intensity. This method provides a theoretical foundation for studying the changing characteristics of underground water systems and predicting water inflow in mines.

### 3.2. Numerical Simulation Method

The numerical simulation method uses the MODFLOW-USG (Unstructured Grid) version released by the United States Geological Survey in 2013. The version supports various structured and unstructured grid types, and its flexibility in refining the density of unstructured grids can achieve precise characterization of the grid in the working face.

The governing equation for groundwater flow and boundary control is as follows:

$$\begin{cases} \frac{\partial}{\partial x}\left(K_L\frac{\partial h}{\partial x}\right)+\frac{\partial}{\partial y}\left(K_L\frac{\partial h}{\partial y}\right)+\frac{\partial}{\partial z}\left(K_z\frac{\partial h}{\partial z}\right)+\varepsilon=0, x,y,z\in\Omega_1 \\ \frac{\partial}{\partial x}\left(K_L(h-h_b)\frac{\partial h}{\partial x}\right)+\frac{\partial}{\partial y}\left(K_L(h-h_b)\frac{\partial h}{\partial y}\right)+K_z\frac{\partial}{\partial z}\left(\frac{\partial h}{\partial z}\right)+\varepsilon=0, x,y,z\in\Omega_2 \\ h(x,y,z)=h_0, x,y,z\in\Omega \\ K_{\vec{\gamma}}\frac{\partial h}{\partial\vec{\gamma}}|\Gamma_1=q, x,y,z\in\Gamma_1 \\ K_{\vec{\gamma}}\frac{\partial h}{\partial\vec{\gamma}}|\Gamma_2=\frac{k}{m}(h_2-h), x,y,z\in\Gamma_2 \end{cases}, \quad (2)$$

$$\begin{cases} \frac{\partial}{\partial x}\left(K_L\frac{\partial h}{\partial x}\right)+\frac{\partial}{\partial y}\left(K_L\frac{\partial h}{\partial y}\right)+\frac{\partial}{\partial z}\left(K_z\frac{\partial h}{\partial z}\right)+\varepsilon=S\frac{\partial h}{\partial t}, x,y,z\in\Omega_1, t\geq 0 \\ \frac{\partial}{\partial x}\left(K_L(h-h_b)\frac{\partial h}{\partial x}\right)+\frac{\partial}{\partial y}\left(K_L(h-h_b)\frac{\partial h}{\partial y}\right)+\frac{\partial}{\partial z}\left(K_z\frac{\partial h}{\partial z}\right)+\varepsilon=u\frac{\partial h}{\partial t}, x,y,z\in\Omega_2, t\geq 0 \\ h(x,y,z,t)|_{t=0}, x,y,z\in\Omega \\ K_{\vec{\gamma}}\frac{\partial h}{\partial\vec{\gamma}}|\Gamma_1=q, x,y,z\in\Gamma_1, t\geq 0 \\ K_{\vec{\gamma}}\frac{\partial h}{\partial\vec{\gamma}}|\Gamma_2=\frac{k}{m}(h_2-h), x,y,z\in\Gamma_2, t\geq 0 \end{cases}, \quad (3)$$

where Equation (2) is the steady flow control equation and Equation (3) is the unsteady flow control equation. The symbols used in the equation are defined as follows:

- $\Omega$—Domain of the porous medium;
- $\Omega_1$—Groundwater aquifer flow zone;
- $\Omega_2$—Confined aquifer flow region;
- $\Gamma_1$—Lateral and bottom boundaries of $\Omega$;
- $h$—Head of the aquifer (L);
- $h_b$—Elevation of the bottom of the confined aquifer;
- $K_L$, $K_z$—Coefficients of horizontal and vertical hydraulic conductivity ($LT^{-1}$);
- $\varepsilon$—Source/sink term of the aquifer ($T^{-1}$);
- $S$—Specific storage coefficient of the aquifer ($L^{-1}$);
- $\mu$—Specific storage of the confined aquifer;
- $h_0$—Initial value of the head of the aquifer (L);
- $h_1$—Initial groundwater level calculated using the steady-state flow model (L);
- $K_n$—Coefficient of permeability in the direction of the normal of the boundary surface ($LT^{-1}$);
- $q$—Flow rate in $\Gamma_1$, positive for inflow and negative for outflow; $q$ is zero for impermeable boundaries ($LT^{-1}$);
- $\gamma$—Normal direction to the boundary surface;
- $\Gamma_2$—Generalized hydraulic head boundary of the flow region;
- $h_2$—Specified hydraulic head outside the $\Gamma_2$ boundary;
- $k$—Hydraulic conductivity of the aquifer between the boundary at $\Gamma_2$ and the specified hydraulic head position at $h_2$;
- $m$—Thickness of the aquifer between the boundary at $\Gamma_2$ and the specified hydraulic head position at $h_2$.

The goaf is generalized as a drain boundary by using the MODFLOW module in the Groundwater Modeling System [29–34]. The principle of The Drain (DRN) Package is as follows:

$$Q_{out}=\begin{cases} CD\left(h_{i,j,k}-HD\right) & h_{i,j,k}>HD \\ Q_{out}=0 & h_{i,j,k}\leq HD \end{cases}, \quad (4)$$

where $Q_{out}$ is the flow from the aquifer into the drain ($L^3T^{-1}$), $CD$ is the drain conductance ($L^2T^{-1}$), $HD$ is the drain elevation (L), and $h_{i,j,k}$ is the head in the cell containing the drain (L).

The following formula is used for fitting the error (Equation (5)):

$$\delta = \frac{\sqrt{\sum_{k=1}^{n}\left(h_k - h_k'\right)^2}}{n\Delta h} \times 100\%, \tag{5}$$

where $\delta$ is the groundwater level fitting error (%), $n$ is the number of stress periods in the model, $\Delta h$ is the maximum measured water level variation during the operation period (L), $h_k$ is the measured groundwater level during stress period $k$ (L), and $h'_k$ is the simulated groundwater level during stress period $k$ (L).

### 3.3. Grey Model (1,1)

The grey theory [35,36] is a mathematical method used to deal with problems involving incomplete data and high uncertainty. It consists of the grey system theory and grey prediction methods, aiming to reveal the inherent patterns of a system through the analysis and prediction of incomplete data. The GM(1,1), or grey model, which is a single-variable time series prediction model based on grey system theory, generates new data sequences by performing first-order accumulations on the original data, and then constructs differential equation models using these new data sequences to achieve the time series prediction [37]. The steps are as follows:

- Establish a first-order accumulation to generate a sequence of numbers.

  Suppose the non-negative original sequence is $x^{(0)}$:

$$x^{(0)} = \left(x^{(0)}(1), x^{(0)}(2), x^{(0)}(3), \ldots, x^{(0)}(n)\right), x^{(0)}(k) > 0, k = 1, 2, \ldots, n. \tag{6}$$

  The accumulation sequence $x^{(1)}$ of $x^{(0)}$ is

$$x^{(1)} = \left(x^{(1)}(1), x^{(1)}(2), x^{(1)}(3), \ldots, x^{(1)}(n)\right), \tag{7}$$

$$x^{(1)}(k) = \sum_{i=1}^{k} x^{(0)}(i). \tag{8}$$

- The assumed first-order model for generating the sequence is

$$\frac{dx^{(1)}}{dt} + ax^{(1)} = b \qquad t \in [0, \infty]. \tag{9}$$

- The dynamic model of the differential equation is given by

$$z^{(1)}(k) = 0.5x^{(1)}(k) + 0.5x^{(1)}(k-1), \tag{10}$$

$$x^{(0)}(k) + az^{(1)}(k) = b, \tag{11}$$

- The least-squares method is used to find the parameters $a$ and $b$:

$$B = \left\{ \begin{array}{cc} -\frac{1}{2}\left[x^{(1)}(1) + x^{(1)}(2)\right], & 1 \\ -\frac{1}{2}\left[x^{(1)}(2) + x^{(1)}(3)\right], & 1 \\ \cdots & \\ -\frac{1}{2}\left[x^{(1)}(n-1) + x^{(1)}(n)\right], & 1 \end{array} \right\}, \tag{12}$$

$$y_n = \left[x^{(0)}(2), x^{(0)}(3), \ldots, x^{(0)}(n)\right]^T, \tag{13}$$

  Identification of parameters $a$ and $b$:

$$\beta = \begin{bmatrix} a \\ b \end{bmatrix} \tag{14}$$

$$\beta = \left(B^T B\right)^{-1} B^T y_n \tag{15}$$

- Solving:

$$\hat{x}^{(1)}(k+1) = \left(x^{(0)}(1) - \frac{b}{a}\right)e^{-ak} + \frac{b}{a}, \ k = 1, 2, \cdots, n-1 \tag{16}$$

### *3.4. Hydrogeological Analog Method*

The hydrogeological analog method is a groundwater exploration method that uses accumulated data from mining wells with similar geological and hydrogeological conditions, and mining methods to predict the water inflow of the designed mining well. The hydrogeological analog method mainly relies on qualitative analysis rather than quantitative data analysis. The following formula is used for calculations [38]:

$$Q = Q_0\left(\sqrt[c]{\frac{F}{F_0}}\right)\left(\sqrt[d]{\frac{S}{S_0}}\right), \tag{17}$$

where $Q$ is the water inflow of the mine ($L^3T^{-1}$), $Q_0$ is the water inflow of the production mine ($L^3T^{-1}$), $F$ is the mining area ($L^2$), $F_0$ is the actual mining area of the production mine ($L^2$), $S$ is the drawdown of the water level (L), $S_0$ is the drawdown of the water level in the production mine (m), $c$ = 1, and $d$ = 2.

### *3.5. Traditional Hydrogeological Conceptual Model*

The traditional hydrogeological conceptual model involves generalizing the boundary properties, internal structure, permeability, hydraulic characteristics, and recharge/discharge conditions of the aquifer into mathematical and physical forms. This process aims to quantify the hydrogeological elements and achieve an accurate numerical simulation. The main aspects of the traditional hydrogeological conceptual model include the following:

- Determining the model domain: The model domain refers to the spatial extent within which the hydrogeological processes will be simulated. It is determined based on the research objectives, the characteristics of the study area, and available observational data.
- Generalization of the aquifer (Aquitard): In numerical modeling, it is common practice to group aquifers (and aquitards) with similar hydraulic properties and close hydraulic connections into a unified aquifer (or aquifer system).
- Generalization of boundary conditions: The model boundaries consist of horizontal boundaries and vertical boundaries. The generalization of boundaries should accurately represent the hydrogeological prototype of the study area, and the generalized boundary conditions should reflect the characteristics of the groundwater flow field in the study area. In the generalization process, natural boundaries should be utilized as much as possible to maintain the representativeness of the model. Natural boundaries refer to physical features such as rivers, lakes, or impermeable formations that naturally limit the flow of groundwater. When determining the properties of artificial boundaries, it is important to consider any relevant adverse factors that may affect the groundwater flow field (e.g., pumping wells, hydraulic barriers, or other anthropogenic influences).

## 4. Results and Discussion

### *4.1. Water Inflow Prediction Using Numerical Simulation Method*

#### 4.1.1. Establishment of Mine Hydrogeological Conceptual Model

The simulation scope of this study is the western area of the Caojiatan coal mine. The surface topography of the 122107 working face mainly comprises crescent-shaped sand dunes with some vegetation. The terrain gradually decreases from the southwest to the northeast. No major faults, igneous rocks, or folding structures are found within the area.

The first step to establish the mine hydrogeological conceptual model of the study area is to analyze the spatial distribution, lithological characteristics, and hydrological properties of the aquifers (aquitards) within the Caotan 122107 working face field.

Based on the hydrological characteristics and hydraulic connections of each aquifer, we classify layers with similar properties into an aquifer group. Therefore, the model is divided into five layers (Figure 3) from top to bottom. The first layer consists of Quaternary porous aquifers, mainly those of the Salawusu Formation aquifer and Lishi Formation. The second layer consists of red clay aquitards of the Baode Formation. The third layer consists of weathered bedrock aquifers. The fourth layer consists of bedrock aquifers, primarily those of the Yanan Formation and Zhiluo Formation. The fifth layer is the $2^{-2}$ coal seam.

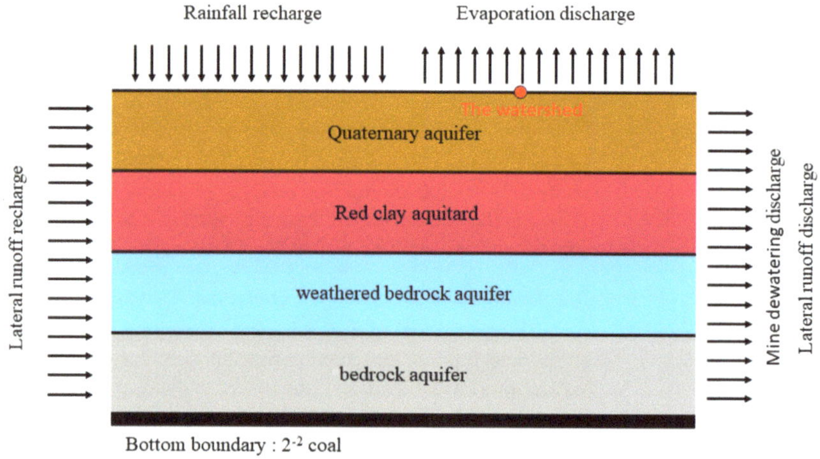


**Figure 3.** Conceptual hydrogeological model.

The second step to establish the mine hydrogeological conceptual model of the study area is to analyze the overlying strata damage after coal mining, predict the development height of the water-conducting fracture zone, and determine the range of water inflow channels.

Formula (1) is used as an empirical formula to determine the development height of water-conducting fractured zones, and the mining height is 11 m. After comparing the results of drilling fluid loss measurements, borehole color TV detection [39], and core data, we find that the normal bedrock layer has a thickness ranging from 167.25 to 254.96 m above the 122107 working face, and the height of the developed water-conducting fracture zone is 183.00 m, most of which has penetrated the weathered rock aquifer and some of which has reached the bottom of the Salawusu Formation aquifer in the red soil extinction area. By analyzing the distribution of the water-conducting fracture zone and the thickness of the bedrock, and calculating the difference between the two, we determined the influence range of the water-conducting fracture zone. On the west side of the 122107 workface, the central area is more affected, and the southern area of the west side is stronger than the northern area.

In the red soil extinction area (Figure 1), there is a possibility of leakage recharge from the Salawusu Formation aquifer to the weathered bedrock. Simultaneously, the water-conducting fracture zone may develop to the bottom of the Salawusu Formation aquifer in the area with thin red soil, leading to a sudden increase in water inflow. In most areas of the working face, the maximum height of the water-conducting fracture zone extends into the weathered bedrock (Figure 4). Therefore, the aquifers of the Zhiluo Formation and the Yanan Formation supply water directly to the $2^{-2}$ coal seam.

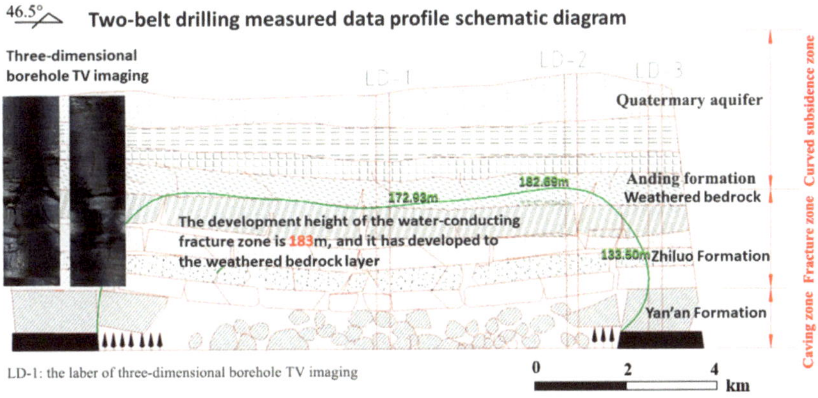


**Figure 4.** Measured profile of the height of the water-bearing fracture zone in the inclined direction of 122107 working face (LD-1, LD-2).

The third step to establish the mine hydrogeological conceptual model of the study area involves analyzing the source of mine water inflow and the variation in water inflow intensity with mining time to determine the water source and intensity characteristics of mine water inflow at different stages. We use a random forest model [13] with 13 physicochemical indicators, including total dissolved solids (TDS), total hardness (TH), pH, total alkalinity (TA), $K^+$, $N_a^+$, $Ca_2^+$, $Mg_2^+$, $Cl^-$, $SO_4^{2-}$, $HCO_3^-$, $CO_3^{2-}$, and $NO_3^-$ to perform water chemistry analysis. The results are shown in Table 1 (higher scores indicate a higher probability of originating from this aquifer), which indicates that during the initial stage of mining at the 122109 working face, the main source of water inflow is from the weathered bedrock aquifer, whereas after stabilization, the water is mainly from the Quaternary aquifer.

**Table 1.** Identification results of eye cutting water source at 122109 working face.

| Time | Score | | | Results |
|---|---|---|---|---|
| | Quaternary Water | Weathered Bedrock Water | Bedrock Water | |
| February 2020 | 0.2033 | 0.4533 | 0.3433 | Weathered bedrock water |
| February 2021 | 0.4200 | 0.2367 | 0.3433 | Quaternary water |

The intensity of mine water inflow is mainly related to several factors: the aquifer productivity of the overlying strata above the coal seam being extracted, the thickness of the bedrock above the coal seam, the degree of communication between water-conducting fracture zones and various aquifers, atmospheric precipitation, the aquitard properties of the post-mining red soil layer, and the mining method and intensity. As mining progresses, the amount of water inflow continues to increase, stabilizing at 650 $m^3/h$ after mining is completed (Figure 5).

The analysis above shows that the water inflow mainly is from the static storage and leaking replenishment of the weathered rock aquifer during the mining, and there is also lateral groundwater recharge from the Quaternary aquifer. After mining, the stable water inflow is from lateral replenishment of the weathered rock aquifer and the Quaternary aquifer.

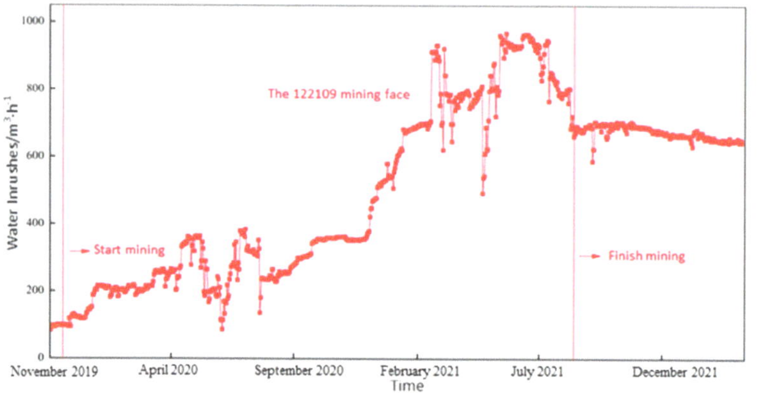


**Figure 5.** Distribution of water inflow curve for the 122109working face (measured data from the working face).

#### 4.1.2. Numerical Simulation Model

The study area is vertically divided into five layers. By collecting borehole and elevation data, a three-dimensional structure of the model (Figure 6a) and the hydrogeological cross-section map (Figure 6b) are generated using the Kriging interpolation method [40].

The model uses unstructured grids to partition each layer. In the horizontal direction, the study area is set as a grid of 200 m $\times$ 200 m, the east and west wings of the 12-panel area are set as a grid of 100 m $\times$ 100 m, and the 122107 working face is set as a grid of 50 m $\times$ 50 m. There is a certain range of missing stratigraphic layers in the red soil aquitard of the second layer in the west wing of the 12-panel area (Figure 7), and other layers are normally distributed.

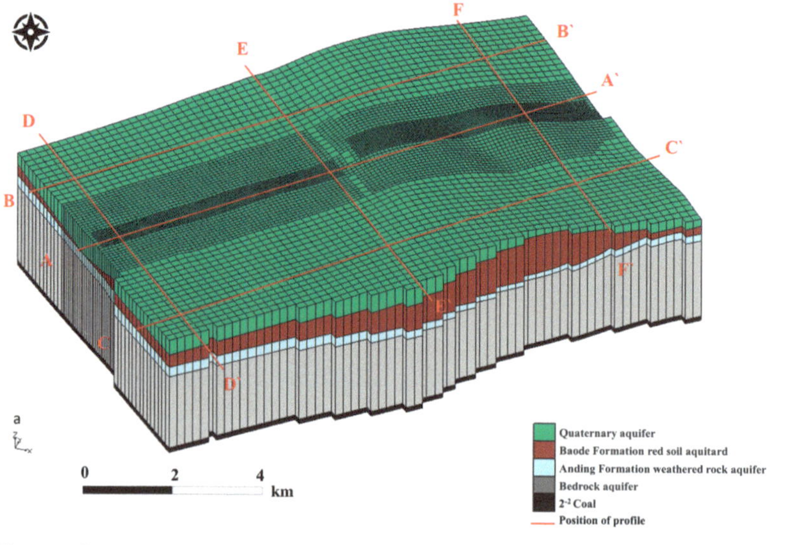


**Figure 6.** *Cont.*

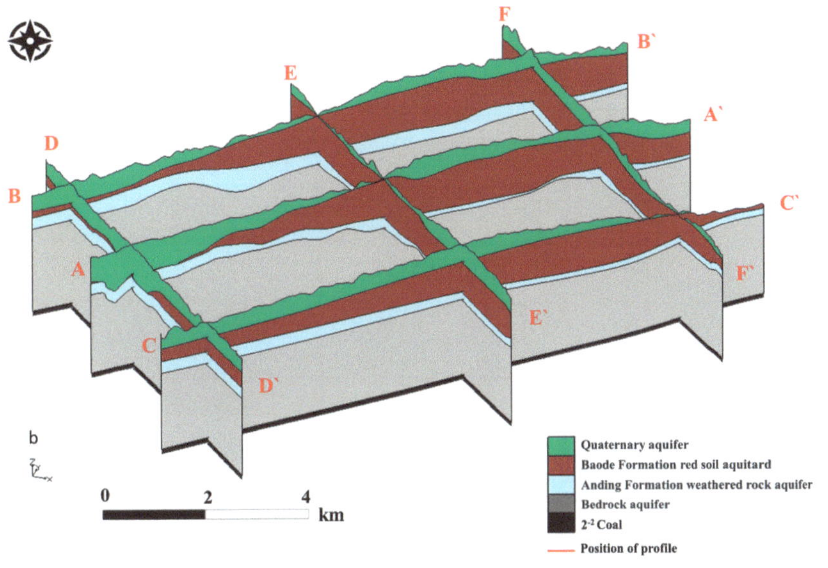


**Figure 6.** Schematic diagram of the hydrogeological conceptual model: three-dimensional structural model; hydrogeological joint profile. (**a**) three-dimensional structure of the model and (**b**) the hydrogeological cross-section map.

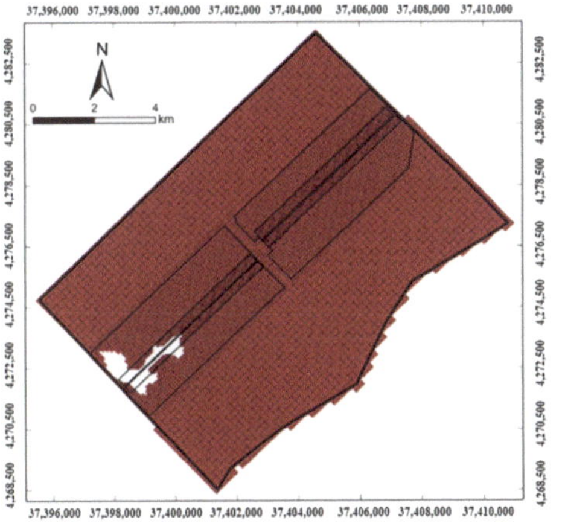


**Figure 7.** Schematic diagram of unstructured grid division of the second layer of laterite aquifer.

Based on the actual hydrogeological conditions, the watershed position of the first layer is set as the aquitard boundary. Other non-natural boundaries are set as general head boundaries in the horizontal direction, and the free water surface of the upper aquifer is set as the upper boundary of the model in the vertical direction. The bottom of the geological layer in the lower part of the model is set as the aquitard boundary. The goaf of the coal seam is generalized as a drain boundary, and the hydrogeological parameters of each aquifer or aquitard are obtained from pumping test data of the hydrological boreholes.

The study area is mainly covered by Quaternary loose sand layers, and atmospheric rainfall is considered an indirect source of water for the working face. The main water sources and sinks are rainfall and evaporation, respectively. The data are obtained from monthly average rainfall and average evaporation data (Figure 8). The amount of water inflow is determined and verified by balancing the flow into and out of the drainage ditch.

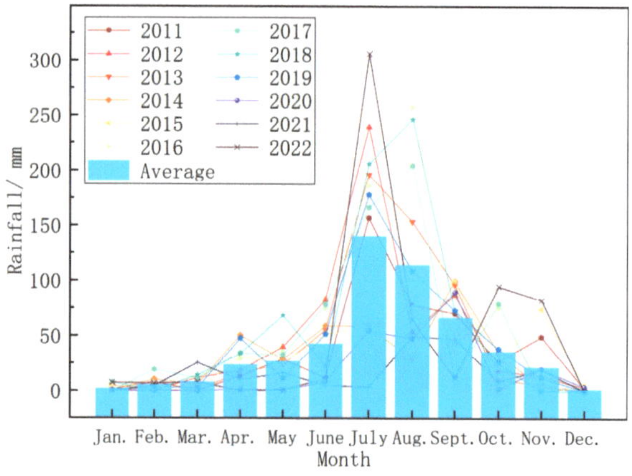


**Figure 8.** Composite monthly average rainfall map of the study area (these data are derived from the local meteorological services).

In this simulation, a steady-state flow model is established after the 122109 working face is mined. Then, based on the initial flow field, the numerical simulation model is used to investigate the changes in groundwater systems, as fitted with the observed water level data and measured water inflow. The stable flow field within the 12th western panel is depicted in Figure 9.

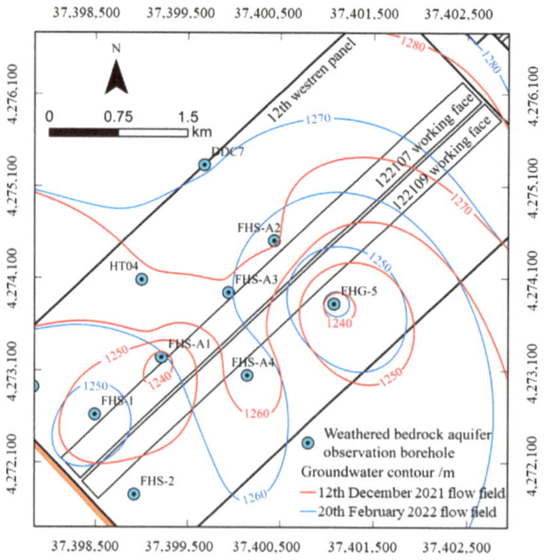


**Figure 9.** Distribution of the stable flow field in the weathered bedrock aquifer of the 12th western panel.

The simulation period of the model is from 1 February 2022 to 31 January 2027. The identification and validation of the model are divided into two parts: stable flow and unsteady flow. In stable flow, a single stable landing funnel ending with the cessation of 122109 workface mining is finally formed by adjusting the boundary conditions and hydrogeological parameters. These flow data are used as the initial flow field of unsteady flow for the simulation of 122107 workface mining. We add pumping wells to simulate the actual pumping of the FHS-1 and FHS-2 wells. By adjusting the parameters, we aim to calibrate the model while fitting the simulated values of the pumping wells to actual values.

In this paper, we reduce the mining damage process by adjusting the permeability coefficient of the overlying rock of the working face, which is a constant due to the limitation of the software and cannot simulate the dynamic change in the coefficient with the mining process. Therefore, we can only obtain the magnitude of water influx within a fixed mining length by equalizing the groundwater over a period of time. The final parameters and their zoning for the Quaternary and weathered bedrock aquifers are shown in Figures 10 and 11, and Table 2, respectively.

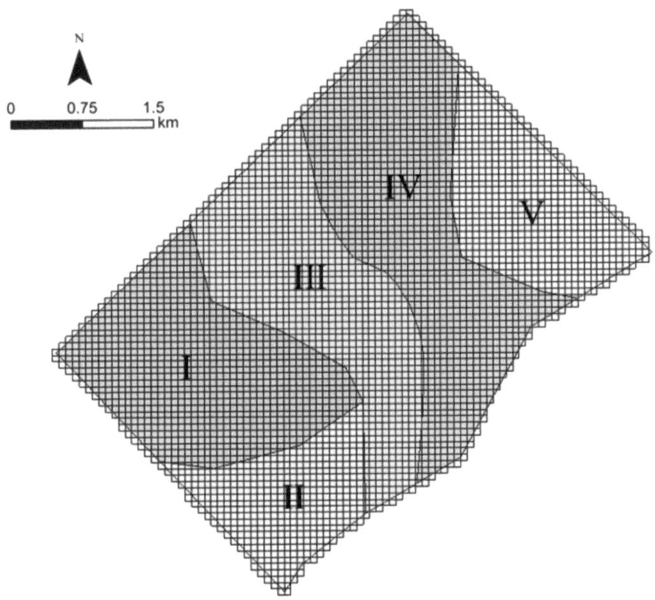


**Figure 10.** Aquifer parameter zoning (Quaternary aquifer).

The fitting is performed for a duration of 48 h (Figures 12 and 13).

The calculated results suggest that the groundwater level fitting errors for boreholes FHS-1 and FHS-2 are 2.37% and 2.03%, respectively, both of which are less than 5%. These results indicate that the fitting is good, and the numerical model is consistent with the actual situation. Therefore, the numerical model can be used to predict the water inflow.

Based on the model, the permeability coefficient of the rock formation above the 122107 working face is adjusted, and the unsteady flow simulation is performed. The simulation results show that the normal water inflow for the 122107 working face is approximately 826 $m^3/h$, and the maximum water inflow, calculated as 1.25 times the normal water inflow, is approximately 1032 $m^3/h$. This value is a safety factor for surge prediction and increases as the mine becomes more water-rich. This safety factor is applied to account for uncertainties and potential variations in the actual groundwater inflow compared to the predicted values. It helps to ensure that the mining operations are adequately prepared to manage and control water inflows for safe and efficient mining activities.

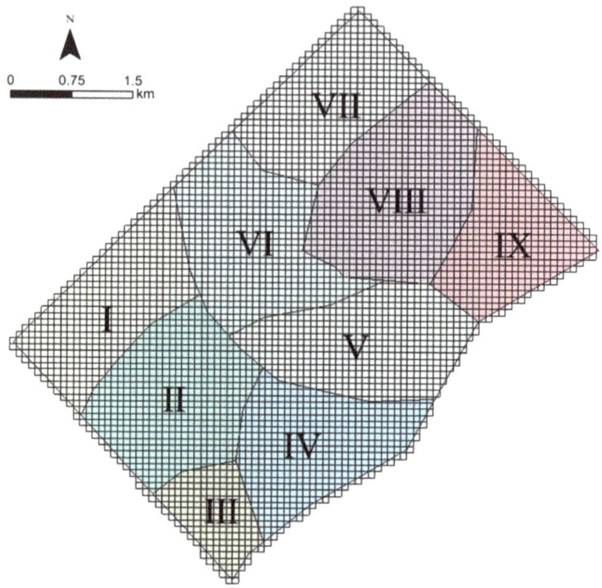


**Figure 11.** Aquifer parameter zoning (weathered bedrock aquifer).

**Table 2.** Aquifer parameter zoning table.

| Layer | Zone | $K_L$/m·d$^{-1}$ | $K_z$/m·d$^{-1}$ | μ | S/m$^{-1}$ |
|---|---|---|---|---|---|
| Quaternary Aquifer | I | 1.84 | 0.18 | 0.12 | / |
| | II | 4.61 | 0.10 | 0.11 | / |
| | III | 1.34 | 0.13 | 0.12 | / |
| | IV | 2.19 | 0.12 | 0.12 | / |
| | V | 3.40 | 0.11 | 0.12 | / |
| Weathered Bedrock Aquifer | I | 0.02 | 0.02 | / | $7.5 \times 10^{-4}$ |
| | II | 0.12 | 0.15 | / | $8.5 \times 10^{-4}$ |
| | III | 0.10 | 0.29 | / | $7.5 \times 10^{-4}$ |
| | IV | 0.03 | 0.23 | / | $7.5 \times 10^{-4}$ |
| | V | 0.22 | 0.11 | / | $7.5 \times 10^{-4}$ |
| | VI | 0.31 | 0.31 | / | $6.5 \times 10^{-4}$ |
| | VII | 0.12 | 0.26 | / | $7.5 \times 10^{-4}$ |
| | VIII | 0.01 | 0.01 | / | $7.5 \times 10^{-4}$ |
| | IX | 0.05 | 0.02 | / | $7.5 \times 10^{-4}$ |

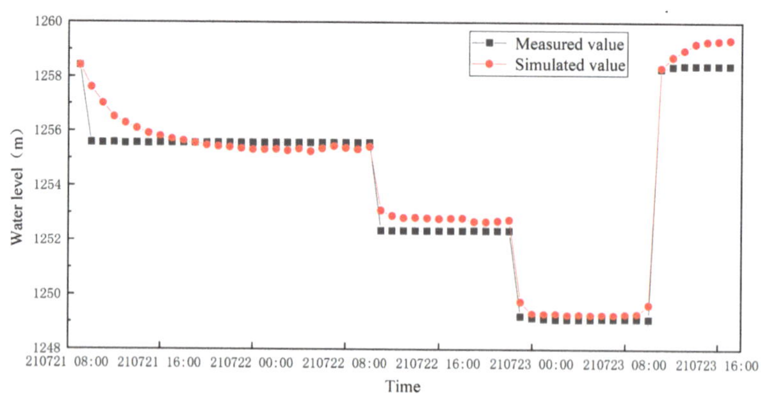


**Figure 12.** Water level fitting curve of FHS-1 hydropore pumping test.

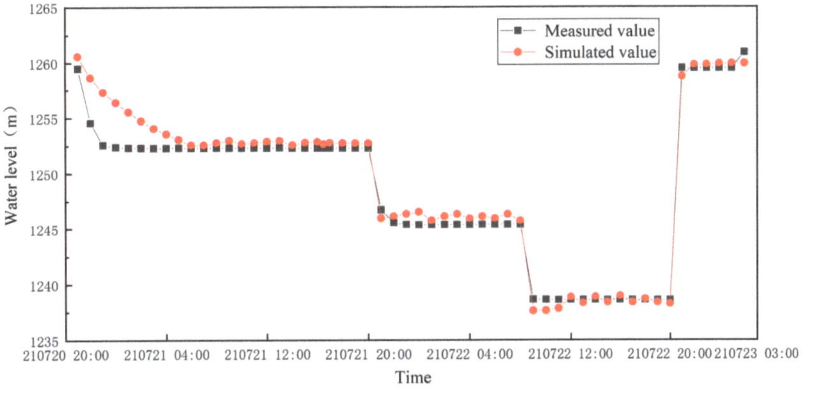


**Figure 13.** Water level fitting curve of FHS-2 hydropore pumping test.

### 4.2. Water Inflow Prediction Using GM(1,1)

Based on the water inflow data of the 122107 working face from March to July 2022, the working face water inflow could be predicted by the GM(1,1) model. The results are shown in Table 3.

**Table 3.** Statistics and forecast results of water inflow at the 122107 working face.

| Time | Measured Value/($m^3 \cdot h^{-1}$) | Predicted Value/($m^3 \cdot h^{-1}$) | Error |
|---|---|---|---|
| March 2022 | 604.8 | 604.8 | 0.0% |
| April 2022 | 654.7 | 591.3 | 9.7% |
| May 2022 | 574.5 | 628.7 | 9.4% |
| June 2022 | 603.7 | 668.5 | 10.7% |
| July 2022 | 768.4 | 710.9 | 7.5% |
| August 2022 | | 755.9 | |
| September 2022 | | 944.9 | |
| October 2022 | | 1181.1 | |

Figure 14 shows that there is some discrepancy between the predicted and measured water inflow values. Overall, the degree of coincidence is relatively high, indicating that the model is only a trend prediction. After calculation, the relative error is far less than 20%. Therefore, it can be used as a predictive model.

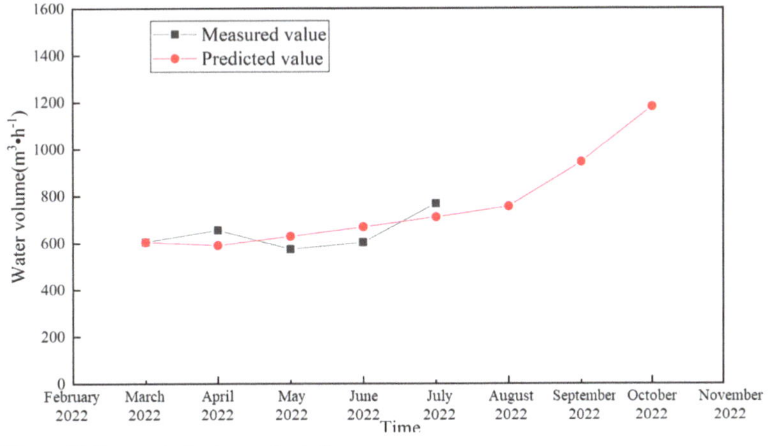


**Figure 14.** Forecast of mine water inflow.

The average normal water inflow for the months of August, September, and October is 960.6 $m^3/h$. A value of 1.25 is selected as the maximum water inflow prediction coef-icient for the mine; thus, the maximum water inflow is 1200.8 $m^3/h$.

### 4.3. Water Inflow Prediction Using Hydrogeological Analog Method

According to the information provided, the 122109 working face, which is in the western wing of the 12th panel area of the Caojiatan coal mine, uses the Longwall Top Coal Caving mining method, with a mining area of approximately 1.56 $km^2$. As the adjacent 122107 working face has similar hydrogeological and mining technical conditions, the 122109 working face data are used to predict the water inflow of the 122107 working face. The area of the 122107 working face is approximately 1.80 $km^2$, and $S = S_0$. In the hydrogeological analogy method, we assume that the hydrogeological conditions of the 122107 working face are similar to those of 122109.

By substituting the selected parameters into Formula (1), the simulation shows that under the condition that the static storage of the water-bearing layer in the roof has been sufficiently pre-drained, the normal water inflow is 923 $m^3/h$ and the maximum water inflow is 1154 $m^3/h$. As seen in Table 4, $S = 526.65$ is the depth of water level drop when pumping the aquifer water level down to the bottom of the coal seam, and the values are average values.

**Table 4.** Parameters for calculation of water inflow using the hydrological analogy method.

| Working Face \ Parameter | F ($10^6$ $m^2$) | S (m) | Q ($m^3$/h) |
|---|---|---|---|
| 122109 | 1.56 | 526.65 | 800 |
| 122107 | 1.80 | 526.65 | 923 |

### 4.4. Comparative Validation and Analysis

Table 5 compares the water inflow prediction accuracy for the 122107 working face obtained using various methods with the actual measured water inflow data.

**Table 5.** List of water inflow prediction results and prediction accuracy of Caojiatan coal mine 122107 working face.

| Results \ Data Sources | GM(1,1) | Hydrogeological Analog Method | Numerical Simulation | Measured Value |
|---|---|---|---|---|
| Normal water inflow Q ($m^3/h$) | 960.6 | 923 | 826 | 807 |
| Maximum water inflowQ ($m^3/h$) | 1200.8 | 1153.8 | 1032.5 | 1008.8 |
| Error (%) | 19.0 | 14.4 | 2.4 | |

Through comparison, we find that the results predicted using conventional methods are larger, and the GM(1.1) results vary with the selection of data. The hydrogeological analog method has obvious shortcomings in restoring the actual hydrogeological characteristics of the working face. However, the numerical simulation method overcomes the shortcomings of both.

We find that the prediction results of the numerical simulation method are closer to the measured values upon comparing them with the water inflow observation ledger, and the accuracy is 12–17% higher than that of traditional methods. Therefore, the numerical model is of great significance to study the impacts of deep coal seam mining on the groundwater environment in the northwestern region of China.

## 5. Conclusions

This paper proposes a method for constructing a mine conceptual hydrogeological model. After fully analyzing the geological characteristics, water sources, water channels, and water inflow intensity, the model can accurately depict the water inflow process in the 122107 working face of the Caojiatan coal mine.

Based on the mine hydrogeological conceptual model, the numerical simulation is used to finely depict the sharp depositional extinction of the red clay layer, and restore the stratigraphic mine–rock–water structure, the response characteristics of the groundwater system under the conditions of high-intensity coal mining, and the water inflow process, achieving the accurate prediction of water inflow.

Numerical simulation modeling through the mine hydrological conceptual modeling method is able to evaluate the process and intensity of mine water inflow and quantitatively predict the size of mine water inflow. This is a new modeling method to analyze the response characteristics of the groundwater system under high-intensity coal seam mining conditions and to predict the amount of mine water inflow.

In particular, the accuracy of the simulation results is related to the number of hydrological boreholes, and at the same time, the numerical simulation method adopts the average value of a period of time for the prediction of water inflow to the working face instead of the instantaneous inflow, and the water control in coal mines should take the numerical feature into full consideration.

**Author Contributions:** Z.H., S.M. and Y.Z. (Yao Zhang): conceptualization, methodology, writing—original draft preparation, writing—review and editing, supervision. Y.Z. (Yifan Zeng): conceptualization, supervision, writing—review and editing. L.W., X.W., Y.L., S.R. and J.L.: data curation. H.B., Y.Z. (Yao Zhang), Z.Z. and L.Z.: visualization, investigation. All authors have read and agreed to the published version of the manuscript.

**Funding:** This research was financially supported by the National Key R&D Program of China (2021YFC2902004) and the China National Natural Science Foundation (42072284, 42372297), Funding Program for Basic Research Operating Expenses of Central Universities (2023ZKPYSH01).

**Data Availability Statement:** The data presented in this study are available on request from the corresponding author. The data are not publicly available due to [Confidentiality principle].

**Conflicts of Interest:** The authors declare that they have no known competing financial interest or personal relationships that could have appeared to influence the work reported in this paper.

## References

1. Mark, C. Overview of ground control research for underground coal mines in the United States. In Proceedings of the 17th International Mining Congress and Exhibition of Turkey (IMCET 2001), Ankara, Turkey, 19–22 June 2001; pp. 3–10.
2. Dudek, M.; Tajduś, k.; Misa, R.; Sroka, A. Predicting of land surface uplift caused by the flooding of underground coal mines—A case study. *Int. J. Rock Mech. Min. Sci.* **2020**, *132*, 104377. [CrossRef]
3. Zeng, Y.F.; Meng, S.H.; Wu, Q.; Mei, A.S.; Bu, W.Y. Ecological water security impact of large coal base development and its protection. *J. Hydrol.* **2023**, *619*, 129319. [CrossRef]
4. Zhou, S.; Liu, T.; Duan, L. Ecological Impact Prediction of Groundwater Change in Phreatic Aquifer under Multi-Mining Conditions. *ISPRS Int. J. Geo Inf.* **2022**, *11*, 359. [CrossRef]
5. Li, B.; Zeng, Y.F.; Zhang, B.B.; Wang, X.Q.; Risk, A. Evaluation Model for Karst Groundwater Pollution Based on Geographic Information System and Artificial Neural Network Applications. *Environ. Earth Sci.* **2018**, *77*, 344.
6. Bicalho, C.C.; Batiot-Guilhe, C.; Taupin, J.D.; Patris, N.; Van Exter, S.; Jourde, H. A Conceptual Model for Groundwater Circulation Using Isotopes and Geochemical Tracers Coupled with Hydrodynamics: A Case Study of the Lez Karst System, France. *Chem. Geol.* **2019**, *528*, 118442. [CrossRef]
7. Lu, C.Y.; He, X.; Zhang, B.; Wang, J.H.; Kidmose, J.; Jarsjö, J. Comparison of Numerical Methods in Simulating Lake–Groundwater Interactions: Lake Hampen, Western Denmark. *Water* **2022**, *14*, 3054. [CrossRef]
8. Zeng, Y.F.; Wu, Q.; Liu, S.Q.; Zhai, Y.L.; Lian, H.Q.; Zhang, W. Evaluation of a Coal Seam Roof Water Inrush: Case Study in the Wangjialing Coal Mine, China. *Mine Water Environ.* **2018**, *37*, 174–184. [CrossRef]
9. Zeng, Y.F.; Li, Z.; Gong, H.J.; Zheng, J.H. Characteristics of Water Abundance in the Aquifer of Weathered Bedrock of 2-2 Coal Roof in Ningtiaota Coal Mine and Prediction on Water Inrush Risk. *Coal Eng.* **2018**, *50*, 100–104.
10. Zeng, Y.F.; Wu, Q.; Liu, S.Q.; Zhai, Y.L.; Zhang, W.; Liu, Y.Z. Vulnerability Assessment of Water Bursting from Ordovician Limestone into Coal Mines of China. *Environ. Earth Sci.* **2016**, *75*, 1431. [CrossRef]
11. Ta, D.; Cao, S.; Steyl, G.; Yang, H.Y.; Li, Y. Prediction of Groundwater Inflow into an Iron Mine: A Case Study of the Thach Khe Iron Mine, Vietnam. *Mine Water Environ.* **2019**, *38*, 310–324. [CrossRef]
12. Duan, L.; Wang, W.K. Water Lsotope Technology for Tracing Groundwater Movement. *Ground Water* **2006**, *2*, 33–36.

13. Wu, Q.; Tu, K.; Zeng, Y.F. Research on China's energy strategic situation under the carbon peaking and carbon neutrality goals. *Chin. Sci. Bull.* **2023**, *68*, 1884–1898. [CrossRef]
14. Wu, Q.; Tu, K.; Zeng, Y.F.; Liu, S.Q. Discussion on the main problems and countermeasures for building an upgrade version of main energy (coal) industry in China. *J. China Coal Soc.* **2019**, *44*, 1625–1636.
15. Li, W.J.; Ren, S.L.; Wu, Q.; Dong, D.L.; Gan, X.Y. Analysis on the dual constraints of energy and environment to the development of China and countermeasures. *Chin. Sci. Bull.* **2019**, *64*, 1535–1544.
16. Wu, C.; Wu, X.; Zhu, G.; Qian, C. Predicting Mine Water Inflow and Groundwater Levels for Coal Mining Operations in the Pangpangta Coalfield, China. *Environ. Earth Sci.* **2019**, *78*, 130. [CrossRef]
17. Zeng, Y.F.; Meng, S.H.; Lü, Y. Advanced Drainage Technology Based on Multi-objective Constraint of Mine Safety and Water Resources Protection. *J. China Coal Soc.* **2022**, *47*, 3091–3100.
18. Chen, S. Mine Water Inrush Prediction Based on Virtual Large Diameter Well Method. *IOP Conf. Ser. Earth Environ. Sci.* **2019**, *300*, 022098.
19. Liu, H.R.; Yan, Y.F.; Yang, H.T. The Prediction of the Water Yield Which Flood from the Pit and the Trait of the Hydrogeology Which Existed in the Mining Area of the Zhujiabaobao Mining in Panzhihua. *Appl. Mech. Mater.* **2013**, *401*, 2155. [CrossRef]
20. Ling, H.; Wang, L.X. A Study on Water Filling Factors and Inflow Prediction in Malin Coal Mine Extended Mining Sector, Lupanshui Mining Area. *Coal Geol. China* **2017**, *29*, 53–56.
21. Mi, J.K.; Wang, T. Application of Analytic Method on Working Face Water Lnflow Prediction in Xinglongzhuang Coal Mine. *Coal Geol. China* **2011**, *23*, 27–30.
22. Wang, F.L. Study on Optimization of Mine Water Inflow Calculation Methods in Different Production Periods. *Coal Sci. Technol.* **2019**, *47*, 177.
23. Luo, Z.J.; Zhao, L.I.; Ren, H.-J. Numerical Simulation Research on Prediction of Mine Inflow. *Coal Sci. Technol.* **2015**, *43*, 33–36.
24. Zhang, B.J. Mine Water Inflow Forecast Based on Visual Modflow in Taigemiao Exploration Area. *Coal Sci. Technol.* **2015**, *43*, 146.
25. Pang, Y.; Li, S.Y. Simulation Prediction of Mine Water Surges Based on Visual MODFLOW. *Sci. Res. Rev.* **2022**, *15*, 126.
26. Wang, G.R.; Wu, Q.; Yan, Z.Z.; Zhao, N.; Duan, C.B.; Cheng, X.; Wang, H. Fine Prediction for Mine Water Inflow on Basis of Visual Modflow. *Int. J. Oil Gas Coal Eng.* **2019**, *7*, 52–59.
27. Bayat, M.; Eslamian, S.; Shams, G.; Hajiannia, A. Groundwater Level Prediction Through GMS Software—Case Study of Karvan Area, Iran. *Quaest. Geogr.* **2020**, *39*, 139–145. [CrossRef]
28. Hu, X.J.; Ling, W.P.; Cao, D.T.; Liu, M.C. Index of Multiple Factors and Expected Height of Fully Mechanized Water Flowing Fractured Zone. *J. China Coal Soc.* **2012**, *37*, 613–620.
29. Zhong, H.; Qu, G.R.; Huang, L.X. Quantitative Simulation of Groundwater by Mathematical Model in Muzhu River Aquifer Using GMS. *IOP Conf. Ser. Earth Environ. Sci.* **2020**, *510*, 042020.
30. Ni, L.D.; Fan, M.Y.; Qu, S.S.; Zheng, Q.Y. Based on GMS Management of Shallow Groundwater Resource in Ningjin, China. *IOP Conf. Ser. Earth Environ. Sci.* **2019**, *237*, 032063. [CrossRef]
31. Liang, Y.M.; Lan, J.K.; Wen, Z.X. Simulation and Prediction of Groundwater Pollution from Planned Feed Additive Project in Nanning City Based on GMS Model. *IOP Conf. Ser. Earth Environ. Sci.* **2018**, *301*, 012159. [CrossRef]
32. Chen, M.J.; Izady, A.; Abdalla, O.A. An Efficient Surrogate-Based Simulation-Optimization Method for Calibrating a Regional MODFLOW Model. *Hydrology* **2017**, *544*, 591–603. [CrossRef]
33. Bo, Q.Y.; Cheng, W.Q.; Sun, T. Groundwater Simulation Model for Baohe River in the Upper Reaches of Baiyangdian Lake Based on Groundwater Simulation Software (GMS). *J. Environ. Prot. Ecol.* **2021**, *22*, 1162–1174.
34. Sartirana, D.; Zanotti, C.; Rotiroti, M.; De Amicis, M.; Caschetto, M.; Redaelli, A.; Fumagalli, L.; Bonomi, T. Quantifying Groundwater Infiltrations into Subway Lines and Underground Car Parks Using MODFLOW-USG. *Water* **2022**, *14*, 4130. [CrossRef]
35. Li, Q. The Grey Elementary Functions and Their Grey Derived Functions. *J. Grey Syst.* **2008**, *20*, 225–245.
36. Islam, M.R.; Kabir, G.; Ng, K.T.W.; Ali, S.M. Yard Waste Prediction from Estimated Municipal Solid Waste Using the Grey Theory to Achieve a Zero-Waste Strategy. *Environ. Sci. Pollut. Res. Int.* **2022**, *29*, 46859. [CrossRef] [PubMed]
37. Wei, W.D.; Wang, G.; Tao, X.; Luo, Q.; Chen, L.X.; Bao, X.L.; Liu, Y.X.; Jiang, J.J.; Liang, H.; Ye, L. Time Series Prediction for the Epidemic Trends of Monkeypox Using the Arima, Exponential Smoothing, GM (1, 1) and LSTM Deep Learning Methods. *J. Gen. Virol.* **2023**, *104*, 001839. [CrossRef]
38. Cao, J.M.; Chi, B.M.; Wang, W.K. *Specialized Hydrogeology*; China Science Publishing: Beijing, China, 2006; pp. 226–241.
39. Xu, N.W.; Dai, F.; Li, B.; Zhu, Y.G.; Zhao, T.; Yang, D.S. Comprehensive Evaluation of Excavation-Damaged Zones in the Deep Underground Caverns of the Houziyan Hydropower Station, Southwest China. *Bull. Eng. Geol. Environ.* **2017**, *76*, 275–293. [CrossRef]
40. Qu, H.; Liu, H.; Tan, K.; Zhang, Q. Geological Feature Modeling and Reserve Estimation of Uranium Deposits Based on Mul-Tiple Interpolation Methods. *Processes* **2022**, *10*, 67. [CrossRef]

*sustainability*

MDPI

*Article*

# Protecting Cape Town's Groundwater from Fuel Stations: An In-Depth Analysis of Regulatory Requirements

**Eden Alexandre Nsimba *, Ntokozo Malaza and Thandazile Marazula**

Department of Environmental and Occupational Studies, Faculty of Applied Sciences, Cape Peninsula University of Technology, Corner of Hanover and Tennant Street, Cape Town 8000, South Africa; malazan@cput.ac.za (N.M.)
* Correspondence: nsimbaeden@gmail.com

**Abstract:** In the face of mounting water supply challenges, Cape Town has increasingly turned to alternative sources, like groundwater. However, the utilisation of groundwater carries inherent risks, particularly the contamination stemming from land-based activities, such as fuel stations. Leaks from underground tanks at these stations represent a major global cause of groundwater pollution, and Cape Town is no exception. To safeguard public health and mitigate potential harm, it is imperative to examine the legal regulations governing fuel station development, assess measures for controlling their environmental impacts and evaluate strategies for managing the associated risks. This study aims to provide an exhaustive review of the regulatory framework concerning the environmental impacts of fuel stations, focusing on groundwater protection in Cape Town. A combination of desk research and interviews was employed to gather and analyse data. The findings show a deficiency in precautionary measures for safeguarding groundwater near fuel stations. Consequently, through this study, the existing legal framework's effectiveness is called into question, with this study suggesting actions to address these identified shortcomings.

**Keywords:** groundwater; contamination; legal framework; environmental regulations; fuel station; land-use; Cape Town

**Citation:** Nsimba, E.A.; Malaza, N.; Marazula, T. Protecting Cape Town's Groundwater from Fuel Stations: An In-Depth Analysis of Regulatory Requirements. *Sustainability* **2023**, *15*, 15135. https://doi.org/10.3390/su152015135

Academic Editors: Zongjun Gao and Jiutan Liu

Received: 14 September 2023
Revised: 12 October 2023
Accepted: 20 October 2023
Published: 22 October 2023



## 1. Introduction

Global water resources are facing unprecedented challenges driven by a combination of factors, prominently including climate change and population growth [1]. In addition to these pressures, the contamination of water resources through anthropogenic activities has emerged as a pressing concern [2]. These multifaceted challenges are progressively diminishing the availability of freshwater reserves accessible to human populations, underscoring the critical need for an effective regulatory system to manage and safeguard these invaluable resources [3].

Several decades ago, policymakers recognised the imperative to enact legislation, regulations and policies aimed at preserving the environment in a holistic sense, with a particular emphasis on water resources [4,5]. However, the efficacy of a legal framework governing environmental concerns can be compromised if it fails to comprehensively address relevant theoretical and practical parameters. These parameters encompass a range of elements, including the implementation of robust mitigation measures to preclude negative environmental impacts [6,7]. The central focus of this study is the examination of fuel stations as potent sources of groundwater contamination. Despite their undeniable socio-economic importance [8,9], fuel stations also engender significant environmental challenges, notably groundwater contamination stemming from leakages in underground storage tanks and associated pipelines [10]. Groundwater pollution originating from fuel stations assumes heightened significance, especially when neighbouring communities rely on groundwater sources for their water supply [11]. This dynamic has been conspicuously observed in the Western Cape province in South Africa, particularly during episodes of water scarcity and drought. In Cape Town, the recurring and severe drought events in the past years have prompted many residents to install private boreholes, and this trend shows no signs of abating in the face of ongoing climate-induced scarcity [12–15]. Therefore, sustained attention must be given to the protection of groundwater to prevent further risks, such as contamination resulting from anthropogenic activities.

Groundwater contamination by petroleum products is a pervasive issue encountered worldwide, with particular prevalence in urban areas [16]. South Africa, including the city of Cape Town, is no exception to this environmental challenge [17]. This is a concern that becomes especially evident when assessing the density and distribution of fuel stations in major South African cities [18]. According to Ahmed et al. [8], fuel stations represent facilities where substantial quantities of petroleum-based fuels, such as petrol, diesel and various refined products, are stored for commercial purposes. However, the inherent technical complexities and environmental vulnerabilities of underground storage tanks (USTs) and associated equipment at fuel stations can lead to leaks, resulting in soil and groundwater contamination [10]. Such contamination incidents often involve the release of hazardous compounds, like Benzene, Toluene, Ethylbenzene, Xylene (BTEX) compounds, polycyclic aromatic hydrocarbons (PAHs), lead and methyl tert-butyl ether (MTBE), an oxygenate blended in unleaded fuel, among others [6].

The human health consequences of exposure to contaminated groundwater can be extremely disastrous [19,20]. Notably, the risk to human health escalates significantly when individuals are regularly exposed to petroleum-contaminated water [21]. Exposure to petroleum hydrocarbons can induce a spectrum of acute health effects, encompassing flu-like symptoms, coughing, headaches, dizziness and eye and skin irritation. With prolonged exposure, chronic health effects can manifest, potentially leading to more severe and enduring issues, such as cancer, cardiotoxicity and endocrine disruptions [22].

Over the years, countries have implemented measures aimed at enhancing regulations and preventing groundwater contamination by fuel stations, particularly addressing concerns arising from ageing USTs [23]. Measures, such as secondary containment for USTs, have emerged particularly as an effective proactive means for controlling the dispersion of fuel released from problematic tanks into soil and groundwater [6]. Ooi et al. [23] also underline the significance of robust monitoring practices and the adoption of state-of-the-art technology for preventing the pollution of water. Furthermore, the role of land-use management cannot be overstated. Precautionary land-use management plays a pivotal role in minimising the environmental risks associated with high-risk activities such as fuel stations [24]. Land-use management assumes critical importance in risk mitigation, particularly concerning safety from risks of human installations, thereby safeguarding communities residing in proximities to major hazard installations [25].

The inherent unpredictability and severity of fuel leakages from USTs necessitate stringent government measures in the form of laws and regulations to avert undesired occurrences such as groundwater contamination [23]. However, it is essential to recognise that laws, regulations and policies may, at times, fall short of achieving their intended effectiveness. Since the advent of modern environmental regulations, the development and implementation of effective regulations have posed one of society's most pressing challenges [26]. Consequently, the continuous refinement of laws and policies is imperative, particularly in the current time earmarked by heightened public awareness of environmental risks [27].

Thus, when existing laws prove inadequate to address burgeoning environmental challenges, reviewing them to develop adequate legal measures to prevent risks becomes imperative [28]. Aven and Renn [27] advise that governments must incorporate clear risk strategies into the formulation of governance policies to enhance risk prevention. Among the several risk strategies that are often discussed in scientific literature, the precautionary approach stands out as a frequently recommended tool in environmental risk governance, especially in the context of activities with a high potential to impact public health and water availability [3,27].

Existing literature and information suggest that fuel stations are among the major contributors to polluters in South Africa [11,17,18,29–31]. This may, therefore, pose a considerable risk to those who rely on groundwater for additional supplies, particularly in cities like Cape Town. Given the escalating reliance on groundwater in the region, a comprehensive understanding of the effectiveness of the current environmental legal framework in safeguarding groundwater resources and the communities dependent upon them is of paramount importance. Currently, only a limited number of studies have endeavoured to assess the effectiveness of the legal frameworks governing the environmental aspects of fuel stations in South Africa [31,32]. However, given the proliferation of fuel stations in densely populated areas of Cape Town, it is reasonable to hypothesise that these facilities pose substantial risks to groundwater users in the city, especially when considering the cases of registered contaminated land cases attributed to fuel stations in the Western Cape Province (Figure 1).

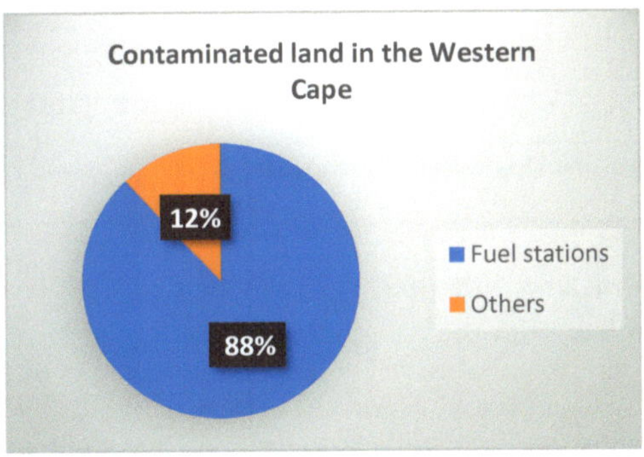


**Figure 1.** This graph shows that fuel stations account for 88% of the contaminated land registered in the National Contaminated Land Register for the Western Cape.

Thus, the primary objective of this study is to meticulously investigate the legal framework governing the environmental impacts of fuel stations, with a specific focus on the risk mitigation measures applied to fuel station developments in Cape Town.

## 2. Research Methodology

The main objective of this study is to investigate the risk prevention measures adopted for fuel stations' impacts on groundwater in Cape Town. The key research question guiding this inquiry is: What regulatory conditions are currently in place to govern the establishment of fuel stations and the control of their impacts on groundwater? To answer this question, an exploratory research methodology was adopted, involving a combination of qualitative research techniques. Data collection included a desk review of policies, legal documents and literature, followed by semi-structured interviews with selected key participants to supplement the review of documents. A total of seven (7) semi-structured interviews were conducted in Cape Town with participants from different layers of the environmental sector locally. The data extracted in response to the research question were then analysed using content analysis.

### 2.1. Desk Review

An initial curation process for legal and policy instruments yielded a list of nine relevant regulatory documents using specific keywords. These regulatory instruments were rigorously analysed alongside literature to address the research question and to find best practices. A summary of the legal documents and policies reviewed to address the research question is presented in Table 1. In South Africa, environmental governance is underpinned by the Constitution, statutes such as the National Environmental Management Act, 107 of 1998, Specific Environmental Management Acts and subordinate legislation at both national and local levels. For this study, an inclusive examination of South African National Standards (SANS) pertaining to the petroleum industry was conducted, with a focus on standards related to the installation of USTs and associated equipment. These standards were scrutinised to extract relevant technical recommendations.

**Table 1.** Reviewed legal and policy documents.

| Document | Level |
| --- | --- |
| The Constitution of the Republic of South Africa Act 108 of 1996 | National |
| National Environmental Management Act 107 of 1998 (NEMA) | National |
| Environmental Impact Assessment Regulations of 2014 (EIA Regulations) | National |
| National Environmental Management: Waste Act 59 of 2008 (NEMWA) | National |
| National Water Act 36 of 1998 (NWA) | National |
| Municipal Planning By-Law, 2015 | Local/Cape Town |
| Framework for the Management of Contaminated Land (2014) | National |
| South African National Standards (SANS 10089-1, SANS 1830 and the SANS 1535) | National |
| EIA Administrative Guideline for Installation and Upgrade of Underground Storage Tank and Associated Equipment | Local/Gauteng |

### 2.2. Semi-Structured Interviews

In tandem with the comprehensive review of legal and policy documents, semi-structured interviews were an additional component of this study, serving as a vital means of gathering insights and perspectives from key stakeholders entrenched in the fields of environmental management and fuel station oversight within Cape Town. The selection criteria for interview participants were stringent, with an emphasis on expertise and substantial knowledge pertaining to environmental regulatory frameworks and management of fuel stations. Rigorous scrutiny of the professional backgrounds of potential participants preceded the interview process to ensure their qualifications met the study's objectives. A total of seven (7) semi-structured interviews were conducted, involving participants from diverse backgrounds and roles (Table 2). The interview discussions were crafted to encompass a range of topics (Appendix A) related to the Environmental Impact Assessment (EIA) process, the potential impacts of fuel stations on groundwater and boreholes, land use management and risk management strategies. A purposive sampling methodology was employed to select the participants for the interviews. It is important to acknowledge that purposive sampling, while invaluable for exploratory studies, presents certain limitations, particularly in terms of the generalisability of findings [33]. However, the use of this sampling method was deemed appropriate to ensure that participants possessed the requisite expertise to provide meaningful insight into the study's specific context.

**Table 2.** Participants who participated in this study.

| Participant | Sector | Function |
| --- | --- | --- |
| Participant 1 | Government | Environmental Officer |
| Participant 2 | Government | Environmental Officer |
| Participant 3 | Research institution | Researcher (Land use planning) |
| Participant 4 | Environmental consulting | Consultant |
| Participant 5 | Fuel station | Site Manager |
| Participant 6 | Fuel station | Site Manager |
| Participant 7 | Fuel station | Fuel station attendant |

## 3. Regulatory Framework for Groundwater Protection and Gaps

This study placed its focus on the key national legislative and regulatory instruments that shape the landscape, including NEMA, NWA and EIA Regulations. Although these three legislations serve complementary functions, NEMA holds a central role in the broader enforcement of environmental regulation.

Cape Town, as the metropolitan municipality of the Western Cape Province, is subject to a regulatory landscape where fuel stations' impacts on groundwater are primarily governed by national environmental legislation. The Western Cape government, acting as the competent authority under NEMA, is tasked with adjudicating environmental impacts associated with fuel stations. Our investigation unveiled several factors that could potentially undermine the efficacy of the legal measures designed to protect groundwater and borehole users. Foremost among these concerns is the inadequate consideration of sensitive land uses in the vicinity of fuel stations, manifesting as the absence of specific criteria for fuel station siting and safe distances between fuel stations and other land uses. Additionally, no explicit legal provisions mandate the installation or upgrading of USTs and associated pipes equipped with groundwater protection mechanisms within fuel stations. Lastly, the legal framework governing the impacts of fuel stations on groundwater is notably fragmented, introducing complexity in its implementation. The outcomes of this study offer valuable insights and may catalyse policy dialogues aimed at formulating additional legal measures to safeguard groundwater and borehole users potentially exposed to groundwater contamination from fuel stations in Cape Town.

### 3.1. *Environmental Authorisation for the Development of Fuel Stations*

The foundational condition for the initiation of activities bearing the potential for substantial environmental impacts is enshrined in Section 24(1) of NEMA, which articulates, "In order to give effects to the general objectives of the integrated environmental management laid down in this chapter, the potential consequences for or impacts on the environment of listed activities or specified activities must be considered, investigated, assessed and reported on to the competent authority or the Minister responsible for the mineral resources, as the case may be, except in respect of those activities that may commence without having to obtain an environmental authorisation in terms of this Act".

Fuel stations, characterised as integrated activities, feature prominently within the ambit of "listed activities" as delineated by the Department in Charge of Environmental Affairs by the EIA Regulations of 2014. Notably, fuel stations are specifically classified under the category of "the construction of facilities for storage or handling of dangerous goods" (Table 3). Consequently, in South Africa, the establishment of fuel stations is contingent upon securing the environmental authorisation as they trigger the legal requisite of an EIA due to the installation of storage tanks for the handling of "dangerous foods". This prerequisite serves as an effective precautionary mechanism against the potential impacts of proposed fuel station developments. However, it is imperative to underscore that the current EIA regulations do not additionally suggest conspicuous provisions addressing groundwater and potential boreholes in the vicinity of fuel stations. In this context, the Gauteng Province stands out among other South African provinces, as they have instituted provincial EIA guidelines, the "EIA Administrative Guideline for Installation and Upgrade of Underground Storage Tank and Associated Equipment (referred to hereafter Gauteng's EIA guideline)", aligning with Section 24 of NEMA.

**Table 3.** EIA-requiring activities triggered by fuel stations (EIA Regulations of 2014).

| Listing and Activity No. | Description of Activity | Assessment |
|---|---|---|
| Listing 1 GNR 983 Activity 14 | The construction of a facility for storage or for the storage and handling of dangerous goods, where storage occurs in containers with a combined capacity of 80 but not exceeding 500 cubic metres. | Basic assessment |
| Listing 1 GNR 983 Activity 51 | The expansion of a facility for storage or for the storage and handling of dangerous goods, where the capacity of such facility will be expanded by 80 cubic metres or more. | |
| Listing 1 GNR 983 Activity 31 | The decommissioning of existing facility or infrastructure for (v) storage or the storage and handling of dangerous goods of more than 80 cubic metres. | |
| Listing 2 GNR 984 Activity 4 | The construction of a facility or infrastructure for storage or for the storage and handling of dangerous goods, where such storage occurs in containers with a combined capacity of 80 exceeding 500 cubic metres. | Full EIA |

### 3.2. Contrasting Regulatory Approaches for Environmental Authorisation

In stark contrast to the Western Cape Province, where the environmental authorisation process for fuel stations is solely governed by the stipulations of Section 24 of NEMA, the Gauteng Province has adopted a more rigorous approach to evaluating the risks posed by fuel stations, as expounded by Kruger [30] and Qonono [18]. Among all the provinces in South Africa, it is the Gauteng Department of Agriculture, Conservation, Environment and Land Affairs (GDAELA) that has taken the initiative to develop a comprehensive provincial EIA guideline. In addition to conforming to the mandates of Section 24 of NEMA, this guideline proffers specific measures to be considered during the application for an environmental authorisation for the construction of a fuel station and upgrading of USTs in existing ones. A distinctive feature of Gauteng's EIA guideline is its central emphasis on land use considerations, prescribing conditions for the siting of fuel stations and establishing safe distances from sensitive land uses [30].

According to Gauteng's EIA guidelines, those seeking environmental authorisation to construct a fuel station must provide detailed information pertaining to the location of boreholes on the proposed site and adjacent properties. Furthermore, developers must determine the extent to which the public depends on groundwater in the areas to ensure that the proposed location of the station does not affect the users of groundwater. The guidelines strictly prohibit new fuel stations that do not meet the stringent criteria established by Gauteng Province's regulatory framework. Additionally, the Gauteng EIA guideline stipulates that no new fuel stations can be approved "within 100 metres of residential properties, schools, or hospitals, unless it can be clearly demonstrated that there will be no significant noise, visual intrusion, safety concerns or fumes and smells."

Information gathered during interviews revealed that the process is less stringent compared to the Gauteng Province. Participant 2, an integral figure involved in the evaluation of applications for environmental authorisation within the Western Cape, provided insights into the prevailing regulatory landscape. The Western Cape lacks a provincial regulatory or policy requirement akin to Gauteng's EIA guideline for USTs. However, Participant 2 elucidated that, in practice, when assessing applications for environmental authorisation for fuel stations, there exists a prevailing ethos to prohibit the construction of fuel stations near residential areas. Furthermore, Participant 2 emphasised the vital role of public participation during the EIA process. This fact of the regulatory demands empowers the public to voice their concern regarding proposed developments, which can, in turn, lead to the rejection of applications for fuel station developments.

Given the argument of Participant 2, it is imperative to recognise that the effectiveness of public participation hinges on the environmental health literacy of the concerned individuals. Not all members of the public may possess the requisite understanding to anticipate the cumulative impacts that might ensue if a fuel station is permitted in close proximity to their properties. Consequently, there arises a compelling need for clearly defined regulations stipulating specific conditions that must be satisfied before the issuance of environmental authorisation.

Drawing inspiration from the Gauteng Province's approach, it becomes evident that establishing a comprehensive guideline in the Western Cape, particularly in Cape Town, is imperative. This guideline should delineate unambiguous requirements aimed at safeguarding sensitive land uses in the vicinity of proposed fuel station developments, thereby mitigating the risks of land-use conflicts. Such a standardised approach would foster consistency in the evaluation of applications for environmental authorisations, departing from the prevalent case-by-case methodology observed in the Western Cape Province. As previously asserted by Qonono [18], the absence of criteria governing the siting of fuel stations in the Western Cape carries inherent risks for the surrounding environment. Given the pervasive nature of petroleum fuels in groundwater, as attested to by [34], it is imperative to prioritise the formulation of precautionary criteria delineating the location of fuel stations and establishing a regulatory safe distance (100 metres in the case of Gauteng's EIA guideline) between fuel station stations and groundwater users. This approach is essential to facilitate the evaluation of environmental authorisation, thereby safeguarding both the environment and public health.

### 3.3. *The Duty of Care and Legal Measures for the Protection of Groundwater*

In the realm of environmental law in South Africa, a fundamental concept is the imposition of a "duty of care" on activities deemed high-risk, such as fuel stations, given their potential to jeopardise the environment. This duty of care is enshrined within Section 28 of NEMA, which statutorily mandates individuals or entities responsible for environmental degradation to assume responsibility for its prevention, minimisation and remediation. While the NEMA establishes this duty, other related acts like the NEMWA and NWA contain provisions that directly address pollution control emanating from high-risk activities, including fuel stations. However, during the review of pertinent legal documents, two legal measures pertaining to the duty of care emerged as critical in the regulation of environmental risks associated with fuel stations, particularly in relation to groundwater protection. These measures are:

(1) Management of Emergency Incidents: Regulated by NEMA Section 30 (also covered in Part 5 of Chapter 3 of the NWA), this measure outlines procedures for handling emergency incidents related to environmental contamination;

(2) Management of Contaminated Land: This is governed by Part 8 of NEMWA and encompasses legal provisions for soil and groundwater monitoring and remediation. Furthermore, the Framework for the Management of Contaminated Land (2010) complement Part 8 of NEMWA, providing guidelines, including acceptable levels of contamination for different land-use types.

While both these measures are pertinent concerning groundwater contamination caused by fuel stations, they are inherently reactive rather than proactive. This means that these provisions are only pivotal for assessing and correcting the consequences of an incident which may potentially affect receptors such as borehole users. Adding to this concern is the absence of legal provisions compelling the installation of fuel station equipment that is up to current standards and adequate groundwater contamination prevention systems. In contrast, the South African Bureau of Standards has formulated several SANSs relevant to groundwater protection, including SANS 10089, SANS 1830 and SANS 1535 [35–37]. These standards offer up-to-date recommendations and best practices for equipment designed to prevent corrosions, leaks and UST failure that could lead to groundwater contamination. However, these SABS-approved standards are currently voluntary, lacking the legal mechanisms necessary for enforcement.

The risk of groundwater contamination remains particularly significant in older fuel stations established before the development of recent groundwater protection technologies. These older stations are problematic due to their antiquated equipment, which lacks effective pollution prevention mechanisms, such as UST double walls, protective coatings, or glass-reinforced polyester (GRP). This issue mirrors challenges faced by China, which, in 2011, enacted the National Groundwater Pollution Control Plan (2011–2020) to compel fuel stations to upgrade underground tanks to state-of-the-art pollution prevention equipment [6].

Given historical incidents of groundwater pollution caused by fuel stations, proactive measures are imperative to avert contamination, such as the case in Beaufort West, a town neighbour to Cape Town in the Western Cape. In Beaufort West, a release of fuel that went unnoticed for several years in three fuel stations in the semi-desert Karoo town, where people are heavily reliant on groundwater, resulted in detrimental consequences, with more than 30 boreholes polluted [11]. Although these incidents are not widely discussed in public discourse in Cape Town, Participant 1 pointed out that it is a serious concern, especially in large cities, such as Cape Town, given the data on fuel station incidents over the years. The Contaminated Land Register of South Africa lists more than 70% of cases resulting from fuel station incidents, alerting on the necessity for proactive legal measures for groundwater protection. According to Participant 1, these measures should encompass mandatory groundwater prevention technology requirements for fuel stations, with urgency for older stations that did not undergo the same EIA scrutiny mandated by NEMA.

Regrettably, the review of selected environmental legal instruments reveals that South Africa lacks an integrated instrument governing the environmental aspects of fuel stations. Pertinent legal provisions addressing the protection of boreholes from fuel station risks are primarily found within the NEMA, NEMWA and NWA. However, these acts, while comprehensive, do not specifically stipulate obligations or preventive measures for the scenario of groundwater abstraction points in proximity to fuel stations. Provisions such as Section 28 of NEMA and Section 19 of NWA, while broadly addressing pollution prevention, lack clarity regarding the regulation of fuel stations' proximity to potential boreholes. Interviews with participants from the fuel retail industry (Participants 5, 6 and 7) indicate that, aside from fire hazards, some fuel stations may not prioritise attention to potential contamination near residential areas. Furthermore, interviews with Participants 5, 6 and 7 revealed that there is a necessity to strengthen requirements for effective groundwater near fuel station operations as this aspect seems to be an aspect that retains little attention from fuel retail operators.

Although the South African Constitution guarantees the right to access sufficient water for everyone, a lack of adequate monitoring provision may result in the public relying upon unsafe and contaminated water. According to Seyler, Witthüser and Sunaitis [38], water pollution stands as a prominent impediment to full access to groundwater in South Africa. Monitoring plays a pivotal role in the protection of groundwater resources [38]. South African water legislation provides mechanisms for monitoring water resources, particularly through Chapter 4 of the NWA, which establishes "National Monitoring Systems". However, small-scale water monitoring, such as groundwater monitoring for fuel station sites, is regulated under Section 24(4) of NEMA. This provision assigns monitoring and impact management responsibilities to those undertaking authorised activities. Notably, the governance of groundwater in South Africa is characterised by insufficient monitoring measures [38]. The deficiency in robust monitoring, especially around fuel stations, poses a significant risk to the public, particularly considering the growing reliance on groundwater in the Cape Town region.

## 4. Land-Use Policies in Cape Town and Groundwater Protection

The crucial task of land use planning and management rests with municipal governments. In Cape Town, the Municipal Planning By-Law of 2015 plays a pivotal role in providing the framework for land-use zoning conditions. This by-law delineates specific zonings under which fuel stations can be established (Table 4). One can identify various zoning specifications pertaining to land-based structures in fuel stations, including aspects such as buildings, land surface and roads, walls and the like. However, a conspicuous omission is the absence of conditions relating to groundwater abstraction points or dictating a safe distance between fuel stations and neighbouring land users. Groundwater abstraction can influence the migration of petroleum fuel plume contaminants. Consequently, the absence of a designated protection zone for groundwater raises the spectre of groundwater users potentially being exposed to the intake of fuel substances [39].

**Table 4.** Zonings for fuel station development (Municipal Planning By-Law, 2015).

| Zoning/Subzoning | No. | Brief Description |
|---|---|---|
| Local Business Zoning 2 | 57 | Provides for low-intensity commercial and mixed-use development, which serves local needs for convenience goods and personal service. |
| General Business subzoning | 61 | Provide for general business activity and mixed-use development of a medium to high intensity. |
| Mixed Use subzoning | 65 | Accommodate a mixture of business and appropriate industrial and residential development. |
| General Industry subzoning | 75 | GI accommodates all forms of industry except noxious trade and risk activity in order to promote the manufacturing sector of the economy. |
| Risk Industry | 77 | Provides for noxious industries which carry high risk in the event of fire or accident. |

This regulatory gap highlights the importance of incorporating protective land-use conditions into local policies governing fuel station establishments. Doing so can bolster preventive measures against the risks posed to borehole users in Cape Town. By integrating protective land-use conditions within local land-use policies, a critical step forward in holistic environmental risk management will be reached. This will ensure that the vulnerabilities of groundwater users are duly considered in the spatial planning of fuel stations. This strategic alignment between land use policies and groundwater protection not only minimises environmental hazards but also upholds the right to an environment that is not harmful, as enshrined in the South African Constitution. Ultimately, this promotes a sustainable and healthy environment for all Cape Town residents.

## 5. Fragmentation Existing Regulatory Instruments and Suggestions for Integration

The issue of fragmentation within South Africa's environmental governance framework is a widespread and well-documented problem. This fragmentation has been a significant hurdle, undermining the effectiveness and efficiency of environmental governance efforts in many instances [40,41]. Consequently, there is a growing recognition of the need to integrate the various elements of the environmental governance framework to enhance implementation efficiency [38].

In the context of regulating the impacts of fuel stations on groundwater, the review of regulatory frameworks and relevant documents highlighted a concerning level of fragmentation within the legal framework. The conditions and requirements for the protection of groundwater and control of the impact of fuel station stations are dispersed across multiple pieces of legislation (NEMA, NEMWA, NWA) and their subordinate documents, often managed by distinct government departments rather than consolidated into a unified document designated aspects related to fuel stations. This fragmentation poses a potential threat to the effectiveness of regulation governing environmental aspects related to fuel stations. Moreover, it introduces complexity in the implementation and compliance with legal measures aimed at safeguarding groundwater.

To address the pervasive issue of fragmentation, the establishment of a specialised regulatory instrument dedicated to governing fuel station aspects in general and groundwater protection, in particular, will be necessary. This is arguably a more pragmatic approach that can circumvent challenges posed by the fragmentation of regulatory documents for fuel stations' environmental aspects. Furthermore, the harmonisation of the disparate legal instruments and requirements currently applicable to fuel stations would culminate in the creation of an integrated and cohesive regulatory process for all aspects of fuel stations (Figure 2), thereby facilitating compliance with legal measures for environmental protection.

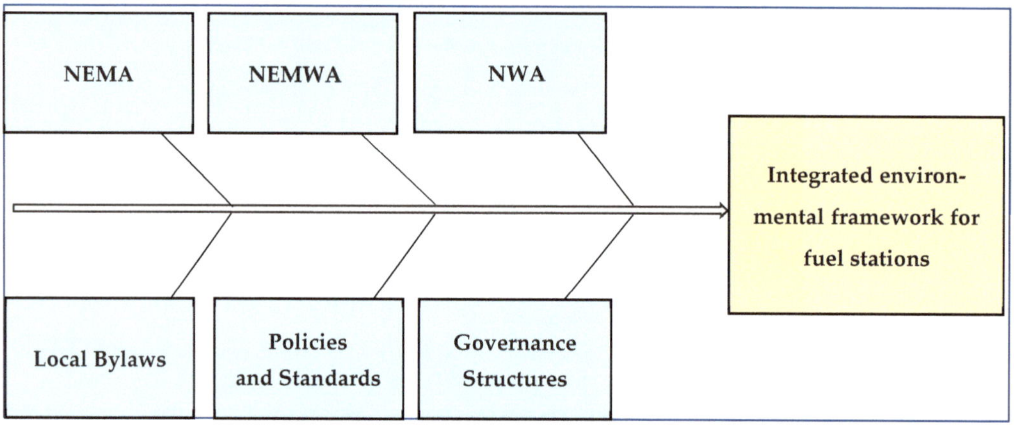


**Figure 2.** Elements to consider for an integrated framework for fuel stations.

## 6. Conclusions

This study undertook a comprehensive review of the legal framework governing the impacts of fuel stations on groundwater, focusing on the city of Cape Town, to illuminate the legal measures adopted for preventing and mitigating these impacts. The regulation of fuel station development and the control of their environmental effects predominantly rely on national environmental legislation, especially NEMA, NEMWA and NWA. Notably, the responsibility for land-use management falls under the purview of local municipal authorities, as mandated by the Constitution of South Africa.

The focal point for controlling fuel development lies in Section 24 of the NEMA, which mandates the requirement for environmental authorisation and EIA for activities classified as "listed activities". This precautionary process serves to uncover potential impacts, such as groundwater contamination, before development proceeds. However, a critical deficiency in the legal framework emerges in the absence of specific provisions at the national, provincial and local levels stipulating safe distance between fuel stations and other land uses. This regulatory gap may have inadvertently facilitated the proliferation of fuel stations in close proximity to residential properties. The study finds that the current legal framework governing the impacts of fuel stations on groundwater exhibits limited adequacy due to several factors. This includes the absence of specific provisions addressing groundwater contamination risks specific to fuel stations, posing a significant challenge. This regulatory gap results in inadequate considerations of potential borehole contamination in the vicinity of fuel stations. Furthermore, the fragmented nature of the instruments regulating fuel stations' impacts on groundwater may impede the effective implementation of existing groundwater protection measures.

To address these shortcomings and enhance the protection of groundwater, several recommendations are put forth. These include mandating the SABS-approved petroleum industry standards as legally enforceable norms. Additionally, it is advisable for authorities responsible for land-use planning in Cape Town to establish specific minimum safe distances (buffers) between fuel stations and sensitive land uses as precautionary measures. This study is expected to contribute significantly to bolstering efforts to prevent groundwater contamination and reduce the vulnerability of residents relying on groundwater near fuel station sites, ultimately fostering greater environmental sustainability.

To ensure the effective implementation of the recommended measures, it is essential to foster collaboration among relevant stakeholders, including government authorities, environmental agencies, industry representatives and local communities. This collaborative effort should focus on developing and implementing necessary policies and regulations. Furthermore, public awareness should be an integral part of this collaboration to educate both communities and fuel station operators about the significance of environmental protection, with a specific emphasis on safeguarding groundwater protection.

**Author Contributions:** E.A.N.: Conceptualisation, methodology, data collection, data analysis, writing—original drafts and project administration. N.M.: Supervision, project administration and writing—review and editing. T.M.: writing—review and editing. All authors have read and agreed to the published version of the manuscript.

**Funding:** This research received no external funding.

**Institutional Review Board Statement:** Ethical approval was obtained from the Cape Peninsula University Technology's Faculty Research Ethics Committee within the Faculty of Applied Sciences.

**Informed Consent Statement:** Informed consent was obtained from all subjects involved in the study.

**Data Availability Statement:** All data are presented in the paper.

**Conflicts of Interest:** The Authors declare no conflict of interest.

## Appendix A

**Table A1.** Semi-structured interview framework.

| |
|---|
| Question 1: In practice, what are the most important pieces of environmental legislation triggered by the development, construction and upgrade of fuel stations in the Western Cape? |
| Question 2: Generally, what are the criteria guiding the review of an environmental impact assessment report with regard to fuel station development applications? |
| Question 3: What are the regulatory requirements for the development of fuel stations and control of the risk they pose in the environment in Cape Town, South Africa? |
| Question 4: Do fuel stations pose risks to groundwater and boreholes in Cape Town? |
| Question 5: During the EIA process, is there any specific consideration given to the proximity of fuel stations and vulnerable land uses, such as residential, hospital, school, hospital, etc.? |
| Question 6: How does "contaminated land" relate to groundwater contamination? |
| Question 7: With regards to land-use management, what are the legal frameworks for the development of fuel stations and control of the risk they pose to the environment? |
| Question 8: Do you see any potential for land use conflicts between fuel stations and sensitive land uses in the Western Cape? |
| Question 9: How would you describe the approach to managing the risk posed by fuel stations in Cape Town? |
| Question 10: Do the legal requirements for the installation and upgrade of underground storage tanks in Cape Town address issues of the vulnerability of boreholes in the surrounding fuel stations? |

## References


1. Malaza, N.; Mabuda, A. Challenges of Integrated Water Resources Management in the Western Cape Province, South Africa. *J. Water Resour. Ocean Sci.* **2019**, *8*, 9–20. [CrossRef]
2. Khatri, N.; Tyagi, S. Influences of natural and anthropogenic factors on surface and groundwater quality in rural and urban areas. *Front. Life Sci.* **2015**, *8*, 23–39. [CrossRef]
3. Gilissen, H.K.; de Jong, E.; van Rijswick, H.F.; van Wezel, A. Towards More Effective Environmental Risk Regulation. *J. Eur. Environ. Plan. Law* **2021**, *18*, 77–102. [CrossRef]
4. Solanes, M. *Water Resources Legislation: A Search for Common Principles*; International Institute for Sustainable Development: London, UK, 2009; Available online: https://www.iisd.org/system/files/publications/water_resources_legislation.pdf (accessed on 20 August 2021).
5. Grönwall, J.; Danert, K. Regarding groundwater and drinking water access through a human rights lens: Self-Supply as a norm. *Water* **2020**, *12*, 419. [CrossRef]
6. Wu, Q.; Zhang, X.; Zhang, Q. Current situation and control measures of groundwater pollution in gas stations. In Proceedings of the IOP Conference Series: Earth and Environmental Science, Singapore, 28–30 July 2017.

7. Maljean-dubois, S. *The Effectiveness of Environmental Law*, 1st ed.; Intersentia: Cambridge, UK, 2017; pp. 1–362.
8. Ahmed, M.M.; Kutty, S.R.M.; Shariff, M.; Khamidi, M.F. Petrol Fuel Station safety and risk assessment framework. In Proceedings of the National Postgraduate Conference—Energy and Sustainability: Exploring the Innovative Minds, Kuala Lumpur, Malaysia, 19–20 September 2011.
9. Ihsan, A.J.; Mustafa, A.E.; Akeel, A.A. The Impact of the Distribution of Fuel Stations on the Urban Environment Case Study Kut-Iraq. In Proceedings of the IOP Conference Series: Materials Science and Engineering, Istanbul, Turkey, 7–9 October 2019.
10. Solecki, T.; Stopa, J. Petroleum substances in soil and groundwater in the urban areas. *AGH Drill. Oil Gas* **2016**, *33*, 135–143. [CrossRef]
11. Gosling, M. Boreholes in Karoo Polluted by Petrol. 2011. Available online: https://www.iol.co.za/capetimes/boreholes-in-karoo-polluted-by-petrol-1098138 (accessed on 30 November 2022).
12. Wright, T.; Jacobs, H.E. Potable water use of residential consumers in the Cape Town metropolitan area with access to groundwater as a supplementary household water source. *Water SA* **2016**, *42*, 144–151. [CrossRef]
13. Visser, W. A perfect storm: The ramifications of Cape Town's drought crisis. *J. Transdiscipl. Res. South. Afr.* **2018**, *14*, 1–10. [CrossRef]
14. Ziervogel, G. *Unpacking the Cape Town Drought: Lessons Learned Report for Cities Support Programme Undertaken by African Centre for Cities*; African Centre for Cities: Cape Town, South Africa, 2019; Available online: https://www.africancentreforcities.net/wp-content/uploads/2019/02/Ziervogel-2019-Lessons-from-Cape-Town-Drought_A.pdf (accessed on 20 September 2022).
15. Jordan, B. Cape Town's Water Plans Go Underground: Number of Boreholes Soars as Cape Town Eyes Groundwater. Sunday Times Website. 2019. Available online: https://www.timeslive.co.za/sunday-times/news/2019-04-07-cape-towns-water-plans-go-underground/ (accessed on 20 June 2020).
16. Azimi, R.; Vaezihir, A.; Lenhard, R.J.; Hassanizadeh, S.M. Evaluation of LNAPL behaviour in water table inter-fluctuate zone under groundwater drawdown condition. *Water* **2020**, *12*, 2337. [CrossRef]
17. Adelana, S.; Xu, Y. Contamination and protection of the Cape Flats Aquifer. In *Groundwater Pollution in Africa*; Xu, Y., Usher, B., Eds.; Taylor & Francis: London, UK, 2006; pp. 265–277.
18. Qonono, K. Analysis of the Fire Hazard Posed by Petrol Stations in Stellenbosch and the Extent to Which Planning Acknowledges Risk. Master's Thesis, Stellenbosch University, Stellenbosch, South Africa, 2019.
19. Michaels, R.A. Legacy Contaminants of Emerging Concern: Lead (Pb), Flint (MI), and Human Health. *Environ. Claims J.* **2019**, *32*, 6–45. [CrossRef]
20. Gibson, J.M.; Fisher, M.; Clonch, A.; MacDonald, J.M. Children drinking private well water have higher blood lead than those with city water. *Proc. Natl. Acad. Sci. USA* **2020**, *117*, 16898–16907. [CrossRef] [PubMed]
21. Kponee, K.Z.; Chiger, A.; Kakulu, I.I.; Vorhees, D.; Heigener-Bernays, W. Petroleum contaminated water and health symptoms: A cross-sectional pilot study in a rural Nigerian community. *Environ. Health* **2015**, *14*, 86. [CrossRef] [PubMed]
22. Bai, X.; Song, K.; Liu, J.; Mohamed, A.K.; Mou, C.; Liu, D. Health risk assessment of groundwater contaminated by oil pollutants based on numerical modelling. *Int. J. Environ. Res. Public Health* **2019**, *16*, 3245. [CrossRef] [PubMed]
23. Ooi, C.S.; Ngui, W.K.; Hoou, H.K.; Lim, M.H.; Leong, M.S. Review of Underground Storage Tank Monitoring Condition Technique. *MATEC Web Conf.* **2019**, *255*, 02009.
24. Cox, L.; Verle, H.; James, A.; John, T.; Ingrid, H.; Nick, F.; Claudia, W.; Scott, A.J.; Yongping, Y.; Anthony, Z.; et al. *Land Use: A Powerful Determinant of Sustainable and Healthy Communities*; US Environmental Protection Agency: Washington, DC, USA, 2013. Available online: https://www.epa.gov/sites/default/files/2016-09/documents/fy13productnheerl4121land_use_synthesis.pdf (accessed on 30 September 2022).
25. Niemand, A.; Jordaan, A.J.; Minnaar, H. Some international perspectives on legislation for the management of human-induced safety risks. *Jamba* **2016**, *8*, 170. [CrossRef] [PubMed]
26. Holley, C. Environmental regulation and governance. In *Regulatory Theory: Foundations and Applications*; Drahos, P., Ed.; ANU Press: Acton, Australia, 2017; pp. 741–758.
27. Aven, T.; Renn, O. Improving government policy on risk: Eight key principles. *Reliab. Eng. Syst. Saf.* **2017**, *76*, 230–241. [CrossRef]
28. Cosens, B.; Craig, R.K.; Hirsh, S.L.; Arnorld, C.A.; Benson, M.H.M.; DeCaro, D.A.; Garmerstani, A.S.; Gosnell, H.; Ruhl, J.B.; Schlager, E. The Role of Law in Adaptive Governance. *Ecol. Soc.* **2017**, *22*, 1–22. [CrossRef] [PubMed]
29. Gomo, M. Site Characterisation of LNAPL-Contaminated Fractured-Rock Aquifer. Master's Thesis, University of the Free State, Bloomfontein, South Africa, 2009.
30. Kruger, R. A Critical Analysis of the Quality of EIA Reports for Filling Stations in South Africa. Master's Thesis, Northwest University, Potchefstroom, South Africa, 2012.
31. Johnston, M. A Critical and Comparative Analysis of the under Regulation of Underground Storage Tanks in South Africa and the Attendant Consequences for Environmental Resources. Master's Thesis, University of Cape Town, Cape Town, South Africa, 2014.
32. Pfotenhauer, T. An Investigation into Factors Increasing Contamination Risk Posed by Fuel Storage Facilities and Concomitant Methods to Mitigate These Risks. Master's Thesis, University of Kwazulu-Natal, Durban, South Africa, 2011.
33. Taherdoost, H. Sampling Methods in Research Methodology; How to Choose a Sampling Technique for Research. *Intern. Journ. of Acad. Res. in Manag.* **2016**, *5*, 18–27. [CrossRef]

34. Logeshwaran, P.; Megharaj, M.; Chadalavada, S.; Bowman, M.; Naidu, R. Petroleum hydrocarbons (PH) in groundwater aquifers: An overview of environmental fate, toxicity, microbial degradation and risk-based remediation approaches. *Environ. Technol. Innov.* **2018**, *10*, 175–193. [CrossRef]
35. *SANS 10089-1*; The Petroleum Industry. South African Bureau of Standard: Pretoria, South Africa, 2008.
36. *SANS 1830*; Flexible Piping for Underground Use at Service Stations and Consumers Installations. South African Bureau of Standard: Pretoria, South Africa, 2006.
37. *SANS 1535*; Glass-Reinforce Polyester-Coated Steel Tanks for the Underground Storage of Hydrocarbons and Oxygenated Solvents and Intended for Burial Horizontally. South African Bureau of Standard: Pretoria, South Africa, 2007.
38. Seyler, H.; Witthüser, K.; Sunaitis, M. *Urban Groundwater Development and Management*; Water Research Commission: Pretoria, South Africa, 2019.
39. California State Water Resources Control Board. Technical Justification for Groundwater Plume Lengths, Indicator Constituents, Concentrations, and Buffer Distances (Separation Distances) to Receptors. 2004. Available online: https://www.waterboards.ca.gov/ust/policy/techjust071211.pdf (accessed on 1 May 2020).
40. Kotze, L. Improving Unsustainable Environmental Governance in South Africa: The Case for Holistic Governance. *Potchefstroom Electron. Law J.* **2016**, *9*, 74–118. [CrossRef]
41. Kotze, L.J.; Nel, J.G.; du Plessus, W.; Snyman, E. Strategies to integrate environmental policy at the operational level: Towards an integrated framework for environmental authorisations. *S. Afr. J. Law Policy* **2007**, *14*, 57–81.

energies

MDPI

Article

# A Comprehensive Study on the Hydrogeochemical and Isotope Characteristics and Genetic Mechanism of Geothermal Water in the Northern Jinan Region

**Zongjun Gao, Mengyuan Hao, Jiutan Liu *, Qiang Li, Menghan Tan and Yiru Niu**

College Of Earth Science and Engineering, Shandong University of Science and Technology, Qingdao 266590, China; gaozongjun@sdust.edu.cn (Z.G.); hmysdust@126.com (M.H.); qiangli6110@126.com (Q.L.); tmhsdust@126.com (M.T.); nyrsdkj@126.com (Y.N.)
* Correspondence: ljtsdust@sdust.edu.cn

**Abstract:** Geothermal water (GW) resources are highly valued as clean, renewable energy sources. In this study, a comprehensive analysis of water chemistry and isotope data from 25 GW samples was conducted to gain insights into the hydrochemical characteristics and formation mechanisms of the GW in the northern Jinan region (NJR). Statistical analysis and hydrochemical methods were employed for relevant analysis. The findings reveal that the GW in the NJR exhibits high salinity, with an average total dissolved solids (TDS) concentration of 9009.00 mg/L. The major ions identified are $Na^+$ and $Cl^-$, with mean concentrations of 2829.73 mg/L and 4425.77 mg/L, respectively, resulting in a hydrochemical type of $Cl^-Na$. The analysis of $\delta^2H$ and $\delta^{18}O$ isotopes indicates that the GW originates from atmospheric precipitation that undergoes deep cycling and interaction with older groundwater. The composition of $^3H$ suggests that the GW in the NJR is a mixture of waters, while radiocarbon dating ($^{14}C$) suggests that the recharge of the GW may have occurred in the late Pleistocene era. The GW in the NJR is classified as partially equilibrated waters. The temperature range of geothermal reservoirs is 57.13 to 99.74 °C. The hydrochemical components primarily result from water–rock interactions, including silicate weathering, cation exchange, as well as carbonate weathering and the dissolution of halite and gypsum. Moreover, taking into account the hydrogeological conditions, hydrochemistry, and isotope analysis, a conceptual model of the geothermal reservoir in the NJR was developed. The research findings serve as a valuable reference and foundation for the development and utilization of geothermal resources in the Jinan region. These originate from the Taiyi mountains in the south or the Taihang mountains in the west, and experience deep circulation and long runoff times. This study provides a reference for the sustainable development and utilization of regional geothermal resources.

**Keywords:** hydrochemical characteristics; isotope composition; genetic mechanism; GW; northern Jinan region

**Citation:** Gao, Z.; Hao, M.; Liu, J.; Li, Q.; Tan, M.; Niu, Y. A Comprehensive Study on the Hydrogeochemical and Isotope Characteristics and Genetic Mechanism of Geothermal Water in the Northern Jinan Region. *Energies* **2023**, *16*, 7658. https://doi.org/10.3390/en16227658

Academic Editors: Benjamin McLellan and Rafael Santos

Received: 26 September 2023
Revised: 13 November 2023
Accepted: 14 November 2023
Published: 19 November 2023



## 1. Introduction

Geothermal energy is a significant natural resource found within the Earth's interior [1]. Global statistics from 2015 to 2019 indicate that a total of 2647 geothermal wells were utilized over that period for geothermal power generation or direct heating in 42 countries [2]. Furthermore, the installed capacity for direct utilization of geothermal resources worldwide in 2019 reached $1.07 \times 10^6$ MWt. In China, the available geothermal resources account for nearly 8% of the global total, representing substantial development potential [3,4]. However, despite this potential, the development and utilization of geothermal resources in China has lagged behind wind and solar energy, primarily due to a lack of comprehensive industrial planning, policy support, and outdated mining techniques [3].

Geothermal fluid serves as the primary medium for the development and utilization of geothermal resources, possessing dual attributes as both water resources and mineral resources. It represents a clean and environmentally friendly source of energy, harboring significant potential for development and utilization [5]. High-temperature GW can be harnessed for power generation, while medium-to-low-temperature GW finds application in heating, bathing, medical care, and other sectors, thus fostering the growth of associated industries [5].

The chemical composition and origin of GW are typically linked to the intricate geological processes and specific hydrogeochemical conditions within a given region [6]. Investigating the hydrochemical characteristics of GW provides insights into the source, origin, age, migration, storage state, reservoir temperature, and other properties of the geothermal fluid. It also enables the examination of information pertaining to geothermal geological conditions through water–rock interactions [7–10]. Currently, statistical methods [5,11], hydrochemical graphing techniques [12,13], and environmental isotope analyses [13,14] are widely employed, offering valuable insights into the formation mechanisms of GW.

Jinan stands as a renowned "spring city" globally, boasting abundant geothermal resources and promising prospects for development and utilization [15]. Previous investigations on geothermal resources in the northern Jinan region (NJR) have primarily concentrated on the region's geothermal geological characteristics, recharge sources, and circulation patterns, yielding fruitful research results [14,16–19]. These studies highlight the way that the formation of GW in NJR is influenced by geological structures and magmatic activities. The primary recharge source for GW emerges from atmospheric precipitation in the mountainous areas of southern Jinan, facilitated by deep runoff. Moreover, carbonate and sulfate mineral dissolution, as well as halite precipitation, play crucial roles in governing the hydrochemical characteristics of groundwater. Precipitation from the southern mountains seeps downward into the ground and is then blocked by the Jejunum intrusive rocks and can only progress downward to greater depths, where the groundwater is heated by a geothermal gradient. The heated water eventually reaches the north and forms GW. The depth of the low-velocity layer under the Jinan intrusive rocks varies with the lateral direction, which may indicate that the depth of groundwater transport under the west and east Jinan intrusive rocks is different. Isotopic data have allowed the origin of thermal water to be determined and have presented different recharge elevations. Reservoir temperature was estimated by chemical geothermometry and validated by fluid–mineral equilibria calculations. The feasibility of different geothermometers was verified, indicating that quartz geothermometers were more suitable for the geothermal system [14,20]. However, the majority of these studies mainly focused on the southern section of the Qihe–Guangrao fault in NJR, limiting geothermal research in the northern segment of this fault. The cationic and $SiO_2$ geothermometers indicate the heat storage temperature in the NJR. The main hydrogeochemical processes of the chemical composition of GWr in the north of Jinan were analyzed by ion ratio. The use of PHREEQC3.1.1 software indicates that GW in the north of Jinan is partially balanced. A conceptual model of geothermal genesis in the northern area of Jinan was established in order to provide conditions for its subsequent rational development and utilization.

Previous studies have focused on the geothermal geologic features, recharge sources and circulation patterns in the southern part of the Qihe–Guangrao rupture, as does this study. Hence, considering the geothermal geological conditions in the NJR, this research employs a comprehensive approach combining hydrochemical and isotopic methods to investigate the hydrochemical characteristics and formation mechanisms of GW. The purpose of this paper is to provide theoretical support for the hydrogeochemical isotope characterization and genesis mechanism of GW in the northern part of the country, and to provide a scientific basis for the development of GW in the northern part of the country.

## 2. Study Area

The NJR (Figure 1a) is situated in the northern part of Jinan city, to the north of the Qihe–Guangrao fault (Figure 1b). It encompasses Shanghe county, the northern part of

Jiyang county, and the northwestern part of Zhangqiu city. The geographical coordinates span from 116°52′00″ E to 117°27′00″ E and 36°57′00″ N to 37°32′00″ N, covering a total area of 1989 km$^2$. The terrain in NJR generally exhibits higher elevation in the southwest and lower elevation in the northeast. The northern yellow floodplain area typically features elevations ranging from 10 to 30 m, with a terrain slope of approximately 0.03‰. The lowest point is located in the northeastern part of Shanghe county, with an elevation of 10.7 m. The landform is primarily characterized by Quaternary alluvial plains, resulting in relatively simple terrain. The topography is predominantly a gently sloping low plain, with some areas transitioning into piedmont sloping plains.

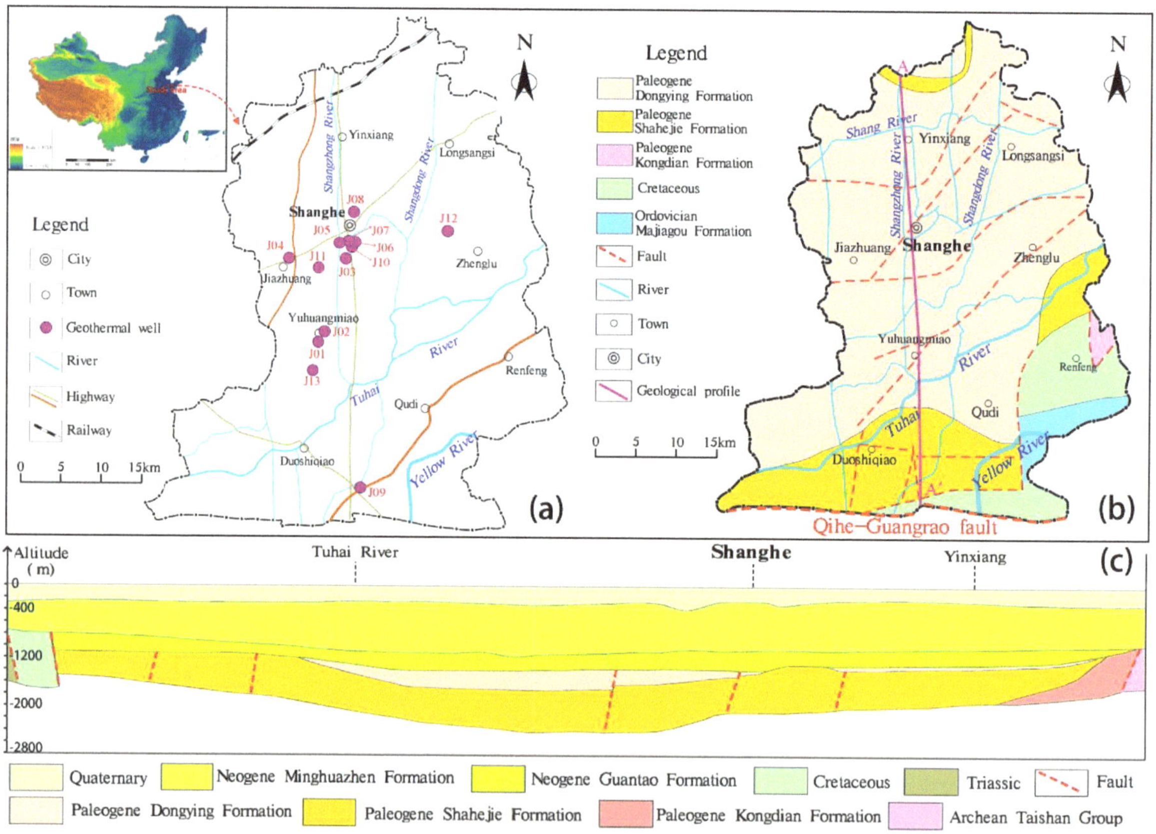


**Figure 1.** Map of Location and distribution of geothermal wells (**a**), pre-Neogene geological (**b**) and geological cross section (**c**).

The NJR experiences a warm temperate continental climate. It is characterized by dry and less rainy springs, hot and rainy summers, cool and breezy autumns, and severely cold and dry winters. The annual average temperature in the region is 13.6 °C (1956 to 2015), and the average annual precipitation amounts to 614.0 mm. More than 70% of the total precipitation is concentrated in the months of July to September, while the period from December to March of the following year witnesses comparatively less rainfall. The NJR falls within the water system areas of both the Yellow River and Tuhai River. Within the Tuhai River water system area, most of the rivers flow in a southwest to northeast direction; the south side of the Tuhai River, however, flows in a southeast to northwest direction before joining the Tuhai River.

In the NJR, the geological formations are arranged in a top-to-bottom sequence, including the Quaternary, Neogene Minghuazhen Formation, Guantao Formation, Paleogene Dongying Formation, Shahejie Formation, Kongdian Formation, and Cretaceous. Among these formations, the Neogene Guantao Formation and the Paleogene Dongying Formation are widely distributed and characterized by substantial thickness. The Cambrian–Ordovician formations, on the other hand, are deeply buried. The primary type of groundwater in the NJR is pore water, which occurs in the upper loose sediments of the Quaternary and Neogene Minghuazhen Formation. The recharge of groundwater is mainly influenced by atmospheric precipitation and infiltration from rivers, both of which are significant sources of groundwater replenishment.

## 3. Materials and Methods

### 3.1. Sampling

In this study, a total of 25 GW samples were collected from 13 geothermal wells in NJR (Figure 1). The geothermal wells were sampled at different frequencies: J02 and J04 were sampled three and four times, respectively, spanning different years. J11, J12, J09, and J01 were sampled once, while the remaining geothermal wells were sampled twice, across different years. All water samples underwent hydrochemical analysis, with eight samples also being tested for hydrogen and oxygen isotopes ($\delta^2H$ and $\delta^{18}O$) as well as tritium ($^3H$) isotopes. Additionally, five water samples were subjected to radiocarbon ($^{14}C$) isotope testing. The locations of the geothermal wells are indicated in Figure 1, and their depths range from 1300 to 2000 m. Temperature measurements were conducted using a deep well detector (SYKJ-6), comprising various components such as the host, measuring line, electronic point thermometer, depth sensor, and display. The temperature measurement range spanned from 0 to 500 m, with a resolution capability of 0.1 °C and an accuracy of $\pm 0.5\%$.

To collect GW samples for the main chemical component test, a 5 L rigid polyethylene plastic bucket was utilized. Prior to sampling, it is recommended to clean the sampling barrel 2–3 times using the GW that will be collected. It is important to fill these bottles without introducing any bubbles, and seal them tightly using parafilm for storage. Once the sampling is completed, it is advisable to send the samples to the laboratory for testing as soon as possible.

### 3.2. Analysis Method

The pH of the GW sample was measured using a pH meter (China Shanghai Yidian Scientific Instrument Co., Shanghai, China, PHS-3C). The levels of total hardness (TH), total dissolved solids (TDS), $K^+$, $Na^+$, $Ca^{2+}$, $Mg^{2+}$, and $H_3SiO_3$ were determined using an inductively coupled plasma emission spectrometer (ICP, PerkinElmer Corporation, Waltham, MA, USA, Optima 7000DV). The concentrations of $SO_4{}^{2-}$, $Cl^-$, and $NO_3{}^-$ were tested using an ion chromatograph (Thermo Fisher Scientific Co., Waltham, MA, USA, ICS-600), while $HCO_3{}^-$ was determined through titration. Stable isotope analysis of $\delta^2H$ and $\delta^{18}O$ was conducted using a stable isotope mass spectrometer (Thermo Fisher Scientific Co., MAT-253), while measurements of 3H and $^{14}C$ were performed using an ultra-low background liquid scintillation spectrometer (Revvity Co., Waltham, MA, USA, Quantulus-220).

The Piper diagram was used to determine the type of water chemistry and analyze geothermal water chemistry, Schoeller diagrams were used to characterize water chemistry, and Na-K-Mg triangles were used to assess water–rock equilibrium and to differentiate between water samples. Gibbers plots were used to determine the primary mechanism controlling water chemistry, ion ratio plots to determine the source of water chemistry, and saturation indices (SI) to determine mineral dissolution or precipitation. SI was calculated using the PHREEQC software based on the mathematical relationship between the ionic activity product and the corresponding equilibrium constant.

## 4. Results and Discussions

### 4.1. Hydrochemical Characteristics of GW

Descriptive statistical analysis proves beneficial in comprehending the general characteristics, enrichment, and variation exhibited by various chemical components present in GW. Table 1 presents the statistical findings concerning the primary physical and chemical components found in GW of NJR. GW temperature falls within the range of 54 °C and 59 °C, with an average of 56.75 °C. The pH values span between 7.15 and 7.55, averaging at 7.42, thereby indicating the weak alkaline nature of the GW in the NJR. TDS concentration in the NJR is relatively high, ranging from 5568.20 mg/L to 12,156.00 mg/L, with an average concentration of 9009.00 mg/L. Dominant cations and anions in GW are $Na^+$ and $Cl^-$, respectively, with concentrations ranging from 1755.00–3755.00 mg/L and 2445.30–6649.00 mg/L. The average values for $Na^+$ and $Cl^-$ are 2829.73 mg/L and 4425.77 mg/L, respectively. Regarding cationic presence, the order is $Na^+ > Ca^{2+} > Mg^{2+} > K^+$. Anionic composition exhibits a hierarchy of $Cl^- > SO_4{}^{2-} > HCO_3{}^- > NO_3{}^-$. Specific data are detailed in Table S1 of the Supplementary Material.

**Table 1.** Statistical characteristics of the main chemical components of GW.

| | Units | Max | Min | Mean | SD |
|---|---|---|---|---|---|
| $K^+$ | mg/L | 67.3 | 3.7 | 23.0 | 13.0 |
| $Na^+$ | mg/L | 3775.0 | 1755.0 | 2829.7 | 548.3 |
| $Ca^{2+}$ | mg/L | 650.4 | 169.8 | 401.7 | 153.9 |
| $Mg^{2+}$ | mg/L | 117.6 | 36.5 | 78.5 | 25.6 |
| $Cl^-$ | mg/L | 6649.0 | 2445.3 | 4425.8 | 1338.1 |
| $SO_4{}^{2-}$ | mg/L | 2903.4 | 280.5 | 1067.8 | 468.4 |
| $HCO_3{}^-$ | mg/L | 231.1 | 26.2 | 155.8 | 53.7 |
| $NO_3{}^-$ | mg/L | 5.9 | 0.02 | 1.6 | 1.5 |
| TH | mg/L | 2087.0 | 603.12 | 1308.3 | 479.2 |
| TDS | mg/L | 12,156.0 | 5568.2 | 9009.0 | 1755.3 |
| pH | / | 7.7 | 7.1 | 7.4 | 0.16 |
| $SiO_2$ | mg/L | 29.0 | 22.6 | 26.3 | 1.3 |
| Temperature | °C | 59.0 | 54.0 | 56.8 | 1.2 |
| $\delta^2H$ | ‰ | −69.1 | −76.0 | −72.2 | 2.6 |
| $\delta^{18}O$ | ‰ | −8.0 | −9.3 | −8.8 | 0.5 |
| $^3H$ | TU | 20.6 ± 1.0 | 2.7 ± 0.1 | / | / |
| $^{14}C$ age | ka BP | 13.4 ± 0.5 | 1.0 ± 0.1 | / | / |

The utilization of the Piper diagram [21] for analyzing water chemistry characteristics is widely employed and serves as a straightforward and efficient tool for hydrochemical classification [22]. As depicted in Figure 2, the GW samples predominantly cluster within the [$Cl^-$] and [$Na^+$] regions of the anion and cation triangle beneath the Piper diagram. Upon projection onto the upper diamond, the water sample points fall within area ②, indicating that the hydrochemical type of deep GW in northern Jinan is uniformly classified as $Cl^-Na$ type.

### 4.2. Isotopic Composition Characteristics

δ2 H/$\delta^{18}O$ and 3H/14C

Isotope methods play a crucial role in determining groundwater origin, recharge elevation, and hydraulic connections between different water bodies [14,19]. In Table 1, the δ2H and δ18O values of the GW in the NJR ranged from −69.96 to −75.98 and −7.94 to −9.25, with mean values of −72.61 and −8.72, respectively. Figure 3 illustrates the relationship between δ2H-δ18O in the GW of NJR. The global meteoric water line (GWML) and the local meteoric water line (LWML) in Figure 3 are derived from previous studies conducted by Craig [23] and Liu et al. [24]. As is evident from Figure 3, the GW sample points can be roughly divided into two groups, namely group (I) and group (II). The water samples in group (I) are closely clustered around the LWML, indicating that they are derived from direct or indirect precipitation in the atmosphere. Due to the influence of the Qihe–Guangrao fault, the depth of groundwater infiltration increases, and the gradually increasing temperature and pressure inside the stratum force the oxygen isotope of the infiltrated atmospheric precipitation to evaporate, or to react with the surrounding rock, and the oxygen isotope exchange of water and minerals, resulting in the oxygen isotope in the GW. Drift, and then the earth's heat flow through the deep circulation forms GW. The extent of this drift depends on factors such as water/rock ratio, interaction strength, and reaction time frame. The water samples belonging to group (II) deviate from the LWML distribution to the right, indicating obvious oxygen isotope drift. This indicates that the NJR GW is less affected by faulting and may originate from older water sources. The water–rock isotope exchange is clear and the cycle period is long. The isotope 3H, a radioactive isotope of hydrogen, serves as a valuable dating technique in hydrogeological research, providing radiometric timing [19,25]. Groundwater typically experiences minimal exchange with surrounding rock media, allowing the 3H content to be solely governed by radioactive decay. As such, it proves useful in investigating hydrogeological and geothermal development issues. When the 3H content is less than 0.8TU, this indicates that the groundwater age is older, and that its source is mainly atmospheric precipitation or formation water before 1950. The range of 0.8–4 TU indicates that atmospheric precipitation after 1950 infiltrated into groundwater, forming a mixture of modern atmospheric precipitation and ancient precipitation or formation water, while the range of 5–15 TU indicates that groundwater is recharged by modern atmospheric precipitation infiltration [26]. In this study, the 3H content is between 2.7~20.6 TU (Table 1), but is slightly biased to 4 TU, indicating that the GW in the NJR is a mixture of modern infiltration water and groundwater or formation water before 1950. Combined with the results of deuterium and oxygen isotope analysis, it can be seen that the GW in the NJR is a mixture of modern infiltration water and formation water before 1950.

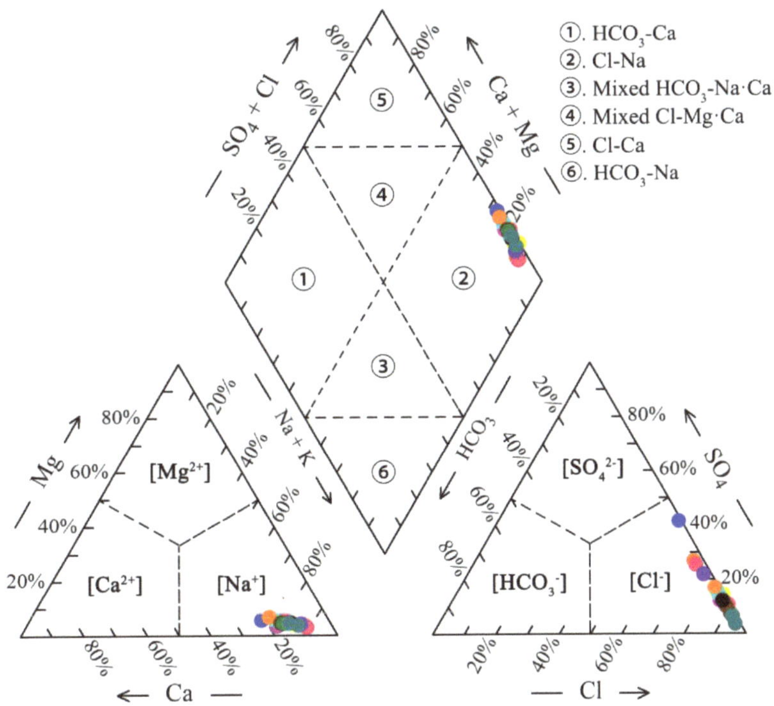


**Figure 2.** Piper diagram of GW sample in NJR.

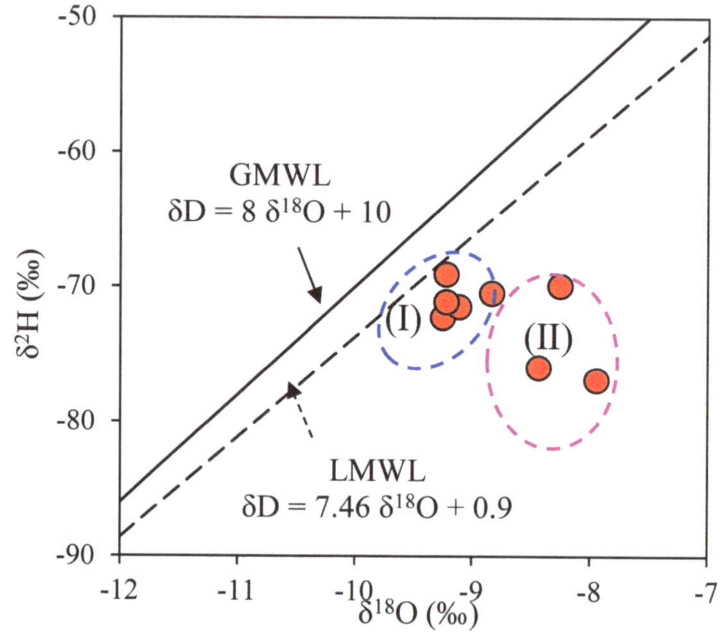


**Figure 3.** Relationship between δD−$\delta^{18}$O in deep GW in NJR.

The 14C dating method is a well-established technique for dating old groundwater and GW and is particularly suited for measuring ages within the range of 100–50,000 years BP [14]. The 14C ages of the GW in the NJR range from 13.36 to 3.85 ka BP (Table 1), suggesting that recharge may have occurred during the late Pleistocene period. Determination of the age of geothermal water provides a reliable analytical basis for later production capacity evaluation, heat preservation and transportation, groundwater recharge, production capacity optimization, intelligent exploitation, and comprehensive utilization.

### *4.3. Geothermal Reservoir Temperature*

Geothermometers serve as a valuable tool in estimating the temperature of geothermal reservoirs. They utilize the relationship between chemical components, isotope values, and temperature in GW. Geothermometers find widespread application in studying the formation mechanisms, predicting the potential, and exploring and developing geothermal resources [14,25]. Typically, four types of geothermometers are utilized: cation geothermometers, $SiO_2$ geothermometers, gas geothermometers, and isotope geothermometers. Each type has its specific application conditions [27]. The above hydrochemical characteristics show that the chemical type of NJR GW is dominated by Na-Cl type water with a high content of $Na^+$, and analysis of the isotopes shows that there is a high amount of silica in the NJR GW, therefore, when estimating the reservoir temperature of GW samples in NJR, cation and silicon chemical geothermometers were employed. The $SiO_2$ geothermal temperature scale was chosen because the abundance of silicon is sufficient and does not change after equilibration. The relationship between cation exchange and temperature can be confirmed in the study area, so a cation temperature scale was also used. The calculated statistical results are presented in Table 2.

**Table 2.** Statistical results of estimating reservoir temperature using the selected chemical geothermometer.

| | | Maximum | Minimum | Mean | Standard Deviation | |
|---|---|---|---|---|---|---|
| (°C) | Quartz | 99.74 | 70.21 | 92.67 | 6.56 | [26] |
| (°C) | Quartz (maximum steam loss) | 81.83 | 57.13 | 75.94 | 5.47 | [26] |
| (°C) | Chalcedony | 46.59 | 17.69 | 39.62 | 6.42 | [26] |
| (°C) | Chalcedony (maximum steam loss) | 49.39 | 21.95 | 42.79 | 6.09 | [28] |
| (°C) | Na-K-Mg | −223.03 | −236.11 | −229.57 | 2.83 | [29] |
| (°C) | Na-K-Ca-Mg | 152.53 | 117.09 | 141.34 | 9.51 | [28] |
| (°C) | Na-K | 96.08 | 18.04 | 32.81 | 26.89 | [29] |
| (°C) | K-Mg | 86.53 | 51.43 | 64.57 | 14.33 | [30] |
| (°C) | Na-K-Ca | 187.29 | 63.93 | 129.26 | 26.99 | [31] |

Generally, a higher concentration of silica in GW correlates with a higher thermal reservoir temperature [25]. However, it is worth noting that the temperature estimated by the silica geothermometer tends to be higher than the measured temperature of the GW. This discrepancy suggests that other factors may be influencing the temperature estimation process. Additionally, the calculated temperatures from the chalcedony geothermometer are lower than both the measured temperatures and the temperatures obtained from quartz geothermometers. This indicates that the chalcedony geothermometer may not be suitable for accurately estimating the temperature of the geothermal reservoir in the NJR. Therefore, the lowest value of the quartz (maximum vapor loss) calculation and the highest value of the quartz calculation were chosen for the thermal storage temperature range.

Silica geothermometers provide temperature estimations based on different silica mineral phases. The calculated temperature ranges for quartz and quartz with maximum steam loss are 70.21 to 99.74 °C and 57.13 to 81.83 °C, with mean values of 92.67 °C and 75.94 °C, respectively. For chalcedony and chalcedony with maximum steam loss, the calculated temperature ranges are 17.69 to 46.59 °C and 21.95 to 49.39 °C, with average values of 39.62 °C and 42.79 °C, respectively. Furthermore, the temperatures obtained using different cation geothermometers are as follows: (1) the calculated temperatures from the Na-K-Mg geothermometer range from −223.03 °C to −236.11 °C, with an average of −229.57 °C; (2) the Na-K-Ca-Mg geothermometer yields temperatures ranging from 117.09 °C to 152.53 °C, with an average of 141.34 °C; (3) the Na-K geothermometer provides temperatures ranging from 18.04 °C to 96.08 °C, with an average of 32.81 °C; (4) the calculated temperature range of the K-Mg geothermometer is 51.43 °C to 86.53 °C, with an average of 64.57 °C; and (5) the Na-K-Ca geothermometer calculates temperatures ranging from 63.93 °C to 187.29 °C, with an average of 129.26 °C. In general, when the reservoir temperature exceeds 150 °C, it is appropriate to utilize the Na-K geothermometer method to estimate the temperature [25]. This method relies on the sodium-to-potassium ratio controlled by minerals such as albite and potassium feldspar present in the geothermal system. However, as shown in Table 2, the estimated temperature range provided by the Na-K geothermometer is relatively large, and some water samples yield temperatures significantly lower than the measured temperature. This suggests that the Na-K geothermometer may not accurately estimate the temperature in all cases. Furthermore, it is worth noting that the calculated temperatures from the Na-K-Mg geothermometer are anomalously lower compared with the measured temperatures (Table 1). This indicates that the Na-K-Mg geothermometer is not suitable for accurately estimating the temperature of the geothermal reservoir.

The Na-K-Mg triangle diagram, proposed by Giggenbach [32] in 1988, serves as a tool to assess the water–rock balance state and to differentiate the various types of water samples. It classifies water samples into three states or types: complete balance, partial balance, mixed waters, and immature water. As depicted in Figure 4, all of the GW samples in the NJR fall within the zone of partially mature and mixed waters. This suggests that water–rock interaction plays a crucial role in shaping the chemical composition of GW. However, it also indicates that the fluid and minerals have not yet reached a state of complete chemical equilibrium. Furthermore, in the NJR, the composition of GW is influenced by the mixing and dilution processes involving groundwater from different recharge sources. This indicates that the GW in the region undergoes interactions with various water sources, resulting in a complex mixture.

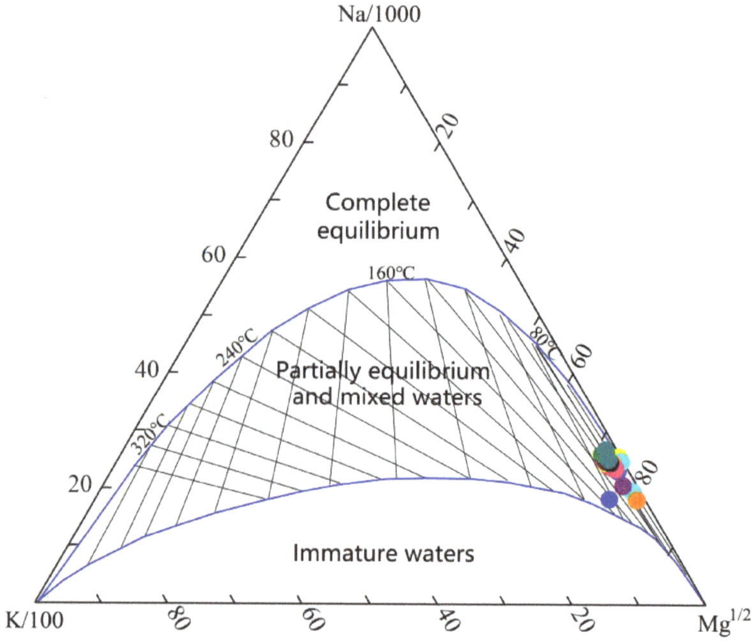


**Figure 4.** Na-K-Mg Giggenbach diagram with GW samples in NJR.

### *4.4. Controlling Factors of Hydrochemical Characteristics of GW*

#### 4.4.1. Schoeller Diagram

The Schoeller diagram, depicted in Figure 5, is extensively utilized in the analysis of water chemistry characteristics. It effectively illustrates fluctuations in the concentration of major ions present in water samples [5,33,34]. When water samples of the same hydrochemical type display parallel lines connecting two ions, it signifies a consistent or similar concentration ratio between those ions. Conversely, if the ion concentrations of two water samples differ, with one positioned above the other on the diagram, it indicates the relative movement direction of groundwater chemical components across different sampling points. This implies a flow of water from regions of lower concentration to those of higher concentration. $SO_4^{2-}$ and $HCO_3^-$ fluctuated greatly, and the other elements fluctuated more consistently (Figure 5). Furthermore, the variation trends observed in the 25 GW samples collected within the NJR can be broadly classified into two groups, consistent with the findings from the δ2H and δ18O analysis. This suggests potential disparities in the recharge sources of GW.

#### 4.4.2. Correlation Analysis

Correlation analysis is commonly employed to investigate the origins of ions, with components from the same source typically displaying notable correlations. Figure 6 depicts the correlation matrix diagram of the principal chemical components found in the GW of northern Jinan. Notably, TDS exhibits a strong correlation with $Cl^-$ (r = 0.95), $Na^+$ (r = 0.95), $Ca^{2+}$ (r = 0.86), and $Mg^{2+}$ (r = 0.84), indicating that these chemical components make significant contributions to TDS. Additionally, $Ca^{2+}$ shows a strong correlation with $Na^+$ (r = 0.95) and $K^+$ (r = 0.54), suggesting a potential common source, likely originating from the dissolution of silicate minerals. Furthermore, a significant correlation exists between

$Ca^{2+}$ and $Mg^{2+}$ (r = 0.89), indicating that the dissolution of carbonate rock minerals may serve as a significant source of $Ca^{2+}$ and $Mg^{2+}$ in GW.

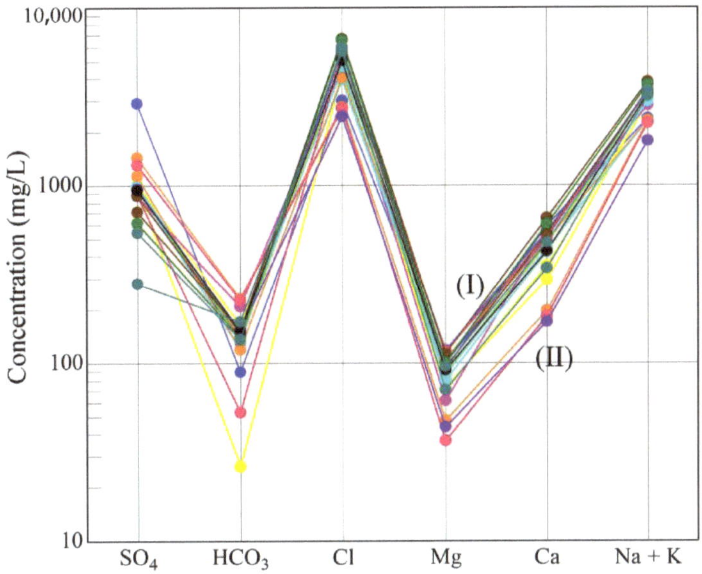


**Figure 5.** Schoeller diagram of GW samples.

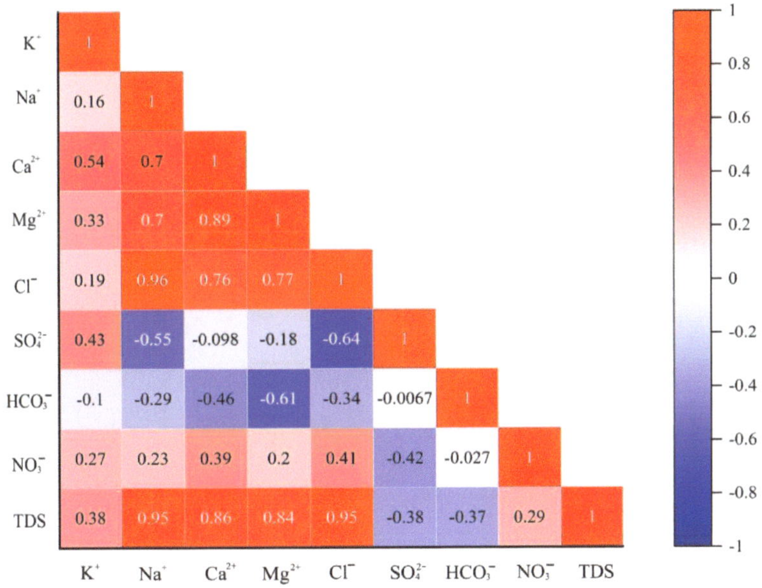

**Figure 6.** Correlation matrix of GW in northern Jinan.

$Cl^-$ is widely considered a conservative parameter and plays a crucial role in analyzing the chemical characteristics of GW due to the limited presence of chlorine-containing minerals in aquifers, apart from halite [5,35,36]. It exhibits a positive correlation with $Na^+$ (r = 0.96), $Ca^{2+}$ (r = 0.76), and $Mg^{2+}$ (r = 0.77), while showing a negative correlation with $SO_4{}^{2-}$ (r = −0.64) and $HCO_3{}^-$ (r = 0.34). Throughout the protracted flow process of

GW, leaching, evaporation, and concentration occur, leading to an increase in $Cl^-$ content and a decrease in $HCO_3^-$ and $SO_4^{2-}$ content. Additionally, the ratios of $HCO_3^-/Cl^-$ (0.006~0.008) and $SO_4^{2-}/Cl^-$ (0.05~0.97) in GW are relatively low, indicating an extended runoff path and a slower deep-water cycle for GW [35].

4.4.3. Ion Ratios Analysis

Typically, the chemical composition of water is influenced by various factors, such as rock weathering, atmospheric precipitation, and evaporation [37]. However, GW follows an extensive circulation path and undergoes continuous leaching, evaporation, and concentration, resulting in a high salinity level. Consequently, evaporation becomes the primary controlling factor for the chemical characteristics of GW. This is further illustrated by the distribution of GW samples in the Gibbs diagram [37] (Figure 7), emphasizing the dominance of evaporation in shaping the chemical properties of GW.

In this study, ion ratio analysis [22,38] was conducted to further investigate the primary hydrogeochemical processes influencing the chemical composition of GW in the NJR. If the dissolution of halite is the sole source of $Na^+$ and $Cl^-$ in groundwater, the $Na^+/Cl^-$ ratio would equal 1, as the dissolution of halite releases equal amounts of $Na^+$ and $Cl^-$ [39]. It can be seen from Figure 8a that the NJR GW samples are distributed on and off the line of $Na^+/Cl^- = 1$, and that four samples fall directly on the line, indicating that rock salt dissolution is a component of GW $Na^+$ and $Cl^-$, but is not the main source. The concentration of $Na^+$ and $Cl^-$ in the GW is high, and 60% of the samples are above the $Na^+/Cl^- = 1$ line, indicating that there is an obvious evaporation and concentration process. In addition, the dissolution or cation exchange of silicate minerals may also be the reason for the presence of $Na^+$ in GW.

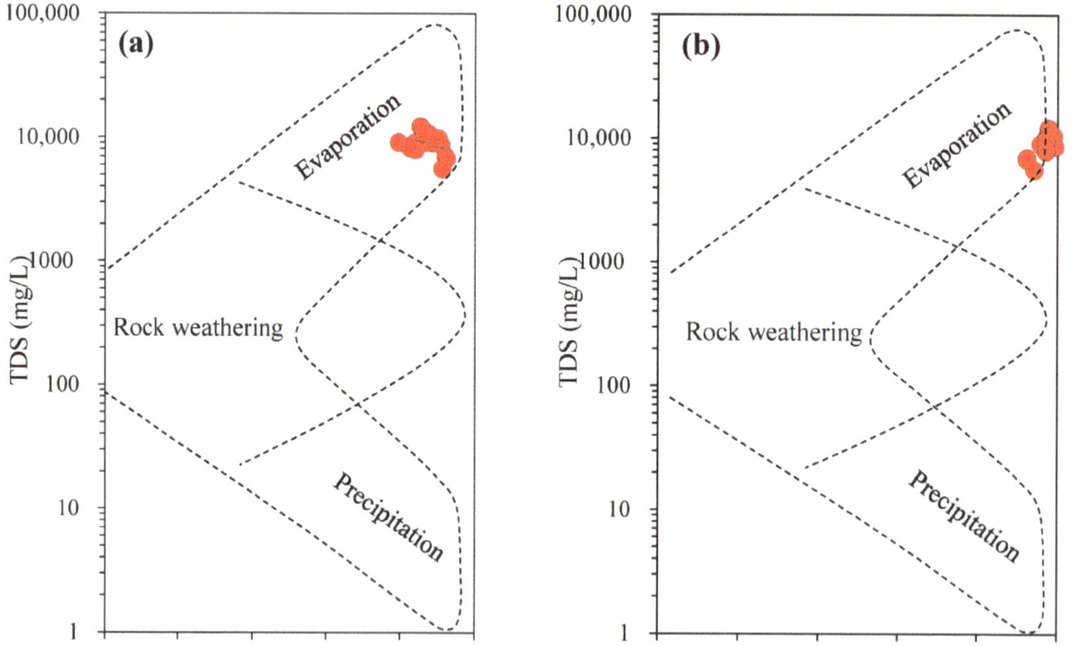


**Figure 7.** Gibbs diagram of deep GW samples in Northern Jinan. (**a**) Relationship between TDS and $Na^+/(Na^++Ca^{2+})$. (**b**) Relationship between TDS and $Cl^-/(Cl^-+HCO3^-)$.

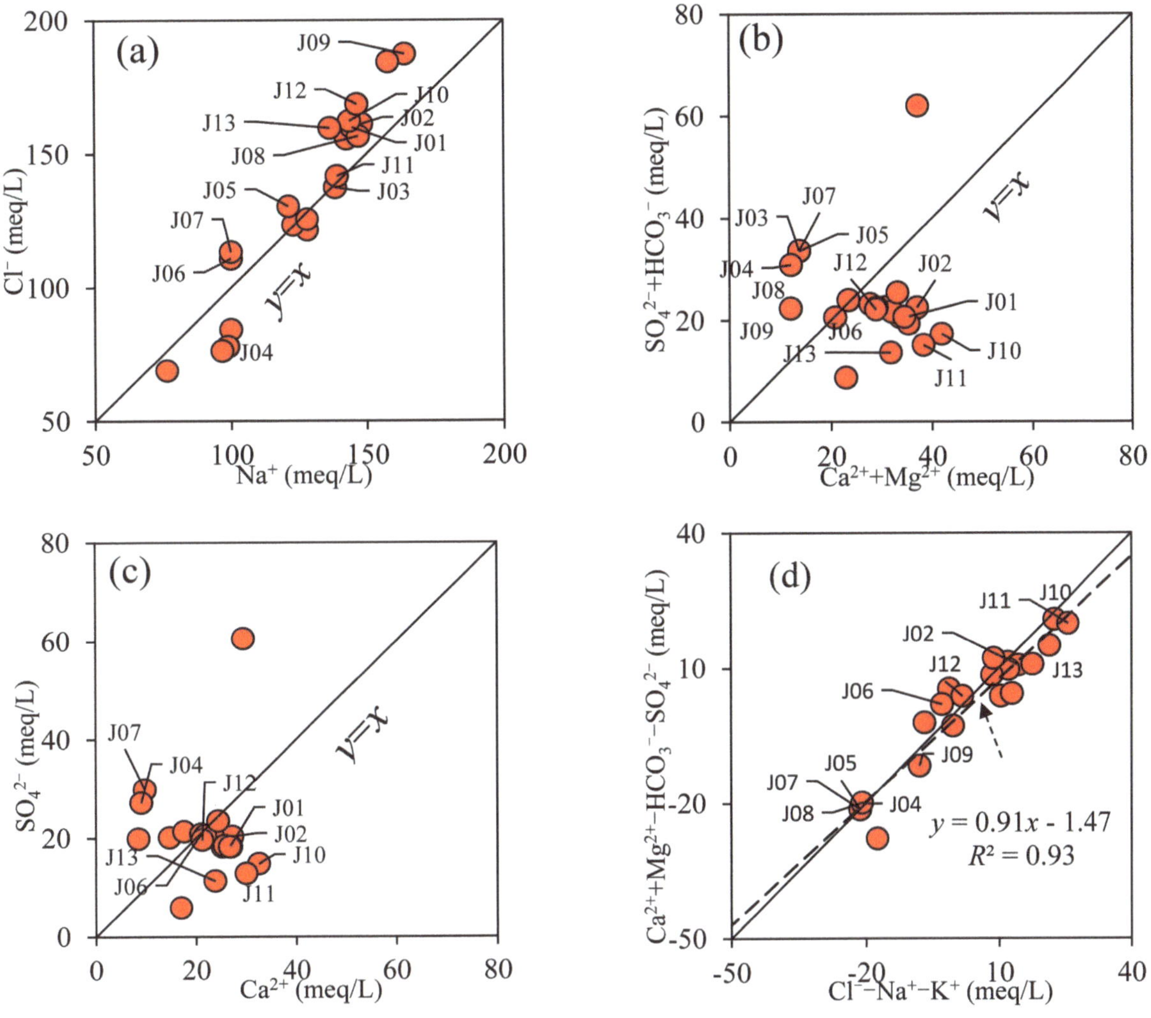


**Figure 8.** Binary scatter plot of major ion ratios in deep GW in northern Jinan. (**a**) Relationship between the ratios of $Na^+$ and $Cl^-$. (**b**) Relationship between the ratios of $Ca^{2+}$+$Mg^{2+}$ and $SO_4^{2-}$+$HCO_3^-$. (**c**) Relationship between the ratios of $Ca^{2+}$ and $SO_4^{2-}$. (**d**) Relationship between the ratios of $Cl^-$-$Na^+$-$K^+$ and $Ca^{2+}$+$Mg^{2+}$-$HCO_3^-$-$SO_4^{2-}$.

If the ratio of $[Ca^{2+} + Mg^{2+}]/[HCO_3^- + SO_4^{2-}]$ is greater than 1 in a water sample, it indicates that silicate weathering plays a significant role in influencing the water chemistry. Conversely, if the ratio is less than 1, it suggests that carbonate weathering is the main controlling factor [22,39]. It can be seen from Figure 8b that 70% of the GW samples of NJR are below the line of $[Ca^{2+} + Mg^{2+}]/[HCO_3^- + SO_4^{2-}] = 1$, indicating that the weathering of silicate rocks is the main controlling factor of the chemical composition of GW, and that the weathering of carbonate rocks also has a certain contribution. In addition, 80% of the GW samples deviate from the distribution of $[Ca^{2+}]/[HCO_3^-] = 1$ line (Figure 8c), and only two samples are close to it, indicating the presence of gypsum dissolution, but it is not the main source of $Ca^{2+}$ and $SO_4^{2-}$ in GW.

The relationship between $[Ca^{2+}+Mg^{2+}-HCO_3^--SO_4^{2-}]$ and $[Cl^--Na^+-K^+]$ is commonly used to identify the occurrence of cation exchange processes in the groundwater system [36,37]. When $[Ca^{2+}+Mg^{2+}-HCO_3^--SO_4^{2-}]/[Cl^--Na^+-K^+]$ equals 1, it signifies that cation exchange is an important hydrogeochemical process influencing the chemical characteristics of water. As depicted in Figure 8d, the GW samples in the NJR are distributed along the $[Ca^{2+}+Mg^{2+}-HCO_3^{-}-SO_4^{2-}]/[Cl^{-}-Na^{+}-K^{+}] = 1$ line, demonstrating a linear relationship between the data points, described by the equation $y = 0.91x - 1.47$ ($R^2 = 0.93$). This confirms that cation exchange is a significant factor impacting the chemical composition of deep GW in the NJR.

#### 4.4.4. Mineral Dissolution Equilibrium

To determine the mineral dissolution and precipitation in GW, the equilibrium state of mineral phases was calculated using PHREEQC software [40], and the saturation index (SI) of the water samples was obtained. Figure 9 presents the statistical results of the SI values for relevant minerals. Overall, the SI values for carbonate minerals such as calcite, dolomite, and aragonite are all above 0, indicating that these minerals are in a supersaturated state and have a tendency to precipitate. On the other hand, the SI values for halite and gypsum are both below 0, indicating that these minerals are not yet in a state of saturation in the GW and will likely continue to dissolve along the water flow path. The SI values for quartz and chalcedony exhibit contrasting characteristics. The SI for quartz is consistently above 0, ranging from 0.16 to 0.28, while the SI for chalcedony is below 0, ranging from −0.18 to −0.06.

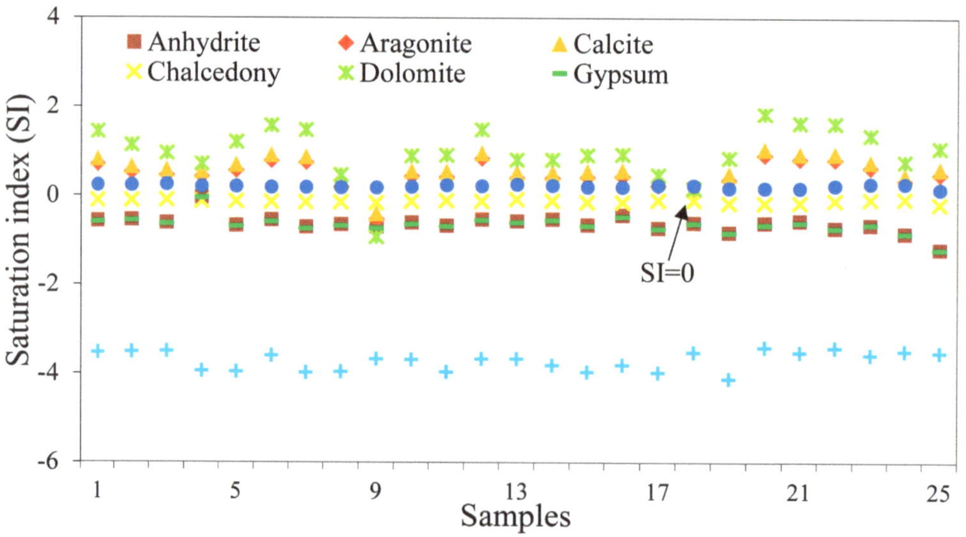


**Figure 9.** SI values for relevant minerals from deep GW in northern Jinan.

### *4.5. Genetic Analysis of Geothermal Field*

The formation of GW is typically associated with various factors, including the cap layer (also known as the insulation layer), thermal storage capacity, heat source, and the source of thermal water recharge.

### *4.6. Geothermal Reservoir*

The primary thermal reservoirs in the NJR consist of the Paleogene Dongying Formation and Guantao Formation. These thermal reservoirs are characterized by lithologies such as coarse sandstone, fine sandstone, and glutenite. The sediment grains exhibit medium roundness and average sorting, with argillaceous cementation. The glutenite shows poor diagenesis and loose properties. Notably, there are well-developed pores and fissures within the reservoir, resulting in a porosity range of 20 to 25%. This configuration provides ample space for efficient heat storage, creating a layered thermal reservoir with a pore-fracture structure. Overall, the thermal reservoir type found throughout the NJR is the Neogene Paleogene porous sandstone reservoir, with a particular focus on the Neogene Guantao Formation for geothermal development in the current stage.

### 4.7. Thermal Reservoir Overlying Strata

The stratum that sits above the thermal reservoir can be referred to as the thermal reservoir overlying strata. In the case of the Dongying Formation, the thermal reservoir overlying strata consists of the Guantao Formation, Minghuazhen Formation, and Quaternary deposits. These strata are primarily characterized by a soft layer composed of cohesive soil and sandy soil. They exhibit low density, considerable thickness, poor thermal conductivity, and significant thermal resistance. As a result, these strata naturally serve as effective thermal reservoir overlying layers.

### 4.8. Heat Source

The NJR primarily derives its heat source from the conduction heat flow of the upper mantle and the normal conduction heat flow of the deep crust. Additionally, the tectonic movement heat source and gravity compression heat source play significant roles and should not be disregarded. During the Mesozoic Yanshan Movement and Cenozoic Himalayan Movement, the region experienced the formation of multiple fault levels, accompanied by periods of magma intrusion. As a result, lava flows emerged along these faults, forming multilayer basic eruptive rocks. Over time, the area accumulated a substantial thickness of Cenozoic strata. Throughout geological history, the Cenozoic strata underwent compaction and diagenesis processes, releasing a substantial amount of heat energy. These various heat sources contribute to the overall heat stored within the thermal reservoir. The overlying thick layers of loose sediment act as thermal insulation, effectively preserving and retaining the heat within the reservoir.

### 4.9. Recharge Source of GW

The GW in the NJR is primarily replenished through lateral runoff from both nearby and distant mountainous areas during a long geological period following sediment deposition. While some sedimentary water and storage water remain trapped during the sedimentation process, the majority of the GW is recharged by runoff from the recharge areas. Based on regional paleogeographic conditions, these recharge areas could be located in the Taiyi mountains to the south or the Taihang mountains to the west. Once atmospheric precipitation infiltrates vertically, it undergoes a deep circulation within the aquifer along the horizontal direction. This recharge process is conducive to the replenishment of GW within the reservoir [41]. The conceptual model diagram illustrating the geothermal genesis for NJR is presented in Figure 10. This conceptual model of geothermal genesis highlights the contribution of lateral runoff from mountainous areas as the primary source of recharge for the GW in the NJR.

### 4.10. Conceptual Model of Geothermal Genesis

The convergence of atmospheric precipitation in the southern and western mountainous areas leads to the formation of surface currents. Over an extended geological timeframe, driven by the energy from head differences, this water migrates deep into the subsurface through fracture zones or the pores of rock formations. As it travels, it undergoes heating from the surrounding rocks and engages in complex water–rock reactions. The heated groundwater, influenced by the density differences caused by the temperature changes, contributes to natural convection. Along with the driving force from the head differences in the recharge area, this convection allows the groundwater to circulate slowly and alternately move (Figure 10). As it moves, the GW is stored within the pores and fissures, forming reservoirs within the geological formations. This process involves a combination of heat transfer, water–rock interactions, and groundwater movement, leading to the accumulation of GW in the area.

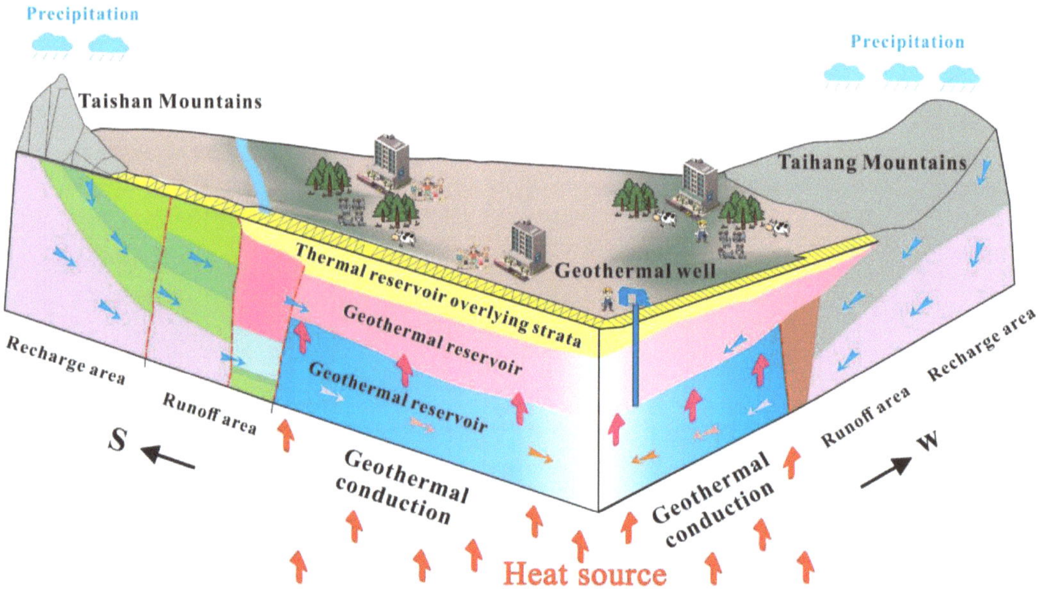


**Figure 10.** Conceptual model of geothermal field in the NJR.

## 5. Conclusions

Jinan possesses abundant geothermal resources, offering significant potential for development and utilization. Previous studies on geothermal resources in Jinan city have mainly focused on the southern section of the Qihe–Guangrao fault. Based on the geothermal geological conditions in the NJR, the chemical characteristics and formation mechanism of GW in the north of Jinan are analyzed by means of hydrochemistry and isotopes. Subsequently, a conceptual model of thermal storage is proposed. This provides a scientific basis and reference for the rational development and utilization of geothermal resources in Jinan city. The main findings and conclusions of this investigation are outlined below:

(1) The GW in the NJR exhibits relatively high salinity, with TDS content ranging from 5586.20 to 12,156.00 mg/L, and an average value of 9009.00 mg/L. The predominant ions in GW are $Na^+$ and $Cl^-$, with mean concentrations of 2829.73 mg/L and 4425.77 mg/L, respectively, resulting in the hydrochemical classification of GW as a $Cl^-$Na type. The cationic composition follows the order of $Na^+ > Ca^{2+} > Mg^{2+} > K^+$, while the anionic composition displays the sequence of $Cl^- > SO_4{}^{2-} > HCO_3{}^- > NO_3{}^-$.

(2) The $\delta^2H$ and $\delta^{18}O$ values of the GW ranged from −69.96 to −75.98 and −7.94 to −9.25, respectively, with mean values of −72.61 and −8.72, respectively. Analysis of the $\delta^2H$ and $\delta^{18}O$ isotopes suggests that the GW in the NJR originates from atmospheric precipitation that has undergone deep circulation and is mixed with older groundwater. Additionally, the composition of $^3H$ supports the notion that the GW is a result of water mixing. Furthermore, based on $^{14}C$ dating, it is inferred that the recharge of the GW likely occurred during the late Pleistocene period.

(3) The Na-K-Mg Giggenbach diagram reveals that all GWs in the NJR are partially equilibrated, indicating that the fluids and minerals have not achieved complete chemical equilibrium. There is significant variation in the temperature estimates obtained from different geothermometers. In this study, the quartz geothermometer is deemed more suitable for estimating the temperature of the geothermal reservoir, which ranges from 57.13 to 99.74 °C.

(4) In the long circulation process, GW undergoes continuous evaporation and leaching, resulting in its high salinity characteristics. The main sources of the chemical compo-

nents of the GW in the NJR are water–rock interactions, such as cation exchange, silicate weathering, carbonate weathering, and gypsum and halite dissolution. Calculations indicate that the SI of carbonate minerals is greater than 0, indicating oversaturation, while the SI values of halite and gypsum are both less than 0, indicating undersaturation.

(5) Based on the hydrogeological conditions, hydrochemistry, and isotope analysis of the NJR, a conceptual model of the geothermal field has been established. The GW is believed to originate from either Taiyi mountain in the south or Taihang mountain in the west, and thus experiences deep circulation and prolonged runoff time, contributing to the high salinity of the GW. These research findings provide valuable references and a foundation for the sustainable development and utilization of the regional geothermal resources in the area.

**Supplementary Materials:** The following supporting information can be downloaded at: https://www.mdpi.com/article/10.3390/en16227658/s1, Table S1: Data.

**Author Contributions:** Conceptualization, Z.G. and J.L.; methodology, M.H. and J.L.; formal analysis, M.H., Q.L., M.T. and Y.N.; writing—original draft preparation, M.H. and J.L.; writing—review and editing, Z.G. and J.L.; resources, Z.G. and M.H. All authors have read and agreed to the published version of the manuscript.

**Funding:** This research was funded by the Natural Science Foundation of Shandong Province (ZR2020MD109, ZR2021QD048).

**Institutional Review Board Statement:** Not applicable.

**Informed Consent Statement:** Not applicable.

**Data Availability Statement:** The data presented in this study are available in Supplementary Material Table S1.

**Conflicts of Interest:** The authors declare no conflict of interest.

## References

1. Zhang, X.; Hu, Q. Development of geothermal resources in China: A Review. *J. Earth Sci.* **2018**, *29*, 452–467. [CrossRef]
2. Lund, J.W.; Toth, A.N. Direct utilization of geothermal energy 2020 worldwide review. *Geothermics* **2020**, *90*, 101915. [CrossRef]
3. Zhao, X.-G.; Wan, G. Current situation and prospect of China hebrew's geothermal resources. *Renew. Sustain. Energy Rev.* **2014**, *32*, 651–661. [CrossRef]
4. Li, X.; Qi, J.; Yi, L.; Xu, M.; Zhang, X.; Zhang, Q.; Tang, Y. Hydrochemical characteristics and evolution of geothermal waters in the eastern Himalayan syntaxis geothermal field, southern Tibet. *Geothermics* **2021**, *97*, 102233. [CrossRef]
5. Duan, R.; Li, P.; Wang, L.; He, X.; Zhang, L. Hydrochemical characteristics, hydrochemical processes and recharge sources of the geothermal systems in Lanzhou City, northwestern China. *Urban Clim.* **2022**, *43*, 101152. [CrossRef]
6. Ersoy, A.F.; Sönmez, S. Hydrogeochemical and isotopic characteristics of the Ilica geothermal system (Erzurum, Turkey). *Environ. Earth Sci.* **2014**, *72*, 4451–4462. [CrossRef]
7. Chenaker, H.; Houha, B.; Vincent, V. Hydrogeochemistry and geothermometry of thermal water from north-eastern Algeria. *Geothermics* **2018**, *75*, 137–145. [CrossRef]
8. Alçiçek, H.; Bülbül, A.; Brogi, A.; Liotta, D.; Ruggieri, G.; Capezzuoli, E.; Meccheri, M.; Yavuzer, I.; Alçiçek, M.C. Origin, evolution and geothermometry of the thermal waters in the Gölemezli Geothermal Field, Denizli Basin (SW Anatolia, Turkey). *J. Volcanol. Geotherm. Res.* **2018**, *349*, 1–30. [CrossRef]
9. Wang, X.; Lu, G.; Hu, B.X. Hydrogeochemical Characteristics and Geothermometry Applications of Thermal Waters in Coastal Xinzhou and Shenzao Geothermal Fields, Guangdong, China. *Geofluids* **2018**, *2018*, 8715080. [CrossRef]
10. Wang, X.; Zhou, X.; Zheng, Y.; Song, C.; Long, M.; Chen, T.; Ren, Z.; Yang, M.; Li, X.; Guo, J. Hydrochemical characteristics and mixing behavior of thermal springs along the Bijiang River in the Lanping basin of China. *Environ. Earth Sci.* **2017**, *76*, 487. [CrossRef]
11. Chiocchini, U.; Castaldi, F.; Barbieri, M.; Eulilli, V. A stratigraphic and geophysical approach to studying the deep-circulating groundwater and thermal springs, and their recharge areas, in Cimini Mountains–Viterbo area, central Italy. *Hydrogeol. J.* **2010**, *18*, 1319–1341. [CrossRef]
12. Belhai, M.; Fujimitsu, Y.; Bouchareb-Haouchine, F.Z.; Haouchine, A.; Nishijima, J. A hydrochemical study of the Hammam Righa geothermal waters in north-central Algeria. *Acta Geochim.* **2016**, *35*, 271–287. [CrossRef]
13. Başaran, C.; Gökgöz, A. Hydrochemical and isotopic properties of Heybeli geothermal area (Afyon, Turkey). *Arab. J. Geosci.* **2016**, *9*, 586. [CrossRef]

14. Wang, J.; Jin, M.; Jia, B.; Kang, F. Hydrochemical characteristics and geothermometry applications of thermal groundwater in northern Jinan, Shandong, China. *Geothermics* **2015**, *57*, 185–195. [CrossRef]
15. Sui, H.; Kang, F.; Li, C.; Han, J.; Xing, L. Relationship between north Ji'nan geothermal water and Ji'nan spring water revealed by hydrogeochemical characteristics. *Carsologica Sin.* **2017**, *36*, 49–58.
16. Kang, F.; Jin, M.; Qin, P. Sustainable yield of a karst aquifer system: A case study of Jinan springs in northern China. *Hydrogeol. J.* **2011**, *19*, 851–863. [CrossRef]
17. Wang, J.; Jin, M.; Lu, G.; Zhang, D.; Kang, F.; Jia, B. Investigation of discharge-area groundwaters for recharge source characterization on different scales: The case of Jinan in northern China. *Hydrogeol. J.* **2016**, *24*, 1723–1737. [CrossRef]
18. Li, C.; Wu, X.; Sun, B.; Sui, H.; Geng, F.; Qi, H.; Ma, X. Hydrochemical characteristics and formation mechanism of geothermal water in northern Ji'nan. *Earth Sci.* **2018**, *43*, 313–325.
19. Kang, F.; Sui, H.; Zheng, T. Heat accumulation and water enrichment mechanism of piedmont karstic geothermal reservoirs: A case study of northern Jinan. *Acta Geol. Sin.* **2020**, *94*, 1606–1624. [CrossRef]
20. Lei, T.; Liang, F.; Han, C.; Wang, Z.; Li, J.; Ma, Y.; Yao, H. Ambient Noise Tomography of Jinan: The Migration of Groundwater and the Formation of Geothermal Water. *Geol. J.* **2022**, *96*, 1716–1728. [CrossRef]
21. Piper, A.M. A graphic procedure in the geochemical interpretation of water-analyses. *Trans.-Am. Geophys. Union* **1944**, *25*, 914–923. [CrossRef]
22. Liu, J.; Peng, Y.; Li, C.; Gao, Z.; Chen, S. Characterization of the hydrochemistry of water resources of the Weibei Plain, Northern China, as well as an assessment of the risk of high groundwater nitrate levels to human health. *Environ. Pollut.* **2021**, *268*, 115947. [CrossRef]
23. Craig, H. Isotopic Variations in Meteoric Waters. *Science* **1961**, *133*, 1702–1703. [CrossRef]
24. Liu, J.; Song, X.; Yuan, G.; Sun, X.; Liu, X.; Wang, S. Characteristics of δ 18O in precipitation over Eastern Monsoon China and the water vapor sources. *Chin. Sci. Bull.* **2010**, *55*, 200–211. [CrossRef]
25. Mao, X.; Zhu, D.; Ndikubwimana, I.; He, Y.; Shi, Z. The mechanism of high-salinity thermal groundwater in Xinzhou geothermal field, South China: Insight from water chemistry and stable isotopes. *J. Hydrol.* **2021**, *593*, 125889. [CrossRef]
26. Fournier, R. Chemical geothermometers and mixing models for geothermal systems. *Geothermics* **1977**, *5*, 41–50. [CrossRef]
27. Clark, I.D.; Fritz, P. *Environmental Isotopes in Hydrology*; Lewis Publishers: Boca Raton, FL, USA, 1997. [CrossRef]
28. Arnórsson, S.; Gunnlaugsson, E.; Svavarsson, H. The chemistry of geothermal waters in Iceland. III. Chemical geothermometry in geothermal investigations. *Geochim. Cosmochim. Acta* **1983**, *47*, 567–577. [CrossRef]
29. Nieva, D.; Nieva, R. Developments in geothermal energy in Mexico—Part twelve. A cationic geothermometer for prospecting of geothermal resources. *Heat Recover. Syst. CHP* **1987**, *7*, 243–258. [CrossRef]
30. Giggenbach, W.; Gonfiantini, R.; Jangi, B.; Truesdell, A. Isotopic and chemical composition of parbati valley geothermal discharges, North-West Himalaya, India. *Geothermics* **1983**, *12*, 199–222. [CrossRef]
31. Fournier, R.; Truesdell, A. An empirical Na-K-MG-Ca geothermometer for natural waters. *Geochim. Cosmochim. Acta* **1973**, *37*, 1255–1275. [CrossRef]
32. Giggenbach, W.F. Geothermal solute equilibria. Derivation of Na-K-Mg-Ca geoindicators. *Geochim. Cosmochim. Acta* **1988**, *52*, 2749–2765. [CrossRef]
33. El Maghraby, M.M. Hydrogeochemical characterization of groundwater aquifer in Al-Madinah Al-Munawarah City, Saudi Arabia. *Arab. J. Geosci.* **2015**, *8*, 4191–4206. [CrossRef]
34. Aksever, F.; Büyükşahin, S. Erratum to: Assessment of variations in water quality using statistical techniques: A case study of Işıklı Lake, Çivril/Denizli, Turkey. *Arab. J. Geosci.* **2017**, *10*, 143. [CrossRef]
35. Han, D.; Liang, X.; Jin, M.; Currell, M.; Song, X.; Liu, C. Evaluation of groundwater hydrochemical characteristics and mixing behavior in the Daying and Qicun geothermal systems, Xinzhou Basin. *J. Volcanol. Geotherm. Res.* **2010**, *189*, 92–104. [CrossRef]
36. Gueroui, Y.; Maoui, A.; Touati, H.; Guettaf, M.; Bousbia, A. Hydrogeochemical and environmental isotopes study of the northeastern Algerian thermal waters. *Environ. Earth Sci.* **2018**, *77*, 747. [CrossRef]
37. Gibbs, R.J. Mechanisms Controlling World Water Chemistry. *Science* **1970**, *170*, 1088–1090. [CrossRef]
38. Zhao, Q.; Luo, Z.; Du, J.; Wang, L.; Li, Z.; Wang, X. Hydrogeochemical characteristics and genesis of geothermal water in Xiaoyangkou, Nantong. *China Environ. Earth Sci.* **2023**, *82*, 456. [CrossRef]
39. Liu, J.; Gao, Z.; Wang, Z.; Xu, X.; Su, Q.; Wang, S.; Qu, W.; Xing, T. Hydrogeochemical processes and suitability assessment of groundwater in the Jiaodong Peninsula, China. *Environ. Monit. Assess.* **2020**, *192*, 384. [CrossRef]
40. Liu, J.; Wang, M.; Gao, Z.; Chen, Q.; Wu, G.; Li, F. Hydrochemical characteristics and water quality assessment of groundwater in the Yishu River basin. *Acta Geophys.* **2020**, *68*, 877–889. [CrossRef]
41. Dallai, L.; Magro, G.; Petrucci, E.; Ruggieri, G. Stable isotope and noble gas isotope compositions of inclusion fluids from Larderello geothermal field (Italy): Constraints to fluid origin and mixing processes. *J. Volcanol. Geotherm. Res.* **2005**, *148*, 152–164. [CrossRef]

*water*

MDPI

*Article*

# Heavy Metal Content Characteristics and Pollution Source Analysis of Shallow Groundwater in Tengzhou Coal Mining Area

**Beibei Yan [1], Qianqian Wei [1,*], Xinfeng Li [1], Xiaoyu Song [2,*], Zongjun Gao [2], Jiutan Liu [2], Ruilin Zhang [2] and Min Wang [2]**

1 Geophysical Prospecting and Surveying Team, Shandong Bureau of Coal Geological, Jinan 250102, China; ybbsdust@126.com (B.Y.); lixinfeng1@126.com (X.L.)
2 College of Earth Science and Engineering, Shandong University of Science and Technology, Qingdao 266590, China; zongjungao1964@163.com (Z.G.); ljtsdust@sdust.edu.cn (J.L.); gltyyan@163.com (R.Z.); brightwangm@163.com (M.W.)
* Correspondence: 18560232967@163.com (Q.W.); 17863961381@163.com (X.S.); Tel.: +86-18560232967 (Q.W.)

**Abstract:** This study analyzed the sources of total metal elements using the positive matrix factorization (PMF) model and conducted human health risk assessment for adults and children using the health risk assessment model recommended by the United States Environmental Protection Agency (USEPA). According to the health risk assessment, As is the main contributor to the non-carcinogenic risk of groundwater in Tengzhou, with drinking water as the main exposure route. Regarding carcinogenic risks (CR), the values of As and Cr for adults and children were higher than $1 \times 10^{-4}$, with drinking water as the main exposure route. Therefore, As is the largest contributor to the CR of groundwater for adults and children and drinking water is the main exposure route in the study area. The primary exposure pathways are oral intake and dermal contact, with oral intake presenting a significant risk. The carcinogenic risks according to principal component analysis (PCA) and PMF analysis showed that the main sources of heavy metals in shallow groundwater in Tengzhou City are agricultural, industrial, natural, and industrial deposition sources, with contribution rates of 21.7%, 27.2%, 31.0%, and 20.1%, respectively. In particular, natural sources are the largest contributor to the accumulation of heavy metals.

**Keywords:** health risk assessment; PMF; shallow ground water; source apportionment

**Citation:** Yan, B.; Wei, Q.; Li, X.; Song, X.; Gao, Z.; Liu, J.; Zhang, R.; Wang, M. Heavy Metal Content Characteristics and Pollution Source Analysis of Shallow Groundwater in Tengzhou Coal Mining Area. *Water* **2023**, *15*, 4091. https://doi.org/10.3390/w15234091

Academic Editor: Andrea G. Capodaglio

Received: 26 October 2023
Revised: 22 November 2023
Accepted: 22 November 2023
Published: 25 November 2023



## 1. Introduction

Groundwater is an important natural resource for human survival and development. At present, China is facing the problem of groundwater pollution. According to the "Report on the State of the Ecology and Environment in China 2020", among the 10,171 groundwater quality monitoring points installed by the natural resources department, Class IV water accounted for 68.8% of the total monitoring points, and Class V water accounted for 17.6% [1]. However, with the rapid development of modern industry and agriculture, the quality of urban groundwater is deteriorating. In particular, heavy metals are the main pollutants in groundwater. Heavy metals such as As, Cr, Hg, and Pb enter the human body through drinking water and skin contact. They can interact strongly with proteins and various enzymes in the human body, making them inactive. They can also be enriched in certain organs of the human body as they are resistant to degradation. When their levels exceed the tolerance limit of the human body, various forms of poisoning, such as acute poisoning, subacute poisoning, and chronic poisoning, will occur, causing great harm to human health [2].

In recent years, with continuous attention being paid to the prevention and control of urban groundwater pollution and the strengthening of people's awareness regarding the health risks of pollutants, the health risk assessment of heavy metals in urban water resources has become a hot topic in urban environmental research. At present, an increasing number of studies are paying attention to the pollution source analysis and health risk assessment of heavy metals in groundwater [3]. The sources of heavy metals in groundwater can be divided into geological background and anthropogenic sources. The geological background sources include the weathering of parent rocks and soil formation, and the anthropogenic sources include metal smelting, sewage irrigation, industrial discharge, and pesticide and fertilizer application [4]. In recent years, the source apportionment of heavy metals in groundwater has primarily been conducted using multivariate statistical analysis and model analysis. Multivariate statistical analysis mainly includes correlation, principal component, and factor analysis, while model analysis mainly includes chemical mass balance (CMB) and absolute principal component. Moreover, multivariate linear regression (APCS-MLR) and positive definite matrix factorization (PMF) have also been applied [5–7].

Tengzhou City is located in the central and southern part of Shandong Province. This city is rich in mineral resources, of which coal is the pillar industry of local economic development. In this study, 24 well points were selected to determine the contents of As, Hg, Cd, Co, Cr, Cu, Ni, Zn, pH, TDS, Th, and COD in the groundwater. The PMF model was used to analyze the sources of total metal elements, and the health risk assessment model recommended by the United States Environmental Protection Agency (USEPA) was used to carry out human health risk assessment for adults and children. The findings are expected to provide a reference for formulating and implementing safety measures regarding groundwater supply and pollution prevention in Tengzhou City [8–10].

## 2. Materials and Methods

### 2.1. Study Area

Tengzhou City is located in the south of Shandong Province (34°50′–35°17′ N, 116°49′–117°24′ E). It is a famous energy base and industrial city in Shandong Province, China, with an area of 1495 km$^2$ and a population of 1.75 million. It borders Zoucheng City in the north, Shanting District in the east, Xuecheng District in the south, Weishan Lake in the west, and Weishan County. The study area is located in the southern part of the warm temperate semi-humid area. The area experiences a prominent monsoon continental climate. Affected by atmospheric circulation, the four seasons are distinct. Springs are windy and rainy, summers are rainy and hot, autumns are sunny and cool, and winters are cold and dry. The annual average temperature is 14.5 °C, average annual precipitation is approximately 695 mm, and average relative humidity is 68%. The terrain is tilted from northeast to southwest, surrounded by mountains on three sides and faces Nansi Lake towards the west. This area features four soil types, among which fluvo-aquic soil and cinnamon soil account for the largest surface area coverage. The study area is divided into six hydrogeological units: Tengxi Plain, Dangshan Fault Terrace, Fushan Fault Block, Jingquan Fault Block, Guanqiao Fault Block, and Yangzhuang Basin. According to the burial characteristics and hydraulic characteristics of groundwater in the study area, the water-bearing rock groups can be divided into three main types. The first type is the Quaternary loose rock group with water-bearing pores, which is distributed throughout the area. The lithology is mainly clay, sub-clay, and gravel, the water type is mainly $HCO_3$–Ca, and the salinity is generally about 0.3 g/L. The second type is carbonate rock and clastic rock group with water-bearing fissures, which is distributed in the northeast of the study area. The third type is the Ordovician and Cambrian carbonate rock group with water-bearing fractures, which is mainly distributed in Yangzhuang, Weizhuang, Houshiwan Village, and Longshantou [11]. Atmospheric precipitation is the main source of supply, and draining is mainly facilitated by artificial mining and natural processes. Tengzhou is an important political, economic, cultural, scientific, educational, and commercial center in southern Shandong Province. Coal and limestone are the main mineral resources of the city. The development and utilization of these mineral resources have made great contributions to the economic and social development of the city.

### 2.2. Sampling and Analysis

According to the hydrogeological conditions of the study area, a monitoring well was set up at the site in February 2022, and a total of 24 groundwater samples were collected. The groundwater sampling points are shown in Figure 1. The sample collection was carried out according to the requirements of the "Technical Guidelines for Sampling Volatile Organic Compounds in Soil and Groundwater" (HJ 1019–2019) [12]. The water samples were collected in a 500 mL polyethylene bottle (Tai'an Xinming Plastic Co., Ltd., Tai'an, China). The sampling bottle was rinsed with distilled water before sampling. After sampling, 2.5 mL nitric acid (1:1) (Shandong Xinhao Chemical Co., Ltd., Zibo, China) was added to the sample bottle to ensure that the pH of the sample was below 2.0, and the sample was then sealed with a white polytetrafluoroethylene bottle cap. After completing sample collection, all samples were sent to the experimental test center of the Shandong Lunan Geological Engineering Exploration Institute for immediate testing. The standard "Water quality-Determination of 65 elements-Inductively coupled plasma-mass spectrometry" (HJ700–2014) was applied for the detection of Cd, Cr, Cu, Ni, Pb, and Zn, and the test instrument was Thermo Fisher iCAP-RQ(Thermo Fisher Scientific Co., Ltd., Waltham, MA, America) [13]. The standard "Water Quality–Determination of Mercury, Arsenic, Selenium, Bismuth and Antimony–Atomic Fluorescence Spectrometry" (HJ694–2014) was applied for the detection of Hg and As, and the test instrument was PF5-2 atomic fluorescence spectrometer(Beijing General Analytical Instrument Co., Ltd., Beijing, China) [14]. The detection limits of As, Cd, Co, Cr, Cu, Hg, Ni, and Zn were 0.12, 0.01, 0.03, 0.11, 0.08, 0.01, 0.06, and 0.67 ug/L, respectively. The standard deviation of the test results for all elements was less than 15%, and the recovery rate was about 90%.

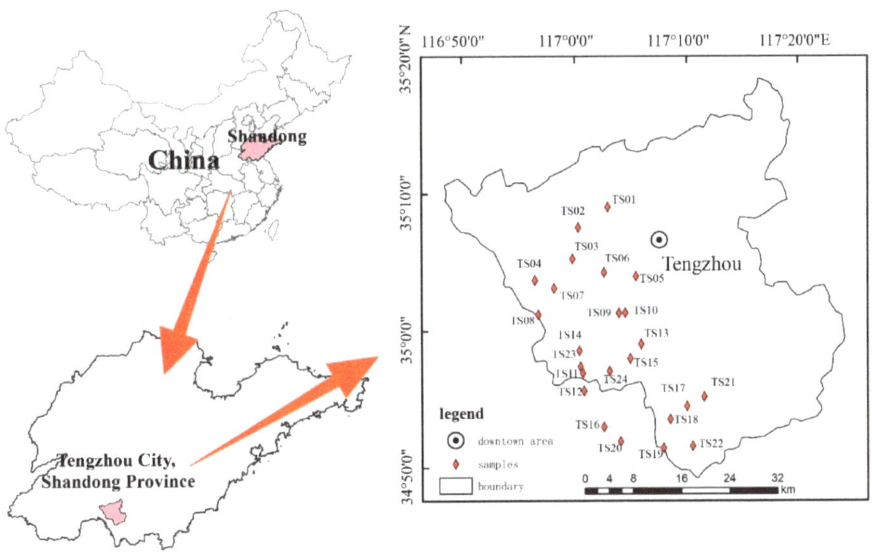


**Figure 1.** Groundwater sampling sites in Tengzhou.

### 2.3. Water Quality Evaluation

The single-factor pollution index and Nemerow index were used to evaluate the heavy metal pollution, and to analyze and discriminate the risk degree of heavy metals in groundwater in Tengzhou City. The Nemerow index method can comprehensively reflect the pollution of each risk element in a water body [15]. Therefore, it has been widely used in China and abroad for calculating the comprehensive pollution index. The calculation formula is as follows:

$$P_i = C_i / H_i \tag{1}$$

$$P_n = \sqrt{\left(P_{imax}^2 + P_{iave}^2\right)/2} \quad (2)$$

In the formula, $P_i$ is the single-factor pollution index of heavy metals in category I; $C_i$ is the measured value of heavy metals in category I; $H_i$ is the weight of type i. The metal water quality standard adopts the Class III water quality standard in the "Standard for Groundwater Quality" (GB14848–2017) as the evaluation standard; $P_{imax}$ is the maximum value of the single-factor pollution index; $P_{iave}$ is the average of the single-factor pollution index; the obtained $P_n$ is classified according to the Nemerow comprehensive pollution classification standard [16,17].

## 2.4. Source Analysis

### 2.4.1. Principal Component Analysis

Principal component analysis (PCA) is a statistical method used to simplify datasets through dimensionality reduction. It can transform a complex dataset containing many variables into a simpler dataset containing a few principal components. These principal components can retain the variation information in the original data as much as possible, such that the data can be analyzed without losing too much information. PCA statistics include the covariance matrix, eigenvalue, principal component score, and cumulative contribution rate. Based on PCA, this study further applied PMF to analyze the sources of heavy metals, thus increasing the reliability of the PMF results [18].

### 2.4.2. Positive Matrix Factorization

Positive matrix factorization (PMF) is a receptor model analysis method. It separates the mixed effects of multiple sources by statistically analyzing the variability among indicators, such that the contribution of each source to the indicators can be determined. The specific formula is as follows:

$$\mathbf{X}_{ij} = \sum_{k=1}^{p} \mathbf{g}_{ik}\mathbf{f}_{kj} + \mathbf{e}_{ij} \quad (3)$$

In the formula, $\mathbf{X}_{ij}$ is the concentration of the jth element in the ith sample; $\mathbf{g}_{ik}$ is the contribution of source k to the first i sample, that is, the contribution rate matrix of the source; $\mathbf{f}_{kj}$ is the concentration of the jth element in the source k, that is, the source component spectrum matrix; $\mathbf{e}$ is the residual matrix; p is the number of pollution data sources.

The factor contribution and distribution are determined by minimizing the objective function Q, and the specific formula is as follows:

$$Q = \sum_{i=1}^{n} \sum_{i=1}^{m} \left(\mathbf{e}_{ij}/\mathbf{u}_{ij}\right) \quad (4)$$

In the formula, i represents the sample serial number; j represents the element serial number; k denotes the sequence number of the contamination data source; n is the total number of samples; m is the total number of elements; $\mathbf{u}_{ij}$ is the uncertainty of heavy metals. The uncertainty formula is as follows:

$$\mathbf{u}_{ij} = \begin{cases} \frac{5}{6}\text{MDL} & C \leq \text{MDL} \\ \sqrt{(\text{EF} \times C)^2 + (0.5 \times \text{MDL})^2}, & C > \text{MDL} \end{cases} \quad (5)$$

In the formula, MDL represents the detection limit of heavy metals in the sample; C is the measured value of heavy metals in the sample (mg/L); EF is the relative uncertainty [19].

## 2.5. Health Risk Assessment

The risks of heavy metals in groundwater to human health are primarily facilitated by oral intake and dermal contact, and 90% of pollutants enter the human body through these two pathways. In this study, the health risk assessment model recommended by the United States Environmental Protection Agency (USEPA) was used to evaluate the health risk of heavy metal elements in the groundwater of the study area. The specific calculation formula of heavy metal exposure dose in soil is as follows:

$$ADI_{oral} = \frac{C \times R_{oral} \times EF \times T}{BW \times AT} \quad (6)$$

$$ADI_{der} = \frac{C \times SA \times AF \times ABS \times EF \times T}{BW \times AT} \quad (7)$$

In the formula, the subscripts oral and der represent oral intake and dermal contact, respectively; C represents the content of heavy metal elements; R represents the intake rate; EF represents the exposure frequency; T represents the exposure time of the population; BW represents the average weight of the population; AT represents the average total exposure time; SA represents the area of skin exposure; AF represents the adhesion factor of heavy metal elements; ABS represents the skin absorption factor. The relevant values are shown in Tables 1 and 2 [20].

**Table 1.** Related parameters of the health risk assessment model [19,20].

| Parameter | Reference | | Unit |
|---|---|---|---|
| | Adults | Children | |
| R | 1.7 | 1.14 | $L \times d^{-1}$ |
| EF | 350 | 350 | $d \times a^{-1}$ |
| T | 30 | 9 | a |
| SA | 16,000 | 9300 | $cm^2$ |
| AF | 0.001 | 0.001 | $L \times cm^{-3}$ |
| BW | 57 | 24 | kg |
| AT | 25,500 | 25,500 | d |

**Table 2.** ABS, RFD, and SF values for the health risk assessment model [19,20].

| Heavy Metals | $ABS \times 10^{-3}/cm \times d^{-1}$ | | $RFD/mg \times (kg \times d)^{-1}$ | | $SF/mg \times (kg \times d)^{-1}$ | |
|---|---|---|---|---|---|---|
| | Adults | Children | Oral | Der | Oral | Der |
| As | 1.13994 | 0.75006 | 0.0003 | 0.0001 | 1.5 | 3.66 |
| Cd | 0.6333 | 0.4167 | 0.0005 | 0.0005 | 6.1 | 0.36 |
| Co | 0.25332 | 0.16668 | 0.02 | 0.016 | | |
| Cr | 1.2666 | 0.8334 | 0.003 | 0.003 | 0.5 | 0.5 |
| Cu | 0.37998 | 0.25002 | 0.04 | 0.012 | | |
| Hg | 1.13994 | 0.75006 | 0.0003 | 0.0013 | | |
| Ni | 0.06333 | 0.04167 | 0.02 | 0.0054 | | |
| Zn | 0.37998 | 0.25002 | 0.3 | 0.01 | | |

According to the pollutants detected in groundwater, the health risks are divided into two categories: non-carcinogenic health risks (HI) and carcinogenic health risks (CR). The calculation formula is as follows:

$$HI = \sum_{i=1}^{n} \left( \frac{ADI_{oral}}{RfD_{oral}} + \frac{ADI_{der}}{RfD_{der}} \right)_i \quad (8)$$

$$CR = \sum_{i=1}^{n} (ADI_{oral} \times SF_{oral} + ADI_{der} \times SF_{der})_i \quad (9)$$

In the formula, RfD represents the reference measurement of each exposure pathway; SF represents the carcinogenic slope factor of each pathway. The relevant values are shown in Table 2.

According to the USEPA classification, for non-carcinogenic risk, HI < 1 indicates a lower non-carcinogenic risk, and HI > 1 indicates potential adverse effects. For carcinogenic risk, if the HI value is lower than $10^{-6}$, the pollutant is considered to have no carcinogenic risk. If the HI value is $10^{-6}$–$10^{-4}$, the pollutant is considered to have a carcinogenic risk tolerable by the human body. If the HI value is greater than $10^{-4}$, the pollutant is considered to pose dangerous carcinogenic risk to the human body [21].

## 3. Results and Discussion

### 3.1. Levels of Major Parameters

In this study, a total of 24 groundwater samples were collected within Tengzhou. Specifically, eight heavy metals were tested, namely As, Cd, Co, Cr, Cu, Hg, Ni, and Zn. The conventional indicators tested included pH, TDS, Th, and COD. The statistical results of the measured data are shown in Table 3. Among the twenty-four samples, the Cd contents of fourteen samples and the Hg contents of five samples were below the detection limits. Nevertheless, heavy metals were detected in the remaining samples. According to the limit values of Class III water in the "Standard for Groundwater Quality" (GB14848-2017) [16], the contents of As, Cd, Co, Cr, Cu, Hg, and Zn were within the permissible limits. Although the average content of Ni was less than the standard limit, the Ni content of one sample exceeded the standard, reflecting some pollution. Among the tested heavy metals, the coefficient of variation of Cd, Cu, and Zn exceeded 1.0, indicating large differences in water quality among the groundwater samples.

**Table 3.** Statistical characteristics of heavy metals and conventional indicators of samples.

| Index | As | Hg | Cr | Cd | Co | Ni | Cu | Zn | pH | TDS | Th | COD |
|---|---|---|---|---|---|---|---|---|---|---|---|---|
| Min | 1.16 | 0.005 | 0.67 | 0.005 | 0.51 | 4.43 | 0.14 | 3.20 | 7.64 | 366 | 269.5 | 0.78 |
| Max | 2.13 | 0.056 | 3.61 | 0.167 | 2.54 | 21.36 | 3.13 | 194.5 | 8.38 | 1972 | 1258.5 | 2.82 |
| Median | 1.54 | 0.025 | 1.545 | 0.005 | 1.18 | 9.735 | 0.37 | 10.1 | 8.04 | 770 | 550.15 | 0.92 |
| Mean | 1.53 | 0.02 | 1.79 | 0.03 | 1.23 | 10.19 | 0.56 | 26.31 | 8.05 | 827.75 | 573.92 | 1.01 |
| SD | 0.18 | 0.02 | 0.75 | 0.05 | 0.50 | 4.22 | 0.70 | 40.10 | 0.20 | 377.97 | 228.63 | 0.40 |
| CV | 0.12 | 0.63 | 0.42 | 1.74 | 0.41 | 0.41 | 1.25 | 1.52 | 0.02 | 0.46 | 0.40 | 0.40 |
| Skewness | 0.99 | 0.37 | 1.05 | 2.29 | 0.93 | 1.03 | 3.12 | 3.50 | −0.22 | 1.80 | 1.61 | 4.26 |
| Kurtosis | 4.80 | −0.65 | 0.54 | 4.06 | 0.82 | 1.08 | 9.41 | 14.16 | −0.73 | 3.41 | 3.04 | 19.59 |
| the III criterion | 10 | 1 | - | 50 | 50 | 20 | 1000 | 1000 | 6.5–8.5 | 1000 | 450 | 3 |

### 3.2. Evaluation of Groundwater Heavy Metal Pollution Risk

The results of the single-factor pollution index analysis and Nemerow comprehensive pollution index are shown in Table 4. The single-factor pollution indexes of As, Cd, Cr, Co, Hg, and Zn in groundwater were 0.12–0.21, 0.001–0.033, 0.013–0.072, 0.01–0.05, 0.005–0.056, and 0.003–0.195, respectively, all of which correspond to the clean grade. Moreover, the single-factor pollution index of Cu was very small, also corresponding to the clean grade. The single-factor pollution index of Ni was 0.22–1.07. It corresponded to the clean grade at most plots except for TS03, at which the value was 1.07, corresponding to mild pollution. The calculation results of the Nemerow comprehensive pollution index showed that the value was below 0.7 at most plots, corresponding to the clean grade, except for TS03, where the value was 0.78. Therefore, this site was classified under the pollution alert grade. In general, under water quality standard III, the groundwater in the area is under good conditions and the water quality is high. The most likely heavy metal pollutants are As and Ni, which requires follow-up attention.

**Table 4.** Results of the Nemerow comprehensive pollution index.

| Number | Single-Factor Pollution Index | | | | | | | | Nemerow Comprehensive Pollution Index |
|---|---|---|---|---|---|---|---|---|---|
| | As | Hg | Cr | Cd | Co | Ni | Cu | Zn | |
| TS01 | 0.12 | 0.005 | 0.027 | 0.003 | 0.020 | 0.42 | $2.8 \times 10^{-4}$ | 0.025 | 0.31 |
| TS02 | 0.16 | 0.005 | 0.026 | 0.001 | 0.024 | 0.48 | $1.4 \times 10^{-4}$ | 0.006 | 0.35 |
| TS03 | 0.12 | 0.023 | 0.053 | 0.033 | 0.051 | 1.07 | $4.4 \times 10^{-4}$ | 0.024 | 0.78 |
| TS04 | 0.13 | 0.017 | 0.041 | 0.006 | 0.010 | 0.23 | $4.9 \times 10^{-4}$ | 0.004 | 0.17 |
| TS05 | 0.16 | 0.013 | 0.045 | 0.033 | 0.027 | 0.53 | $3.4 \times 10^{-4}$ | 0.005 | 0.39 |
| TS06 | 0.16 | 0.031 | 0.072 | 0.001 | 0.019 | 0.34 | $2.1 \times 10^{-4}$ | 0.015 | 0.25 |
| TS07 | 0.17 | 0.034 | 0.030 | 0.025 | 0.040 | 0.47 | $2.4 \times 10^{-3}$ | 0.049 | 0.35 |
| TS08 | 0.16 | 0.056 | 0.037 | 0.001 | 0.017 | 0.37 | $2.5 \times 10^{-4}$ | 0.009 | 0.27 |
| TS09 | 0.14 | 0.043 | 0.026 | 0.001 | 0.028 | 0.62 | $1.8 \times 10^{-4}$ | 0.195 | 0.46 |
| TS10 | 0.15 | 0.010 | 0.025 | 0.003 | 0.020 | 0.47 | $3.1 \times 10^{-3}$ | 0.059 | 0.34 |
| TS11 | 0.15 | 0.005 | 0.037 | 0.001 | 0.025 | 0.50 | $2.3 \times 10^{-4}$ | 0.018 | 0.37 |
| TS12 | 0.16 | 0.021 | 0.027 | 0.001 | 0.014 | 0.29 | $2.1 \times 10^{-4}$ | 0.009 | 0.21 |
| TS13 | 0.16 | 0.029 | 0.034 | 0.001 | 0.027 | 0.73 | $3.0 \times 10^{-4}$ | 0.014 | 0.53 |
| TS14 | 0.15 | 0.021 | 0.013 | 0.001 | 0.016 | 0.32 | $3.4 \times 10^{-4}$ | 0.010 | 0.24 |
| TS15 | 0.15 | 0.030 | 0.023 | 0.001 | 0.023 | 0.53 | $2.7 \times 10^{-4}$ | 0.007 | 0.39 |
| TS16 | 0.15 | 0.005 | 0.024 | 0.001 | 0.022 | 0.50 | $1.8 \times 10^{-4}$ | 0.006 | 0.37 |
| TS17 | 0.15 | 0.030 | 0.066 | 0.002 | 0.040 | 0.85 | $4.4 \times 10^{-4}$ | 0.010 | 0.62 |
| TS18 | 0.15 | 0.053 | 0.031 | 0.002 | 0.028 | 0.58 | $4.5 \times 10^{-4}$ | 0.005 | 0.42 |
| TS19 | 0.16 | 0.037 | 0.031 | 0.001 | 0.029 | 0.60 | $5.4 \times 10^{-4}$ | 0.003 | 0.44 |
| TS20 | 0.17 | 0.046 | 0.035 | 0.003 | 0.013 | 0.28 | $5.9 \times 10^{-4}$ | 0.061 | 0.21 |
| TS21 | 0.21 | 0.032 | 0.049 | 0.001 | 0.025 | 0.51 | $6.3 \times 10^{-4}$ | 0.040 | 0.38 |
| TS22 | 0.14 | 0.027 | 0.059 | 0.001 | 0.041 | 0.91 | $6.0 \times 10^{-4}$ | 0.044 | 0.66 |
| TS23 | 0.15 | 0.014 | 0.023 | 0.001 | 0.012 | 0.22 | $4.0 \times 10^{-4}$ | 0.010 | 0.16 |
| TS24 | 0.16 | 0.005 | 0.021 | 0.011 | 0.022 | 0.42 | $4.7 \times 10^{-4}$ | 0.005 | 0.31 |

### 3.3. Groundwater Health Risk Assessment

According to the health risk assessment model recommended by the USEPA, the daily average exposure of the eight heavy metals was analyzed under two exposure pathways (oral intake and dermal contact). Moreover, the health risk assessment was carried out according to the content of heavy metal elements in groundwater. The health risk assessment model was used to calculate the single non-carcinogenic health risk index and carcinogenic health risk index of heavy metals in adults and children under the two exposure pathways. The results are shown in Table 5.

**Table 5.** HI and CR of heavy metals in groundwater of Tengzhou.

| Index | HM | Adults | | | Children | | |
|---|---|---|---|---|---|---|---|
| | | Oral | Dermal | Total | Oral | Dermal | Total |
| Non-carcinogenic Risk | As | 1.51 | $4.85 \times 10^{-2}$ | 1.56 | 0.72 | $1.32 \times 10^{-2}$ | 0.73 |
| | Cd | $1.67 \times 10^{-2}$ | $9.97 \times 10^{-5}$ | $1.68 \times 10^{-2}$ | $7.99 \times 10^{-3}$ | $2.72 \times 10^{-5}$ | $8.02 \times 10^{-3}$ |
| | Co | $1.82 \times 10^{-2}$ | $5.42 \times 10^{-5}$ | $1.82 \times 10^{-2}$ | $8.68 \times 10^{-3}$ | $1.48 \times 10^{-5}$ | $8.70 \times 10^{-3}$ |
| | Cr | 0.18 | $2.09 \times 10^{-3}$ | 0.18 | $8.38 \times 10^{-2}$ | $5.70 \times 10^{-4}$ | $8.44 \times 10^{-2}$ |
| | Cu | $4.14 \times 10^{-3}$ | $4.94 \times 10^{-5}$ | $4.19 \times 10^{-3}$ | $1.98 \times 10^{-3}$ | $1.35 \times 10^{-5}$ | $1.99 \times 10^{-3}$ |
| | Hg | $2.42 \times 10^{-2}$ | $6.00 \times 10^{-5}$ | $2.43 \times 10^{-2}$ | $1.16 \times 10^{-2}$ | $1.63 \times 10^{-5}$ | $1.16 \times 10^{-2}$ |
| | Ni | 0.15 | $3.31 \times 10^{-4}$ | 0.15 | $7.17 \times 10^{-2}$ | $9.03 \times 10^{-5}$ | $7.18 \times 10^{-2}$ |
| | Zn | $2.59 \times 10^{-2}$ | $2.77 \times 10^{-3}$ | $2.86 \times 10^{-2}$ | $1.24 \times 10^{-2}$ | $7.56 \times 10^{-4}$ | $1.31 \times 10^{-2}$ |
| | HI | 1.92 | $5.40 \times 10^{-2}$ | 1.98 | 0.92 | $1.47 \times 10^{-2}$ | 0.93 |
| Carcinogenic Risk | As | $6.78 \times 10^{-4}$ | $1.78 \times 10^{-5}$ | $6.96 \times 10^{-4}$ | $3.24 \times 10^{-4}$ | $4.84 \times 10^{-6}$ | $3.29 \times 10^{-4}$ |
| | Cd | $5.1 \times 10^{-5}$ | $1.79 \times 10^{-8}$ | $5.10 \times 10^{-5}$ | $2.44 \times 10^{-5}$ | $4.89 \times 10^{-9}$ | $2.44 \times 10^{-5}$ |
| | Cr | $2.63 \times 10^{-4}$ | $3.14 \times 10^{-6}$ | $2.66 \times 10^{-4}$ | $1.26 \times 10^{-4}$ | $8.55 \times 10^{-7}$ | $1.27 \times 10^{-4}$ |
| | CR | $9.93 \times 10^{-4}$ | $2.09 \times 10^{-5}$ | $1.01 \times 10^{-3}$ | $4.74 \times 10^{-4}$ | $5.70 \times 10^{-6}$ | $4.80 \times 10^{-4}$ |

The order of magnitude of the non-carcinogenic health risks of heavy metals through oral intake and dermal contact was between 1 and $10^{-5}$. For adults, the HI values followed the order As > Cr > Ni > Zn > Hg > Co > Cd > Cu. In particular, the HI value of As was 1.51, and the other heavy metals were classified as posing low risk. Regarding oral intake, the HI value for adults was 1.92, which is close to twice the value recommended by the USEPA.

Regarding dermal contact, the HI value for adults was smaller than the recommended value by two orders of magnitude. The HI values of heavy metals for children were much lower than the USEPA-recommended values, except for As with a value of 0.72 under the oral intake pathway. In general, As was the main contributor to the non-carcinogenic risk of groundwater in Tengzhou, and oral intake is the main pathway of exposure.

In terms of carcinogenic risk, the values of As and Cr exceeded the maximum acceptable value of $1 \times 10^{-4}$ for adults and children recommended by the USEPA, and the main exposure pathway is oral intake. The CR of As to adults and children through the two pathways was the largest, and the CR value of oral intake exceeded the acceptable level by two orders of magnitude, reflecting a significant risk. The CR values of As through dermal contact were $1.78 \times 10^{-5}$ and $4.84 \times 10^{-6}$ for adults and children, respectively, which were classified as possible risks. Cr presented a significant carcinogenic risk to both adults and children through drinking water, and Cd may pose a carcinogenic risk to both adults and children through drinking water. In general, As was the primary contributor to the carcinogenic health risk to the population in the region, with oral intake as the main exposure pathway.

Pearson correlation analysis was conducted to analyze the correlation of heavy metals in the groundwater in Shijiazhuang. The results show that the correlation coefficients between heavy metals Co–Cd, Co–Cr, Co–Ni, and Cr–Ni were 0.5, 0.43, 0.91, and 0.44, respectively, indicating significant positive correlations between them ($p < 0.05$). They may have common material sources and similar hydrogeochemical behaviors. There was no significant correlation between Zn and As and other metal elements, indicating different material sources or migration and transformation pathways of these metal elements(Figure 2).

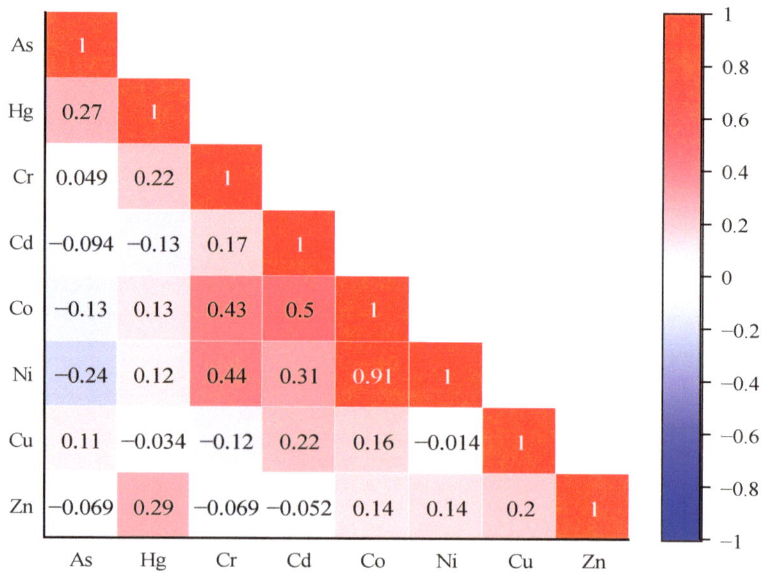


**Figure 2.** Correlation coefficients between metal elements in groundwater.

SPSS26.0 was used to standardize the original data. Thereafter, the KMO test and Bartlett's test of sphericity were performed on the standardized data. The KMO statistic value was 0.507 (greater than 0.5), and Bartlett's test of sphericity yielded a value less than 0.05, indicating that the data are suitable for PCA. Four principal components, PC1, PC2, PC3, and PC4, were extracted using the Kaiser standardized orthogonal rotation method. By multiplying the linear combination coefficient with the quotient of the variance interpretation rate and the cumulative variance interpretation rate, the loadings of the principal components were obtained, and the results are presented in Table 6.

Table 6. Composition matrix table of principal component analysis.

| HM | PC1 | PC2 | PC3 | PC4 |
|---|---|---|---|---|
| As | −0.082 | 0.396 | −0.006 | 0.618 |
| Hg | 0.072 | 0.585 | −0.082 | −0.055 |
| Cr | 0.233 | 0.142 | −0.378 | 0.195 |
| Cd | 0.224 | −0.249 | 0.214 | 0.348 |
| Co | 0.376 | −0.016 | 0.039 | 0.024 |
| Ni | 0.359 | −0.019 | −0.089 | −0.158 |
| Cu | 0.053 | 0.037 | 0.644 | 0.289 |
| Zn | 0.072 | 0.332 | 0.402 | −0.504 |

### 3.4. Source Apportionment of Heavy Metal Components in Groundwater

Combined with the PCA results, the PMF model was applied using EPA PMF 5.0 software to quantitatively analyze the sources of heavy metal elements in the study area. In the calculation of the uncertainty of heavy metal elements, the signal-to-noise ratios of Cd, Cu, and Zn were found to be small ($S/N < 0.5$). Therefore, these three heavy metals were classified as "weak", and the remaining heavy metals were classified as "strong". The number of factors was set to 3–6 and calculations were conducted 20 times. When the number of factors was 4, $Q_{Robust}/Q_{true}$ decreased rapidly, and the residuals of each sample were between − 3 and 3, indicating that the PCA model and the PMF model are reasonable. The fitting degree $R^2$ of As, Cr, Co, Hg, and Ni was greater than 0.9, indicating good fitting of the source apportionment of the PMF model for As, Cr, Co, Hg, and Ni. In contrast, the fitting degree $R^2$ of Cd, Cu, and Zn was below 0.7, indicating poor fitting for these three elements. This may be related to the large coefficient of variation among Cd, Cu, and Zn, which needs to be explained by combining the PCA and PMF models. The PMF model yielded four factors named PMF1, PMF2, PMF3, and PMF4, accounting for 21.7%, 27.2%, 31.0%, and 20.1% of the total contribution, respectively.

According to Figure 3, the weight of PMF1 is mainly attributable to Cr (54%), followed by Zn (39%). As Zn showed a strong coefficient of variation, its distribution in groundwater may be affected by humans. In the correlation analysis, no significant correlation was observed between Cr and Zn. Studies have shown that the main sources of Zn and Cu are automobile exhaust, livestock and poultry manure, and pesticides. In local farms with livestock, feed additives, pesticides, and fertilizers are frequently used. In the principal component analysis, Zn and Cu in PC3 showed strong positive loads. It can be considered that PMF1 and PC3 are attributable to agricultural sources. Considering the poor fitting of Zn and Cu in the PMF model, these two elements were excluded in other PMF factors [22,23].

The weights of As, Cd, Co, and Ni accounted for approximately 40% in PMF2. The study area includes mineral mining and smelting activities. Studies have shown that As, Cu, and Cd are mainly affected by human activities. For example, the development of metal minerals has caused serious pollution of heavy metals Cd and As in the mining area. PMF2 may be attributable to a variety of anthropogenic industrial sources, including industrial and agricultural activities and traffic emissions [24].

In PMF3, the weights of Cr, Co, and Ni contribute 42%, 48%, and 49%, respectively. A strong homology was observed between Cr and Co in soil, and the average values of Cr, Co, and Ni were lower than the background values of groundwater in Tengzhou City. In the correlation analysis, a certain correlation was found between Cr, Co, and Ni and other elements. In addition to industrial activities, Cr and Co are related to the mineral composition of the parent rock. Therefore, PMF3 can be primarily attributed to natural sources [25,26].

Hg presented high loading in PMF4. In urban areas, the main sources of soil Hg are atmospheric deposition and surface water. Chemical enterprises use oil and coal as the main fuel, resulting in the emission of Hg-containing gases. Studies have shown that Hg enrichment is mainly caused by the accumulation of microparticles released from surrounding metallurgy and coal combustion processes into the soil through atmospheric deposition. Therefore, PMF4 is considered to be attributable to industrial deposition sources [27].

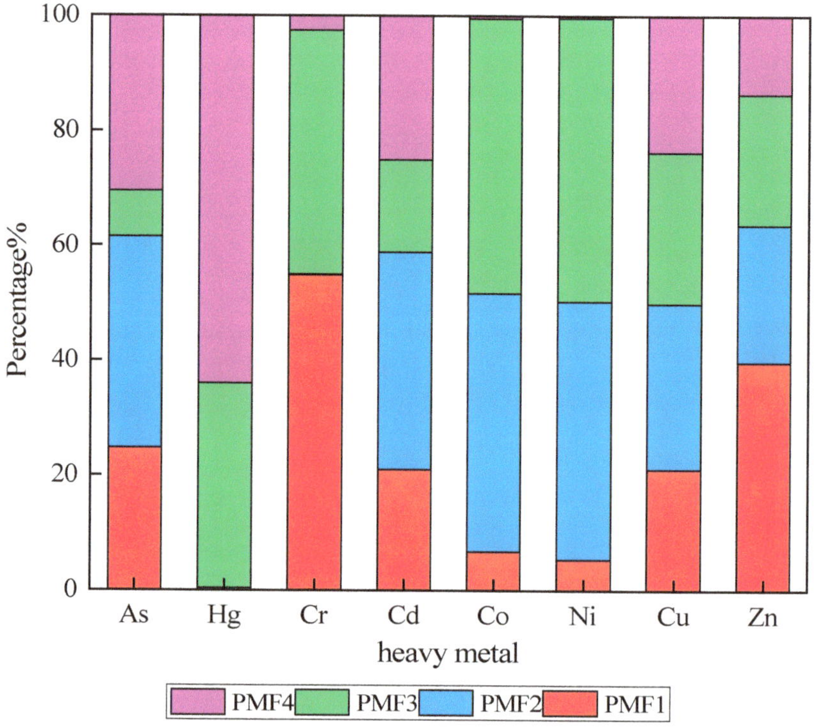


**Figure 3.** Contribution rates of heavy metal pollution sources in PMF.

## 4. Conclusions

Under water quality standard III, the single-factor pollution index of Ni was 1.07 at plot TS03, reflecting mild pollution, whereas the values of other heavy metal elements corresponded to the clean grade. The calculation results of the Nemerow comprehensive pollution index showed that the value reached 0.78 only at TS03, which corresponds to the pollution alert state. The Nemerow comprehensive pollution index of other plots was less than 0.7. Therefore, the groundwater in this area is in good condition and the water quality is high.

The non-carcinogenic risk values of heavy metals for adults followed the order As > Cr > Ni > Zn > Hg > Co > Cd > Cu. The non-carcinogenic risk value of As was 1.51, but the other heavy metals posed low risk. Under the oral intake pathway, the HI value for adults was 1.92. The non-carcinogenic risk values of heavy metals for children were much lower than the USEPA-recommended values, except for As (0.72) under the oral intake pathway.

As and Cr are the main contributors to carcinogenic health risks for adults and children, with oral intake as the main exposure pathway. As poses the greatest carcinogenic health risk to adults and children through both pathways. Cr poses a significant carcinogenic risk to both adults and children through drinking water, and Cd may pose carcinogenic risks to both adults and children through drinking water.

The main sources of heavy metals in shallow groundwater in Tengzhou City are agriculture, industrial activities, natural processes, and industrial deposition, with contribution rates of 21.7%, 27.2%, 31.0%, and 20.1%, respectively. Among them, natural sources contribute the most to the accumulation of heavy metals.

**Author Contributions:** Conceptualization, B.Y., Q.W., X.L. and X.S.; Data curation, B.Y., Q.W., X.L., X.S. and R.Z.; Formal analysis, X.S., Z.G., J.L., R.Z. and M.W.; Funding acquisition, Q.W.; Investigation, B.Y., Q.W., X.L., X.S. and Z.G.; Methodology, B.Y., Q.W. and X.S.; Project administration, Z.G.; Resources, B.Y., Q.W., X.S., Z.G. and J.L.; Software, Z.G., J.L., R.Z. and M.W.; Supervision, X.L., X.S., Z.G., J.L. and M.W.; Validation, B.Y., Q.W., X.L. and Z.G.; Visualization, Z.G., J.L. and M.W.; Writing—original draft, Q.W., Z.G. and M.W.; Writing—review and editing, B.Y., Q.W., X.S. and Z.G. All authors have read and agreed to the published version of the manuscript.

**Funding:** This research was supported by Bureau Research Special Project of Shandong Provincial Bureau of Coal Geological (LMDK(2022)47, LMDK(2023)10).

**Data Availability Statement:** Data are contained within the article.

**Acknowledgments:** This research was supported by Bureau Research Special Project of Shandong. We thank the editors and anonymous reviewers for their valuable suggestions regarding the manuscript.

**Conflicts of Interest:** The authors declare no conflict of interest.

## References

1. Ministry of Ecology and Environment of the People's Republic of China. Report on the State of the Ecology and Environment in China 2020. Available online: https://www.mee.gov.cn/hjzl/sthjzk/zghjzkgb (accessed on 26 May 2021).
2. Swaine, D.J. Why trace elements are important. *Fuel Process. Technol.* **2000**, *65–66*, 21–33. [CrossRef]
3. Chowdhury, S.; Mazumder, M.J.; Al-Attas, O.; Husain, T. Heavy metals in drinking water: Occurrences, implications, and future needs in developing countries. *Sci. Total Environ.* **2016**, *569–570*, 476–488. [CrossRef]
4. Wang, S.-L.; Xu, X.-R.; Sun, Y.-X.; Liu, J.-L.; Li, H.-B. Heavy metal pollution in coastal areas of South China: A review. *Mar. Pollut. Bull.* **2013**, *76*, 7–15. [CrossRef]
5. Ahmed, S.; Khurshid, S.; Qureshi, F.; Hussain, A.; Bhattacharya, A. Heavy metals and geo-accumulation index development for groundwater of Mathura City, Uttar Pradesh. *Desalin. Water Treat.* **2019**, *138*, 291–300. [CrossRef]
6. Wang, G.; Yinglan, A.; Jiang, H.; Fu, Q.; Zheng, B. Modeling the source contribution of heavy metals in surficial sediment and analysis of their historical changes in the vertical sediments of a drinking water reservoir. *J. Hydrol.* **2015**, *520*, 37–51. [CrossRef]
7. Dwivedi, A.K.; Vankar, P.S. Source identification study of heavy metal contamination in the industrial hub of Unnao, India. *Environ. Monit. Assess.* **2014**, *186*, 3531–3539. [CrossRef] [PubMed]
8. Deda, A.; Alushllari, M.; Mico, S. Measurement of heavy metal concentrations in groundwater. In Proceedings of the WOMEN IN PHYSICS: 6th IUPAP International Conference on Women in Physics, Birmingham, UK, 16–20 July 2017.
9. Abdelhalim, A.; Howard, G.; Howden, N.J.K.; Ahmed, M.; Ismail, E. Carcinogenic and non-carcinogenic health risk assessment of heavy metals contamination in groundwater in the west of Minia area, Egypt. *Hum. Ecol. Risk Assess. Int. J.* **2023**, *29*, 571–596. [CrossRef]
10. Fang, H.; Wang, X.; Xia, D.; Zhu, J.; Yu, W.; Su, Y.; Zeng, J.; Zhang, Y.; Lin, X.; Lei, Y.; et al. Improvement of Ecological Risk Considering Heavy Metal in Soil and Groundwater Surrounding Electroplating Factories. *Processes* **2022**, *10*, 1267. [CrossRef]
11. Zhao, W.X. The Geochemical Characteristics of the Shallow Groundwater in Nansihu-Lake Basin: The Contribution of Water-Rock Interact and Anthropogenic Activities. Master's Thesis, China University of Geosciences (Beijing), Beijing, China, 2023. [CrossRef]
12. Ministry of Ecology and Environment of the People's Republic of China. *HJ 1019–2019 Technical Guidelines for Sampling Volatile Organic Compounds in Soil and Groundwater*; Ministry of Ecology and Environment of the People's Republic of China: Beijing, China, 2019.
13. Ministry of Ecology and Environment of the People's Republic of China. *HJ 700–2014 Water Quality-Determination of 65 Elements-Inductively Coupled Plasma-Mass Spectrometry*; Ministry of Ecology and Environment of the People's Republic of China: Beijing, China, 2014.
14. Ministry of Ecology and Environment of the People's Republic of China. *HJ 694–2014 Water Quality–Determination of Mercury, Arsenic, Selenium, Bismuth and Antimony–Atomic Fluorescence Spectrometry*; Ministry of Ecology and Environment of the People's Republic of China: Beijing, China, 2014.
15. Li, M.-T.; Shen, C.; Wu, J.; Huang, S.-F.; Li, D.-Y.; Wang, M. Content and Ecological Risks of Heavy Metals in Soil with Different Land Uses in a Rapidly Urbanizing Area. *Huan Jing Ke Xue* **2021**, *42*, 4889–4896. [CrossRef] [PubMed]
16. *GB14848–2017*; Standard for Groundwater Quality. Standardization Administration of the People's Republic of China: Beijing, China, 2017.
17. Maskooni, E.K.; Naseri-Rad, M.; Berndtsson, R.; Nakagawa, K. Use of Heavy Metal Content and Modified Water Quality Index to Assess Groundwater Quality in a Semiarid Area. *Water* **2020**, *12*, 1115. [CrossRef]
18. Singaraja, C.; Chidambaram, S.; Srinivasamoorthy, K.; Anandhan, P.; Selvam, S. A Study on Assessment of Credible Sources of Heavy Metal Pollution Vulnerability in Groundwater of Thoothukudi Districts, Tamilnadu, India. *Water Qual. Expo. Health* **2015**, *7*, 459–467. [CrossRef]
19. Wen, C.; Zhu, S.; Li, N.; Luo, X. Source apportionment and risk assessment of metal pollution in natural biofilms and surface water along the Lancang River, China. *Sci. Total Environ.* **2022**, *843*, 156977. [CrossRef] [PubMed]

20. Chen, H.; Zhao, X.-Y.; Chang, S.; Song, Y.-M.; Lu, M.-Q.; Zhao, B.; Chen, H.-D.; Gao, S.; Wang, L.-J.; Cui, J.-S.; et al. Source Analysis and Health Risk Assessment of Heavy Metals in the Groundwater of Shijiazhuang, a Typical City in North China Plain. *Huan Jing Ke Xue* **2023**, *44*, 4884–4895. [CrossRef] [PubMed]
21. Li, W.; Wang, M.-Y.; Liu, L.-Y.; Wang, H.-F.; Yu, S. Groundwater heavy metal levels and associated human health risk in the North China Plain. *Arab. J. Geosci.* **2015**, *8*, 10389–10398. [CrossRef]
22. Belon, E.; Boisson, M.; Deportes, I.; Eglin, T.; Feix, I.; Bispo, A.; Galsomies, L.; Leblond, S.; Guellier, C. An inventory of trace elements inputs to French agricultural soils. *Sci. Total Environ.* **2012**, *439*, 87–95. [CrossRef] [PubMed]
23. Qu, M.; Wang, Y.; Huang, B.; Zhao, Y. Source apportionment of soil heavy metals using robust absolute principal component scores-robust geographically weighted regression (RAPCS-RGWR) receptor model. *Sci. Total Environ.* **2018**, *626*, 203–210. [CrossRef] [PubMed]
24. Sheng, D.; Wu, J.; Wen, X.; Wu, M.; Zhang, C. Contamination and ecological health risks of heavy metals in groundwater of a typical agricultural area in NW China. *Geochem. Explor. Environ. Anal.* **2020**, *20*, 440–450. [CrossRef]
25. Nanos, N.; Martín, J.A.R. Multiscale analysis of heavy metal contents in soils: Spatial variability in the Duero river basin (Spain). *Geoderma* **2012**, *189–190*, 554–562. [CrossRef]
26. Fei, X.; Lou, Z.; Xiao, R.; Ren, Z.; Lv, X. Contamination assessment and source apportionment of heavy metals in agricultural soil through the synthesis of PMF and GeogDetector models. *Sci. Total Environ.* **2020**, *747*, 141293. [CrossRef]
27. Zhang, H.; Zhang, F.; Song, J.; Tan, M.L.; Kung, H.-T.; Johnson, V.C. Pollutant source, ecological and human health risks assessment of heavy metals in soils from coal mining areas in Xinjiang, China. *Environ. Res.* **2021**, *202*, 111702. [CrossRef] [PubMed]

*hydrology*

MDPI

*Article*

# Estimating Thermal Impact on Groundwater Systems from Heat Pump Technologies: A Simplified Method for High Flow Rates

**David Krcmar [1,*], Tibor Kovacs [2], Matej Molnar [2], Kamila Hodasova [1] and Martin Zatlakovic [1]**

1 Department of Engineering Geology, Hydrogeology and Applied Geophysics, Faculty of Natural Science, Comenius University in Bratislava, Ilkovicova 6, 84215 Bratislava, Slovakia; hodasova11@uniba.sk (K.H.)
2 NuSi, Svatovojtesska 7, 83103 Bratislava, Slovakia
* Correspondence: david.krcmar@uniba.sk

**Abstract:** This research delves into the potential thermal effects on underground water systems caused by the use of thermal technologies involving extraction and injection wells. We developed a unique approach that combines straightforward calculations with computer-based modeling to evaluate thermal impacts when water flow rates exceed 2 L/s. Our model, based on a system with two wells and a steady water flow, was used to pinpoint the area around the thermal technology where the temperature varied by more than 1 °C. Our findings suggest that the data-based relationships we derived from our model calculations provide a cautious estimate of the size of the affected area, or 'thermal cloud'. However, it is important to note that our model's assumptions might not fully account for the complex variables present in real-world underground water systems. This highlights a need for more research and testing. A key contribution of our study is the development of a new method to assess the thermal impact of operations involving heat pumps. In conclusion, while our proposed method needs more fine-tuning, it shows promise in estimating temperature changes within water-bearing rock layers, or aquifers. This is crucial in the effective use of thermal technologies while also ensuring the protection and sustainable management of our underground water resources.

**Keywords:** groundwater; heat pump; thermal feedback; well doublet; analytical solutions

**Citation:** Krcmar, D.; Kovacs, T.; Molnar, M.; Hodasova, K.; Zatlakovic, M. Estimating Thermal Impact on Groundwater Systems from Heat Pump Technologies: A Simplified Method for High Flow Rates. *Hydrology* **2023**, *10*, 225. https://doi.org/10.3390/hydrology10120225

Academic Editors: Zongjun Gao and Jiutan Liu

Received: 30 October 2023
Revised: 25 November 2023
Accepted: 27 November 2023
Published: 29 November 2023



## 1. Introduction

Groundwater heat pump (GWHP) systems have emerged as a sustainable solution for heating and cooling in various settings, based on the thermal stability of the subsurface environment [1,2]. However, these systems are not without complexity. The design and operation of GWHP systems require a thorough understanding of the thermal and hydraulic dynamics of the subsurface, which can be complex and very specific to the site [3,4].

Historically, researchers have attempted to address these complexities through a variety of analytical and numerical methods [5–7]. However, these methods often depend on simplifying assumptions, such as steady-state conditions and advection-dominated transport, which may not accurately represent the dynamic nature of real-world GWHP systems [8,9].

The development of advanced computational capabilities and modeling techniques has launched a new era of possibilities for more accurate and detailed assessments of GWHP systems [10,11]. For instance, numerical models can now account for the transient nature of thermal loads and the spatial variability in subsurface properties, providing a more realistic representation of system dynamics [12–14].

One of the critical areas of research in this field is the estimation of the thermal impact on groundwater systems from GWHPs [5,15,16]. A comprehensive understanding of the thermal effects on groundwater systems is crucial for the design, implementation, and management of these technologies. Pioneering works in this area, such as [15], have developed mathematical models to investigate the transient temperature behavior of production wells during the reinjection of heat-depleted water into aquifers with a uniform regional flow.

However, a significant challenge identified in the literature is to estimate the thermal impact on groundwater systems at high flow (pumping) rates, particularly those exceeding 2 L/s [5,17]. This article seeks to address these challenges and knowledge gaps. Our goal is to develop a novel method based on analytical calculations that accelerates the process of assessing thermal impacts on groundwater systems for flow rates higher than those currently estimated. This paper, based on [5,17,18], exhaustively examines these theoretical proposals and, through rigorous testing and experimentation, aims to extend the scope of these models and propose a more reliable, comprehensive approach.

By utilizing simplified methods with improved accuracy, the adoption and application of environmentally friendly heating and cooling technologies are applicable while ensuring that their operation has minimal adverse impacts on our crucial groundwater resources.

## 2. Materials and Methods

### *2.1. Hydraulics of a Two-Well Heat Pump System*

In order to utilize analytical formulas, specific conditions and limitations were considered. A two-well system (pumped and injected) was used in the direction of groundwater flow. The water-bearing layer was an infinite plane of constant thickness with a stressed surface and a uniform water flow. The schematic situation is depicted in Figure 1.

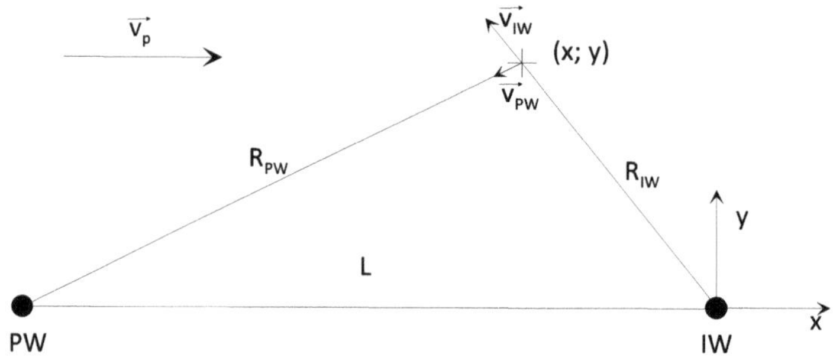


**Figure 1.** Schematic diagram of the situation of the heat pump well.

The coordinate system was positioned in place of the injected well, with the orientation of the $x$-axis in the direction of an undisturbed medium flow. In this case, the well coordinates were $(0;\ 0)$ for the pumped well and $(-L;\ 0)$ for the injected well, and the vector of the natural fluid flow had the form $\vec{v_p} = (v_x; 0)$.

First, the hydraulic aspect of the problem was addressed. From the principle of superposition, it follows that the velocity at any point is determined as the vector sum of the natural velocities resulting from the pumping and injecting wells in the following form:

$$\vec{v_c} = \vec{v_p} + \vec{v_{PW}} + \vec{v_{IW}} \tag{1}$$

In the case of a confined head, it is possible to express the steady velocity field in the vicinity of a well in an infinite water layer with thickness $b$ from the continuity equation. The velocity field was determined based on the position and amount of water being pumped $Q$:

$$\vec{v_c}(x,y,Q) = k\,\vec{\imath} + \frac{-Q}{2\pi R_{PW} b}\,\vec{J}_{cs} + \frac{Q}{2\pi R_{IW} b}\,\vec{J}_{IW} \tag{2}$$

where

$\vec{J}_{PW}$ is the unit vector in the direction of $\vec{v}_{PW}$;
$\vec{J}_{IW}$ is the unit vector in the direction of $\vec{v}_{IW}$;
$R(x;y)$ are the distances to the relevant well, as shown in Figure 1;
$\vec{v}_C$ expresses the Darcy velocity of the steady fluid flow at each point in space;
$b$ is the thickness of the water layer;

$k$ is the hydraulic conductivity;
$x$ *and* $y$ are the spatial coordinates;
$Q$ is the quantity of pumped water;
$\vec{i}$ is the slope vector of the groundwater level with amplitude $i$.

For the purposes of the task, it was essential to distinguish between two situations. In the first case, some of the water from the injection well returned to the pump well (the wells mutually influenced each other hydraulically). In the second case, the water from the injected well did not flow back into the pump well (the wells were hydraulically separated). Hydraulic separation of the wells therefore indicates that the water in the pumped well will not be affected by the injection of thermally influenced water into the injected well. Hydraulically separated wells ensure temperature stability and thus the functionality of the technology. However, in practice, most technologies are not actually hydraulically separated. Even lower amounts of pumped and injected water are usually associated with the flow of some of the injected water back into the pumped well in natural groundwater flows. Despite this situation, technology can still be functional during the utilized seasons. Hydraulic separation of wells occurs exactly when the component of the total velocity in the $x$ direction on the line is greater than 0.

$$\left[\vec{v_c}\left(-\frac{L}{2};0;Q\right)\right]_X > 0 \tag{3}$$

This condition is met exactly when [19]

$$L > \frac{2Q}{T\pi i} \tag{4}$$

where $T = kb$ is the transmissivity coefficient of the medium, $L$ is the distance between the wells, and $Q$ is the quantity of pumped water. Absolute non-influence of the wells occurs only in the case of a very large distance between the wells or when the pumped amounts are small.

The same conditions for the hydraulic separation of wells in the heat pump system are also stated in [5,18].

With the existence of backward flow between wells, it was necessary to quantify its magnitude. For this purpose, the speed of groundwater flow was expressed on the axis between the wells (a line in our chosen coordinate system). From symmetry $\vec{v_c}$, it follows that the velocity component in the direction of the "$y$" axis on this axis is always zero. It represents $\left[\vec{v_c}\left(\frac{L}{2};y\right)\right]_x = \left|\vec{v_c}\left(\frac{L}{2};y\right)\right|$. This speed on the axis was expressed as follows:

$$\left|\vec{v_c}\left(\frac{L}{2};y\right)\right| = ki - \frac{QL}{2\pi b}\frac{1}{y^2 + \left(\frac{L}{2}\right)^2} \tag{5}$$

For the backward flow of water into the pumped well, there is an area for the values of the parameter $y$ in the interval $(-Y;\ Y)$ where the speed $[v_c]_x < 0$. The entire backward flow to the well will pass through this area on the well axis (Figure 2). The magnitude of the backward flow from the injection well to the pumping well was expressed as follows:

$$\begin{aligned} Q_{back} &= \int_{-Y}^{Y} -\left|v_c\left(\tfrac{L}{2};y\right)\right| b\,dy \\ Q_{back} &= -2kibY + \tfrac{2Q}{\pi}\arctan\tfrac{2Y}{L} \end{aligned} \tag{6}$$

$Y$ is the distance on the $x$ axis at $L/2$ where $|vc| = 0$ (Figure 2). After substitution into Equation (5), the following equation was obtained for the distance $Y$:

$$Y = \sqrt{\frac{QL}{2\pi bki} - \left(\frac{L}{2}\right)^2} \tag{7}$$

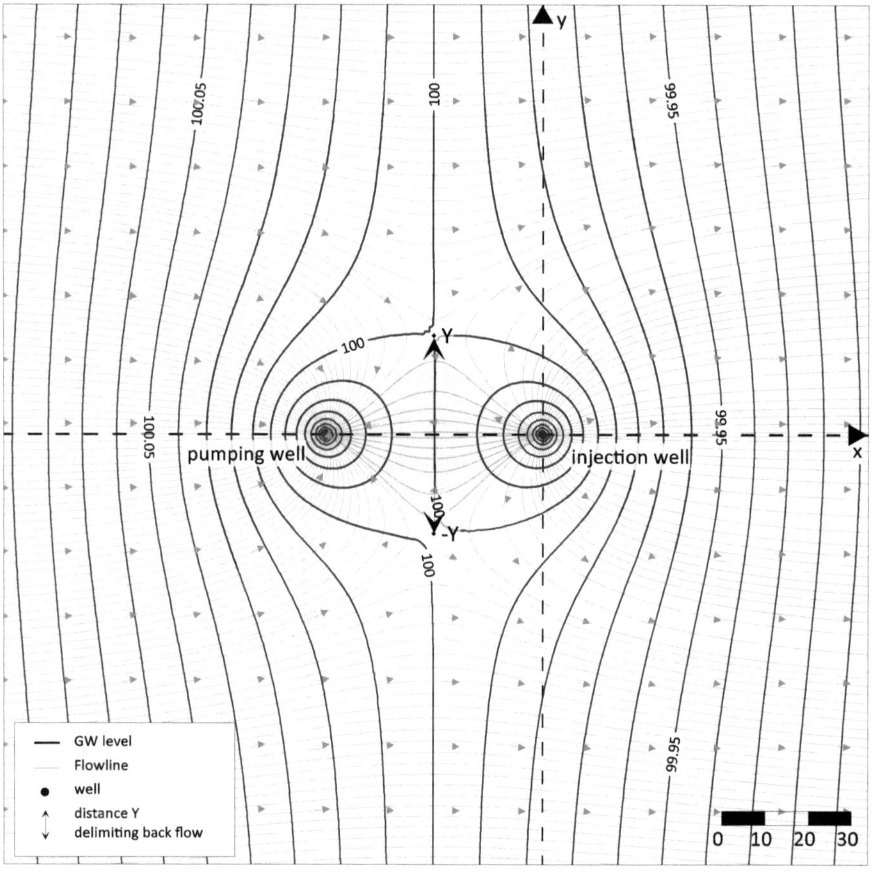


**Figure 2.** Context of groundwater level and flow lines around the heat pump with backward water flow between wells.

Similar analytical equations evaluating these parameters can be found in [20].

### 2.1.1. Characteristics of the Heat Pump System

In open-loop water-to-water heat pump systems, the degree of influence of the temperature of the pumped water on the temperature of the water flowing back from the injected well to the pumped one is significant. This effect of backflow reduces the system's efficiency, or leads to the same cooling output, leading to increased water extraction (which worsens the back influence). If there is a natural water flow, part of the heat-affected water exits the area of use, even when the same amount is pumped and infiltrated. Hence, water-to-water heat pump systems thermally impact the groundwater in their surroundings, following the flow of groundwater. During the technology approval process, it is necessary to limit this thermal influence on the geological environment due to possible negative impacts on the quality of groundwater. It is also crucial to evaluate whether the technology explored does not hinder the functionality of existing water-to-water systems nearby.

### 2.1.2. Determining the Backflow to the Pumping Well

The ideal case for a heat pump occurs when no heated water returns to the pumping well in a natural flow. In the preceding chapter, the condition of no backwater flow between wells was introduced by Equation (4). The same condition is also stated in [5,18]. If this condition is met, the wells will not thermally influence each other. However, if a portion of the water from the injected well flows back to the pumped well, the water in the pumped well will be thermally affected. This thermal effect will change over time and will depend on the hydraulic and transport characteristics of the groundwater flow. Even in a fairly simple natural flow system with a pumped well and an injection well, there is no straightforward analytical solution. However, it is possible to determine the water arrival time from the injection well to the pump well and use this interval to conservatively restrict the possible use of heat pumps during the cooling or heating season. If, however, this determined time indicates the non-functionality of the technology, the calculations will need to be refined via numerical modeling, which is more accurate.

Even when complete hydrodynamic separation of the pumped and injected wells is not detected, it is possible to use heat pumps. As mentioned, the arrival time of the heat to the injected well is an essential factor. For the heat flow, one must know the time of arrival of the water time between the two wells and the retardation factor $R$.

Based on Equation (4), we chose the parameter $\beta$ in the following form:

$$\beta = \frac{2Q}{kb\ \pi i L} \quad (8)$$

Well interference occurs when $\beta > 1$. The published relationship for the groundwater flow time $t_{hyd}$ between the wells [5] in the following form was used:

$$t_{hyd} = \frac{Ln}{Ki}\left[\frac{\beta}{\sqrt{\beta-1}}\arctan\left(\frac{1}{\sqrt{\beta-1}}\right) - 1\right] \quad (9)$$

where $n$ is the porosity of the rock environment.

For the calculation, a quantity called the retardation coefficient $R$ was needed, which expresses the deceleration of heat propagation due to heat exchange between water and the rock environment. The retardation coefficient is determined by the following equation [21]:

$$\frac{1}{R} = \frac{v_{the}}{v_{hyd}} = \frac{nC_W}{C_{aq}} = \frac{nc_w\rho_w}{nc_w\rho_w + (1-n)c_s\rho_s} \quad (10)$$

$v_{hyd}$ is the water flow speed in a porous environment;
$v_{the}$ is the heat propagation speed in a porous environment;
$c_w$ is the heat capacity of the water;
$c_s$ is the heat capacity of the rock;
$\rho_w$ is the water density;
$\rho_s$ is the rock density;
$C_w$ is the volume heat capacity of the water;
$C_{aq}$ is the volume heat capacity of the water environment;
$t_{the}$ is the time of the heat change transport which, because of the heat capacity of the environment, is greater than the groundwater flow time. From Equation (10), it follows that the ratio between $t_{hyd}$ and $t_{the}$ is

$$t_{the} = R\ t_{hyd} \quad (11)$$

From Equations (8) and (10) together, the formula for the heat input into the pumping well follows.

$$t_{the} = \frac{C_{aq}L}{C_W ki}\left[\frac{\beta}{\sqrt{\beta-1}}\arctan\left(\frac{1}{\sqrt{\beta-1}}\right) - 1\right] \quad (12)$$

The calculation for time does not include diffusion and dispersion effects, so the temperature in the pumping well will gradually increase. The time $t_{the}$ represents a boundary at which the thermal influence disrupts the usability of the heat pump.

### 2.2. Thermal Impact on Surroundings

To estimate the extent of thermal impact, the procedure published in [17] was first verified. This article describes the following three analytical solutions for heat transport in an environment:

- The radial transport model;
- The linear advective model of heat transport;
- The planar advective model of heat transport.

Taking into account the thermal influence on the surroundings of a two-well system under natural groundwater flow, the planar advective model was selected as the most suitable approximation for the heat transport of the problem because it considers the impact of the local hydraulic gradient on the heat distribution around the well. In the following text, the equations according to which the method determines the thermal influence on the surroundings and the variables that appear in them are described.

Table 1 presents a list of physical quantities with the labels used in the calculations.

**Table 1.** Variables included in the analytical calculation.

| Variable | Symb | Unit |
|---|---|---|
| Temperature gradient of poured water | $\Delta T_{inj}$ | K |
| Temperature gradient in the source term | $\Delta T_0$ | K |
| Width of the source term | $Y$ | m |
| Poured amount of water | $Q$ | $m^3/s$ |
| Seepage velocity | $v_a$ | m/s |
| Specific heat capacity of substance x (w—water; s—solid) | $c_x$ | J/(kg·K) |
| Volumetric heat capacity of substance x (w—water; s—solid) | $C_x$ | $J/(m^3 \cdot K)$ |
| Density of substance x (w—water; s—solid substance) | $\rho_x$ | $kg/m^3$ |
| Thermal conductivity of substance x (w—water; s—solid) | $\lambda_x$ | W/(m·K) |
| Dispersion (transverse longitudinal) | $\alpha_{T,L}$ | m |
| Retardation coefficient | $R$ | --- |

The source term was an area perpendicular to the direction of groundwater flow in the environment. The area had a height equal to the depth of the aquifer. In [17], the authors introduced two models for the width of the source area:

$$Y_0 = \frac{Q}{2bki} \tag{13}$$

$$Y_{max} = \frac{Q}{bki} \tag{14}$$

The thermal power supplied to the environment was determined based on the volumetric heat capacity of water $C_w$, the amount of poured water $Q$, and the temperature gradient between the poured and pumped water $\Delta T_{inj}$ according to the following relationship:

$$q_h = \Delta T_{inj} C_w Q \tag{15}$$

From the thermal power, it was possible to determine the thermal gradient on the surface according to the following equation:

$$\Delta T_0 = \frac{q_h}{b v_a n C_w Y} \tag{16}$$

The properties of the environment related to the thermal flow were characterized for further needs by the quantity $D_{x,y}$, which includes heat diffusion via thermal conductivity and the expansion of the thermal cloud based on dispersion (transverse and longitudinal), according to the following relationship:

$$D_{x,\,y} = \frac{\lambda_m}{n C_w} + \alpha_{L,T} v_a \tag{17}$$

The equation of temperature change propagation in the environment using a natural flow with diffusion and dispersion was analytically expressed under the given conditions as follows:

$$\Delta T(x,\, y,\, t) = \left(\frac{\Delta T_0}{4}\right) erfc\left(\frac{Rx - v_a t}{2\,\sqrt{D_x R t}}\right)\left\{ \mathrm{erf}\left[\frac{y + \frac{Y}{2}}{2\sqrt{\frac{D_y x}{v_a}}}\right] - \mathrm{erf}\left[\frac{y - \frac{Y}{2}}{2\sqrt{\frac{D_y x}{v_a}}}\right]\right\} \quad (18)$$

The equation describes the spread of the temperature gradient and does not depend on the background temperature of the environment.

In Equation (18), there is also a unit of time. Heat pumps can be used for both cooling and heating. In the calculations, a period of one heating or cooling season of 6 months was assumed.

The method provides a sufficiently accurate solution only for small amounts pumped at up to 2 L/s and with the conditions described in the given article [17]. In attempts to expand this method to larger amounts of pumped water, it was not possible to obtain sufficiently representative results. The main problem with this method lies in the assumption of an almost undisturbed fluid flow at a certain distance from the well. However, this condition is not met for pumping larger amounts. In the wider vicinity of the well, the natural flow in the environment is affected by pumping. The influence increases with the amount of pumped water and the reduction in the amount of the natural flow of groundwater.

## 3. Results

### 3.1. Empirical Methods

In the next part of this work, an approximate conservative dependency was determined for larger amounts of pumped water, estimating the size of the cloud from the pumping parameters and the hydraulic and transport parameters of the environment. A conservative dependency, means a safe (overestimated) estimate to determine the range of temperature changes. A numerical model of heat propagation was compiled in a water-saturated environment under simplified conditions, namely neglecting heat loss to the surrounding environment. This is a conservative limitation since heat exchange with the surface and substrate reduces the reach of the thermal impact on the surroundings. MODFLOW [22] and MT3D [23,24] software were used to create the model.

The numerical model was developed based on the assumptions described in the analytical solution mentioned earlier for a homogeneous, isotropic, infinite aquifer with constant thickness, a hydrostatic pressure level, and a uniform flow of groundwater. The boundary conditions of a constant level were chosen in the model far enough so that they had no impact on the actual flow around the wells. The computational grid had the densest area around the infiltration well of a size of 150 × 150 m, with a cell size of 2 × 2 m. Toward the edges of the model, the cell size gradually increased.

For a conservative possibility of verifying the impact of thermal technology on surrounding groundwater, typical curves of temperature change propagation in groundwater were empirically determined by analyzing the modeling results with a larger pumped amount.

Model solutions were sought for the case of the typical hydraulic and thermal properties of gravel–sand aquifers (Table 2). The selected parameters of the system and groundwater flow were changed (Table 3). The values in Tables 2 and 3 are typical values for alluvial sediments in Slovakia. The environmental parameters related to heat capacity and conductivity remained unchanged since they do not change much for aquifers usable with the technology of water–water heat pump systems.

**Table 2.** Used parameters of the gravel–sand aquifer.

| Parameter | Unit | Value |
|---|---|---|
| Water density | kg/m$^3$ | 1000 |
| Specific heat capacity of water | J/(kg·K) | 4180 |
| Rock density—quartz | kg/m$^3$ | 2600 |
| Specific heat capacity—quartz | J/(kg·K) | 820 |
| Thermal conductivity of porous environment | W/(m·K) | 2.5 |
| Longitudinal dispersion | m | 1.8 |
| Transverse dispersion | m | 0.18 |

**Table 3.** Range of varying system parameters for individual series of calculations.

| Parameter | Unit | Value Range |
|---|---|---|
| Thickness of aquifer | m | 4–20 |
| Pumped amount of water | L/s | 0–10 |
| Distance of wells | m | 30–100 |
| Groundwater gradient | m/m | 0.0005–0.002 |
| Hydraulic conductivity | m/s | $1 \times 10^{-3}$–$5 \times 10^{-3}$ |

A series of calculations with varying parameters was performed in which the maximum width ($Y_{cloud}$) and length ($X_{cloud}$) of the thermal cloud were determined (Figure 3). For all cases, the heat transport was modeled for a period of 6 months, which represents an estimate of the period of using heat pump technology in cooling or heating mode during the year. The undisturbed temperature of the groundwater was 12 °C in the calculations, and the temperature of the heated water poured into the infiltration well was 24 °C. The usual temperature of groundwater is 12 °C, and a temperature of 24 °C is the highest permitted temperature of water infiltrating the rock environment in the Slovak legislation. With the model, the area around the thermal technology in which the temperature changed by more than 1 °C was determined. Temperature changes in space do not have a linear dependency. However, it can be stated that the determined typical curve will represent the place where the temperature change will be 1/12 of the difference between the temperatures of the poured water and the surrounding water (it is the ratio of the determined change in the groundwater temperature and the modeled maximum temperature change at the location of the infiltration well).

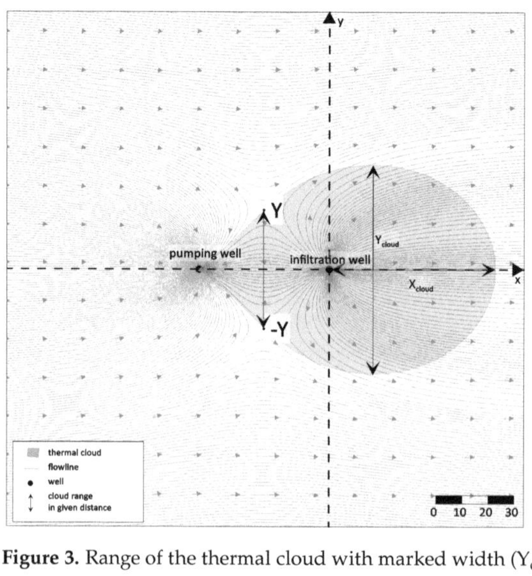


**Figure 3.** Range of the thermal cloud with marked width ($Y_{cloud}$) and length ($X_{cloud}$).

From the results of modeling individual cases of heat transport, curves were determined that safely delimit the calculated values of the extent of the thermal change area. The curves obtained represent a conservative estimate of the dimensions of the thermal cloud after half a year of using thermal technology.

From a time perspective, it is not possible to examine the results of the calculations for the entire range of input parameter values in Table 3 within this task. The series of calculations listed in Table 4 were performed. In each series, the pumped amount varied in the range of 0–30 L/s.

**Table 4.** Overview of the parameters of individual series of calculations.

| Parameter | k | b | n | L | I |
|---|---|---|---|---|---|
| Unit | m/s | m | | m | m/m |
| Series 1 | 0.00125 | 10 | 0.3 | 50 | 0.001 |
| Series 2 | 0.00125 | 10 | 0.3 | 50 | 0.002 |
| Series 3 | 0.00125 | 10 | 0.3 | 50 | 0.0005 |
| Series 4 | 0.00125 | 2.5 | 0.3 | 50 | 0.001 |
| Series 5 | 0.00125 | 10 | 0.22 | 50 | 0.001 |
| Series 6 | 0.00125 | 10 | 0.3 | 20 | 0.001 |
| Series 7 | 0.004 | 10 | 0.3 | 50 | 0.001 |
| Series 8 | 0.001 | 4 | 0.3 | 50 | 0.001 |

#### 3.1.1. Cloud Length

The length of the thermal cloud, starting from the infiltration well, was determined as the distance of heat transport in the natural flow and the additional distance of heat transport due to pumping and pouring into the wells. The radius of temperature change from the heat source in the natural flow of groundwater for t = 182 days is expressed as follows:

$$x_{natural} = \frac{ki}{Rn}\,t \tag{19}$$

$R$ is the retardation factor, which expresses the ratio between the speed of propagation of water and the speed of propagation of temperature in a porous environment. It is caused by heat exchange between water and the rock environment. It depends on the thermal parameters of the water and the rock environment. A typical value of $R = 3$ was used for the calculation.

Based on the results of the modeling, a graph was compiled (Figure 4) of the dependence of the additional distance $x'$ on the parameter $p$, expressed as

$$p = \frac{Q(1 - Q_{back})ki}{nb} \tag{20}$$

The graph shows the modeled distances from the change in groundwater temperature by 1 °C depending on the parameter $p$. A curve delimiting the calculated values was determined, assuming that pumped amounts larger than 10 L/s do not necessarily fall under this curve. However, the shape of the curve is also influenced by the results of the calculations, with yields greater than 10 L/s. The conservative delimiting curve obtained in this way is shown in Figure 4. The curve is described by the following function:

$$x' = 46944.56\,p^{0.33} + 9.71 \tag{21}$$

The final length of the cloud will be limited by the data from the delimiting curve:

$$X_{cloud} = \frac{ki}{Rn}\,t + 46944.56\left(\frac{Q(1 - Q_{back})ki}{nb}\right)^{0.33} + 9.71 \tag{22}$$

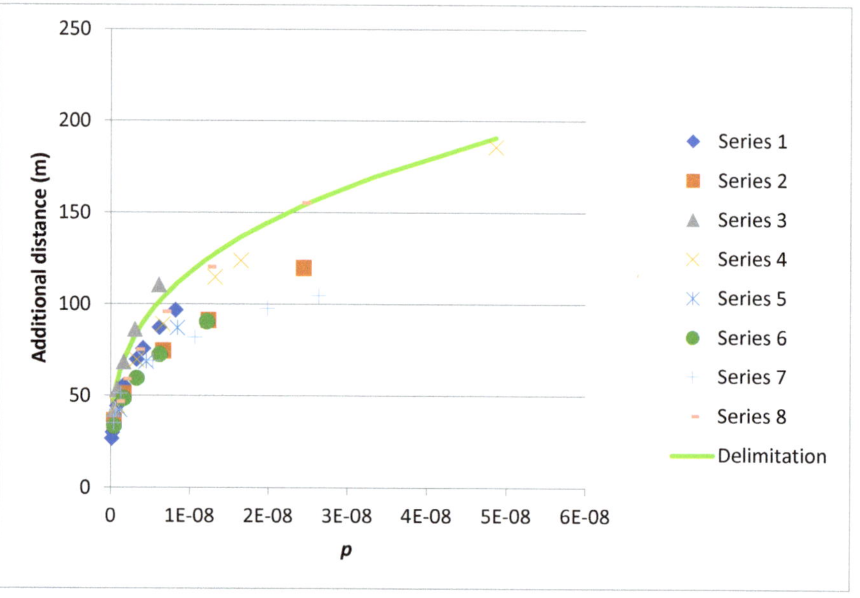


**Figure 4.** Graph of the dependence of the additional distance on environmental parameters.

3.1.2. Cloud Width

For the maximum width of the thermal cloud, its dependence on the width of the return flow between the wells was calculated. The width of the entire return flow is $2Y$ (Equation (7)). In the case of hydraulically isolated wells, the distance $Y$ was replaced by zero. In the graph shown in Figure 5, the resulting values of the width of the area of the temperature change of the groundwater by 1 °C depending on the value of $2Y$ are displayed.

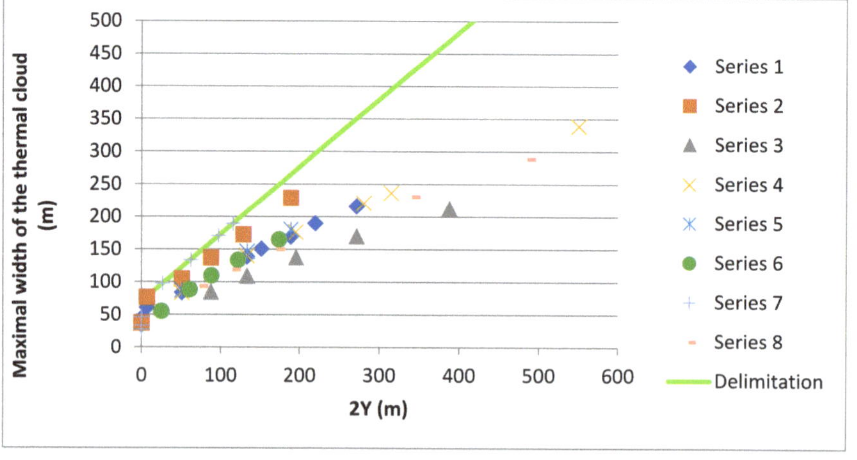


**Figure 5.** Graph of the dependence of the thermal cloud width on the return flow width between the wells.

This display represents a linear course of the observed dependence. Since the width of the return flow is zero for hydraulically separated wells, it is not possible to express the width of the thermal cloud for wells that do not influence each other. The determined delimitated half-line marked in Figure 5 is higher for all pumped amounts of hydraulically separated wells than the value of the intersecting curve at the zero point. From the evaluated data, it is possible to propose to set the boundary width of the cloud as data from the delimiting curve as follows:

$$Y_{cloud} = 1.031\ (2Y) + 70.35\ (23) \tag{23}$$

The delimiting curves (23) and (24) will need to be updated with further calculations.

#### 3.1.3. Simplified Methodology for the Assessment of Heat Pump Technology

The assessment of heat pump technology comprises three stages:

1. The evaluation of the hydraulic separation of the wells. If the wells are hydraulically separated, there is no increase in temperature in the pumped well due to the inflow of already heated water. If the condition from Equation (4) is met, the wells are hydraulically separated, and one can advance to step 3. If not, progression to step 2 is required.
2. The determination of the time of temperature backflow. If the wells are not hydraulically separated, the resulting system can still be used until the thermally adjusted water flows back into the pumped well. From the Equation (9), the water flow duration ($t_{hyd}$) is calculated, and subsequently, the heat transport time ($t_{the}$) is derived from Equation (11). If the heat transport time is less than the number of days in the heating/cooling seasons, the wells thermally influence each other.
3. The estimation of the thermal cloud's extent. This involves determining the estimated dimensions of the thermal cloud that will spread into the surroundings according to Figure 3. The length of the cloud ($X_{cloud}$) is calculated from Equation (22), and the width of the cloud ($Y_{cloud}$) is calculated from Equation (23). It should be noted that the dimensions of the thermal cloud are ascertained based on the numerical modeling of an idealized scenario. These should be overestimated values, determined at a thermal gradient of 12 °C. The boundaries are selected lines at which the thermal change compared to the original state is 1 °C. If the gradient of the poured water is smaller, we cannot predict how the boundaries will diminish or expand, only that the temperature change on the given line will be directly proportionally smaller. Such calculated values are never superior to more precise calculations as this is a very rough approximation.

An Excel sheet, prepared for these calculations, is included in the Supplementary Material.

## 4. Discussion and Conclusions

In this study, a comprehensive series of calculations and modeling efforts were conducted to estimate the thermal impact on groundwater systems resulting from the implementation of heat pump technologies, with a specific focus on high flow rates in gravel–sand aquifers. The primary objective was to determine the maximum width and length of the thermal cloud, which represents the extent of temperature changes within the aquifer. The calculations were performed over a period of 6 months, corresponding to the typical duration of heat pump usage for cooling or heating purposes throughout the year. The undisturbed groundwater temperature was assumed to be 12 °C, while the temperature of the heated water introduced into the infiltration well was set at 24 °C, adhering to the highest permitted temperature according to Slovak legislation.

The modeling results provided valuable insights into the spatial distribution of temperature changes within the aquifer. By identifying the area where the temperature changed by more than 1 °C, the study established a reference point for assessing the thermal impact of heat pump technologies. Although temperature changes in space do not exhibit a linear dependency, a typical curve was identified to approximate the location where the temperature change is approximately 1/12 of the difference between the temperatures of the injected water and the surrounding groundwater. This ratio serves as an estimate of the change in groundwater temperature based on the modeled maximum temperature change at the infiltration well location.

Similar studies in the literature have also addressed the estimation of thermal impacts on groundwater systems resulting from heat pump technologies. Notably, [5,17] explored theoretical proposals that align with the objectives of this study. Ref. [5] emphasized the significance of the accurate estimation of thermal impacts, while the authors of [17] specifically focused on high flow rates. These studies, along with our research, contribute to the broader understanding of thermal impacts and provide valuable insights into the design and operation of heat pump systems.

Equations (22) and (23) safely estimate the extent of temperature changes but cannot assess the efficiency of the technology or the possibilities of its implementation in terms of functionality. In this study, a condition was defined in which part of the water from the injection well would flow back to the extraction well, i.e., the wells would affect each other hydraulically and thus thermally. The return flow to the extraction well reduces the efficiency of the technology because after the time needed for the flow from the injection to the extraction well, the temperature in the extraction well will change. This time is determined in Equation (12). If the technology is not used for longer than this time during the season, its efficiency will not be reduced. Otherwise, to determine the realistically usable amount of heat from the aquifer, more detailed solutions of numerical modeling will be needed under the conditions of the given technology.

Despite the advancements made in this study, it is important to acknowledge certain limitations. The proposed method relies on simplified analytical calculations and numerical modeling which may not fully capture the complexities inherent in real-world groundwater systems. Assumptions made in the model, such as steady-state conditions and advection-dominated transport, may not entirely represent the dynamic nature of heat pump systems. Additionally, the model assumes a two-well system and a uniform water flow which may not be universally applicable to all scenarios. Further research and testing are necessary to refine and validate the proposed method, considering a wider range of system configurations and aquifer characteristics.

In conclusion, this study successfully proposed and verified a method for estimating the extent of temperature changes within aquifers resulting from the implementation of thermal technologies utilizing extraction and injection wells. The empirical dependencies derived from the model calculations provide a conservative estimate of the dimensions of the thermal cloud. For flow rates up to 10 L/s in gravel–sand aquifers, these dependencies are deemed safe and reliable for determining the dimensions of thermal clouds. However, if the calculated influence in a given area raises concerns, more sophisticated numerical modeling techniques are required to accurately determine the actual extent of temperature changes within an aquifer. By addressing these limitations and conducting further research, the accuracy and applicability of the proposed method can be enhanced, facilitating the implementation of thermal technologies while ensuring the protection and sustainable management of groundwater resources.

**Supplementary Materials:** The following supporting information can be downloaded at: https://www.mdpi.com/article/10.3390/hydrology10120225/s1.

**Author Contributions:** D.K.: supervision, formal analysis, funding acquisition, and writing—review and editing; T.K. and M.M.: investigation, conceptualization, methodology, modeling, and writing—original draft preparation; K.H. and M.Z.: formal analysis and writing—review and editing. All authors have read and agreed to the published version of the manuscript.

**Funding:** This research was funded by the Slovak Research and Development Agency under contract No. APVV–14–0174, as well as the Ministry of Education, Science, Research and Sport of the Slovak Republic under contract No. VEGA 1/0302/21.

**Data Availability Statement:** The data presented in this study are available on request from the corresponding author.

**Conflicts of Interest:** The authors declare no conflict of interest. The funders had no role in the design of the study; in the collection, analyses, or interpretation of data; in the writing of the manuscript; or in the decision to publish the results.

## References


1. Ampofo, F.; Maidment, G.; Missenden, J. Review of groundwater cooling systems in London. *Appl. Therm. Eng.* **2006**, *26*, 2055–2062. [CrossRef]
2. Andersson, O. Heat pump supported ATES applications in Sweden. *IEA Heat Pump Cent. Newsl.* **1998**, *16*, 20–21.
3. Bakema, G.; Snijders, A. ATES and ground-source heat pumps in the Netherlands. *IEA Heat Pump Cent. Newsl.* **1998**, *16*, 15–17.
4. Ball, D.A.; Fischer, R.D.; Talbert, S.G.; Hodgett, D.; Auer, F. *State of-the-Art Survey of Existing Knowledge for the Design of Ground Source Heat Pump Systems*; Report ORNL/Sub 80-7800/2; Battelle Columbus Laboratories: Columbus, OH, USA, 1983; 75p.
5. Banks, D. Thermogeological Assessment of Open-Loop Well-Doublet Schemes: A Review and Synthesis of Analytical Approaches. *Hydrogeol. J.* **2009**, *17*, 1149–1155. [CrossRef]
6. Bodvarsson, G. Thermal problems in the siting of reinjection wells. *Geothermics* **1972**, *1*, 63–66. [CrossRef]
7. Clyde, C.G.; Madabhushi, G.V. Spacing of wells for heat pumps. *J. Water Resour. Plan. Manag.* **1983**, *109*, 203–212. [CrossRef]
8. De Marsily, G. *Quantitative Hydrogeology: Groundwater Hydrology for Engineers*; Academic: Orlando, FL, USA, 1986; pp. 277–283.
9. Domenico, P.A.; Schwartz, F.W. *Physical and Chemical Hydrogeology*; Wiley: New York, NY, USA, 1990; 824p.
10. Clauser, C. *Numerical Simulation of Reactive Flow in Hot Aquifers: SHEMAT and Processing SHEMAT*; Springer: Berlin/Heidelberg, Germany, 2003; 332p.
11. Ferguson, G.; Woodbury, A.D. Thermal sustainability of groundwater-source cooling in Winnipeg, Manitoba. *Can. Geotech. J.* **2005**, *42*, 1290–1301. [CrossRef]
12. Gringarten, A.C. Reservoir lifetime and heat recovery factor in geothermal aquifers used for urban heating. *Pure Appl. Geophys.* **1978**, *117*, 297–308. [CrossRef]
13. Liu, G.; Pu, H.; Zhao, Z.; Liu, Y. Coupled thermo-hydro-mechanical modeling on well pairs in heterogeneous porous geothermal reservoirs. *Energy* **2019**, *171*, 631–653. [CrossRef]
14. Di Dato, M.; D'angelo, C.; Casasso, A.; Zarlenga, A. The impact of porous medium heterogeneity on the thermal feedback of open-loop shallow geothermal systems. *J. Hydrol.* **2022**, *604*, 127205. [CrossRef]
15. Gringarten, A.C.; Sauty, J.P. A theoretical study of heat extraction from aquifers with uniform regional flow. *J. Geophys. Res.* **1975**, *80*, 4956–4962. [CrossRef]
16. Tsang, C.F.; Lippmann, M.J.; Witherspoon, P.A. Production and reinjection in geothermal reservoirs. *Trans. Geotherm. Resour. Counc.* **1977**, *1*, 301–303.
17. Pophillat, W.; Attard, G.; Bayer, P.; Hecht-Méndez, J.; Blum, P. Analytical solutions for predicting thermal plumes of groundwater heat pump systems. *Renew. Energy* **2020**, *147*, 2696–2707. [CrossRef]
18. Banks, D. *An Introduction to Thermogeology: Ground Source Heating and Cooling*, 2nd ed.; Wiley: Chichester, UK, 2012; 544p.
19. Luo, J.; Kitanidis, P.K. Fluid residence times within a recirculation zone created by an extraction–injection well pair. *J. Hydrol.* **2004**, *295*, 149–162. [CrossRef]
20. Milnes, E.; Perrochet, P. Assessing the impact of thermal feedback and recycling in open-loop groundwater heat pump (GWHP) systems: A complementary design tool. *Hydrogeol. J.* **2013**, *21*, 505–514. [CrossRef]
21. Piga, B.; Casasso, A.; Pace, F.; Godio, A.; Sethi, R. Thermal Impact Assessment of Groundwater Heat Pumps (GWHPs): Rigorous vs. Simplified Models. *Energies* **2017**, *10*, 1385. [CrossRef]
22. McDonald, M.G.; Harbaugh, A.W. *A Modular Three-Dimensional Finite-Difference Ground-Water Flow Model*; Holcomb University: Indianapolis, IN, USA, 1988. [CrossRef]
23. Zheng, C.; Wang, P.P. *MT3DMS: A Modular 3-D Multispecies Model for Simulation of Advection, Dispersion and Chemical Reactions of Contaminants in Groundwater Systems. Documentation and User's Guide*; Contract Report SERDP-99-1; Army Engineer Research and Development Center: Vicksburg, MS, USA, 1999.
24. Hetch-Mendéz, J. Implementation and Verification of the USGS Solute Transport Code MT3DMS for Groundwater Heat Transport Modeling. Master's Thesis, University of Tübingen, Tübingen, Germany, 2008.

water

MDPI

*Article*

# Evaluation of Groundwater Vulnerability of Yishu River Basin Based on DRASTIC-GIS Model

**Jiaqi Hu [1], Peng Yang [1,*], Qiang Li [2], Min Wang [2], Jianguo Feng [2], Zongjun Gao [2] and Jiutan Liu [2]**

1 No. 8 Institute of Geology and Mineral Resources Exploration of Shandong Province, Rizhao 276826, China; dkj8ycwk@shandong.cn

2 College of Earth Science and Engineering, Shandong University of Science and Technology, Qingdao 266590, China; qiangli6110@126.com (Q.L.); brightwangm@163.com (M.W.); fengjianguo20316@sohu.com (J.F.); zongjungao1964@163.com (Z.G.); ljtsdust@sdust.edu.cn (J.L.)

* Correspondence: pyang@email.cugb.edu.cn

**Abstract:** The evaluation of vulnerability is a crucial aspect in the sustainable development, utilization, and preservation of groundwater resources. This study utilizes a comprehensive approach, integrating systematic analysis of hydrogeological conditions and the utilization of observed and collected data. The evaluation of groundwater vulnerability in the Yishu River Basin (YRB) was conducted by employing the DRASTIC model, along with the zone overlay function of GIS software. Seven evaluation indicators were considered in this assessment. The findings demonstrate that the groundwater vulnerability in the YRB can be categorized into five divisions: excellent, good, medium, poor, and very poor, accounting for 14.5%, 42.3%, 27.9%, 14.0%, and 1.3% respectively. The areas with low vulnerability are predominantly located in the eastern part of the study area, covering the largest proportion of the total area. Conversely, areas with high vulnerability are found alongside both banks of the Shu River, forming narrow strips. Although these areas have smaller overall coverage, they contain dispersed water sources that require careful attention. These research findings provide valuable scientific insights and serve as a reference for urban planning, land use management, and groundwater resource protection in the YRB. The formulation and adoption of targeted protection measures in accordance with different groundwater vulnerability zoning, the formulation of scientific groundwater resource development and utilization programs, and execution of land resource planning are of great significance from the perspective of groundwater resource protection.

**Keywords:** groundwater; DRASTIC-GIS model; vulnerability; Yishu River Basin

**Citation:** Hu, J.; Yang, P.; Li, Q.; Wang, M.; Feng, J.; Gao, Z.; Liu, J. Evaluation of Groundwater Vulnerability of Yishu River Basin Based on DRASTIC-GIS Model. *Water* **2024**, *16*, 429. https://doi.org/10.3390/w16030429

Academic Editor: Lucila Candela

Received: 8 December 2023
Revised: 22 January 2024
Accepted: 23 January 2024
Published: 29 January 2024



## 1. Introduction

Groundwater is a precious and widely available resource that serves as a vital water source for various purposes, including industrial production, human consumption, socio-economic development, and ecological preservation [1–3]. In arid and semi-arid regions, it often stands as the sole water source. Currently, global concern centers around changes in both the quantity and quality of groundwater [4–6]. On one hand, industrial activities contribute to groundwater pollution through the discharge of wastewater, exhaust gases, and solid waste [1,7]. Similarly, agricultural practices, such as excessive fertilizer application and pesticide residue from spraying, further deteriorate groundwater quality [8,9]. On the other hand, the escalating demands for agricultural and irrigation water, burgeoning population, and excessive groundwater extraction due to urban and industrial expansion have resulted in detrimental effects on the geological and ecological landscape [10,11]. Impacts include land subsidence, shrinking lakes and wetlands, and the degradation of vegetation, ultimately posing significant threats to the environment.

Groundwater has many advantages over surface water, yet it possesses characteristics of slow renewable speed and poor regulation ability [12]. Once polluted, treatment and regulation of groundwater pose significant challenges. Heavy metals are cumulative and persistent, and once they enter the groundwater body they will remain for a long time and gradually accumulate, leading to groundwater pollution, and contaminated groundwater poses a serious threat to the ecological environment and human health [13]. Firstly, contaminated groundwater affects the source of drinking water and directly jeopardizes human health. Secondly, heavy metals further diffuse into the water system through the interaction between groundwater and surface water, endangering aquatic organisms and ecosystems. Finally, contaminated groundwater in agricultural irrigation and industrial water use can also lead to the contamination of agricultural and industrial products, affecting food safety and industrial development. The results of single-parameter sensitivity analysis and map-removal sensitivity analysis of groundwater vulnerability indicators in Shanghai city by combining the hydrogeological characteristics of the plains river network illustrate the reasonableness of the application of the DRAHC model in the plains river network area [14]. Tang Xuefang et al. evaluated the antifouling performance of groundwater in Chengdu city, and the results showed that the higher zones were mainly distributed in the northwestern terrace area and its vicinity, the medium antifouling performance zones were mainly distributed in the secondary terraces, and the lower antifouling performance zones were mainly distributed along the rivers and their branch channels of diffuse flats, the primary terraces and other zones in the form of a strip [15]. Due to the complexity of groundwater systems, there are considerable difficulties in managing contaminated groundwater [16,17]. On the one hand, governance projects incur huge costs; on the other hand, their effects are slow to take root. Compounded by various pathways of groundwater supply, runoff, and discharge, it becomes a problematic undertaking in a complex geological environment [18]. Furthermore, each aquifer is hydraulic, making it challenging to delineate and limit the pollution range once groundwater is contaminated. To effectively prevent and control groundwater pollution, it is crucial to strengthen the protection and management of groundwater, as well as to evaluate its vulnerability [18–21]. Reasonable use of groundwater is imperative.

The groundwater's ability to withstand external pollution under various hydrogeological conditions, without considering the influence of human activities and pollution sources, and solely focusing on internal hydrogeological factors, plays a crucial role in evaluating its vulnerability [22–24]. The DRASTIC model has its limitations in assigning weights and ratios to different thematic layers and their subcategories. Different researchers have used other techniques to address these limitations, such as calculating recharge using the SWAT model and modifying the weights and ratios of the layers and their subclasses using Analytical Hierarchy Process (AHP) and Frequency Ratio (FR) techniques [13]. There are many methods for evaluating the vulnerability of groundwater, such as the iterative index method, fuzzy mathematical comprehensive evaluation method, and numerical simulation method. Currently, the most commonly utilized approach to assess groundwater pollution prevention performance, both nationally and internationally, is the DRASTIC model [22,23,25,26]. The GOD model, AVI model, ISIS model, etc. [27], use the sum of the scores of each parameter. The GOD method is only applicable to areas with large differences in vulnerability and has a large error when used in areas with small changes in vulnerability. The DRASTIC model, Sintacs model, Siga model, Epik model, etc., of the counting system models use the weight-scoring method [28], which is mostly applied to the shallow groundwater vulnerability assessment. The main evaluation models for porous media aquifers are DRASTIC, AVI, Siga, and Epik models. Among the DRASTIC, AVI, Sintacs, ISIS, and Galdit methods [28], the Galdit model considers seawater intrusion in coastal aquifers and has been widely used to assess coastal aquifer vulnerability [29,30]. The DRASTIC model has been extensively applied in groundwater vulnerability assessments in the United States and the European Union. The notable advantage of this model lies in its simplicity, ease of implementation, and ability to quantitatively analyze hydrogeological factors affecting pollutant transport [13]. In view of the fact that there are few studies on the YRB [31], in order to determine the regional groundwater environment situation, in this work, the DRASTIC evaluation model was adopted, the groundwater vulnerability evaluation index system of the YRB was constructed, and a specific study was conducted, which provides a scientific example for other regions to carry out the evaluation of groundwater vulnerability. In addition, Geographic Information System (GIS) technology facilitates a wide range of functions such as data input, storage, management, retrieval, processing, and comprehensive analysis of geospatial data and information. Accordingly, it has been widely applied across various domains. In the context of aquifer vulnerability evaluation, the vector or grid units in GIS graphics can correspond to the evaluation units employed in evaluation methods [26,32]. Furthermore, the classification and overlay analysis process of GIS software aligns precisely with the factor scoring process and weighted sum analysis of diverse factors in evaluating groundwater vulnerability [33,34]. The utilization of GIS technology's attribute database and spatial overlay analysis for evaluating the vulnerability of aquifers proves to be both straightforward and prevalent.

Groundwater serves as the primary water supply for industrial and agricultural activities, urban development, and domestic consumption in the Yishu River Basin (YRB). Given its crucial role in the region's sustainable economic development, it is indispensable [3]. To date, many scholars have conducted a large amount of research on the characteristics and quality of groundwater pollution in the YRB, but there is a lack of research on the vulnerability of groundwater in the YRB, and there is no relevant model to simulate the vulnerability of groundwater. Hence, this study employed the DRASTIC model in conjunction with the GIS to assess the groundwater's vulnerability. The purpose of this study is to determine the vulnerability of groundwater in the area by collecting and organizing hydrogeological data and modeling to assess the groundwater fouling resistance. The findings will furnish essential foundational information for the rational protection of groundwater resources. Furthermore, the results will aid environmental management departments in formulating plans for the safeguarding of groundwater, as well as in industrial and agricultural planning, environmental monitoring, and other related endeavors. Ultimately, this research will serve as a theoretical basis for the prevention and control of groundwater pollution.

## 2. Materials and Methods

### 2.1. Study Area

The YRB is situated in the central and southern parts of Shandong Province, covering a vast area of 17,911 km$^2$ (Figure 1). The region experiences a typical northern warm temperate monsoon climate. It exhibits four distinct seasons, ample sunlight, and a mild climate. Precipitation in the study area exhibits significant year-to-year variability, leading to an uneven annual distribution. Approximately 75% of the annual precipitation occurs between June and September. The yearly average precipitation is 824.8 mm. The distribution of the surface water system in the study area and its vicinity is influenced by four primary faults within the Yishu fault zone. The rivers' courses generally align with the main fault orientation, flowing from north to south. Key rivers in the area include the Yi River, Shu River, Dongwen River, Meng River, and Beng River. The Yi River and Shu River (Figure 1) and their extensive alluvial plain, running from north to south, comprise the crucial runoff formations stemming from alluvial and diluvial accumulations. As such, the region boasts abundant groundwater resources. The study collected data from 132 sampling sites in the YRB during the dry season in May 2017 and during the rich season in September 2017, for a total of 264 datasets in both the dry and wet seasons (Figure 1).

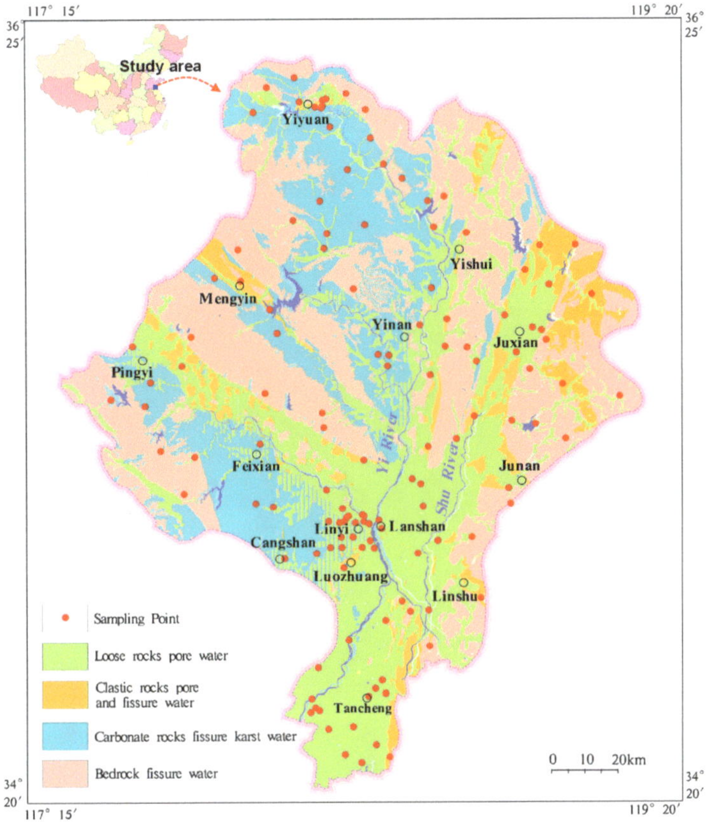


**Figure 1.** Hydrogeological map and distribution map of monitoring points in the YRB.

The topography of the YRB is high in the north and low in the south, and the west side belongs to the middle and low hilly area in the south of Shandong. Geomorphology is mainly divided into erosion landforms, erosion denudation landforms, denudation landforms, denudation accumulation landforms, and accumulation landforms, as shown in Figure 2. The erosion landforms, located in the Yishan and Mengshan mountains on the west side of the YRB, have active tectonic movement and strong water erosion. The erosion denudation landforms are distributed in the middle and low mountainous areas, such as Menglianggu uplift and Mamuchi Dome fault in the northwest of the YRB, and in the northern mountainous area of Junan. The accumulation landforms are distributed in the plains on both sides of the Yi River and the Shu River and in the Lintancang plain area.

Based on the stratigraphy, structure, and groundwater occurrence conditions of the YRB, as well as the physical properties of water in rocks, groundwater can be classified into four categories: unconsolidated rock pore water, clastic rock pore and fissure water, carbonate rock fissure karst water, and bedrock fissure water (Figure 1). Unconsolidated rock pore water is primarily found in the first terrace of rivers, southern alluvial plains, residual hills, basin edges, peripheries of alluvial layers, and mountain valley areas. The aquifer is composed of medium to fine sand, coarse sand gravel, sandy clay mixed with ginger stone, and thin layers of fine sand. Clastic rock pore and fissure water is mainly distributed in the central graben of the study area. The aquifer consists of Cretaceous Qingshan Group tuffaceous sandstone, Dasheng Group conglomerate, and sandstone. Carbonate rock fissure karst water exists as both groundwater and confined water in the cracks and karst of Cambrian and Ordovician limestone, as well as limestone interbedded with shale and sandstone. Bedrock fissure water primarily occurs in weathering fissures and structural fracture zones of the Middle Archean Yishui Group, the New Archean Mount Taishan Group metamorphic rocks, magmatic rocks of various stages, and the Cretaceous Qingshan Group extrusive rocks.

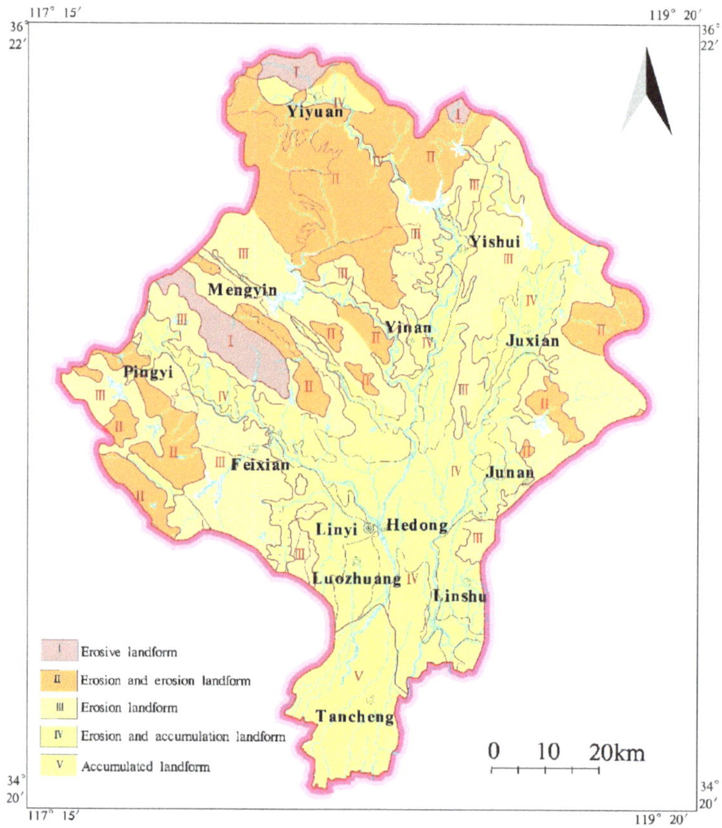


**Figure 2.** Outline map of landform zoning in the study area.

### 2.2. Data Collection

There are 24 groundwater sources in the YRB, mainly located in urban areas, including Yiyuan urban area, Yishui urban area, Ju County urban area, Mengyin urban area, Tancheng urban area, Yinan urban area, and Pingyi urban area. According to the hydrogeological subdivision, there are 4 water sources located in the southeast of the Laiwu fault basin groundwater resources section, 3 water sources located in the Yishui fault valley groundwater resources section, 3 water sources located in the Ju County hilly valley groundwater resources section, 2 water sources located in the Yinan hilly valley groundwater resources section, 4 water sources located in the Mengyin-Tambu fault valley groundwater resources section, 5 water sources located in the Pingyi-Feixian hilly valley groundwater resources section, and 3 water sources located in the middle reaches of the YRB hilly valley groundwater resources section. There are two types of water sources, namely karst fissure water sources and pore water sources, of which there are 17 pore water sources and 7 karst water sources; there are 4 small-sized water sources, 19 medium-sized water sources, and 1 large-sized water source according to the scale of water sources.

### 2.3. Thematic Layer Preparation

This study examined the vulnerability of groundwater in the YRB using the DRASTIC assessment method (Figure 3). Firstly, based on the regional hydrogeological data, the indicators were subdivided to evaluate the vulnerability; subsequently, the ArcGIS 10.2 software was utilized to superimpose the regions according to the weighting values of the indicators, the raw data input was carried out using vector data (points, lines, and surfaces), and the input vector data were transformed into raster data and spatially analyzed accordingly through the GIS software so that the relationships between each data layer and the spatial distribution status of each data layer could be quantified and understood. The input vector data were transformed into raster data using GIS software and analyzed accordingly so as to quantitatively grasp the relationship between each data layer and the spatial distribution of each data layer. Taking the GIS as a working platform, the raw data of groundwater vulnerability energy influencing factors were scanned, aligned, edited with vectorization, converted into the appropriate format, and topologically formed into zones through the process of map scanning, and vectorization editing. Finally, the spatial layers of the influencing factors were stored in the format of vector data. After the completion of vector data input, all the vector data formats were converted to raster data format with the vector raster conversion tool, and then the layers were divided into standard grids. The size of the standard grids was decided after considering the spatial distribution of each item, the characteristics of the changes, and the accuracy of the collected data so as to delineate the vulnerability zones in the YRB. The DRASTIC model, a well-established framework for assessing regional groundwater pollution potential, quantifies the impact of each assessment index on groundwater pollution prevention through scoring values [33,34]. Each evaluation index is classified into grades, and its score is determined based on its influence on groundwater pollution. The fundamental concept underlying the DRASTIC model asserts that the pollution prevention performance of an aquifer is primarily governed by its essential hydrogeological characteristics [35–37]. Consequently, hydrogeological parameters pertaining to pollution prevention performance were comprehensively evaluated to assess the aquifer's ability to prevent pollution.

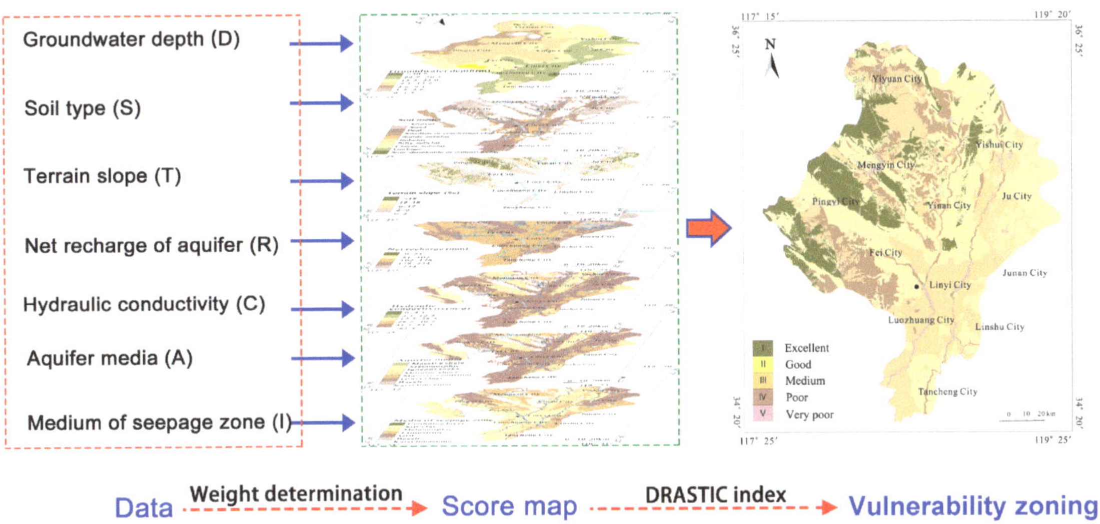


Figure 3. Flowchart of the DRASTIC methodology.

The DRASTIC model comprises seven primary evaluation indexes that influence the vulnerability of groundwater. These indexes are the groundwater depth (D), net recharge of aquifers (R), aquifer media (A), soil type (S), topography (T), medium of seepage zone (I), and hydraulic conductivity of aquifer (C). Each of these indicators represents a specific hydrogeological condition that contributes to the overall vulnerability of groundwater [33,34,36].

The groundwater depth (D) plays a crucial role in determining the vulnerability of pollutants reaching the aquifer. In general, a deeper underground water depth is associated with lower vulnerability. Conversely, a shallower depth indicates high vulnerability. Net recharge of aquifers (R) refers to the amount of water that percolates from the surface to groundwater. A higher net recharge leads to increased transport of polluting substances in the infiltrating water, thereby reducing the groundwater's vulnerability. The aquifer medium (A) is an intrinsic property of the aquifer and influences its vulnerability. Fine particles in the medium result in more primary cracks and pores, enhancing the barrier, adsorption, and degradation of pollutants. Thus, a smaller particle size indicates lower vulnerability ability, while a larger particle size implies weaker performance. Soil medium (S) refers to the upper part of the vadose zone where biological activities occur. Clay properties and particle size in the soil significantly impact vulnerability. Soils with smaller clay particles and minimal expansion and shrinkage demonstrate stronger groundwater pollution prevention performance, while soils with larger particles and higher variability exhibit weaker performance. Topographic slope (T) refers to a surface slope or its changes. A larger topographic slope results in stronger surface runoff and less groundwater infiltration recharge, indirectly affecting regional net recharge. Regions with steep terrain slopes tend to exhibit stronger groundwater pollution prevention performance, while those with gentler slopes show weaker performance. The medium of the seepage zone (I) is an intrinsic characteristic that primarily reflects the particle thickness and fracture development degree. Coarse particles and well-developed cracks indicate higher vulnerability, while finer particles correspond to stronger performance. When evaluating a confined aquifer, the impact of the vadose zone is expanded to include both the vadose zone and any saturated zones which overlie the aquifer. In the case of a confined aquifer, the significantly restrictive zone above the aquifer which forms the confining layer is used as the type of media which has the most significant impact [38]. According to the range of scores given by the DRASTIC model for each medium, the assigned value for a pressurized aquifer is always 1. Hydraulic conductivity (C) relates to the aquifer medium's hydraulic transmission capacity. A lower hydraulic conductivity coefficient indicates a lower vulnerability, whereas a higher coefficient suggests weaker performance. The assessment indicators in the DRASTIC pollution susceptibility scoring system were determined as shown in Table 1.

### 2.4. Evaluation Index Weight Determination

The seven index factors in the DRASTIC model exhibit different degrees of influence on the prevention of groundwater pollution. To determine the relative importance of each factor in evaluating groundwater pollution prevention performance, corresponding weights were assigned to each score within a defined range [39]. These weights range from 1 to 5, with higher weights indicating a greater impact of the index factor on the evaluation of groundwater pollution prevention performance. Table 2 presents the evaluation index weights for groundwater vulnerability.

**Table 1.** Fragility evaluation index scoring system.

| Index | Range/Type | Mark |
|---|---|---|
| Groundwater depth (D) | Range (m) | Mark |
| | 0~1.5 | 10 |
| | 1.5~4.6 | 9 |
| | 4.6~9.1 | 7 |
| | 9.1~15.2 | 5 |
| | 15.2~22.9 | 3 |
| | 22.9~30.5 | 2 |
| | >30.5 | 1 |
| Soil type (S) | Type | Mark |
| | Thin layer or none | 10 |
| | Gravel | 10 |
| | Sand | 9 |
| | Peat | 8 |
| | Swollen or condensed clay | 7 |
| | Sandy subclay | 6 |
| | Subclay | 5 |
| | Silty subclay | 4 |
| | Clayey subclay | 3 |
| | Garbage | 2 |
| | Non-shrinkable or cohesive clay | 1 |
| Terrain slope (T) | Range (%) | Mark |
| | 0~2 | 10 |
| | 2~6 | 9 |
| | 6~12 | 5 |
| | 12~18 | 3 |
| | >18 | 1 |
| Net recharge of aquifer (R) | Type (mm) | Mark |
| | 0~51 | 1 |
| | 51~102 | 3 |
| | 102~178 | 6 |
| | 178~254 | 8 |
| | >254 | 9 |
| Hydraulic conductivity (C) | Type (m/d) | Mark |
| | 0~4.1 | 1 |
| | 4.1~12.2 | 2 |
| | 12.2~28.5 | 4 |
| | 28.5~40.7 | 6 |
| | 40.7~81.5 | 8 |
| | >81.5 | 10 |
| Aquifer media (A) | Type | Mark |
| | Massive shale | 2 |
| | Metamorphic/igneous rocks | 3 |
| | Weathered metamorphic/igneous rocks | 4 |
| | Moraine sheet | 5 |
| | Stratified sandstone, limestone, and shale sequences | 6 |
| | Massive sandstone | 6 |
| | Massive limestone | 6 |
| | Gravel layer | 8 |
| | Basalt | 9 |
| | Karst limestone | 10 |
| Medium of seepage zone (I) | Type | Mark |
| | Confined layer | 1 |
| | Silt/clay | 3 |
| | Shale | 3 |
| | Limestone | 6 |
| | Sandstone | 6 |
| | Stratified limestone, sandstone, shale | 6 |
| | Gravel containing more silt and clay | 6 |
| | Metamorphic/igneous rocks | 4 |
| | Grit | 8 |
| | Basalt | 9 |
| | Karst limestone | 10 |

**Table 2.** Weights of evaluation indicators for groundwater vulnerability.

| Evaluation Index | D | R | A | S | T | I | C |
|---|---|---|---|---|---|---|---|
| Weight | 5 | 4 | 3 | 2 | 1 | 5 | 3 |

### 2.5. Calculation of DRASTIC Index

The assessment of the vulnerability of groundwater in a given area is determined by the linear weighted sum of each index factor's scores. This approach enables the evaluation of the relative strength of groundwater vulnerability based on its scores [40,41]. The calculation of the DRASTIC Index (DI) is achieved through Equation (1):

$$\mathrm{DI} = \mathrm{D}w \times \mathrm{D}r + \mathrm{R}w \times \mathrm{R}r + \mathrm{A}w \times \mathrm{A}r + \mathrm{S}w \times \mathrm{S}r + \mathrm{T}w \times \mathrm{T}r + \mathrm{I}w \times \mathrm{I}r + \mathrm{C}w \times \mathrm{C}r \quad (1)$$

The formula incorporates two variables: '$w$' and '$r$', where '$w$' represents the weight assigned to each evaluation indicator, while 'r' represents the corresponding score value. Seven evaluation indicators ('D', 'R', 'A', 'S', 'T', 'I', and 'C') are denoted to assess the groundwater quality. The DRASTIC Index (DI) is determined as the weighted sum of the scores attributed to each evaluation indicator. A lower DI value indicates a lower level of vulnerability in groundwater. Consequently, the likelihood of pollution decreases, while, conversely, higher vulnerability is associated with higher DI values [42,43].

## 3. Results and Discussion

### 3.1. Single Evaluation of Vulnerability

#### 3.1.1. Groundwater Depth

A total of 264 datasets were collected from the YRB during the dry and abundant water periods in May and September 2017, respectively. The depth to the water table has a minimum value of 0.5 m and a maximum value of 25 m. Seasonal variations in groundwater levels were explained by calculating average groundwater level values at each point using data from both periods. Using ArcGIS interpolation, a contour map of groundwater depth was generated. Furthermore, a subregion map depicting the groundwater depth in the YRB was created based on the evaluation index's subregion values (Table 1), as illustrated in Figure 4. Notably, the YRB's northeast and southwest regions exhibited groundwater depth scores below 3, whereas the midwest had comparatively higher scores. Overall, there was a decreasing trend in the scores from northwest to southeast. This suggests that the vulnerability of groundwater in the YRB improved gradually from northwest to southeast, as indicated by the groundwater depth.

#### 3.1.2. Net Recharge of Aquifers

In this study, the average annual rainfall of the YRB was multiplied by the corresponding infiltration coefficient to derive the index value. Subsequently, the index value was scored based on the rating table presented in Table 1, and a regional scoring map was generated. The results indicate that the aquifer net recharge scores in the study area predominantly range from 3 to 6, with higher scores of 8 observed in the eastern Juxian, southwestern Mengyin, and Yiyuan regions. Figure 5 depicts the score map illustrating the net recharge of aquifers in the YRB.

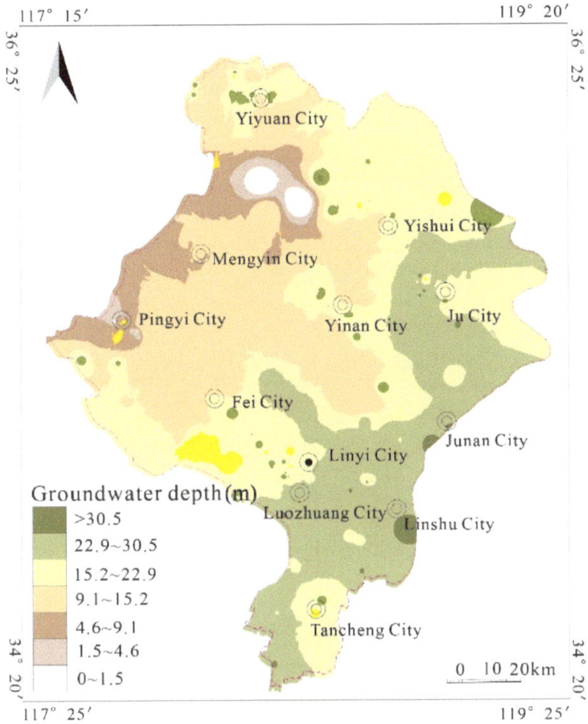


**Figure 4.** Scoring map for groundwater depth (D).

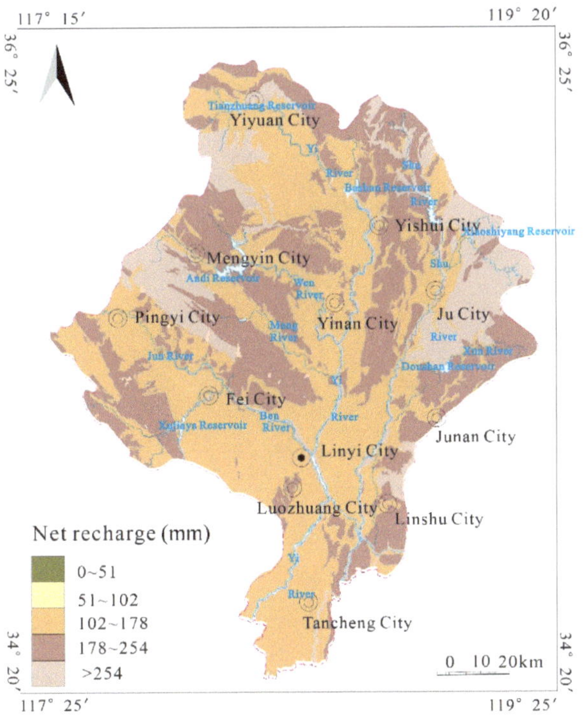


**Figure 5.** Scoring map for net recharge (R).

### 3.1.3. Types of Aquifer Media

Based on the hydrogeological map (Figure 1) and the topographic and geomorphological map (Figure 2) of the YRB, ArcGIS was utilized for projection and grading purposes. The topography of the basin showcases a gradual decrease in elevation from the north to the south. Various landform types, such as erosion, denudation, and accumulation, can be observed. The gravel layer accounts for about 50% of the entire study area and is mainly distributed along the river, being most widely distributed in Tancheng City and Linyi City. The distribution of karst limestone accounts for about 15% of the total, mainly in Yiyuan City and Fei City, with a small amount of distribution in the upper reaches of the river. Massive shale accounts for about 5% of the total and is mostly found in the cities of Jounan and Linshu. Metamorphic/igneous rocks, Moraine sheet, and other water-bearing media account for about 30% of the total and are interspersed with the gravel layer. The YRB exhibits diverse groundwater sources, encompassing pore water found in unconsolidated rock, fractured water in clastic rock, fractured karst water in carbonate rock, and fractured water in bedrock. The larger the particle size of the aquifer medium or the more fissured river karst pipes, the greater the permeability and the less capable the aquifer medium is of weakening pollutants. Less permeable cohesive soils have a barrier to pollutant infiltration while possibly causing range spreading on the face of the contaminated area due to the development of surface runoff. Figure 6 illustrates the scoring map that portrays the distribution of aquifer media across the YRB.

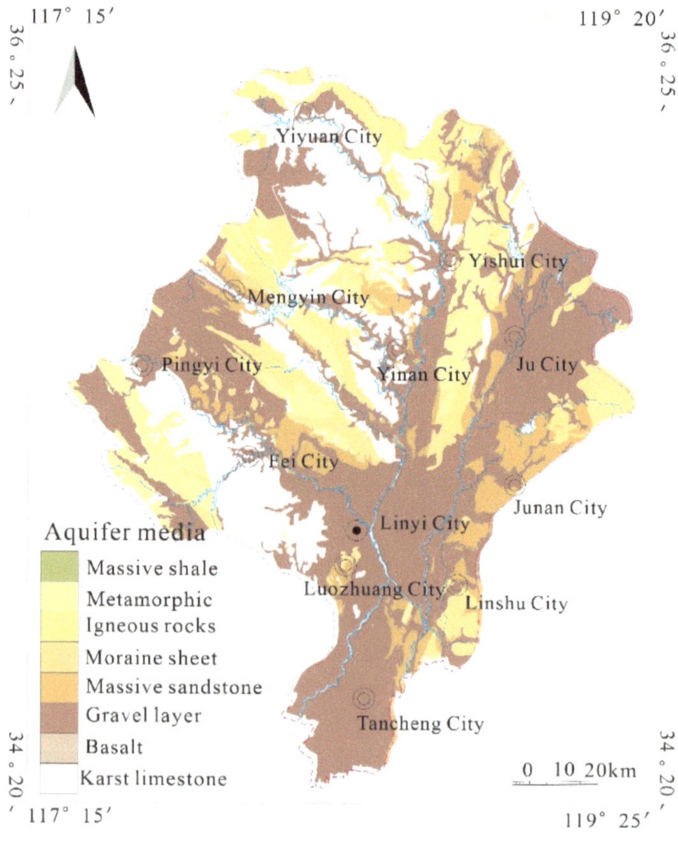


**Figure 6.** Scoring map for aquifer media (A).

3.1.4. Soil Type

A geological map and soil type zoning map of the YRB was utilized to analyze and score the soil types within the study area. These maps were projected utilizing ArcGIS 10.2 software, as illustrated in Figure 7. The thin layer or none and gravel account for about 70% of the total and are mainly distributed in the northern part of the cities of Mengyin, Yishui, Yiyuan, and Fei. Peat, swollen or condensed clay, sandy subclay, and other soil media accounted for about 30% of the total, mainly distributed in the southern cities of Linyi, Tancheng, and Linshu.

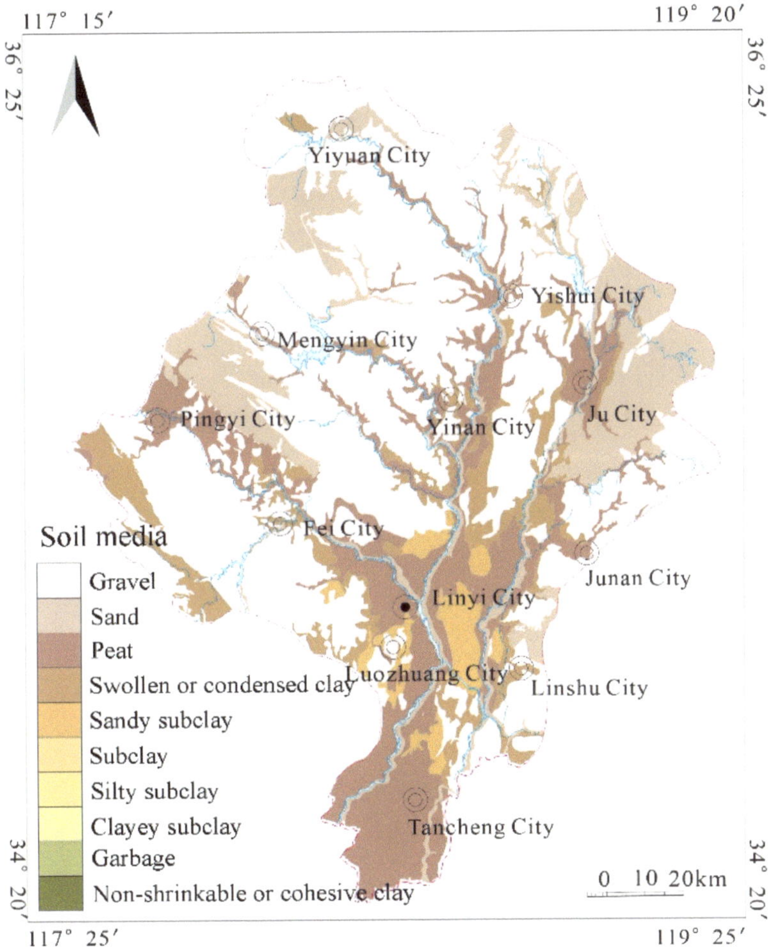


**Figure 7.** Scoring map for soil media (S).

3.1.5. Topographic Slope

Initially, the digital elevation model (DEM) data for the YRB were downloaded, followed by the use of ArcGIS software's slope extraction function to obtain the topographic slope of the study area. This information was then utilized to divide the region into distinct topographic slope divisions. Consequently, a scoring map for the topography of the YRB was produced, as depicted in Figure 8.

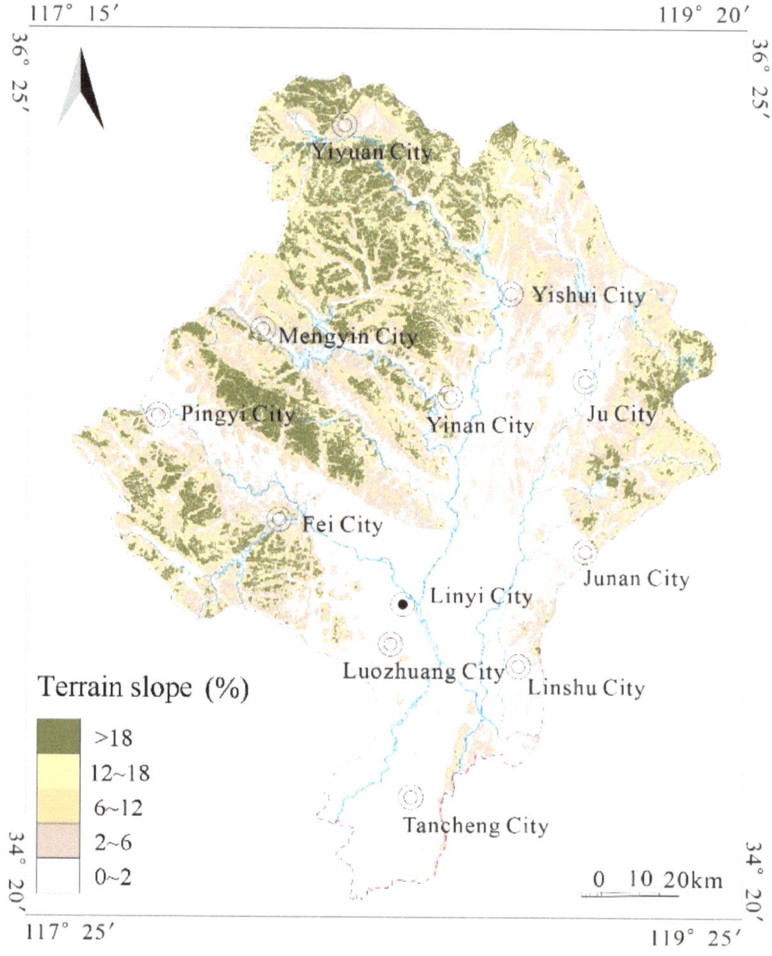


**Figure 8.** Scoring map for topography (T).

#### 3.1.6. Medium in the Seepage Area

The study area was classified using the geological map and soil type distribution map prepared by the Hydrology Bureau of Shandong Province. ArcGIS software was subsequently employed for projection and scoring, as demonstrated in Figure 9.

#### 3.1.7. Hydraulic Conductivity of Aquifer

Based on the geological map and hydrogeological map of the YRB, the study area's aquifer medium lithology map was divided, resulting in the aquifer medium lithology map specific to the YRB. The water conservancy conductivity coefficient was assigned based on the aquifer lithology, leading to the creation of a scoring map pertaining to the hydraulic conductivity of the YRB, as illustrated in Figure 10.

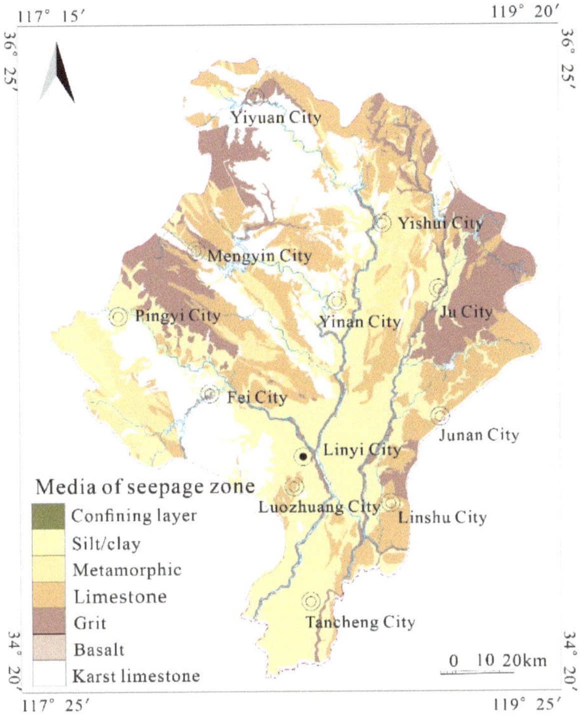


**Figure 9.** Scoring map for medium in seepage area (I).

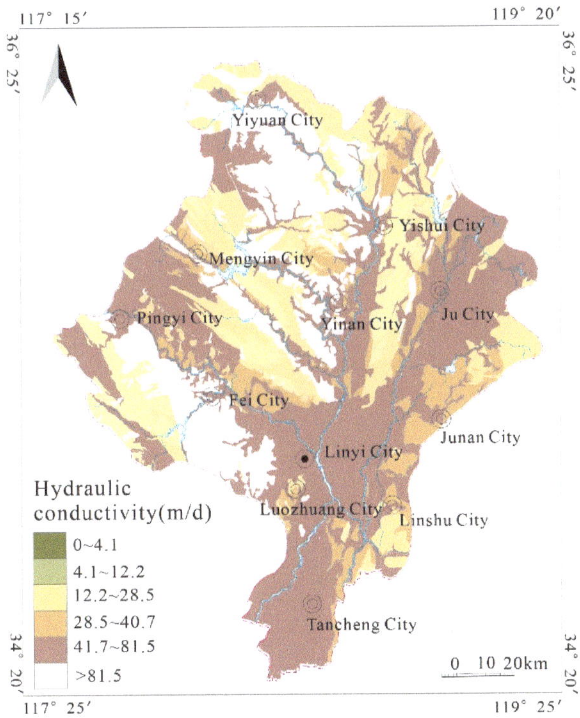


**Figure 10.** Scoring map for hydraulic conductivity (C).

### 3.2. Comprehensive Evaluation of Vulnerability

The vulnerability of the entire YRB was evaluated based on single evaluation results using a weighted approach. Each evaluation index was multiplied by its corresponding weight and aggregated to obtain the overall evaluation results. The resultant DI value ranged from 80 to 221, with higher values indicating higher groundwater vulnerability and greater vulnerability to pollution. The assessment index analysis and evaluation results, combined with the original DRASTIC vulnerability classification principle, were used to divide the basin into five levels corresponding to different vulnerability zones: excellent, good, medium, poor, and very poor. These zones are outlined in Table 3.

**Table 3.** Vulnerability assessment degree division.

| DI | Vulnerability | Rank |
|---|---|---|
| <108 | Excellent | I |
| 108~136 | Good | II |
| 136~164 | Medium | III |
| 164~192 | Poor | IV |
| >192 | Very poor | V |

The vulnerability evaluation of the YRB was derived based on the established classification criteria, as depicted in Figure 11.

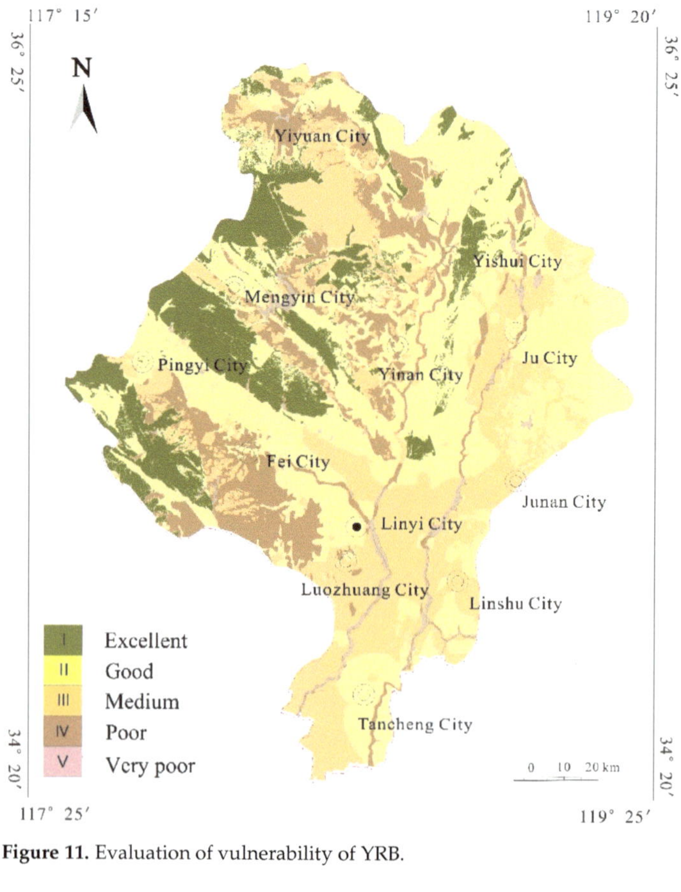


**Figure 11.** Evaluation of vulnerability of YRB.

The classification criteria are described below: (1) Area of low vulnerability: This area exhibits a DI value below 108 and is predominantly situated in the western and southwestern portions of the YRB, covering an area of 2602.28 km$^2$, which represents 14.5% of the total study area. The groundwater in this region is found at significant depths and is primarily composed of water in massive rock fissures, with minimal development of rock cracks and poor permeability. Additionally, the surface is overlaid with a substantial Quaternary system, characterized by predominantly silty clay lithology, which effectively safeguards the groundwater.

(2) Area of low vulnerability performance: This region demonstrates DI values ranging from 108 to 136 and is primarily distributed in the eastern part of the study area, encompassing an area of 7583 km$^2$, accounting for 42.3% of the total area. The groundwater in this area mostly comprises fissure water in massive rock formations and unconsolidated rock pore water, with limited development of rock fissures and low permeability. The Quaternary system in this area consists of a clay layer, providing an effective barrier against the penetration of pollutants into the underground.

(3) Area of moderate vulnerability: This area exhibits DI values ranging from 136 to 164 and is mainly located in the southern and northern regions of the study area, covering an area of 5000.59 km$^2$, which constitutes 27.9% of the total area. It lies within the transitional zone between karst water and Quaternary pore water. Although the surface is covered by the Quaternary system, the system is generally thin, resulting in limited groundwater protection. Consequently, this area is classified as a zone of moderate vulnerability based on various evaluation indicators.

(4) Area of high vulnerability: The DI values in this particular region range from 164 to 192 and are primarily concentrated in the southwest, central, and northern parts of the YRB, covering an extent of 2502.2 km$^2$, which accounts for approximately 14.0% of the total area. The dominant groundwater in this region is characterized as karst fissure water, with limited Quaternary protection on the surface. As a result, pollutants easily permeate through the surface and contaminate the groundwater. Moreover, the aquifer primarily consists of unconsolidated rock pore water, characterized by high permeability, leading to further vulnerability to groundwater pollution. Based on diverse evaluation indicators, this area is classified as having high vulnerability.

(5) Area of very high vulnerability: The DI value in this area exceeds 192 and is primarily observed in surface water bodies such as reservoirs and rivers within the YRB, spanning an area of 22.98 km$^2$, which corresponds to approximately 1.3% of the total area. Given the direct hydraulic connection between this area and the underlying groundwater system, any pollution in the surface water directly impacts and contaminates the groundwater. Consequently, this region is classified as a zone with high vulnerability based on various evaluation indicators.

The groundwater vulnerability in the YRB can be mainly classified into two categories: good and medium (Figure 11). The areas with low vulnerability are the most extensive and are geographically widespread, spanning both the eastern and western regions of the study area. The medium vulnerability areas are primarily concentrated in the central region. Conversely, the areas with high vulnerability are limited in extent and are predominantly located along the banks of the Shu River, forming a narrow strip. Although their spatial coverage is relatively small, these areas coincide with the first-level protected zones of crucial water sources. Consequently, the aquifers in these areas are characterized by high inherent vulnerability and pollution hazard indices. As a result, the safeguarding of groundwater within the protected zones becomes particularly crucial to prevent pollution from infiltrating the water sources.

## 4. Conclusions and Suggestions

The ability of groundwater to resist external pollution can be evaluated according to its vulnerability. In this study, the hydrogeological conditions in the YRB were analyzed, and seven indicators were selected to evaluate the vulnerability of groundwater using the

DRASTIC model combined with GIS technology. This approach improves the reliability of regional vulnerability evaluations and provides guidance for regional groundwater pollution prevention, water source selection, and rational utilization of groundwater resources. The groundwater vulnerability in the YRB was divided into five sub-regions, with 14.5%, 42.3%, 27.9%, 14.0%, and 1.3% of the total area classified as excellent, good, medium, poor, and very poor, respectively. The area with low vulnerability accounted for the largest proportion of the total area and was mainly distributed in the eastern part of the study area. The areas with high vulnerability were mainly distributed along the banks of the Shu River in a strip with a small area but with water source distribution that requires attention. The evaluation results are of some guiding significance for the formulation of groundwater resource protection planning and groundwater pollution prevention planning in the YRB and the rational development and utilization of groundwater resources and land resources. According to different groundwater vulnerability zoning, formulating and adopting targeted protection measures, formulating scientific groundwater resource development and utilization programs, and carrying out land resource planning from the perspective of groundwater resource protection are of great significance for alleviating the shortage of groundwater resources, the continuous decline of groundwater level, and maintaining the sustainable development and utilization of groundwater resources. Based on the evaluation results, the study suggests strict control of groundwater pollution sources, rational planning of industrial layout, optimization of groundwater exploitation mode, and strengthening of groundwater dynamic monitoring. These measures are necessary for the rational development, utilization, and protection of regional groundwater resources.

**Author Contributions:** Conceptualization, J.H., P.Y., Z.G., Q.L., M.W., Q.L. and J.F.; validation, J.H.; formal analysis, J.H. and P.Y.; writing—original draft preparation, J.H.; writing—review and editing, P.Y., Q.L. and J.L.; super-vision, Z.G., J.H., M.W., J.F. and J.L. All authors have read and agreed to the published version of the manuscript.

**Funding:** This project was supported by the Rizhao Ecological Geological Environment Monitoring (Lu Di Zi (2023) No. 4).

**Data Availability Statement:** The data presented in this study are available on request from the corresponding author.

**Conflicts of Interest:** The authors declare no conflicts of interest.

## References

1. Liu, J.; Peng, Y.; Li, C.; Gao, Z.; Chen, S. Characterization of the hydrochemistry of water resources of the Weibei Plain, Northern China, as well as an assessment of the risk of high groundwater nitrate levels to human health. *Environ. Pollut.* **2021**, *268*, 115947. [CrossRef]
2. Ayuba, R.; Tijani, M.N. Hydrochemical characterization of groundwater in Lokoja, North-Central Nigeria. *Sustain. Water Resour. Manag.* **2021**, *7*, 61. [CrossRef]
3. Liu, J.; Wang, M.; Gao, Z.; Chen, Q.; Wu, G.; Li, F. Hydrochemical characteristics and water quality assessment of groundwater in the Yishu River basin. *Acta Geophys.* **2020**, *68*, 877–889. [CrossRef]
4. Liu, J.; Gao, Z.; Wang, M.; Li, Y.; Ma, Y.; Shi, M.; Zhang, H. Study on the dynamic characteristics of groundwater in the valley plain of Lhasa City. *Environ. Earth Sci.* **2018**, *77*, 646. [CrossRef]
5. Mohammadi Arasteh, S.; Shoaei, S.M. Simulation of groundwater resource quantity and quality and assessment of the effects of alluvial material dissolution on the changes of qualitative parameters of the Zanjan Plain, Iran. *Arab. J. Geosci.* **2023**, *16*, 60. [CrossRef]
6. Van Rooyen, J.D.; Watson, A.P.; Miller, J.A. Combining quantity and quality controls to determine groundwater vulnerability to depletion and deterioration throughout South Africa. *Environ. Earth Sci.* **2020**, *79*, 255. [CrossRef]
7. Abdulsalam, A.; Ramli, M.F.; Jamil, N.R.; Ashaari, Z.H.; Umar, D.U.A. Hydrochemical characteristics and identification of groundwater pollution sources in tropical savanna. *Environ. Sci. Pollut. Res.* **2022**, *29*, 37384–37398. [CrossRef] [PubMed]
8. Gugulothu, S.; Subbarao, N.; Das, R.; Dhakate, R. Geochemical evaluation of groundwater and suitability of groundwater quality for irrigation purpose in an agricultural region of South India. *Appl. Water Sci.* **2022**, *12*, 142. [CrossRef]
9. Mukonazwothe, M.; Munyai, L.F.; Mutoti, M.I. Groundwater quality evaluation for domestic and irrigation purposes for the Nwanedi Agricultural Community, Limpopo Province, South Africa. *Heliyon* **2022**, *8*, e09203. [CrossRef] [PubMed]

10. Akbar, H.; Nilsalab, P.; Silalertruksa, T.; Gheewala, S.H. Comprehensive review of groundwater scarcity, stress and sustainability index-based assessment. *Groundw. Sustain. Dev.* **2022**, *18*, 100782. [CrossRef]
11. Dhaouadi, L.; Besser, H.; Karbout, N.; Wassar, F.; Alomrane, A.R. Assessment of natural resources in tunisian Oases: Degradation of irrigation water quality and continued overexploitation of groundwater. *Euro-Mediterr. J. Environ. Integr.* **2021**, *6*, 36. [CrossRef]
12. Li, C.; Gao, Z.; Chen, H.; Wang, J.; Liu, J.; Li, C.; Teng, Y.; Liu, C.; Xu, C. Hydrochemical analysis and quality assessment of groundwater in southeast North China Plain using hydrochemical, entropy-weight water quality index, and GIS techniques. *Environ. Earth Sci.* **2021**, *80*, 523. [CrossRef]
13. Umar, M.; Khan, S.; Arshad, A.; Aslam, R.; Khan, H.; Rashid, H. A modified approach to quantify aquifer vulnerability to pollution towards sustainable groundwater management in irrigated indus basin. *Environ. Sci. Pollut. Res.* **2022**, *29*, 27257–27278. [CrossRef]
14. Zhu, F.; Xiong, L.; Wu, J.; Zhao, Y.; Huang, J.; Wei, J.; Lin, K. Groundwater Vulnerability Assessment in Plain River Network Areas Based on the Improved DRASTIC Model. *Environ. Sci. Technol.* **2020**, *43*, 187–193.
15. Tang, X.; Wu, Y.; Chen, J.; Deng, D. DRASTIC-GIS Model for Assessing Groundwater Vulnerability in Typical Area of Chengdu. *Environ. Manag. Monit. Technol.* **2020**, *32*, 28–32.
16. Zhu, Z.; Wang, J.; Hu, M.; Jia, L. Geographical detection of groundwater pollution vulnerability and hazard in karst areas of Guangxi Province, China. *Environ. Pollut.* **2019**, *245*, 627–633. [CrossRef] [PubMed]
17. Qiu, J. Safeguarding China's water resources. *Natl. Sci. Rev.* **2018**, *5*, 102–107. [CrossRef]
18. Dizaji, A.R.; Hosseini, S.A.; Rezaverdinejad, V.; Sharafati, A. Groundwater contamination vulnerability assessment using DRASTIC method, GSA, and uncertainty analysis. *Arab. J. Geosci.* **2020**, *13*, 645. [CrossRef]
19. Shakoor, A.; Khan, Z.M.; Farid, H.U.; Sultan, M.; Ahmad, I.; Ahmad, N.; Mahmood, M.H.; Ali, M.U. Delineation of regional groundwater vulnerability using DRASTIC model for agricultural application in Pakistan. *Arab. J. Geosci.* **2020**, *13*, 195. [CrossRef]
20. Rajput, H.; Goyal, R.; Brighu, U. Modification and optimization of DRASTIC model for groundwater vulnerability and contamination risk assessment for Bhiwadi region of Rajasthan, India. *Environ. Earth Sci.* **2020**, *79*, 136. [CrossRef]
21. Asfaw, D.; Mengistu, D. Modeling megech watershed aquifer vulnerability to pollution using modified DRASTIC model for sustainable groundwater management, Northwestern Ethiopia. *Groundw. Sustain. Dev.* **2020**, *11*, 100375. [CrossRef]
22. Janipella, R.; Quamar, R.; Sanam, R.; Jangam, C.; Pandurang Balwant Jyothi, V.; Padmakar, C.; Pujari, P.R. Evaluation of Groundwater Vulnerability to Pollution using GIS Based DRASTIC Method in Koradi, India—A Case Study. *J. Geol. Soc. India* **2020**, *96*, 292–297. [CrossRef]
23. Arya, S.; Subramani, T.; Vennila, G.; Roy, P.D. Groundwater vulnerability to pollution in the semi-arid Vattamalaikarai River Basin of south India thorough DRASTIC index evaluation. *Geochemistry* **2020**, *80*, 125635. [CrossRef]
24. Mali, S.C.; Thabaj, K.A.; Purandara, B.K. Evaluation of Geochemical Characteristics of Groundwater in Parts of Ghataprabha Sub-basin Using DRASTIC Indices. *J. Geol. Soc. India* **2020**, *95*, 513–519. [CrossRef]
25. Rahman, M.; Haque, M.M.; Tareq, S.M. Appraisal of groundwater vulnerability in south-central part of Bangladesh using DRASTIC model: An approach towards groundwater protection and health safety. *Environ. Chall.* **2021**, *5*, 100391. [CrossRef]
26. Shamsuddin, A.S.; Ismail, S.N.S.; Abidin, E.Z.; Bin, H.Y.; Juahir, H.; Bakar, W.A.M.A. Application of GIS-based DRASTIC model approaches in assessing groundwater vulnerability for shallow alluvial aquifer deposited. *Arab. J. Geosci.* **2021**, *14*, 2693. [CrossRef]
27. Antonakos, A.K.; Lambrakis, N.J. Development and testing of three hybrid methods for the assessment of aquifer vulnerability to nitrates, based on the drastic model, an example from ne korinthia, greece. *J. Hydrol.* **2007**, *333*, 288–304. [CrossRef]
28. Goyal, D.; Haritash, A.K.; Singh, S.K. A comprehensive review of groundwater vulnerability assessment using index-based, modelling and coupling methods. *J. Environ. Manag.* **2021**, *296*, 113161. [CrossRef] [PubMed]
29. Moazamnia, M.; Hassanzadeh, Y.; Nadiri, A.A.; Sadeghfam, S. Vulnerability indexing to saltwater intrusion from models at two levels using artificial intelligence multiple model (AIMM). *J. Environ. Manag.* **2020**, *255*, 109871.1–109871.11. [CrossRef]
30. Mojgan, B.; Aminreza, N.; Saman, J. A new hybrid framework for optimization and modification of groundwater vulnerability in coastal aquifer. *Environ. Sci. Pollut. Res.* **2019**, *26*, 21808–21827.
31. Celia, S.F.; Marina, C.B.; Rubén, L.L.; Fátima, S.G. Multiresponse Performance evaluation and life cycle assessment for the optimal elimination of pb (ii) from industrial wastewater by adsorption using vine shoot activated carbon. *Sustainability* **2023**, *15*, 11007.
32. Chakraborty, B.; Roy, S.; Bera, A.; Adhikary, P.P.; Bera, B.; Sengupta, D.; Bhunia, G.S.; Shit, P.K. Groundwater vulnerability assessment using GIS-based DRASTIC model in the upper catchment of Dwarakeshwar river basin, West Bengal, India. *Environ. Earth Sci.* **2021**, *81*, 2. [CrossRef]
33. Koon, A.B.; Anornu, G.K.; Dekongmen, B.W.; Sunkari, E.D.; Agyare, A.; Gyamfi, C. Evaluation of groundwater vulnerability using GIS-based DRASTIC model in Greater Monrovia, Montserrado County, Liberia. *Urban Clim.* **2023**, *48*, 101427. [CrossRef]
34. Patel, P.; Mehta, D.J.; Sharma, N.D. A GIS-based DRASTIC Model for Assessing Groundwater Quality Vulnerability: Case Study of Surat and its Surroundings. *J. Geol. Soc. India* **2023**, *99*, 578–582. [CrossRef]
35. Patel, P.; Mehta, D.; Sharma, N. A review on the application of the DRASTIC method in the assessment of groundwater vulnerability. *Water Supply* **2022**, *22*, 5190–5205. [CrossRef]
36. Patle, D.; Nema, S.; Awasthi, M.K.; Sharma, S.K.; Tiwari, Y.K. Groundwater vulnerability assessment using DRASTIC model in Niwari District of Bundelkhand Region, Madhya Pradesh, India. *Arab. J. Geosci.* **2022**, *15*, 1590. [CrossRef]

37. Smail, R.Q.S.; Dişli, E. Assessment and validation of groundwater vulnerability to nitrate and TDS using based on a modified DRASTIC model: A case study in the Erbil Central Sub-Basin, Iraq. *Environ. Monit. Assess.* **2023**, *195*, 567. [CrossRef] [PubMed]
38. Aller, L.T.; Bennett, T.; Lehr, J.H.; Petty, R.J.; Hackett, G. Drastic: A standardized system for evaluating ground water pollution potential using hidrogeologic settings. *J. Geol. Soc. India* **1987**, *29*.
39. Venkatesan, G.; Pitchaikani, S.; Saravanan, S. Assessment of Groundwater Vulnerability Using GIS and DRASTIC for Upper Palar River Basin, Tamil Nadu. *J. Geol. Soc. India* **2019**, *94*, 387–394. [CrossRef]
40. Aslam, B.; Ismail, S.; Ali, I. A GIS-based DRASTIC model for assessing aquifer susceptibility of Safdarabad Tehsil, Sheikhupura District, Punjab Province, Pakistan. *Model. Earth Syst. Environ.* **2020**, *6*, 995–1005. [CrossRef]
41. Hasan, M.; Islam, M.A.; Aziz Hasan, M.; Alam, M.J.; Peas, M.H. Groundwater vulnerability assessment in Savar upazila of Dhaka district, Bangladesh—A GIS-based DRASTIC modeling. *Groundw. Sustain. Dev.* **2019**, *9*, 100220. [CrossRef]
42. Kaliraj, S.; Chandrasekar, N.; Peter, T.S.; Selvakumar, S.; Magesh, N.S. Mapping of coastal aquifer vulnerable zone in the south west coast of Kanyakumari, South India, using GIS-based DRASTIC model. *Environ. Monit. Assess.* **2014**, *187*, 4073. [CrossRef] [PubMed]
43. Karimzadeh Motlagh, Z.; Derakhshani, R.; Sayadi, M.H. Groundwater vulnerability assessment in central Iran: Integration of GIS-based DRASTIC model and a machine learning approach. *Groundw. Sustain. Dev.* **2023**, *23*, 101037. [CrossRef]

water

Article

# Hydrochemical Characteristics, Controlling Factors and Strontium Enrichment Sources of Groundwater in the Northwest Plain of Shandong Province, China

Jingpeng Chen [1], Xiaohua Wu [1,*], Jichu Zhao [1], Shuai Liu [1], Yuqi Zhang [2], Jiutan Liu [2] and Zongjun Gao [2]

1 The Second Institute of Hydrogeology and Engineering Geology, Shandong Provincial Bureau of Geology & Mineral Resources (Lubei Geo-Engineering Exploration Institute of Shandong Province), Dezhou 253072, China; chenjingpeng@126.com (J.C.); zhaojichu_1@163.com (J.Z.); vileesy@126.com (S.L.)
2 College of Earth Science and Engineering, Shandong University of Science and Technology, Qingdao 266590, China; zyqsdust@163.com (Y.Z.); ljtsdust@sdust.edu.cn (J.L.); zongjungao1964@163.com (Z.G.)
* Correspondence: private_official@163.com

**Abstract:** To elucidate the hydrochemical characteristics, controlling factors, sources and mechanisms of strontium ion enrichment in groundwater in the northwest plain of Shandong Province, China, 88 groundwater samples were collected, including 51 shallow pore groundwater samples, 29 deep pore groundwater samples and 8 karst groundwater samples. The hydrochemical characteristics of the different types of groundwater were quite different. The karst groundwater samples were all fresh water with a single hydrochemical type, either $HCO_3$-Ca or $HCO_3$-Ca·Mg. The deep pore groundwater samples were mainly brackish water, and the shallow pore groundwater samples were brackish water–salt water, which has complex hydrochemical types. The hydrochemical characteristics of all the types of groundwater were controlled by mineral dissolution and active positive cation exchange. In shallow pore groundwater, deep pore groundwater and karst groundwater, the dissolution of silicate, evaporite and carbonate minerals dominated the hydrogeochemical process. The strontium in groundwater was derived from the dissolution of minerals with strontium isomorphism. The average contents of strontium in shallow, deep and karst groundwater were 1.59 mg/L, 0.58 mg/L and 0.50 mg/L, respectively. The strontium in shallow pore groundwater was mainly derived from the enrichment of groundwater runoff, and its sources are abundant, with silicic rock being the main source. The deep pore groundwater mainly derived from the evaporative minerals containing strontium, and the karst water mainly derived from carbonate rock dissolution with similar characteristics.

**Keywords:** groundwater; hydrochemical characteristics; strontium; northwest plain of Shandong Province

**Citation:** Chen, J.; Wu, X.; Zhao, J.; Liu, S.; Zhang, Y.; Liu, J.; Gao, Z. Hydrochemical Characteristics, Controlling Factors and Strontium Enrichment Sources of Groundwater in the Northwest Plain of Shandong Province, China. *Water* **2024**, *16*, 550. https://doi.org/10.3390/w16040550

Academic Editors: Chin H Wu and Dimitrios E. Alexakis

Received: 10 January 2024
Revised: 24 January 2024
Accepted: 7 February 2024
Published: 10 February 2024



## 1. Introduction

Groundwater is an important resource supplying human industry, agricultural production and domestic water [1–5]. The hydrochemical characteristics of groundwater depend on its formation process, storage conditions, flow path and interaction with the surrounding environment and are influenced by human activities [6–8]. In its natural state, the hydrochemical composition of groundwater is mainly controlled by water–rock interaction, including rock dissolution, evaporation and concentration, cation exchange and other factors [9,10]. However, with the economic growth and social development of human beings, human activities are increasingly impacting the Earth's natural environment. Emissions from human production and life have greatly changed the hydrochemistry of groundwater [11–13].

Strontium (Sr) is a necessary trace element for human health, maintaining the normal function of the human body [14,15]. Humans consume strontium through food and drinking water. The main source of strontium in groundwater is the Earth's crust. Suitable hydrogeological conditions are the basic conditions required for strontium enrichment in groundwater [16]. Groundwater reacts with minerals in surrounding rock continuously and dissolves them via processes such as leaching, ion exchange, etc., while runoff and other conditions promote the increase in strontium content in groundwater. The abundance of strontium in the surrounding rock determines the content of strontium in groundwater [17]. There have been many studies on the formation mechanism and enrichment factors of strontium in groundwater [6,14]. By means of the hydrochemical ion ratio, statistics, hydrogeochemistry, geological tectonics and other methods, scholars have studied the causes of the enrichment of relevant groundwater chemical components [18–20].

The northwest plain of Shandong Province is a typical alluvial plain area located in the lower reaches of the Yellow River Basin. Pore groundwater is the most widely distributed type of groundwater in the region, including shallow pore groundwater and deep pore groundwater. In addition, karst groundwater is distributed in a small area. In this area, the quality of pore water is generally poor, the groundwater contains more hydrochemical components, and the distribution area of high-quality groundwater is small, thus leading to a water shortage due to quality in northern China. At the same time, the plain in the lower reaches of the Yellow River Basin has a large population and is a major human gathering place with prosperous social and economic activities. The shortage of water resources restricts the development of this region to a certain extent. In this study, the groundwater sources in the northwest plain of Shandong Province (NPSP) were selected to carry out research with a view to determining the origin of this groundwater and conserving it, and the phenomenon of strontium enrichment in groundwater was elucidated during this investigation.

There have been many studies on groundwater in this region but few on the cause of strontium enrichment in high-quality groundwater. We studied the hydrochemical characteristics, controlling factors and cause of strontium enrichment in groundwater on the basis of hydrogeological conditions, descriptive analysis, a Piper trilinear diagram, a Gibbs diagram, the ion ratio and correlation analysis, and suggestions for the development and utilization of high-quality strontium-rich groundwater are put forward.

## 2. Study Area

### 2.1. Natural Conditions

The northwest plain of Shandong Province (NPSP) is located in the north of the lower Yellow River plain (Figure 1), which is located in a warm temperate climate zone. The average annual temperature is 12 °C. The average annual rainfall is about 574 mm. The terrain in the NPSP is flat with a ground elevation of 2–30 m, and the overall terrain slopes downward from the southwest to the northeast coast. The rivers in the NPSP include the Yellow River, the Tuhai River and the Majia River. Rivers within the territory are rain-fed rivers. Their direction and terrain are largely consistent.

There are more than 17 million people living in this area. Dezhou and Liaocheng are major agricultural cities in Shandong Province, China. Binzhou and Dongying are rich in petroleum, with a well-developed refining industry chain and a relatively high level of industrial activities. These activities profoundly affect the groundwater environment.

### 2.2. Hydrological and Geological Conditions

The majority of the NPSP is covered by thick Quaternary loose rocks and mainly loose rock sediments, and Cambrian–Ordovician limestone occurs under the Quaternary layer (about 260 m) along the Yellow River near Donga-Qihe. The groundwater in the NPSP is mainly pore groundwater, and the shallow pore groundwater aquifer is mainly found in fine sand, medium-coarse sand and clay at a depth of 0–60 m. The deep pore groundwater occurs at a depth of 260–800 m, where the aquifer is dominated by fine sand and medium-fine sand. At 60–260 m, the middle-deep loose-rock pore groundwater is mainly brackish water, which is not utilized and is not discussed in this study. Only in some areas is there karst water, which occurs in Ordovician limestone at a depth of 250–520 m.

The samples in this study were collected from a mining water source, reserve water source and other places with relatively good water quality, including shallow, deep and karst groundwater sources.

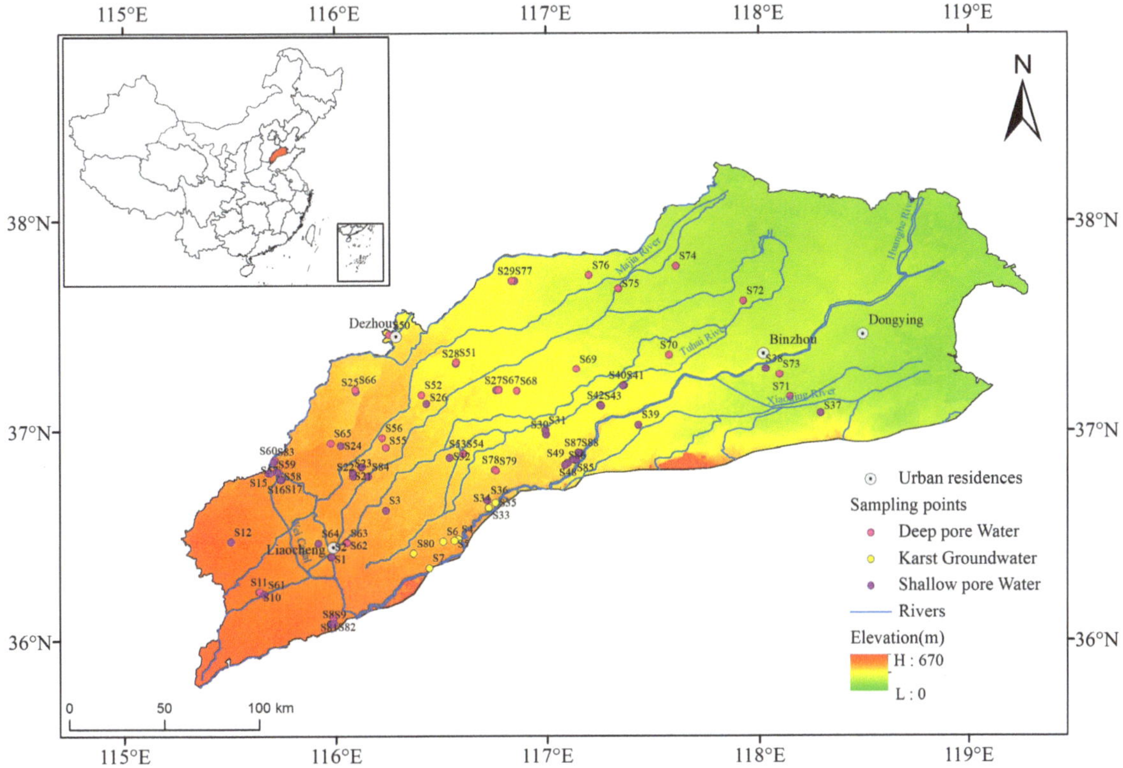


**Figure 1.** Location of the sampling points in the northwest plain of Shandong Province.

## 3. Materials and Methods

### 3.1. Field Sampling

In this study, 88 groundwater samples were collected from the NPSP from June to August 2017, including 51 shallow pore groundwater samples, 29 deep pore groundwater samples and 8 karst groundwater samples. The sampling points were located at groundwater sources, or alternate, smaller sources, and 1~2 groundwater samples were collected from each source. The depth range of shallow pore wells is 15–60 m, that of deep pore wells is 120–400 m, that of individual wells is 600 m, and that of most karst wells is 350–520 m. Figure 1 shows the locations of the groundwater samples. Before sample collection, groundwater was pumped for more than 30 min to ensure that the groundwater was fresh from the aquifer. pH and TDS were tested on site using a portable water quality analyzer. The samples were stored at 4 °C during transportation.

### 3.2. Laboratory Sample Analysis

The laboratory of the Shandong Lubei Geological Engineering Investigation Institute was used for this test. Cations ($Ca^{2+}$, $Mg^{2+}$) were tested by titration, $K^+$ and $Na^+$ were tested by flame spectroscopy, anions ($HCO_3^-$, $Cl^-$, $SO_4^{2-}$) were tested by titration, and $NO_3^-$ was tested by ultraviolet spectrophotometry. The test was completed according to the 'Methods for analysis of groundwater quality' [21] and complied with the relevant specification requirements. Table 1 shows the accuracy and detection limits for the chemical analysis of individual ions.

**Table 1.** The accuracy and precision of the chemical analysis.

| Index | Detection Limit (mg/L) |
|---|---|
| pH | / |
| TH | 3 mg/L ($CaCO_3$) |
| TDS | / |
| $K^+$ | 0.5 |
| $Na^+$ | 0.6 |
| $Ca^{2+}$ | 4 |
| $Mg^{2+}$ | 3 |
| $Cl^-$ | 3 |
| $SO_4^{2-}$ | 10 |
| $HCO_3^-$ | 5 |
| $NO_3^-$ | 0.2 |

### 3.3. Data Analysis and Methods

The obtained water chemical data were statistically analyzed, and the main water chemical components were counted using Origin2022 SR1 software [22]. A Gibbs diagram and Piper three-line diagram were drawn based on the above methods. Descriptive analysis, the ion ratio and correlation analysis were used to analyze and study the hydrochemical characteristics of high-quality strontium-rich groundwater in the NPSP, China.

## 4. Results

### 4.1. Hydrochemical Characteristics

#### 4.1.1. General Hydrochemistry of Groundwater

A hydrochemical statistical analysis method was used to gain a more intuitive understanding of groundwater hydrochemical characteristics in the region [9]. The main hydrochemical data of the samples were statistically analyzed (Tables 2 and A1). The pH of the three groundwaters ranged between 7.00 and 8.60, all of which were weakly alkaline. The shallow pore groundwater (SPGW) was predominantly brackish, with a TDS ranging from 556.5 to 3403.03 mg/L, while the TDS of deep pore groundwater (DPGW) ranged from 461.80 to 2671.00 mg/L, with an average of 1069.72 mg/L, and the karst groundwater (KGW) was freshwater with a TDS ranging from 314.51 to 600.54 mg/L. Nitrate is mainly derived from anthropogenic emissions, including agricultural cultivation and domestic wastewater discharge [23]. Nearly all the groundwater samples were low in nitrate, indicating that the groundwater in the area is less affected by human nitrate emissions.

The cation found in the highest concentration in SPGW was $Na^+$, followed by $Mg^{2+}$ and then $Ca^{2+}$; their average mass concentrations were 242.80 mg/L, 113.95 mg/L and 101.18 mg/L, respectively. The anion found in the highest concentration was $HCO_3^-$, followed by $SO_4^{2-}$ and then $Cl^-$; their average mass concentrations were 706.64 mg/L, 326.37 mg/L and 224.73 mg/L, respectively.

The cation found in the highest concentration in DPGW was $Na^+$, and the anion found in the highest concentration was $HCO_3^-$.

$Ca^{2+}$ and $HCO_3^-$ were the anion and cation, respectively, with the highest concentrations in KGW. Their average concentrations were 77.25 mg/L and 283.74 mg/L, respectively. The content of other ions was relatively low.

The coefficient of variation and standard deviation can reflect the differences in each component to some extent. Table 2 shows that the coefficient of variation of cations in SPGW followed the order $Na^+ > K^+ > Mg^{2+} > Ca^{2+}$, the coefficient of variation of anions followed the order $NO_3^- > SO_4^{2-} > Cl^- > HCO_3^-$ and $Mg^{2+}$, $Ca^{2+}$ and $HCO_3^-$ were relatively stable. The remaining ions had large spatial differences, following the order $Mg^{2+} > Ca^{2+} > K^{+} > Na^{+}$, $NO_3^{-} > Cl^{-} > SO_4^{2-} > HCO_3^{-}$ in DPGW. The coefficient of variation of KGW was small. The coefficient of variation of $Cl^{-}$ was 0.66, which is larger than that found for other ions, indicating that the spatial distribution of each ion in KGW was relatively stable.

**Table 2.** Mass concentration statistics of main hydrochemical indexes in the northwest plain of Shandong Province.

| Type | Index | $K^+$ | $Na^+$ | $Ca^{2+}$ | $Mg^{2+}$ | $Cl^-$ | $SO_4^{2-}$ | $HCO_3^-$ | $NO_3^-$ | TH | TDS | pH | $Sr^{2+}$ |
|---|---|---|---|---|---|---|---|---|---|---|---|---|---|
| SPGW | AV | 2.24 | 242.80 | 101.18 | 113.95 | 224.73 | 326.37 | 706.64 | 8.14 | 721.95 | 1382.29 | 7.65 | 1.59 |
| | CV | 2.50 | 0.63 | 0.48 | 0.53 | 0.70 | 0.89 | 0.23 | 2.87 | 0.41 | 0.50 | 0.04 | 0.36 |
| | Min | 0.10 | 53.00 | 10.02 | 37.67 | 31.91 | 28.82 | 390.53 | 0.00 | 280.22 | 556.86 | 7.00 | 0.60 |
| | MD | 1.20 | 200.00 | 90.18 | 95.99 | 191.43 | 206.53 | 695.63 | 0.68 | 660.53 | 1135.85 | 7.70 | 1.60 |
| | Max | 40.00 | 690.00 | 230.46 | 336.56 | 645.19 | 1104.69 | 1263.11 | 153.00 | 1491.19 | 3403.03 | 8.20 | 3.00 |
| DPGW | AV | 1.05 | 332.76 | 30.41 | 26.19 | 191.19 | 244.13 | 440.82 | 0.94 | 183.77 | 1069.72 | 8.02 | 0.58 |
| | CV | 0.82 | 0.42 | 1.06 | 1.09 | 0.74 | 0.67 | 0.33 | 1.27 | 1.02 | 0.42 | 0.05 | 0.71 |
| | Min | 0.25 | 84.00 | 4.01 | 1.22 | 46.09 | 52.83 | 237.98 | 0.00 | 15.01 | 461.80 | 7.20 | 0.10 |
| | MD | 1.00 | 340.00 | 16.03 | 14.58 | 173.71 | 206.53 | 414.94 | 0.58 | 110.09 | 983.58 | 8.10 | 0.40 |
| | Max | 5.00 | 670.00 | 116.23 | 128.79 | 659.37 | 691.63 | 811.57 | 5.01 | 820.66 | 2671.04 | 8.60 | 1.50 |
| KGW | AV | 0.68 | 27.50 | 77.15 | 25.52 | 34.12 | 64.24 | 283.74 | 11.41 | 297.74 | 390.07 | 7.83 | 0.50 |
| | CV | 0.32 | 0.36 | 0.27 | 0.48 | 0.66 | 0.35 | 0.17 | 0.46 | 0.23 | 0.25 | 0.03 | 0.39 |
| | Min | 0.40 | 18.00 | 50.10 | 15.80 | 21.27 | 38.42 | 237.98 | 3.52 | 240.19 | 314.51 | 7.40 | 0.30 |
| | MD | 0.75 | 24.00 | 75.15 | 20.66 | 24.82 | 55.24 | 274.59 | 13.15 | 272.72 | 347.61 | 7.80 | 0.50 |
| | Max | 1.00 | 46.00 | 124.25 | 52.25 | 85.08 | 96.06 | 378.32 | 17.67 | 450.36 | 600.54 | 8.10 | 0.80 |

Note: MD (median), AV (average), CV (coefficient of variation). Unit: mg/L, except for pH and CV.

The main ion hydrochemical box line diagram (Figure 2) shows that the content of different hydrochemical components in SPGW, DPGW and KGW was quite different, indicating that the composition of DPGW and KGW was relatively stable. The source of SPGW was rich, and the three material sources were different.

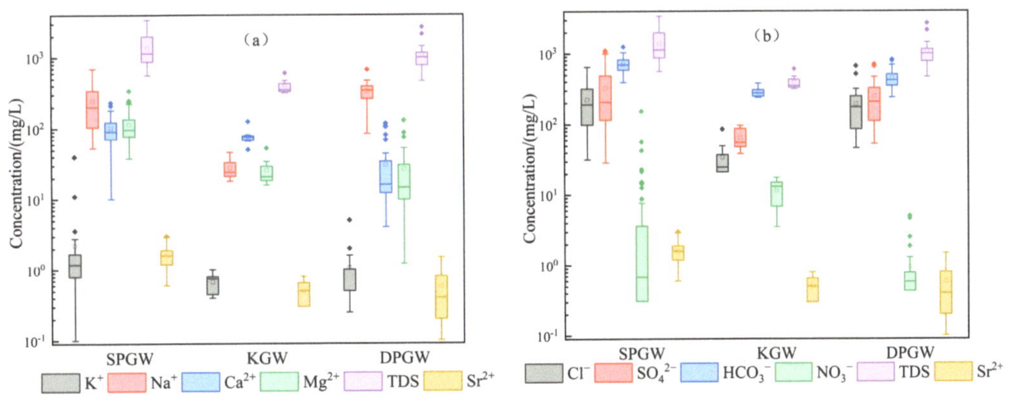


**Figure 2.** Box diagram of hydrochemical components of different groundwater types (**a**,**b**) in the northwest plain of Shandong Province.

There were some differences in $Sr^{2+}$ content in the different types of groundwater (Table 2). The concentration of $Sr^{2+}$ in SPGW fell between 0.60 and 3.00 mg/L, with an average concentration of 1.59 mg/L and a coefficient of variation of 0.36. The concentration of $Sr^{2+}$ in DPGW fell between 0.10 and 1.50 mg/L, with an average of 0.58 mg/L and a coefficient of variation of 0.41. The concentration of $Sr^{2+}$ in KGW fell between 0.30 and 0.80 mg/L, with an average of 0.50 mg/L and a coefficient of variation of 0.39. It can be seen that $Sr^{2+}$ had obvious differences in different sources of groundwater. The content of $Sr^{2+}$ in SPGW was significantly higher than that in DPGW and KGW, and the coefficient of variation was small. The spatial distribution of $Sr^{2+}$ content across samples of the same type of groundwater was not obvious, and there was a certain continuity.

#### 4.1.2. Hydrochemical Type

The Piper diagram is a simple and effective water type classification method in hydrochemical research [2,11,24]. As can be observed in Figure 3, the hydrochemical type of SPGW was complex, and the cation content was dominated by Na·Mg, Ca·Na and Ca·Mg, while the anion content was dominated by $HCO_3^-$, followed by $HCO_3$·Cl and $SO_4$-$HCO_3$·Cl. The DPGW cations were dominated by the Na type, followed by the Ca·Na·Mg type, while the anion types were relatively complex, with the $HCO_3$·Cl, $HCO_3$ and $HCO_3$·Cl·$SO_4$ types dominating. KGW's chemistry was relatively unique, with the cations being mainly Ca, Mg and Ca types and the anion type and $HCO_3^-$ being absolutely dominant, with the $HCO_3$·Ca-Mg type and $HCO_3$·Ca type largely present. Moreover, Figure 3 shows that there was no significant correlation between $Sr^{2+}$ content and groundwater chemistry.

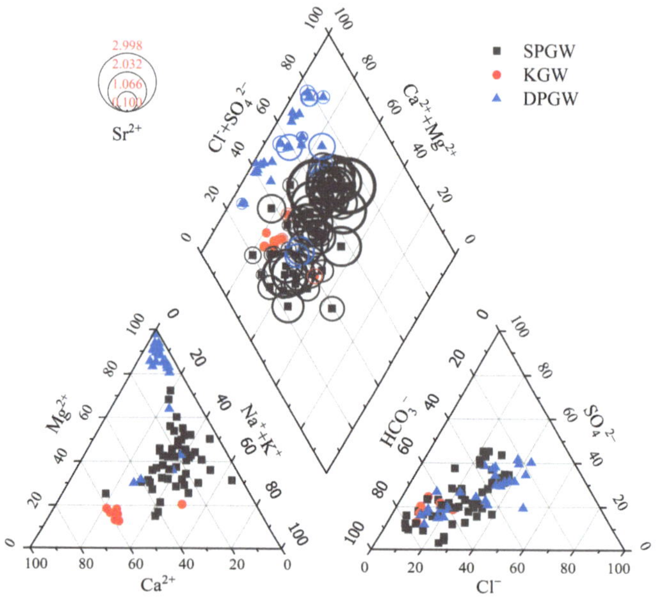


**Figure 3.** Hydrochemical Piper diagram of groundwater source water in the northwest plain of Shandong Province.

### *4.2. Controlling Factors of Groundwater Hydrochemistry*

The chemical components of groundwater are determined by physical and chemical changes under complex geological and climatic conditions, and the main factors controlling their formation include rock weathering, evaporation and concentration and atmospheric precipitation [25,26]. Gibbs established a semi-logarithmic graph with ordinate and abscissa representations of TDS (unit: mg·$L^{-1}$) and $Na^+/(Na^+ + Ca^{2+})$ or $Cl^-/(Cl^- + HCO_3^-)$, respectively. We used it to determine the main factors that control the chemical characteristics of water bodies.

The water sample points were in the rock weathering and evaporation–concentration control areas (Figure 4), and the KGW was derived from the rock weathering area, indicating that rock weathering heavily influences the formation of KGW's chemical components. The

SPGW was mainly collected from the rock weathering and evaporation–concentration areas. The DPGW was collected from the evaporation–concentration and rock weathering areas.

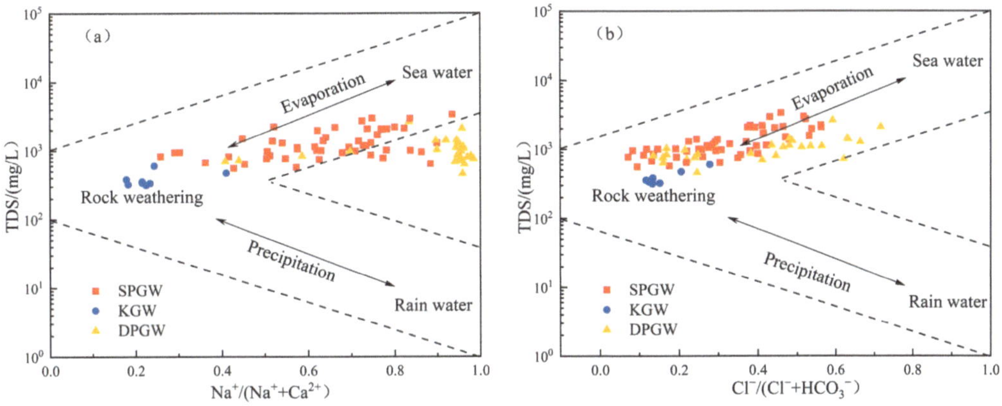


**Figure 4.** Gibbs diagram of hydrochemistry in groundwater source water (**a**,**b**) in northwest plain of Shandong Province.

### 4.2.1. Dissolution

Groundwater dissolves rock minerals during runoff. Gaillardet constructed ion ratio diagrams of $Ca^{2+}$, $Mg^{2+}$, $Na^{+}$ and $HCO_3^-$ to determine the effects of water–rock interactions between three main rock types (silicate, carbonate and evaporate) on the hydrochemical properties of a region [27]. As shown in Figure 5a,b, SPGW was mainly found in the silicate area, DPGW in the evaporite area, and KGW in the carbonate and silicate areas, indicating that the formation of the three types of groundwater is heavily influenced by the weathering and dissolution of these three minerals.

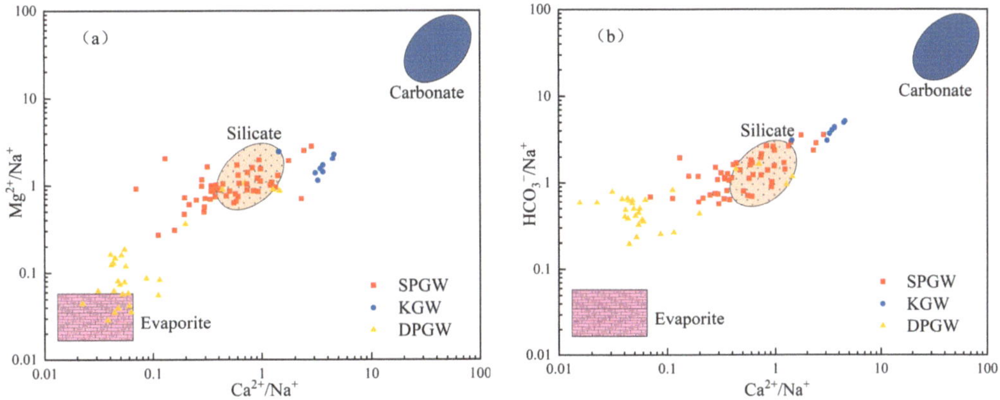


**Figure 5.** Relative contribution of rock weathering (**a**) and dissolution (**b**) to groundwater composition in northwest plain of Shandong Province.

### 4.2.2. Alternating Adsorption of Cations

The chemical characteristics of groundwater in a region are affected by cation exchange. The ratio of ($Mg^{2+}$ + $Ca^{2+}$-$HCO_3^-$-$SO_4^{2-}$) to ($Na^+$ + $K^+$-$Cl^-$) is generally used to identify cation exchange processes [9,28]. Figure 6a shows that the groundwater samples taken from almost all of the study areas were affected by cation exchange. We further introduce the chlor-alkali index to indicate the direction of cation exchange. As can be seen in Figure 6b, the chlor-alkali index of most the groundwater samples was negative, indicating that positive cation exchange occurs in groundwater in this area, the concentration of $Na^+$ and $K^+$ increases, and the concentration of $Ca^{2+}$ and $Mg^{2+}$ decreases.

$$\text{CAI-I} = (Cl^- - Na^+ - K^+)/(Cl^-), \quad (1)$$

$$\text{CAI-II} = (Cl^- - Na^+ - K^+)/(HCO_3^- + SO_4^{2-} + CO_3^{2-} + NO_3^-) \quad (2)$$

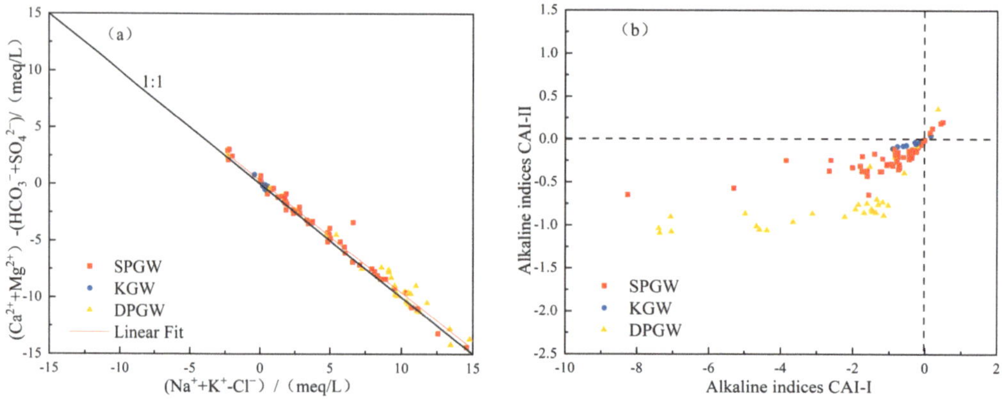


**Figure 6.** Alternate cation adsorption (**a**), and chlor-alkali index (**b**) of groundwater source in northwest plain of Shandong Province.

### 4.2.3. Impact of Human Activities

Nitrate is an important indicator of human activities affecting the groundwater environment [2,9]. Table 2 shows the average content of nitrate in groundwater. The higher the ratios of $Cl^-/Na^+$ and $NO_3^-/Na^+$, the more obvious the influence of human activities on groundwater [29,30]. Since the water samples in the NPSP were taken near the Y = 0 line (Figure 7), far from agricultural activities, the groundwater was mainly affected by the dissolution of minerals, and the impact of anthropogenic nitrate emissions was weak.

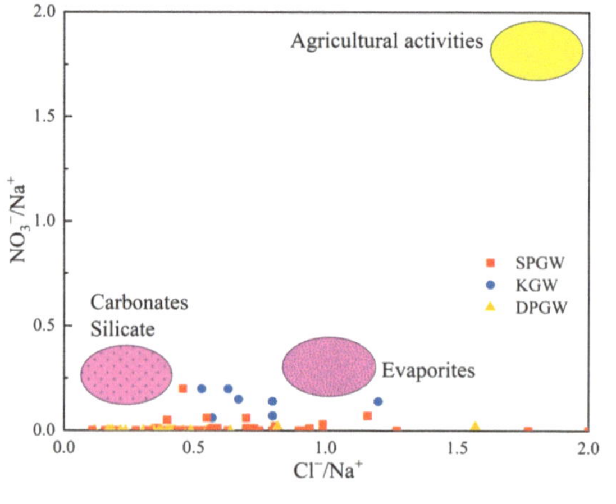


**Figure 7.** Relationship between hydrochemistry and human activities in the northwest plain of Shandong Province.

#### 4.2.4. Ion Ratio

The ratio of major ions in water can be used to study the source of major ions and the evolution of water chemistry [31]. In order to further explore the water–rock interaction process in groundwater, we adopted the ratio method to analyze the major ions. The concentration ratio of $Na^+$ and $Cl^-$ dissolved in salt was 1:1 (Equation (3)), while the ratio in seawater was 0.86 [32–34]. The ratio of silicate dissolution was greater than 1 [35]. It can be seen from Figure 8a that the water sample points above y = x mainly derive from the dissolution of silicate rocks, and the water sample points below y = x may be affected by cation exchange or human activities. The effect of the dissolution of silicate rock minerals existed in different types of groundwater. The content of $Na^+$ and $Cl^-$ ions in KGW was low (Figure 8b), and $HCO_3^-$ was the main ion present, which indicates that carbonate rocks are the main source of its hydrochemical composition. The $Na^+$ content in DPGW was greater than that of $HCO_3^-$, indicating that the dissolution of evaporative salt rock (such as salt rock) in DPGW is increased, releasing more $Na^+$ and $Cl^-$ content. Other mineral components contributed less.

$$NaCl \rightarrow Na^+ + Cl^-, \quad (3)$$

$$2NaAlSi_3O_8 + 2CO_2 + 11H_2O \rightarrow Al_2Si_2O_5(OH)_4 + 2Na^+ + 4H_4SiO_4 + 2HCO_3^-, \quad (4)$$

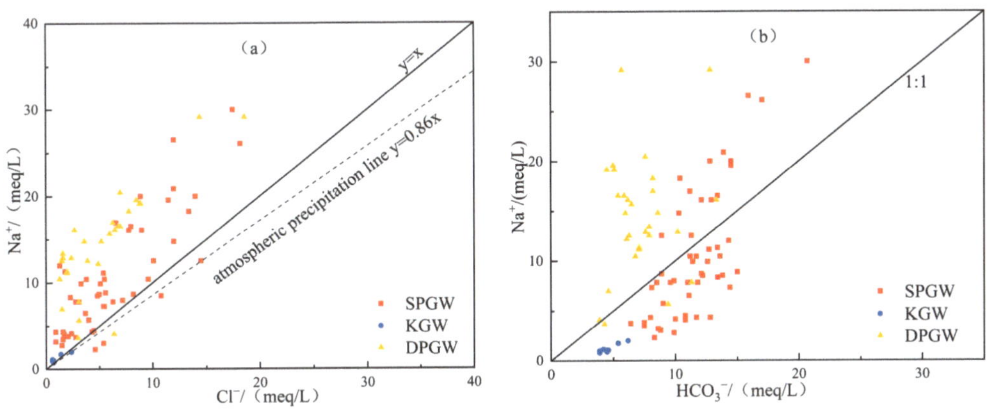


**Figure 8.** The relationship between $Na^+$ and $Cl^-$ (**a**), Na+ and $HCO_3^-$ (**b**) in groundwater source water in the northwest plain of Shandong Province.

Studies have shown that $Ca^{2+}$ and $Mg^{2+}$ in groundwater are mainly derived from evaporative rock and carbonate rock, and $SO_4^{2-}$ is mainly derived from evaporative rock. The ratio relationship between ($Ca^{2+}$ + $Mg^{2+}$) and ($HCO_3^-$ + $SO_4^{2-}$) can be used to characterize the sources of $Ca^{2+}$, $Mg^{2+}$ and $SO_4^{2-}$ [36].

All KGS samples were taken around the 1:1 straight line (Figure 9a). Both the $Ca^{2+}$ and $Mg^{2+}$ ions in KGW come from carbonate minerals. The SPGW samples were taken around or below the line. The $Ca^{2+}$ and $Mg^{2+}$ contents in groundwater are not enough to balance the $HCO_3$- and $SO_4^{2-}$ ions in the water, and more $Na^+$ and $K^+$ ions are needed to balance it, indicating that SPGW mainly comes from carbonate and silicate minerals. The DPGW samples were taken far away from this line, and combined with the above, this indicates that the $Ca^{2+}$ and $Mg^{2+}$ contents in DPGW are mostly derived from evaporative minerals.

The concentration ratio relationship between ($Ca^{2+}$ + $Mg^{2+}$-$HCO_3^-$) and ($SO_4^{2-}$-$Na^+$-$Cl^-$) can further determine whether $SO_4^{2-}$ is derived from the dissolution of gypsum. $SO_4^{2-}$-($Na^+$-$Cl^-$) represents the dissolution rate of $Ca^{2+}$ in gypsum, while $SO_4^{2-}$-($Na^+$+$Cl^-$) represents the dissolution rate of $SO_4^{2-}$ in gypsum. ($Ca^{2+}$ + $Mg^{2+}$-$HCO_3^-$) represents the dissolution rate of $Ca^{2+}$ in gypsum, and $SO_4^{2-}$-($Na^+$ + $Cl^-$) represents the dissolution rate of $SO_4^{2-}$ in gypsum [37]. Figure 9b shows that all water sample points fell near the 1:1 line, indicating that $SO_4^{2-}$ in different types of groundwater in the NPSP is mostly derived from the dissolution of gypsum, which is consistent with the presence of gypsum in the Neogene strata in the work area.

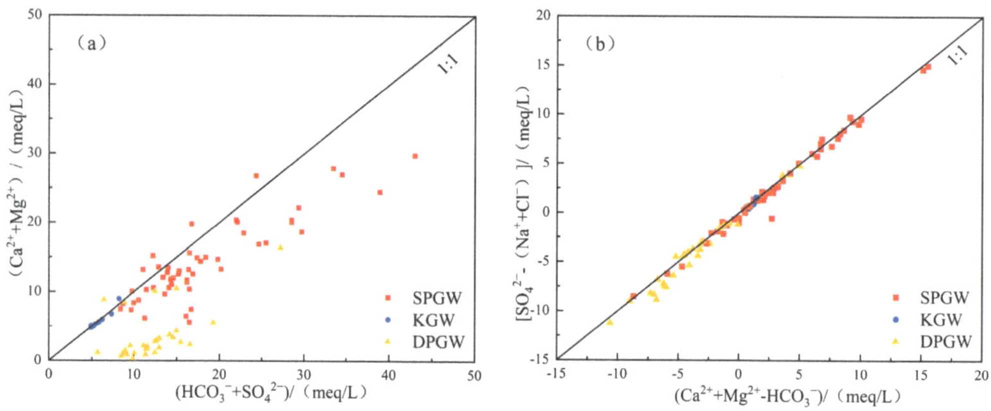


**Figure 9.** Relationships between ($Ca^{2+}$ + $Mg^{2+}$) and ($HCO_3^-$ + $SO_4^{2-}$) (**a**), ($Ca^{2+}$ + $Mg^{2+}$-$HCO_3^-$) and ($SO_4^{2-}$-$Na^+$-$Cl^-$) (**b**) in the northwest plain of Shandong Province.

## 5. Discussion

### 5.1. Controlling Factors of Groundwater's Hydrochemical Characteristics

The above findings show that the strontium content in the three groundwater types was 1.59 mg/L, 0.58 mg/L and 0.50 mg/L, respectively. The hydrochemical characteristics were mainly controlled by mineral dissolution and cation exchange in the NPSP. Silicate, evaporative minerals and carbonate mineral dissolution are the dominant factors influencing the hydrogeochemical processes in SPGW, DPGW and KGW, respectively, and the cation exchange was found to be a mainly positive exchange. The quality of KGW was relatively good, indicating it is suitable for use as drinking water, and SPGW is suitable for farmland irrigation.

### 5.2. Origin of Strontium Ions in Groundwater

Human activities have relatively little influence on the NPSP, which can be regarded as a natural study site for groundwater chemical evolution analysis. On the basis of the above research, the model of the regional water chemical field was established (Figure 10). In the process of groundwater formation, minerals are first dissolved in groundwater, and the evaporation and cation exchange occurring during runoff affect the hydrochemical components.

The above indicates that the composition of groundwater in the NPSP is mainly derived from rock weathering, with the strontium in rocks being the source of elements in groundwater. Strontium is an alkaline earth metal group with active chemical properties and a strong migration ability. It belongs to the same group as Ca and Mg and has similar chemical properties. It often coexists with Ca and Mg in minerals in the form of homomorphism and is abundant in granite, diorite, carbonate rocks and clay minerals [38].

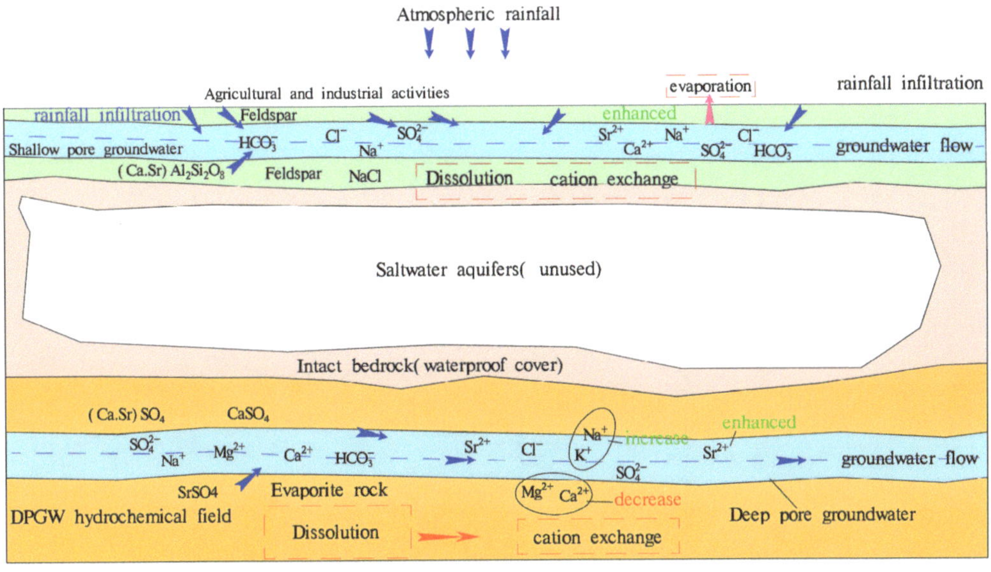


**Figure 10.** Hydrochemical field evolution model of the northwest plain of Shandong Province.

The SPGW recharge area is relatively close, but shallow burial, loose strata and rock minerals such as silicate rock and groundwater contact areas are highly present, and so it is easily affected by evaporation, concentration, ion exchange and other factors. The mineral solubility is remarkable; the $Sr^{2+}$ in SPGW mainly comes from strontium minerals in silicate rocks, such as anorthite (Equation (5)). The contact area between the minerals and water is sufficient, and the degree of dissolution is high, so the content of strontium is high in shallow groundwater.

$$(Ca.Sr)Al_2Si_2O_8 + 2CO_2 + 3H_2O \rightarrow AlSi_2O_5(OH)_4 + 2HCO_3^- + Sr^{2+}(Ca^{2+}) \quad (5)$$

In contrast, DPGW has the characteristics of deep burial, a slow runoff, a far-off recharge area, long formation, evaporation and concentration times and an ion exchange effect; the $Sr^{2+}$ in DPGW mainly comes from evaporative salt rocks, including gypsum (Equation (6)).

$$(Ca.Sr)SO_4 \cdot 2H_2O \rightarrow Ca^{2+}Sr^{2+} + SO_4^{2-} + 2H_2O \quad (6)$$

KGW originates from atmospheric precipitation, and the recharge comes from the upstream mountainous area. Groundwater immediately enters the Ordovician limestone strata after rainfall, with a short runoff distance. Meanwhile, the rocks in the Ordovician limestone strata are hard and dominated by solution pores and caves, and there are few contact surfaces between the groundwater and bedrock, which do not fully come into contact. However, its dissolution is closely related to the content of $CO_2$, so the content of all the ions in karst water is low, and the upper layer is covered by a thick Quaternary layer, and the effect of evaporation and concentration is weak (Figure 11). The $Sr^{2+}$ in KGW is mainly derived from strontium-containing minerals (such as calcite, dolomite, etc.) in carbonate rocks (Equations (7) and (8)) [39]. Its runoff distance is short, and the contact area between rock and water is small. Therefore, mineral dissolution is relatively small, and the content of each ion is low. Compared with DPGW, the dissolution of $Sr^{2+}$ is relatively greater, indicating that the content of $Sr^{2+}$ in carbonate minerals is higher than that in gypsum-based evaporite, and the solubility of strontium-containing minerals is greater than that of gypsum-based evaporite minerals.

$$(Ca.Sr)CO_3 + CO_2 + H_2O \rightarrow Ca^{2+}Sr^{2+} + 2HCO_3^-, \quad (7)$$

$$(Ca.Sr)Mg(CO_3)_2 + CO_2 + H_2O \rightarrow Ca^{2+}Sr^{2+} + Mg^{2+} + 2HCO_3^- + CO_3^{2-} \quad (8)$$

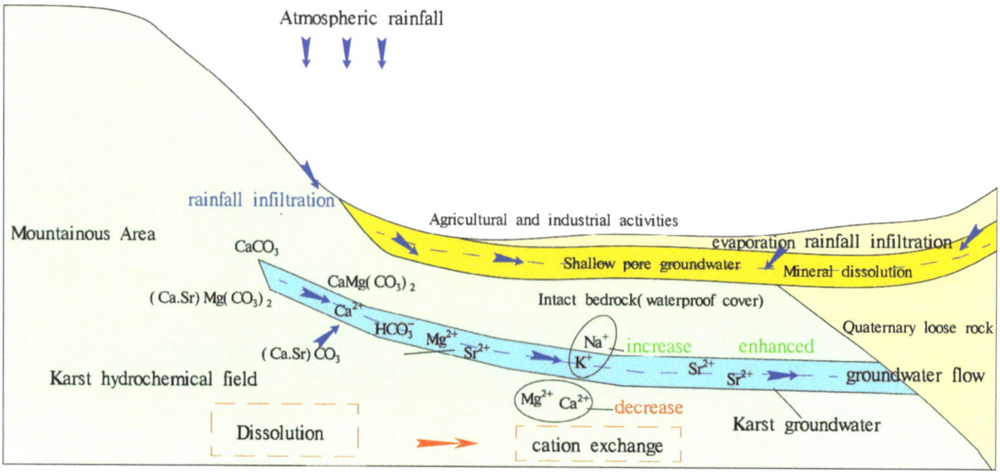


**Figure 11.** Strontium enrichment model of karst groundwater in the northwest plain of Shandong Province.

### *5.3. Suggestions for Groundwater Development and Utilization*

According to the chemical characteristics of groundwater in this area, SPGW contains component concentrations that exceed the standard concentrations, and thus can be used for agricultural irrigation after treatment. All the ion targets of KGW are in line with the characteristics of natural strontium-rich groundwater. It is suggested that the exploitation of strontium-rich karst water be appropriately increased and that DPGW be rationally exploited within the scope permitted by law without causing other geological environmental problems.

## 6. Conclusions

Three types of groundwater samples were collected in the northwest plain of Shandong Province, China. including shallow pore groundwater, deep pore groundwater and karst-soluble groundwater. The hydrochemical characteristics and controlling factors of different groundwater types were studied, and the source and enrichment of strontium in groundwater were analyzed. Our conclusions are as follows:

(1) There are significant differences in different groundwater types in the northwest plain of Shandong Province, China. Shallow groundwater is mainly brackish–saline water, deep groundwater is mainly brackish water, with some fresh water, and karst groundwater is fresh water. The shallow pore water is rich in sources and diverse in hydrochemistry. The cation content mainly comprises the Na·Mg type, Ca·Na type and Ca·Mg type and the anion content mainly comprises the $HCO_3$ type, followed by the $HCO_3$·Cl type and $SO_4$·$HCO_3$·Cl type. The cations in deep pore water are mainly of the Na type, followed by the Ca·Na·Mg type, and the anionic types are relatively complex, mainly including the $HCO_3$·Cl and $HCO_3$·Cl·$SO_4$ types. The hydrochemical types of karst are relatively simple, mainly including the $HCO_3$-Ca·Mg type and $HCO_3$-Ca type.

(2) The main components of groundwater sources in the NPSP are derived from water–rock interaction, and rock mineral leaching plays a dominant role, which is more obvious in karst water, and the influencing factors of evaporation and concentration are highly present in shallow and deep pore water. Shallow pore water mainly

comprises dissolved silicate minerals, deep pore water mainly contains dissolved evaporative minerals, largely gypsum, and karst water is mainly composed of dissolved carbonate minerals. Cation exchange exists in these three kinds of groundwater, and a positive exchange is the main effect. Natural geological conditions and good water source protection measures mean that human activities have relatively little influence on the water chemical formation process.

(3) The distribution of strontium-rich groundwater is the highest in shallow pore water, where the average $Sr^{2+}$ content reaches 1.59 mg/L, while the average $Sr^{2+}$ content in deep pore water and karst water is 0.58 mg/L and 0.50 mg/L, respectively. The strontium in shallow pore groundwater mainly comes from the enrichment of groundwater runoff, and its sources are abundant, though it mainly derives from silicic rock. Deep pore groundwater mainly comprises evaporative minerals containing strontium, and karst water is mainly derived from carbonate rock dissolution with similar characteristics.

Finally, on the basis of this study, we suggest strengthening the development of strontium-rich karst groundwater as drinking water resources and doing a good job of water source protection, which can produce good economic and social benefits. Shallow groundwater can be used for farmland irrigation.

**Author Contributions:** Conceptualization, J.C. and X.W.; methodology, J.C. and J.L.; software, J.C. and Y.Z.; validation, J.C. and J.Z.; formal analysis, J.C. and Y.Z.; investigation, J.C. and S.L.; resources, X.W.; data curation, J.C. and X.W.; writing—original draft preparation, J.C.; writing—review and editing, X.W.; visualization, J.Z.; supervision, Z.G. and X.W.; project administration, J.Z.; funding acquisition, X.W. All authors have read and agreed to the published version of the manuscript.

**Funding:** This research received no external funding.

**Data Availability Statement:** Data available on request due to restrictions eg privacy or ethical.

**Conflicts of Interest:** The authors declare no conflicts of interest.

## Appendix A

**Table A1.** Complete chemical analysis of all study samples.

| Sample Number | Type | $K^+$ | $Na^+$ | $Ca^{2+}$ | $Mg^{2+}$ | $Cl^-$ | $SO_4^{2-}$ | $HCO_3^-$ | $NO_3^-$ | TH | TDS | pH | $Sr^{2+}$ |
|---|---|---|---|---|---|---|---|---|---|---|---|---|---|
| S1 | SPGW | 1.60 | 86.00 | 90.18 | 46.17 | 99.26 | 110.47 | 390.53 | 0.52 | 415.33 | 639.83 | 7.4 | 1.1 |
| S2 | SPGW | 1.30 | 132.00 | 112.22 | 78.98 | 141.80 | 211.33 | 549.18 | 21.55 | 605.48 | 981.27 | 7.4 | 1.7 |
| S3 | SPGW | 3.60 | 196.00 | 210.42 | 113.00 | 382.86 | 218.54 | 744.44 | 0.55 | 990.79 | 1508.78 | 7.0 | 2.3 |
| S4 | KGW | 0.80 | 24.00 | 76.15 | 18.23 | 24.82 | 38.42 | 280.69 | 9.68 | 265.21 | 339.94 | 7.8 | 0.5 |
| S5 | KGW | 0.80 | 24.00 | 76.15 | 21.87 | 21.27 | 62.44 | 274.59 | 3.95 | 280.22 | 355.28 | 7.8 | 0.5 |
| S6 | KGW | 1.00 | 20.00 | 80.16 | 24.30 | 24.82 | 81.65 | 274.59 | 3.52 | 300.24 | 380.89 | 8.0 | 0.6 |
| S7 | KGW | 0.80 | 46.00 | 124.25 | 34.02 | 85.08 | 96.06 | 378.32 | 17.67 | 450.36 | 600.54 | 7.6 | 0.7 |
| S8 | SPGW | 0.40 | 80.00 | 52.10 | 59.54 | 56.72 | 43.23 | 457.65 | 42.70 | 375.30 | 574.01 | 8.0 | 1.2 |
| S9 | SPGW | 1.80 | 70.00 | 148.30 | 94.77 | 191.43 | 165.70 | 536.98 | 0.86 | 760.61 | 949.17 | 7.4 | 1.8 |
| S10 | SPGW | 1.00 | 240.00 | 78.16 | 94.77 | 194.98 | 240.15 | 683.42 | 0.45 | 585.47 | 1200.58 | 8.1 | 2.1 |
| S11 | DPGW | 0.75 | 440.00 | 20.04 | 13.37 | 315.51 | 341.01 | 274.59 | <0.20 | 105.08 | 1288.69 | 8.3 | 0.8 |
| S12 | SPGW | 2.20 | 200.00 | 148.30 | 92.34 | 177.25 | 302.59 | 738.34 | 0.75 | 750.60 | 1304.84 | 7.4 | 1.1 |
| S13 | SPGW | 1.20 | 150.00 | 80.16 | 104.49 | 131.17 | 196.92 | 677.32 | 3.53 | 630.50 | 1013.63 | 7.7 | 1.6 |
| S14 | SPGW | 2.80 | 450.00 | 116.23 | 134.87 | 407.68 | 487.50 | 884.79 | 12.73 | 845.68 | 2061.57 | 7.4 | 3.0 |
| S15 | SPGW | 1.70 | 100.00 | 122.24 | 69.26 | 56.72 | 72.05 | 781.06 | 0.35 | 590.47 | 822.08 | 7.2 | 1.3 |
| S16 | SPGW | 2.80 | 204.00 | 78.16 | 113.00 | 198.52 | 67.24 | 915.30 | 0.86 | 660.53 | 1135.85 | 7.4 | 2.3 |
| S17 | SPGW | 1.20 | 196.00 | 70.14 | 92.34 | 170.16 | 28.82 | 842.08 | 1.61 | 555.44 | 993.40 | 7.8 | 2.4 |
| S18 | SPGW | 0.80 | 228.00 | 72.14 | 109.35 | 180.80 | 206.53 | 768.85 | 0.55 | 630.50 | 1191.84 | 7.8 | 1.8 |
| S19 | SPGW | 1.50 | 180.00 | 116.23 | 85.05 | 223.34 | 206.53 | 585.79 | 1.43 | 640.51 | 1120.71 | 7.5 | 1.8 |
| S20 | SPGW | 0.80 | 240.00 | 120.24 | 102.06 | 134.71 | 292.98 | 713.93 | 153.00 | 720.58 | 1408.26 | 7.8 | 1.1 |
| S21 | SPGW | 11.00 | 180.00 | 74.15 | 72.90 | 109.90 | 124.88 | 671.22 | 23.16 | 485.39 | 939.09 | 7.8 | 0.8 |
| S22 | SPGW | 0.80 | 260.00 | 44.09 | 64.40 | 63.81 | 158.50 | 817.67 | 0.91 | 375.30 | 1010.42 | 8.0 | 0.8 |
| S23 | SPGW | 0.60 | 390.00 | 38.08 | 55.89 | 233.97 | 235.35 | 683.42 | 0.20 | 325.26 | 1304.30 | 8.2 | 0.8 |
| S24 | SPGW | 0.50 | 180.00 | 44.09 | 111.78 | 99.26 | 211.33 | 720.04 | 3.64 | 570.46 | 1019.80 | 8.2 | 1.5 |
| S25 | SPGW | 1.00 | 480.00 | 90.18 | 153.09 | 425.40 | 557.15 | 848.18 | 0.44 | 855.68 | 2138.29 | 7.7 | 1.8 |

**Table A1.** *Cont.*

| Sample Number | Type | $K^+$ | $Na^+$ | $Ca^{2+}$ | $Mg^{2+}$ | $Cl^-$ | $SO_4{}^{2-}$ | $HCO_3{}^-$ | $NO_3{}^-$ | TH | TDS | pH | $Sr^{2+}$ |
|---|---|---|---|---|---|---|---|---|---|---|---|---|---|
| S26 | SPGW | 1.00 | 184.00 | 100.20 | 103.28 | 255.24 | 196.92 | 604.10 | 0.40 | 675.54 | 1150.59 | 7.8 | 1.5 |
| S27 | SPGW | 1.40 | 340.00 | 180.36 | 134.87 | 425.40 | 569.16 | 628.51 | 14.88 | 1005.80 | 1988.49 | 7.4 | 1.6 |
| S28 | SPGW | 1.00 | 420.00 | 110.22 | 157.95 | 475.03 | 600.38 | 634.61 | 15.33 | 925.74 | 2104.90 | 7.6 | 1.9 |
| S29 | SPGW | 2.50 | 600.00 | 176.35 | 221.13 | 645.19 | 840.53 | 1037.34 | 1.11 | 1351.08 | 3018.28 | 7.5 | 2.5 |
| S30 | SPGW | 1.80 | 192.00 | 98.20 | 66.83 | 81.54 | 148.89 | 817.67 | 0.28 | 520.42 | 1009.21 | 7.8 | 0.8 |
| S31 | SPGW | 1.60 | 96.00 | 82.16 | 78.98 | 85.08 | 110.47 | 610.20 | <0.20 | 530.42 | 768.59 | 7.8 | 1.3 |
| S32 | SPGW | 1.00 | 74.00 | 86.17 | 37.67 | 31.91 | 52.83 | 524.77 | <0.20 | 370.30 | 556.86 | 7.6 | 0.6 |
| S33 | KGW | 0.50 | 26.00 | 74.15 | 15.80 | 21.27 | 48.03 | 256.28 | 14.33 | 250.20 | 335.72 | 8.1 | 0.3 |
| S34 | KGW | 0.40 | 18.00 | 70.14 | 19.44 | 24.82 | 48.03 | 237.98 | 14.89 | 255.20 | 322.20 | 7.8 | 0.3 |
| S35 | KGW | 0.40 | 22.00 | 66.13 | 18.23 | 21.27 | 48.03 | 237.98 | 11.96 | 240.19 | 314.51 | 8.1 | 0.3 |
| S36 | SPGW | 0.10 | 64.00 | 98.20 | 65.61 | 53.18 | 76.85 | 604.10 | 0.25 | 515.41 | 669.73 | 7.2 | 0.8 |
| S37 | SPGW | 0.30 | 100.00 | 52.10 | 91.13 | 152.44 | 86.45 | 488.16 | 8.80 | 505.40 | 742.89 | 7.9 | 1.2 |
| S38 | SPGW | 0.40 | 180.00 | 46.09 | 47.39 | 106.35 | 134.48 | 518.67 | 0.26 | 310.25 | 782.13 | 8.1 | 0.6 |
| S39 | SPGW | 40.00 | 240.00 | 66.13 | 207.77 | 340.32 | 403.45 | 829.87 | 0.53 | 1020.82 | 1721.66 | 7.4 | 2.1 |
| S40 | SPGW | 0.40 | 288.00 | 230.46 | 185.90 | 514.03 | 626.79 | 689.53 | 56.75 | 1341.07 | 2254.58 | 7.3 | 2.1 |
| S41 | SPGW | 2.50 | 690.00 | 42.08 | 336.56 | 620.38 | 1071.07 | 1263.11 | 0.52 | 1491.19 | 3403.03 | 7.4 | 2.4 |
| S42 | SPGW | 1.00 | 200.00 | 162.32 | 91.13 | 290.69 | 365.03 | 543.08 | 7.64 | 780.62 | 1396.85 | 7.8 | 1.0 |
| S43 | SPGW | 1.50 | 53.00 | 132.26 | 80.19 | 163.07 | 134.48 | 506.47 | <0.20 | 660.53 | 831.04 | 7.4 | 1.6 |
| S44 | SPGW | 1.50 | 370.00 | 90.18 | 193.19 | 276.51 | 787.69 | 738.34 | 0.70 | 1020.82 | 2097.56 | 7.8 | 1.7 |
| S45 | SPGW | 0.60 | 256.00 | 88.18 | 125.15 | 191.43 | 341.01 | 774.95 | 0.38 | 735.59 | 1398.72 | 7.7 | 1.3 |
| S46 | SPGW | 0.50 | 94.00 | 78.16 | 98.42 | 60.27 | 182.51 | 659.02 | <0.20 | 600.48 | 853.86 | 8.2 | 1.3 |
| S47 | SPGW | 0.50 | 100.00 | 72.14 | 85.05 | 31.91 | 115.27 | 713.93 | <0.20 | 530.42 | 770.66 | 7.8 | 1.7 |
| S48 | SPGW | 1.50 | 380.00 | 118.24 | 172.53 | 283.60 | 725.25 | 817.67 | 2.60 | 1005.80 | 2101.05 | 7.8 | 1.9 |
| S49 | SPGW | 1.50 | 460.00 | 152.30 | 246.65 | 496.30 | 991.82 | 781.06 | 14.49 | 1396.12 | 2761.24 | 7.5 | 2.9 |
| S50 | DPGW | 0.80 | 296.00 | 4.01 | 1.22 | 85.08 | 110.47 | 463.75 | 0.63 | 15.01 | 760.40 | 8.4 | 0.2 |
| S51 | DPGW | 0.50 | 390.00 | 14.03 | 25.52 | 223.34 | 206.53 | 500.36 | 0.43 | 140.11 | 1151.73 | 8.3 | 1.4 |
| S52 | DPGW | 0.50 | 296.00 | 8.02 | 9.72 | 56.72 | 110.47 | 622.40 | 0.57 | 60.05 | 803.78 | 8.2 | 0.4 |
| S53 | DPGW | 1.60 | 180.00 | 68.14 | 86.27 | 109.90 | 177.71 | 689.53 | <0.20 | 525.42 | 978.39 | 7.8 | 1.3 |
| S54 | DPGW | 1.00 | 130.00 | 80.16 | 74.12 | 109.90 | 148.89 | 573.59 | 0.43 | 505.40 | 841.29 | 7.4 | 1.5 |
| S55 | DPGW | 1.00 | 370.00 | 36.07 | 10.94 | 95.72 | 105.67 | 811.57 | 0.58 | 135.11 | 1036.19 | 7.2 | 0.6 |
| S56 | DPGW | 0.50 | 420.00 | 20.04 | 41.31 | 276.51 | 326.60 | 500.36 | 0.42 | 220.18 | 1345.94 | 7.8 | 0.4 |
| S57 | DPGW | 0.50 | 280.00 | 12.02 | 10.94 | 173.71 | 139.29 | 372.22 | 0.60 | 75.06 | 812.49 | 8.0 | 0.9 |
| S58 | SPGW | 2.60 | 168.00 | 108.22 | 92.34 | 194.98 | 43.23 | 878.69 | 0.45 | 650.52 | 1062.70 | 7.9 | 1.8 |
| S59 | DPGW | 0.25 | 240.00 | 8.02 | 3.65 | 46.09 | 81.65 | 414.94 | 0.29 | 35.03 | 638.68 | 8.4 | 0.2 |
| S60 | DPGW | 0.50 | 288.00 | 14.03 | 18.23 | 138.26 | 158.50 | 384.43 | 0.43 | 110.09 | 869.89 | 8.3 | 0.4 |
| S61 | SPGW | 0.70 | 168.00 | 50.10 | 76.55 | 191.43 | 120.08 | 494.26 | <0.20 | 440.35 | 862.56 | 7.8 | 1.7 |
| S62 | DPGW | 1.25 | 450.00 | 34.07 | 20.66 | 301.33 | 461.09 | 305.10 | 0.58 | 170.14 | 1430.34 | 7.8 | 0.9 |
| S63 | DPGW | 1.25 | 440.00 | 44.09 | 19.44 | 311.96 | 437.07 | 311.20 | 0.33 | 190.15 | 1419.05 | 7.8 | 0.9 |
| S64 | SPGW | 2.40 | 288.00 | 140.28 | 95.99 | 358.05 | 408.26 | 543.08 | 0.36 | 745.60 | 1577.36 | 7.8 | 1.5 |
| S65 | DPGW | 0.50 | 260.00 | 10.02 | 4.86 | 70.90 | 100.86 | 433.24 | 0.67 | 45.04 | 711.60 | 8.3 | 0.3 |
| S66 | DPGW | 0.50 | 288.00 | 12.02 | 6.08 | 53.18 | 105.67 | 482.06 | 0.80 | 55.04 | 762.21 | 8.4 | 0.2 |
| S67 | DPGW | 1.50 | 370.00 | 14.03 | 12.15 | 230.43 | 254.56 | 378.32 | 0.86 | 85.07 | 1082.60 | 8.0 | 0.2 |
| S68 | DPGW | 1.00 | 380.00 | 20.04 | 8.51 | 241.06 | 273.77 | 360.02 | 0.58 | 85.07 | 1113.71 | 8.1 | 0.2 |
| S69 | DPGW | 1.00 | 380.00 | 18.04 | 15.80 | 248.15 | 278.57 | 329.51 | 0.49 | 110.09 | 1116.79 | 8.0 | 0.3 |
| S70 | DPGW | 1.00 | 360.00 | 16.03 | 30.38 | 212.70 | 312.20 | 396.63 | 0.77 | 165.13 | 1142.48 | 8.2 | 0.6 |
| S71 | DPGW | 0.70 | 160.00 | 6.01 | 10.94 | 53.18 | 52.83 | 280.69 | 0.30 | 60.05 | 461.80 | 8.4 | 0.2 |
| S72 | DPGW | 2.00 | 670.00 | 26.05 | 52.25 | 510.48 | 653.21 | 347.81 | 2.52 | 280.22 | 2129.59 | 8.5 | 0.7 |
| S73 | DPGW | 0.60 | 256.00 | 14.03 | 4.86 | 70.90 | 76.85 | 433.24 | 0.54 | 55.04 | 687.82 | 8.6 | 0.2 |
| S74 | DPGW | 5.00 | 670.00 | 116.23 | 128.79 | 659.37 | 691.63 | 781.06 | 0.78 | 820.66 | 2671.04 | 7.4 | 1.4 |
| S75 | DPGW | 1.50 | 470.00 | 24.05 | 14.58 | 248.15 | 432.27 | 463.75 | 1.87 | 120.10 | 1435.39 | 7.8 | 0.4 |
| S76 | DPGW | 1.00 | 340.00 | 14.03 | 14.58 | 127.62 | 211.33 | 524.77 | 0.78 | 95.08 | 983.58 | 7.8 | 0.2 |
| S77 | DPGW | 1.00 | 308.00 | 6.01 | 7.29 | 56.72 | 168.11 | 482.06 | 1.30 | 45.04 | 838.31 | 8.3 | 0.1 |
| S78 | DPGW | 0.80 | 94.00 | 104.21 | 44.96 | 226.88 | 120.08 | 237.98 | 5.01 | 445.36 | 724.92 | 7.4 | 0.7 |
| S79 | DPGW | 0.90 | 84.00 | 106.21 | 38.88 | 106.35 | 220.94 | 262.39 | 4.61 | 425.34 | 702.58 | 7.5 | 0.7 |
| S80 | KGW | 0.70 | 40.00 | 50.10 | 52.25 | 49.63 | 91.26 | 329.51 | 15.29 | 340.27 | 471.48 | 7.4 | 0.8 |
| S81 | DPGW | 1.00 | 340.00 | 12.02 | 29.16 | 184.34 | 321.80 | 366.12 | <0.20 | 150.12 | 1080.73 | 8.2 | 0.6 |
| S82 | SPGW | 0.80 | 88.00 | 10.02 | 95.99 | 74.45 | 120.08 | 457.65 | 15.11 | 420.34 | 642.01 | 8.0 | 1.3 |
| S83 | SPGW | 0.80 | 276.00 | 38.08 | 44.96 | 46.09 | 105.67 | 872.59 | 0.27 | 280.22 | 957.77 | 7.8 | 0.9 |
| S84 | SPGW | 1.70 | 104.00 | 208.42 | 38.88 | 159.53 | 100.86 | 659.02 | 2.03 | 680.54 | 954.04 | 7.0 | 1.5 |
| S85 | SPGW | 0.60 | 228.00 | 68.14 | 120.29 | 116.99 | 422.66 | 695.63 | 0.87 | 665.53 | 1315.20 | 8.0 | 1.3 |
| S86 | SPGW | 1.00 | 370.00 | 164.33 | 170.10 | 319.05 | 790.09 | 787.16 | 0.31 | 1110.89 | 2218.25 | 7.5 | 2.1 |
| S87 | SPGW | 1.00 | 460.00 | 100.20 | 166.46 | 315.51 | 730.06 | 884.79 | 0.68 | 935.75 | 2228.15 | 7.3 | 1.7 |
| S88 | SPGW | 1.50 | 610.00 | 104.21 | 234.50 | 425.40 | 1104.69 | 970.22 | 0.26 | 1225.98 | 2978.98 | 7.2 | 2.0 |

## References


1. Gao, M.; Li, X.; Qian, J.; Wang, Z.; Hou, X.; Gui, C.; Bai, Z.; Fu, C.; Li, J.; Zuo, X. Hydrochemical Characteristics and Controlling Factors of Shallow and Deep Groundwater in the Heilongdong Spring Basin, Northern China. *Sustainability* **2023**, *15*, 15447. [CrossRef]
2. Liu, J.; Peng, Y.; Li, C.; Gao, Z.; Chen, S. An investigation into the hydrochemistry, quality and risk to human health of groundwater in the central region of Shandong Province, North China. *J. Clean. Prod.* **2021**, *282*, 125416. [CrossRef]
3. Li, Z.; Coles, A.; Xiao, J. Groundwater and streamflow sources in China's Loess Plateau on catchment scale. *CATENA* **2019**, *181*, 104075. [CrossRef]
4. Qu, S.; Duan, L.; Mao, H.; Wang, C.; Liang, X.; Luo, A.; Huang, L.; Yu, R.; Miao, P.; Zhao, Y. Hydrochemical and isotopic fingerprints of groundwater origin and evolution in the Urangulan River basin, China's Loess Plateau. *Sci. Total Environ.* **2023**, *866*, 161377. [CrossRef] [PubMed]
5. Lee, K.; Yun, S.; Yu, S.; Kim, K.; Lee, J.; Lee, S. The combined use of self-organizing map technique and fuzzy c-means clustering to evaluate urban groundwater quality in Seoul metropolitan city, South Korea. *J. Hydrol.* **2019**, *569*, 685–697. [CrossRef]
6. Alvarez, M.; Funes, D.; Dapeña, C.; Bouza, P. Origin and hydrochemical characteristics of groundwater in the Northeastern Patagonia, Argentina: The relationship with geomorphology and soils. *Environ. Earth Sci.* **2020**, *79*, 503. [CrossRef]
7. Liang, C.; Wang, W.; Ke, X.; Ou, A.; Wang, D. Hydrochemical Characteristics and Formation Mechanism of Strontium-Rich Groundwater in Tianjiazhai, Fugu, China. *Water* **2022**, *14*, 1874. [CrossRef]
8. Li, X.; Wu, H.; Qian, H.; Gao, Y. Groundwater Chemistry Regulated by Hydrochemical Processes and Geological Structures: A Case Study in Tongchuan, China. *Water* **2018**, *10*, 338. [CrossRef]
9. Liu, J.; Gao, Z.; Wang, M.; Li, Y.; Shi, M.; Zhang, H.; Ma, Y. Hydrochemical characteristics and possible controls in the groundwater of the Yarlung Zangbo River Valley, China. *Environ. Earth Sci.* **2019**, *78*, 76. [CrossRef]
10. Liu, J.; Wang, F.; Cai, W.; Wang, Z.; Li, C. Numerical investigation on the effects of geological parameters and layered subsurface on the thermal performance of medium-deep borehole heat exchanger. *Renew. Energy* **2020**, *149*, 384–399. [CrossRef]
11. Zhai, Y.; Zheng, F.; Zhao, X.; Xia, X.; Teng, Y. Identification of hydrochemical genesis and screening of typical groundwater pollutants impacting human health: A case study in Northeast China. *Environ. Pollut.* **2019**, *252*, 1202–1215. [CrossRef] [PubMed]
12. Li, S.; Lu, X.; Bush, R. Chemical weathering and $CO_2$ consumption in the Lower Mekong River. *Sci. Total Environ.* **2014**, *472*, 162–177. [CrossRef] [PubMed]
13. Yin, Z.; Luo, Q.; Wu, J.; Xu, S.; Wu, J. Identification of the long-term variations of groundwater and their governing factors based on hydrochemical and isotopic data in a river basin. *J. Hydrol.* **2021**, *592*, 125604. [CrossRef]
14. Dai, L.; Mei, M.; Chu, C.; Lo, E. Remineralizing effect of a new strontium-doped bioactive glass and fluoride on demineralized enamel and dentine. *J. Dent.* **2021**, *108*, 103633. [CrossRef] [PubMed]
15. Li, D.; Gan, S.; Li, J.; Dong, Z.; Long, Q.; Qiu, S.; Zhou, Y.; Lu, C. Groundwater Chemistry and Pollution in a Changing Environment: Monitoring, Mechanisms, and Remediations. *J. Chem.* **2021**, *2021*, 5547924. [CrossRef]
16. Wang, R.; Wu, X.; Zhai, Y.; Su, Y.; Liu, C. An Experimental Study on the Sources of Strontium in Mineral Water and General Rules of Its Dissolution—A Case Study of Chengde, Hebei. *Water* **2021**, *13*, 699. [CrossRef]
17. Zuo, R.; Teng, Y.; Wang, J. Modeling migration of strontium in sand and gravel aquifer in the candidate VLLW disposal site. *J. Radioanal. Nucl. Chem.* **2009**, *281*, 653–662. [CrossRef]
18. Zhang, B.; Zhang, R.; Han, P.-F.; Wang, J.; Wan, L. Evolution of Hydrogeochemistry in the Haolebaojinao Watershed of the Ordos Basin, China. *Sustainability* **2023**, *15*, 5091. [CrossRef]
19. Peng, H.; Zou, P.; Ma, C.; Xiong, S.; Lu, T. Elements in potable groundwater in Rugao longevity area, China: Hydrogeochemical characteristics, enrichment patterns and health assessments. *Ecotoxicol. Environ. Saf.* **2021**, *218*, 112279. [CrossRef]
20. Athamena, A.; Gaagai, A.; Aouissi, H.A.; Burlakovs, J.; Bencedira, S.; Zekker, I.; Krauklis, A.E. Chemometrics of the Environment: Hydrochemical Characterization of Groundwater in Lioua Plain (North Africa) Using Time Series and Multivariate Statistical Analysis. *Sustainability* **2023**, *15*, 20. [CrossRef]
21. *DZ/T 0064*; Methods for Analysis of Groundwater Quality. Ministry of Natural Resources, China: Beijing, China.
22. Peng, H.; Hou, Q.; Zen, M.; Huang, C.; Shi, H.; Pi, P.; Pan, Y. Hydrochemical Characteristics and Controlling Factors of Groundwater in the Leizhou Peninsula. *Environ. Sci.* **2021**, *42*, 5374–5384.
23. Xiao, Y.; Liu, K.; Hao, Q.; Xiao, D.; Zhu, Y.; Yin, S.; Zhang, Y. Hydrogeochemical insights into the signatures, genesis and sustainable perspective of nitrate enriched groundwater in the piedmont of Hutuo watershed, China. *Catena* **2022**, *212*, 106020. [CrossRef]
24. Piper, A.M. A graphic procedure in the geochemical interpretation of water-analyses. EOS Trans. *Am. Geophys. Union* **1944**, *25*, 914–928. [CrossRef]
25. Wei, H.; Liang, X.; Liu, S.; Liu, M.; Xiao, C. Hydrochemical Evolution of Groundwater in Dehui, China. *Water* **2020**, *12*, 3378. [CrossRef]
26. Gibbs, R. Mechanisms controlling world water chemistry. *Science* **1970**, *170*, 1088–1890. [CrossRef] [PubMed]
27. Gaillardet, J.; Dupre, B.; Louvat, P.; Allegre, C. Global silicate weathering and $CO_2$ consumption rates deduced from the chemistry of large rivers. *Chem. Geol.* **1999**, *159*, 3–30. [CrossRef]
28. Yuan, H.; Yang, S.; Wang, B. Hydrochemistry characteristics of groundwater with the influence of spatial variability and water flow in Hetao Irrigation District, China. *Environ. Sci. Pollut. Res.* **2022**, *29*, 71150–71164. [CrossRef] [PubMed]

29. Fan, B.; Zhao, Z.; Tao, F.; Liu, B.; Tao, Z.; Gao, S.; Zhang, L. Characteristics of carbonate, evaporite and silicate weathering in Huanghe River basin: A comparison among the upstream, midstream and downstream. *J. Asian Earth Sci.* **2014**, *96*, 17–26. [CrossRef]
30. Ren, C.; Zhang, Q. Groundwater Chemical Characteristics and Controlling Factors in a Region of Northern China with Intensive Human Activity. *Int. J. Environ. Res. Public Health* **2020**, *17*, 9126. [CrossRef]
31. Peng, H.; Yang, W.; Ferrer, A.; Xiong, S.; Li, X.; Niu, G.; Lu, T. Hydrochemical characteristics and health risk assessment of groundwater in karst areas of southwest China: A case study of Bama, Guangxi. *J. Clean. Prod.* **2022**, *341*, 130872. [CrossRef]
32. Magaritz, M.; Nadler, A.; Koyumdjisky, H.; Dan, J. The use of Na/Cl ratios to trace solute sources in a semiarid zone. *Water Resour. Res.* **1981**, *17*, 602–608. [CrossRef]
33. Sami, K. Recharge mechanisms and geochemical processes in a semi-arid sedimentary basin, Eastern Cape, South Africa. *J. Hydrol.* **1992**, *139*, 27–48. [CrossRef]
34. Thimonier, A.; Schmitt, M.; Waldner, P.; Schleppi, P. Seasonality of the Na/Cl ratio in precipitation and implications of canopy leaching in validating chemical analyses of throughfall samples. *Atmos. Env.* **2008**, *42*, 9106–9117. [CrossRef]
35. Chen, X.; Jiang, C.; Zheng, L.; Zhang, L.; Fu, X.; Chen, S.; Chen, Y.; Hu, J. Evaluating the genesis and dominant processes of groundwater salinization by using hydrochemistry and multiple isotopes in a mining city. *Environ. Pollut.* **2021**, *283*, 117381. [CrossRef]
36. Li, C.; Gao, X.; Wang, Y. Hydrogeochemistry of high-fluoride groundwater at Yuncheng Basin, northern China. *Sci. Total Environ.* **2015**, *508*, 155–165. [CrossRef]
37. Zhang, T.; Wang, M.; Zhang, Z.; Liu, T.; He, J. Hydrochemical characteristics and possible controls of the surface water in ranwu lake basin. *Environ. Sci.* **2020**, *41*, 4003–4010. (In Chinese)
38. Su, C. Formation Mechanism of Strontium-Rich Groundwater in Xintian County, Hunan Province. Ph.D. Thesis, China University of Geosciences, Shenzhen, China, 2022. (In Chinese)
39. Chen, C. Distribution Characteristics and Formation Mechanism of Strontium Rich and Metasilicic Acid Rich Groundwater in Muwen River Groundwater System. Master's Thesis, Shandong University of Science and Technology, Qingdao, China, 2020. (In Chinese)

*sustainability*

MDPI

*Review*

# Influencing Factors and Evaluation of Groundwater Ecological Function in Arid/Semiarid Regions of China: A Review

**Haohao Cui [1,2], Mingjiang Yan [1], Qian Wang [1,2,*], Guanghui Zhang [1], Huimin Feng [3] and Xujuan Lang [4,5,6]**

1 Institute of Hydrogeology and Environmental Geology, Chinese Academy of Geological Sciences, Shijiazhuang 050061, China; cuihaohao@mail.cgs.gov.cn (H.C.); yanmingjiang@mail.cgs.gov.cn (M.Y.); zhangguanghui@mail.cgs.gov.cn (G.Z.)

2 Key Laboratory of Groundwater Sciences and Engineering, Ministry of Natural Resources, Shijiazhuang 050061, China

3 College of Urban and Rural Construction, Shanxi Agricultural University, Jinzhong 030801, China; fenghuimin1997@163.com

4 School of Water Resources and Environment, Hebei GEO University, Shijiazhuang 050031, China; langlan77@163.com

5 Hebei Province Key Laboratory of Sustained Utilization and Development of Water Resources, Shijiazhuang 050031, China

6 Hebei Province Collaborative Innovation Center for Sustainable Utilization of Water Resources and Optimization of Industrial Structure, Shijiazhuang 050031, China

* Correspondence: wqianqian@mail.cgs.gov.cn

**Abstract:** In arid and semi-arid areas, due to drought climate and shortage of water resources, groundwater is crucial for natural ecological protection and economic development. It serves a dual role as a resource function and an ecological function. However, with the continuous improvement of the exploitation and utilization of groundwater by human activities during rapid economic development, the phenomenon of groundwater overexploitation is becoming more and more serious, which has destroyed the natural balance of groundwater recharge and discharge. As a result, natural vegetation has lost the maintenance of the ecological function of groundwater, and a series of ecological and environmental problems have occurred, such as natural vegetation degradation, land desertification, sandstorms, and so on. In recent years, scholars have carried out research on groundwater resource management and optimization of water resource allocation, trying to solve the problem of water balance in arid regions. However, there is still a lack of comprehensive understanding and systematization regarding influencing factors and degeneration mechanisms related to groundwater's ecological function. By summarizing and analyzing the previous research results, this paper summarizes the influencing factors, evaluation methods, existing problems and future directions of groundwater ecological function research in China to provide a reference for rational exploitation and utilization of groundwater and ecological protection. This paper is divided into four main contents. The first part introduces the definition of groundwater ecological function (GEF); the second part summarizes the research status of influencing factors of GEF, including the groundwater table depth, vegetation root system and lithologic structure of vadose zone, etc.; the third part analyzes the evaluation of groundwater ecological function; the fourth part discusses the existing problems in the study of groundwater ecological functions, and based on the above research the evaluation framework of GEF is proposed with the Shiyang River basin as a case study; and finally, it highlights the future research directions about GEF.

**Keywords:** groundwater; natural vegetation; northwest China; vadose zone; ecological water level

**Citation:** Cui, H.; Yan, M.; Wang, Q.; Zhang, G.; Feng, H.; Lang, X. Influencing Factors and Evaluation of Groundwater Ecological Function in Arid/Semiarid Regions of China: A Review. *Sustainability* **2024**, *16*, 1631. https://doi.org/10.3390/su16041631

Academic Editors: Zongjun Gao and Jiutan Liu

Received: 30 November 2023
Revised: 4 February 2024
Accepted: 9 February 2024
Published: 16 February 2024



## 1. Introduction

In arid and semi-arid areas of China, due to the arid climate and scarcity of rainfall, the ecological status of the local natural oases is closely linked to the groundwater table buried depth (GD). Consequently, the natural oases in the middle and lower reaches of each river basin exhibit a strong reliance on the GD. When GD exceeds the "limit ecological water level" depth for wetland or natural vegetation, there is an imbalance or failure in groundwater's ecological function (GEF). The prolonged absence of artificial water supply will lead to significant shrinkage or even complete drying up of natural wetland areas, as well as severe degradation or death of natural vegetation. In the past 50 years, with the development of society and economy, the utilization rate of water resources in the inland basins of northwest China has exceeded 65%, far surpassing the average water utilization rate of 30% in global arid areas. Shiyang River Basin is a typical area in Northwest China; due to the reuse of water resources and over-extraction of groundwater, its water resources utilization has reached 172%, which has exceeded the renewable water resources in the basin [1,2]. With the increasing exploitation of groundwater, the groundwater table in Northwest China has dropped sharply, and the problem of groundwater over-extraction is very serious; in some areas, the water resource depletion phenomenon has occurred. This unsustainable use and irrational exploitation have caused deterioration in groundwater's ecological function while triggering significant hydrological changes and environmental disruptions within arid regions, such as drastic reductions in water volume of downstream rivers or even cut-off flow, shrinkage or drying of terminal lakes, soil desertification and salinization, vegetation degradation, biodiversity reduction, frequent sandstorms, etc. It has seriously threatened the local natural ecological security [3].

In order to solve the problem of rational allocation of water resources between economic society and ecological protection and promote the harmonious and sustainable development of humans and water in arid and semi-arid areas, it is necessary to further understand the important role of groundwater in maintaining natural oasis ecology in arid areas. And a thorough study of the characteristics, influencing factors, and formation mechanisms of GEF's degeneration crisis is necessary to provide a scientific basis for rational groundwater exploitation and ecological protection in arid regions. In view of the above contents, scholars have conducted numerous studies. By summarizing and analyzing the previous research results, this paper has outlined the influencing factors, evaluation methods, existing problems and research trends of groundwater ecological function research in China. And it aims to provide a valuable reference for the rational exploitation and utilization of groundwater resources as well as ecological preservation efforts. This paper is divided into four main contents. The first part introduces the definition of GEF; the second part offers a comprehensive overview of the current research status regarding the influencing factors of GEF, including the GD, vegetation root system and lithologic structure of vadose zone, etc.; the third part puts forward the evaluation framework of GEF based on the above research; and finally, it highlights the existing challenges and future research directions about GEF.

## 2. Definition of Groundwater Ecological Function

The groundwater function refers to the impact or effect on human society and the environment caused by the changes in groundwater quality or quantity in time or space. This mainly encompasses the functions of groundwater as a resource supply for maintaining the stability of the ecological and geological environments [4]. The function of groundwater in maintaining an ecological environment, known as groundwater ecological function (GEF), refers to the positive role or effect that the groundwater system plays in maintaining the surface vegetation, lake, wetland or land quality. It is one of the important components of groundwater function. Groundwater can transport water to the surface of the vadose zone by supporting capillary action, thereby supporting natural vegetation and natural wetlands, etc. This process is intimately linked to the lithologic structure of the vadose zone and the depth of the root development within the vegetation community. Any changes to the groundwater system will have corresponding effects on the ecological environment. In arid and semi-arid areas, the ecological environment is closely related to groundwater, and further research on the GEF can provide a scientific foundation for the exploitation and utilization of groundwater resources and ecological protection in the basin [5].

## 3. Influencing Factors of Groundwater Ecological Function

It can be inferred from the definition of GEF that the factors influencing GEF mainly include GD, vegetation root system and lithologic structure of the vadose zone. The key index to measure the intensity of GEF is the threshold of limit ecological water level [6], which denotes the minimum GD at which groundwater must rise through capillary action to sustain surface moisture in the vadose zone, thereby ensuring ecological security of natural vegetation or natural wetland [7]. It is closely linked to the lithologic structure of the vadose zone and the depth of vegetation root development. Once the buried depth of the groundwater table is deeper than the threshold of limited ecological water level for a long time, the natural vegetation ecosystem will lose the water supplied by the capillary water from groundwater [8], which will inevitably lead to the degradation and even extinction of natural vegetation and thus land desertification [9,10].

### *3.1. The Influence of GD on GEF in Arid Region*

The variation of water content in the root layer of natural vegetation is closely related to the dynamics of GD. Suitable GD varies across different ecological environments, consequently giving rise to diverse ecological functions associated with groundwater. Therefore, the ecological water level is composed of a series of GD that meet the requirements of the ecological environment [11,12]. Based on distinct growth characteristics exhibited by vegetation under varying GD conditions, the ecological water level can be further categorized into marsh water level, salinization water level, suitable ecological water level, ecological warning water level, desertification water level and so on (Figure 1) [13–15].

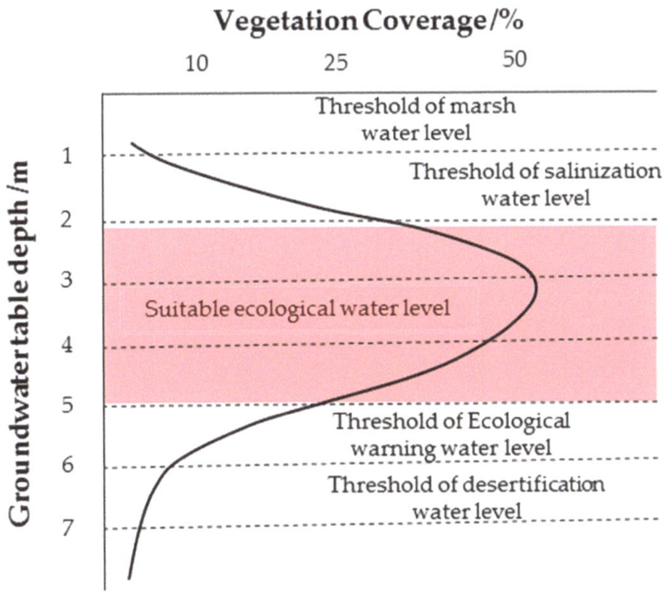


**Figure 1.** Coverage of natural vegetation under different ecological water levels in arid regions.

In the arid area of northwest China, the suitable GD of natural vegetation communities ranges from 2 to 5 m [16]. When the GD is less than 2 m, surface salinization will occur, while when the GD is more than 5 m, the natural vegetation will be degraded to a different degree. However, the suitable GD can vary across different basins. For example, in Yanqi Basin, Wang Shuixian et al. [17] reported that the minimum GD for soil salinization control in an oasis irrigation area is about 2.5 m. In the Minqin Basin, the suitable GD of natural vegetation in the oasis is 2.5–3.9 m [18], and when the GD is less than 4 m, the normalized difference vegetation index (NDVI) of natural vegetation in Minqin Oasis is significantly negatively correlated with the GD. In Hunshendak Sandy Land, Inner Mongolia, Zhang Gaoqiang et al. [19] analyzed the relationship between the enhanced vegetation index (ENI) of vegetation and groundwater and found that the local natural vegetation coverage reached its peak at a GD of 1.5 m when the GD exceeds 8.0 m, the vegetation coverage does not change with the dynamic change of the GD, while when the GD is 1.5–8.0 m, the vegetation coverage decreases exponentially with the increase of the GD, but the response of vegetation growth state to the dynamic change of the GD has a certain lag effect [20]. In the Yerqiang River basin in the middle of the Tianshan Mountains, Xinjiang, based on the different guarantee rates (45–90%) of the ecological water demand of groundwater, the GD ranges from 2.89 to 4.65 m [21]. Zhang Guanghui et al. [22] studied the relationship between GD and vegetation growth in the Shiyang River Basin and showed that the GD had a significant impact on the growth of natural vegetation when it was less than 10.3 m. Depending on the different growth states of vegetation, the GD is divided into different sections (Table 1).

**Table 1.** Relationship between the status of GEF and GD in northwest China.

| GD (m) | Status of GEF | Response Characteristics of Vegetation Key Indicators |
|---|---|---|
| 1.0~3.2 | Good | Both ecological NDVI index and vegetation coverage increased significantly and are positively correlated with GD. |
| 3.2~5.3 | Normal | Ecological NDVI index and vegetation coverage decreased significantly and are negatively correlated with GD. Most natural vegetation still shows a good growth state, but some moist vegetation, such as reed, is in poor ecological condition. |
| 5.3~7.4 | Gradual degeneration | The ecological NDVI index and vegetation coverage show a continuous decrease, and they are negatively correlated with GD. Some vegetation, such as licorice and reed, begin to wither and die. |
| 7.4~10.3 | Gradual degeneration | The ecological NDVI index and vegetation coverage show a slight decrease, and they are still negatively correlated with GD. The natural vegetation system shows a poor growth state on the whole. |
| >10.3 | Catastrophic degeneration | Ecological NDVI index and vegetation coverage are not related to GD. Most of the desert vegetation is dead, the coverage is very low, and the surface has been desertification or desertification. |

### 3.2. *Effects of Vegetation Water Use Sources on GEF in Arid Regions*

The research on water use sources of vegetation in arid regions has become a pivotal research topic both domestically and internationally. The characteristics of the root system, community, and water demand of different dominant vegetation all play a significant role in determining groundwater ecological level, which leads to differences in groundwater ecological functions.

In the natural environment, the water used by vegetation growth mainly includes atmospheric precipitation, surface water (rivers or lakes), soil water and groundwater (Figure 2) [23,24]. The depth and overlap of root distribution determine the dependence of vegetation on groundwater, especially for annual vegetation [25–28]. With the gradual increase in groundwater depth, the vegetation community in the ecosystem will transition from mesophytic series to xerophytic series, deep-rooted and xerophytic vegetation will become the dominant species in the ecosystem, and the community association will gradually become more monolithic [29]. The correlation between vegetation coverage, NDVI and GD is often used to reveal the succession process of vegetation community [30–32].

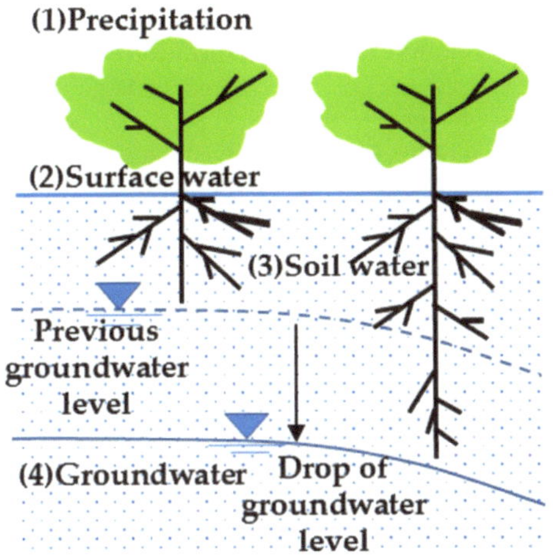


**Figure 2.** Water use sources and strategy of vegetation.

The water consumption characteristic of vegetation is also an important factor in determining the suitable ecological water level, especially under the action of water stress; the absorption and utilization capacities of groundwater by roots are different [33–35], leading to different GEF. In arid regions, natural vegetation will choose the most suitable water source for absorption and utilization according to different external hydrological conditions and meet its own growth and development needs [36,37]. There are several methods for natural vegetation utilization of water sources, including the root excavation method [38], fluid flow technique method [39] and stable isotope method [40–42]. For example, Nie Yunpeng et al. [43] used liquid flow technology to qualitatively identify the source of water absorbed by vegetation through the difference of leaf stomatal conductance and vegetation water potential change. Wang Yanli et al. [44] emphasized that stable isotope method has the advantages of non-destructive and high accuracy. For most terrestrial vegetation, the isotopic composition of water does not change during the transport of water from roots to unsuberized vegetation stems. But, the deuterium–oxygen isotopes of water from different sources exhibit significant differences, supporting the use of stable isotope information to determine the water sources of vegetation.

The water sources utilized by different plants can vary depending on their species and growth stage. Yang Guang et al. [45] compared the water sources of two-year *Haloxylon ammodendron* and *Tamarix* in the lower reaches of Manas Basin. They found that the main water source of *Haloxylon ammodendron* is groundwater, followed by soil water at a depth of 80~150 cm, while the main water source of *Tamarix* is soil water at a depth of 60~150 cm, followed by groundwater. Wan Yanbo et al. [46] studied the water sources of young, mature and overripe *Populus euphratica* and found that at 2 m of GD, the surface water was the main source of young *Populus euphratica*, while groundwater and deep soil water were the main sources of mature and overripe *Populus euphratica*. And at 4.2 m of GD, the main water sources of *Populus euphratica* at different ages were deep soil water and groundwater. Gao Ya et al. [47] studied the water use sources of *Caragana* of different tree ages in the Gonghe Basin, Qinghai Province. Before rainfall, young *Caragana* mainly used shallow soil water, and mature *Caragana* mainly used middle and deep soil water. After rainfall, *Caragana* of all ages preferentially used shallow soil water supplied by precipitation. Wang Yuyang et al. [48] studied the typical desert vegetation oasis in the lower reaches of the Tarim River and found that the shallow soil water utilization was the main feature of *Alhagi* in the early growing season, while in the peak and the end of the growing season, the water used by *Alhagi* was mainly deep soil water and groundwater. Eggemeyer et al. [49] found that *Pondus pine* mainly used deep soil water below 0.9 m in winter and shallow soil water between 0.05 and 0.5 m in spring. Tian Lihui et al. [50] showed that in northwest China, due to the difference in soil texture and soil water content on the windward slope, dune top and leeward slope, the depth of soil water utilization by *Salix cheilophila* was greatly different.

The difference in water utilization by vegetation in the above areas is mainly controlled by the distribution of vegetation roots, soil water content and GD [51–53]. With the dynamic change of GD, vegetation will adjust the water use strategy. For the same vegetation species, when the GD is shallow, the roots are concentrated on the surface. However, with the increase in GD, the root distribution will show a deepening trend (Figure 2), and the root distribution will be more dispersed in the vertical direction so as to obtain survival water from a larger spatial range. Therefore, the water use strategy characteristic of vegetation with growth determines that the ecological function of groundwater is not static but changes with the growth of vegetation.

### 3.3. Influence of Vadose Zone on GEF in Arid Region

The vadose zone is the main water supply for the growth of natural vegetation, and it is also the link of water mutual transformation in the groundwater–soil–vegetation–atmosphere continuum, which is closely related to groundwater [54]. Vegetation roots can absorb soil water from the vadose zone. Therefore, the characteristics of water transport and lithologic structure of the vadose zone not only determine the distribution and redistribution process of soil water but also directly affect the growth of vegetation [55]. Furthermore, under extreme arid climate conditions, soil moisture stored in the vadose zone has the ecological function of maintaining vegetation survival [56]. Therefore, it is of great significance to further understand the influence mechanism of the vadose zone on GEF.

Xu Yuanzhi et al. [57] proposed that the characteristics and ecological effects of water transport in the vadose zone are related to the initial water content and lithologic structure of the vadose zone and are affected by rainfall, ground temperature and vegetation growth. Precipitation intensity, rainfall, duration of precipitation, and ground slope all affect the infiltration rate of precipitation and thus affect the characteristics of water transport in the vadose [58–60]. In arid regions, the significant change of soil temperature in the vertical soil profile can also affect the water distribution and transport in the vadose zone [61]. Studies have shown that the above variations are the result of changes in soil water potential [62]. Additionally, vegetation growth also has a certain impact on the change of water in the vadose zone. For example, groundwater evaporation in vegetated areas is higher than that in bare land, which is related to the vegetation type, vegetation growth stage and vegetation transpiration rate [63]. The lithologic structure of the vadose zone affects the infiltration process of soil water. Finer lithologic particles within the vadose zone result in lower infiltration rates. When the lithologic structure of the vadose zone presents different structural characteristics, such as interlayers of coarse and fine particles, the influence on water migration in the vadose zone is obviously different. Furthermore, the stratified structure in the lithologic structure of the vadose zone may hinder (or accelerate) vertical water migration [64], thereby affecting the ecological function of groundwater to maintain natural vegetation [65,66].

Simultaneously, the support capillary rise height of the vadose zone varies depending on its different lithologic structure, thereby affecting the groundwater ecological level and leading to different ecological responses of the natural vegetation (Figure 3) [67]. Through the layered soil column test, Zhang Ping et al. [68] found that when the thickness of the lower soil layer is less than the maximum height of the capillary water rise height within that layer, the capillary water can migrate through the lower soil layer to the upper soil layer. In such cases, the height of the capillary water rise is determined by the properties of the upper soil layer. Shi Wenjuan et al. [69] conducted a systematic study on the maximum rise height of capillary water in layered soil under different conditions. Their findings revealed that the maximum rise height of capillary water in layered soil depends on factors such as the texture, thickness, stratification and their relationship with each other. Conversely, in the process of water release, the unsynchronized release of water by pore channels in the soil delays gravitational drainage, leading to different water retention performance and water retention capacity of the vadose zone (Figure 3) [70,71]. Zhang Renquan et al. [72] discussed the water release mechanism of homogeneous soil and heterogeneous soil, pointing out that delayed water release is a common occurrence in both types, and the delayed water release effect of layered soil is more significant. The complex lithologic interlayer structure, more layers of fine-grained soil and large drop rate of water level will lead to a significant water release delay effect. Moreover, the varying thicknesses of each layer across layered soil impact water release differently, leading to diverse delayed water release processes in various vadose lithologic structures.

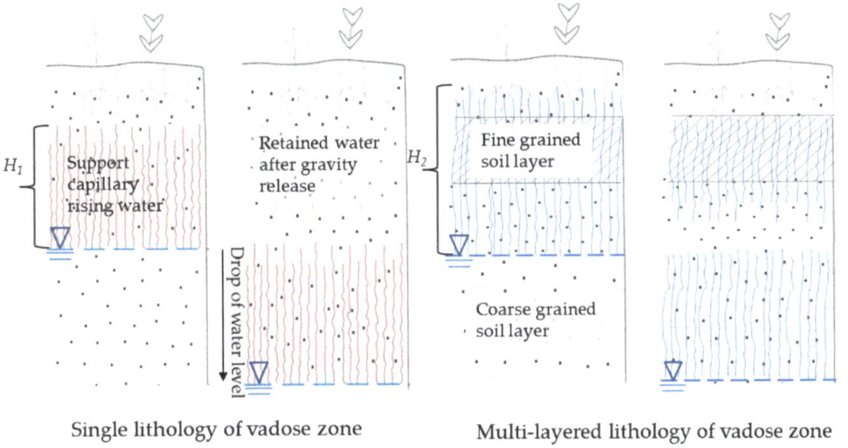


**Figure 3.** Capillary water rise height and retained water after the gravity release of different lithologic structures in the vadose zone. Note: $H_1$ is the height of support capillary water rise in a single lithology vadose zone, and $H_2$ is the height of support capillary water rise in a multi-layered lithology vadose zone. Due to the influence of the fine-grained layer in the multilayer structure, $H_2 > H_1$. The arrow indicates a groundwater table drop event. Both vadose zone structures will retain water in the former capillary zone after the groundwater table drops, but the vadose zone with a multilayer structure retains more water.

Currently, there is a certain understanding of the ecological effects of the vadose zone in arid regions, and various research methods have been widely applied in the study of water transport and ecological effects of the vadose zone in arid regions, such as capillary potential theory, wetting front infiltration theory and preferential flow theory [73], the model test method [74], zero flux surface method [75], tracer experiment method [76], lysimeter experiment method [77] and water balance method [78]. Different methods have their own advantages and applicable conditions. For example, the preferential flow theory has solved the heterogeneity of water transport in vadose zones, while the lysimeter experiment method is well-suited for field-scale application [79]. The combination of various methods [80], such as in-situ monitoring using the zero flux surface method and the numerical simulation techniques, can better describe the water transport process in the vadose zone. Nevertheless, the research on the transformation mechanism of water flux between vegetation-vadose zone-groundwater remains limited. In particular, the study of the influence of multi-layered vadose zone on the GEF is still an important research topic in the future.

## 4. The Evaluation of Groundwater Ecological Function

The method of groundwater function evaluation has been successfully studied and applied in different areas of China. The main idea of this method is to comprehensively consider the regional groundwater resource function, ecological environment function and geological environment function. It involves identifying appropriate indices for evaluating groundwater function based on driving, state, and response factors within the groundwater system. The analytic hierarchy process, comprehensive index method, and GIS platform are then utilized to evaluate the groundwater function. Subsequently, functional zones at different levels are delineated based on the evaluation results. However, scholars differ in the selection of indicators and the division of functional areas. For example, Huang Pengfei et al. [81] conducted a study on groundwater function evaluation in Minqin Oasis, pointing out that some indicators are only applicable to individual zones and suggesting that grade standards should be established for these indicators and "multi-indicator" evaluation should be carried out for individual zones by direct assignment method. Yan Chengyun et al. [82] constructed the groundwater function evaluation system of the middle and lower reaches of the Shule River Basin, divided and evaluated individual functions and comprehensive functions at five levels, and pointed out the application prospect of groundwater. During the functional characteristics and evaluation of groundwater in the North China Plain, Zhang Guanghui et al. [83] emphasized that the selected groundwater functional evaluation indexes should not only objectively express the natural properties of groundwater but also take into account the relevant social properties. If there are too many indicators, it will bring unnecessary basic data processing work. On the contrary, it will obviously affect the accuracy and reliability of the evaluation results. Sun Caizhi et al. [84] evaluated the groundwater function in the Lower Liaohe Plain, establishing a new groundwater function evaluation index system from the perspective of the supply and demand of groundwater function. By combining this with the existing evaluation index system, the research results have important reference significance for the normal maintenance of groundwater function and the scientific management of groundwater resources in the study area. Aiming at the need for rational groundwater development and ecological protection in arid and semi-arid areas, Wang Jinzhe et al. [85] developed a groundwater function evaluation and regionalization method suitable for these areas. This method fully considers the threshold correlation and spatial distribution characteristics between different types of ecological environment degeneration and groundwater table buried depth (Figure 4). However, it does not account for the characteristics of vegetation roots and the heterogeneity of lithologic structure in the vadose zone.

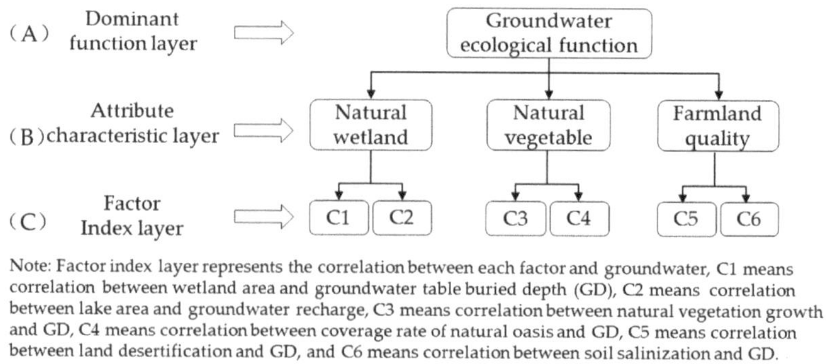


**Figure 4.** Groundwater function evaluation and zoning system in arid region.

In other countries, the research on GEF mainly focuses on the identification and regionalization of groundwater-dependent ecosystems. The main research methods used include field ecological survey [86], remote sensing [87] and geographic information technology [88]. Through the scientific zoning of groundwater ecological functions on the basin scale, the interaction between groundwater and the ecosystem, as well as the rational allocation of regional water resources, are deeply understood. For example, Australia carried out a study on the regionalization of watershed-scale groundwater-dependent ecosystems in 2016, which provided a scientific basis for natural resource management [89]. Jeanette

Howard et al. [90] clarified the interconnections between groundwater and surface ecosystems by identifying and mapping the distribution of groundwater-dependent ecosystems in California. Cruz et al. [91] identified and regionalized the ecological functions of groundwater in the Azores Volcanic Islands and divided the dependence of the ecosystem on groundwater into five levels. Liu et al. [92] carried out the regionalization of groundwater-dependent ecosystems in Central Asia by remote sensing and verified the accuracy of the regionalization results by superposition analysis, correlation analysis and water balance method. The verification results were consistent. Based on field ecological surveys, remote sensing and geographic information technology methods, some scholars have developed a time-comprehensive method for regionalization of groundwater-dependent ecosystems, which has been applied and verified in Chile. The results showed that this method could regionalize and analyze the spatio-temporal changes of groundwater-dependent ecosystems on an annual scale [93].

## 5. Discussion and Case Study

In the arid inland region of Northwest China, the population and its living and production water consumption have far exceeded the local catchment-scale water resources carrying capacity. Addressing the conflict between economic and social water consumption and natural ecological water demand is one of the most critical challenges. Ensuring orderly and reasonable restoration, as well as effective protection of groundwater ecological functions in natural oases, while achieving "harmonious" development between humans and nature, remains a significant challenge. In order to solve the above problems, many scholars at home and abroad have carried out scientific research on efficient agricultural water-saving and water source replacement, optimal allocation of water resources in river basins, artificial planting of xeromorphic vegetation to control desertification and artificial external ecological water transfer, evaluation and regionalization of groundwater ecological functions in arid regions [94–96] and so on. In the northwest inland river basin, the key to the rational allocation of water resources is to determine the proportion of socio-economic water and ecological water under the premise of meeting the minimum ecological water consumption [97–100]. A research finding from the Chinese Academy of Engineering [101] pointed out that groundwater and surface water come from the same source in the northwest inland river basin. Therefore, in the allocation of water resources, it is not only necessary to perform overall planning for the whole basin but also to fully consider the mutual transformation of groundwater and surface water so as to maintain the sustainable use of water resources while ensuring ecological and social economic water use.

The groundwater ecological function evaluation and regionalization technology is the basis of the above research work. The rational allocation of water resources can only be achieved through detailed regionalization and evaluation of the groundwater ecological function. Although scholars have conducted many studies on GEF and proposed theories and research methods for groundwater function evaluation and regionalization, or groundwater-dependent ecosystem evaluation and regionalization, which have been widely applied and verified in different regions, the quantitative evaluation of the GEF in arid and semi-arid areas needs to be further studied.

Taking the Shiyang River basin as a case study, this paper put forward a theoretical formula to express the quantitative expression of groundwater ecological function based on the above literature review and comprehensive consideration of the influencing factors of groundwater ecological function [102–104].

$$A_{gw} = \frac{W_s(h_{\mathrm{gw}}, H_{\mathrm{mx}}, D_{zb})}{W_p(E_{\mathrm{zs}}, \theta_{\mathrm{sm}})} \times 100\% \quad (1)$$

The meanings of each parameter in the formula are as follows.

$A_{gw}$ is the ecological function intensity of groundwater, which represents the percentage value of groundwater water supply capacity and vegetation water demand. Under the premise of a certain amount of vegetation water demand, the stronger the water supply capacity of groundwater, the stronger its ecological function.

$W_s$ is the amount of water supplied by groundwater to natural vegetation, which is the function of the buried depth of the groundwater table ($h_{gw}$), supporting capillary water rise height ($H_{mx}$) and root development depth ($D_{zb}$) of natural vegetation community in the area.

(1) Under the same buried depth of groundwater table: The deeper the root development, or the higher the support capillary water rise height, the stronger the groundwater supply capacity, showing the stronger groundwater ecological function;
(2) Due to the hydrotaxis of vegetation root development in arid regions, the slower the groundwater table decline, or the more significant the annual fluctuation of the groundwater table, the more conducive to the vertical development of vegetation root system, thus absorbing more water from groundwater, showing a stronger groundwater ecological function. Furthermore, when the buried depth of the groundwater table exceeds the limited ecological water level threshold, the delayed water release effect of the vadose zone will store part of the water and continue to supply water to vegetation, extending the role of groundwater ecological function. The more complex the lithologic structure of the vadose zone, the more obvious this effect is.

$W_p$ is the maximum value of water required by a natural vegetation community for normal growth, which is the function of vegetation transpiration intensity ($E_{zs}$) and soil moisture content ($\theta_{sm}$) in the root layer, and is related to vegetation type, depth of root layer development and meteorological (light, temperature, etc.) conditions.

According to the proposed formula, the quantitative changes in groundwater ecological function intensity from 2018 to 2021 were calculated for the Shiyang River Basin (Figure 5). The results indicated that groundwater ecological function plays a significant role from April to September each year, while it reaches its minimal state from October to March of the following year, corresponding to the end of the vegetation growth period. It is evident from the figure that there is a strong correlation between the strength of GEF and the GD, and it exhibits an inter-annual periodic change. However, the existing research often provides a static assessment of the current status of groundwater ecological function, which does not reflect its dynamic change characteristics, and cannot formulate and revise groundwater management strategies in real-time. Therefore, the combination of groundwater ecological function zoning and dynamic real-time evaluation is one of the important contents to be further studied so as to guide the fine optimal allocation of basin water resources.

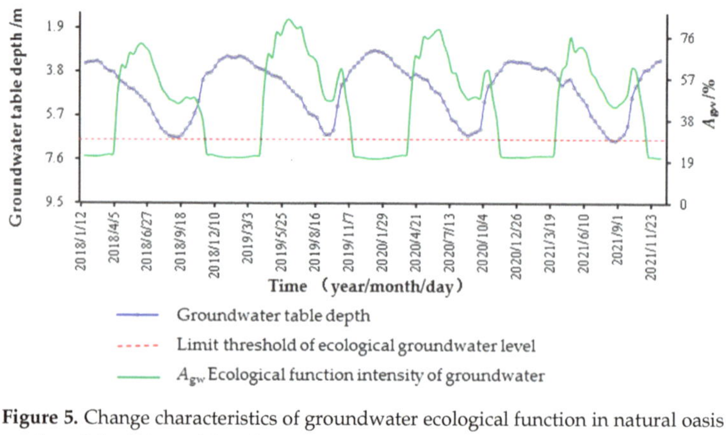


**Figure 5.** Change characteristics of groundwater ecological function in natural oasis area in the lower reaches of the Shiyang River basin.

## 6. Conclusions and Future Directions

This paper presents the definition of groundwater ecological function (GEF), provides an overview of the research status of influencing factors of GEF, analyzes the evaluation of groundwater ecological function, discusses the existing challenges in the study of GEF, and proposes an evaluation framework of GEF with Shiyang River basin as a case study. However, the current research on groundwater ecological function primarily focuses on assessing the present situation and lacks real-time dynamic monitoring, prediction and early warning capabilities for future trends. Therefore, the research on groundwater ecological function should be strengthened in the following aspects in the future:

(1) Quantitative evaluation of groundwater ecological function.

All the factors that affect the ecological function of groundwater should be considered comprehensively, including the changes in groundwater depth, the characteristics of vegetation root development, the lithologic structure of the vadose zone and meteorological conditions. However, there is limited understanding regarding how vadose zone lithologic structure affects groundwater ecological function, particularly concerning delayed release water's ecological effects under drought stress, which requires further exploration;

(2) Cyclical variation of groundwater ecological function.

Revealing the annual and interannual cyclical variation of groundwater ecological function can enhance the understanding of the degeneration mechanism of groundwater ecological function and provide accurate guidance for sustainable groundwater exploitation and oasis ecological protection;

(3) Strengthen the dynamic monitoring of groundwater ecological function.

With the advent of cutting-edge technologies such as cloud services, the Internet of Things, big data and artificial intelligence, there is a growing need to establish a sophisticated technology platform (or system) for real-time dynamic monitoring, early warning and management that integrates air, sky, earth and water [105]. On the basis of a deep understanding of evolving patterns of groundwater ecological functions, real-time dynamic monitoring of groundwater ecological functions can be carried out to guide the allocation of basin water resources by zoning and grading. This approach allows for continuous optimization and adjustment, ultimately preventing and mitigating groundwater overexploitation and natural ecological degradation.

**Author Contributions:** Conceptualization, G.Z. and H.C.; methodology, H.C.; software, M.Y.; validation, X.L., H.C. and Q.W.; formal analysis, Q.W. and H.F.; investigation, X.L.; resources, X.L.; data curation, H.C.; writing—original draft preparation, H.C.; writing—review and editing, X.L.; visualization, Q.W.; supervision, Q.W.; project administration, G.Z.; funding acquisition, M.Y., Q.W. and H.F. All authors have read and agreed to the published version of the manuscript.

**Funding:** This research was funded by Fundamental Scientific Research Funds from the Chinese Academy of Geological Sciences of China (SK202313), the Science and Technology Innovation Program of Shanxi Agricultural University, China (2015ZZ04), and the Geological Survey Project of China Geological Survey, grant number DD20230557.

**Institutional Review Board Statement:** Not applicable.

**Informed Consent Statement:** Not applicable.

**Data Availability Statement:** Data used for analysis were extracted from the list referenced. Metadata is available upon request from the corresponding author.

**Acknowledgments:** The authors thank the editor and anonymous reviewers for providing helpful advice.

**Conflicts of Interest:** The authors declare no conflicts of interest.

## References


1. Wang, G.X.; Cheng, G.D.; Xu, Z.M. The Utilization of Water Resource and Its Influence on Eco-Environment in the Northwest Arid Area of China. *J. Nat. Resour.* **1999**, *14*, 110–116.
2. Wang, L.; Chen, F. Change in water-use efficiency of irrigated areas before and after integrated management in Shiyang River Basin. *Acta Ecol. Sin.* **2018**, *38*, 3692–3704.
3. Gou, T.Y.; Tong, L.; Kang, D.K.; He, Y.J.; Wang, W.Z.; Kang, S.Z.; Fu, J. Evaluating the comprehensive effects of the Key Master Plan of the Shiyang River Basin in arid areas of northwest China. *Trans. Chin. Soc. Agric. Eng.* **2022**, *38*, 74–84.
4. Zhang, G.H.; Shen, J.M.; Nie, Z.L.; Wang, J.Z.; Yan, M.J.; Hao, M.L. Theory and methodology of regional groundwater function and sustainable utilization assessment in China. *Hydrogeol. Eng. Geol.* **2006**, *33*, 62–66+71.
5. Xu, W.J.; Kong, F.H.; Mao, R.C.; Song, J.X.; Sun, H.T.; Wu, Q.; Liang, D.; Bai, H.F. Identifying and mapping potential groundwater-dependent ecosystems for a semi-arid and semi-humid area in the Weihe River, China. *J. Hydrol.* **2022**, *609*, 127789. [CrossRef]
6. Hu, S.; Ma, R.; Sun, Z.Y.; Ge, M.Y.; Zeng, L.L.; Huang, F.; Bu, J.W.; Wang, Z. Determination of the optimal ecological water conveyance volume for vegetation restoration in an arid inland river basin, northwestern China. *Sci. Total Environ.* **2021**, *788*, 147775. [CrossRef]
7. Wang, J.Z.; Zhang, G.H.; Wang, Q.; Cui, H.H.; Liu, P.F. Construction and application of evaluation index system of groundwater ecological function in northwest arid area. *Acta Geol. Sin. (Chin. Ed.)* **2021**, *95*, 1573–1581. [CrossRef]
8. Qi, Z.W.; Xiao, C.L.; Wang, G.; Liang, X.J. Study on ecological threshold of groundwater in typical salinization area of Qian′an county. *Water* **2021**, *13*, 856. [CrossRef]
9. Wang, W.K.; Gong, C.C.; Zhang, Z.Y.; Chen, L. Research Status and Prospect of the Subsurface Hydrology and Ecological Effect in Arid Regions. *Adv. Earth Sci.* **2018**, *33*, 702–718.
10. Huang, F.; Ochoa, C.G.; Chen, X.; Zhang, D.R. Modeling oasis dynamics driven by ecological water diversion and implications for oasis restoration in arid endorheic basins. *J. Hydrol.* **2021**, *593*, 125774. [CrossRef]
11. Jia, L.M.; Guo, Z.X.; Long, Y.H.; Guo, K.; Liao, Z.L. Research advances in ecological groundwater level in arid areas. *Ecol. Sci.* **2015**, *34*, 187–193. [CrossRef]
12. Zhang, R.; Wu, J.W.; Yang, Y.; Peng, X.; Li, C.S.; Zhao, Q. A method to determine optimum ecological groundwater table depth in semi-arid areas. *Ecol. Indic.* **2022**, *139*, 108915. [CrossRef]
13. Huang, F.; Zhang, Y.D.; Zhang, D.R.; Chen, X. Environmental groundwater depth for groundwater-dependent terrestrial ecosystems in arid/semiarid regions: A review. *Int. J. Environ. Res. Public Health* **2019**, *16*, 763. [CrossRef] [PubMed]
14. Xin, P.; Kong, J.; Li, L.; Barry, D.A. Modeling of groundwater and vegetation interactions in a tidal marsh. *Adv. Water Resour.* **2013**, *57*, 52–68. [CrossRef]
15. Fan, Z.L.; Ma, Y.J.; Zhang, H.; Wang, R.H.; Zhao, Y.J.; Zhou, H.F. Research of eco-water table and rational depth of groundwater of Tarim River drainage basin. *Arid Land Geogr.* **2004**, *27*, 8–13. [CrossRef]
16. Zhang, Y.Y.; Chen, X.; Gao, M.; Liu, X.Q. Meta-analysis of ecological depth to groundwater table and its influencing factors in aird region of northwest China. *South-to-North Water Divers. Water Sci. Technol.* **2020**, *18*, 57–65. [CrossRef]
17. Wang, S.; Wu, B.; Yang, P.; Li, X.; Dong, X. Determination of the Ecological Groundwater Depth Considering Ecological Integrity over Oasis Irrigation Areas in the Yanqi Basin. *Resour. Sci.* **2011**, *33*, 422–430.
18. Cao, L.; Nie, Z.L.; Liu, M.; Lu, H.X.; Wang, L.F. Changes in natural vegetation growth and groundwater depth and their relationship in the Minqin oasis in the Shiyang River Basin. *Hydrogeol. Eng. Geol.* **2020**, *47*, 25–33. [CrossRef]
19. Zhang, G.Q.; Nie, H.F.; Xiao, C.L.; Xue, H.; Zhao, X.Q.; Li, T.Q.; Zhu, Z.Z. Effect of groundwater depth on vegetation coverage in southeastern margin of Otindag Sandy Land. *J. Arid Land Resour. Environ.* **2022**, *36*, 147–153. [CrossRef]
20. Kai, W.M.; Liang, X.J.; Xiao, C.L.; Qi, Z.W.; Jia, L. Determination and analysis of groundwater ecological level in Changling county. *China Rural Water Hydropower* **2022**, *1*, 32–38.
21. Aziz, M. Determination and control measures of groundwater ecological level in Yarkand River Basin. *Groundwater* **2023**, *45*, 83–84+105. [CrossRef]
22. Zhang, G.H.; Nie, Z.L.; Cui, H.H.; Wang, Q.; Ma, R.; Xu, Z.X.; Chen, X.; Shao, J.L.; Zou, S.Z.; Fei, Y.H.; et al. *Theory and Practice of Groundwater Rational Development and Ecological Function Protection in Northwest Inland Arid Area*; Science Press: Beijing, China, 2022; pp. 191–193.
23. Miguez-Macho, G.; Fan, Y. Spatiotemporal origin of soil water taken up by vegetation. *Nature* **2021**, *598*, 624–628. [CrossRef] [PubMed]
24. Hao, S.; Li, F.D. Water sources of the typical desert vegetation in Ebinur Lake basin. *Acta Geogr. Sin.* **2021**, *76*, 1649–1661. [CrossRef]
25. Midgley, G.F. Ecology. Biodiversity and ecosystem function. *Science* **2012**, *335*, 174–175. [CrossRef] [PubMed]
26. Oerter, E.; Slessarev, E.; Visser, A.; Min, K.; Kan, M.; McFarlane, K.J.; Saha, M.C.; Berhe, A.A.; Pett-Ridge, J.; Nuccio, E. Hydraulic redistribution by deeply rooted grasses and its ecohydrologic implications in the southern Great Plains of North America. *Hydrol. Process.* **2021**, *35*, e14366. [CrossRef]
27. Yu, K.L.; Odorico, P.D. Climate, vegetation and soil controls on hydraulic redistribution in shallow tree roots. *Adv. Water Resour.* **2014**, *66*, 70–80. [CrossRef]
28. Marchionni, V.; Daly, E.; Manoli, G.; Tapper, N.J.; Walker, J.P.; Fatichi, S. Groundwater buffers drought effects and climate variability in urban reserves. *Water Resour. Res.* **2020**, *56*, e2019WR026192. [CrossRef]

29. Wan, L.; Cao, W.B.; Hu, F.S.; Jin, X.M.; Chen, J.S.; Gong, B. *Ecohydrogeology*; Geology Press: Beijing, China, 2005; pp. 122–131.
30. Zhang, H.; Wang, X.S. The impact of groundwater depth on the spatial variance of vegetation index in the Ordos Plateau, China: A semivariogram analysis. *J. Hydrol.* **2020**, *588*, 125096. [CrossRef]
31. Phiri, M.; Shiferaw, Y.A.; Tesfamichael, S.G. Modelling the relationship between groundwater depth and NDVI using time series regression with Distributed Lag M. *S. Afr. J. Geomat.* **2018**, *7*, 147–163. [CrossRef]
32. Song, G.; Huang, J.T.; Ning, B.H.; Wang, J.W.; Zeng, L. Effects of groundwater level on vegetation in the arid area of western China. *China Geol.* **2021**, *4*, 527–535. [CrossRef]
33. Lee, E.; Kumar, P.; Knowles, J.F.; Minor, R.L.; Tran, N.; Barron-Gafford, G.A.; Scott, R.L. Convergent hydraulic redistribution and groundwater access supported facilitative dependency between trees and grasses in a semi-arid environment. *Water Resour. Res.* **2021**, *57*, e2020WR028103. [CrossRef]
34. Chen, X.L.; Chen, Y.P.; Li, W.H.; Wang, Y.Y. Spatial distribution characteristics of fine roots of Populus euphratica Oliv. under different groundwater depths in arid regions. *Plant Sci. J.* **2018**, *36*, 45–53.
35. Liu, S.S.; Xu, G.Q.; Li, Y.; Wu, X.; Liu, J.; Mi, X.J. Difference and consistency of responses of five sandy shrubs to changes in groundwater level in the Hailiutu River Basin. *Acta Ecol. Sin.* **2021**, *41*, 615–625.
36. Zhu, Y.J.; Cui, Q.G.; Du, J.; Xu, S.H.; Liu, Z.L. Water use of three shrub communities in Mu Us Sandy Land. *Acta Ecol. Sin.* **2020**, *40*, 4470–4478.
37. Barbeta, A.; Mejía-Chang, C.; Ogaya, R.; Voltas, J.; Dawson, T.E.; Penuelas, J. The combined effects of a long-term experimental drought and an extreme drought on the use of plant-water sources in a Mediterranean forest. *Glob. Chang. Biol.* **2015**, *21*, 1213–1225. [CrossRef] [PubMed]
38. Xu, S.Q.; Ji, X.B.; Jin, B.W.; Zhang, J.L. Root distribution of three dominant desert shrubs and their water uptake dynamics. *J. Plant Ecol.* **2017**, *10*, 780–790. [CrossRef]
39. Niu, Y.M.; Jia, G.D.; Liu, Z.H.; Wang, X.; Liu, Z.Q. Soil moisture absorption and utilization of Quercus variabilis in Beijing mountain area. *J. Beijing For. Univ.* **2022**, *44*, 16–24.
40. de la Casa, J.; Barbeta, A.; Rodríguez-Uña, A.; Wingate, L.; Ogée, J.; Gimeno, T.E. Isotopic offsets between bulk plant water and its sources are larger in cool and wet environments. *Hydrol. Earth Syst. Sci.* **2022**, *26*, 4125–4146. [CrossRef]
41. Dai, J.J.; Zhang, X.P.; Luo, Z.D.; Wang, R.; Liu, Z.L.; He, X.G.; Rao, Z.G.; Guan, H.D. Variation of the stable isotopes of water in the soil-plant-atmosphere continuum of a Cinnamomum camphora woodland in the East Asian monsoon region. *J. Hydrol.* **2020**, *589*, 125199. [CrossRef]
42. Han, L.; Han, Y.G.; Huang, X.Y.; Gao, Y.; Sun, Z.J. Hydrogen and oxygen stable isotope tracer plant water source and water use strategy in arid region. *J. Anhui Agric. Univ.* **2020**, *47*, 435–441. [CrossRef]
43. Nie, Y.P.; Chen, H.S.; Wang, K.L. Methods for determining plant water source in thin soil region: A review. *Chin. J. Appl. Ecol.* **2010**, *21*, 2427–2433. [CrossRef]
44. Wang, Y.L.; Liu, L.C.; Gao, Y.H.; Li, G.; Zhao, J.C.; Xie, M. Analysis of water sources of plants in artificial sand-fixation vegetation area based on large rainfall events. *Chin. J. Appl. Ecol.* **2016**, *27*, 1053–1060. [CrossRef]
45. Yang, G.; Li, X.L.; Chen, S.; He, X.L.; Yang, M.J.; Long, A.H.; Xue, L.Q. A Study of Water Use Sources of Typical Desert Vegetations in Manasi River Basin Based on $^{18}$O Isotope. *China Rural Water Hydropower* **2017**, *11*, 94–97+103.
46. Wan, Y.F.; Shi, Q.D.; Dai, Y.; Li, T.; Peng, L.; Li, H. Water sources of Populus euphratica with different tree ages in the oasis of desert hinterland. *Chin. J. Appl. Ecol.* **2022**, *33*, 353–359. [CrossRef]
47. Gao, Y.; He, L.X.Z.; Jia, Z.Q.; Li, Q.X.; Dai, J. Effects of precipitation on water use characteristics of Caragana intermedia plantations with different stand ages in alpine sandy land. *Chin. J. Appl. Ecol.* **2021**, *32*, 1935–1942. [CrossRef]
48. Wang, Y.Y.; Chen, Y.P.; Li, W.H.; Wang, R.Z.; Zhou, Y.Y.; Zhang, J.P. Water sources of typical desert riparian plants in the lower reaches of Tarim River. *J. Desert Res.* **2017**, *37*, 1150–1157.
49. Eggemeyer, K.D.; Awada, T.; Harvey, F.E.; Wedin, D.A.; Zhou, X.H.; Zanner, C.W. Seasonal changes in depth of water uptake for encroaching trees Juniperus virginiana and Pinus ponderosa and two dominant C4 grasses in a semiarid grassland. *Tree Physiol.* **2009**, *29*, 157–169. [CrossRef]
50. Tian, L.H.; Wang, H.J.; Zhang, D.S.; Wang, Q.Y.; Liu, R.N. Water use patterns for a typical afforested shrub among topographic positions in an alpine desert of Qinghai-Tibet Plateau using stable isotopes tracers. *Acta Ecol. Sin.* **2021**, *41*, 6215–6226.
51. Pan, Y.X.; Wang, X.P.; Ma, X.Z.; Zhang, Y.F.; Hu, R. The stable isotopic composition variation characteristics of desert plants and water sources in an artificial revegetation ecosystem in Northwest China. *Catena* **2020**, *189*, 104499. [CrossRef]
52. Tiemuerbieke, B.; Min, X.J.; Zang, Y.X.; Xing, P.; Ma, J.Y.; Sun, W. Water use patterns of co-occurring C3 and C4 shrubs in the Gurbantonggut desert in northwestern China. *Sci. Total Environ.* **2018**, *634*, 341–354. [CrossRef]
53. Zhou, H.; Zhao, W.Z.; He, Z.B.; Yan, J.L.; Zhang, G.F. Variation in depth of water uptake for Pinus sylvestris var. mongolica along a precipitation gradient in sandy regions. *J. Hydrol.* **2019**, *577*, 123921. [CrossRef]
54. Zhou, H. Review of studies on the relationship between soil water movement and energy and their driving forces in the vadose zone of arid regions. *Acta Ecol. Sin.* **2019**, *39*, 6586–6597. [CrossRef]
55. Kroes, J.; Supit, I.; van Dam, J.; van Walsum, P.; Mulder, M. Impact of capillary rise and recirculation on simulated crop yields. *Hydrol. Earth Syst. Sci.* **2018**, *22*, 2937–2952. [CrossRef]
56. Zhao, C.L.; Jia, X.X.; Shao, M.A.; Zhu, Y.J. Regional variations in plant-available soil water storage and related driving factors in the middle reaches of the Yellow River Basin, China. *Agric. Water Manag.* **2021**, *257*, 107131. [CrossRef]

57. Xu, Y.Z.; Zhao, G.Z.; Mu, N.S.; Wang, L.L.; Peng, H.X. Review on factors affecting the process of water movement in vadose zone. *J. North China Univ. Water Resour. Electr. Power Nat. Sci. Ed.* **2019**, *40*, 37–41. [CrossRef]
58. Wang, C.S.; Gao, F.; Sun, W.Y.; Mu, X.M.; Gao, P.; Zhao, G.J.; Song, X.Y. Rainfall-infiltration process and characteristics of slope-gully system in the hilly and gully region of the Loess Plateau. *Acta Ecol. Sin.* **2021**, *41*, 3111–3122.
59. Chen, J.; Song, N.P.; Chen, L.; Wang, X.; Wang, Q.X. Soil moisture dynamics and its response to precipitation in different cover types of desert steppe. *J. Soil Water Conserv.* **2021**, *35*, 198–206. [CrossRef]
60. Shi, M.M.; Wang, X.M.; Chen, Q.; Han, B.H.; Zhou, B.R.; Xiao, J.S.; Xiao, H.B. Responses of soil moisture to precipitation and infiltration in dry and wet alpine grassland ecosystems. *Acta Pratacult. Sin.* **2021**, *30*, 49–58.
61. Ma, Z.T.; Wang, W.K.; Zhang, Z.Y.; Zhao, M.; Chen, L.; Song, H.; Yu, P.Y. Simulation study on diurnal distribution variation of soil water in shallow vadose zone. *J. Water Resour. Water Eng.* **2019**, *30*, 245–251+260.
62. Ren, T.; Jia, Z.F.; Wang, Z.; Zhong, Z. Dynamics of soil water potential under plastic membrane and geotextile cloth mulching in Mu Us Desert Land. *J. Soil Water Conserv.* **2020**, *34*, 78–84. [CrossRef]
63. Zhang, W.J.; Wang, N.A.; Yu, X.R.; Niu, Z.M.; Zhao, L.Q. Magnitude of groundwater evapotranspiration in the Badain Jaran Desert based on groundwater dynamics method and empirical model: A case study of the Sumujilin Lake Area. *Arid Zone Res.* **2020**, *37*, 1215–1222. [CrossRef]
64. Yu, P.Y.; Wang, W.K.; Wang, Z.F.; Gong, C.C.; Zhang, Z.Y.; Chen, L. Influence of lithologic structure of vadose zone on rainfall infiltration capacity. *Water Resour. Hydropower Eng.* **2019**, *50*, 25–33. [CrossRef]
65. Pandey, P.K.; Pandey, V. Estimation of infiltration rate from readily available soil properties (RASPs) in fallow cultivated land. *Sustain. Water Resour. Manag.* **2019**, *5*, 921–934. [CrossRef]
66. Zornberg, J.G.; Bouazza, A.; Mccartney, J.S. Geosynthetic capillary barriers: Current state of knowledge. *Geosynth. Int.* **2015**, *17*, 273–300. [CrossRef]
67. Zhou, H.; Zhao, W.Z. Soil physical characteristics of shallow vadose zone and modeling its effects on upward capillary rise of groundwater in an arid-desert area. *Chin. J. Appl. Ecol.* **2019**, *30*, 2999–3009. [CrossRef]
68. Zhang, P.; Wu, H.; Yin, H.J.; Li, B.G.; Wang, H.K. Research on influence of construction of soil layer on height of capillary water upward movement and evaporation of groundwater. *Water Sav. Irrig.* **2011**, *3*, 6–8.
69. Shi, W.J.; Shen, B.; Wang, Z.R.; Wang, W.Y. Maximum height of upward capillary water movement in layered soil. *Agric. Res. Arid Area* **2007**, *1*, 94–97.
70. Cui, H.H.; Zhang, B.; Feng, X.; Gao, Y.X. Soil water-holding properties of different soil body configuration. *Agric. Res. Arid Areas* **2016**, *34*, 1–5. [CrossRef]
71. Ge, J.; Huang, D.W.; Gao, X.; Tang, J.; Shen, H. Water Retention Capacity of Drained Soil Columns with Grained Layers. Southwest China. *J. Agric. Sci.* **2019**, *32*, 2126–2132. [CrossRef]
72. Zhang, R.Q.; Gao, Y.F.; Wang, P.Y. A Preliminary Study on The Mechanism of Water Release From Saturated Layered Soils. *Earth Sci.* **1985**, *1*, 21–27.
73. Wang, Q.J.; Shao, M.A.; Zheng, J.Y. *Water Movement and Solute Migration in Soil*; China Water Power Press: Beijing, China, 2007.
74. Liu, H.H.; Wang, G.L.; Yin, H.S.; He, R.P. Characteristics of soil water infiltration under different vegetation cover in jujube economic forest. *J. Irrig. Drain.* **2020**, *39*, 52–58. [CrossRef]
75. Yang, L.H.; Xu, Y.C.; Ma, Y.X.; Song, X.F.; Zhao, Z.P.; Gong, L.; Liu, H.Y.; Song, Q.Y. Water movement in the unsaturated zone and groundwater recharge in the Yellow River irrigation area. *Res. Soil Water Conserv.* **2021**, *28*, 168–174. [CrossRef]
76. Lin, D.; Jin, M.G.; Liang, X.; Zhang, H.B. Estimating groundwater recharge beneath irrigated farmland using environmental tracers' fluoride, chloride and sulfate. *Hydrogeol. J.* **2013**, *21*, 1469–1480. [CrossRef]
77. Xu, C.C.; Xu, G.Q.; Chen, Y.P.; Qiu, S.; Li, Y. Application of a new ground medium vaporizing instrument in laboratory simulated evaporation precipitation infiltration test. *Groundwater* **2018**, *40*, 127–129.
78. You, C.Y.; Wang, J.Z.; Fang, G.Q.; Zhang, J.J. Application of water balance method in groundwater resources calculation for agriculture area in Arong County. *Water Resour. Hydropower Northeast China* **2019**, *37*, 19–21+71. [CrossRef]
79. Lai, H.A.; Wang, Y.Y. Review of research methods on soil water transport in vadose zone. *Agric. Technol.* **2022**, *42*, 37–40. [CrossRef]
80. Gulati, D.; Satpute, S.; Kaur, S.; Aggarwal, R. Estimation of potential recharge through direct seeded and transplanted rice fields in semi-arid regions of Punjab using HYDRUS-1D. *Paddy Water Environ.* **2022**, *20*, 79–92. [CrossRef]
81. Huang, P.F.; Ma, D.H.; Wang, Z.J.; Gong, H.B. The application of administrative analysis method in groundwater function evaluation of Minqin Oasis. *China Environ. Manag.* **2006**, *2*, 2–5.
82. Yan, C.Y.; Nie, Z.L.; Zhang, G.H.; Shen, J.M.; Cheng, X.X.; Yu, S.B. Assessment of groundwater function in the middle and lower reaches of the Shulehe River Basin. *Hydrogeol. Eng. Geol.* **2007**, *4*, 62–66+71.
83. Zhang, G.H.; Yang, L.Z.; Nie, Z.L.; Shen, J.M.; Wang, J.Z.; Yan, M.J. Assessment of Groundwater Function in North China Plain. *Resour. Sci.* **2009**, *31*, 368–374.
84. Sun, C.Z.; Yang, L.; Hu, D.L. Groundwater ecological sensitivity assessment in the lower Liaohe River Plain based on GIS technique. *Acta Ecol. Sin.* **2011**, *31*, 7428–7440.
85. Wang, J.Z.; Zhang, G.H.; Yan, M.J.; Tian, Y.L.; Wang, Q. Index weight analysis of groundwater function evaluation and zoning system in arid areas. *Trans. Chin. Soc. Agric. Eng.* **2020**, *36*, 133–143. [CrossRef]

86. Pérez Hoyos, I.C.; Krakauer, N.Y.; Khanbilvardi, R.; Armstrong, R.A. A review of advances in the identification and characterization of groundwater dependent ecosystems using geospatial technologies. *Geosciences* **2016**, *6*, 17. [CrossRef]
87. Martínez-Santos, P.; Díaz-Alcaide, S.; de la Hera-Portillo, A.; Gomez-Escalonilla, V. Mapping groundwater-dependent ecosystems by means of multi-layer supervised classification. *J. Hydrol.* **2021**, *603*, 126873. [CrossRef]
88. Páscoa, P.; Gouveia, C.M.; Kurz-Besson, C. A simple method to identify potential groundwater-dependent vegetation using NDVI MODIS. *Forests* **2020**, *11*, 147. [CrossRef]
89. Glanville, K.; Ryan, T.; Tomlinson, M.; Muriuki, G.; Ronan, M.; Pollett, A. A Method for Catchment Scale Mapping of Groundwater-Dependent Ecosystems to Support Natural Resource Management (Queensland, Australia). *Environ. Manag.* **2016**, *57*, 432–449. [CrossRef] [PubMed]
90. Howard, J.; Merrifield, M. Mapping Groundwater Dependent Ecosystems in California. *PLoS ONE* **2020**, *5*, e11249. [CrossRef] [PubMed]
91. Cruz, J.V.; Coutinho, R.; Andrade, C.; Medeiros, D.; Cymbron, R. Identification and Mapping of Groundwater Dependent Ecosystems in the AZORES Volcanic Archipelago (Portugal). *Water* **2022**, *14*, 1126. [CrossRef]
92. Liu, C.; Liu, H.; Yu, Y.; Zhao, W.Z.; Guo, L.; Yetemen, O. Mapping groundwater-dependent ecosystems in arid Central Asia: Implications for controlling regional land degradation. *Sci. Total Environ.* **2021**, *797*, 14902. [CrossRef] [PubMed]
93. Duran-Llacer, I.; Luis Arumí, J.; Arriagada, L.; Aguayo, M.; Rojas, O.; Gonzalez-Rodriguez, L.; Rodriguez-Lopez, L.; Martinez-Retureta, R.; Oyarzun, R.; Singh, S.K. A new method to map groundwater-dependent ecosystem zones in semi-arid environments: A case study in Chile. *Sci. Total Environ.* **2022**, *816*, 151528. [CrossRef]
94. Huang, L.; Cao, W.; Wu, D.; Gong, G.L.; Zhao, G.S. Assessment on the changing conditions of ecosystems in key ecological function zones in China. *Chin. J. Appl. Ecol.* **2015**, *26*, 2758–2766. [CrossRef]
95. Deng, M.J.; Fan, Z.L.; Xu, H.L.; Zhou, H.Y. Ecological function regionalization of Tarim River Basin. *Arid Land Geogr.* **2017**, *40*, 705–717. [CrossRef]
96. Wang, L.X.; Zhang, M.S.; Sui, L.C.; Zhang, S.C.; Yang, Y. Ecological Function Regionalization in the Weihe River Basin. *Arid Zone Res.* **2020**, *37*, 236–243. [CrossRef]
97. Wang, H.; You, J.J. Progress of water resources allocation during the past 30 years in China. *J. Hydraul. Eng.* **2016**, *47*, 265–271+282. [CrossRef]
98. Yang, X.X.; Guo, P.; Li, M. A fuzzy multi-objective optimal allocation model of water resource oriented ecology in the middle reaches of Heihe River. *Water Sav. Irrig.* **2016**, *5*, 65–70.
99. Song, S.H.; Nie, Z.L.; Geng, X.X.; Shen, X.; Wang, Z.; Zhu, P.C. Response of runoff to climate change in the area of runoff yield in upstream Shiyang River Basin, Northwest China: A case study of the Xiying River. *J. Groundwater Sci. Eng.* **2023**, *11*, 89–96. [CrossRef]
100. Guo, R.Q. Scientific allocation of water resources and promotion of ecological management of dongjuyanhai: Study on ecological protection and high quality development scheme in the lower reaches of Heihe River. *J. Irrig. Drain.* **2020**, *39* (Suppl. 2), 136–139. [CrossRef]
101. Water Resources in Northwest Region Task Group of Chinese Academy of Engineering (Chinese Academy of Engineering, Beijing\100038, China). Strategic study on allocation of water resources, conservation and upgrading of eco-environment and sustainable development in North-west China. *Strateg. Study CAE* **2003**, *4*, 1–26.
102. Liu, M.; Nie, Z.L.; Cao, L.; Wang, L.F.; Lu, H.X.; Wang, Z.; Zhu, P.C. Comprehensive evaluation on the ecological function of groundwater in the Shiyang River watershed. *J. Groundwater Sci. Eng.* **2021**, *9*, 326–340. [CrossRef]
103. Cui, H.H.; Zhang, G.H.; Wang, Q.; Yan, M.J.; Cao, L.; Liu, P.F. Periodicity and mechanism of groundwater ecological function in natural oasis in the lower reaches of Shiyang River Basin. *J. Hydraul. Eng.* **2023**, *54*, 199–207+219. [CrossRef]
104. Wang, J.Z.; Zhang, G.H.; Cui, H.H.; Wang, Q.; Dong, H.B.; Hao, J. System index attribute and application of groundwater function zoning in northwest inland area of China. *J. Hydraul. Eng.* **2020**, *51*, 796–804.
105. Wang, L.L.; Zhu, Y.M.; Liu, H.Y.; Wang, Q.; Wang, K. Design and implementation of an integrated platform for smart watersheds. *Water Wastewater Eng.* **2021**, *57* (Suppl. 2), 456–460. [CrossRef]

*water*

MDPI

*Article*

# Source Apportionment and Health Risk Assessment of Groundwater Potentially Toxic Elements (PTEs) Pollution Characteristics in an Accident Site in Zhangqiu, China

**Min Wang [1], Xiaoyu Song [1], Yu Han [2,3,4,*], Guantao Ding [2,3,5], Ruilin Zhang [1], Shanming Wei [2,3], Shuai Gao [2] and Yuxiang Liu [2,3]**

1 College of Earth Science and Engineering, Shandong University of Science and Technology, Qingdao 266590, China; brightwangm@163.com (M.W.); 17863961381@163.com (X.S.); gltyyan@163.com (R.Z.)
2 801 Institute of Hydrogeology and Engineering Geology, Shandong Provincial Bureau of Geology & Mineral Resources, Ji'nan 250014, China; m15275183308@163.com (G.D.); wsm0810@163.com (S.W.); shuaigaohome@163.com (S.G.); 15966066381@163.com (Y.L.)
3 Shandong Engineering Research Center for Environmental Protection and Remediation on Groundwater, Ji'nan 250014, China
4 School of Management, Shandong University, Ji'nan 250100, China
5 College of Environmental Science and Engineering, Ocean University of China, Qingdao 266100, China
* Correspondence: 18615268830@163.com; Tel.: +86-18615268830

**Abstract:** In order to understand the pollution degree and source of potentially toxic elements (PTEs) in groundwater around the accident site and evaluate their harm to human health, 22 groundwater samples were collected around the accident well, and the contents of As, Cd, Cr, Cu, Hg, Ni, Pb, Zn, $CH_2Cl_2$ and $C_2H_4Cl_2$ were determined. On the basis of water quality evaluation, the source apportionment method combining qualitative and quantitative analysis was used to determine the main sources of PTEs in the region, and the health risk assessment model was used to evaluate the health risk of PTEs to the human body. The results show that pH, TDS, Th and COD all exceed the standard to varying degrees, among which TH is the index with the largest number exceeding the standard. The quality of the groundwater environment in the study area is at a very poor level, and the F value is between 7.25 and 8.49. The exposure results model showed that there was no non-carcinogenic risk of PTEs in the study area, and the health risk of oral intake in the exposed population was greater than that of skin contact. Compared with adults, children were more vulnerable to the health risk stress of PTEs in groundwater. The total carcinogenic risk is higher than the total non-carcinogenic risk. As, Cd and Cr are the primary factors causing carcinogenic health risks in this area. Principal component analysis (PCA) was used to analyze the sources of PTEs in groundwater, and three principal components were extracted. It was preliminarily determined that PTE pollution was mainly related to agricultural sources, anthropogenic industrial sources and industrial sedimentation sources. The results of positive definite factor matrix analysis (PMF) were basically similar to those of PCA, but PMF further clarified the contribution rate of three pollution sources, among which agricultural sources contributed the most to the accumulation of PTEs.

**Keywords:** health risk assessment; groundwater; PTEs; PMF; source analysis

**Citation:** Wang, M.; Song, X.; Han, Y.; Ding, G.; Zhang, R.; Wei, S.; Gao, S.; Liu, Y. Source Apportionment and Health Risk Assessment of Groundwater Potentially Toxic Elements (PTEs) Pollution Characteristics in an Accident Site in Zhangqiu, China. *Water* **2024**, *16*, 768. https://doi.org/10.3390/w16050768

Academic Editor: Lucila Candela

Received: 4 February 2024
Revised: 28 February 2024
Accepted: 2 March 2024
Published: 4 March 2024



## 1. Introduction

Groundwater is a kind of extensive and fragile natural resource that is very important for human survival and development. According to statistics, the total amount of water resources in China in 2022 was 2708.81 billion $m^3$, of which 792.44 billion $m^3$ was underground water resources, accounting for about 30% of the total water resources [1]. Groundwater resource storage and water security play an important role in regional economic development and residents' lives. Especially in northern China, most of the water used in human daily life comes from groundwater [2].

However, in recent years, with the rapid growth of the economy and population, urbanization and industrialization have rapidly improved. At the same time, due to the large discharge of industrial waste, agricultural waste and municipal solid waste, the quality of the groundwater environment has been seriously threatened [3–7], especially groundwater pollution with potentially toxic element (PTE) pollution as an important source of pollution. According to the 2006 report of the World Health Organization, nearly 80% of human diseases worldwide are caused by groundwater pollution. Heavy metal pollution and organic pollution in groundwater are some of the causes of biological and ecosystem toxicity [8].

These typical PTEs have strong neurotoxicity, long-distance migration and difficult degradation [9–11]. PTEs bioaccumulate in the process of transferring along the food chain, which has a more serious impact on the ecological environment and human health [12–15]. For example, large amounts of toxic elemental mercury have been found in fish or other animals in remote Arctic regions [16]. In recent years, scholars at home and abroad have focused on different areas and different types of soil around coal-fired power plants. Among the many evaluation methods, the Nemerow pollution index method, principal component analysis method, gray system method and fuzzy comprehensive evaluation method, which add the uncertainty theory of pollution to the research method, are the most frequently used. The health risk model was used to carry out research on the human health risk assessment of PTEs [17–20]. There are also some studies on the qualitative source identification and quantitative source analysis of pollutant source analysis methods such as joint correlation analysis, cluster analysis and the positive matrix factorization model [21–23]. However, there are few reports on the quantitative source apportionment of PTEs in groundwater, especially the combination of the dual quantitative source apportionment receptor model for source apportionment and mutual comparison.

Due to the existence of factories around the study area, there are unknown people dumping waste liquid in the accident well, resulting in environmental pollution incidents. Therefore, this paper selects 22 groundwater sampling points at an accident site in Zhangqiu City, Shandong Province, China. According to the detection results of various indicators of groundwater in this area, based on the analysis and determination of PTE content in groundwater, the single-factor pollution index and Nemerow comprehensive pollution index method were used to comprehensively evaluate the PTE pollution characteristics of groundwater. Based on the IBM SPSS Statistics 26.0 model of correlation analysis combined with principal component analysis and the positive matrix factorization model, the PTE pollution source of groundwater was quantitatively explained, and the health risk system proposed by USEPA was used to evaluate the health risk of PTEs in groundwater to the human body in two ways. It provides a scientific basis for maintaining the safety of the groundwater environment and human health.

## 2. Materials and Methods

### 2.1. Study Area

Zhangqiu District is located in the eastern region of Jinan City, with an area of 1719 km$^2$. It is connected to the Zhoucun District of Zibo City in the east and the Daiyue District of Tai an City in the south. The traffic in the area is convenient. Zhangqiu District is located on the northern foot of Taiyi Mountain, bordering on the North China Plain [24]. The Great Wall Ridge stretches to the south, and Changbai Mountain stands to the east. The terrain is tilted from southeast to northwest and from south to north, followed by mountains, hills, plains and depressions. It is a continental climate in the warm temperate monsoon region. The four seasons are distinct, and the rain and heat occur in the same season. Spring is dry and windy, summer rainfall is concentrated, autumn is mild and cool and winter snow is less dry and cold [25]. The accident site is located in the surrounding area of Shanggao Village Coal Mine in Zhangqiu City, Shandong Province.

The southern part of the Tuozhangqiu belongs to the Luxi uplift, and the northern part belongs to the Jiyang sag. The geological structure is dominated by monoclinic strata, with a strike of 60° NW and a dip angle of 7° to 13°. The strata exposed in the study area are mainly Quaternary and Ordovician, and the remaining strata are less or not exposed [26]. The sedimentary sequence of each geological age is the Archean metamorphic rock series, the Paleozoic sedimentary rock series, the Mesozoic sedimentary rock series and the Cenozoic sedimentary rock series [27]. The larger synclines in the study area include the Wenzu–Bucun syncline and the Xilanggou syncline. The faults and fissures are mostly NNW and NNE, and the larger faults are the Wenzu fault, Yuwangshan fault, Ganquan fault, Songjiazhuang fault and so on.

The hydrogeological unit of the study area belongs to the upstream runoff area of the Mingshui spring basin. The piedmont zone in the northern part of the study area is mainly loose rock pore water. The main aquifer of carbonate rock fissure karst water is the Wuyangshan Formation, which is mainly located in the south and exposed to the surface. The karst fissure water of clastic rock and carbonate rock is mainly coal-bearing strata. The bedrock weathering fissure water is mainly distributed in the northern part of the study area. Artificial mining discharge is an important discharge form of karst water in the area. Because the area is mainly a monoclinic structure, confined groundwater is often formed in the continuous distribution section of sandstone.

### *2.2. Sampling and Analysis*

Combined with the geological structure, environment and human activities, according to 'the technical guidelines for the sampling of volatile organic compounds in soil and groundwater' (HJ 1019-2019), 'Groundwater environment investigation and evaluation work guide', 'Groundwater environment monitoring technical specification' (HJ 164-2020), the point layout is required, 22 monitoring wells are set up and a total of 22 groups of groundwater samples are collected. The groundwater sampling points are shown in Figure 1. Taking the accident well as the center, 22 monitoring wells were taken around the accident well. Clean, dry polyethylene plastic bottles were used to take water samples, and the sample pretreatment was completed within 24 h.

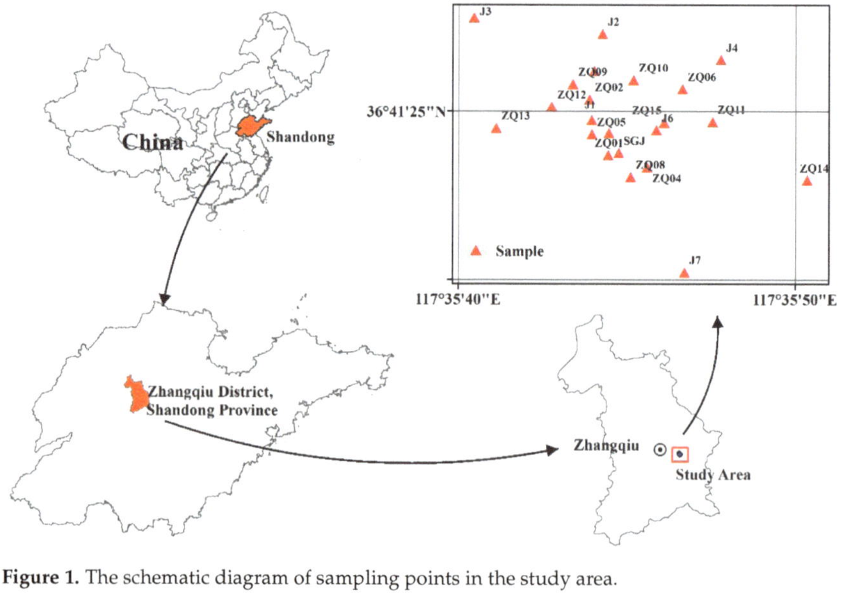


**Figure 1.** The schematic diagram of sampling points in the study area.

Inductively coupled plasma mass spectrometry (ICP-MS) (ICP-MS, Thermo X Series 2, Waltham, MA, USA) is used to detect heavy metal ions in groundwater samples. Volatile organic compounds in water samples are detected according to 'Determination of volatile organic compounds in water quality-Purge and trap/gas chromatography-mass spectrometry' (HJ 639-2012). Quality assurance and control are carried out by means of method blank and matrix addition. We use a purge and trap device (ATOMX-XYZ 1090L0429, Mason, OH, USA), and the chromatographic mass spectrometer is Agilent 8890-5977B (Agilent, Santa Clara, CA, USA) [28]. All tests are performed in triplicate to ensure the accuracy of the analysis. The error of parameter analysis is detected by the cation–anion concentration balance method, and the measurement error is less than 5%.

### 2.3. Groundwater Quality Assessment

The comprehensive evaluation method of groundwater quality is the evaluation method recommended by the 'Groundwater Quality Standard' (GB/T 14848-2017) [29]. The specific steps are as follows: Firstly, the element indexes were determined according to 'Groundwater Quality Standard' [29], and the pollution indexes of single component and multiple components were calculated with the limitation of class III water quality standard as the evaluation standard. Then, the pollution grade of sampling points was determined by combining the classification standards of multiple indexes [30,31].

#### 2.3.1. Single-Factor Evaluation Method

The measured concentration value is compared with the limit value in the evaluation standard to determine the pollution level of the pollutant. The formula is as follows:

$$\mathrm{P_{ki}} = \frac{\mathrm{c_{ki}}}{\mathrm{S_{ij}}} \tag{1}$$

In the formula, $\mathrm{P_{ki}}$ is the pollution index of the pollution index i in the water quality sample of the k monitoring well; $\mathrm{c_{ki}}$ is the measured concentration of the pollution index i in the water quality sample of the k monitoring well; $\mathrm{s_{ij}}$ is the limit value of pollution index i in GB/T 14848-2017 j water quality standard.

The pollution level is divided into six grades, and the degree of pollution has never reached serious pollution. The specific evaluation and grading standards of single-factor evaluation are shown in Table 1.

**Table 1.** Single-factor pollution score levels.

| Pki | Class of Pollution | Pollution Level |
|---|---|---|
| ≤0 | I | uncontaminated |
| 0~1 | II | light pollution |
| 1~3 | III | moderate pollution |
| 3~6 | IV | more serious pollution. |
| 6~10 | V | heavy pollution |
| >10 | VI | extremely heavy pollution |

#### 2.3.2. Nemerow Comprehensive Evaluation Method

The comprehensive evaluation method of the Nemerow index is used to obtain the environmental pollution index by comparing the measured value of the water quality monitoring index with the environmental standard value and then calculating the weight on the basis of the obtained environmental pollution index, which can accurately take into account the influence of the maximum value and avoid the influence of personal subjective factors in the process of weighted calculation. The formula is as follows:

$$F_i = \frac{c_{ki}}{S_{ij}} \tag{2}$$

$$\overline{F} = \frac{1}{n}\sum\nolimits_{i=1}^{n} F_i \tag{3}$$

$$F = \sqrt{\frac{F_{max}^2 + \overline{F}^2}{2}} \quad (4)$$

In the formula, $c_{ki}$ is the measured concentration of the pollution index $i$ in the water quality sample of the k monitoring well; $s_{ij}$ is the limit value of pollution index $i$ in GB/T 14848-2017 $j$ water quality standards; $F_{max}$ is the maximum value of each pollution index $F_i$, which is dimensionless; $\overline{F}$ is the average value of each pollution index $F_i$, which is dimensionless. The classification of the Nemerow comprehensive pollution index is shown in Table 2. The pollution level is divided into five grades, and the pollution degree is superior to the range.

**Table 2.** Nemerow comprehensive evaluation score levels.

| F | Class of Pollution | Pollution Level |
|---|---|---|
| ≤0.8 | I | superior |
| 0.8~2.5 | II | better |
| 2.5~425 | III | good |
| 4.25~7.2 | IV | worse |
| >7.2 | V | range |

### 2.4. Health Risk Assessment Model

PTEs in groundwater mainly harm human health through oral intake (including direct drinking water and indirect drinking water) and skin contact (including bathing and swimming). These two methods account for more than 90% of the pollutants ingested in or in contact with the human body. PTEs in groundwater in the study area can be divided into chemical carcinogenic elements (As, Cd, Cr, $CH_2Cl_2$ and $C_2H_4Cl_2$) and chemical non-carcinogenic elements (Cu, Zn, Ni, Hg and Pb). Their risk characterization is mainly carried out in two aspects, namely non-carcinogenic risk assessment and carcinogenic risk assessment [32].

Recipients exposed to groundwater in and around industrial areas can be divided into adults and children. The US Environmental Protection Agency (USEPA) has developed a human health risk assessment model. Based on this model, different calculation methods are used to calculate the health risks caused by oral intake and skin contact [32]. The groundwater health risks in China are mostly based on foreign models and parameters. The actual evaluation results will be affected by uncertainties such as system models, scenarios and human activities. Therefore, the exposure parameters of Shandong Province are used as much as possible in the specific calculation process to make the research results more in line with the actual situation of local people.

(1) Risk exposure through oral intake route: $ADD_i$, mg $(kg \times d)^{-1}$

$$ADD_i = \frac{c_w \cdot IR \cdot ED \cdot EF}{BW \cdot AT} \quad (5)$$

In the formula, $c_w$ is the pollutant concentration, mg $\times$ $L^{-1}$; $IR$ is drinking water intake, L $\times$ $d^{-1}$; $EF$ is drinking water exposure frequency, d $\times$ $a^{-1}$; $ED$ is the exposure period of drinking water, a; $BW$ is the average body weight of the population in Shandong Province (the two groups are divided into adults and children), kg; $AT$ is exposure time (non-carcinogenic exposure time and carcinogenic exposure time), d. Reference values of model parameters are in Table 3.

(2) Risk exposure of skin contact pathway, $ADD_d$, mg $(kg \times d)^{-1}$

$$ADD_d = \frac{c_w \cdot SA \cdot PC \cdot ET \cdot ED \cdot EF \cdot CF}{BW \cdot AT} \quad (6)$$

In the formula, $c_w$ is the pollutant concentration, mg × $L^{-1}$; *SA* is the skin contact surface area, $cm^2$; *PC* is skin permeation constant, cm × $h^{-1}$; *ET* is the exposure time, h × $d^{-1}$; *EF* is the frequency of skin contact exposure, d × $a^{-1}$; *ED* is the duration of skin contact exposure, a; *CF* is the volume conversion factor, L × $cm^{-3}$. The reference values and units of model parameters are shown in Table 3.

(3) Risk characterization

The PTEs detected in the samples were divided into two categories, non-carcinogenic substances and carcinogenic substances, and their health risks are calculated according to Equations (7) and (8), respectively [33]. The values of $RfD_i$ and $SF_i$ are shown in Table 4.

$$R_i^n = \frac{ADD_i}{RfD_i} \times 10^{-6} \quad (7)$$

$$R_i^c = SF_i \times LADD_i \quad (8)$$

In the formula, $R_i^c$ is the lifetime excess carcinogenic risk of carcinogenic pollutant *i* to the population for 70 years, which is dimensionless, $LADD_i$ is the daily average exposure dose of threshold chemical pollutants, mg (kg × d)$^{-1}$, and the calculation method is the same as *ADD*.

**Table 3.** Exposure measurement calculation parameters.

| Exposure Pathways | Parameter | Unit | Adults | Children | Reference |
|---|---|---|---|---|---|
| oral ingestion | IR | L × $d^{-1}$ | 2.54 | 0.92 | [33,34] |
| | EF | D × $a^{-1}$ | 365 | 365 | [35] |
| | ED | a | 30 | 6 | [35,36] |
| | BW | kg | 62.8 | 22 | [33,34] |
| skin exposure | SA | $cm^2$ | 1.7 × $10^4$ | 8100 | [33,34] |
| | EF | D × $a^{-1}$ | 200 | 200 | [35,37] |
| | ET | H × $d^{-1}$ | 0.173 | 0.145 | [33,34] |
| | CF | L × $cm^{-3}$ | 0.001 | 0.001 | [38] |
| | AT (non-carcinogenic) | d | ED × 365 | ED × 365 | [35] |
| | AT (carcinogenic) | d | 74.68 × 365 | 10 × 365 | [35] |

**Table 4.** Skin permeation constant and reference dose of PTEs for oral and transdermal delivery [34–38].

| PTEs | PC cm × $h^{-1}$ | SF/mg × (kg × d)$^{-1}$ | | RFD/mg × (kg × d)$^{-1}$ | |
|---|---|---|---|---|---|
| | | Oral Ingestion | Skin Exposure | Oral Ingestion | Skin Exposure |
| As | 1.8 × $10^{-3}$ | 1.5 | 1.5 | 3.0 × $10^{-4}$ | 3.0 × $10^{-4}$ |
| Cr | 2.0 × $10^{-3}$ | 0.501 | 20.0 | 3.0 × $10^{-3}$ | 6.0 × $10^{-5}$ |
| Cd | 1.0 × $10^{-3}$ | 6.1 | 6.1 | 1.0 × $10^{-3}$ | 1.0 × $10^{-5}$ |
| Ni | 1.0 × $10^{-4}$ | - | - | 2.0 × $10^{-2}$ | 5.4 × $10^{-3}$ |
| Zn | 6.0 × $10^{-4}$ | - | - | 3.0 × $10^{-1}$ | 1.0 × $10^{-2}$ |
| Pb | 4.0 × $10^{-6}$ | - | - | 1.4 × $10^{-3}$ | 1.4 × $10^{-3}$ |
| Hg | 6.0 × $10^{-4}$ | - | - | 3.0 × $10^{-4}$ | 2.1 × $10^{-5}$ |
| Cu | 6.0 × $10^{-4}$ | - | - | 4.0 × $10^{-2}$ | 1.2 × $10^{-2}$ |
| $CH_2Cl_2$ | 1.0 × $10^{-3}$ | 0.002 | - | 6.0 × $10^{-3}$ | - |
| $C_2H_4Cl_2$ | 1.0 × $10^{-3}$ | 0.091 | - | 6.0 × $10^{-3}$ | - |

Note: - means that there is no correlation coefficient.

Assuming that PTEs in water have a cumulative relationship with human health risks, the health risk $R_T$ caused by a single element to two groups of people can be expressed as in Equation (9), and the total health risk HI caused by all elements in two ways to two groups of people can be expressed as in Equation (10).

$$R_T = R_i^n + R_i^c \tag{9}$$

$$H_i = \sum R \tag{10}$$

In the formula, $R_i^n$ is the lifetime risk of health hazards caused by non-carcinogenic pollutants; $ADD_i$ is the average daily exposure dose of threshold chemical pollutants, mg (kg × d)$^{-1}$; $RfD_i$ is the reference dose for a certain exposure pathway of chemical pollutants, mg (kg × d)$^{-1}$; $10^{-6}$ is the acceptable risk level of the hypothesis corresponding to $RfD$.

### 2.5. Analytical Method of Pollution Source

#### 2.5.1. Correlation Analysis

Correlation analysis is a statistical method for studying the uncertain relationship between variables that can be used to analyze the degree of correlation between two or more correlated variable elements. In general, if there is a positive correlation between elements, it indicates that there may be the same or similar behavior and source between elements; if it is negatively correlated, it indicates that there are different sources or no correlation between elements [39].

#### 2.5.2. Principal Component Analysis

Principal component analysis (PCA) can use dimensionality reduction to transform the original complex and mutually contained environmental features into a few main original feature indicators that are closely related to each other. In the process of induction, as much information in the data sample should be retained as possible to ensure the reliability of the analysis results. As the basic basis for evaluating the importance of alternative indicators, the public degree of the selected indicator is closer to 1, which means that the indicator is more difficult to replace. According to Caesar's criterion, the impact factor index with an eigenvalue greater than 1 is retained as the principal component [40].

#### 2.5.3. Positive Definite Matrix Factorization Model

The positive matrix factorization (PMF) model is a receptor model that can determine the contribution value of the pollution source to the receptor. Its application premise is that there is no value less than 0 in the source contribution matrix and the pollution source matrix, and the source contribution matrix ($g_{ik}$) and the source component spectrum matrix ($f_{kj}$) are obtained [41]. The specific calculation formula is

$$X_{ij} = \sum_{k=1}^{p} g_{ik} f_{kj} + e_{ij} \tag{11}$$

$$Q = \sum_{i=1}^{n} \sum_{j=1}^{m} \left( \frac{X_{ij} - \sum_{k=1}^{p} g_{ik} f_{kj}}{u_{ij}} \right) \tag{12}$$

In the formula, $X_{ij}$ represents the content of the $j$th element in the $i$th sample; $p$ is the number of pollution sources; $g_{ik}$ represents the contribution rate of source $k$ to the first $i$ sample; $f_{kj}$ denotes the content of the $j$th element in the source $k$; $e_{ij}$ is the residual matrix; $Q$ is the objective function; $n$ is the number of samples; $m$ is the number of element types; $u_{ij}$ is the uncertainty value of $j$ element content in the $i$th sample.

When the element content is less than or equal to the method detection limit (MDL), $u_{ij}$ can be calculated by Formula (13):

$$u_{ij} = \frac{5}{6} MDL \tag{13}$$

When the element content is higher than *MDL*, it is calculated from Formula (14) [42]:

$$u_{ij} = \sqrt{(EF \times c)^2 + \left(\frac{MDL}{2}\right)^2} \tag{14}$$

In the formula, *EF* is the percentage of uncertainty, and *c* is the measured value of the element.

### 2.6. Data Analysis and Mapping Tools

Microsoft Excel 2019 was used for statistical analysis of element data. Principal component analysis and correlation analysis were carried out by SPSS 26.0 and Origin2021 software, respectively, and PMF5.0 was used to analyze the source of elemental pollution. We used Origin2021 and Arcgis10.7 software to complete the drawing and editing of the map.

## 3. Results and Discussion

### 3.1. Levels of Major Parameters

A total of 22 groundwater samples were collected in this study. The PTE indicators tested included 10 indicators of As, Cd, Cr, Cu, Hg, Ni, Pb, Zn, $CH_2Cl_2$ and $C_2H_4Cl_2$. The statistical results of conventional indicators of groundwater quality are shown in Table 5. According to the class III standard limit specified in the 'Groundwater Quality Standard' (GB/T 14848-2017) [29], the pH range is 3.10–7.70, with an average value of 6.54, indicating that the pH of groundwater in the study area is neutral and acidic. The content of TDS is 513.00–34,400.00 mg × $L^{-1}$, which indicates that the groundwater in this area is fresh water and brackish water. The variation range of TH is 342.00 mg × $L^{-1}$–2930.00 mg × $L^{-1}$, which indicates that there is a large area of high-hardness water in the groundwater in this area. The indexes of groundwater quality exceed the national class III water quality standard to varying degrees, and the most excessive index is TH.

**Table 5.** Statistical results of groundwater quality indicators.

| Index | Detection Limit | Min | Average | Max | Standard Deviation | Coefficient of Variation |
|---|---|---|---|---|---|---|
| pH | - | 3.10 | 6.54 | 7.7 | 1.19 | 0.18 |
| TH (mg × $L^{-1}$) | 5 | 342.00 | 1235.19 | 2930 | 771.30 | 0.62 |
| TDS (mg × $L^{-1}$) | 5 | 513.00 | 4471.59 | 34,400 | 8159 | 1.82 |
| COD (mg × $L^{-1}$) | 0.5 | 0.95 | 70.9 | 504.6 | 153.04 | 2.48 |
| As (mg × $L^{-1}$) | 0.005 | 0.00 | 0.02 | 0.05 | 0.02 | 0.85 |
| Cd (mg × $L^{-1}$) | 0.0001 | 0.00 | 0.01 | 0.07 | 0.02 | 2.88 |
| Cr (mg × $L^{-1}$) | 0.001 | 0.00 | 0.1 | 0.94 | 0.25 | 2.33 |
| Cu (mg × $L^{-1}$) | 0.001 | 0.00 | 0.24 | 2.46 | 0.62 | 2.48 |
| Hg (mg × $L^{-1}$) | 0.0005 | 0.00 | 0.00292 | 0.0049 | 0.00126 | 0.43 |
| Ni (mg × $L^{-1}$) | 0.001 | 0.01 | 1.4 | 18.25 | 3.99 | 2.78 |
| Pb (mg × $L^{-1}$) | 0.001 | 0.00 | 0.04 | 0.33 | 0.08 | 0.89 |
| Zn (mg × $L^{-1}$) | 0.005 | 0.01 | 1.57 | 15.24 | 4.02 | 2.48 |
| $CH_2Cl_2$ (mg × $L^{-1}$) | 0.005 | 0.00807 | 0.2808 | 1.39 | 0.35122 | 1.36 |
| $C_2H_4Cl_2$ (mg × $L^{-1}$) | 0.0005 | 0.00058 | 0.00283 | 0.00911 | 0.00282 | 0.99 |

The distribution of PTE content in groundwater is shown in Figure 2. It can be seen from the diagram that various indicators have high outliers to varying degrees. Ni and Zn at some sites were also significantly higher. The study area is seriously polluted, and the spatial difference is large. It is necessary to evaluate the human health risk of groundwater in this area. Among the groundwater detection indicators, the coefficient of variation for Cd, Cr, Cu, Ni and Zn exceeded 1.0, indicating that there was a large difference in water quality between groundwater samples.

## 3.2. Groundwater Quality Evaluation Results

From the perspective of single-factor pollution in Table 6, all eight heavy metal indicators have water samples at moderate to severe pollution levels. Among them, the contribution of Ni pollution was the highest, the number of samples with severe pollution reached 14, accounting for 51.8%, and the highest value of the single-factor pollution index reached 915. Followed by Pb, four samples were at the level of severe pollution, and the highest value of the single-factor pollution index reached 32.6%. All eight indexes of ZQ01 points in all samples exceeded the standard and were at the level of heavy pollution. Except for Hg, the single-factor pollution index of the other seven indexes was at the maximum value in all samples. The second is the J1 point, which has Cr, Ni and As indicators of a heavy pollution level. From the comprehensive evaluation of groundwater indicators in Table 6, it can be seen that the quality of the groundwater environment in the study area is at a very poor level, the F value is between 7.25 and 8.49, and the drinking water wells around the accident wells are at a poor level.

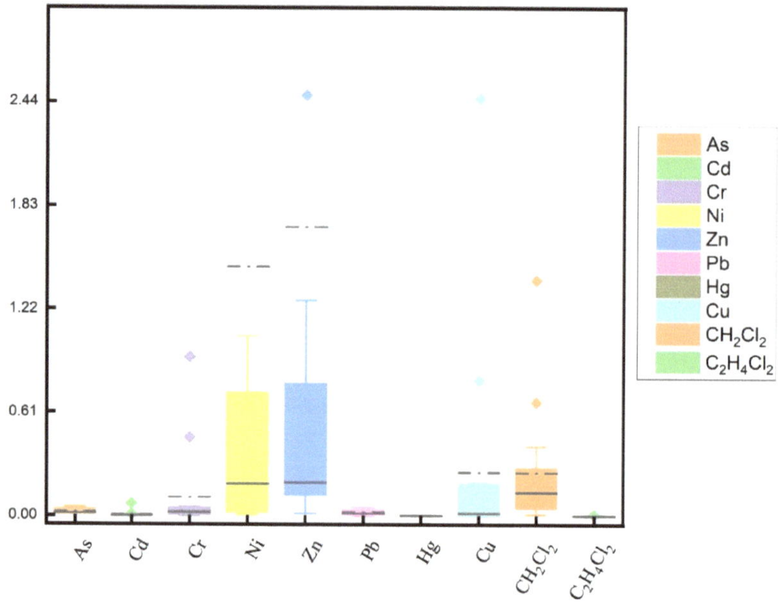


**Figure 2.** Distribution map of PTE content in groundwater.

**Table 6.** Groundwater quality evaluation table.

| Analysis Index | As | Cd | Cr | Cu | Hg | Ni | Pb | Zn | F |
|---|---|---|---|---|---|---|---|---|---|
| ZQ01 | 10 | 10 | 10 | 10 | 10 | 10 | 10 | 10 | 8.49 |
| ZQ02 | 1 | 3 | 3 | 6 | 10 | 10 | 3 | 3 | 7.88 |
| ZQ03 | 0 | 1 | 0 | 0 | 0 | 6 | 3 | 1 | 7.36 |
| ZQ04 | 1 | 1 | 3 | 1 | 10 | 1 | 3 | 1 | 7.61 |
| ZQ05 | 0 | 0 | 0 | 0 | 0 | 0 | 1 | 0 | 7.33 |
| ZQ06 | 0 | 0 | 0 | 0 | 0 | 1 | 0 | 0 | 7.24 |
| ZQ07 | 1 | 3 | 3 | 6 | 0 | 10 | 1 | 10 | 7.85 |
| ZQ08 | 0 | 1 | 0 | 0 | 10 | 10 | 0 | 1 | 7.62 |
| ZQ09 | 0 | 1 | 0 | 0 | 0 | 10 | 0 | 1 | 7.51 |
| ZQ10 | 0 | 0 | 0 | 0 | 0 | 1 | 0 | 0 | 725 |
| ZQ11 | 0 | 0 | 0 | 0 | 0 | 1 | 0 | 0 | 7.25 |
| ZQ12 | 0 | 3 | 1 | 1 | 10 | 10 | 3 | 1 | 7.59 |
| ZQ13 | 0 | 3 | 1 | 6 | 0 | 10 | 3 | 6 | 7.73 |
| ZQ14 | 0 | 1 | 0 | 0 | 0 | 1 | 0 | 1 | 7.27 |
| ZQ15 | 0 | 3 | 0 | 0 | 0 | 10 | 0 | 1 | 7.59 |
| J1 | 3 | 1 | 10 | 0 | 0 | 10 | 3 | 6 | 7.87 |
| J2 | 0 | 0 | 3 | 1 | 0 | 10 | 0 | 1 | 7.44 |
| J3 | 0 | 0 | 0 | 0 | 0 | 6 | 0 | 1 | 7.30 |
| J4 | 3 | 0 | 3 | 0 | 0 | 10 | 0 | 1 | 7.39 |
| J5 | 0 | 0 | 0 | 0 | 0 | 1 | 0 | 1 | 7.30 |
| J6 | 3 | 3 | 3 | 3 | 0 | 10 | 3 | 6 | 7.72 |
| J7 | 0 | 0 | 6 | 1 | 0 | 6 | 3 | 1 | 7.49 |

### *3.3. Groundwater Health Risk Assessment*

Based on the health risk assessment model, the health risks of PTEs in groundwater in the study area to adults and children through oral intake and dermal contact were calculated as shown in Table 7. The total health risks of the five non-carcinogenic PTEs in the study area through oral intake and skin contact with the population did not exceed the maximum acceptable level recommended by the International Commission on Radiological Protection (ICRP) of $5.0 \times 10^{-5}$ $a^{-1}$. The total health risks of the three carcinogenic heavy metal elements in the PTEs through oral intake to the population exceeded the maximum acceptable level, which should be paid attention to. From Table 7, it can be seen that Cd has the greatest carcinogenic health risk to children through oral intake, which is $9.20 \times 10^{-4}$, and Cr has the greatest carcinogenic health risk to children through skin contact, which is $7.44 \times 10^{-5}$. The maximum non-carcinogenic health risk of Ni to children through oral intake was $3.07 \times 10^{-6}$, and the maximum non-carcinogenic health risk of Zn to children through skin contact was $2.99 \times 10^{-9}$. The total non-carcinogenic risks of PTEs in groundwater to the two groups of people under the oral intake route and the skin contact exposure route were Ni > Pb > Hg > Cu > Zn and Zn > Hg > Ni > Cu > Pb, respectively. The non-carcinogenic risk of oral intake was consistent with the trend of total non-carcinogenic risk. The total non-carcinogenic risk is 1~2 orders of magnitude lower than the maximum acceptable risk level ($5.0 \times 10^{-5}$ $a^{-1}$), and the total carcinogenic risk value is 2~4 orders of magnitude higher than the total non-carcinogenic risk value, indicating that the non-carcinogenic health risk of groundwater in the study area is small, and the daily drinking water, bathing, swimming and other activities of the population will not cause significant harm.

**Table 7.** Health risk assessment results.

| | Oral Route | | Skin Exposure | | Comprehensive Risks | |
|---|---|---|---|---|---|---|
| | **Children** | **Adults** | **Children** | **Adults** | **Children** | **Adults** |
| As | $7.85 \times 10^{-4}$ | $5.08 \times 10^{-4}$ | $9.88 \times 10^{-7}$ | $5.81 \times 10^{-7}$ | $7.86 \times 10^{-4}$ | $5.09 \times 10^{-4}$ |
| Cd | $9.20 \times 10^{-4}$ | $5.96 \times 10^{-4}$ | $6.44 \times 10^{-7}$ | $3.78 \times 10^{-7}$ | $9.21 \times 10^{-4}$ | $5.96 \times 10^{-4}$ |
| Cr | $1.39 \times 10^{-3}$ | $8.97 \times 10^{-4}$ | $7.74 \times 10-5$ | $4.54 \times 10^{-5}$ | $1.46 \times 10^{-3}$ | $9.42 \times 10^{-4}$ |
| Cu | $2.71 \times 10^{-7}$ | $2.62 \times 10^{-7}$ | $3.79 \times 10^{-10}$ | $3.32 \times 10^{-10}$ | $2.71 \times 10^{-7}$ | $2.62 \times 10^{-7}$ |
| Hg | $4.10 \times 10^{-7}$ | $3.96 \times 10^{-7}$ | $2.46 \times 10^{-9}$ | $2.16 \times 10^{-9}$ | $4.12 \times 10^{-7}$ | $3.99 \times 10^{-7}$ |
| Ni | $3.07 \times 10^{-6}$ | $2.97 \times 10^{-6}$ | $7.96 \times 10^{-10}$ | $6.98 \times 10^{-10}$ | $3.07 \times 10^{-6}$ | $2.97 \times 10^{-6}$ |
| Pb | $1.33 \times 10^{-6}$ | $1.29 \times 10^{-6}$ | $3.72 \times 10^{-12}$ | $3.27 \times 10^{-12}$ | $1.33 \times 10^{-6}$ | $1.29 \times 10^{-6}$ |
| Zn | $2.37 \times 10^{-7}$ | $2.30 \times 10^{-7}$ | $2.99 \times 10^{-9}$ | $2.62 \times 10^{-9}$ | $2.40 \times 10^{-7}$ | $2.32 \times 10^{-7}$ |
| $CH_2Cl_2$ | $1.29 \times 10^{-5}$ | $8.38 \times 10^{-6}$ | — | — | $1.29 \times 10^{-5}$ | $8.35 \times 10^{-6}$ |
| $C_2H_4Cl_2$ | $6.46 \times 10^{-6}$ | $4.18 \times 10^{-6}$ | — | — | $6.46 \times 10^{-6}$ | $4.18 \times 10^{-6}$ |

The EPA's acceptable risk level for carcinogens is in the range of $10^{-6}$–$10^{-4}$. Less than $10^{-6}$ indicates that the risk is not obvious, $10^{-6}$–$10^{-4}$ indicates that there is a risk, and more than $10^{-4}$ indicates that there is a significant risk. From Table 7, it can be seen that Cr has the greatest health risk of carcinogenicity for children through both routes. The carcinogenic risk of oral intake is $1.39 \times 10^{-3}$, which exceeds the acceptable risk level by 1–3 orders of magnitude, with a significant risk. The carcinogenic risk of skin contact is $7.74 \times 10^{-5}$. The carcinogenic risks of $CH_2Cl_2$ and $C_2H_4Cl_2$ in PTEs to children through oral intake were $1.29 \times 10^{-5}$ and $6.46 \times 10^{-6}$, respectively, which were risky but not significant. As, Cd and Cr have a significant risk to both groups under the exposure route of oral intake, and the risk is not obvious under the exposure route of skin contact. The carcinogenic risk of oral intake is 2–4 orders of magnitude larger than that of skin contact. Therefore, As, Cd and Cr are the three primary factors that cause the carcinogenic health risk of the population (especially children) in this area. We should pay attention to the pollution of As, Cd and Cr in the groundwater in the study area.

In general, by comparing the health risks of groundwater PTEs to the two groups of people in different exposure routes, it can be seen that the health risks of children and adults are in the same order of magnitude, whether through oral intake or skin contact, but children are always slightly higher than adults. Therefore, we should focus on the health risks of groundwater PTE toxicity to children.

### *3.4. Source Apportionment of Elements in Groundwater*

#### 3.4.1. Correlation Analysis

Correlation analysis has a basic distinguishing effect on exploring the sources of different PTEs, and can simply identify the correlation between different PTEs. In general, correlation analysis can be used to analyze whether different PTEs have a significant positive correlation, so as to determine whether different PTEs may come from similar sources or the same source [43].

In this paper, the correlation between groundwater PTEs in the study area was analyzed by Origin2021 software (Figure 3).

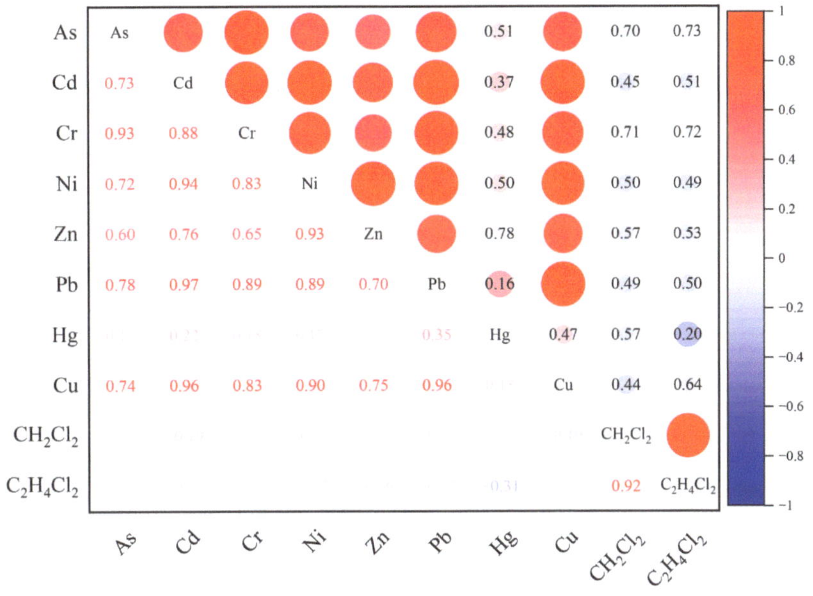


Siginficant level: 0.05

**Figure 3.** Correlation analysis matrix.

The results show that there is a significant positive correlation between the seven heavy metal elements except Hg in the study area, and the correlation is pairwise, indicating that there may be an information overlap between As, Cd, Cr, Cu, Ni, Pb and Zn elements in groundwater in the study area, and the correlation coefficients are all > 0.5, so they may undergo the same environmental geochemistry or have a compound pollution relationship. However, the correlation coefficient between $C_2H_4Cl_2$ and $CH_2Cl_2$ is 0.92, which is a very significant correlation level but has no significant relationship with other PTEs. Therefore, $CH_2Cl_2$ and $C_2H_4Cl_2$ are more likely to come from the same pollution source. In addition, there is no obvious correlation between Hg and other PTEs in the study area, and there may be independent pollution sources.

3.4.2. PCA

The results of the KMO and Bartlett sphericity tests showed that the KMO test value was 0.652, the Bartlett sphericity test statistic was 0, and the significance test value of the Bartlett sphericity test was $p < 0.05$, indicating that this group of data was suitable for PCA. PCA was carried out by IBM SPSS Statistics 26.0 software. Three principal components were extracted by the Kaiser standardized orthogonal rotation method, which were F1, F2 and F3.

Three principal components (F1–F3) were extracted by principal component analysis and could explain 89.454% of the total variance. The principal component loads contained in each component are shown in Table 8. The contribution of As, Cd, Cr, Cu, Ni, Pb and Zn in F1 is higher, and the absolute value of the factor load exceeds 0.8. The principal component contains seven heavy metal elements. The long-term input of livestock and poultry manure and the application of chemical fertilizers may be the main sources of Cu, Zn, Cd and Cr in the study area. Livestock feed additives, pesticides and fertilizers are used more frequently in the local area [44,45], indicating that they are mainly affected by agricultural sources. The contribution of $CH_2Cl_2$ and $C_2H_4Cl_2$ in F2 is higher, and the factor load is greater than 0.9. Since some indicators of groundwater have changed significantly due to the dumping of chemical waste by three enterprises in the study area, F2 explains the impact of this pollution incident on groundwater quality, indicating that it may be affected by anthropogenic industrial sources. The factor load of Hg in F3 is greater than 0.8, and Hg is seriously polluted in some sample points. However, due to the volatility of Hg, its content has a large degree of spatial dispersion. It is preliminarily judged that Hg enrichment is mainly caused by the accumulation of micro-particles released by surrounding coal combustion through atmospheric deposition, which is mainly an industrial sedimentation source.

**Table 8.** The main calculation results of PTEs' principal component analysis.

| HM | F1 | F2 | F3 |
|---|---|---|---|
| As | 0.843 | 0.135 | 0.037 |
| Cd | 0.969 | 0.038 | −0.02 |
| Cr | 0.924 | 0.13 | 0.009 |
| Cu | 0.952 | 0.061 | −0.052 |
| Hg | 0.281 | −0.298 | 0.899 |
| Ni | 0.959 | 0.055 | −0.109 |
| Pb | 0.968 | 0.022 | 0.214 |
| Zn | 0.827 | 0.06 | −0.219 |
| $CH_2Cl_2$ | −0.185 | 0.92 | 0.26 |
| $C_2H_4Cl_2$ | −0.209 | 0.949 | 0.045 |
| Eigenvalue | 6.104 | 1.884 | 0.957 |
| Variance explained rate (%) | 61.045 | 18.838 | 9.572 |

3.4.3. PMF

Principal component analysis combined with correlation analysis basically explained the possible sources of 10 PTEs in the study area. On this basis, the PMF method can be used to quantitatively analyze the source of PTEs in the study area. When calculating the uncertainty of PTEs, the signal-to-noise ratio of Ni and Zn is small (S/N = 0), which is classified as 'Bad', the signal-to-noise ratio of Cu is small (S/N < 0.5), which is classified as 'weak', and the remaining PTEs are classified as 'strong'. After trial calculation, when the number of setting factors is 2–7, the number of operations is 25 times and the number of factors is 3, $Q_{Robust}/Q_{true}$ decreases rapidly, and the residual value is between −3 and 3. The proportion is the largest, indicating that the correlation between the measured value and the predicted value of the PMF model is high [46], and it is inferred that there are three main sources of PTEs in this area.

According to the source composition spectrum (Figure 4) and source contribution diagram (Figure 5), it was found that As, Cd, Cr, Cu and Pb had higher loads on PMF1, which were 34%, 88%, 54.7%, 93.1% and 38.2%, respectively, which were consistent with the results of F1. Cd, as a representative element of PMF1, is an iconic element of agricultural activities [47]. Agricultural activities such as industrial waste, pesticides, fertilizers and plastic film use will lead to its high content. The excessive point of the Cd element is distributed near villages and towns in the study area, which also confirms this point. In the study area, there is the Minggao No. 2 coal mine, which is located in the east of the Jidong coalfield. There are many roadways filled with water. Some studies have pointed out that As, Ni and Cr are relatively stable in nature. They are usually regarded as crustal elements, mainly derived from soil characteristics [48]. Therefore, PMF1 is mainly affected by agricultural sources.

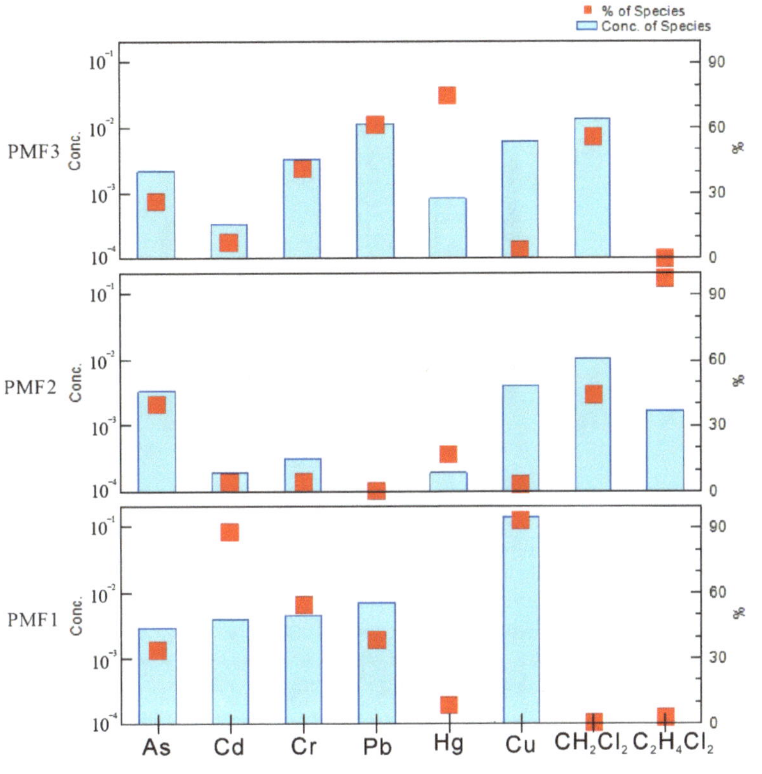


**Figure 4.** Analytical results of PMF model.

The contribution rates of $CH_2Cl_2$ and $C_2H_4Cl_2$ in PMF2 were 43.9% and 97.6%, respectively, indicating that the load of similar elements was higher, which was due to the fact that the surrounding companies poured waste samples into the accident well, mainly due to the influence of artificial industrial sources.

PMF3 has higher Hg and Pb element loads, with contribution rates of 74.7% and 61.8%. Atmospheric deposition and surface water are the main sources of Hg. Chemical enterprises use oil and coal as the main fuel, and coal combustion leads to the emission of Hg-containing gases. Pb is the iconic element of traffic activity [49], the main source of automobile exhaust and coal combustion, so it suggests that Hg enrichment is mainly due to its release into the surrounding metallurgy and coal, mainly from industrial sedimentation sources.

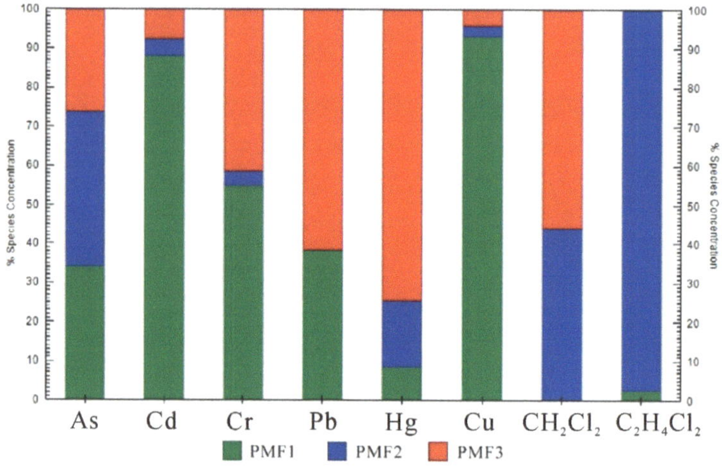


**Figure 5.** The proportion of PTE factors in groundwater.

From the source apportionment results, the two analyses have similar identification results in the qualitative identification of potential pollution sources and generally represent the same potential pollution sources. In addition, the two models are slightly different in the distribution of principal component composition but do not affect the overall analytical results. From the perspective of the contribution rate of pollution sources, the trend and contribution rate of pollution sources analyzed by the two models are different. The reason for this difference is that the factorization process of the two models is different, that is, although both of them are based on the calculation of the least squares method, they are different in the factor extraction model [50,51].

According to the factor profile of PMF, the average contribution of sample sources to the total mass was calculated (Figure 6). The results showed that As, Cd, Cr, Cu and Pb are mainly from agricultural sources, with a contribution rate of 39.84%. $CH_2Cl_2$ and $C_2H_4Cl_2$ are mainly affected by anthropogenic industrial sources, with a contribution rate of 26.13%. Hg is from an industrial sedimentation source, with a contribution rate of 34.04%.

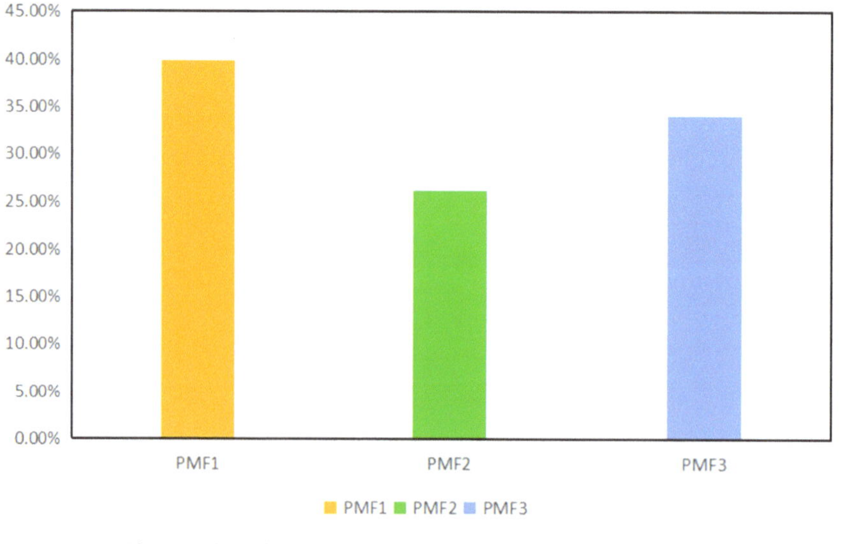


**Figure 6.** Contribution rate distribution diagram of sources.

## 4. Conclusions

This paper took an accident site in Zhangqiu City, Shandong Province, as the main research area, determined the pollution situation and pollution source of the research area, made a pollution assessment and health risk assessment of the research area, and provided information for the sustainable development and utilization of groundwater resources in the research area and on the protection of and improvement in the living environment. It is of great significance. The comprehensive indexes of pH, TDS, COD, TH and PTEs in groundwater at the accident site are significantly high, exceeding the class III standard limit to varying degrees, and the spatial dispersion of PTE content is large. The single-factor pollution index values of Ni and Pb in the PTE index are high. According to the single-factor pollution index method, most of the PTEs are at a safe or alert level. According to the Nemerow comprehensive pollution index, there are heavy pollution levels. The sample points are ZQ01 and ZQ07, and Ni is the main contribution factor in the region. In view of the comprehensive consideration of PTEs, the groundwater in the study area is seriously polluted. According to the non-carcinogenic risk assessment, the total non-carcinogenic risk index of groundwater around the accident site for nearby children and adults does not exceed the non-carcinogenic safety risk value, but the risk value for children is higher than that for adults. According to the carcinogenic risk assessment, the carcinogenic health risk values of adults and children exposed to As, Cd and Cr through oral intake in the study area are all high, and the maximum acceptable risk value is $1.0 \times 10^{-4}$. The control of As, Cd and Cr in groundwater in the study area should be strengthened. Among the two exposure routes, the oral intake route has the greatest impact. On the basis of correlation analysis and principal component analysis, combined with the PMF model, the main pollution sources of PTEs in the study area were determined. As, Cd, Cr, Cu, Ni, Pb and Zn are from agricultural sources (PMF1), $CH_2Cl_2$ and $C_2H_4Cl_2$ are from human-made industrial sources (PMF2) and the accumulation of Hg is mainly from industrial sedimentation sources (PMF3). The contribution rates of the three pollution sources were 39.84%, 26.13% and 34.04%, respectively.

**Author Contributions:** M.W.: conceptualization, investigation, methodology, project administration, writing—original draft; X.S.: conceptualization, methodology, project administration, writing—review and editing; Y.H.: conceptualization, investigation, writing—original draft; G.D.: methodology, writing—original draft; R.Z.: methodology, software, writing—review and editing; S.W.: methodology, data curation; S.G.: supervision; Y.L.: validation. All authors have read and agreed to the published version of the manuscript.

**Funding:** The study was supported by the National Natural Science Foundation of China (No. 42202294).

**Data Availability Statement:** Data are available upon request from the corresponding author. The data are not publicly available due to privacy reasons.

**Conflicts of Interest:** The authors declare no conflicts of interest.

## References

1. 2022 China Water Resources Bulletin. *Water Resour. Dev. Manag.* **2023**, *9*, 2. Available online: http://www.mwr.gov.cn/xw/slyw/202306/t20230629_1672395.html (accessed on 10 January 2024).
2. Ministry of Ecology and Environment of the People's Republic of China National Groundwater Pollution Prevention and Control Plan (2011–2020) [EB]. 2011. Available online: https://www.mee.gov.cn/gkml/hbb/bwj/201111/W020111109376922920938.pdf (accessed on 28 October 2011).
3. Xiao, H.; Shahab, A.; Li, J.Y.; Xi, B.; Sun, X.; He, H.; Yu, G. Distribution, ecological risk assessment and source identification of heavy metals in surface sediments of Huixian karst wetland, China. *Ecotoxicol. Environ. Saf.* **2019**, *185*, 109700. [CrossRef]
4. Ahmed, S.; Khurshid, S.; Madan, R.; Bassam, A.A.; Naushad, M. Water quality assessment of shallow aquifer based on Canadian Council of Ministers of the environment index and its impact on irrigation of Mathura District, Uttar Pradesh. *J. King Saud Univ.-Sci.* **2020**, *32*, 1218–1225. [CrossRef]
5. Ahmedd, S.; Akhtar, N.; Rahman, A.; Mondal, N.; Khurshid, S.; Sacsh, S.; Khan, M.; Kamboj, V. Evaluating groundwater pollution with emphasizing heavy metal hotspots in an urbanized alluvium watershed of Yamuna River, northern India. *Environ. Nanotechnol. Monit. Manag.* **2022**, *18*, 100744. [CrossRef]

6. Pazand, K.; Khosrvai, D.; Ghaderi, M.R.; Rezvaniyanzadeh, M. Hydrogeochemistry and lead contamination of groundwater in the north part of Esfahan province, Iran. *J. Water Health* **2018**, *16*, 622–634. [CrossRef]
7. Baskaran, P.; Abraham, M. Evaluation of groundwater quality and heavy metal pollution index of the industrial area, Chennai. *Phys. Chem. Earth Parts A/B/C* **2022**, *128*, 103259. [CrossRef]
8. Belkhiri, L.; Mouni, L.; Sheikhy Narany, T.; Tiri, A. Evaluation of potential health risk of heavy metals in groundwater using the integration of indicator kriging and multivariate statistical methods. *Ground Water Sustain. Dev.* **2017**, *4*, 12–22. [CrossRef]
9. Wang, S.; Gu, Y.; Yang, J.; Jiang, Y.; Yu, X.; Zhang, Z.; Wang, J.; Wang, T.; Zhang, Y. Emissions and distribution characteristics of heavy metal in 1000MW near-zero emission coal-fired units. *China Environ. Sci.* **2022**, *42*, 2060–2069.
10. Zhao, L.Y.; Kong, L.H.; Zhao, Z.G.; Zhou, W.; Qiu, J.; Huang, J.; Geng, B.; Yang, M.; Han, X.; Li, Y.; et al. Pollution Characteristics, Source Analysis and Risk Assessment of Heavy Metals in Soil Around a Gold Mine in Jiaodong Peninsula. *Geol. China* **2023**, 1–28. Available online: http://kns.cnki.net/kcms/detail/11.1167.P.20231106.1329.002.html (accessed on 26 February 2024).
11. Ljung, K.; Otabbong, E.; Selinus, O. Natural and anthropogenic metal inputs to soils in urban uppsala, sweden. *Environ. Geochem. Health* **2006**, *28*, 353–364. [CrossRef]
12. Huang, R.J.; Cheng, R.; Jing, M.; Yang, L.; Li, Y.; Chen, Q.; Chen, Y.; Yan, J.; Lin, C.; Wu, Y.; et al. Source-specific health risk analysis on particulate trace elements: Coal combustion and traffic emission as major contributors in wintertime Beijing. *Environ. Sci. Technol.* **2018**, *52*, 10967–10974. [CrossRef]
13. Gu, C.; Zhao, Y. Research progress of air pollutant emissions of Chinese coal-fired power plants. *J. China Coal Soc.* **2022**, 1–12. Available online: https://kns.cnki.net/kcms/detail/11.2190.TD.20220816.1038.003.html (accessed on 9 January 2024).
14. Xie, T.Y.; Shi, Z.M.; Gao, Y.W.; Tan, L.; Meng, L. Modeling analysis of the characteristics of selenium-rich soil in heavy metal high background area and its impact on main crops. *Ecol. Inform.* **2021**, *66*, 101420. [CrossRef]
15. Wang, J.; Zhang, X.; Chen, A.; Wang, B.; Zhao, Q.; Liu, G.; Xiao, X.; Cao, J. Source analysis and risk evaluation of heavy metal in the river sediment of polymetallic mining area: Taking the Tonglüshan skarn type Cu-Fe-Au deposit as an example, Hubei section of the Yangtze River Basin, China. *China Geol.* **2022**, *5*, 649–661.
16. Outridge, P.M.; Mason, R.P.; Wang, F.; Guerrero, S.; Heimburger, L. Updated global and oceanic mercury budgets for the United Nations global mercury assessment. *Environ. Sci. Technol.* **2018**, *52*, 11466–11477. [CrossRef]
17. Huang, X.; Hu, J.; Qin, F.; Quan, W.; Cao, R.; Fan, M.; Wu, X. Heavy metal pollution and ecological assessment around the Jinsha coal-fired power plant (China). *Int. J. Environ. Res. Public Health* **2017**, *14*, 1589. [CrossRef]
18. Minkina, T.; Konstantinova, E.; Bauer, T.; Mandzhieva, S.; Sushkova, S.; Chaplygin, V.; Burachevskaya, M.; Nazarenko, O.; Kizilkaya, R.; Gülser, C.; et al. Environmental and human health risk assessment of potentially toxic elements in soils around the largest coal-fired power station in Southern Russia. *Environ. Geochem. Health* **2020**, *43*, 2285–2300. [CrossRef]
19. Linnik, V.G.; Minkina, T.M.; Bauer, T.V.; Saceliev, A.; Mandzhieva, S. Geochemical assessment and spatial analysis of heavy metals pollution around coalfired power station. *Environ. Geochem. Health* **2021**, *42*, 4087–4100. [CrossRef]
20. Liu, Z.; Fang, Z.; Ding, X. Heavy metal pollution and health risk assessment of soil in ash yard of coal-fired power plant. *Ecol. Environ.* **2021**, *30*, 1916–1922. [CrossRef]
21. Yang, Z.; Xiao, R.; Chen, Y.; Deng, Y.; Han, C.; Liu, C.; Gao, Z.; Huang, S.; Dai, W. Heavy metal distribution, risk assessment and source analysis of soil around a typical coalfired power plant in South China. *Acta Ecol. Sin.* **2020**, *40*, 4823–4835.
22. Yang, M.; Li, X.; Wang, Y.; Zhang, J. Spatial distribution, source analysis and health risk assessment of heavy metals in surface dust around a coal-fired power plant. *J. Arid Land Resour. Environ.* **2022**, *36*, 176–182. [CrossRef]
23. Che, K.; Chen, C.; Zheng, Q.; Fan, H.; Wei, M.; Luo, P.; Yu, J. Heavy metal emissions from coal-fired power plants and heavy metal pollution characteristics and health risks in surrounding soils. *Environ. Sci.* **2022**, *43*, 4578–4589. [CrossRef]
24. Chen, H.; Wu, W.; Cao, L.; Zhou, X.; Guo, R.; Nie, L.; Shang, W. Source Analysis and Contamination Assessment of Potentially Toxic Element in Soil of Small Watershed in Mountainous Area of Southern Henan, China. *Int. J. Environ. Res. Public Health* **2022**, *19*, 13324. [CrossRef]
25. Han, Y.; Liu, Y.; Wei, S.; Wang, M.; Ding, G.; Song, X.; Shen, D.; Gao, S.; Tang, C.; Ma, G. Source Apportionment and Health Risk Assessment of Heavy Metals in Karst Water from Abandoned Mines in Zhangqiu, China. *Water* **2023**, *15*, 3440. [CrossRef]
26. Gao, D.; Yan, Y.; Yao, R.; Shi, H.; Chen, Q. Spatial analysis and GIS mapping of regional hydrochemistry, groundwater quality, and probabilistic health risk in western Sichuan Basin, southwestern China. *Environ. Earth Sci.* **2023**, *83*, 41. [CrossRef]
27. Cai, X.L.; Cao, J.M.; Zhu, J.S.; Cheng, X. System of Crust-Mantle Ductile Shear Zone in the Continental Lithosphere in China. *Earth Sci. Front.* **2008**, *15*, 36–54. [CrossRef]
28. Elizabeth, J.; Mary, V.; Michelle, L. Effect of Fe(III) on 1,1,2,2-Tetrachloroethane Degradation and Vinyl Chloride Accumulation in Wetland Sediments of the Aberdeen Proving Ground. *Bioremediat. J.* **2004**, *8*, 31–45. [CrossRef]
29. *GB/T 14848-2017*; Natural Resources and Territory Spatial Planning. Standard for Groundwater Quality. Natural Resources and Territory Spatial Planning: Beijing, China, 2014.
30. Li, M.; Zhang, J.; Liang, N.; Lin, L.; Li, Q.; Wen, X. Comparisons of some common methods for water environmental quality assessment. *Prog. Geogr.* **2012**, *31*, 617–624. [CrossRef]
31. Wang, W.; Ji, M.; Su, Y. Reviews of the progress in the research of water quality evaluation and the methods for water quality evaluation. *Sci-Tech Inf. Dev. Econ.* **2012**, *22*, 129–131.
32. Wu, W.H.; Zou, H.; Zhu, G.; Liao, Y.; Pan, H.; Xiao, C.; Fan, J.; Li, L. Characteristics and health risk assessment of heavy metal pollution in groundwater of a mining area. *J. Ecol. Rural Environ.* **2018**, *34*, 1027–1033.

33. USEPA. *Risk Assessment Guidance for Superfund Volume I: Human Health Evaluation Muanual (Part E, Supplemental Guidance for Dermal Risk Assessment)*; USEPA: Washington, DC, USA, 2004.
34. Duan, X.L. *Research Methods of Exposure Parameters and Their Application in Environmental Health Risk Assessment*; Science Press: Beijing, China, 2012.
35. Liu, B.; Cui, X.; Wang, X.; Hu, Q. Source identification and health risk assessment of heavy metals in groundwater of Yongqing County, Hebei. *J. Ecol. Rural. Environ.* **2022**, *39*, 741–749. [CrossRef]
36. Ma, C.; Zhou, J.; Zeng, Y.; Bai, F.; Yan, Z. Source analysis and health risk assessment of heavy metals in groundwater in the oasis belt of Ruoqiang County, Xinjiang. *Acta Sci. Circumstantiae* **2023**, *43*, 266–277. [CrossRef]
37. Cao, Z.; Gou, J.; Qiu, X.; Zhou, Q.; Wan, Y.; Zhao, Z. Spatial and temporal distribution characteristics and health risk assessment of heavy metals in water of Ningxia section of the Yellow River. *Environ. Monit. Manag. Technol.* **2022**, *34*, 33–38. [CrossRef]
38. An, W.; Sun, L.; Ma, L.; Zhang, S. Assessment of heavy metals and human health risk in surface soils in a legacy land of typical industrial agglomeration area. *J. Nat. Sci. Hunan Norm. Univ.* **2022**, *45*, 108–116.
39. Rao, K.; Tang, T.; Zhang, X.; Wang, M.; Liu, J.; Wu, B.; Wang, P.; Ma, Y. Spatial-temporal dynamics, ecological risk assessment, source identification and interactions with internal nutrients release of heavy metals in surface sediments from a large Chinese shallow lake. *Chemosphere* **2021**, *282*, 131041. [CrossRef]
40. Singaraja, C.; Chidambaram, S.; Srinivasamoorthy, K.; Anandhan, P.; Selvam, S. A study on assessment of credible sources of heavy metal pollution vulnerability in groundwater of thoothukudi districts, Tamilnadu, India. *Water Qual. Expo. Health* **2015**, *7*, 459–467. [CrossRef]
41. Paatero, P.; Tapper, U. Positive matrix factorization: A non-negative factor model with optimal utilization of error estimates of data values. *Environmetrics* **1994**, *5*, 111–126.
42. Wen, C.; Zhu, S.; Li, N.; Luo, X. Source apportionment and risk assessment of metal pollution in natural biofilms and surface water along the Lancang River, China. *Sci. Total Environ.* **2022**, *843*, 156977. [CrossRef]
43. Wang, Z.; Han, R.; Muhammad, A.; Guan, D.; Zama, E.; Li, G. Correlative distribution of DOM and heavy metals in the soils of the Zhangxi watershed in Ningbo city, East of China. *Environ. Pollut.* **2022**, *299*, 118811.
44. Belon, E.; Boisson, M.; Deportes, I.Z.; Eglin, T.K.; Feix, I.; Bispo, A.O.; Galsomies, L.; Leblond, S.; Guellier, C.R. An inventory of trace elements inputs to French agricultural soils. *Sci. Total Environ.* **2012**, *439*, 87–95. [CrossRef]
45. Qu, M.; Wang, Y.; Huang, B.; Zhao, Y. Source apportionment of soil heavy metals using robust absolute principal component scores-robust geographically weighted regression (RAPCS-RGWR) receptor model. *Sci. Total Environ.* **2018**, *626*, 203–210. [CrossRef]
46. Li, J.; Teng, Y.; Wu, J.; Jiang, J.; Huang, Y. Source Apportionment of Soil Heavy Metal in the Middle and Upper Reaches of Le'an River based on PMF Model and Geostatistics. *Res. Environ. Sci.* **2019**, *32*, 984–992. [CrossRef]
47. Jiang, Y. Distribution Characteristics and Environmental Health Risk Assessment of Dissolved Trace Elements in River Flow under the Influence of Human Activities. Ph.D. Thesis, Guizhou University, Guiyang, China, 2021. [CrossRef]
48. Ye, P.; Abdugheni, A.; Sun, X.; Halidan, A. E Source analysis and pollution assessment of soil heavy metals in the economic belt on the northern slope of Tianshan. *China Environ. Sci.* **2022**, *42*, 4704–4712. [CrossRef]
49. Cai, L.; Ma, J.; Zhou, Y.; Huang, L.; Dou, L.; Zhang, C.; Fu, S. Multivariate Geostatistics and GIS-based Approach to Study the Spatial Distribution and Sources of Heavy Metals in Agricultural Soil in the Pearl River Delta, China. *Environ. Sci.* **2008**, *29*, 3496–3502. [CrossRef]
50. Ma, J.; Shen, Z.; Zhang, P.; Liu, P.; Liu, J.; Sun, J.; Wang, L. Pollution characteristics and source apportionment of heavy metals in farmland soils around the gangue heap of coal mine based on APCS-MLR and PMF receptor model. *Environ. Sci.* **2023**, *44*, 2192–2203. [CrossRef]
51. Zhao, D.; Jiang, C.; Zhao, Q.; Chen, X.; Li, C.; Zheng, L.; Chen, Y. Distribution characteristics and source apportionment of polycyclic aromatic hydrocarbons in groundwater of a coal mining area based on PMF and PCA-APCS-MLR model. *Earth Environ.* **2022**, *50*, 721–732. [CrossRef]

sustainability

MDPI

*Article*

# The Challenge Posed by Emerging Environmental Contaminants: An Assessment of the Effectiveness of Phenoxyethanol Biological Removal from Groundwater through Mesocosm Experiments

**Laura Ducci [1], Pietro Rizzo [1,*], Antonio Bucci [2], Riccardo Pinardi [1], Pamela Monaco [2] and Fulvio Celico [1]**

[1] Department of Chemistry, Life Sciences and Environmental Sustainability, University of Parma, Parco Area delle Scienze 157/A, 43124 Parma, Italy; laura.ducci@unipr.it (L.D.); riccardo.pinardi@unipr.it (R.P.)

[2] Department of Biosciences and Territory, University of Molise, Contrada Fonte Lappone, 86090 Pesche, Italy; antonio.bucci@unimol.it (A.B.)

* Correspondence: pietro.rizzo@unipr.it; Tel.: +39-0521-905330

**Abstract:** The occurrence of emerging pollutants (EPs) such as pharmaceuticals and personal care products (PPCPs) has raised serious concerns about the possible adverse effects on ecosystem integrity and human health. Wastewater treatment facilities appear to be the main sources of PPCPs released in aquatic environments. This research examines the effectiveness of groundwater microbial community activities to remove phenoxyethanol (Phy-Et), currently exploited as a preservative in many cosmetic formulations at a maximum concentration of 1% but which has shown, at higher levels of exposure, adverse systemic effects on animals. Mesocosm experiments were carried out for 28 days using two different concentrations of the substance (5.2 mg/L and 27.4 mg/L). The main results obtained through chemical and microbiological investigations revealed a significant Phy-Et reduction (≈100% when added at a concentration of 5.2 mg/L and ≈84% when added at a concentration of 27.4 mg/L), demonstrating that some autochthonous microorganisms in the analyzed samples played a "key role" in removing this compound, despite its proven antimicrobial activity. Nevertheless, the decrease in the "natural attenuation" efficacy (≈16%) when using higher concentrations of the chemical suggests the existence of a "dose-dependent effect" of Phy-Et on the process of biodegradation. Biomolecular investigations carried out through next-generation sequencing (NGS) revealed (i) the presence of a significant fraction of hidden microbial diversity to unravel, (ii) variations of the composition and species abundance of the groundwater microbial communities induced by Phy-Et, and (iii) a biodiversity reduction trend correlated to the increase of Phy-Et concentrations. Overall, the preliminary information obtained from the experiments carried out at the laboratory scale appears encouraging, although it reflects only partially the complexity of the phenomena that occur in natural environments and influences their "auto-purification capability". Accordingly, this research paves the way for more in-depth investigations to develop appropriate tools and protocols to evaluate the occurrence and fate of Phy-Et in nature and assess the impact of its release and the effects of long-term exposure (even at low concentrations) on ecosystems and health.

**Keywords:** emerging environmental contaminants; phenoxyethanol; mesocosms; natural attenuation; groundwater microbial communities

**Citation:** Ducci, L.; Rizzo, P.; Bucci, A.; Pinardi, R.; Monaco, P.; Celico, F. The Challenge Posed by Emerging Environmental Contaminants: An Assessment of the Effectiveness of Phenoxyethanol Biological Removal from Groundwater through Mesocosm Experiments. *Sustainability* **2024**, *16*, 2183. https://doi.org/10.3390/su16052183

Academic Editors: Zongjun Gao and Jiutan Liu

Received: 19 January 2024
Revised: 16 February 2024
Accepted: 5 March 2024
Published: 6 March 2024



## 1. Introduction

Environmental pollution threatens human and planetary health, jeopardizing modern societies' sustainability [1]. In 2015, diseases attributable to pollution caused 9 million premature deaths, accounting for 16% of all deaths worldwide, a number three times higher compared to deaths for AIDS, tuberculosis, and malaria combined and 15 times higher compared to deaths for all wars and other forms of violence [2]. Unfortunately, estimates updated to 2019 still confirmed this number of deaths per year [1]. Overall, ambient air pollution and toxic chemical pollution, which are consequences of industrialization and urbanization, have caused many deaths, rising by 7% since 2015 and over 66% since 2000 [1]. Among the various categories of pollutants, there has been a growing interest in the so-called emerging pollutants (EPs) in the last decades. This category includes all the substances that could be pollutants that are not currently covered by controls but that could be covered in the future; this will depend on future ecotoxicity studies and concentrations found [3]. More than 700 compounds grouped in 20 classes of EPs are present in the NORMAN database (www.norman-network.net) [4] and include surfactants, antibiotics and other pharmaceuticals, steroid hormones and other endocrine-disrupting compounds (EDCs), fire retardants, sunscreens, disinfection by-products, new pesticides and pesticide metabolites, naturally occurring algal toxins, etc. [5,6]. In addition, another group of pollutants denoted as "contaminants of emerging concern" (CECs) chemical substances used for long periods in various anthropic sectors and released into the environment in the absence of control, and by-products of these, which are now detected by analysis of soils, surface waters, and groundwater. The term "contaminants of emerging concern" is usually used when very little information is available about the magnitude and frequency of risks posed by this category of substances in the environment and human health [7–9].

Contamination caused by urbanization, fast population growth, agricultural activities, and industrial development has drastically affected water resource quality [10]. The discovery of numerous new compounds in drinking, ground, and surface water has alarmed the public, mainly when human health-based guidelines are unavailable [11,12]. One of the largest sources of EPs is represented by wastewater treatment plants (WWTPs) [13] since wastewater treatments are not designed to treat these compounds and, thus, high amounts are released into the environment through effluents [14].

The main aim of this research was to investigate the effectiveness of biological removal of (Phy-Et) from groundwater through mesocosm experiments to assess the contribution of autochthonous microbial communities in the natural attenuation process of this chemical substance.

Phy-Et is an aromatic glycol ether that occurs naturally in green tea or can be produced in the laboratory due to its commercial importance [15] It is widely found in a large range of leave-on and rinse-off cosmetic products such as moisturizers, hand disinfectants, soaps, serums, sunscreen creams, mascaras, eyeliners, eye balms, and perfumes due to its pure chemical form, pleasant smell, and colorless appearance [16,17]. Other uses include shampoos, shaving creams, ultrasound gels, insect repellents, antiseptics, solvents, anesthetics, cellulose acetate solvents, dyes, and ink and ink manufacture [15]. This compound has a large spectrum of antimicrobial activity and is effective against Gram-negative and Gram-positive bacteria, such as *Escherichia coli* and *Staphylococcus aureus*, and yeasts, such as *Candida* [15,18,19], although it has only a weak inhibitory effect on resident skin flora [20].

The European Scientific Committee on Consumer Safety considers Phy-Et safe for all consumers (including children of all ages) when used as a preservative in cosmetic products up to a concentration of 1%. However, the French National Agency for the Safety of Medicines and Health Products (ANSM) advised against using Phy-Et as a preservative in cosmetic products intended for application to the nappy area of infants and children under three years [20]. Nevertheless, toxicological studies revealed adverse systemic effects in animals at levels of exposure many magnitudes higher (around 200-fold higher) than those to which consumers are exposed [20]. Accordingly, in light of the vast increase in personal care product consumption and their significant release in aquatic ecosystems, it is important to assess the fate of Phy-Et in the environment and the microorganisms' ability to degrade.

## 2. Materials and Methods

### 2.1. Study Area

Groundwater samples for mesocosm experiments have been collected from a piezometer (Pz3) located within the alluvial Parma aquifer (northern Italy; Figure 1). Parma is the second town by population of the Emilia Romagna region and represents a critical contamination source of surface waters and groundwater from leaks in the urban sewerage system [21]. The area is characterized by large industrial districts and intensive agriculture practices both in urban and suburban areas [21]. The piezometer Pz3 is very close to a pipeline of a wastewater treatment plant called "Parma Ovest", in whose waters has been ascertained a continuous and constant leak from the local sewers demonstrated by the detection of fecal coliforms and enterococci [21]. Accordingly, for assessing the biological removal of Phy-Et, groundwater microbial communities have most likely been shaped by a selective pressure caused by frequent contact with sewage and, thus, emerging environmental contaminants.

The aquifer system of the Parma urban area is represented by an alternation of fine-grained (clays and silts) and coarse-grained bodies (gravels and sands) belonging to the alluvial geological units (or synthems) that filled the Po River basin during the Pleistocene–Holocene period (Figure 1). Specifically, the sedimentation processes in this area primarily result from the sedimentary dynamics originating from Apennine streams [22–25]. Groundwater in this aquifer system exhibits a predominant southwest-to-northeast flow at the basin scale, from the Apennine to the central Po Plain and consequently toward the Adriatic Sea [25]. The hydraulic conductivity of the studied aquifer is frequently variable in the three dimensions due to the wide heterogeneity of the geological medium. Zanini et al. [26] calculated a hydraulic conductivity varying from $1.2 \times 10^{-5}$ to $4.9 \times 10^{-5}$ m/s (mean $2.3 \times 10^{-5}$ m/s; median $1.7 \times 10^{-5}$ m/s) in coarse-grained horizons and from $9.3 \times 10^{-9}$ to $1.3 \times 10^{-7}$ m/s (mean $1.6 \times 10^{-7}$ m/s; median $9.7 \times 10^{-8}$ m/s) in fine-grained layers.

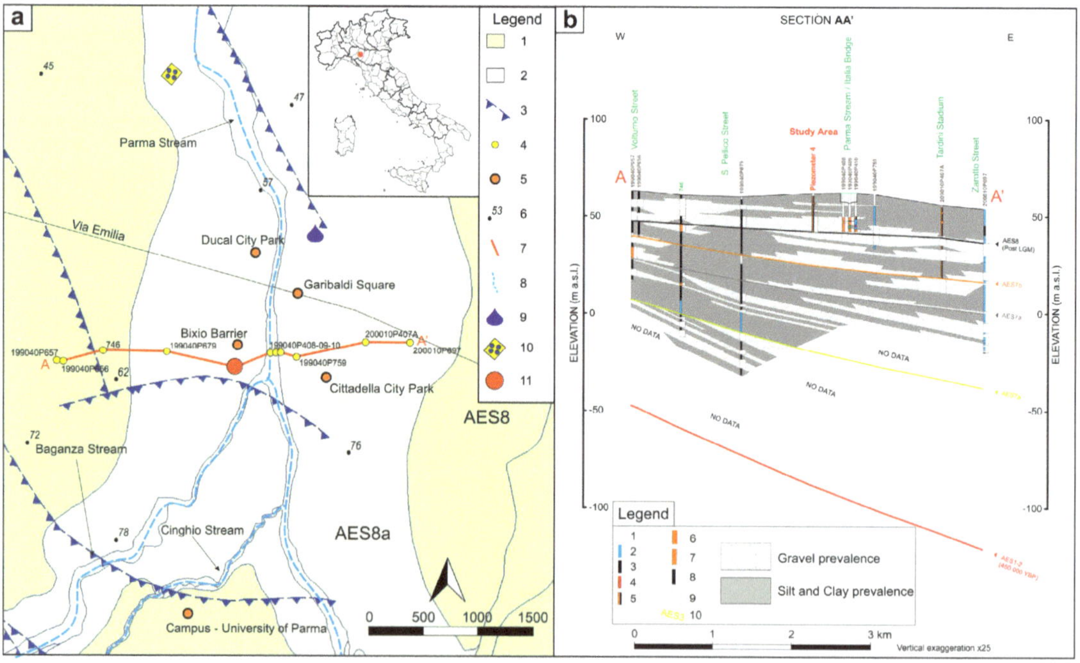


**Figure 1.** (**a**) Geological map and section at basin scale (from Ducci et al., 2022, modified [21]); (1) AES8: Emiliano–Romagnolo superiore synthem—Ravenna subsynthem; (2) AES8a: Emiliano–Romagnolo

superiore synthem—Modena subsynthem; (3) main thrust; (4) borehole; (5) toponym; (6) quoted point; (7) geological section's trace; (8) stream; (9) meteorological station; (10) wastewaters treatment plant "Parma Ovest"; (11) study area; (from Ducci et al., 2022, modified [21]). (**b**) Geological section at basin scale (the trace is shown in (**a**)): (1) fill material, (2) grey/light blue clay and silt; (3) yellow/brown clay and silt; (4) red clay and silt; (5) silt; (6) sand; (7) gravel and sand; (8) gravel and clay/silt; (9) gravel; (10) Code of Geological Unit (sensu Di Dio et al., 2005 [27]): AES1: Emiliano–Romagnolo superiore synthem—Monterlinzana subsynthem; AES2: Emiliano–Romagnolo superiore synthem—Maiatico subsynthem; AES7a: Emiliano–Romagnolo superiore synthem—Villa Verucchio subsynthem—Niviano unit; AES7b: Emiliano–Romagnolo superiore synthem—Villa Verucchio subsynthem—Vignola unit; AES8: Emiliano–Romagnolo synthem—Ravenna subsynthem.

The presence of samples derived from 30 core borings made it possible to reconstruct the local stratigraphy at the site scale [21]. Excluding the anthropogenic material (gravel and bricks) that fill the entire study area about 1 m thick, founded succession can be summarized as follows: (i) the first layer consists of silt and clay with a thickness ranging from 5 to 7 m; (ii) the second layer is primarily composed of gravel and sand, with a thickness varying between 1 and 5 m. Particularly, it exhibits a clear separation into two distinct elements as it extends northward, near Pz3; (iii) the third layer comprises silt and clay, with a thickness ranging from 2 to 3 m; and (iv) the fourth layer represents the shallowest alluvial aquifer at the test site and is composed of gravel and sand (Figure 2). These sedimentary bodies result from ancient alluvial dynamics associated with the paleo-river systems of the Parma and Baganza streams.

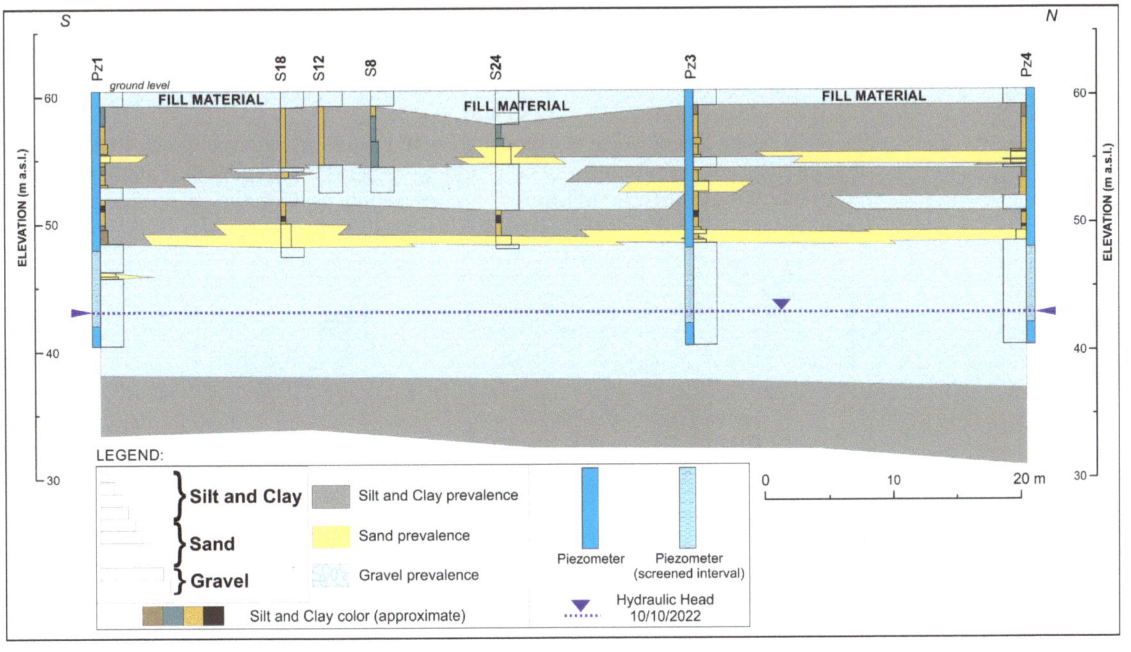


**Figure 2.** Geological section at site scale (from Ducci et al., 2022, modified [21]).

The groundwater flow net reconstructed in the study area is characterized by seasonal variations during the hydrologic year, even though the overall flow is always directed from South to North (Figure 3).

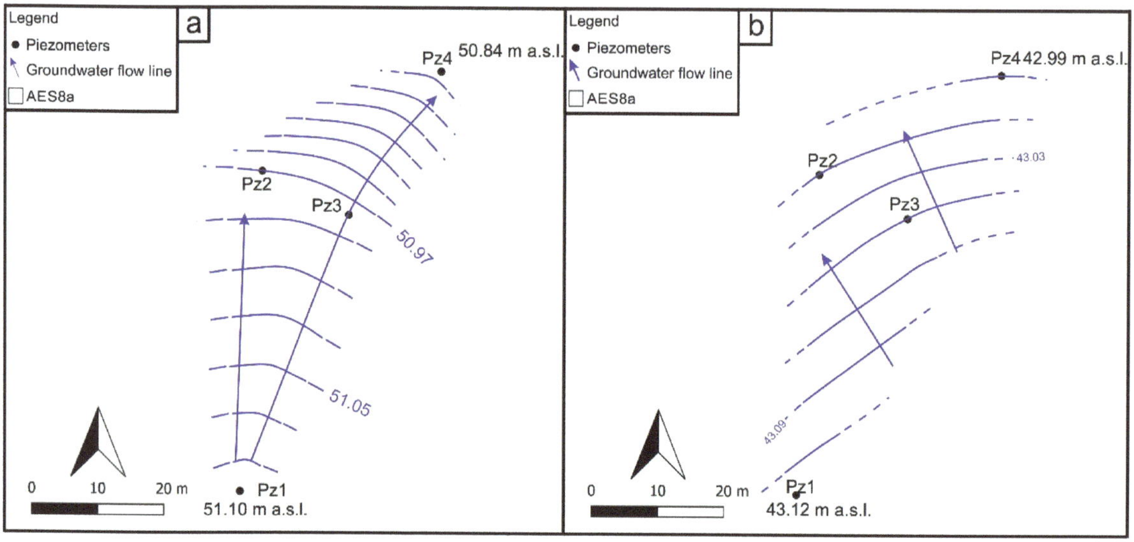


**Figure 3.** (**a**) Groundwater flow net in May 2022; (**b**) groundwater flow net in October 2022 (from Ducci et al., 2022, modified [21]).

## 2.2. *Groundwater Sampling and Mesocosm Setup*

Groundwater samples were collected from the piezometer Pz3, immediately below the hydraulic head, with a sterile bailer in October 2022.

The groundwater temperature has been monitored monthly at the piezometer Pz3 from March 2022 to October 2022 using the multiparameter HANNA probe (model HI9828, HANNA Instruments, Villafranca Padovana, Italy).

In detail, for mesocosm experiments, 16 L of water was sampled. In addition, another 2 L of water was used to carry out biomolecular and chemical analyses aimed at characterizing autochthonous microbial communities and determining the "background contamination" by Phy-Et (Figure 4a).

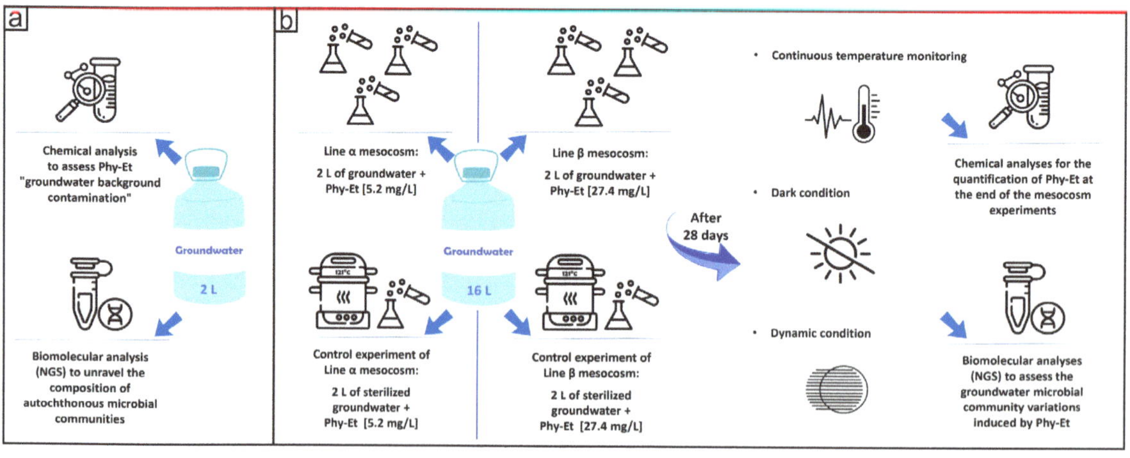


**Figure 4.** (**a**) Groundwater microbial community characterization and assessment of Phy-Et background contamination; (**b**) experimental protocol setup of mesocosms.

Mesocosms were prepared by transferring 2 L of groundwater in sterile glass Erlenmeyer flasks and by adding Phy-Et (BiOrigins, Fordingbridge, United Kingdom) at two different concentrations: 5.2 mg/L and 27.4 mg/L. The experiments carried out with Phy-Et at a concentration of 5.2 mg/L will be henceforth also indicated as Line α, whereas the ones carried out with Phy-Et at a concentration of 27.4 mg/L will be reported as Line β. For each Line, three different mesocosms were set up (Figure 4b). The Phy-Et concentrations arbitrarily chosen for the experiments were more than 20 and 100 times higher than the instrumental detection limit for this compound (0.2 mg/L).

Afterward, the mesocosms were sealed and incubated at room temperature for 28 days, in the dark, with agitation (140 rpm), using the OrbitalShaker—OHAUS® (Model: SHHD6825DG, S/N: 210331001, Nänikon, Zürich, Switzerland).

In addition, the experimental plan has included two controls, consisting of sterilized groundwater samples added with Phy-Et at the concentrations of 5.2 mg/L and 27.4 mg/L, to highlight a possible removal of this compound from groundwater not attributable to the microbial activities (Figure 4b).

### *2.3. Temperature Monitoring and Chemical Analyses for Phy-Et Detection*

For the entire duration of the mesocosm experiments (28 days), the temperature was monitored every 5 min through an immersion thermometer with a data logger (Elitech model: GSP-6 Serial No. EFG218100242, London, UK).

Chemical analyses for Phy-Et detection were performed both on the original groundwater and samples collected at the end of the mesocosm experiments. A total of 50 mL of water was poured into amber glass bottles and transported to the laboratory in a refrigerated box. The analyses were performed at Abich S.r.l., an Italian company with UNI EN ISO 9001:2015 [28] and UNI CEI EN ISO 13485:2016 [29] accreditations. According to an accredited in-house method, they were based on the ultra-high-performance liquid chromatographic-diode array detection method (UHPLC/DAD) (Rif. HPLC n.0018).

### *2.4. Next-Generation Sequencing (NGS) for Bacterial Community Analyses*

Next-generation sequencing (NGS) analyses were carried out firstly to characterize autochthonous groundwater microbial communities in groundwater samples collected at the piezometer (Pz3) and then to assess those variations in their composition after the addition of Phy-Et to the mesocosms at two different concentrations (5.2 mg/L and 27.4 mg/L).

Groundwater samples (1.95 L) were filtered through sterile mixed esters of cellulose filters (S-Pak™ Membrane Filters, 47 mm diameter, 0.22 μm pore size, Millipore Corporation, Billerica, MA, USA) within 24 h from the collection. Bacterial DNA extraction from filters was performed using the commercial kit FastDNA SPIN Kit for soil (MP Biomedicals, LLC, Solon, OH, USA) and FastPrep® Instrument (MP Biomedicals, LLC, Solon, OH, USA). After the extraction, DNA integrity and quantity were evaluated using electrophoresis in 0.8% agarose gel containing 1 μg/mL of Gel-Red™ (Biotium, Inc., Fremont, CA, USA). Next-generation sequencing (NGS) technologies generated the bacterial community profiles in the samples at the Genprobio Srl Laboratory following the protocol reported by Ducci et al. [21].

The 16S rRNA gene sequences obtained in this study were deposited in the National Center for Biotechnology Information (NCBI) Sequence Read Archive under the accession number PRJNA977861.

## 3. Results

### *3.1. Microbiological and Physico-Chemical Characterization of Groundwater*

Autochthonous bacterial communities of groundwater collected from the piezometer Pz3 were characterized by a predominance of *Proteobacteria* (relative abundance 64.14%), followed by *Bacteroidetes* (relative abundance 8.34%), *Chloroflexi* (relative abundance 5.23%), *Verrucomicrobia* (relative abundance 3.91%), *Actinobacteria* (relative abundance 3.38%), *Aci-*

*dobacteria* (relative abundance 3.07%), and other phyla, which were detected with relative abundance values lower than 3% (Figure 5). At the genus level, the five major taxa were *Methylotenera*, *Pseudomonas*, unclassified microorganisms of the *Burkholderiaceae* family, *Massilia*, and *Cavicella*, which were retrieved at percentages of 8.71%, 7.27%, 4.36%, 2.18%, and 2.00%, respectively (Figure 5).

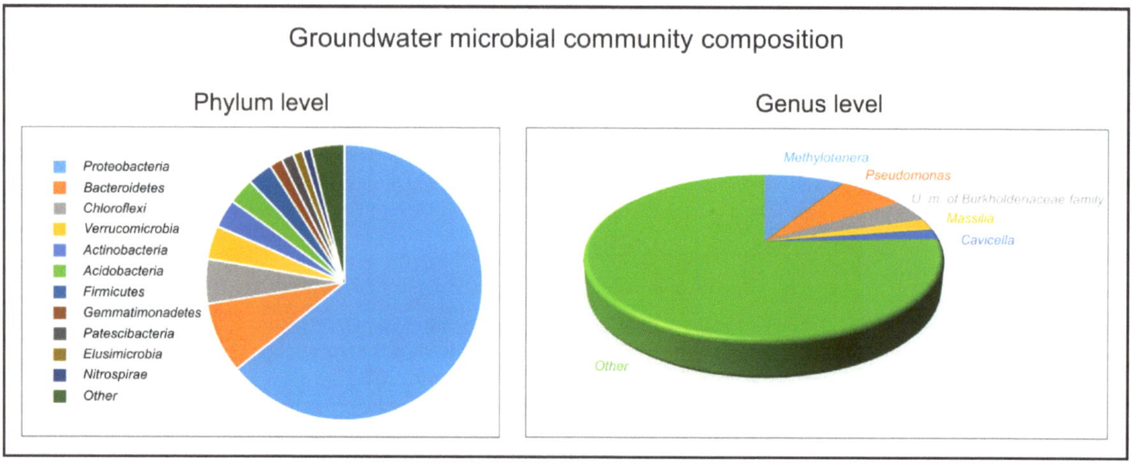


**Figure 5.** Microbial community composition at phylum (**on the left**) and genus (**on the right**) levels of groundwater collected from the piezometer Pz3. Taxa with relative abundance values below 1% and 2% for phylum and genus levels, respectively, are labeled "Other". U. m. refers to unclassified microorganisms.

Chemical analyses for determining Phy-Et revealed that its concentration was lower than the instrumental detection limit (0.2 mg/L). Groundwater temperature during the observation period ranged from 13.2 °C to 16.1 °C, with an average of 14.4 °C.

### *3.2. Results of Physico-Chemical and Microbiological Analyses for the Mesocosm Experiments*

At the end of the 28-day experimental period, water samples were collected from the mesocosms in which Phy-Et was initially added at two different concentrations and were subjected to chemical analyses to assess whether there had been a reduction.

The same analyses were also performed on the two controls, consisting of sterilized groundwater added with Phy-Et, to detect any losses of this compound from the systems, not due to biodegradation processes.

The results obtained clearly showed the removal of the compound of about 100% in Line $\alpha$ mesocosms (from 5.2 mg/L to below the detection limit [0.2 mg/L]). On the other hand, in the Line $\beta$ mesocosms, a reduction of about 84% was detected (from 27.4 mg/L to concentrations ranging between $3.21 \pm 0.01$ mg/L and $5.24 \pm 0.02$ mg/L). No significant variations in the concentration of the chemical were recorded in the controls, evidence which demonstrates that microorganisms were the "key player" in the degradation process.

The continuous monitoring of water temperature in the mesocosms allowed the detection of fluctuations varying from a minimum of 17.9 °C to a maximum of 24.2 °C, with an average of 21.0 °C.

On the whole, the results of the biomolecular investigations carried out to unravel the composition of bacterial communities exposed to the two concentrations of Phy-Et in the mesocosm experiments revealed, at the phylum level, the predominance of *Proteobacteria*, followed by *Bacteroidetes*, *Verrucomicrobia*, *Acidobacteria*, *Actinobacteria*, and *Patescibacteria* (Figure 6).

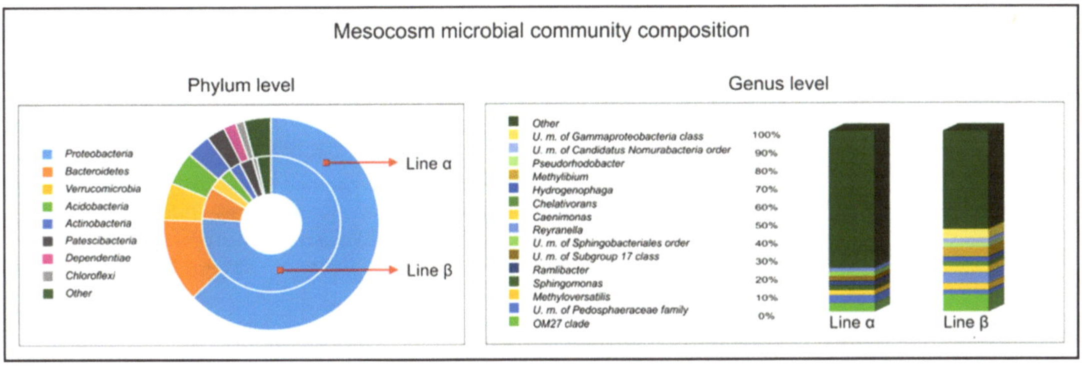


**Figure 6.** Mesocosm microbial community composition at phylum (**on the left**) and genus (**on the right**) levels of Lines α and β. Taxa with relative abundance values below 1% and 2% for phylum and genus levels, respectively, are labeled "Other". U. m. refers to unclassified microorganisms. The relative abundance of each taxon has been calculated as the mean of values of the triplicate experiments.

However, differences emerged when comparing the percentages of these taxa with the relative abundance of the same phyla retrieved in the original groundwater sample. For example, the abundance of *Proteobacteria* went from 64.14% to 62.92% (on average) in Line α mesocosms and 76.36% (on average) in Line β mesocosms. Moreover, the same goes for the other phyla, as reported in Table 1.

**Table 1.** Variations of the relative abundance average values of the main phyla in groundwater samples.

| Taxonomy | Pz3 | Line α | Line β |
|---|---|---|---|
| *Proteobacteria* | 64.14% | 62.92% | 76.36% |
| *Bacteroidetes* | 8.34% | 12.71% | 7.51% |
| *Chloroflexi* | 5.23% | 1.27% | 1.02% |
| *Verrucomicrobia* | 3.91% | 5.68% | 3.03% |
| *Actinobacteria* | 3.38% | 3.69% | 2.70% |
| *Acidobacteria* | 3.07% | 5.08% | 2.92% |
| *Firmicutes* | 2.90% | 0.22% | 0.32% |
| *Gemmatimonadetes* | 1.53% | 0.65% | 0.23% |
| *Patescibacteria* | 1.49% | 2.78% | 3.08% |

In addition, the phylum *Chloroflexi*, which was among the major three phyla in the original groundwater sample (5.23%), showed a decrease in the mesocosms to percentages of 1.27% and 1.02% when Phy-Et was added at 5.2 mg/L and 27.4 mg/L, respectively.

When analyzing the mesocosm microbial communities at the genus taxonomic level (Figure 6), the top six taxa retrieved in the Line α experiments were: microorganisms of the clade *OM27* (average relative abundance 4.77%), unclassified microorganisms of *Pedosphaeraceae* family (average relative abundance 4.46%), *Methyloversatilis* (average relative abundance 2.72%), *Sphingomonas* (average relative abundance 2.64%), *Ramlibacter* (average relative abundance 2.61%), and unclassified microorganisms of Subgroup 17 class (*Acidobacteria* phylum; average relative abundance 2.51%). On the other hand, the top six genera found in the Line β mesocosms were microorganisms of the clade *OM27* (average relative abundance 9.49%), *Reyranella* (average relative abundance 6.14%), unclassified microorganisms of *Gammaproteobacteria* class (average relative abundance 5.35%), *Methylibium* (average relative abundance 5.18%), *Methyloversatilis* (average relative abundance 3.57%), and *Caenimonas* (average relative abundance 3.47%).

These results suggest that Phy-Et has modified the composition and species abundance of the groundwater microbial communities, most likely inhibiting the growth of cells sensitive to its action and favoring the survival of those tolerant or able to metabolize it.

In addition, another important aspect that emerged is the presence of diverse unclassified microorganisms at the genus level, accounting on average for 41.58% and 29.14% in Line α Line β mesocosms, respectively. This observation demonstrates that a large portion of microbial diversity in the analyzed systems remained hidden.

The rarefaction analysis, a measure used to estimate the alpha diversity in samples and gauge whether sequencing efforts captured the microbial diversity (Supplementary Materials Figure S1), revealed a biodiversity reduction trend correlated to the increase of Phy-Et concentrations.

## 4. Discussion

On a global scale, pharmaceuticals and personal care products (PPCPs) have risen in the last decade due to advances in research and development, increased world population, and accessibility to healthcare [30,31]. Given their known or suspected adverse ecological or human health effects, PPCPs are considered two of the most abundant classes of emerging contaminants in the environment [32,33]. Among the most dramatic effects on living organisms are the feminization of fishes [10,34], alterations in the reproduction and development of some fish species caused by very low concentrations of environmental estrogens [35], and increased antibiotic resistance of pathogenic microorganisms [36]. Their presence in aquatic and terrestrial ecosystems is associated with their occurrence at influents and effluents of wastewater treatment plants [33], whose removal capacity strongly depends on the features of the pollutant and the technologies used for wastewater treatments.

In this research, the authors have focused on Phy-Et, a compound present as a preservative in many cosmetics. According to the currently available safety data, this substance is considered safe when used at a concentration of up to 1%, is a rare sensitizer, and is one of the most well-tolerated preservatives [20]. Nevertheless, animal studies revealed different adverse systemic effects at higher levels of exposure than consumers when using Phy-Et-containing products [20]. In addition, other authors who analyzed the toxic effects of this compound in an eukaryotic model organism, *Allium cepa*, claimed that, in the light of their data, "it should be ensured that the use of Phy-Et should be limited, if not preferred, or if it is absolutely necessary to use it in doses that do not have toxic effects on the organisms" [15]. Accordingly, the main aim of this work was to analyze the capacity of microbial communities to remove Phy-Et (supplied at two different concentrations) from groundwater through mesocosm experiments designed to represent and simulate, on a laboratory scale, the conditions of shallow groundwater flowing in the daily heterothermic zone, characterized by significant and frequent temperature fluctuations closely tied to atmospheric conditions. Additionally, in the piezometer Pz3, the recorded temperature values ranged from 13.2 °C to 16.1 °C, with an average value of 14.4 °C whereas the temperature of the mesocosms ranged from a minimum of 17.9 °C to a maximum of 24.2 °C, with an average value of 21.0 °C. Even though the water temperatures in the lab-scale experiments were higher than those measured in the field during the observation period, it is worth noting that many of the genera detected in the mesocosms include species characterized by a wide range of growth temperatures. Accordingly, the authors are convinced that the observed microbial community variations, induced by the presence of Phy-Et, nevertheless provide an accurate representation of the events that would occur in nature.

However, in a wider context, similar temperatures to those measured in the mesocosms have also been found in other hydrogeological systems [37], with values reaching 24.0 °C during summer. In a future perspective, further studies will be carried out with the use of a refrigerated thermostat (e.g., [38]) to control the temperatures to which these systems are exposed, to subject the autochthonous microbial communities to the same fluctuations observed during the hydrological year in the field.

In the frame time considered, autochthonous microorganisms (i) were able to reduce by ≈100% and ≈84% the amount of Phy-Et present in the systems with higher removal effectiveness when the concentration of the contaminant was lower (5.2 mg/L), and (ii) represented the major contributors to the natural attenuation process. Additionally, these particularly encouraging data also suggest the existence of a "dose-dependent effect" of Phy-Et on the process of biodegradation, given the observed decrease of the "natural attenuation efficacy" (≈16%) with the increasing contaminant concentrations. The analysis of bacterial communities through NGS technologies revealed drastic changes in their composition determined by the presence of Phy-Et, evidenced by a reduction in biodiversity linked to increasing Phy-Et concentrations. It is known that Phy-Et exerts its antimicrobial activity by uncoupling oxidative phosphorylation from respiration and by competitively inhibiting malate dehydrogenase [39]. Moreover, it also acts as a bactericidal agent by increasing the permeability of the cell membrane to potassium ions and exerts a direct inhibitory effect on microbial DNA and RNA synthesis [39].

Therefore, it is not surprising that the addition of the contaminant has shaped microbial communities so that the more sensitive microorganisms were under-represented in the mesocosm experiments compared to the original groundwater.

On the other hand, the presence of microorganisms of the clade *OM27* as dominant components of the "contaminated" samples is noteworthy. *OM27* clade is a cluster of unculturable bacteria, phylogenetically related to the predatory deltaproteobacterial genus *Bdellovibrio* [40,41], which have a geographically wide distribution [40] but of which very little is known [42]. Their potential predatory "attitude" could confer an advantage under the selective pressure of Phy-Et. Conversely, the biodegradation potentials of some members belonging to the bacterial genera such as, for example, *Methyloversatilis*, *Methylibium*, *Ramlibacter*, *Reyranella*, and *Sphingomonas*, are known, and this could explain why they were found as the main representatives in the mesocosm experiments.

Interestingly, there is a large portion of microbial biodiversity that remains hidden, as demonstrated by the relatively high percentages of unclassified microorganisms at the genus level, and that should be uncovered through the development of appropriate research tools, given the importance of acquiring knowledge on the main processes underlying the bacterial biodegradation of contaminants in nature.

Several thousands of PPCPs are produced globally every year, and their release into the environment remains an unavoidable by-product of a modernized lifestyle [43–46]. PPCPs may enter the environment as components of human or animal waste after incomplete absorption and excretion from the body or may result from medical, industrial, agricultural, or household waste emissions [47–50]. Trace amounts of PPCP-related compounds have been found in waste, aquatic ecosystems, or finished drinking water [48,51–58], suggesting widespread contamination and highlighting the importance of monitoring these compounds once they are released into the environment.

This research has demonstrated that microorganisms have a high potential for removing the largely used preservative Phy-Et from groundwater, at least in an artificial laboratory system. However, knowledge about the extent of their degradative capabilities in natural contexts still remains scarce. Accordingly, this topic should be further deepened, for this and many other contaminants of emerging concern, through the improvement of tools and protocols to evaluate their occurrence and fate in nature (for example, by promoting the execution of several large-scale monitoring campaigns in different environmental contexts and developing of different analytical methods for CEC detection, together with extensive analyses aimed at unraveling microbe potentiality for biodegradation, based on traditional and molecular microbiological methods) and the assessment of the impact of their release and the effects of long-term exposure (even at low concentrations) on ecosystems and health.

## 5. Conclusions

In light of the extensive global production of pharmaceuticals and personal care products (PPCPs), their inevitable release into the environment represents a salient implication of contemporary lifestyles. Consequently, persistent endeavors are required to evaluate these compounds' occurrence and subsequent fate in the natural environment and understand the potential long-term effects on ecosystems and human health.

This research's focal point revolved around examining Phy-Et, a common preservative in cosmetic products. Although considered safe at concentrations up to 1% based on existing data, studies have raised concerns about its adverse systemic effects at higher exposure levels.

This work sought to understand the potential of microbial communities in removing Phy-Et from groundwater through specifically designed mesocosm experiments, which aimed to simulate shallow groundwater conditions.

While this research has demonstrated encouraging results, including significant reductions in Phy-Et concentrations and the pivotal role played by autochthonous microorganisms, it has also revealed the existence of a dose-dependent effect. The composition of bacterial communities was notably altered by Phy-Et, which determined a decrease in biodiversity. These findings underscore the significance of Phy-Et's impact on microbial ecosystems.

In addition, the presence of both unculturable microorganisms, such as the *OM27* clade, and a significant fraction of "hidden" microbial diversity to unravel raises intriguing possibilities and challenges for future explorations.

**Supplementary Materials:** The following supporting information can be downloaded at: https://www.mdpi.com/article/10.3390/su16052183/s1, Figure S1: Rarefaction curves of the samples. The alpha diversity plots were obtained by using the Shannon index.

**Author Contributions:** Conceptualization, L.D., P.R. and F.C.; methodology, L.D. and P.R.; software, R.P. and P.M.; validation, A.B. and F.C.; formal analysis, L.D., P.R., A.B., R.P. and F.C.; investigation, L.D. and P.R.; resources, F.C.; data curation, L.D., P.R., A.B., R.P. and P.M.; writing—original draft preparation, L.D., P.R., A.B., R.P. and F.C.; writing—review and editing, L.D., P.R., A.B. and F.C.; visualization, L.D., P.R., R.P. and P.M.; supervision, A.B. and F.C.; project administration, F.C.; funding acquisition, F.C. All authors have read and agreed to the published version of the manuscript.

**Funding:** This research received no external funding.

**Institutional Review Board Statement:** Not applicable.

**Informed Consent Statement:** Not applicable.

**Data Availability Statement:** The data presented in this study are available on request from the corresponding author.

**Acknowledgments:** This research benefited from the equipment and framework of the COMP-R Initiative, funded by the "Departments of Excellence" program of the Italian Ministry for University and Research (MUR, 2023–2027). We thank Nicola Bianco for contributing to executing some technical-operational activities. We are grateful to the reviewers for their constructive comments and valuable suggestions.

**Conflicts of Interest:** The authors declare that they have no known competing financial interests or personal relationships that could have appeared to influence the work reported in this paper.

## References

1. Fuller, R.; Landrigan, P.J.; Balakrishnan, K.; Bathan, G.; Bose-O'Reilly, S.; Brauer, M.; Caravanos, J.; Chiles, T.; Cohen, A.; Corra, L.; et al. Pollution and health: A progress update. *Lancet Planet. Health* **2022**, *6*, e535–e547. [CrossRef] [PubMed]
2. Landrigan, P.J.; Fuller, R.; Acosta, N.J.; Adeyi, O.; Arnold, R.; Basu, N.; Baldé, A.B.; Bertollini, R.; Bose-O'Reilly, S.; Boufford, J.I.; et al. The Lancet Commission on pollution and health. *Lancet* **2018**, *391*, 462–512. [CrossRef] [PubMed]
3. Yadav, D.; Rangabhashiyam, S.; Verma, P.; Singh, P.; Devi, P.; Kumar, P.; Hussain, C.M.; Gaurav, G.K.; Kumar, K.S. Environmental and health impacts of contaminants of emerging concerns: Recent treatment challenges and approaches. *Chemosphere* **2021**, *272*, 129492. [CrossRef]

4. Vasilachi, I.C.; Asiminicesei, D.M.; Fertu, D.I.; Gavrilescu, M. Occurrence and fate of emerging pollutants in water environment and options for their removal. *Water* **2021**, *13*, 181. [CrossRef]
5. Geissen, V.; Mol, H.; Klumpp, E.; Umlauf, G.; Nadal, M.; van der Ploeg, M.; van de Zee, S.E.A.T.M.; Ritsema, C.J. Emerging pollutants in the environment: A challenge for water resource management. *Int. Soil Water Conserv. Res.* **2015**, *3*, 57–65. [CrossRef]
6. Snow, D.D.; Cassada, D.A.; Larsen, M.L.; Mware, N.A.; Li, X.; D'Alessio, M.; Zhang, Y.; Sallach, J.B. Detection, occurrence and fate of emerging contaminants in agricultural environments. *Water Environ. Res.* **2017**, *89*, 897–920. [CrossRef] [PubMed]
7. Gomes, A.R.; Justino, C.; Rocha-Santos, T.; Freitas, A.C.; Duarte, A.C.; Pereira, R. Review of the ecotoxicological effects of emerging contaminants to soil biota. *J. Environ. Sci. Health Part A* **2017**, *52*, 992–1007. [CrossRef]
8. Risk Assessment Guidance for Superfund Volume 1. Human Health Evaluation Manual (Part E, Supplemental Guidance for Dermal Risk Assessment), EPA/540/R/99/005 Office of Superfund Remediation and Technology Innovation. U.S. Environmental Protection Agency: Washington, DC, USA. Available online: https://www.epa.gov/sites/production/files/2015-09/documents/part_e_final_revision_10-03-07.pdf (accessed on 12 October 2020).
9. United States Environmental Protection Agency (USEPA). Contaminants of Emerging Concern Including Pharmaceuticals and Personal Care Products. Washington, DC, USA. Available online: https://www.epa.gov/wqc/contaminants-emerging-concern-including-Pharmaceuticals-and-personal-care-products (accessed on 12 October 2020).
10. Patel, M.; Kumar, R.; Kishor, K.; Mlsna, T.; Pittman, C.U., Jr.; Mohan, D. Pharmaceuticals of emerging concern in aquatic systems: Chemistry, occurrence, effects, and removal methods. *Chem. Rev.* **2019**, *119*, 3510–3673. [CrossRef]
11. Baken, K.A.; Sjerps, R.M.; Schriks, M.; van Wezel, A.P. Toxicological risk assessment and prioritization of drinking water relevant contaminants of emerging concern. *Environ. Int.* **2018**, *118*, 293–303. [CrossRef]
12. Khatib, J.M.; Baydoun, S.; ElKordi, A.A. Water pollution and urbanisation trends in Lebanon: Litani River basin case study. In *Urban Pollution: Science and Management*; John Wiley & Sons Ltd.: Hoboken, NJ, USA, 2018; pp. 397–415. [CrossRef]
13. Petrie, B.; Barden, R.; Kasprzyk-Hordern, B. A review on emerging contaminants in wastewaters and the environment: Current knowledge, understudied areas and recommendations for future monitoring. *Water Res.* **2015**, *72*, 3–27. [CrossRef]
14. Park, J.; Yamashita, N.; Park, C.; Shimono, T.; Takeuchi, D.M.; Tanaka, H. Removal characteristics of Pharmaceuticals and Personal Care Products: Comparison between membrane bioreactor and various biological treatment processes. *Chemosphere* **2017**, *179*, 347–358. [CrossRef]
15. Akgündüz, M.Ç.; Çavuşoğlu, K.; Yalçın, E. The potential risk assessment of Phenoxyethanol with a versatile model system. *Sci. Rep.* **2020**, *10*, 1209. [CrossRef]
16. Wang, J.; Liu, Y.; Kam, W.R.; Li, Y.; Sullivan, D.A. Toxicity of the cosmetic preservatives parabens, phenoxyethanol and chlorphenesin on human meibomian gland epithelial cells. *Exp. Eye Res.* **2020**, *196*, 108057. [CrossRef]
17. Panico, A.; Serio, F.; Bagordo, F.; Grassi, T.; Idolo, A.; De Giorgi, M.; Guido, M.; Congedo, M.; De Donno, A. Skin safety and health prevention: An overview of chemicals in cosmetic products. *J. Prev. Med. Hyg.* **2019**, *60*, E50.
18. Gilbert, P.; Beveridge, E.G.; Crone, P.B. The action of Phenoxyethanol upon respiration and dehydrogenase enzyme systems in Escherichia coli. *J. Pharm. Pharmacol. Suppl.* **1976**, *28*, 51.
19. Lowe, I.; Southern, J. The antimicrobial activity of Phenoxyethanol in vaccines. *Lett. Appl. Microbiol.* **1994**, *18*, 115–116. [CrossRef] [PubMed]
20. Dréno, B.; Zuberbier, T.; Gelmetti, C.; Gontijo, G.; Marinovich, M. Safety review of Phenoxyethanol when used as a preservative in cosmetics. *J. Eur. Acad. Dermatol. Venereol.* **2019**, *33*, 15–24. [CrossRef]
21. Ducci, L.; Rizzo, P.; Pinardi, R.; Solfrini, A.; Maggiali, A.; Pizzati, M.; Balsamo, F.; Celico, F. What Is the Impact of Leaky Sewers on Groundwater Contamination in Urban Semi-Confined Aquifers? A Test Study Related to Fecal Matter and Personal Care Products (PCPs). *Hydrology* **2022**, *10*, 3. [CrossRef]
22. Di Dio, G.; Lasagna, S.; Preti, D.; Sagne, M. Carta geologica dei depositi quaternari della provincia di Parma. *Il Quat.* **1997**, *10*, 443–450.
23. Di Dio, G. *Applicazione di Concetti e Metodi della Stratigrafia Fisica alla Ricerca di Risorse Idriche nel Sottosuolo della Pianura Emiliano-Romagnola*; Serie 3a; Giornale di Geologia: Bologna, Italy, 1998; Volume 60, pp. 35–39.
24. Regione Emilia-Romagna; ENI-AGIP. *Riserve Idriche Sotterranee Della Regione Emilia-Romagna*; Cura di, A., Di Dio, G.M., Eds.; Regione Emilia-Romagna, ENI Agip Divisione Esplorazione e Produzione; S.EL.CA: Firenze, Italy, 1998; p. 120.
25. Pinardi, R.; Feo, A.; Ruffini, A.; Celico, F. Purpose-Designed Hydrogeological Maps for Wide Interconnected Surface–Groundwater Systems: The Test Example of Parma Alluvial Aquifer and Taro River Basin (Northern Italy). *Hydrology* **2023**, *10*, 127. [CrossRef]
26. Zanini, A.; Ghirardi, M.; Emiliani, R. A multidisciplinary approach to evaluate the effectiveness of natural attenuation at a contaminated site. *Hydrology* **2021**, *8*, 101. [CrossRef]
27. Di Dio, G.; Martini, A.; Lasagna, S.; Zanzucchi, G. *Illustrative Notes of the Geological Map of Italy at 1:50,000 Scale Sheet 199 Parma Sud. Geological, Seismic, and Soil Service of the Emilia-Romagna Region, APAT-Geological Service of Italy*; S.EL.CA: Florence, Italy, 2005.
28. *UNI EN ISO 9001:2015*; Quality Management Systems Requirements. International Organization for Standardization: Geneva, Switzerland, 2015.
29. *UNI EN ISO 13485:2016*; Medical Devices—Quality Management Systems—Requirements for Regulatory Purposes. International Organization for Standardization: Geneva, Switzerland, 2016.
30. Van Boeckel, T.P.; Gandra, S.; Ashok, A.; Caudron, Q.; Grenfell, B.T.; Levin, S.A.; Laxminarayan, R. Global antibiotic consumption 2000 to 2010: An analysis of national Pharmaceutical sales data. *Lancet Infect. Dis.* **2014**, *14*, 742–750. [CrossRef]

31. Adeleye, A.S.; Xue, J.; Zhao, Y.; Taylor, A.A.; Zenobio, J.E.; Sun, Y.; Han, Z.; Salawu, O.A.; Zhu, Y. Abundance, fate, and effects of Pharmaceuticals and Personal Care Products in aquatic environments. *J. Hazard. Mater.* **2022**, *424*, 127284. [CrossRef]
32. Khan, S.; Naushad, M.; Govarthanan, M.; Iqbal, J.; Alfadul, S.M. Emerging contaminants of high concern for the environment: Current trends and future research. *Environ. Res.* **2022**, *207*, 112609. [CrossRef]
33. Lozano, I.; Pérez-Guzmán, C.J.; Mora, A.; Mahlknecht, J.; Aguilar, C.L.; Cervantes-Avilés, P. Pharmaceuticals and Personal Care Products in water streams: Occurrence, detection, and removal by electrochemical advanced oxidation processes. *Sci. Total Environ.* **2022**, *827*, 154348. [CrossRef]
34. Jarque, S.; Quirós, L.; Grimalt, J.O.; Gallego, E.; Catalan, J.; Lackner, R.; Piña, B. Background fish feminization effects in European remote sites. *Sci. Rep.* **2015**, *5*, 11292. [CrossRef]
35. Kalia, V.C. Pharmaceutical and personal care product contamination: A global scenario. In *Pharmaceuticals and Personal Care Products: Waste Management and Treatment Technology*; Butterworth-Heinemann: Oxford, UK, 2019; pp. 27–61. [CrossRef]
36. Zhang, X.; Yan, S.; Chen, J.; Tyagi, R.D.; Li, J. Physical, chemical, and biological impact (hazard) of hospital wastewater on environment: Presence of Pharmaceuticals, pathogens, and antibiotic-resistance genes. In *Current Developments in Biotechnology and Bioengineering*; Elsevier: Amsterdam, The Netherlands, 2020; pp. 79–102. [CrossRef]
37. Segadelli, S.; Vescovi, P.; Ogata, K.; Chelli, A.; Zanini, A.; Boschetti, T.; Petrella, E.; Toscani, L.; Gargini, A.; Celico, F. A conceptual hydrogeological model of ophiolitic aquifers (serpentinised peridotite): The test example of Mt. Prinzera (North. Italy). *Hydrol. Process.* **2017**, *31*, 1058–1073. [CrossRef]
38. Allocca, V.; Celico, F.; Petrella, E.; Marzullo, G.; Naclerio, G. The role of land use and environmental factors on microbial pollution of mountainous limestone aquifers. *Environ. Geol.* **2008**, *55*, 277–283. [CrossRef]
39. CIR (Cosmetic Ingredient Review). Final report on the safety assessment of Phenoxyethanol. *J. Am. Coll. Toxicol.* **1990**, *9*, 259–278. [CrossRef]
40. Orsi, W.D.; Smith, J.M.; Liu, S.; Liu, Z.; Sakamoto, C.M.; Wilken, S.; Poirier, C.; Richards, T.A.; Keeling, P.J.; Worden, A.Z.; et al. Diverse, uncultivated bacteria and archaea underlying the cycling of dissolved protein in the ocean. *ISME J.* **2016**, *10*, 2158–2173. [CrossRef] [PubMed]
41. Gorokhova, E.; Motiei, A.; El-Shehawy, R. Understanding biofilm formation in ecotoxicological assays with natural and anthropogenic particulates. *Front. Microbiol.* **2021**, *12*, 632947. [CrossRef]
42. Yilmaz, P.; Yarza, P.; Rapp, J.Z.; Glöckner, F.O. Expanding the world of marine bacterial and archaeal clades. *Front. Microbiol.* **2016**, *6*, 1524. [CrossRef]
43. Caldwell, D.J.; Mastrocco, F.; Margiotta-Casaluci, L.; Brooks, B.W. An integrated approach for prioritizing Pharmaceuticals found in the environment for risk assessment, monitoring and advanced research. *Chemosphere* **2014**, *115*, 4–12. [CrossRef]
44. Fatta-Kassinos, D.; Meric, S.; Nikolaou, A. Pharmaceutical residues in environmental waters and wastewater: Current state of knowledge and future research. *Anal. Bioanal. Chem.* **2011**, *399*, 251–275. [CrossRef] [PubMed]
45. Khetan, S.K.; Collins, T.J. Human Pharmaceuticals in the aquatic environment: A challenge to green chemistry. *Chem. Rev.* **2007**, *107*, 2319–2364. [CrossRef] [PubMed]
46. Tran, N.; Drogui, P.; Brar, S.K. Sonochemical techniques to degrade Pharmaceutical organic pollutants. *Environ. Chem. Lett.* **2015**, *13*, 251–268. [CrossRef]
47. Li, W.C. Occurrence, sources, and fate of Pharmaceuticals in aquatic environment and soil. *Environ. Pollut.* **2014**, *187*, 193–201. [CrossRef] [PubMed]
48. Michael, I.; Rizzo, L.; McArdell, C.S.; Manaia, C.M.; Merlin, C.; Schwartz, T.; Dagot, C.; Fatta-Kassinos, D. Urban wastewater treatment plants as hotspots for the release of antibiotics in the environment: A review. *Water Res.* **2013**, *47*, 957–995. [CrossRef] [PubMed]
49. Sharma, V.K.; Anquandah, G.A.; Nesnas, N. Kinetics of the oxidation of endocrine disruptor nonylphenol by ferrate (VI). *Environ. Chem. Lett.* **2009**, *7*, 115–119. [CrossRef]
50. Taylor, D.; Senac, T. Human Pharmaceutical products in the environment—The "problem" in perspective. *Chemosphere* **2014**, *115*, 95–99. [CrossRef] [PubMed]
51. Anquandah, G.A.; Sharma, V.K.; Panditi, V.R.; Gardinali, P.R.; Kim, H.; Oturan, M.A. Ferrate (VI) oxidation of propranolol: Kinetics and products. *Chemosphere* **2013**, *91*, 105–109. [CrossRef] [PubMed]
52. Kuzmanović, M.; Ginebreda, A.; Petrović, M.; Barceló, D. Risk assessment based prioritization of 200 organic micropollutants in 4 Iberian rivers. *Sci. Total Environ.* **2015**, *503*, 289–299. [CrossRef] [PubMed]
53. Petrović, M.; Škrbić, B.; Živančev, J.; Ferrando-Climent, L.; Barcelo, D. Determination of 81 Pharmaceutical drugs by high performance liquid chromatography coupled to mass spectrometry with hybrid triple quadrupole–linear ion trap in different types of water in Serbia. *Sci. Total Environ.* **2014**, *468*, 415–428. [CrossRef] [PubMed]
54. Qin, Q.; Chen, X.; Zhuang, J. The fate and impact of Pharmaceuticals and Personal Care Products in agricultural soils irrigated with reclaimed water. *Crit. Rev. Environ. Sci. Technol.* **2015**, *45*, 1379–1408. [CrossRef]
55. Sharma, V.K.; Anquandah, G.A.; Yngard, R.A.; Kim, H.; Fekete, J.; Bouzek, K.; Ray, A.K.; Golovko, D. Nonylphenol, octylphenol, and bisphenol-A in the aquatic environment: A review on occurrence, fate, and treatment. *J. Environ. Sci. Health Part A* **2009**, *44*, 423–442. [CrossRef]
56. Sharma, V.K.; Liu, F.; Tolan, S.; Sohn, M.; Kim, H.; Oturan, M.A. Oxidation of β-lactam antibiotics by ferrate (VI). *Chem. Eng. J.* **2013**, *221*, 446–451. [CrossRef]

57. Tölgyesi, Á.; Verebey, Z.; Sharma, V.K.; Kovacsics, L.; Fekete, J. Simultaneous determination of corticosteroids, androgens, and progesterone in river water by liquid chromatography–tandem mass spectrometry. *Chemosphere* **2010**, *78*, 972–979. [CrossRef]
58. Verlicchi, P.; Galletti, A.; Petrovic, M.; Barceló, D. Hospital effluents as a source of emerging pollutants: An overview of micropollutants and sustainable treatment options. *J. Hydrol.* **2010**, *389*, 416–428. [CrossRef]

water

MDPI

Article

# Influence of Structural Symmetry of Fault Zones on Fluid-Induced Fault Slips and Earthquakes

Zhiyong Niu [1,*] and Bing Bai [2]

1 School of Resources and Civil Engineering, Suzhou University, Suzhou 234000, China
2 Hubei Key Laboratory of Environmental Geotechnology, Institute of Rock and Soil Mechanics, Chinese Academy of Sciences, Wuhan 430071, China; bbai@whrsm.ac.cn
* Correspondence: niuzhiyong88@ahszu.edu.cn

**Abstract:** Subsurface fluid injection and extraction can reactivate faults and induce earthquakes. In current research, faults are typically described as symmetrical structures and the presence of asymmetric structures is often overlooked. The reality is that numerous asymmetric faults exist within the Earth's crust. The architectural and permeability characteristics of fault zones differ significantly between symmetrical and asymmetrical faults. These differences may have a great influence on fault stability during fluid injection or extraction. In this study, the impact of fault zone structures on fluid-induced slips and seismic activity were investigated through numerical analysis. The findings indicated that symmetrical faults were more likely to induce larger slips and earthquakes during various subsurface fluid operations. For asymmetric faults, larger induced slips occurred when fluid was operated in a hanging wall reservoir than in a footwall reservoir. In symmetrical faults, the opposite was true. When evaluating the stability of a fault in subsurface fluid engineering, the fault structure and fluid pattern and their combined effects must be considered comprehensively.

**Keywords:** fault zone; fault slip; fluid injection–production; asymmetric fault; symmetrical fault; induced earthquake

**Citation:** Niu, Z.; Bai, B. Influence of Structural Symmetry of Fault Zones on Fluid-Induced Fault Slips and Earthquakes. *Water* **2024**, *16*, 1118. https://doi.org/10.3390/w16081118

Academic Editors: Giuseppe Pezzinga, Zongjun Gao and Jiutan Liu

Received: 9 March 2024
Revised: 9 April 2024
Accepted: 10 April 2024
Published: 15 April 2024



## 1. Introduction

Numerous subsurface fluid injection and extraction activities have been conducted by humans for a variety of purposes, including resource extraction, energy generation, and environmental management, among others. Fluid extraction engineering is employed in the production of oil [1], natural gas [2], shale gas [3], coalbed methane [4], tight sandstone gas [5], geothermal energy [6], and so on. Fluid injection engineering is applied in wastewater injection into deep wells [7], $CO_2$ geological storage [8], hydraulic fracturing for resource extraction [9], and so on. Many engineering projects involve both subsurface fluid injection and extraction, for example, enhanced oil recovery by water or gas (N2, $CO_2$, etc.) [10–12]; enhanced geothermal systems (EGSs), in which cold fluid is injected into geothermal reservoirs to capture heat [13–15]; enhanced water system recovery (EWR), where gas is injected into deep saline formations to obtain water and liquid minerals [16,17]; and the extraction of shale gas (or coalbed methane) via hydraulic fracking [18–20]. Fluid injection or extraction will directly lead to changes in the effective stress within the target area, disrupting the original stress balance. The stress perturbation may cause faults to slip and even induce earthquakes.

Fault activation and earthquakes induced by subsurface fluid injection or extraction have become significant factors influencing the efficacy and safety of the aforementioned engineering activities and have attracted considerable attention. Research has indicated that numerous large earthquakes in Oklahoma were induced by oil production activities [21]. Gas extraction in the Netherlands led to a series of earthquakes, causing substantial damage to numerous buildings [22]. In 2011, the extraction of groundwater in Lorca, Spain, induced a 5.1-magnitude earthquake, leading to heavy casualties and extensive property damage [23]. In 2012, a 4.8-magnitude earthquake was induced by wastewater injection in Timpson, East Texas, which was followed by a sequence of smaller earthquakes [24]. During the period from 2 December to 8 December 2006, fluid injection in Basel, Switzerland, triggered over 10,000 seismic events. Among these, four earthquakes with magnitudes exceeding 3.0 caused extensive damage and widespread public panic [25]. In 2017, South Korea experienced a 5.5 magnitude earthquake triggered by an EGS development. The event led to significant damage and prompted the government to halt the project [26]. Hydraulic fracturing in western Canada has been associated with fault activation and earthquakes, sparking broad public concern [27]. Overall, it can be seen that both subsurface fluid injection and extraction can induce fault activation and earthquakes (see Figure 1). This issue significantly hinders the advancement of subsurface engineering projects involving fluid injection or extraction. Consequently, assessing the risks of induced seismic events and managing fault stability have emerged as critical concerns in recent years [28–33]. The key issue is to reveal the induced slip behavior of faults in the complex fluid–solid coupling process. Numerical simulation has become an indispensable method, offering several key advantages: (1) the capability to model complex geological structures [34,35]; (2) the flexibility to simulate varied fluid flow conditions [36,37]; and (3) the ability to solve multi-field coupling [38]. In this work, our investigation focused on analyzing the behavior of induced fault slips and earthquakes using numerical modeling techniques.

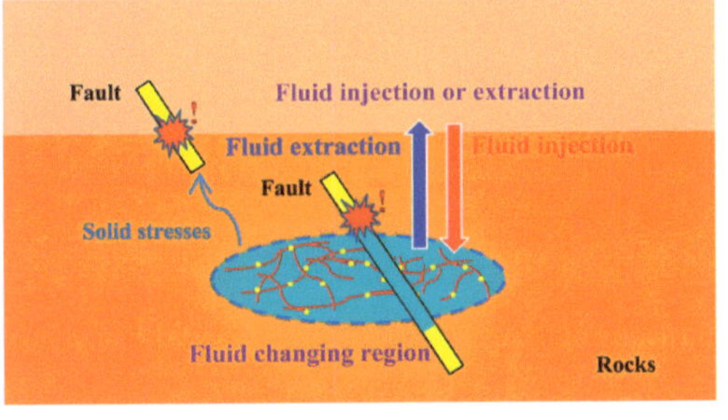


**Figure 1.** Schematic diagram of fluid-induced earthquakes.

Faults are geological structures that are the result of tectonic movements. Rock layers fracture under crustal stress and glide along a plane, resulting in a fault. Faults typically develop over a long period, creating a complex zone that consists of a primary fault plane and surrounding fractured rock masses. This zone also includes an array of secondary faults and fracture planes called a fault zone. Faults are complex geological entities, and it is difficult to solve fluid–solid coupling problems through theoretical calculations. Therefore, numerical simulation has become an indispensable research method. In numerical simulations, the geometry of faults, material properties, and boundary conditions must be appropriately simplified to make the calculations feasible and ensure the reliability of the simulation results. At present, the widely accepted fault structure model is the symmetrical dual structure shown in Figure 2a. In this model, the fault zone consists of a high-permeability fault core and a low-permeability damage zone [39–41]. There is also a model that further subdivides fault zones based on this symmetrical dual structure [42]. Another frequently employed model in numerical simulations simplifies the fault as a zone of homogeneous rock layers [43–45]. In summary, the representation of faults in numerical simulations is predominantly characterized by symmetrical structures.

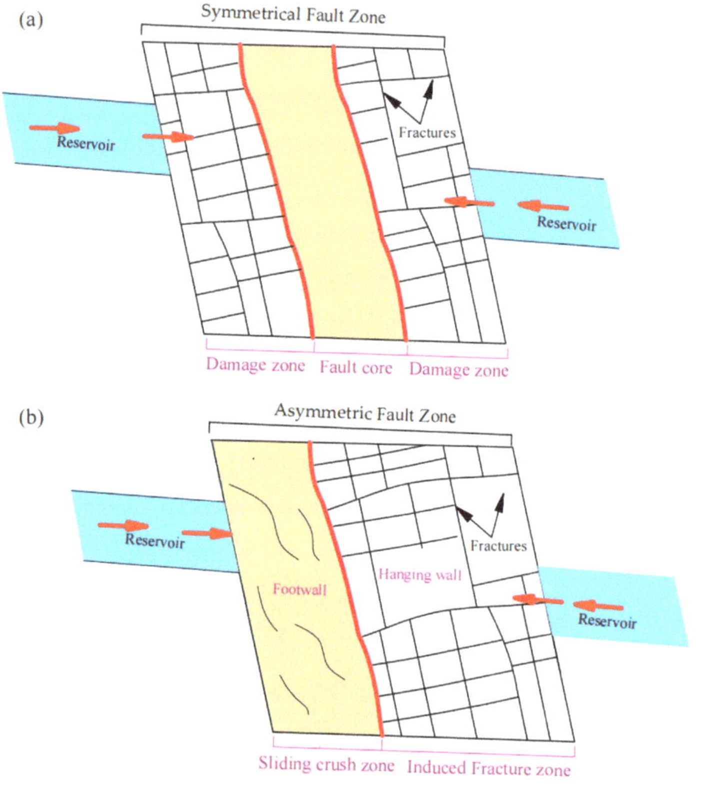


**Figure 2.** Structure of the symmetrical fault (**a**) and the asymmetric fault (**b**).

However, the final forms of fault zones are not always symmetrical structures. Numerous asymmetric faults exist in the Earth's crust. A large number of studies have confirmed that inclined faults often present an asymmetric structure [46–50]. The asymmetric fault zone is primarily evident through the structural and permeability differences between the hanging wall and the footwall, as illustrated in Figure 2b. During the growth of a fault, it usually forms an active wall and a passive wall [51]. The active wall—commonly referred to as the hanging wall—typically undergoes more severe slippage, forming an induced fracture zone. On the other hand, the footwall remains relatively motionless, forming a sliding crush zone. As a consequence, the active wall experiences a higher level of fragmentation than the passive wall [51]. Furthermore, during the violent sliding of the hanging wall, fault gouges and loose sediment are more likely to be forced into the fractures and pores of the footwall—a phenomenon known as clay smear. Clay smear is recognized as a critical mechanism in fault sealing [52–54]. Consequently, the porosity and permeability of the sliding crush zone are generally found to be significantly lower than those of the induced fracture zone.

To fully reveal the behavior of induced fault slips under the complex fluid–solid coupling process, considering the impact of the fault zone's structural symmetry is crucial. In this paper, we studied it by numerical simulation. We considered two categories of fault zones: one with a symmetrical structure and another with an asymmetric structure. To make the study more comprehensive and convincing, we considered six different fluid patterns which can represent all the operations involved in the various subsurface fluid injection and production engineering techniques. The slip characteristics of the faults under the different fluid patterns were then studied.

## 2. Theoretical Background

### 2.1. Sliding Model

The software ABAQUS 6.14 was used to accomplish the numerical simulation. Due to the potential for significant fault slips under conditions of excess pore pressure, the sliding formulation employed in this study was chosen to be of the finite-sliding type. It allows for arbitrary separation, sliding, and rotation of contact surfaces. As shown in Figure 3, slave node S1 may come into contact anywhere along the master surface. While in contact, it is constrained to slide along the master surface. Figure 3 also shows the possible evolution of the contact between node S1 and its master surface. Node S1 is in contact with the element face with end nodes M2 and M3 at time $t_1$. The load transfer at this time occurs between node S1 and nodes M2 and M3 only. Later on, at time $t_2$, node S1 may find itself in contact with the element face with end nodes M5 and M6. Then, the load transfer will occur between node S1 and nodes M5 and M6.

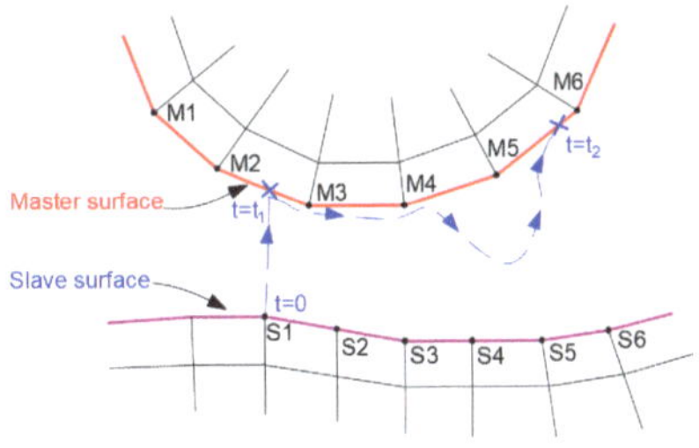


**Figure 3.** Contacting bodies and the trajectory of node S1 in finite-sliding contact.

We used the master–slave contact method to calculate the slip under the finite-sliding interaction. The surface of the surrounding rock was chosen as the master surface, and the surface of the fault was chosen as the slave surface, due to the greater stiffness of the rock mass compared to the fault. Figure 4 shows the slave surface (node $n_1$) and the master surface (node $n_2$, node $n_3$, and node $n_4$). $\boldsymbol{n}$ and $\boldsymbol{t}$ are, respectively, the normal and the tangent. The variable $h$ is the overclosure of the surfaces, where $nh = x - x_1$. The slip between the surfaces can be solved as follows:

$$d\delta h = -\delta \boldsymbol{x}_i \cdot \left( \boldsymbol{n} \frac{dN_i}{ds} N_j \boldsymbol{t} + \boldsymbol{t} N_i \frac{dN_j}{ds} \boldsymbol{n} + \boldsymbol{t} N_i \rho_n N_j \boldsymbol{t} \right) d\boldsymbol{x}_j \tag{1}$$

$$d\delta s = \delta \boldsymbol{x}_i \left( \boldsymbol{t} \frac{dN_i}{ds} N_j \boldsymbol{t} - \boldsymbol{n} N_i \frac{dN_j}{ds} \boldsymbol{n} - \boldsymbol{n} N_i \rho_n N_j \boldsymbol{t} \right) \cdot d\boldsymbol{x}_j \tag{2}$$

where $s$ is the slip; $N_i$ is the interpolation function; and $\rho_n$ is the segment curvature, defined as $\rho_n \overset{\text{def}}{=} -n \cdot \frac{d^2 x}{dg^2} / \left| \frac{dx}{dg} \right|^2$.

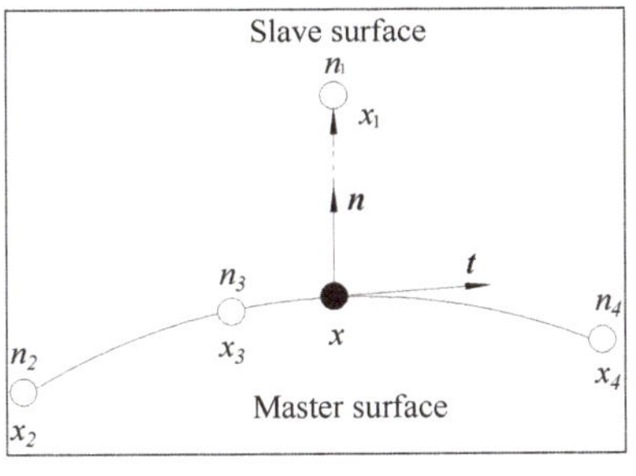


**Figure 4.** Slide line segment.

The ABAQUS theory manual [55] can be referred to for a detailed derivation.

## 2.2. Material Model

The reservoir and fault were set as porous elastic material. The seepage behavior was controlled by Darcy's law:

$$f = -\frac{\kappa}{\mu}\nabla p \tag{3}$$

where $f$ is the volumetric flux vector, $\mu$ is the viscosity, $\kappa$ is the intrinsic permeability of the medium, and $\nabla p$ is the pressure gradient vector.

The governing equation of fluid mass ($K$) conservation is as follows [56]:

$$\frac{dK}{dt} + \nabla(\rho_0 \cdot f) = 0 \tag{4}$$

The mechanical behavior of the reservoir was controlled by a porous elastic material model [57,58]:

$$2G\varepsilon_{ij} = \sigma_{ij} - \frac{v}{(1+v)}\sigma_{kk}\delta_{ij} + \frac{3(v_u - v)}{B(1+v)(1+v_u)}p\delta_{ij} \tag{5}$$

$$\Delta m = \frac{3\rho_0(v_u - v)}{2GB(1+v)(1+v_u)}(\sigma_{kk} + 3p/B) \tag{6}$$

where $G$ is the shear modulus, $\varepsilon_{ij}$ is the strain, $\sigma_{ij}$ is the stress, $\sigma_{kk}$ is the mean stress deviation, $p$ is the pore pressure, $v_u$ is the undrained Poisson's ratio, $v$ is the drained Poisson's ratio, $\delta_{ij}$ is the Kronecker delta, $\Delta m$ is the change in fluid mass per unit volume, and $\rho_0$ is the density of the pore fluid in the reference state. $B$ is Skempton's coefficient:

$$B = \frac{1/K - 1/K_s}{1/K - 1/K_S + (1/K_f - 1/K_n)} \tag{7}$$

where $K$ is the drained bulk modulus, $K_S$ is the solid grain's bulk modulus, $K_f$ is the pore fluid bulk modulus, and $1/K_n$ is the unjacketed pore compressibility.

## 2.3. Seismic Computational Model

The seismic moment model (Aki, 1967) was used to estimate the induced earthquake [59]:

$$M_0 = \mu DS \tag{8}$$

where $M_0$ is the seismic moment, $\mu$ is the rigidity of the fault (Pa), $D$ is the rupture area of the fault (m$^2$), and $S$ is the average slip displacement of the fault (m).

Then, the moment magnitude scale ($M_w$) of the induced earthquake can be obtained by the following formulation [60]:

$$M_w = \frac{2}{3}(logM_0 - 9.1) \tag{9}$$

# 3. Numerical Modelling

## 3.1. Model Setup

The finite-element software ABAQUS has significant advantages in nonlinear analysis, which encompasses material, geometric, and boundary nonlinearities. Contact analysis belongs to a typical nonlinear problem of boundary nonlinearity. The boundary conditions cannot be fully given at the beginning of the calculation. During the calculation process, the contact area and the pressure between contact bodies will vary with load and deformation. Computing the value of a fault slip along the surrounding rock is a typical boundary nonlinear contact issue. So, ABAQUS was selected to accomplish the simulation of the fault.

A basic model (Figure 5) was designed to simulate the subsurface fluid injection and extraction engineering at the fault. As shown, the length and height of the model were, respectively, 3200 m and 1000 m. The thickness of the reservoir was 60 m. The fault was normal, with a length, thickness, and inclined angle of 700 m, 30 m, and 60°. The distance from the top of the fault to the top of the model fault was 260 m. The distances from the top of the fault to the left and right boundaries were 1140 m and 2050 m, respectively.

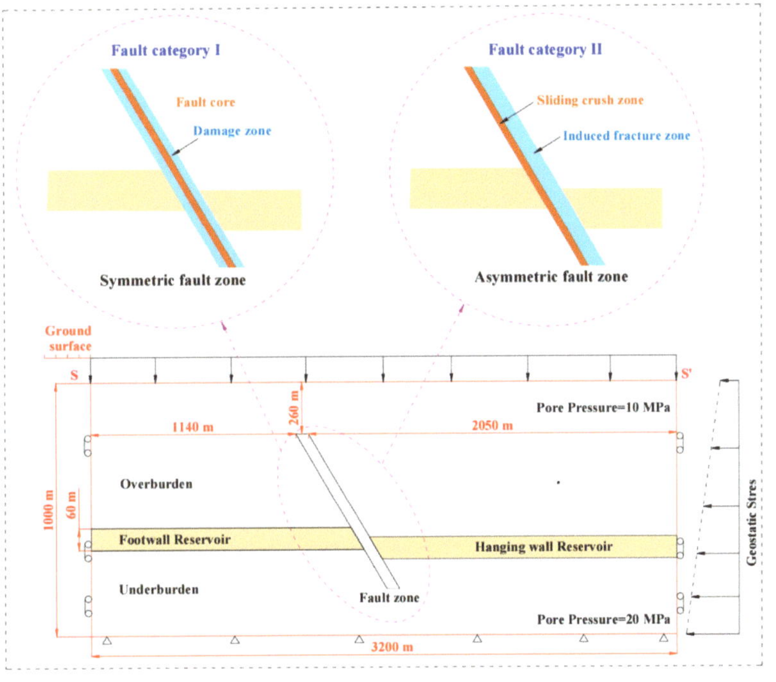


**Figure 5.** Basic model.

In order to reveal the influence of fault structure on the behavior of fluid-induced slips, we designed two categories of fault zone: a symmetric fault (Fault Category I) and an asymmetric fault (Fault Category II), both with the same length, thickness, and angle. Fault Category I consisted of the fault core (low permeability) and damage zone (high permeability). The thickness of the fault core was 6 m. Fault Category II consisted of a sliding crush zone (low permeability) and an induced fracture zone (high permeability). The thickness of the sliding crush zone was 12 m. The physical and mechanical properties of the fault core and the sliding crush zone were also the same (Table 1). The model's representation of fault dimensions and subsurface property values was conceptual. For this information, we relied on observed variations in the caprock, fault, and reservoir characteristics within the Qiabuqia Geothermal Field in Qinghai Province's Gonghe Basin, China. The unmeasurable parameters were largely estimated using established selection guidelines, and data from additional scholarly sources were referred to [42,61–63].

**Table 1.** Physical parameters of the rock mass.

| Property | Density ($kg/m^3$) | Young's Modulus (GPa) | Poisson's Ratio (-) | Porosity (-) | Permeability (m/s) |
|---|---|---|---|---|---|
| Overburden | 2600 | 20 | 0.37 | 0.01 | $1 \times 10^{-16}$ |
| Reservoir | 2200 | 25 | 0.30 | 0.2 | $1 \times 10^{-7}$ |
| Underburden | 2600 | 30 | 0.33 | 0.36 | $1 \times 10^{-16}$ |
| Fault core | 2500 | 10 | 0.3 | 0.05 | $1 \times 10^{-15}$ |
| Damage zone | 2000 | 10 | 0.3 | 0.2 | $1 \times 10^{-7}$ |
| Sliding crush zone | 2500 | 10 | 0.3 | 0.05 | $1 \times 10^{-15}$ |
| Induced fracture zone | 2000 | 10 | 0.3 | 0.2 | $1 \times 10^{-7}$ |

It is the fluid changes in a reservoir that directly induce fault activation during subsurface fluid engineering. In order to reveal the influence of the relative spatial position between the reservoir and the fault, two categories of reservoirs were designed. As shown in Figure 6, they were located at the upper and lower parts of the fault, respectively.

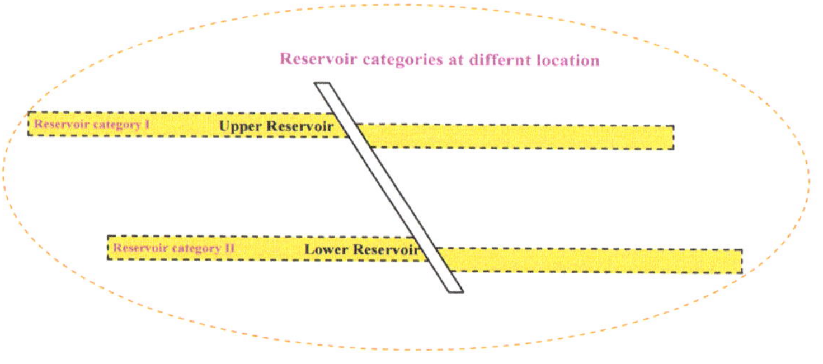


**Figure 6.** Reservoir categories.

The boundary conditions were as shown in Figure 5. The bottom boundary was fixed, such that vertical and horizontal displacement was not allowed. The left and right boundaries were only constrained horizontally. The top boundary was free, and both vertical and horizontal displacement were allowed. To simulate the actual depth of the fault, a surface pressure of $1.0 \times 10^7$ MPa was loaded on the top boundary. The initial pore pressure was hydrostatic, with 10 MPa on the top and 20 MPa on the bottom.

### *3.2. Fluid Injecting and Extracting*

The initial stresses were geostatic. After initial geostress balancing, the fluid injection and extraction were simulated. To thoroughly elucidate the characteristics of the induced fault slip behavior by various subsurface fluid injection and extraction operations, we considered six distinct fluid flow patterns. As shown in Figure 7, Patterns I and II represented fluid injection engineering, such as wastewater injection into deep wells. Pattern I represented fluid injection into the footwall reservoir, while Pattern II illustrated fluid injection into the hanging wall reservoir. Pattern III and Pattern IV represented fluid extraction engineering, such as oil exploitation. Pattern III represented fluid extraction from the footwall reservoir, while Pattern IV represented fluid extraction from the hanging wall reservoir. Pattern V and Pattern VI represented fluid injection and extraction engineering, such as EOR and EWR, which were described in the introduction. The six different fluid patterns can represent all the operations in the various subsurface fluid injection and production engineering methods.

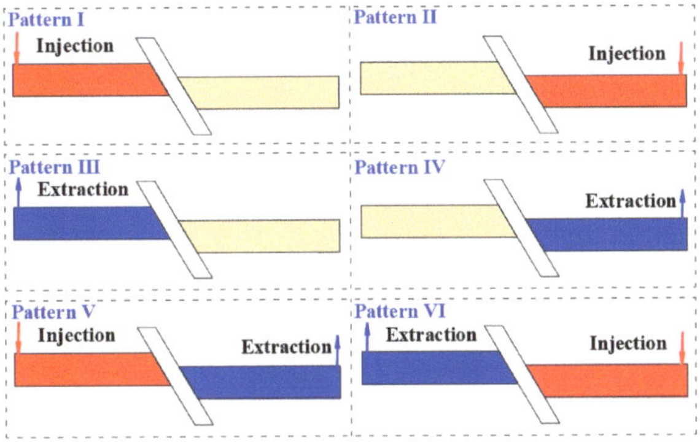


**Figure 7.** Diagram of the six different injection and extraction patterns.

In all six fluid patterns, the injection pressure and the extraction pressure were under 10 MPa. After the simulation, the induced fault slips and earthquakes could be investigated to assess the influence of different fluid operations on the stability of faults with different structures.

## 4. Results and Discussion

### *4.1. Upper Reservoir*

#### 4.1.1. Fluid Pattern I

Figure 8 displays the slip displacement of the symmetrical fault (represented by the black line) and the asymmetric fault (represented by the red line) under Fluid Pattern I when the reservoir was located in the upper part of the fault. The results indicated that the slip behaviors of the two categories of fault were nearly the same. When fluid was injected into the footwall reservoir, the two faults both slid down along the fault plane. The asymmetric fault exhibited a maximum slip displacement of just 15.77 mm. In contrast, the symmetrical fault's maximum slip displacement reached 30.35 mm, which is nearly double that of the asymmetric fault. The average slip displacement of the asymmetric fault was 4.40 mm, while that of the symmetrical fault was 16.47 mm. The average slip of the symmetrical fault was nearly four times that of the asymmetric fault. According to the seismic computational model in chapter 2, the $M_w$ of the induced earthquake of the symmetrical fault was 3.15 and that of the asymmetric fault was 2.77. The results revealed that the induced slip and earthquake of the asymmetric fault were significantly larger than those of the symmetric fault when fluid was injected into the footwall reservoir.

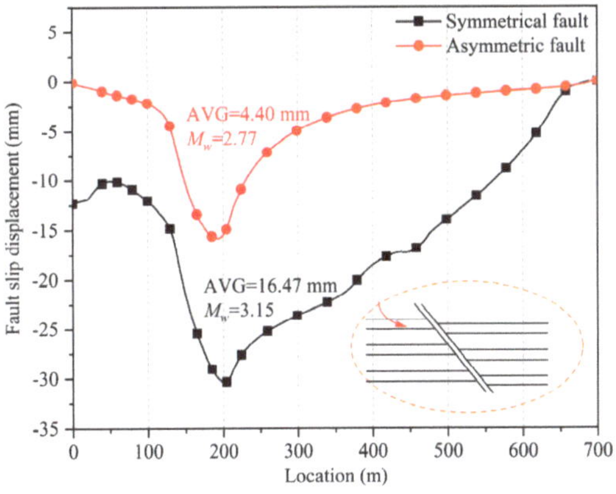


**Figure 8.** The induced fault slip displacement of the upper reservoir—Fluid Pattern I.

Figure 9 illustrates the pore-pressure distribution of the symmetrical fault and the asymmetric fault after fluid injection into the footwall reservoir. When the fault was symmetrical, there was a noticeable increase in pore pressure within the left damage zone (as shown in Figure 9a). In contrast, for the asymmetric fault, the pore pressure within the fault remained largely unchanged (refer to Figure 9b). The symmetrical fault comprised a low-permeability fault core surrounded by a high-permeability damage zone, which prevented the formation of lateral sealing. As depicted in Figure 9a, upon fluid injection into the footwall reservoir, the fluid directly flowed into the left damage zone of the fault. The fluid could not flow into the fault core due to its low permeability. Therefore, the pore pressure in the footwall reservoir and the left damage zone of the fault was obviously increased. For the asymmetric fault, the low-permeability sliding crush zone was docked with the injected reservoir. Fluid could not flow into the fault. As can be seen from Figure 9b, the pore pressure of the fault was almost unchanged. According to the principles of effective stress proposed by Biot [57] and Rice [58], the effective stress is reduced when pore pressure increases. According to the Mohr–Coulomb strength criterion, a decrease in effective stress leads to a reduction in the shear strength of the fault plane, which may in turn cause shear failure and slip. In the numerical model, the Coulomb friction model was also used to describe the tangential behavior of the contact pair of the fault and its surrounding rock. This model incorporates a critical shear stress ($\tau_{cirt}$) between the two contacting surfaces. When the shear stress exceeds the critical shear stress, the contact pair may experience instability and slip. The increase in pore pressure within the fault can directly reduce the critical shear stress, leading to fault sliding. The greater the increase in pore pressure, the more pronounced the induced fault slippage became. Consequently, when fluid was injected into the footwall reservoir, the slippage of the symmetrical fault was significantly greater than that of the asymmetric fault.

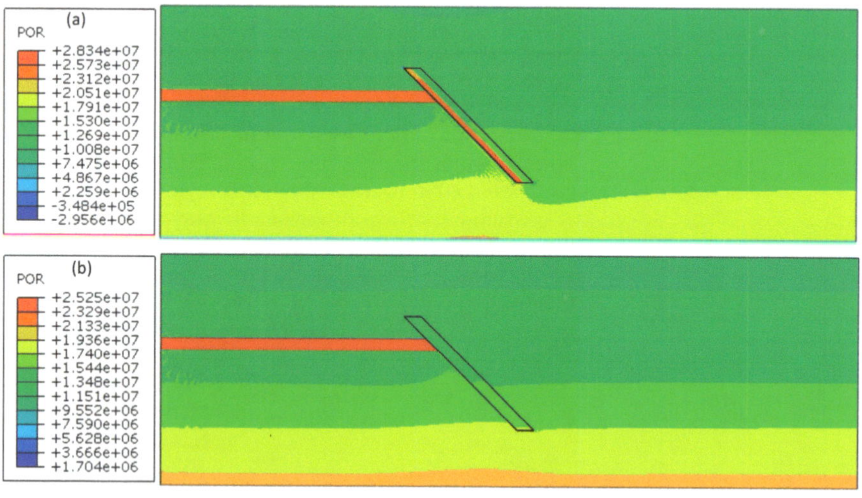


**Figure 9.** The pore−pressure distribution of the symmetrical fault (**a**) and the asymmetric fault (**b**) under Fluid Pattern I.

Figure 10 shows the distribution of shearing stress for the symmetrical fault (Figure 10a) and the asymmetric fault (Figure 10b). The shearing stress of the symmetrical fault was obviously larger than that of the asymmetric fault. We set a path on the footwall fault plane to obtain the shear stress data, as shown in Figure 11. As can be seen, the shear stress of the symmetrical fault was greater than that of the asymmetric fault. By calculation, the average shear stress along the fault plane of the symmetrical fault was 1.55 MPa, while that of the asymmetric fault was only 0.80 MPa. This further confirmed that the structure of a fault might greatly affect the distribution of fault stress. When the fluid was injected into the footwall reservoir, the asymmetric fault would be more stable than the symmetrical fault.

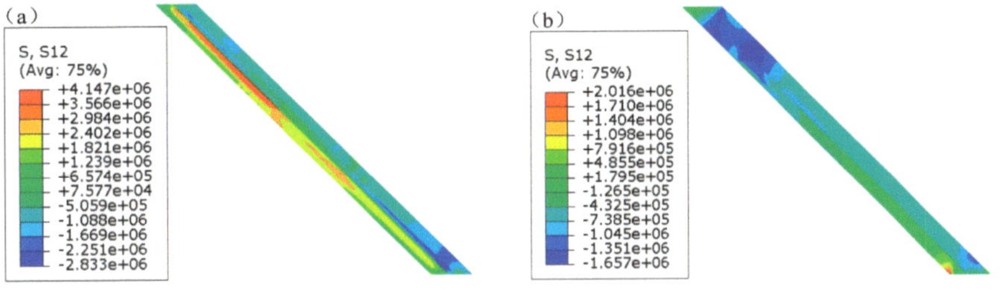


**Figure 10.** The shear stress distribution of the symmetrical fault (**a**) and the asymmetric fault (**b**) under Fluid Pattern I.

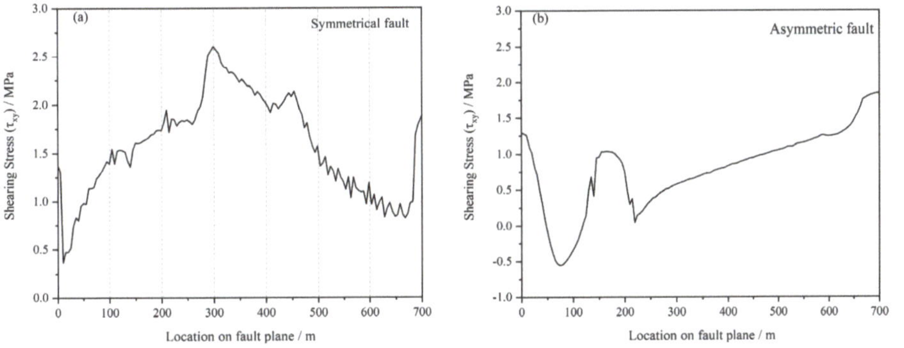


**Figure 11.** The shear stress along the fault plane of the symmetrical fault (**a**) and the asymmetric fault (**b**) under Fluid Pattern I.

4.1.2. Fluid Pattern II

Figure 12 shows the slip displacement of the symmetrical fault and the asymmetric fault when fluid was injected into the hanging wall reservoir. The slip behaviors of the two categories of faults were also similar. The symmetrical fault exhibited the same varying tendency as the asymmetrical fault, with the upper part sliding downward and the middle and lower parts sliding upward. The maximum slip displacement of the symmetrical fault was 14.74 mm, and that of the asymmetric fault was 15.11 mm. The average slip displacement for the asymmetric fault was 8.96 mm, while the symmetrical fault exhibited an average of 7.85 mm. The results for the two faults were quite close to each other. Based on calculations, the induced earthquake moment ($M_w$) for the symmetrical fault was 2.94, and for the asymmetric fault, it was 2.98. Although the asymmetric fault induced a slightly larger earthquake, the difference was not statistically significant. The reason was that both the induced fracture zone of the asymmetric fault and the right damage zone of the symmetrical fault had a high permeability. When fluid was injected into hanging wall reservoir, it could flow into both the induced fracture zone and the damage zone. Figure 13 displays the pore-pressure distribution of the two faults after fluid injection into the hanging wall reservoir. Overall, the pore pressure within the asymmetric fault (Figure 13a) was higher and had a wider distribution range than that within the symmetrical fault (Figure 13b). The reason was that the asymmetric fault's induced fracture zone was thicker than the symmetrical fault's right damage zone. Fluid injection resulted in a greater reduction in effective stress for the asymmetric fault. As a result, it experienced a slightly higher level of induced slip and seismic activity.

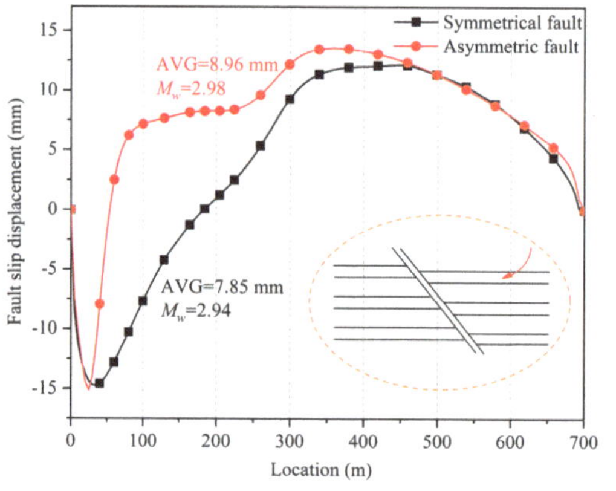


**Figure 12.** The induced fault slip displacement of the upper reservoir—Fluid Pattern II.

**Figure 13.** The pore-pressure distribution of the symmetrical fault (**a**) and the asymmetric fault (**b**) under Fluid Pattern II.

### 4.1.3. Fluid Pattern III

Figure 14 shows the results under Fluid Pattern III. The slip trends of the two faults were also similar and exhibited a reserved trend consistent with that observed under Pattern I. On the one hand, the slip displacement of the symmetrical fault was larger than that of the asymmetric fault. The asymmetric fault had a maximum slip displacement of 11.85 mm, with an average of 4.08 mm. In contrast, the symmetrical fault's maximum slip displacement was 13.32 mm, with an average of 5.18 mm. The induced earthquake ($M_w$) of the symmetrical fault was 2.82. In comparison, the asymmetric fault had an $M_w$ of 2.73. On the other hand, compared with Fluid Pattern I, the difference between the results for the two faults under Pattern III was less significant.

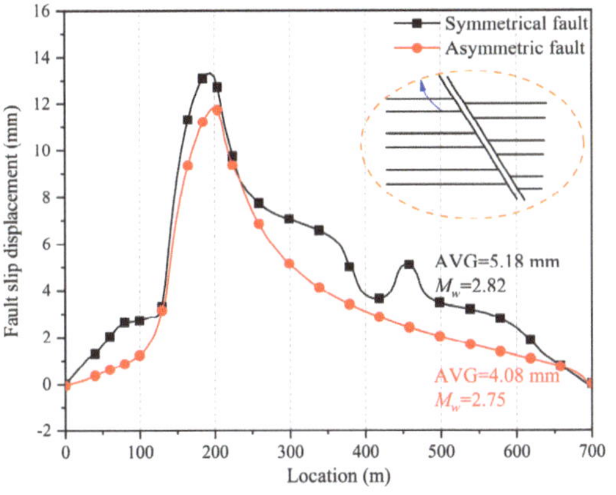


**Figure 14.** The induced fault slip displacement of the upper reservoir—Fluid Pattern III.

When fluid was extracted from the footwall reservoir, the induced slippage and earthquake of the symmetrical fault were only slightly larger than those of the asymmetric fault. The difference between the two faults was not significant, which differed from the results observed for Pattern I. When fluid was extracted from the reservoir, the primary cause of fault sliding shifted from the Mohr–Coulomb strength criterion to the compressive deformation of the reservoir. Since both the reservoir and the fault consisted of porous elastic materials, compressive deformation occurred in response to increasing load. Fluid extraction directly reduced pore pressure, which was equivalent to applying load. Therefore, when fluid was extracted, the region experienced a decrease in pore pressure and underwent compressive deformation. Excessive deformation can also lead to fault slipping. Since the sliding crush zone of the asymmetric fault was connected to the footwall reservoir, the pore pressure in the fault remained almost unchanged when fluid was extracted from the reservoir. The fault slip was primarily caused by the compressive deformation of the reservoir. For the symmetrical fault, the slip was caused by a combination of the compressive deformation of the reservoir and the damage zone on the fault's left side. As a result, the induced slippage and earthquake of the symmetrical fault were larger than those of the asymmetric fault.

Figure 15 presents nephograms of the vertical displacement for the symmetrical and asymmetric faults. From these images, it is evident that fluid extraction led to significant compression of the rock formations. We selected the fault and its surrounding area for analysis. When the fault was symmetrical, both the pore pressure in the reservoir and the left fault damage zone decreased. This reduction led to compressive deformation in the fault and its surrounding areas (Figure 15a). As indicated by the purple dashed ellipse, the vertical displacement in the upper part of the fault and its top-right surrounding area was downward, while the vertical displacement in the bottom-left area of the fault was upward. When the fault was asymmetric, the area indicated by the purple dashed ellipse was obviously small (Figure 15b). This was because the sliding crush zone of the fault was impermeable. When fluid was extracted, only the reservoir underwent compression and induced the surrounding fault areas to deform. Therefore, the slip region and degree of the asymmetric fault were smaller than for the symmetrical fault. Next, we compared the outcomes of Fluid Pattern III with those of Fluid Pattern I. Under Pattern I, the induced average slip of the symmetrical fault was 12 mm greater than that of the asymmetric fault. Under Pattern III, the difference was only 1.1 mm. The induced slippage differences of the two faults under Pattern III were less significant than those under Pattern I. It can be concluded that the degree of fault slippage caused by compressive deformation is lower than that caused by shear slippage.

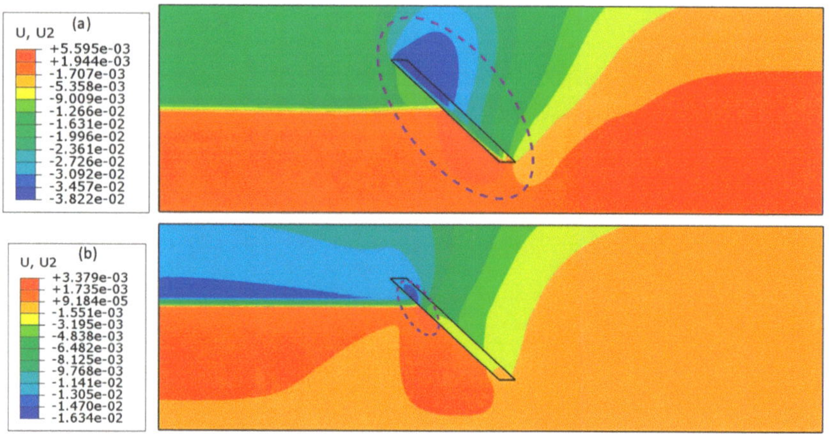


**Figure 15.** The vertical displacement distribution of the symmetrical fault (**a**) and the asymmetric fault (**b**) under the upper reservoir—Fluid Pattern III.

4.1.4. Fluid Pattern IV

Figure 16 shows the slip displacement of the symmetrical fault and the asymmetric fault under Fluid Pattern IV. When fluid was extracted from the hanging reservoir, the slips of the symmetrical fault and the asymmetric fault exhibited similar varying tendencies and quantities. The maximum slip displacements of the symmetrical fault and the asymmetric fault were 14.76 mm and 18.43 mm, respectively. The average slip displacements were 5.07 mm for the symmetrical fault and 5.97 mm for the asymmetric fault. The induced earthquake ($M_w$) of the symmetrical fault was 2.81, while that for the asymmetric fault was 2.86. The induced slippage and earthquakes of the symmetrical fault were slightly larger. The right damage zone of the symmetrical fault and the induced fractured zone of the asymmetric fault were connected to the hanging wall reservoir. Both zones had high permeability. When fluid was extracted from the hanging wall reservoir, the pore pressure in both zones decreased. The compressive deformation of the reservoir and the fault zone led fault to a loss of stability and to sliding. The induced fracture zone of the symmetrical fault was thicker than its right damage zone, resulting in greater induced slippage and seismic activity for the symmetrical fault. However, the impact was not significant. The difference in slippage between the two faults was minimal.

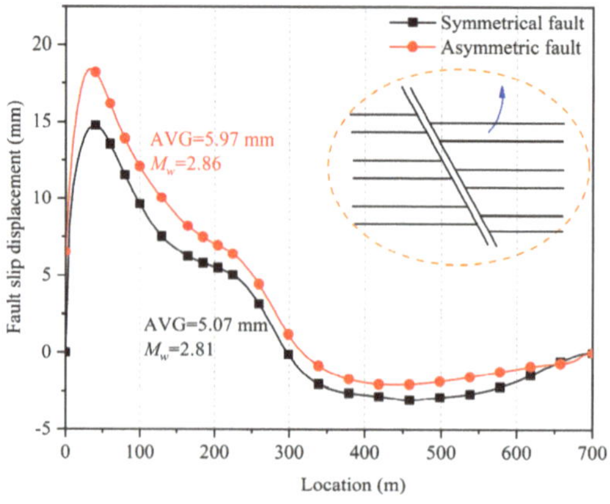


**Figure 16.** The induced fault slip displacement of the upper reservoir—Fluid Pattern IV.

#### 4.1.5. Fluid Pattern V

Fluid Pattern VI represented fluid being injected into the footwall reservoir and simultaneously extracted from the hanging wall reservoir. Figure 17 shows the slip displacement of the two faults under Fluid Pattern V. The slip tendencies of the symmetrical fault and the asymmetric fault were similar, with the upper part sliding upward and the lower part sliding downward. The maximum slip displacement of the asymmetric fault was 17.78 mm, with an average of 6.84 mm. In contrast, the symmetrical fault exhibited a maximum slip displacement of 23.28 mm, with an average of 13.83 mm. The $M_w$ values of the asymmetric fault and the symmetrical fault were 2.90 and 3.10, respectively. The average displacement of the symmetrical fault was more than double that of the asymmetric fault, with the induced earthquake magnitude being larger by 0.2. The differences in the results between the two faults were still quite pronounced.

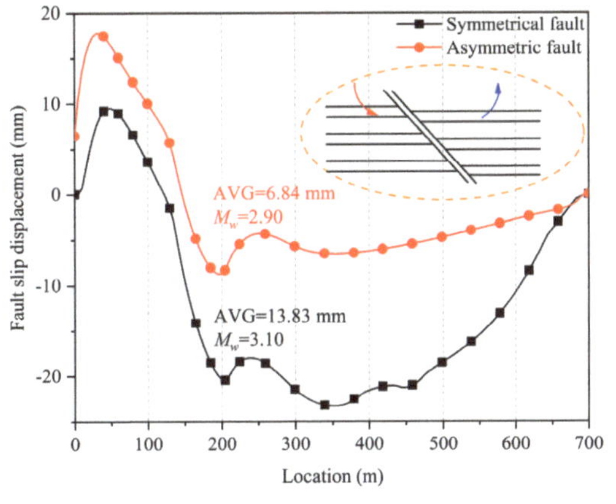


**Figure 17.** The induced fault slip displacement of the upper reservoir—Fluid Pattern V.

#### 4.1.6. Fluid Pattern VI

Fluid Pattern VI represented the fluid being injected into the hanging wall reservoir and simultaneously extracted from the footwall reservoir. Figure 18 shows the slip displacement of the symmetrical fault and the asymmetric fault. The slip tendency of the faults under Fluid Pattern VI was the opposite of that under Pattern V. The maximum slip displacement of the asymmetric fault was 17.37 mm, with an average of 6.46 mm. The maximum slip displacement of the symmetrical fault was 10.75 mm, with an average of 10.85 mm. The $M_w$ of the induced earthquake of the asymmetric fault was 3.03. The $M_w$ of the induced earthquake of the symmetrical fault was 2.88. The induced slippage and earthquake of the symmetrical fault were larger than those of the asymmetric fault, which was the opposite of what occurred under Pattern V.

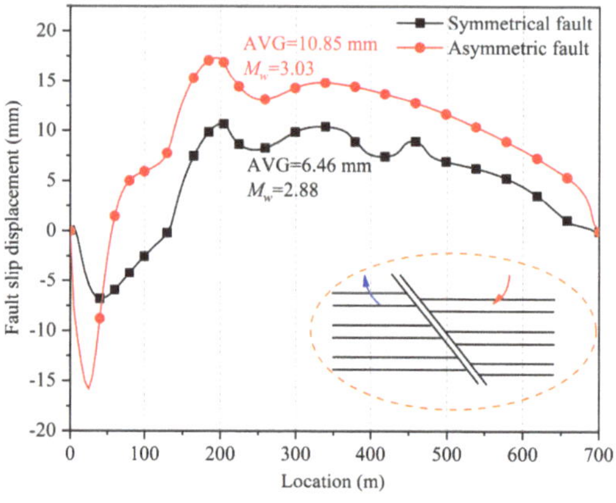


**Figure 18.** The induced fault slip displacement of the upper reservoir—Fluid Pattern VI.

### 4.2. Lower Reservoir

Figure 19 illustrates the fault slip displacement for both the symmetrical and asymmetric faults under the six fluid patterns in the lower reservoir.

Figure 19a shows the slip of the faults under Fluid Pattern I. The asymmetric fault exhibited a maximum slip displacement of 13.40 mm and an average slip displacement of 2.03 mm. In comparison, the symmetrical fault had a maximum slip displacement of 28.21 mm, with an average of 19.44 mm. The $M_w$ of the induced earthquake of the asymmetric fault was 2.55. In contrast, the $M_w$ of the induced earthquake for the symmetrical fault was as high as 3.20. The induced slippage and seismic activity of the symmetrical fault were considerably greater than those of the asymmetric fault, with an $M_w$ difference of 0.65. The induced slippage and earthquake of the symmetrical fault were much larger than those of the asymmetric fault, with the $M_w$ being larger by 0.65.

Figure 19b shows the slip of the faults under Fluid Pattern II. The maximum slip displacement of the asymmetric fault was 14.26 mm, with an average slip displacement of 9.40 mm. The maximum slip displacement of the symmetrical fault was 15.58 mm, with an average slip displacement of 7.92 mm. The induced earthquake of the asymmetric fault had an $M_w$ of 2.99, while the symmetrical fault's induced earthquake registered an $M_w$ of 2.94. Consequently, the slippage and seismic activity induced by the symmetrical fault were slightly less than those of the asymmetric fault. Compared to the results from the upper reservoir, both the asymmetric and symmetrical faults showed a slight increase in their results when fluid was injected into the lower reservoir.

Figure 19c showed the slip of the faults under Fluid Pattern III. The maximum slip displacement of the asymmetric fault was 9.97 mm, with an average slip displacement of 1.59 mm. The maximum slip displacement of the symmetrical fault was 15.25 mm, with an average slip displacement of 9.35 mm. The $M_w$ of the induced earthquake of the asymmetric fault was 2.47, while that of the induced earthquake of the symmetrical fault was 2.99. The induced slippage of the symmetrical fault was obvious larger than that of the asymmetric one. The earthquake of the symmetrical fault was larger than that of the asymmetric fault by 0.52.

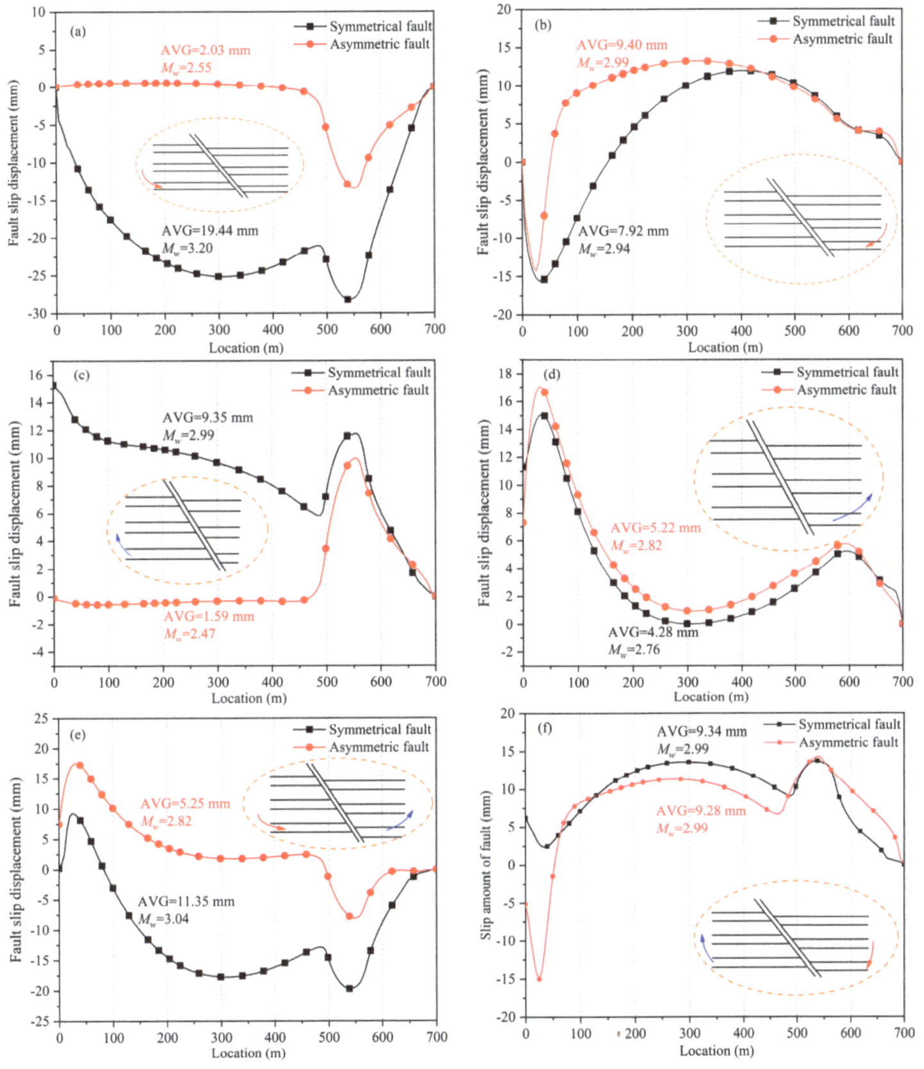


**Figure 19.** The induced fault slip displacements of the lower reservoir under the six fluid patterns: Pattern I (**a**), Pattern II (**b**), Pattern III (**c**), Pattern IV (**d**), Pattern V (**e**), and Pattern VI (**f**).

Figure 19d shows the slip of the faults under Fluid Pattern IV. The maximum slip displacement of the asymmetric fault was 17.04 mm, with an average slip displacement of 5.22 mm. The maximum slip displacement of the symmetrical fault was 15.08 mm, with an average slip displacement of 4.28 mm. The $M_w$ of the induced earthquake of the asymmetric fault was 2.82, while that of the induced earthquake of the symmetrical fault was 2.76. The induced slippage and earthquake of the symmetrical fault were slightly smaller than those of the asymmetric fault.

Figure 19e shows the slip of the faults under Fluid Pattern V. The maximum slip displacement of the asymmetric fault was 17.59 mm, with an average slip displacement of 5.25 mm. The maximum slip displacement of the symmetrical fault was 19.72 mm, with an average slip displacement of 11.35 mm. The $M_w$ of the induced earthquake of the asymmetric fault was 2.82, while that of the induced earthquake of the symmetrical fault was 3.04. The induced slippage and earthquake of the symmetrical fault were larger than those of the asymmetric fault.

Figure 19f shows the slip of the faults under Fluid Pattern VI. The maximum slip displacement of the asymmetric fault was 14.98 mm, with an average slip displacement of 9.28 mm. The maximum slip displacement of the symmetrical fault was 13.72 mm, with an average slip displacement of 9.34 mm. The $M_w$ values of the induced earthquake of the asymmetric fault and the symmetrical fault were both 2.99. The induced slippage and earthquake of the symmetrical fault were almost the same as those of the asymmetric fault.

We thus obtained results for fault slips and induced earthquakes for the three categories of reservoir and six fluid injection–production patterns (as shown in Table 2).

**Table 2.** Induced slip of asymmetric and symmetrical faults under different fluid patterns and reservoir categories.

| Reservoir Category | Fluid Pattern | Asymmetric Fault | | Symmetrical Fault | |
|---|---|---|---|---|---|
| | | Average Slip (mm) | Induced Earthquake ($M_w$) | Average Slip (mm) | Induced Earthquake ($M_w$) |
| Upper reservoir | I | 4.40 | 2.77 | 16.47 | 3.15 |
| | II | 8.96 | 2.98 | 7.85 | 2.94 |
| | III | 4.08 | 2.75 | 5.18 | 2.82 |
| | IV | 5.97 | 2.86 | 5.07 | 2.81 |
| | V | 6.84 | 2.90 | 13.83 | 3.10 |
| | VI | 10.85 | 3.03 | 6.46 | 2.88 |
| Lower reservoir | I | 2.03 | 2.55 | 19.44 | 3.20 |
| | II | 9.40 | 2.99 | 7.92 | 2.94 |
| | III | 1.59 | 2.47 | 9.35 | 2.99 |
| | IV | 5.22 | 2.82 | 4.28 | 2.76 |
| | V | 5.25 | 2.82 | 11.35 | 3.04 |
| | VI | 9.28 | 2.99 | 9.34 | 2.99 |

First, we compared the outcomes of Fluid Pattern I with those of Pattern II. For both categories of reservoirs, the induced average slippage and earthquake of the asymmetric fault under Pattern II were greater than those under Pattern I. Taking the upper reservoir as an example, the average slip of Pattern I was only 4.40 mm and that of Pattern II was as high as 8.96 mm. This revealed that for the asymmetric fault, injecting fluid into the hanging wall reservoir was more dangerous than injecting fluid into the footwall reservoir. For the symmetrical fault, however, the opposite was the case. Taking the lower reservoir as an example, the average slip of Pattern I was 19.44 mm and that of Pattern II was only 7.92 mm. Therefore, for the symmetrical fault, injecting fluid into the footwall reservoir was more dangerous. Similarly, comparing the results for Fluid Pattern III with those for Pattern IV, we found that extracting fluid from the hanging wall reservoir was more dangerous when the fault was asymmetric. When the fault was symmetrical, extracting fluid from the hanging wall reservoir was more dangerous than extracting fluid from the footwall reservoir.

Then, we compared the results of induced slippage and earthquake for the symmetrical fault and the asymmetric fault for the two categories of reservoir. It was found that the induced slip of the asymmetric fault was smaller than that of the symmetrical fault for Fluid Patterns I, III, and V. For Fluid Patterns II, IV, and VI, however, the induced slip of the asymmetric fault was larger than that of the asymmetric fault. For the same fault, the fault slip results under different fluid patterns were significantly different. Furthermore, under the same fluid patterns, the slip results for the faults with different structures also varied. This tells us that fault stability is the result of various factors. When evaluating the stability of faults in subsurface fluid engineering, it is essential to consider the fault structure, fluid patterns, and their combined effects comprehensively.

Among all the average slip results, the maximum obtained for the asymmetric fault was 1085 mm. This was much larger than that obtained for the symmetrical fault (19.44 mm). Among all the results for the induced earthquake, there were five situations under which $M_w$ was larger than 3.0. These situations were as follows: upper reservoir—Fluid Pattern VI—asymmetric fault; upper reservoir—Fluid Pattern I—symmetrical fault; upper reservoir—Fluid Pattern V—symmetrical fault; lower reservoir—Fluid Pattern I—symmetrical fault; lower reservoir—Fluid Pattern V—symmetrical fault. Among these situations, four concerned the symmetrical fault and only one concerned the asymmetric fault. It was thus revealed that in various subsurface fluid injection and extraction engineering projects, symmetrical faults may pose a greater risk than asymmetric faults. The reason in the cases considered here is that both the left and right damage zones of the symmetrical fault were highly permeable. In contrast, for the asymmetric fault, only the induced fracture zone in the hanging wall exhibited high permeability.

## 5. Conclusions

Faults are geological structures formed as a result of tectonic movements. Numerous asymmetric faults exist in the Earth's crust. Despite this, current research on the risk assessment of induced earthquakes and the control of fault stability often describes fault zones as if they were symmetrical structures. In this paper, we concentrate on examining the impact of the structural symmetry of fault zones on fluid-induced slip and the potential for seismic activity. The results of the study can provide some useful references. The main conclusions were as follows:

(1) The stability of a fault was the result of a variety of factors. For the same fault, the fault slip results under different fluid patterns were significantly different. Simultaneously, under the same fluid pattern, the slip results for faults with different structures also varied. When evaluating the stability of faults in subsurface fluid engineering, the fault structure, the fluid pattern, and their combined action must be considered comprehensively.

(2) When fluids were injected into a subsurface reservoir, the primary cause of fault slippage was the reduction in shear strength due to a decrease in effective stress. When fluid was extracted from the reservoir, the main cause of fault sliding was often the increase in effective stress, which could lead to compressive deformation of the reservoir rock. When fluid was injected into the footwall reservoir, the asymmetric fault was more stable than the symmetrical fault. When fluid was injected into the hanging wall reservoir, in contrast, the symmetrical fault was more stable.

(3) When the fault was asymmetric, it was more dangerous if fluid was injected into or extracted from the hanging wall reservoir than from the footwall reservoir. When the fault was symmetrical, injecting fluid into or extracting fluid from the footwall reservoir posed a greater danger.

(4) In numerous subsurface fluid injection and extraction projects, symmetrical faults are generally regarded as posing a higher risk compared to asymmetric faults. The reason is that symmetrical faults typically have high permeability in both the left and right damage zones. For asymmetric faults, however, only the induced fracture zone exhibits high permeability. Comprehensive investigation of the effect of fault zone symmetry on fault slips is essential for accurately assessing the risks associated with subsurface fluid injection and extraction.

**Author Contributions:** Conceptualization, Data curation, Formal analysis, Investigation, Methodology, Visualization, Writing—original draft, Funding acquisition: Z.N.; Software, Writing—review and editing: B.B. All authors have read and agreed to the published version of the manuscript.

**Funding:** This research work has been made possible thanks to financing from Key Projects of Natural Science Research in Colleges and Universities funded by Anhui Provincial Department of Education (no. KJ2021A1112), provincial quality engineering projects for institutions of higher education funded by Anhui Provincial Department of Education (no. 2022zygzts116), and the Civil Engineering Dual Capable Teaching Team funded by Suzhou University (no. 2023XJSN07).

**Data Availability Statement:** The original contributions presented in the study are included in the article; further inquiries can be directed to the corresponding author.

**Conflicts of Interest:** The authors declare no conflicts of interest.

## References

1. Ite, A.E.; Ibok, U.J.; Ite, M.U.; Petters, S.W. Petroleum Exploration and Production: Past and Present Environmental Issues in the Nigeria's Niger Delta. *Am. J. Environ. Prot.* **2013**, *1*, 78–90. [CrossRef]
2. Leather, D.T.B.; Bahadori, A.; Nwaoha, C.; Wood, D.A. Engineering, A review of Australia's natural gas resources and their exploitation. *J. Nat. Gas Sci. Eng.* **2013**, *10*, 68–88. [CrossRef]
3. Luo, X.; Wang, S.; Jing, Z.; Lv, M.; Wang, Z.; Zhai, Z.; Zhang, Y. A Review of Shale Gas Exploitation Technology. In Proceedings of the International Conference on Renewable Energy and Environmental Technology, Bali, Indonesia, 28 September 2014.
4. Neelu, R.; Upadhyay, B.; Parida, A. New Approach to Evaluate and Predict the Reservoir Performance of a Two-Phase Flow Coalbed Methane Well with Fully Coupled Time-Dependent Desorption and Pressure-Dependent Diffusivity Effects. *Energy Fuel* **2022**, *36*, 5284–5306. [CrossRef]
5. Huang, J.; Liao, Z.; Hu, Q.; Song, Z.; Wang, X. Fracture mechanism of tight sandstone under high and complex 3-D stress compression: Insights from acoustic emission. *J. Pet. Sci. Eng.* **2022**, *208*, 109635. [CrossRef]
6. Yadav, K.; Sircar, A. Geothermal energy provinces in India: A renewable heritage. *TIDEE TERI Inf. Dig. Energy Environ.* **2022**, *21*, 175. [CrossRef]
7. Grant, F.J. Groundwater, Deep Injection of Waste Water in the Western Canada Sedimentary Basin. *Groundwater* **2014**, *53*, 187–194.
8. van der Zwaan, B.; Gerlagh, R. Economics of geological $CO_2$ storage and leakage. *Clim. Change* **2009**, *93*, 285–309. [CrossRef]
9. Wang, J.; Xie, H.P.; Kmatthai, S.; Hu, J.J.; Li, C.B. The role of natural fracture activation in hydraulic fracturing for deep unconventional geo-energy reservoir stimulation. *Pet. Sci.* **2023**, *20*, 2141–2164. [CrossRef]
10. Thomas, S. Enhanced Oil Recovery—An Overview. *Oil Gas Sci. Technol. Rev. L'ifp* **2008**, *63*, 9–19. [CrossRef]
11. Graham, R.A.; Cox, R.J.; Stadlwieser, J.A.; Stinn, R. *Enhanced Oil Recovery by Gas Injection: Proposed Screening Criteria*; The Petroleum Society: Richardson, TX, USA, 1996; pp. 96–119.
12. Islam, M.R.; Chakma, A.; Jha, K.N. Engineering, Heavy oil recovery by inert gas injection with horizontal wells. *J. Pet. Sci. Eng.* **1994**, *11*, 213–226. [CrossRef]
13. Bayer, P.; Rybach, L.; Blum, P.; Brauchler, R. Review on life cycle environmental effects of geothermal power generation. *Renew. Sustain. Energy Rev.* **2013**, *26*, 446–463. [CrossRef]
14. Arman, A.; Christian, B. From hot rock to useful energy: A global estimate of enhanced geothermal systems potential. *Appl. Energy* **2020**, *279*, 115769.
15. Feng, Z.; Zhao, Y.; Zhou, A.; Zhang, N. Development program of hot dry rock geothermal resource in the Yangbajing Basin of China. *Renew. Energy* **2012**, *39*, 490–495. [CrossRef]
16. Li, Q.; Wei, Y.N.; Liu, G.; Shi, H. $CO_2$-EWR: A cleaner solution for coal chemical industry in China. *J. Clean. Prod.* **2015**, *103*, 330–337. [CrossRef]
17. Wei, N.; Liu, S.; Li, X. Evaluation on potential of $CO_2$ enhanced water recovery deployment in China's coal chemical industry. *Clim. Change Res.* **2021**, *17*, 70–78.
18. Tong, L.; Wei, X.; Liu, X.; Liang, L.; Wang, X.; Chen, J.; Lei, H. A Criterion for a Hydraulic Fracture Crossing a Natural Fracture in Toughness Dominant Regime and Viscosity Dominant Regime. *Eng. Fract. Mech.* **2023**, *289*, 109421. [CrossRef]
19. Jones, P.; Hillier, D.; Comfort, D. Fracking for Shale Gas: Planning Policy and Practice. *Town Ctry. Plan.* **2016**, *85*, 145–150.
20. Wu, B.; Wei, X.; Wang, W.; Li, J.; Liu, T.; Wang, X. Effect of stress and material barriers on hydraulic fracture height containment in layered formations. *Environ. Earth Sci.* **2022**, *81*, 255. [CrossRef]
21. Hough, S.E.; Page, M. A Century of Induced Earthquakes in Oklahoma? *Bull. Seismol. Soc. Am.* **2015**, *105*, 2863–2870. [CrossRef]
22. Vlek, C. Induced earthquakes from long-term gas extraction in Groningen, the Netherlands: Statistical analysis and prognosis for acceptable-risk regulation. *Risk Anal.* **2018**, *38*, 1455–1473. [CrossRef]
23. González, P.J.; Tiampo, K.F.; Palano, M.; Cannavó, F.; Fernández, J. The 2011 Lorca earthquake slip distribution controlled by groundwater crustal unloading. *Nat. Geosci.* **2012**, *5*, 821–825. [CrossRef]
24. Fan, Z.; Eichhubl, P.; Gale, J.F.W. Geomechanical analysis of fluid injection and seismic fault slip for the Mw 4.8 Timpson, Texas, earthquake sequence. *J. Geophys. Res. Solid Earth* **2016**, *121*, 2798–2812. [CrossRef]
25. Deichmann, N.; Giardini, D. Earthquakes Induced by the Stimulation of an Enhanced Geothermal System below Basel (Switzerland). *Seismol. Res. Lett.* **2009**, *80*, 784–798. [CrossRef]
26. Ellsworth, W.L.; Giardini, D.; Townend, J.; Ge, S.; Shimamoto, T. Triggering of the Pohang, Korea, Earthquake (Mw 5.5) by Enhanced Geothermal System Stimulation. *Seismol. Res. Lett.* **2019**, *5*, 1844–1858. [CrossRef]
27. Bao, X.; Eaton, D.W. Fault activation by hydraulic fracturing in western Canada. *Science* **2016**, *354*, 1406–1409. [CrossRef]
28. Watts, D.R.; Hauser, E.C.; Mohshin, M.; Dominic, D.F. Induced seismicity risk management: The problem of disappearing faults. *AAPG Bull.* **2021**, *105*, 265–273. [CrossRef]

29. White, J.A.; Foxall, W. A Phased Approach to Induced Seismicity Risk Management. *Energy Procedia* **2014**, *63*, 4841–4849. [CrossRef]
30. Gaucher, E.; Schoenball, M.; Heidbach, O.; Zang, A.; Fokker, P.A.; van Wees, J.D.; Kohl, T. Induced seismicity in geothermal reservoirs: A review of forecasting approaches. *Renew. Sustain. Energy Rev.* **2015**, *52*, 1473–1490. [CrossRef]
31. Goertz-Allmann, B.P.; Wiemer, S. Geomechanical modeling of induced seismicity source parameters and implications for seismic hazard assessment. *Geophysics* **2012**, *78*, KS25–KS39. [CrossRef]
32. Wang, L.; Kwiatek, G.; Bohnhoff, M.; Rybacki, E.; Dresen, G. Injection-induced fault slip and associated seismicity in the lab: Insights from source mechanisms, local stress states and fault geometry. *Earth Planet. Sci. Lett.* **2024**, *626*, 118515. [CrossRef]
33. Wang, L.; Kwiatek, G.; Rybacki, E.; Bonnelye, A.; Bohnhoff, M.; Dresen, G. Laboratory Study on Fluid-Induced Fault Slip Behavior: The Role of Fluid Pressurization Rate. *Geophys. Res. Lett.* **2020**, *47*, e2019GL086627. [CrossRef]
34. Jia, Z.; Liu, S.; Cheng, S.; Zhao, X.; Zhang, G. Modeling of Complex Geological Body and Computation of Geomagnetic Anomaly. *Math. Probl. Eng.* **2020**, *2020*, 9839606. [CrossRef]
35. Zhang, Q.; Zhang, X.-P.; Ji, P.-Q. Numerical study of interaction between a hydraulic fracture and a weak plane using the bonded-particle model based on moment tensors. *Comput. Geotech.* **2019**, *105*, 79–93. [CrossRef]
36. Sainoki, A.; Schwartzkopff, A.K.; Jiang, L.; Mitri, H.S. Numerical Modeling of Complex Stress State in a Fault Damage Zone and Its Implication on Near-Fault Seismic Activity. *J. Geophys. Res. Solid Earth* **2021**, *126*, e2021JB021784. [CrossRef]
37. Foley, L.; Daltaban, T.S.; Wang, J.T. Numerical simulation of fluid flow in complex faulted regions. *Geol. Soc. Lond. Spec. Publ.* **1998**, *127*, 121–132. [CrossRef]
38. Chen, S.; Ding, B.; Gong, L.; Huang, Z.; Yu, B.; Sun, S. Comparison of multi-field coupling numerical simulation in hot dry rock thermal exploitation of enhanced geothermal systems. *Adv. Geo-Energy Res.* **2019**, *3*, 396–409. [CrossRef]
39. Gan, Q.; Elsworth, D. Thermal drawdown and late-stage seismic-slip fault reactivation in enhanced geothermal reservoirs. *J. Geophys. Res. Solid Earth* **2014**, *119*, 8936–8949. [CrossRef]
40. Yoon, S.; Lee, H.; Kim, J. The modeling of fault activation, slip, and induced seismicity for geological $CO_2$ storage at a pilot-scale site in the Janggi Basin, South Korea. *Int. J. Rock Mech. Min. Sci.* **2023**, *170*, 105441. [CrossRef]
41. Yoon, J.S.; Zang, A.; Stephansson, O.; Zimmermann, G. Modelling of Fluid-injection-induced Fault Reactivation Using a 2D Discrete Element Based Hydro-mechanical Coupled Dynamic Simulator. *Energy Procedia* **2016**, *97*, 454–461. [CrossRef]
42. Treffeisen, T.; Henk, A. How pore pressure changes affect stress and strain in fault zones with complex internal structure—Insights from hydromechanical modelling. *J. Struct. Geol.* **2023**, *170*, 104847. [CrossRef]
43. Li, C.; Xu, Z. Numerical Modeling and Investigation of Fault-Induced Water Inrush Hazard under Different Mining Advancing Directions. *Mathematics* **2022**, *10*, 1561. [CrossRef]
44. Park, J.-W.; Guglielmi, Y.; Graupner, B.; Rutqvist, J.; Park, E.-S. Numerical modelling of Fault Reactivation Experiment at Mont Terri Underground Research Laboratory in Switzerland: DECOVALEX-2019 TASK B (Step 2). *Tunn. Undergr. Space* **2019**, *29*, 197–213.
45. Chang, K.W.; Yoon, H.; Kim, Y.; Lee, M.Y. Operational and geological controls of coupled poroelastic stressing and pore-pressure accumulation along faults: Induced earthquakes in Pohang, South Korea. *Sci. Rep.* **2020**, *10*, 2073. [CrossRef]
46. Flodin, E.; Aydin, A. Faults with asymmetric damage zones in sandstone, Valley of Fire State Park, southern Nevada. *J. Struct. Geol.* **2004**, *26*, 983–988. [CrossRef]
47. Brogi, A. Fault zone architecture and permeability features in siliceous sedimentary rocks: Insights from the Rapolano geothermal area (Northern Apennines, Italy). *J. Struct. Geol.* **2008**, *30*, 237–256. [CrossRef]
48. Berg, S.S.; Skar, T. Controls on damage zone asymmetry of a normal fault zone: Outcrop analyses of a segment of the Moab fault, SE Utah. *J. Struct. Geol.* **2005**, *27*, 1803–1822. [CrossRef]
49. Ben-Zion, Y.; Sammis, C.G. Characterization of Fault Zones. *Pure Appl. Geophys.* **2003**, *160*, 677–715. [CrossRef]
50. Wang, L.; Kwiatek, G.; Renard, F.; Guérin-Marthe, S.; Rybacki, E.; Bohnhoff, M.; Naumann, M.; Dresen, G. Fault roughness controls injection-induced seismicity. *Proc. Natl. Acad. Sci. USA* **2024**, *121*, e2310039121. [CrossRef]
51. Wu, Z.; Chen, W.; Xue, Y.; Song, G.; Liu, H. Structural Characteristics of Faulting Zone and Its Ability in Transporting and Sealing Oil and Gas. *Acta Geol. Sin.* **2010**, *84*, 570–578.
52. Egholm, D.L.; Clausen, O.R.; Sandiford, M.; Kristensen, M.B.; Korstgård, J.A. The mechanics of clay smearing along faults. *Geology* **2008**, *36*, 787–790. [CrossRef]
53. Xiao-Fei, F.U.; Xue, G.; Li-Xu, Z.; Yan-Fang, L.V. Formation and evolution of clay smear and hydrocarbon migration and sealing. *J. China Univ. Min.* **2012**, *41*, 52–63.
54. Allan, U.S. Model for hydrocarbon migration and entrapment within faulted structures. *AAPG Bull.* **1989**, *73*, 803–811.
55. Hibbitt, Karlsson & Sorensen. *ABAQUS Theory Manual, Version 6.2*; Hibbitt, Karlsson & Sorensen: Providence, RI, USA, 2001.
56. Zhou, X.; Burbey, T.J. Deformation characteristics of a clayey interbed during fluid injection. *Eng. Geol.* **2014**, *183*, 185–192. [CrossRef]
57. Biot, M.A. General Theory of Three-Dimensional Consolidation. *J. Appl. Phys.* **1941**, *12*, 155–164. [CrossRef]
58. Rice, J.R.; Cleary, M.P. Some basic stress diffusion solutions for fluid-saturated elastic porous media with compressible constituents. *Rev. Geophys. Space Phys.* **1976**, *14*, 227–241. [CrossRef]
59. Aki, K. Scaling Law of Seismic Spectrum. *J. Geophys. Res. Atmos.* **1967**, *72*, 1217–1231. [CrossRef]
60. Hanks, T.C.; Kanamori, H. A moment magnitude scale. *J. Geophys. Res.* **1979**, *84*, 2348. [CrossRef]

61. Mazzoldi, A.; Rinaldi, A.P.; Borgia, A.; Rutqvist, J. Induced seismicity within geological carbon sequestration projects: Maximum earthquake magnitude and leakage potential from undetected faults. *Int. J. Greenh. Gas Control.* **2012**, *10*, 434–442. [CrossRef]
62. Rinaldi, A.P.; Rutqvist, J. Modeling of deep fracture zone opening and transient ground surface uplift at KB-502 $CO_2$ injection well, in Salah, Algeria. *Int. J. Greenh. Gas Control.* **2013**, *12*, 155–167. [CrossRef]
63. Guo, M.; Ji, C.; Liu, Z.; Zhang, P. Artificial hydraulic fracture propagation law of hot dry rock reservoir in Gonghe Basin, Qinghai province. *J. Xi'an Univ. Sci. Technol.* **2023**, *43*, 514–522.

*water*

*Article*

# Predicting Arsenic Contamination in Groundwater: A Comparative Analysis of Machine Learning Models in Coastal Floodplains and Inland Basins

**Zhenjie Zhao [1,2], Amit Kumar [1,2] and Hongyan Wang [1,2,3,*]**

1 Key Laboratory of Hydrometeorological Disaster Mechanism and Warning, Ministry of Water Resources, Nanjing University of Information Science and Technology, Nanjing 210044, China
2 School of Hydrology and Water Resources, Nanjing University of Information Science and Technology, Nanjing 210044, China
3 State Key Lab of Urban and Regional Ecology, Research Center for Eco-Environmental Sciences, Chinese Academy of Sciences, Beijing 100085, China
* Correspondence: why_cugb@163.com

**Abstract:** Arsenic (As) contamination in groundwater represents a major global health threat, potentially impacting billions of individuals. Elevated As concentrations are found in river floodplains across south and southeast Asia, as well as in the inland basins of China, despite varying sedimentological and hydrogeochemical conditions. The specific mechanisms responsible for these high As levels remain poorly understood, complicating efforts to predict and manage the contamination. Applying hydro-chemical, geological, and soil parameters as explanatory variables, this study employs multiple linear regression (MLIR) and random forest regression (RFR) models to estimate groundwater As concentrations in these regions. Additionally, random forest classification (RFC) and multivariate logistic regression (MLOR) models are applied to predict the probability of As levels exceeding 10 μg/L in the Hetao Basin (China) and Bangladesh. Model validation reveals that RFR explains 80% and 70% of spatial variability of As concentration in the Hetao Basin and Bangladesh, respectively, outperforming MLIR, which accounts for only 35% and 32%. Similarly, RFC outperforms MLOR in predicting high As probability, achieving correct classification rates of 98.70% (Hetao Basin) and 98.25% (Bangladesh) on training datasets, and 82.76% (Hetao Basin) and 91.20% (Bangladesh) on validation datasets. The performance of the MLOR model on the validation set yields accuracy rates of 81.60% and 72.18%, respectively. In the Hetao Basin, $Ca^{2+}$, redox potential (Eh), Fe, pH, $SO_4{}^{2-}$, and $Cl^-$ are key predictors of As contamination, while in Bangladesh, soil organic carbon (SOC), pH, and $SO_4{}^{2-}$ are significant predictors. This study underscores the potential of random forest (RF) models as robust tools for predicting groundwater As contamination.

**Keywords:** arsenic; groundwater; floodplain; inland basins; multiple linear regression; random forest

**Citation:** Zhao, Z.; Kumar, A.; Wang, H. Predicting Arsenic Contamination in Groundwater: A Comparative Analysis of Machine Learning Models in Coastal Floodplains and Inland Basins. *Water* **2024**, *16*, 2291. https://doi.org/10.3390/w16162291

Academic Editors: Zongjun Gao and Jiutan Liu

Received: 20 July 2024
Revised: 13 August 2024
Accepted: 13 August 2024
Published: 14 August 2024



## 1. Introduction

Geogenic arsenic (As) contamination in groundwater is a pressing global environmental health issue, particularly affecting south and southeast Asian countries [1–3]. Despite extensive research [4–6], the mechanisms driving As mobilization and the prediction of its concentrations remain elusive due to irregular distribution of As and the limited availability of data. Consequently, there is a critical need for the development of reliable prediction methods [7,8]. Elevated As levels are typically found in flat, low-lying sedimentary aquifers from the Late Pleistocene to Holocene periods [9,10]. Potential geochemical processes responsible for high As concentrations include reductive dissolution of As-bearing iron oxides under reducing conditions, oxidative dissolution of As-bearing pyrite in oxic environments, and As desorption driven by pH changes and competitive ion interactions [11]. These primary release mechanisms are influenced by various hydrological and biogeochemical factors [12,13], with geographical variations further complicating As contamination patterns. Consequently, predicting groundwater As contamination is challenging due to the interplay of multiple factors such as pH, redox potential (Eh), ionic composition, and mineral type [14] and is difficult to predict using a single parameter.

Inland basins, typically the Datong, the Hetao, and the Huhhot Basins in North China, exhibit widespread As contamination. Previous studies demonstrate that elevated As concentrations occurred resulting from Fe oxide dissolution under reducing conditions [15], sulfide mineral weathering, and cation exchange processes [16]. Additionally, historical mining activities may have contributed to As transport downstream [17]. Similarly, floodplains, especially in Bangladesh, Cambodia, and Pakistan, have high-As groundwater. Previous studies indicate that elevated As concentrations occurred resulting from redox changes involving the reduction of Fe-oxyhydroxides and subsequent oxidation [18–20]. Furthermore, $HCO_3^-$ presence and elevated pH levels can mobilize As from minerals and sediments [21,22]. Overall, inland basins and floodplains exhibit different yet overlapping As contamination mechanisms. Previous studies [23,24] typically relied on hydro-chemical or geological data for predicting As contamination in a single area, either in inland basins or flood Basins. A combination of hydrogeochemical and geological data and simultaneously predicting groundwater As contamination in those two regions could provide insight for understanding mechanisms to control As contaminations.

Accurate and precise estimates of groundwater As contamination is essential for effective pollution controls. Various modeling techniques, such as artificial neural networks (ANN), multiple linear regression (MLIR), random forest (RF), and principal component regression (PCR), have been widely used for contamination forecasting [25–28]. While MLIR is commonly employed for predicting As concentrations, its linear nature and difficulty in capturing outliers often result in lower accuracy [29,30]. In this context, MLIR and MLOR models are now commonly employed as benchmarks to assess the performance of alternative models. These models set the baseline for the minimum acceptable accuracy in various studies [31]. For example, RF models generally demonstrate superior performance compared to MLIR and MLOR models [32,33]. Despite this, MLIR remains a valuable tool; for instance, it has shown better performance than ANN in predicting water quality indices in Nigeria [34]. Additionally, research has indicated that decision trees (DT), used in isolation, can achieve greater accuracy and efficiency compared to PCR, with RF representing an ensemble of such decision trees, demonstrating even greater efficacy over DT [35,36]. In contrast, random forest regression (RFR) offers several advantages, including the ability to establish nonlinear relationships, handle categorical and continuous variables, prevent overfitting, and provide unbiased error rate measurements [37]. These characteristics make RFR particularly suitable for complex environmental datasets. Few studies have demonstrated the effectiveness of RFR in predicting As contamination in groundwater [38,39]. Furthermore, implementing binary classification could offer valuable insights into the presence or absence of contamination, thus supporting more targeted and efficient mitigation strategies. However, binary classification methods have not been extensively explored in As contamination predictions. In this study, MLIR and RFR models were utilized to predict As concentrations in groundwater. Additionally, multivariate logistic regression (MLOR) and random forest classification (RFC) methods were applied to assess the probability of high As contamination (>10 μg/L) in the Hetao Basin and Bangladesh. The objectives were to: (i) model and predict As contamination in the Hetao Basin and Bangladesh using MLIR, RFR, MLOR, and RFC by incorporating hydro-chemical, soil, and geological data; (ii) evaluate and compare the performance of these models in forecasting groundwater As concentrations; and (iii) identify and analyze differences in As contamination patterns between the Hetao Basin and Bangladesh.

## 2. Materials and Methods

### 2.1. Study Area

The Hetao Basin is located between the Yellow River and the Langshan Mountains (Figure 1). The Langshan Mountains, which rise up to 2365 m above mean sea level (m asl), bound the basin to the northwest. These mountains were uplifted by a normal fault along the mountain front, which originated at the end of the Jurassic period and remained active until the end of the Late Pleistocene, with a fault slope of 60–70°. The Langshan Mountains consist primarily of a metamorphic complex [40], encompassing Jurassic to Cretaceous metamorphic sedimentary rocks (sandstone, mudstone, and shale) and Mesoproterozoic metamorphic rocks (quartz slate, quartzite, phyllite, marble, schist, and two-mica schist), as well as intrusive rocks (granite and diorite). Tectonic movements have folded and fractured these complex Cenozoic sedimentary rocks, including conglomerate and sandstone, which outcrop in the valleys. During the Tertiary period, red or deep brown sedimentary rocks formed under oxic conditions [41], serving as the primary source of sediments for the adjacent flat plain.

Alluvial fans have developed near the mountain valleys due to the deposition of pluvial sediments, extending from the mountain ranges to a broad plain with gradients between 1/100 and 1/150. The flat plain, covering 75% of the study area, features gradients between 1/1000 and 1/5000 and contains 1500 to 8000 m of sediments. Pluvial sediments, primarily gravel and coarse to medium sand, are found in the alluvial fans, while inland lacustrine sediments, composed of silt and fine sand, were locally deposited in the plain during the Quaternary period [42]. Groundwater is mainly found in Quaternary alluvial, alluvial-pluvial, and alluvial-lacustrine aquifers. The alluvial-pluvial unconfined aquifers are predominant in the belt of alluvial fans, whereas fluvial-lacustrine leaky-confined aquifers are common in the flat plain [43].

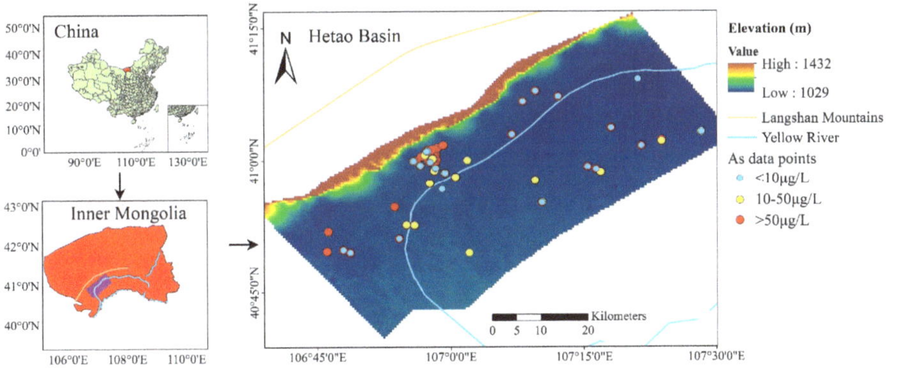


**Figure 1.** Spatial distribution of groundwater As concentrations in the Hetao Basin (data from [16,44]).

The Bengal delta plain, one of the largest deltas globally, has been shaped by sediment deposition from the Ganges–Meghna–Brahmaputra (GMB) river system (Figure 2). The Bangladesh encompasses most of the Bengal delta plain, with a portion extending into West Bengal. The Pleistocene uplands in northwest Bangladesh are known as the Barind Tract, while the central region contains the Madhupur Tract. These areas feature the characteristic brown Madhupur Clay and Dupi Tila sands, remnants of uplifted blocks from periods of low sea level. The Ganges–Brahmaputra–Meghna rivers transport an estimated annual sediment load of several $10^{12}$ kg, with about one-third deposited in the subaerial delta and floodplain [45]. The substantial influx of detritus from the eroded ultramafic, granitic, and high-grade metamorphic rocks of the Himalayas' northern, central, and southern regions is the primary source of dominant minerals such as quartz, biotite, and feldspar [46]. In the Bengal Delta, alluvial deposits of silt, clay, sand, and gravel, typically 30–80 m thick, have accumulated over the past 5–7 thousand years of the Holocene, unconformably overlying the Pleistocene fluvio-deltaic sediments. These Pleistocene sediments, comprising oxidized mud and sand layers, were tectonically uplifted and are associated with several active NE–SW faults. Sediment types deposited in the delta have been significantly influenced by global sea level changes during Pleistocene glaciations [45,47].

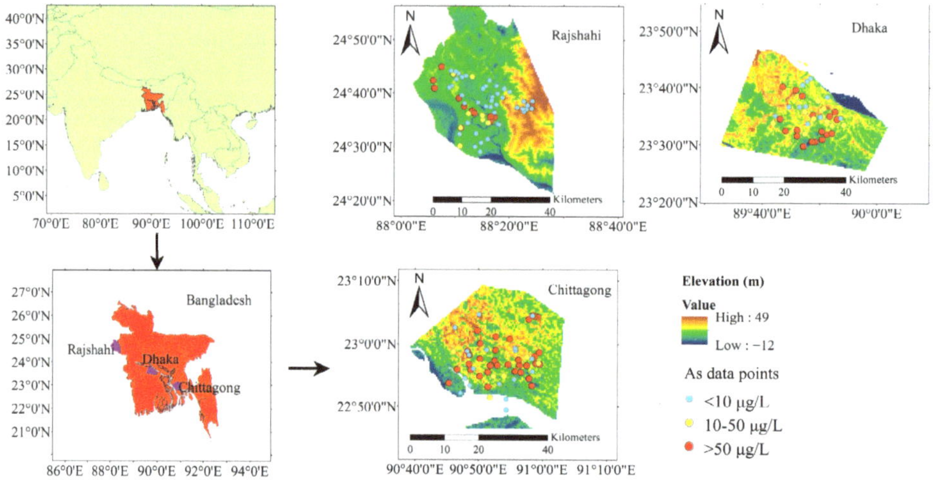


**Figure 2.** Spatial distribution of groundwater arsenic concentrations in Bangladesh (data from [48]).

## 2.2. Hydrogeochemical Data Collecting and Processing

Studies indicate that high As concentrations in groundwater are generally influenced by hydrogeochemical, geological, and soil factors [11]. The interaction between surface water and groundwater, precipitation, and human activities can also influence As concentration [46,49–51]. The selection of hydrogeochemical and soil-related parameters is based on their direct influence on the concentration and behavior of As in groundwater. Hydrogeochemical parameters, such as pH, oxidation-reduction potential (Eh), and the presence of specific ions like calcium ($Ca^{2+}$), chloride ($Cl^-$), and sulfate ($SO_4{}^{2-}$) are crucial as they directly affect the mobility and speciation of arsenic. Soil properties, including organic carbon density (OCD), clay content (CC), soil organic carbon (SOC), bulk density (BD), silt content (SC), and cation exchange capacity (CEC), play significant roles in arsenic retention and release in the soil-water system. These parameters determine the adsorption capacity of soils, influencing how arsenic is transported through the subsurface environment.

By considering these hydrogeochemical and soil-related parameters, we aim to capture the primary factors that control arsenic distribution in groundwater, providing a robust basis for predicting arsenic concentrations. While other factors like surface water interactions, precipitation, and human activities also play roles, our focus on these parameters ensures a comprehensive understanding of the natural geochemical processes governing arsenic behavior in groundwater systems.

To predict groundwater As concentrations, we utilized As concentration data from groundwater wells in the Hetao Basin and three regions in Bangladesh (Rajshahi, Dhaka, and Chittagong). The As datasets for Bangladesh were sourced from [48], encompassing data collected from 1997 to 2001. Data for the Hetao Basin were obtained from [44] and [16], with As concentrations measured during field campaigns in July 2007 and July 2017. The study area in the Hetao Basin is divided into three zones: the alluvial fan, the transitional zone, and the plain zone. Water table levels in the alluvial fan are approximately 20 m below the land surface (BLS), while in the plain zone, groundwater levels drop to about 1 m BLS [44]. Ref. [16] reported groundwater levels ranging from 13 to 32 m BLS, with an average of 23 m BLS. In comparison, Ref. [48] documented groundwater levels ranging from 3.6 to 37.5 m BLS, with an average of 24.5 m BLS [48]. Regarding the concern about the comparability of hydrochemical data collected at different times, we argue that groundwater in these regions is highly stable over time. This stability ensures that using data from different periods is reasonable and reliable for predicting As concentrations. This inherent stability in groundwater characteristics allows for robust comparisons and analyses across datasets collected at different times. Hydrogeochemical data including $Ca^{2+}$, $Cl^{-}$, DOC, Eh, Mg, pH, and $SO_4^{2-}$ were included within these As datasets. Additional geological and soil properties, including OCD, CC, SOC, BD, SC, and CEC, were acquired from SoilGrids (https://soilgrids.org/). The spatial resolution of the SoilGrids data is 250 m. The geological cross sections of the two areas are shown in Figure S1 and Figure 2.

After collecting the data, the first step involves removing any missing or blank entries to ensure the dataset's integrity and reliability. This preprocessing step is crucial for eliminating noise and potential biases that could affect subsequent analyses. Once the dataset is cleaned, it is then processed and analyzed according to the three sigma ($3\sigma$) rule [52]. After applying the $3\sigma$ rule, a comprehensive statistical analysis is performed on the refined dataset. This analysis includes calculating key statistical metrics such as the maximum value, minimum value, mean, and variance. These metrics provide a detailed summary of the data distribution and variability, offering insights into the central tendency and dispersion of the data.

With the statistical groundwork laid, the next phase involves conducting modeling analysis. This process is systematic and methodical, adhering to a predefined sequence of methods as illustrated in Figure 3. Each step in this sequence builds upon the previous one, ensuring a robust and thorough evaluation of the data. The modeling methods are chosen based on their relevance and applicability to the specific characteristics of the dataset, aiming to extract meaningful patterns and relationships.

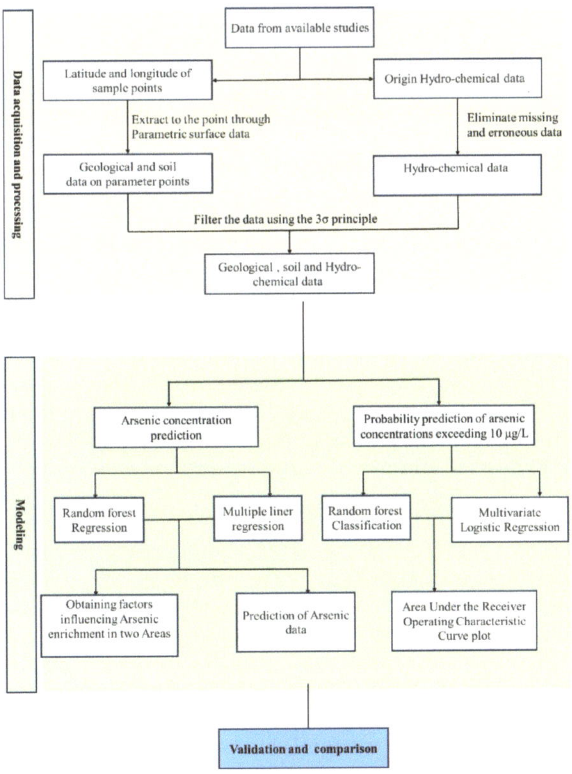


**Figure 3.** Flow diagram of methodologies adopted for this study.

## 2.3. Multicollinearity Assessment

In this study, two methods were employed to assess the relationships among the selected factors: Variance Inflation Factor (VIF) with a threshold of less than 10, and Pearson's Correlation Coefficients (r) not exceeding ±0.7 [53]. If the VIF and r values are >10 and >±0.7, respectively, this indicates significant collinearity issues. The VIF analysis revealed that the highest value was 9.4 (Table S1), indicating that all values were within the acceptable range (VIF < 10). This suggests that multicollinearity is not a significant concern, allowing us to confidently include all examined features in the model without risking inflated standard errors or misleading parameter estimates. The Pearson Correlation plot, shown in Figure 4, and the corresponding values listed in Tables S2 and S3 demonstrate a maximum correlation of 0.70 and a minimum of −0.63, which are within the acceptable standard of $r \leq \pm 0.7$. These results indicate that none of the predictors are highly correlated with each other, further reducing the likelihood of multicollinearity affecting the model's performance.

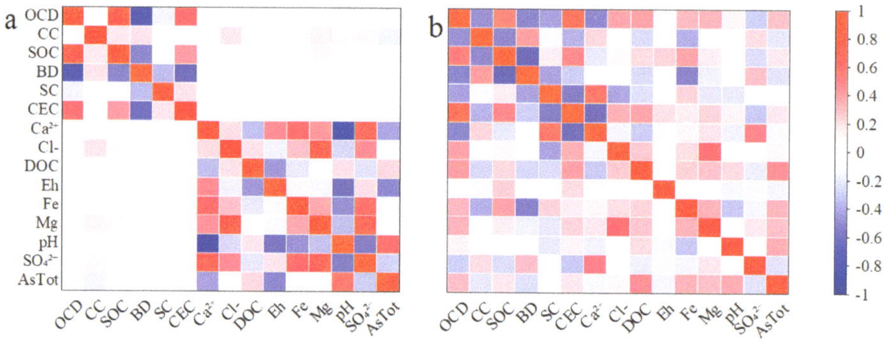


**Figure 4.** Graphical presentation of Pearson's Correlation Coefficient. (**a**): Hetao Basin; (**b**): Bangladesh. Abbreviations: OCD: organic carbon density; CC: clay content; SOC: soil organic carbon; BD: bulk density; SC: silt content; CEC: cation exchange capacity; DOC: dissolved organic carbon; AsTot: total arsenic.

These assessments influenced the final model selection by ensuring that only variables with acceptable levels of collinearity were included. By confirming that both VIF and r values are within the acceptable ranges, we established a robust foundation for the model. This careful selection process enhances the reliability and validity of the model's predictions, ensuring that the estimated coefficients are unbiased and that the model's interpretability is maintained. Consequently, this approach eliminates the possibility of false significance. For instance, if Feature 1 is significant for As concentration but also significantly correlated with Feature 2, leading to the incorrect conclusion that Feature 2 is also significant for As concentration, our method prevents such misleading outcomes [54].

### 2.4. Feature Selections for Models

For MLIR, we assessed the significance of independent variables using *p*-values, considering a threshold of 0.05 to indicate significant correlations. Variables identified as significant ($p < 0.05$) parameters in Table 1 were used in the MLIR model to predict As contamination. In the Hetao Basin, the significant predictors were pH, Fe, and $SO_4^{2-}$, while significant predictors were pH, Mg, Fe, and DOC in Bangladesh. For the RFR model, we employed feature combination optimization to minimize the root mean square error (RMSE) and determine the optimal combination of predictive variables [55]. Various combinations of hydro-chemical, geological, and soil parameters were tested, and the RMSE value in the validation phase was used to identify the best model. The optimal predictive variables for the Hetao Basin were determined to be $Ca^{2+}$, Eh, Fe, pH, $SO_4^{2-}$, and $Cl^-$, whereas for Bangladesh, the optimal predictors were SOC, pH, and DOC.

**Table 1.** Results of multiple linear regression (MLIR) model.

| **Hetao Basin** | | | | |
|---|---|---|---|---|
| **Predictor Variables** | **Coefficients** | **Standard Error** | **Standardized Coefficients** | ***p* Value** |
| intercept | −2312.778 | 836.709 | / | 0.007 |
| OCD | −0.887 | 1.291 | −0.137 | 0.493 |
| CC | −0.852 | 0.649 | −0.094 | 0.192 |
| SOC | 1.062 | 0.808 | 0.162 | 0.191 |
| BD | −1.031 | 3.225 | −0.047 | 0.75 |
| SC | −0.052 | 0.521 | −0.011 | 0.92 |
| CEC | −0.461 | 0.742 | −0.054 | 0.536 |
| $Ca^{2+}$ | 0.718 | 0.578 | 0.18 | 0.216 |
| $Cl^-$ | −0.031 | 0.071 | −0.051 | 0.657 |
| DOC | 6.54 | 3.934 | 0.134 | 0.099 |
| Eh | −0.25 | 0.244 | −0.101 | 0.307 |
| Fe | 85.617 | 31.68 | 0.254 | **0.008** |
| Mg | 0.68 | 0.511 | 0.179 | 0.186 |
| pH | 349.246 | 67.566 | 0.679 | **<0.001** |
| $SO_4{}^{2-}$ | −0.201 | 0.1 | −0.24 | **0.047** |
| **Bangladesh** | | | | |
| **Predictor Variables** | **Coefficients** | **Standard Error** | **Standardized Coefficients** | ***p* Value** |
| intercept | −877.873 | 598.590 | / | 0.145 |
| OCD | 0.102 | 0.383 | 0.033 | 0.790 |
| CC | −1.427 | 2.995 | −0.037 | 0.634 |
| SOC | −0.486 | 2.850 | −0.015 | 0.865 |
| BD | 0.075 | 0.342 | 0.026 | 0.826 |
| SC | −0.034 | 0.215 | −0.024 | 0.874 |
| CEC | −0.981 | 0.596 | −0.170 | 0.102 |
| $Ca^{2+}$ | −0.225 | 0.281 | −0.092 | 0.424 |
| $Cl^-$ | −0.156 | 0.090 | −0.141 | 0.086 |
| DOC | 10.117 | 2.981 | 0.244 | **0.001** |
| Eh | −0.052 | 0.101 | −0.034 | 0.605 |
| Fe | 10.207 | 2.554 | 0.338 | **<0.001** |
| Mg | 2.005 | 0.669 | 0.275 | **0.003** |
| pH | 173.960 | 32.274 | 0.379 | **<0.001** |
| $SO_4{}^{2-}$ | −0.784 | 0.427 | −0.144 | 0.069 |

Furthermore, for MLOR and RFC, we utilized a variable selection method based on r [56,57]. This method involves: (i) starting with the variables with the highest positive or negative r from the available dataset, and (ii) gradually adding variables with lower r until the variable with the lowest r is included. The goal was to determine whether the variable with the highest r alone could accurately predict the outcome or if additional variables were necessary for optimal prediction. Models were evaluated based on RMSE values in the validation phase to determine the best variable combination. Ultimately, MLOR identified pH, Fe, and $SO_4{}^{2-}$ as the optimal predictors for the Hetao Basin, highlighting the significant role of these factors in the region's environmental conditions. For Bangladesh, pH, Mg, Fe, and DOC were the significant predictors, indicating the diverse influences of geologic conditions and water chemistry in this area. On the other hand, RFC determined $Ca^{2+}$, Fe, pH, $SO_4{}^{2-}$, and $Cl^-$ as the best predictors for the Hetao Basin. This suggests that the combination of these variables captures the complexity of the environmental interactions more effectively for this region. For Bangladesh, RFC determined SOC, pH, and DOC as the optimal predictors, underlining the importance of soil organic content and chemical properties in this context.

The feature importance plots (Figure S3) visually represent the significance of each variable in the RF models. These plots clearly show the dominance of certain variables, with pH and DOC consistently emerging as crucial factors across different regions and models. The consistency of pH as a significant predictor underscores its fundamental role in influencing environmental processes. Similarly, the importance of DOC in both regions indicates its critical function in soil and water chemistry dynamics.

RF is a machine learning algorithm that creates multiple decision trees and aggregates their results to produce a final output [58]. Generally, a higher number of decision trees improves model performance. However, excessive decision trees can lead to overfitting, diminishing the model's reliability [59]. To prevent this, we restricted the number of decision trees in the RFR and RFC models to a maximum of 100. The optimal number of trees is determined by selecting the quantity that maximizes the $R^2$ coefficient for RFR and the accuracy rate for RFC.

### 2.5. Adopted Modeling Approach and Validation

This study employed two predictive models, RFR and MLIR, to forecast As contamination. The RFC and MLOR approaches were used to develop a predictive model for assessing probability of high or low risk based on a specified As threshold. To date, these methods have been widely used in research on groundwater As contamination [39,60]. The models utilized raw As concentration values for RFR and MLIR, while RFC and MLOR used binary-coded data points: contaminated (As $\geq$ 10 µg/L, coded as 1) or uncontaminated (As < 10 µg/L, coded as 0). These analyses were conducted using the Python programming language [61]. For model construction and validation, sampling points were randomly split into two subsets: 80% of the samples were used to build the model, and the remaining 20% were used for validation. The MLOR model uses a set of $x$ independent variables to predict the response variable Y, which in this study represents the As concentration. This model is formulated as shown in Equation (1),

$$Y = \beta_0 + \beta_1 x_1 + \beta_2 x_2 + \beta_3 x_3 + \beta_4 x_4 + \epsilon \quad (1)$$

where Y is a dependent variable; $x_1$, $x_2$, $x_3$, and $x_4$ are the independent variables; $\beta_0$ is the y intercept; $\beta_1$, $\beta_2$, $\beta_3$, and $\beta_4$ are the slope; and $\epsilon$ is a random error.

The sigmoid function is represented by Equation (2) as follows:

$$P = 1/1 + e^{-y} \quad (2)$$

Apply the sigmoid function in Equation (3) as follows:

$$P = 1/1 + e^{-(\beta_0 + \beta_1 x_1 + \beta_2 x_2 + \beta_3 x_3 + \beta_4 x_4)} \quad (3)$$

For the evaluation of models, RFR and MLIR use the coefficient of determination ($R^2$) of the regression model, and RFC and MLOR employ three statistical validation techniques and one graphical technique: operating characteristics (ROC) curve, Correct Classification Rate, False Negative Rate, and False Positive Rate. The AUC (area under the ROC curve) ranges from 0.5 to 1, with values closer to 1 indicating better model performance [62]. These metrics are used to assess accuracy [63], calculated as follows:

$$AUC = \frac{(\Sigma TP + \Sigma TN)}{(P + N)} \quad (4)$$

$$Correct\ Classification\ Rate = \frac{TP + TN}{TP + TN + FP + FN} \quad (5)$$

$$False\ Negative\ Rate = \frac{FN}{TP + FN} \quad (6)$$

$$False\ Positive\ Rate = \frac{FP}{TN + FP} \quad (7)$$

where *TP*, *TN*, *FP*, and *FN* interpret the true positive, true negative, false positive, and false-negative accordingly, and *N* is the total number of samples.

## 3. Results and Discussion

### *3.1. Hydro-Chemical and Geological Characteristics of Groundwater in the Hetao Basin and Bangladesh*

This study compares groundwater hydrogeochemical and geological characteristics between the Hetao Basin in China and three regions in Bangladesh (Rajshahi, Dhaka, and Chittagong). Key geological parameters include OCD, CC, SOC, BD, SC, CEC (Table 2), while key hydrogeochemical characteristics include $Ca^{2+}$, $Cl^{-}$, DOC, Eh, Fe, Mg, pH, $SO_4^{2-}$, and As (Table 2).

**Table 2.** Descriptive statistics of the variables in models.

| | | Maximum | | Minimum | | Average ± Standard Deviation | |
|---|---|---|---|---|---|---|---|
| **Dataset** | **Unit** | **Hetao Basin** | **Bangladesh** | **Hetao Basin** | **Bangladesh** | **Hetao Basin** | **Bangladesh** |
| OCD | hg/m$^3$ | 205 | 364 | 32 | 247 | 81.6 ± 31.3 | 300.4 ± 30.7 |
| CC | g/kg | 246 | 131 | 117 | 121 | 166.5 ± 22.3 | 127.1 ± 2.5 |
| SOC | t/ha | 221 | 51 | 24 | 35 | 65.6 ± 30.9 | 42.3 ± 3 |
| BD | cg/cm$^3$ | 154 | 384 | 119 | 243 | 141 ± 9.3 | 321.2 ± 32.9 |
| SC | g/kg | 277 | 549 | 85 | 313 | 149.2 ± 40.8 | 422.8 ± 65 |
| CEC | mmol/kg | 252 | 227 | 123 | 156 | 163.2 ± 23.8 | 188.4 ± 16.4 |
| $Ca^{2+}$ | mg/L | 219.8 | 183 | 3.6 | 16.6 | 67.2 ± 51 | 78.9 ± 38.8 |
| $Cl^{-}$ | mg/L | 1645 | 637 | 31.8 | 1 | 340 ± 325.8 | 46.3 ± 85.5 |
| DOC | mg/L | 34 | 14 | 1 | 0.1 | 5 ± 4.2 | 2.4 ± 2.3 |
| Eh | mv | 143.6 | 278 | −246 | −105 | −102.6 ± 81.8 | 76.5 ± 62.2 |
| Fe | mg/L | 2.7 | 12.1 | 0 | 0.007 | 0.4 ± 0.6 | 2.9 ± 3.1 |
| Mg | mg/L | 264.7 | 79.9 | 12.6 | 8.9 | 64.7 ± 53.3 | 30.5 ± 13 |
| pH | unitless | 8.8 | 7.55 | 7 | 6.53 | 7.9 ± 0.4 | 7 ± 0.2 |
| $SO_4^{2-}$ | mg/L | 1123 | 115 | 0.4 | 0.2 | 237.6 ± 242.5 | 8.2 ± 17.4 |
| As | μg/L | 946.2 | 409 | 0.3 | 0.5 | 189.3 ± 202.8 | 68.9 ± 94.8 |

Groundwater in the Hetao Basin is characterized by high salinity, indicated by elevated $Cl^{-}$ concentrations, with a maximum of 1645 mg/L and an average of 340 mg/L. $SO_4^{2-}$ levels are also high, with a maximum of 1123 mg/L and an average of 237.6 mg/L. These values reflect significant evaporative effects, leading to high total dissolved solids (TDS) and an alkaline pH of 7.9. The groundwater (Figure S4) in the Hetao Basin is primarily classified as Na-Cl and Ca-$SO_4$ type due to these high ion concentrations, alkaline conditions, and paleolake sedimentation environment [64]. In contrast, groundwater in Bangladesh demonstrates lower salinity, with $Cl^{-}$ averaging 46.3 mg/L and $SO_4^{2-}$ at 8.2 mg/L. The average pH in Bangladesh is 7.0, indicating more neutral conditions and less evaporative influence. This results in groundwater that is typically classified as Ca-$HCO_3$ and Mg-$HCO_3$ type, reflecting the significant recharge from rainfall and lower concentrations of dissolved ions (Figure S5).

The concentrations of redox elements including Fe, $SO_4^{2-}$, and DOC in the Hetao Basin and Bangladesh groundwater systems show notable differences. In the Hetao Basin, Fe predominantly adsorbed onto sediments, which are less soluble. This could be a result of the alkaline pH in the Hetao Basin [65].

The OCD is higher in Bangladesh, with a maximum of 364 hg/m$^3$ and an average of 300.4 hg/m$^3$, compared to the Hetao Basin's maximum of 205 hg/m$^3$ and an average of 81.6 hg/m$^3$. This higher organic content, along with reducing conditions, should impact As mobility [66,67]. The SOC content in the Hetao Basin is 65.6 t/ha, higher than Bangladesh's 42.3 t/ha, indicating different organic matter sources and decomposition rates, which probably results from the lacustrine deposition environment in the Hetao Basin [42]. The Hetao Basin has higher CC and CEC, with maximum values of 246 g/kg and 252 mmol/kg, respectively. This contributes to the soil's ability to retain cations, which would influence contaminant mobilities [68]. BD is higher in Bangladesh, with a maximum of 384 cg/cm$^3$

and an average of 321.2 cg/cm$^3$, compared to the Hetao Basin's maximum of 154 cg/cm$^3$ and average of 141 cg/cm$^3$, indicating a denser soil structure.

These distinct hydro-chemical conditions and soil parameters influence As behavior differently in the two regions. In the Hetao Basin, high ionic strength and alkaline pH promote As desorption from mineral surfaces. Reduction of high $SO_4^{2-}$ concentrations may also interact with iron minerals, affecting As release and transport. In Bangladesh, reductive dissolution under anoxic conditions, facilitated by high organic carbon content, could be the dominant mechanism for As release. Understanding these differences is crucial for developing targeted groundwater management and As mitigation strategies in these regions.

### 3.2. Performance of Estimation Models

Model validation results (Figures 5 and 6) indicate that the RFR model demonstrates superior performance in estimating arsenic (As) concentrations, capturing 80% and 70% of the spatial variability in the Hetao Basin and Bangladesh, respectively. This performance underscores the RFR model's high accuracy and robustness, especially in managing the intricacies of complex environmental datasets. The RFR model's ability to handle non-linear relationships and high-dimensional data attributes significantly to its effectiveness, making it a highly suitable tool for environmental science research.

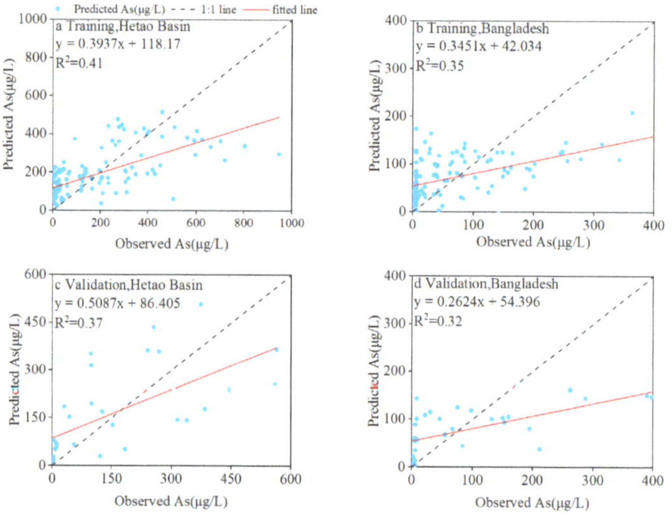


**Figure 5.** Relationship between predicted and measured As contamination for MLIR model. (**a**): Training data for Hetao Basin; (**b**): Training data for Bangladesh; (**c**): Validation data for Hetao Basin; (**d**): Validation data for Bangladesh.

In contrast, the MLIR model explains only 35% and 32% of the spatial variability in the same regions. This substantial discrepancy highlights the limitations of MLIR in capturing the complex, non-linear interactions that characterize environmental data. The inferior performance of MLIR underscores its inadequacy in contexts where data complexity and dimensionality are paramount.

The findings clearly advocate for the preferential use of sophisticated and flexible regression methods like RFR in environmental studies. The RFR model's superior handling of non-linearity and high-dimensional datasets makes it a valuable tool for researchers aiming to achieve accurate and reliable predictions. Moreover, the robustness of RFR in diverse geographical settings, as evidenced by its performance in both the Hetao Basin and Bangladesh, further solidifies its applicability and utility in global environmental science endeavors.

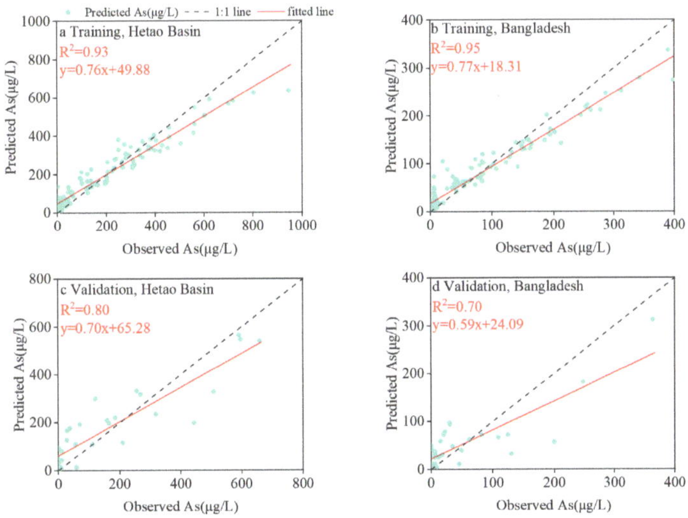


**Figure 6.** Relationship between predicted and measured As contamination for RFR model. (**a**): Training data for Hetao Basin; (**b**): Training data for Bangladesh; (**c**): Validation data for Hetao Basin; (**d**): Validation data for Bangladesh.

In terms of computational efficiency and scalability, while the RFR model offers superior predictive accuracy, it is computationally intensive, particularly with large datasets, due to its ensemble nature and the need for multiple tree evaluations. This can pose challenges in scalability and real-time application. Conversely, MLIR is more computationally efficient and easier to scale given its simpler structure, which involves solving linear equations. Regarding interpretability, MLIR is inherently more interpretable due to its straightforward linear relationships, which facilitate understanding and communication of the influence of each predictor variable. In contrast, the RFR model, despite its high accuracy, suffers from lower interpretability as it functions as a "black box" [36], making it difficult to discern the individual contributions of predictors to the overall model output.

In order to make full use of the advantages of both models, optimization techniques such as hyperparameter tuning, feature selection, and dimensionality reduction can be applied. Furthermore, hybrid approaches can improve predictive accuracy. For instance, combining RFR with MLIR through ensemble methods like stacking or blending can enhance model performance. In such hybrid models, MLIR can provide baseline predictions, while RFR can capture complex patterns and residuals, leading to a more robust and accurate predictive framework.

In conclusion, for environmental studies that necessitate dealing with complex data structures and seeking high predictive accuracy, the random forest regression model stands out as the preferred methodological approach. Future research should prioritize the implementation of such advanced regression techniques to enhance the precision and reliability of environmental assessments and predictions.

### *3.3. Application of the Binary Classification Models for Groundwater Arsenic Contamination Probability in Hetao Basin and Bangladesh*

Based on the aforementioned results from MLIR and RFR, we developed models to predict the probability of high As contamination in groundwater in the Hetao Basin and Bangladesh using RFC and MLOR. The classification results for our training dataset, along with the AUC values, demonstrate strong predictive capabilities. Specifically, the AUC for the RFC model is 0.97 for the Hetao Basin and 0.99 for Bangladesh, significantly outperforming the MLOR model, which has an AUC of 0.88 for the Hetao Basin and 0.83 for Bangladesh (Figure 7a,b). For the training dataset (Table 3), the MLOR model achieved a correct classification rate of 82.76% for the Hetao Basin, with a false positive rate of 10.35% and a false negative rate of 6.90%. For Bangladesh, the MLOR model had a correct classification rate of 79.41%, a false positive rate of 17.65%, and a false negative rate of 2.94%. In comparison, the RFC model showed significantly higher performance on the training dataset, with a correct classification rate of 98.70% for the Hetao Basin and 98.25% for Bangladesh. The false positive and false negative rates for the RFC model were substantially lower, at 0.88% and 0.42% for the Hetao Basin and 0.75% and 1.00% for Bangladesh, respectively. The classification results of the predictive model for Hetao Basin and Bangladesh are given in Table 3.

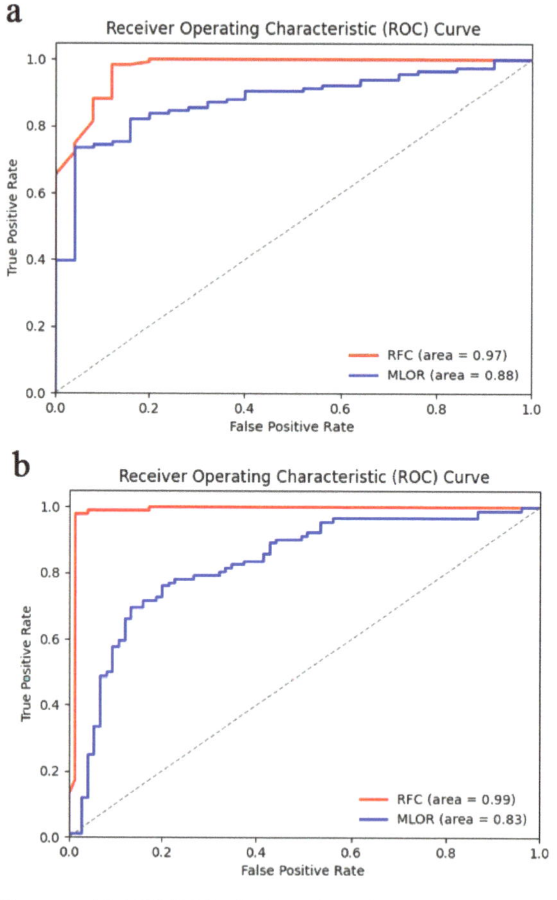


**Figure 7.** (**a**) AUC-ROC of adopted models in Hetao Basin. (**b**) AUC-ROC of adopted models in Bangladesh.

In the validation dataset, the MLOR model's performance decreased slightly, achieving a correct classification rate of 81.60% for the Hetao Basin with a false positive rate of 17.50% and a false negative rate of 0.90%. In Bangladesh, the MLOR model's correct classification rate dropped to 72.18%, with false positive and false negative rates of 19.55% and 8.27%, respectively. The RFC model maintained a better performance on the validation dataset as well, with a correct classification rate of 82.76% for the Hetao Basin and 91.20% for Bangladesh. The false positive and false negative rates for the RFC model were 15.24% and 2.00% for the Hetao Basin, and 2.94% and 5.88% for Bangladesh, respectively. These observations highlight the RFC model's enhanced robustness and predictive accuracy. The superior performance of the RFC model in validation datasets underscores its capability to generalize better across different geographic contexts, maintaining lower error rates and higher classification accuracy. This robustness suggests that the RFC model is more effective in adapting to variations in validation data compared to the MLOR model. This improved performance demonstrates the RFC model's reliability and strong predictive power for identifying high arsenic contamination in groundwater.

**Table 3.** Classification results for the predictive models in Hetao Basin and Bangladesh.

| | Correct Classification | False Positive | False Negative |
|---|---|---|---|
| | Training Dataset | | |
| MLOR (Hetao Basin) | 82.76% | 10.35% | 6.90% |
| MLOR (Bangladesh) | 79.41% | 17.65% | 2.94% |
| RFC (Hetao Basin) | 98.70% | 0.88% | 0.42% |
| RFC (Bangladesh) | 98.25% | 0.75% | 1.00% |
| | Validation Dataset | | |
| MLOR (Hetao Basin) | 81.60% | 17.50% | 0.90% |
| MLOR (Bangladesh) | 72.18% | 19.55% | 8.27% |
| RFC (Hetao Basin) | 82.76% | 15.24% | 2.00% |
| RFC (Bangladesh) | 91.20% | 2.94% | 5.88% |

Note: The cutoff value is 0.48.

### *3.4. Mechanisms Controlling Geogenic Groundwater as Contamination in Hetao Basin and Bangladesh*

According to the results of the RFR and MLIR models, key indicators for the occurrence of geogenic high As groundwater in the Hetao Basin include concentrations of $SO_4^{2-}$, Fe, pH, $Cl^-$, Eh, and $Ca^{2+}$, whereas in Bangladesh, the important indicators are $SO_4^{2-}$, Fe, pH, and DOC. Studies have demonstrated that the Holocene strata in the Hetao Basin are generally rich in organic carbon [69]. During the degradation of organic matter (OM), Fe can be utilized as an electron acceptor, leading to the reduction of Fe oxides/oxyhydroxides and the subsequent release of coexisting As [70,71]. The dissolution of Fe oxides is further exacerbated by low redox conditions, which are often indicated by low Eh values. As Fe oxides dissolve, As is released into the groundwater. In addition to dissolution of Fe oxides, competitions between anions and As can contribute to As mobilization in the Hetao Basin. Sulfate ($SO_4^{2-}$) can undergo microbial reduction to sulfide ($S^{2-}$) under anaerobic conditions, which can then react with Fe to form iron sulfide minerals, further releasing As previously adsorbed onto Fe oxides. Chloride ($Cl^-$) can also compete with As for adsorption sites, reducing As adsorption and increasing its mobility in groundwater. The positive correlation between As and pH suggests that higher pH levels are often associated with As enrichment in groundwater. Arsenate, the predominant As species in the Hetao Basin's groundwater, exhibits the greatest adsorption at low pH, with adsorption decreasing as pH increases [72]. This decrease in adsorption at higher pH levels results in higher concentrations of As in the groundwater. Furthermore, under alkaline conditions, $Ca^{2+}$ can co-precipitate with As species to form compounds such as $CaH_2AsO_4^+$, $CaHAsO_4$, and $CaAsO_4^-$ [73], which may temporarily reduce As mobility but can dissolve under changing geochemical conditions.

Compared to the Hetao Basin, soil organic carbon content was more pronounced in influencing groundwater As concentration. The aquifer sediments in Bangladesh contain OM and peat, as reported in several studies [74,75]. The degradation of OM within the aquifer generates DOC [76]. Under reducing conditions, higher DOC facilitates the Fe reduction and releases As from younger sediments [77,78]. This process is supported by microbial activity that utilizes DOC as an energy source, leading to the reduction of Fe(III) to Fe(II) and the concomitant release of As adsorbed onto Fe oxides. Furthermore, a negative correlation between As and sulfate ions suggests that As dissolution occurs under reducing conditions. Sulfate reduction can lead to the formation of sulfide, which can react with Fe to form iron sulfides. This process can either sequester As within newly formed minerals or release it from Fe oxide surfaces, depending on the specific geochemical conditions. The presence of DOC further enhances this process by providing the necessary organic substrate for sulfate-reducing bacteria.

These mechanisms illustrate how the interplay of redox conditions, organic matter degradation, and competitive adsorption processes influence the concentration and mobility of As in groundwater in both the Hetao Basin and Bangladesh.

## 4. Conclusions and Recommendations

This study demonstrates that groundwater As enrichment in both inland basins and floodplains is predominantly influenced by hydro-chemical factors, with geological features such as topography contributing to regional variations. Water-rock interactions are fundamental to the mobilization of As, with specific hydro-chemical and geological data providing valuable indicators for predicting As levels in groundwater. The RFR model was better than MLIR in predicting groundwater As contamination, offering more accurate predictions closely aligned with actual values in both training and validation datasets. Additionally, the RFC model outperformed the MLOR model in predicting high As groundwater probability, capturing nuanced variations and demonstrating higher accuracy. These findings highlight the potential of the random forest approach as a robust tool for predicting global As exposure risks. The practical implications of our findings for environmental management and policy-making are significant. Integrating predictive models like RFR and RFC into existing monitoring frameworks can vastly improve the management and mitigation of arsenic contamination. Predictive models can guide the allocation of resources for monitoring, enabling targeted sampling in high-risk areas and thereby optimizing efforts to mitigate As contamination. Based on the results, the following recommendations have been enlisted below:

- Continual calibration and validation: Regularly calibrate and validate prediction models using diverse datasets to enhance the accuracy and reliability of groundwater As predictions. This iterative process ensures that models remain accurate as new data become available.
- Implement comprehensive monitoring programs: Implement comprehensive monitoring programs that include regular sampling and analysis of hydro-chemical and geological parameters to provide up-to-date data for model inputs. This approach helps in maintaining the accuracy of predictive models and allows for early detection of potential contamination events.
- Strategic management plans: Develop and implement strategic management plans based on predictive model outcomes to mitigate As contamination in high-risk areas. Policies should focus on sustainable groundwater management and remediation efforts tailored to the specific conditions indicated by model predictions.
- Academic research integration: Foster continuous academic research that integrates hydrogeology, geochemistry, and data science to comprehensively address the complex challenge of groundwater As contamination. This multidisciplinary approach can lead to innovative solutions and improvements in prediction models.
- Public health initiatives: Strengthen public health initiatives by disseminating information on As risks and providing resources for safe water alternatives in affected regions. Public awareness and education are crucial for minimizing exposure and protecting health.
- International collaboration: Encourage international collaboration to share knowledge, data, and resources, enhancing the global capacity to effectively predict and manage groundwater As contamination. Collaboration can lead to the development of more robust models and shared strategies for mitigation.

By adopting these recommendations, we can improve the predictive accuracy of groundwater As models and develop effective strategies to safeguard aquifers and protect human health from the adverse effects of As contamination.

Despite the robust findings, there are potential limitations and confounding factors that could influence groundwater As contamination. These include data quality and availability, as the accuracy of predictive models depends on the quality and availability of input data; inconsistent or sparse data, which can reduce the model's reliability; temporal variations, as groundwater As concentrations can vary seasonally and annually, and so models must account for these temporal variations to provide accurate predictions over time; complex hydrogeological interactions, as interactions between various hydro-chemical and geological factors are complex and not fully understood, and simplifications and assumptions in the models might overlook some critical interactions; and human activities, since agricultural practices, industrial activities, and land use changes can significantly impact groundwater As levels. These anthropogenic factors should be integrated into predictive models to improve their accuracy.

The findings of this study underscore the importance of hydro-chemical factors in influencing groundwater As contamination and highlight the effectiveness of random forest models in predicting As exposure risks. By integrating these predictive models into environmental management frameworks, we can enhance the precision of As contamination predictions and develop effective strategies to safeguard aquifers and protect public health. Adopting the outlined recommendations will improve model accuracy and reliability, foster interdisciplinary research, and promote international collaboration, ultimately leading to better management and mitigation of groundwater As contamination on a global scale.

**Supplementary Materials:** The following supporting information can be downloaded at: https://www.mdpi.com/article/10.3390/w16162291/s1, Figure S1 Hydrogeological cross-section from north to south across Hetao Basin. Figure S2 Hydrogeological cross-section from north to south across Bangladesh. Figure S3 Feature Importance of Different Variables (a: Hetao Basin; b: Bangladesh). Figure S4 Piper plot of groundwaters in Hetao Basin. Figure S5 Piper plot of groundwaters in Bangladesh. Table S1 Multicollinearity analysis of different factors. Table S2 Pearson's correlation coefficient between factors in Hetao Basin (n = 143). Table S3 Pearson's correlation coefficient between factors in Bangladesh (n = 167). Reference [79] is cited in the Supplementary Materials.

**Author Contributions:** Conceptualization: H.W.; Writing—original draft preparation: Z.Z.; Writing—review & editing: A.K. and H.W.; Supervision: H.W. All authors have read and agreed to the published version of the manuscript.

**Funding:** This study was supported by the National Natural Science Foundation of China (No. 42207268).

**Data Availability Statement:** The raw data supporting the conclusions of this article will be made available by the authors on request.

**Conflicts of Interest:** The authors declare there to be no conflicts of interest.

## References

1. Argos, M.; Kalra, T.; Rathouz, P.J.; Chen, Y.; Pierce, B.; Parvez, F.; Islam, T.; Ahmed, A.; Rakibuz-Zaman, M.; Hasan, R.; et al. Arsenic Exposure from Drinking Water, and All-Cause and Chronic-Disease Mortalities in Bangladesh (HEALS): A Prospective Cohort Study. *Lancet* **2010**, *376*, 252–258. [CrossRef] [PubMed]
2. Chowdhury, U.K.; Biswas, B.K.; Chowdhury, T.R.; Samanta, G.; Mandal, B.K.; Basu, G.C.; Chanda, C.R.; Lodh, D.; Saha, K.C.; Mukherjee, S.K.; et al. Groundwater Arsenic Contamination in Bangladesh and West Bengal, India. *Environ. Health Perspect.* **2000**, *108*, 393–397. [CrossRef] [PubMed]
3. Karim, M.M. Arsenic in Groundwater and Health Problems in Bangladesh. *Water Res.* **2000**, *34*, 304–310. [CrossRef]
4. Roberts, L.C.; Hug, S.J.; Dittmar, J.; Voegelin, A.; Saha, G.C.; Ali, M.A.; Badruzzaman, A.B.M.; Kretzschmar, R. Spatial Distribution and Temporal Variability of Arsenic in Irrigated Rice Fields in Bangladesh. 1. Irrigation Water. *Environ. Sci. Technol.* **2007**, *41*, 5960–5966. [CrossRef] [PubMed]
5. Concha, G.; Nermell, B.; Vahter, M. Spatial and Temporal Variations in Arsenic Exposure via Drinking-Water in Northern Argentina. *J. Health Popul. Nutr.* **2006**, *24*, 317–326. [PubMed]
6. Dittmar, J.; Voegelin, A.; Roberts, L.C.; Hug, S.J.; Saha, G.C.; Ali, M.A.; Badruzzaman, A.B.M.; Kretzschmar, R. Spatial Distribution and Temporal Variability of Arsenic in Irrigated Rice Fields in Bangladesh. 2. Paddy Soil. *Environ. Sci. Technol.* **2007**, *41*, 5967–5972. [CrossRef] [PubMed]
7. Mahimairaja, S.; Bolan, N.S.; Adriano, D.C.; Robinson, B. Arsenic Contamination and Its Risk Management in Complex Environmental Settings. *Adv. Agron.* **2005**, *86*, 1–82.

8. Ayotte, J.D.; Nolan, B.T.; Nuckols, J.R.; Cantor, K.P.; Robinson, G.R.; Baris, D.; Hayes, L.; Karagas, M.; Bress, W.; Silverman, D.T.; et al. Modeling the Probability of Arsenic in Groundwater in New England as a Tool for Exposure Assessment. *Environ. Sci. Technol.* **2006**, *40*, 3578–3585. [CrossRef] [PubMed]
9. Nordstrom, D.K. Arsenic in the Geosphere Meets the Anthroposphere. In Proceedings of the Understanding the Geological and Medical Interface of Arsenic, As 2012—4th International Congress: Arsenic in the Environment, Cairns, Australia, 22–27 July 2012.
10. Smedley, P.; Zhang, M.; Zhang, G.; Luo, Z. Mobilisation of Arsenic and Other Trace Elements in Fluviolacustrine Aquifers of the Huhhot Basin, Inner Mongolia. *Appl. Geochem.* **2003**, *18*, 1453–1477. [CrossRef]
11. Wang, Y.; Pi, K.; Fendorf, S.; Deng, Y.; Xie, X. Sedimentogenesis and Hydrobiogeochemistry of High Arsenic Late Pleistocene-Holocene Aquifer Systems. *Earth-Sci. Rev.* **2019**, *189*, 79–98. [CrossRef]
12. Harvey, C.F.; Swartz, C.H.; Badruzzaman, A.B.M.; Keon-Blute, N.; Yu, W.; Ali, M.A.; Jay, J.; Beckie, R.; Niedan, V.; Brabander, D.; et al. Arsenic Mobility and Groundwater Extraction in Bangladesh. *Science* **2002**, *298*, 1602–1606. [CrossRef] [PubMed]
13. van Geen, A.; Bostick, B.C.; Trang, P.T.K.; Lan, V.M.; Mai, N.-N.; Manh, P.D.; Viet, P.H.; Radloff, K.; Aziz, Z.; Mey, J.L.; et al. Retardation of Arsenic Transport through a Pleistocene Aquifer. *Nature* **2013**, *501*, 204–207. [CrossRef]
14. Masscheleyn, P.H.; Delaune, R.D.; Patrick, W.H. Effect of Redox Potential and pH on Arsenic Speciation and Solubility in a Contaminated Soil. *Environ. Sci. Technol.* **1991**, *25*, 1414–1419. [CrossRef]
15. Guo, H.; Liu, C.; Lu, H.; Wanty, R.B.; Wang, J.; Zhou, Y. Pathways of Coupled Arsenic and Iron Cycling in High Arsenic Groundwater of the Hetao Basin, Inner Mongolia, China: An Iron Isotope Approach. *Geochim. Cosmochim. Acta* **2013**, *112*, 130–145. [CrossRef]
16. Luo, T.; Hu, S.; Cui, J.; Tian, H.; Jing, C. Comparison of Arsenic Geochemical Evolution in the Datong Basin (Shanxi) and Hetao Basin (Inner Mongolia), China. *Appl. Geochem.* **2012**, *27*, 2315–2323. [CrossRef]
17. Zhang, H.; Ma, D.; Hu, X. Arsenic Pollution in Groundwater from Hetao Area, China. *Environ. Geol.* **2002**, *41*, 638–643. [CrossRef]
18. Zheng, Y.; Stute, M.; van Geen, A.; Gavrieli, I.; Dhar, R.; Simpson, H.; Schlosser, P.; Ahmed, K. Redox Control of Arsenic Mobilization in Bangladesh Groundwater. *Appl. Geochem.* **2004**, *19*, 201–214. [CrossRef]
19. Zheng, Y.; van Geen, A.; Stute, M.; Dhar, R.; Mo, Z.; Cheng, Z.; Horneman, A.; Gavrieli, I.; Simpson, H.; Versteeg, R.; et al. Geochemical and Hydrogeological Contrasts between Shallow and Deeper Aquifers in Two Villages of Araihazar, Bangladesh: Implications for Deeper Aquifers as Drinking Water Sources. *Geochim. Cosmochim. Acta* **2005**, *69*, 5203–5218. [CrossRef]
20. Buschmann, J.; Berg, M.; Stengel, C.; Sampson, M.L. Arsenic and Manganese Contamination of Drinking Water Resources in Cambodia: Coincidence of Risk Areas with Low Relief Topography. *Environ. Sci. Technol.* **2007**, *41*, 2146–2152. [CrossRef] [PubMed]
21. Anawar, H.M.; Akai, J.; Komaki, K.; Terao, H.; Yoshioka, T.; Ishizuka, T.; Safiullah, S.; Kato, K. Geochemical Occurrence of Arsenic in Groundwater of Bangladesh: Sources and Mobilization Processes. *J. Geochem. Explor.* **2003**, *77*, 109–131. [CrossRef]
22. Baig, J.A.; Kazi, T.G.; Arain, M.B.; Afridi, H.I.; Kandhro, G.A.; Sarfraz, R.A.; Jamal, M.K.; Shah, A.Q. Evaluation of Arsenic and Other Physico-Chemical Parameters of Surface and Ground Water of Jamshoro, Pakistan. *J. Hazard. Mater.* **2009**, *166*, 662–669. [CrossRef] [PubMed]
23. Podgorski, J.E.; Eqani, S.A.M.A.S.; Khanam, T.; Ullah, R.; Shen, H.; Berg, M. Extensive Arsenic Contamination in High-pH Unconfined Aquifers in the Indus Valley. *Sci. Adv.* **2017**, *3*, e1700935. [CrossRef] [PubMed]
24. Rodríguez-lado, L.; Sun, G.; Berg, M.; Zhang, Q.; Xue, H.; Zheng, Q.; Johnson, C.A. Groundwater Arsenic Contamination throughout China. *Science* **2013**, *341*, 866–868. [CrossRef] [PubMed]
25. Chakraborty, M.; Sarkar, S.; Mukherjee, A.; Shamsudduha, M.; Ahmed, K.M.; Bhattacharya, A.; Mitra, A. Modeling Regional-Scale Groundwater Arsenic Hazard in the Transboundary Ganges River Delta, India and Bangladesh: Infusing Physically-Based Model with Machine Learning. *Sci. Total Environ.* **2020**, *748*, 141107. [CrossRef] [PubMed]
26. Tan, K.; Ye, Y.; Cao, Q.; Du, P.; Dong, J. Estimation of Arsenic Contamination in Reclaimed Agricultural Soils Using Reflectance Spectroscopy and ANFIS Model. *IEEE J. Sel. Top. Appl. Earth Obs. Remote Sens.* **2014**, *7*, 2540–2546. [CrossRef]
27. Podgorski, J.; Araya, D.; Berg, M. Geogenic Manganese and Iron in Groundwater of Southeast Asia and Bangladesh—Machine Learning Spatial Prediction Modeling and Comparison with Arsenic. *Sci. Total Environ.* **2022**, *833*, 155131. [CrossRef] [PubMed]
28. Cho, K.H.; Sthiannopkao, S.; Pachepsky, Y.A.; Kim, K.-W.; Kim, J.H. Prediction of Contamination Potential of Groundwater Arsenic in Cambodia, Laos, and Thailand Using Artificial Neural Network. *Water Res.* **2011**, *45*, 5535–5544. [CrossRef] [PubMed]
29. Charulatha, G.; Srinivasalu, S.; Maheswari, O.U.; Venugopal, T.; Giridharan, L. Evaluation of Ground Water Quality Contaminants Using Linear Regression and Artificial Neural Network Models. *Arab. J. Geosci.* **2017**, *10*, 128. [CrossRef]
30. Sayegh, A.S.; Munir, S.; Habeebullah, T.M. Comparing the Performance of Statistical Models for Predicting PM10 Concentrations. *Aerosol Air Qual. Res.* **2014**, *14*, 653–665. [CrossRef]
31. Haggerty, R.; Sun, J.; Yu, H.; Li, Y. Application of Machine Learning in Groundwater Quality Modeling—A Comprehensive Review. *Water Res.* **2023**, *233*, 119745. [CrossRef] [PubMed]
32. Ouedraogo, I.; Defourny, P.; Vanclooster, M. Application of Random Forest Regression and Comparison of Its Performance to Multiple Linear Regression in Modeling Groundwater Nitrate Concentration at the African Continent Scale. *Hydrogeol. J.* **2019**, *27*, 1081–1098. [CrossRef]
33. Tesoriero, A.J.; Gronberg, J.A.; Juckem, P.F.; Miller, M.P.; Austin, B.P. Predicting Redox-Sensitive Contaminant Concentrations in Groundwater Using Random Forest Classification. *Water Resour. Res.* **2017**, *53*, 7316–7331. [CrossRef]

34. Akakuru, O.C.; Akaolisa, C.C.Z.; Aigbadon, G.O.; Eyankware, M.O.; Opara, A.I.; Obasi, P.N.; Ofoh, I.J.; Njoku, A.O.; Akudinobi, B.E.B. Integrating Machine Learning and Multi-Linear Regression Modeling Approaches in Groundwater Quality Assessment around Obosi, Se Nigeria. *Environ. Dev. Sustain.* **2023**, *25*, 14567–14606. [CrossRef]
35. Saghebian, S.M.; Sattari, M.T.; Mirabbasi, R.; Pal, M. Ground Water Quality Classification by Decision Tree Method in Ardebil Region, Iran. *Arab. J. Geosci.* **2014**, *7*, 4767–4777. [CrossRef]
36. Biau, G.; Scornet, E. A Random Forest Guided Tour. *Test* **2016**, *25*, 197–227. [CrossRef]
37. Liaw, A.; Wiener, M. Classification and Regression by RandomForest. *R News* **2002**, *2*, 18–22.
38. Tan, Z.; Yang, Q.; Zheng, Y. Machine Learning Models of Groundwater Arsenic Spatial Distribution in Bangladesh: Influence of Holocene Sediment Depositional History. *Environ. Sci. Technol.* **2020**, *54*, 9454–9463. [CrossRef] [PubMed]
39. Bindal, S.; Singh, C.K. Predicting Groundwater Arsenic Contamination: Regions at Risk in Highest Populated State of India. *Water Res.* **2019**, *159*, 65–76. [CrossRef]
40. Feng, L.-X.; Brown, R.W.; Han, B.-F.; Wang, Z.-Z.; Łuszczak, K.; Liu, B.; Zhang, Z.-C.; Ji, J.-Q. Thrusting and Exhumation of the Southern Mongolian Plateau: Joint Thermochronological Constraints from the Langshan Mountains, Western Inner Mongolia, China. *J. Asian Earth Sci.* **2017**, *144*, 287–302. [CrossRef]
41. He, J.; Ma, T.; Deng, Y.; Yang, H.; Wang, Y. Environmental Geochemistry of High Arsenic Groundwater at Western Hetao Plain, Inner Mongolia. *Front. Earth Sci. China* **2009**, *3*, 63–72. [CrossRef]
42. Guo, H.; Yang, S.; Tang, X.; Li, Y.; Shen, Z. Groundwater Geochemistry and Its Implications for Arsenic Mobilization in Shallow Aquifers of the Hetao Basin, Inner Mongolia. *Sci. Total Environ.* **2008**, *393*, 131–144. [CrossRef] [PubMed]
43. Deng, Y.; Wang, Y.; Ma, T. Isotope and Minor Element Geochemistry of High Arsenic Groundwater from Hangjinhouqi, the Hetao Plain, Inner Mongolia. *Appl. Geochem.* **2009**, *24*, 587–599. [CrossRef]
44. Gao, Z.; Weng, H.; Guo, H. Unraveling Influences of Nitrogen Cycling on Arsenic Enrichment in Groundwater from the Hetao Basin Using Geochemical and Multi-Isotopic Approaches. *J. Hydrol.* **2021**, *595*, 125981. [CrossRef]
45. Goodbred, S.L.; Kuehl, S.A. The Significance of Large Sediment Supply, Active Tectonism, and Eustasy on Margin Sequence Development: Late Quaternary Stratigraphy and Evolution of the Ganges–Brahmaputra Delta. *Sediment. Geol.* **2000**, *133*, 227–248. [CrossRef]
46. Karim, M.M.; Safiuddin, M. *Arsenic Contamination of Groundwater in Bangladesh*; British Geological Survey Technical Report WC/00/19; British Geological Survey: Nottingham, UK, 2001; Volume 4, pp. 162–165.
47. Acharyya, S.K.; Lahiri, S.; Raymahashay, B.C.; Bhowmik, A. Arsenic Toxicity of Groundwater in Parts of the Bengal Basin in India and Bangladesh: The Role of Quaternary Stratigraphy and Holocene Sea-Level Fluctuation. *Environ. Geol.* **2000**, *39*, 1127–1137. [CrossRef]
48. Huang, G.; Song, J.; Han, D.; Liu, R.; Liu, C.; Hou, Q. Assessing Natural Background Levels of Geogenic Contaminants in Groundwater of an Urbanized Delta through Removal of Groundwaters Impacted by Anthropogenic Inputs: New Insights into Driving Factors. *Sci. Total Environ.* **2023**, *857*, 159527. [CrossRef] [PubMed]
49. Rodríguez, R.; Ramos, J.; Armienta, A. Groundwater Arsenic Variations: The Role of Local Geology and Rainfall. *Appl. Geochem.* **2004**, *19*, 245–250. [CrossRef]
50. Huang, G.; Zhang, M.; Liu, C.; Li, L.; Chen, Z. Heavy Metal(loid)s and Organic Contaminants in Groundwater in the Pearl River Delta That Has Undergone Three Decades of Urbanization and Industrialization: Distributions, Sources, and Driving Forces. *Sci. Total Environ.* **2018**, *635*, 913–925. [CrossRef] [PubMed]
51. Huang, G.; Chen, Z.; Liu, F.; Sun, J.; Wang, J. Impact of Human Activity and Natural Processes on Groundwater Arsenic in an Urbanized Area (South China) Using Multivariate Statistical Techniques. *Environ. Sci. Pollut. Res.* **2014**, *21*, 13043–13054. [CrossRef] [PubMed]
52. Pukelsheim, F. The Three Sigma Rule. *Am. Stat.* **1994**, *48*, 88–91. [CrossRef]
53. Ahmad, S.; Ahmad, I.; Umar, R.; Farooq, S.H. Spatio-Temporal Variation and Health Risk Associated with Trace Element Concentrations in Groundwater of Mathura City Using Modified Indexing Approach. *Arab. J. Geosci.* **2022**, *15*, 318. [CrossRef]
54. Li, X.; Ge, J.; Liu, Z.; Yang, S.; Wang, L.; Liu, Y. Estimating the Methane Flux of the Dajiuhu Subalpine Peatland Using Machine Learning Algorithms and the Maximal Information Coefficient Technique. *Sci. Total Environ.* **2024**, *916*, 170241. [CrossRef] [PubMed]
55. Rodriguez-Galiano, V.; Mendes, M.P.; Garcia-Soldado, M.J.; Chica-Olmo, M.; Ribeiro, L. Predictive Modeling of Groundwater Nitrate Pollution Using Random Forest and Multisource Variables Related to Intrinsic and Specific Vulnerability: A Case Study in an Agricultural Setting (Southern Spain). *Sci. Total Environ.* **2014**, *476–477*, 189–206. [CrossRef] [PubMed]
56. Khosravi, K.; Mao, L.; Kisi, O.; Yaseen, Z.M.; Shahid, S. Quantifying Hourly Suspended Sediment Load Using Data Mining Models: Case Study of a Glacierized Andean Catchment in Chile. *J. Hydrol.* **2018**, *567*, 165–179. [CrossRef]
57. Sharafati, A.; Khosravi, K.; Khosravinia, P.; Ahmed, K.; Salman, S.A.; Yaseen, Z.M.; Shahid, S. The Potential of Novel Data Mining Models for Global Solar Radiation Prediction. *Int. J. Environ. Sci. Technol.* **2019**, *16*, 7147–7164. [CrossRef]
58. Jin, Z.; Shang, J.; Zhu, Q.; Ling, C.; Xie, W.; Qiang, B. RFRSF: Employee Turnover Prediction Based on Random Forests and Survival Analysis. In *Web Information Systems Engineering—WISE 2020*; Lecture Notes in Computer Science; Springer: Cham, Swizerland, 2020; Volume 12343, pp. 503–515. [CrossRef]
59. Guo, W.; Gao, Z.; Guo, H.; Cao, W. Hydrogeochemical and Sediment Parameters Improve Predication Accuracy of Arsenic-Prone Groundwater in Random Forest Machine-Learning Models. *Sci. Total Environ.* **2023**, *897*, 165511. [CrossRef] [PubMed]

60. Zhang, Q.; Rodríguez-Lado, L.; Johnson, C.A.; Xue, H.; Shi, J.; Zheng, Q.; Sun, G. Predicting the Risk of Arsenic Contaminated Groundwater in Shanxi Province, Northern China. *Environ. Pollut.* **2012**, *165*, 118–123. [CrossRef] [PubMed]
61. Raschka, S. *Python Machine Learning*, 1st ed.; Equation Reference; Packt Publishing: Birmingham, UK, 2015; Volume 2015, pp. 1–71.
62. Chen, W.; Tsangaratos, P.; Ilia, I.; Duan, Z.; Chen, X. Groundwater Spring Potential Mapping Using Population-Based Evolutionary Algorithms and Data Mining Methods. *Sci. Total Environ.* **2019**, *684*, 31–49. [CrossRef] [PubMed]
63. Bellu, A.; Fernandes, L.F.S.; Cortes, R.M.V.; Pacheco, F.A.L. A Framework Model for the Dimensioning and Allocation of a Detention Basin System: The Case of a Flood-Prone Mountainous Watershed. *J. Hydrol.* **2016**, *533*, 567–580. [CrossRef]
64. Wang, H.; Eiche, E.; Guo, H.; Norra, S. Impact of Sedimentation History for As Distribution in Late Pleistocene-Holocene Sediments in the Hetao Basin, China. *J. Soils Sediments* **2020**, *20*, 4070–4082. [CrossRef]
65. Guo, H.; Zhang, B.; Li, Y.; Berner, Z.; Tang, X.; Norra, S.; Stüben, D. Hydrogeological and Biogeochemical Constrains of Arsenic Mobilization in Shallow Aquifers from the Hetao Basin, Inner Mongolia. *Environ. Pollut.* **2011**, *159*, 876–883. [CrossRef] [PubMed]
66. Bauer, M.; Blodau, C. Mobilization of Arsenic by Dissolved Organic Matter from Iron Oxides, Soils and Sediments. *Sci. Total Environ.* **2006**, *354*, 179–190. [CrossRef] [PubMed]
67. Guo, H.; Li, X.; Xiu, W.; He, W.; Cao, Y.; Zhang, D.; Wang, A. Controls of Organic Matter Bioreactivity on Arsenic Mobility in Shallow Aquifers of the Hetao Basin, P.R. China. *J. Hydrol.* **2019**, *571*, 448–459. [CrossRef]
68. Khalid, S.; Shahid, M.; Niazi, N.K.; Rafiq, M.; Bakhat, H.F.; Imran, M.; Abbas, T.; Bibi, I.; Dumat, C. Arsenic Behaviour in Soil-Plant System: Biogeochemical Reactions and Chemical Speciation Influences. In *Enhancing Cleanup of Environmental Pollutants*; Springer: Cham, Switzerland, 2017; Volume 2.
69. Deng, Y.; Wang, Y.; Ma, T.; Gan, Y. Speciation and Enrichment of Arsenic in Strongly Reducing Shallow Aquifers at Western Hetao Plain, Northern China. *Environ. Geol.* **2009**, *56*, 1467–1477. [CrossRef]
70. Postma, D.; Larsen, F.; Thai, N.T.; Trang, P.T.K.; Jakobsen, R.; Nhan, P.Q.; Long, T.V.; Viet, P.H.; Murray, A.S. Groundwater Arsenic Concentrations in Vietnam Controlled by Sediment Age. *Nat. Geosci.* **2012**, *5*, 656–661. [CrossRef]
71. Rowland, H.A.L.; Pederick, R.L.; Polya, D.A.; Pancost, R.D.; Van Dongen, B.E.; Gault, A.G.; Vaughan, D.J.; Bryant, C.; Anderson, B.; Lloyd, J.R. The Control of Organic Matter on Microbially Mediated Iron Reduction and Arsenic Release in Shallow Alluvial Aquifers, Cambodia. *Geobiology* **2007**, *5*, 281–292. [CrossRef]
72. Stollenwerk, K.G. Geochemical Processes Controlling Transport of Arsenic in Groundwater: A Review of Adsorption. In *Arsenic in Ground Water*; Springer: Boston, MA, USA, 2005.
73. Hafeznezami, S.; Zimmer-Faust, A.G.; Jun, D.; Rugh, M.B.; Haro, H.L.; Park, A.; Suh, J.; Najm, T.; Reynolds, M.D.; Davis, J.A.; et al. Remediation of Groundwater Contaminated with Arsenic through Enhanced Natural Attenuation: Batch and Column Studies. *Water Res.* **2017**, *122*, 545–556. [CrossRef]
74. Nickson, R.T.; McArthur, J.M.; Ravenscroft, P.; Burgess, W.G.; Ahmed, K.M. Mechanism of Arsenic Release to Groundwater, Bangladesh and West Bengal. *Appl. Geochem.* **2000**, *15*, 403–413. [CrossRef]
75. McArthur, J.M.; Ravenscroft, P.; Safiulla, S.; Thirlwall, M.F. Arsenic in Groundwater: Testing pollution Mechanisms for Sedimentary Aquifers in Bangladesh. *Water Resour. Res.* **2001**, *37*, 109–117. [CrossRef]
76. Aiken, G.; Kuniansky, E.L. *U.S. Geological Survey Artificial Recharge Workshop Proceedings, April 2-4, 2002, Sacramento, California*; U.S. Geological Survey: Reston, VA, USA, 2002; pp. 47–50.
77. Bhattacharya, P.; Welch, A.H.; Ahmed, K.M.; Jacks, G.; Naidu, R. Arsenic in Groundwater of Sedimentary Aquifers. *Appl. Geochem.* **2004**, *19*, 163–167. [CrossRef]
78. Akai, J.; Izumi, K.; Fukuhara, H.; Masuda, H.; Nakano, S.; Yoshimura, T.; Ohfuji, H.; Anawar, H.M.; Akai, K. Mineralogical and Geomicrobiological Investigations on Groundwater Arsenic Enrichment in Bangladesh. *Appl. Geochem.* **2004**, *19*, 215–230. [CrossRef]
79. Cao, W.; Guo, H.; Zhang, Y.; Ma, R.; Li, Y.; Dong, Q.; Li, Y.; Zhao, R. Controls of Paleochannels on Groundwater Arsenic Distribution in Shallow Aquifers of Alluvial Plain in the Hetao Basin, China. *Sci. Total Environ.* **2018**, *613–614*, 958–968. [CrossRef] [PubMed]

*water*

*Article*

# Characterization of Groundwater Hydrochemistry and Temporal Dynamics of Water Quality in the Northern Baiquan Spring Basin

**Di Wu [1,2], Bo Li [1,2], Yuxing Li [1,2], Qingbin Li [1,2], Chen Sheng [1,2], Jiutan Liu [3], Min Wang [3], Yangyang Min [1,2], Jianguo Feng [3], Yuqi Zhang [3] and Jieqing Yu [4,*]**

1 801 Institute of Hydrogeology and Engineering Geology, Shandong Provincial Bureau of Geology & Mineral Resources, Jinan 250014, China; sjzwoody@163.com (D.W.); libo05048888@163.com (B.L.); dkkcy801@163.com (Y.L.); lqb7904@163.com (Q.L.); m15264753886@163.com (C.S.); litinamin@yahoo.com (Y.M.)

2 Shandong Engineering Research Center for Environmental Protection and Remediation on Groundwater, Jinan 250014, China

3 College of Earth Science and Engineering, Shandong University of Science and Technology, Qingdao 266590, China; ljtsdust@sdust.edu.cn (J.L.); brightwangm@163.com (M.W.); fengjianguo20316@163.com (J.F.); zyqsdust@163.com (Y.Z.)

4 Department of Geology and Engineering Survey, Hebei Geological Workers' University, Shijiazhuang 050081, China

* Correspondence: yjq880326@163.com

**Abstract:** Groundwater represents a critical resource for sustaining the livelihoods of both urban and rural populations, facilitating economic and social development, and preserving ecological equilibrium. This study leverages groundwater quality monitoring data from the northern Baiquan spring basin (NBSB) to elucidate groundwater hydrochemical characteristics and decipher the temporal variability in water quality. Findings suggest that groundwater within the NBSB is predominantly weakly alkaline and characterized as hard-fresh, with $HCO_3^-$ and $Ca^{2+}$ as the predominant ions, which collectively demarcate the hydrochemical type as predominantly $HCO_3$-Ca. The principal constituents of NBSB groundwater are influenced predominantly by the weathering of carbonates and silicates alongside the dissolution of gypsum and halite. Moreover, agricultural operations and similar human activities have exerted an impact on the hydrochemical attributes of NBSB's groundwater. Generally, fluctuations in groundwater anion concentrations over time are more pronounced than those of cations, exemplified by a significant upward trend in the major ion concentrations at the BQ03 monitoring site in the later stages. While the general groundwater quality within the NBSB is deemed satisfactory, most monitoring sites have experienced an escalation in water quality indices over time, notably at BQ03, which warrants serious attention. The findings of this research contribute to the efficient management and sustainable utilization of groundwater resources in the NBSB.

**Keywords:** groundwater; hydrochemical characteristics; dynamic changes; water quality assessment; northern Baiquan spring basin

**Citation:** Wu, D.; Li, B.; Li, Y.; Li, Q.; Sheng, C.; Liu, J.; Wang, M.; Min, Y.; Feng, J.; Zhang, Y.; et al. Characterization of Groundwater Hydrochemistry and Temporal Dynamics of Water Quality in the Northern Baiquan Spring Basin. *Water* **2024**, *16*, 2519. https://doi.org/10.3390/w16172519

Academic Editor: Pankaj Kumar

Received: 5 August 2024
Revised: 30 August 2024
Accepted: 2 September 2024
Published: 5 September 2024



## 1. Introduction

Groundwater, as a natural water storage system, is widely distributed, stable, and reliable, with substantial reserves constituting approximately 99% of the Earth's liquid freshwater [1–3]. It plays a critical role in providing drinking water, supporting sanitation systems, agriculture, industry, and ecosystems [4–8]. Globally, groundwater supplies 50% of residential water and approximately 25% of agricultural irrigation water, which supports 38% of the world's irrigated land. However, rapid population growth, accelerated urbanization, and changing patterns in water usage are posing significant threats and challenges to groundwater resources [3,9–11]. Many underground aquifers are being exploited at unsustainable rates, and groundwater pollution is becoming increasingly problematic [12–14]. Consequently, the rational development and management of groundwater resources has emerged as an urgent issue that must be addressed.

The formation of groundwater is a prolonged hydrological process influenced by environmental factors such as rainfall, evaporation, infiltration, and seepage [1,13–16]. Groundwater undergoes intricate geological processes and geochemical reactions within underground aquifers, interacting with subterranean rocks to create distinct geochemical characteristics [2,5,17–19]. Analyzing dynamic changes in groundwater's chemical composition and evaluating water quality can enhance our understanding of groundwater's composition, trends, and potential issues, thereby guiding the scientific management and protection of groundwater resources. For example, Liu et al. [7] investigated groundwater level and quality fluctuations in the valley plain of Lhasa City on the Qinghai-Tibet Plateau, identifying influencing factors. Mishra et al. [20] examined spatiotemporal variations in groundwater quality in Madhya Pradesh, India, assessed its suitability for drinking, and offered recommendations for groundwater management. Ismail et al. [21] analyzed groundwater quality in Lahore from 2005 to 2021, revealing a general decline associated with urbanization, and providing insights for decision-making on groundwater management and urban expansion in Lahore. Pacheco Castro et al. [22] employed hierarchical clustering analysis to study spatiotemporal changes in groundwater quality of karst aquifers in Yucatan, Mexico, enhancing understanding of how precipitation and human activities impact groundwater quality. These studies collectively contribute valuable insights for the rational development and management of groundwater resources in their respective regions.

Groundwater serves as a vital water source and a key component of the ecological environment in the Baiquan spring basin (BSB), significantly impacting the lives and economic progress of local communities. The groundwater in the BSB mainly consists of carbonate rock fissure karst water and Quaternary loose rock pore water. In previous studies, the groundwater in typical water sources in the study area was classified as Class III water, with the main indicators exceeding the standard being nitrate, sulfate, TDS, and TH. The concentrations of each indicator have significantly increased compared to decades ago, indicating that the groundwater has been polluted to a certain extent. In order to prevent the continuous deterioration of groundwater quality, certain groundwater protection measures need to be taken [23,24]. Despite its importance, there is a notable dearth of research exploring the hydrochemistry and water quality of BSB groundwater. To address this gap, our study capitalized on long-term monitoring data from groundwater sites in the northern Baiquan spring basin (NBSB). Employing a combination of hydrochemical and statistical techniques, we aimed to: (1) delve into the hydrochemical traits and primary ion sources of NBSB groundwater, (2) elucidate the temporal shifts in groundwater's main ion concentrations, and (3) assess groundwater quality and its evolving trends. The insights gained from this research offer valuable guidance for the sustainable utilization, management, and conservation of groundwater resources in the NBSB region.

*Study Area*

The BSB (Figure 1), one of the four major spring areas in Jinan, constitutes a relatively independent hydrogeological unit with a total area of 783.48 km$^2$. This research area is situated in a mid-latitude region within the warm temperate semi-humid monsoon climate zone. It experiences an average annual temperature of 12.8 °C, average annual precipitation of 671.2 mm, and average annual evaporation of 1804 mm [25]. The region features a well-developed water system, including lakes and reservoirs, and is part of the Yellow River basin. Notable rivers in the area include the Juye River, Xibalou River, and Ganggou River.

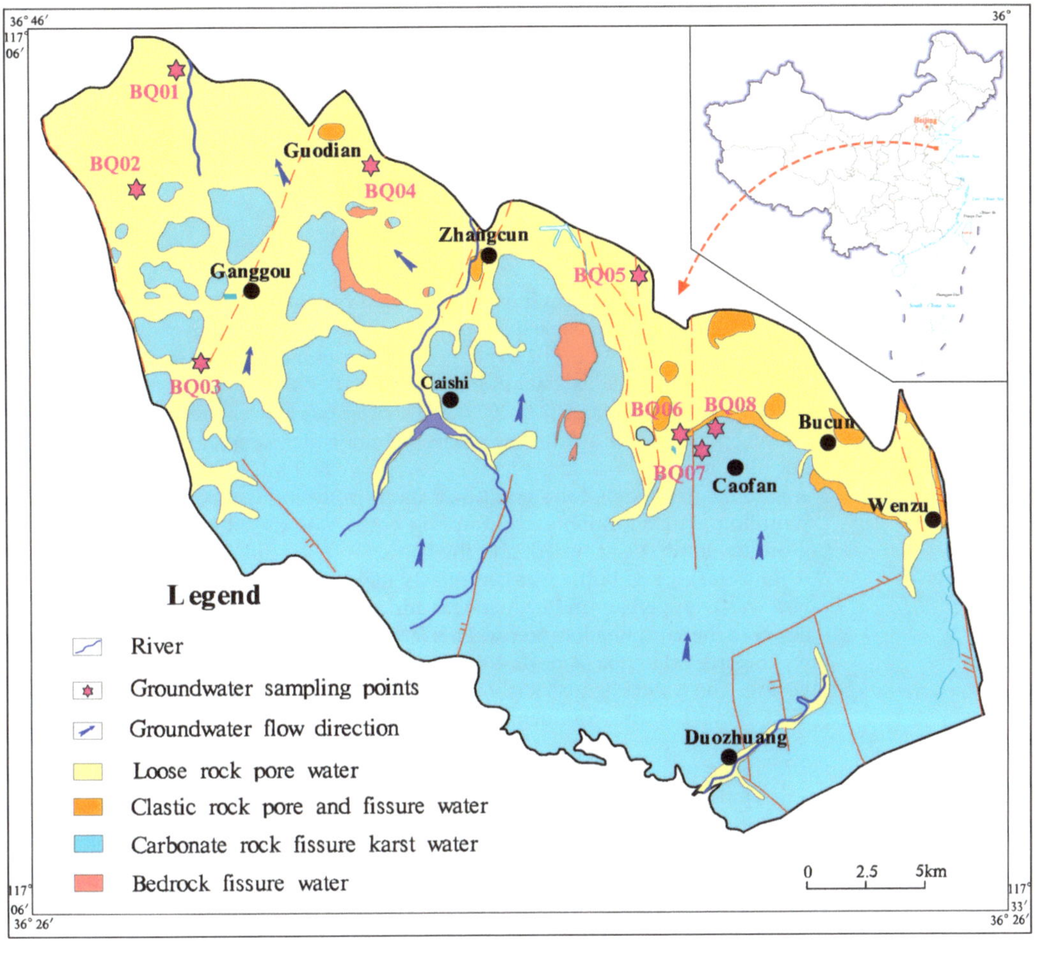


**Figure 1.** Location map of BSB and groundwater monitoring points.

The southern boundary of the BSB is delineated by a surface watershed, while the northern boundary corresponds to the contact zone between the Carboniferous and Permian strata and the associated magmatic rocks. The eastern and western limits are defined by the Wenzu fault and Dongwu fault, respectively, both of which exhibit relative water resistance. The altitude of the research area decreases gradually from south to north, with the southern region characterized by hilly terrain and low mountains, transitioning to a plain area in the north. From south to north, the exposed geological formations include metamorphic rocks from the Mount Taishan Group, Cambrian limestone, and Ordovician limestone in succession [23,25].

The primary water types found in the BSB include Quaternary loose rock pore water, clastic rock pore and fissure water, and carbonate rock fissure and karst water. Quaternary loose rock pore water is predominantly located in the central and northern plain areas of the BSB. Clastic rock pore and fissure water mainly occur within Carboniferous–Permian sandstone, sandy shale, and Tertiary strata. Carbonate fissure and karst water are present in thick layers of limestone aquifers, thin layers of limestone, and interbedded shale aquifers [23]. Additionally, bedrock fissure water is sporadically distributed within the BSB (Figure 1). Groundwater recharge, runoff, and discharge in the BSB are influenced by factors such as meteorology, hydrology, stratigraphy, lithology, artificial exploitation, and irrigation practices. The principal source of groundwater recharge is atmospheric precipitation, followed by infiltration from field irrigation. Groundwater runoff generally follows the same direction as surface water, flowing from southeast to northwest. Under natural conditions, discharge occurs primarily through lateral runoff and groundwater evaporation [26]. Currently, the main discharge methods are manual extraction and downstream runoff discharge, with significant variation in water level depths observed from south to north.

## 2. Materials and Methods

### *2.1. Sampling*

A total of 8 long-term dynamic observation points were set up in the research area, with continuous and reliable data sources, so these 8 points were chosen for sampling, with a time span from 2018 to 2023.The monitoring points, as illustrated in Figure 1, yielded a total of 70 water quality data points. Notably, BQ02 (monitored from 2020 to 2023) and BQ08 (monitored from 2022 to 2023) had shorter monitoring durations, and there are some missing data at other monitoring points. Groundwater samples were collected twice yearly, once in June and once in September, using two 500 mL polyethylene plastic bottles. To ensure the freshness of the samples, a stable pumping was conducted for 5 min prior to sampling. Furthermore, each sampling bottle was rinsed at least three times with the groundwater to be sampled, and the bottles were completely filled, leaving no bubbles. After the sampling bottle was completely filled, the bottle cap was tightened and sealing film was wrapped around it. Acid was added to pH < 2 for detecting $NO_3^-$. The distance between the sampling location and the laboratory was relatively close. After the sampling was completed on the same day, the samples were sent to the laboratory for water quality testing. The water chemistry laboratory of Shandong Provincial Geo-mineral Engineering Exploration Institute conducted the water sample testing.

### *2.2. Testing*

Hydrochemical composition analysis of samples in the laboratory was conducted, including major anions ($Cl^-$, $SO_4^{2-}$, $HCO_3^-$, $NO_3^-$, $F^-$), cations ($Ca^{2+}$, $Mg^{2+}$, $K^+$, $Na^+$), pH, total hardness (TH), and total dissolved solids (TDS). The pH value was measured using an acidity meter (PHS-3C, Shanghai Yidian Scientific Instrument Co., Ltd., Shanghai, China), while cations, TDS, and TH were tested using an inductively coupled plasma emission spectrometer (ICP, Optima 7000DV, Thermo Fisher Scientifc, Waltham, MA, USA). No other ions were detected, so they are concentrated in the following main ions. Anions, except for $HCO_3^-$ (which was determined by titration), were measured using an ion chromatograph (ICS-600, Thermo Fisher Scientifc, Waltham, MA, USA). Additionally, to ensure the accuracy of the water quality data, charge balance error (CBE) calculations were performed on all water samples. The results indicated that the CBE values for all samples were within ±5%, confirming the reliability of the water quality monitoring conducted in this study.

### *2.3. Groundwater Quality Evaluation*

Groundwater quality evaluation is a comprehensive process that assesses multiple water quality parameters. It involves comparing the actual concentrations of each parameter in groundwater with standard concentrations to determine an evaluation grade that reflects water quality directly [2,6,27,28]. In this study, groundwater quality at the NBSB was evaluated using the entropy weight water quality index (EWQI), based on China's "Groundwater Quality Standards". Eight indicators were selected for this evaluation: pH, total hardness (TH), total dissolved solids (TDS), $Na^+$, $Cl^-$, $SO_4^{2-}$, $NO_3^-$, and $F^-$. The calculation process for the EWQI is as follows:

Firstly, construct the initial matrix $X_{ij}$ where $i = 1, 2, \ldots, m$ and $j = 1, 2, \ldots, n$. Here, m represents the total number of water samples, and n denotes the number of hydrochemical parameters for each sample.

$$x_{ij} = \begin{bmatrix} x_{11} & \cdots & x_{1n} \\ \vdots & \ddots & \vdots \\ x_{m1} & \cdots & x_{mn} \end{bmatrix} \quad (1)$$

$$y_{ij} = \frac{x_{ij} - (x_{ij})_{min}}{(x_{ij})_{max} - (x_{ij})_{min}} \quad (2)$$

Secondly, normalize the initial matrix $X_{ij}$ to obtain the judgment matrix $Y$.

$$Y = \begin{bmatrix} y_{11} & \cdots & y_{1n} \\ \vdots & \ddots & \vdots \\ y_{m1} & \cdots & y_{mn} \end{bmatrix} \quad (3)$$

In the formula, $(x_{ij})$min and $(x_{ij})$max represent the minimum and maximum values of the same hydrochemical parameters across all samples, respectively.

Then, calculate the information entropy ($e_j$), entropy weight ($w_j$), and quality evaluation quantity ($q_j$) using the following equations:

$$\begin{aligned} e_j &= -\frac{1}{\ln m} \sum_{i=1}^{m} P_{ij} \ln P_{ij} \\ P_{ij} &= 1 + \frac{(1+y_{ij})}{\sum_{i=1}^{m}(1+y_{ij})} \end{aligned} \quad (4)$$

$$w_j = \frac{1 - e_j}{\sum_{j=1}^{n}(1 - e_j)} \quad (5)$$

$$q_j = \frac{C_j}{S_j} \times 100 \quad (6)$$

In the formula, $c_j$ represents the concentration of evaluation indicator $j$ (mg/L), and $s_j$ represents the allowable limit value of indicator $j$ (mg/L).

Finally, calculate the EWQI using the following equation:

$$\text{EWQI} = \sum_{i=1}^{m} w_j q_j \quad (7)$$

Based on the classification criteria for groundwater quality:

EWQI < 25: Groundwater quality is classified as Excellent (Class I).

25 ≤ EWQI < 50: Groundwater quality is classified as Good (Class II).

50 ≤ EWQI < 100: Groundwater quality is classified as Medium (Class III).

100 ≤ EWQI < 150: Groundwater quality is classified as Poor (Class IV).

EWQI > 150: Groundwater quality is classified as Extremely Poor (Class V).

These classifications help in assessing the suitability of groundwater for various uses and determining the need for treatment or remediation.

## 3. Results

### 3.1. Overall Characteristics of Groundwater Hydrochemistry

Table 1 and Figure 2a present the statistical results of groundwater quality monitoring data for the NBSB from 2018 to 2023. The pH values of the groundwater range from 7.1 to 8.1, with a mean of 7.59, indicating a weakly alkaline environment. TDS and TH are key indicators of groundwater quality [2,29,30]. In the NBSB, groundwater TH ranges from 298 mg/L to 950 mg/L, and TDS ranges from 371 mg/L to 1215 mg/L, with average values of 527.80 mg/L and 665.63 mg/L, respectively. Except for pH and $HCO_3^-$, the coefficient of variation of other ions is greater than 30%, indicating that these ions have a high degree of dispersion and instability, which may be influenced by supply sources or human activities. According to the TDS and TH water quality classification (Figure 2b), the groundwater in the NBSB is predominantly classified as hard-fresh water (82.9%), with a smaller proportion categorized as hard-brackish water (17.1%).

**Table 1.** Statistical results of groundwater hydrochemical composition of NBSB.

| | Maximum | Minimum | Mean | Standard Deviation | Coefficient of Variation |
|---|---|---|---|---|---|
| pH | 8.1 | 7.1 | 7.59 | 0.23 | 3.03 |
| TH (mg/L) | 950 | 298 | 527.80 | 187.66 | 35.56 |
| TDS (mg/L) | 1215 | 371 | 665.63 | 248.05 | 37.27 |
| $Ca^{2+}$ (mg/L) | 301 | 89.6 | 160.62 | 58.62 | 36.50 |
| $Mg^{2+}$ (mg/L) | 66.6 | 18.1 | 30.77 | 11.29 | 36.69 |
| $K^+$ (mg/L) | 3.62 | 0.46 | 1.39 | 0.92 | 66.19 |
| $Na^+$ (mg/L) | 44.8 | 2.93 | 19.77 | 13.75 | 69.55 |
| $Cl^-$ (mg/L) | 165 | 8.25 | 53.39 | 47.17 | 88.35 |
| $SO_4^{2-}$ (mg/L) | 583 | 58.7 | 184.39 | 113.34 | 61.47 |
| $HCO_3^-$ (mg/L) | 578 | 187 | 329.79 | 93.94 | 28.48 |
| $NO_3^-$ (mg/L) | 131 | 12.2 | 37.76 | 17.62 | 46.66 |
| $F^-$ (mg/L) | 0.57 | 0.06 | 0.26 | 0.11 | 42.31 |

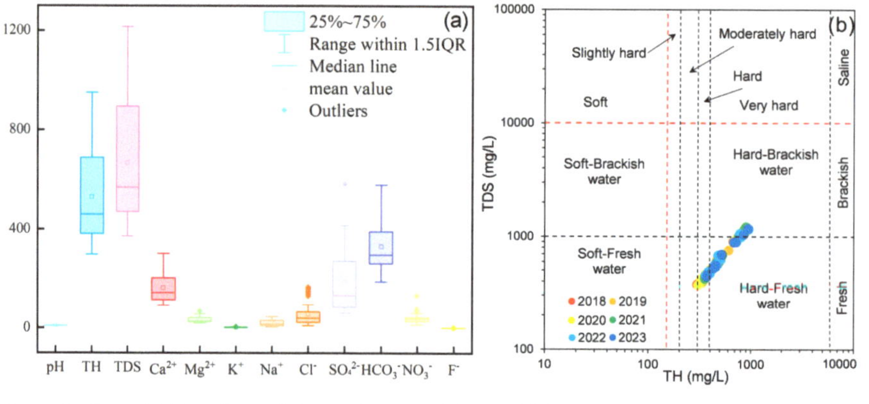


**Figure 2.** Plots of (**a**) bar chart of chemical composition and (**b**) TH vs. TDS water type.

The dominant cation in the groundwater of the NBSB is $Ca^{2+}$, with concentrations ranging from 89.6 mg/L to 301 mg/L and a mean value of 160.62 mg/L. $Mg^{2+}$ is the second most prevalent cation, varying from 18.1 mg/L to 66.6 mg/L and averaging at 30.77 mg/L. Among anions, $HCO_3^-$ is the most abundant, with levels between 187 mg/L and 578 mg/L, averaging at 329.97 mg/L. $SO_4^{2-}$ is the next most common anion, ranging from 58.7 mg/L to 583 mg/L and averaging 184.39 mg/L. In general, the cation concentration follows the order $Ca^{2+} > Mg^{2+} > Na^+ > K^+$, while the anion concentration follows $HCO_3^- > SO_4^{2-} > Cl^- > NO_3^- > F^-$ (Figure 2a). The $NO_3^-$ content, which serves as an indicator of human activity intensity and its impact on the groundwater environment [10,31,32], ranges from 12.2 mg/L to 131 mg/L in NBSB groundwater from 2018 to 2023, with an average concentration of 37.76 mg/L. Notably, the average $NO_3^-$ concentration in NBSB groundwater remains below the standard limit of 88.57 mg/L.

### 3.2. Hydrochemical Types

The Durov plot [33] is a widely utilized method for characterizing groundwater chemistry. Compared to the Piper diagram, the Durov plot not only depicts the hydrochemical type but also simultaneously represents the TDS and pH values of water samples, which facilitates a more comprehensive understanding of hydrochemical characteristics [2,14,34,35]. Figure 3 illustrates the projection of groundwater monitoring data from the NBSB for the period 2018 to 2023 onto the Durov plot. It is evident that, with the exception of one groundwater sample located in area ①, the majority of the samples are distributed in area ②. Therefore, according to the Durov plot, the predominant hydrochemical type of groundwater in the NBSB is classified as Ca-$HCO_3$ type.

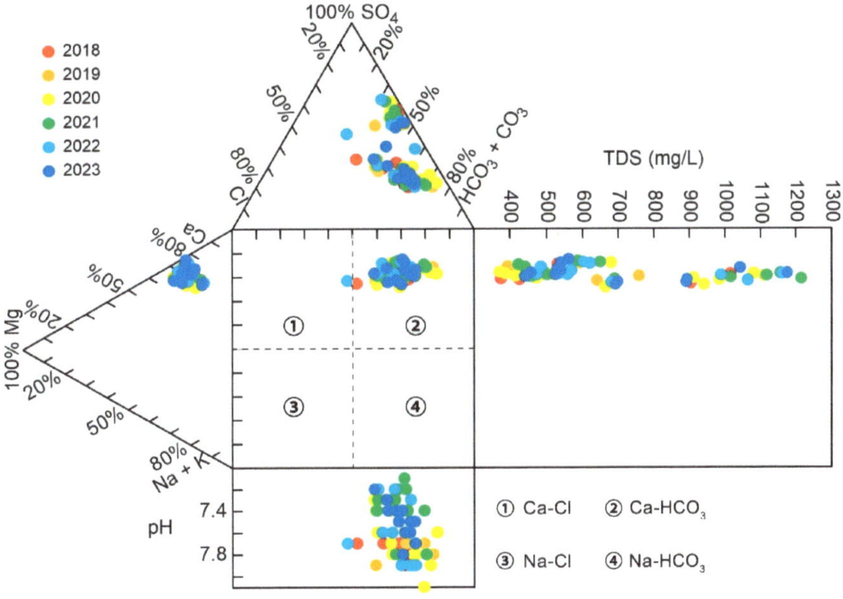


**Figure 3.** Durov plot of groundwater hydrochemistry.

### 3.3. Water Quality Evaluation Results

This study evaluated groundwater quality at five monitoring points (BQ01, BQ03, BQ04, BQ06, and BQ07) within the NBSB region, utilizing the EWQI based on comprehensive groundwater dynamic monitoring data. The results indicated that EWQI values ranged from 19.18 to 76.09, with a mean value of 36.65. Specifically, 23.64% of the samples were classified as Class I water (13 samples), 52.73% as Class II water (29 samples), and 13.64% as Class III water (13 samples). Overall, groundwater quality in the NBSB is rated as excellent, making it an ideal water supply source.

## 4. Discussion

### 4.1. Sources of Major Ions

The hydrochemical composition of groundwater in a region is primarily influenced by rock weathering, evaporation, and precipitation. Among these, rock weathering—particularly of silicates, carbonates, and evaporites—is often the most significant factor [5,17,27,36]. In coastal areas, seawater intrusion can also play a substantial role [1,37,38]. Accelerated urbanization and population growth in many parts of the world have exacerbated groundwater pollution, mainly through the misuse of groundwater assets and the discharge of domestic and industrial wastewater into groundwater systems. Groundwater pollution that includes radiation, wastewater, heavy metal pollution, microbial contamination and more, can also have an impact on hydrochemical characteristics [39]. Additionally, in regions with significant human activity, anthropogenic factors often significantly impact hydrochemical characteristics and should not be overlooked [4,18,30,32]. Conversely, analyzing the characteristics and interrelationships of major ions in groundwater can help identify their sources.

One of the primary sources of $Na^+$ and chloride $Cl^-$ in groundwater is the dissolution of halite (NaCl) in sedimentary rocks. Therefore, if halite dissolution is the main source of $Na^+$ and $Cl^-$ in groundwater, their ratio should theoretically approach 1 [17,32,36]. Figure 4a illustrates the relationship between $Na^+$ and $Cl^-$ in the groundwater of the NBSB region. The data points are predominantly located above the Na/Cl = 1 line, with some points situated close to both sides of this line. This suggests that while halite dissolution contributes to the $Na^+$ and $Cl^-$ concentrations, other hydrogeochemical processes—such as cation exchange or anthropogenic inputs—also influence these ions [2,36]. The relationship between $Ca^{2+}$ + $Mg^{2+}$ and $HCO_3^-$ + $SO_4^{2-}$ is used to assess the impact of carbonate and silicate rock weathering on groundwater chemistry [6,29,40]. As depicted in Figure 4b, the water sample points are mainly distributed on either side of the $(Ca^{2+} + Mg^{2+})/(HCO_3^- + SO_4^{2-}) = 1$ line, with a concentration in the upper region. This indicates that groundwater in the NBSB area is influenced by the dissolution of carbonate and silicate minerals, with a more pronounced effect from carbonates.

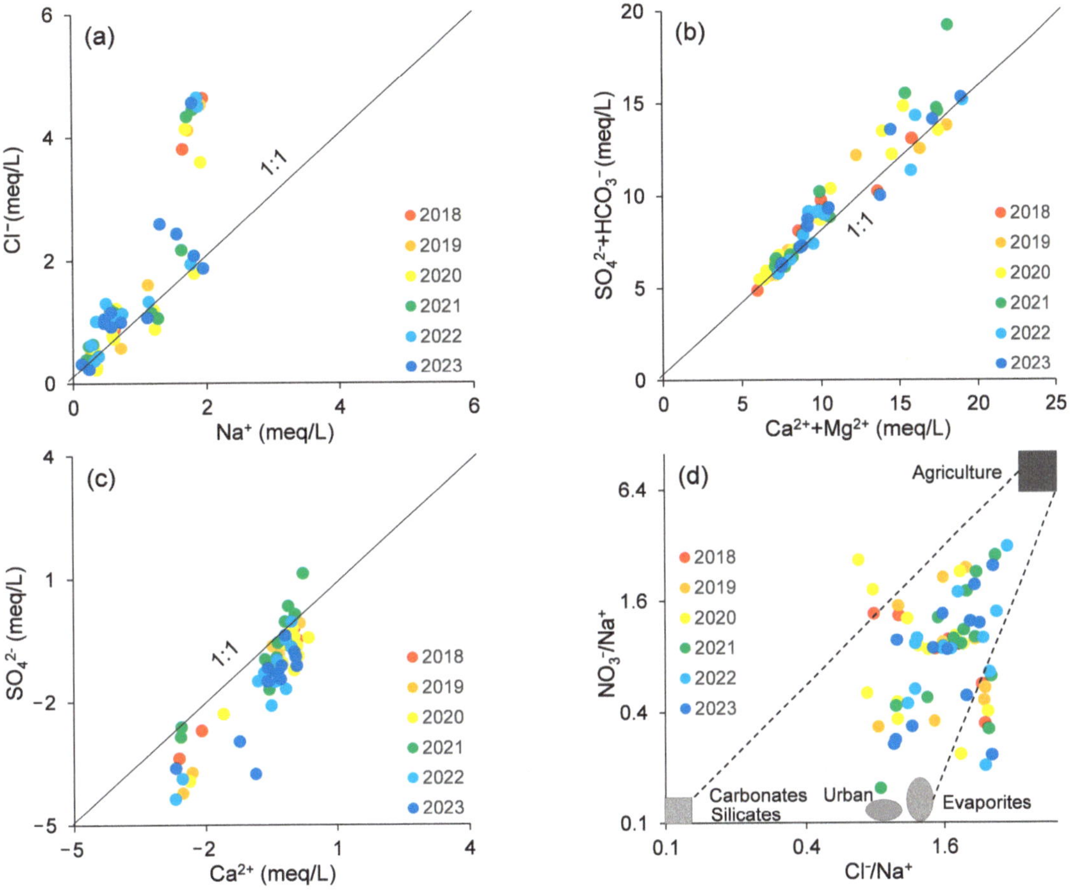


**Figure 4.** Major ion ratios diagram in groundwater of NBSB. (**a**) Ratio of $Na^+$ to $Cl^-$; (**b**) The ratio of $Ca^{2+}$ + $Mg^{2+}$ to $SO_4^{2-}$ + $HCO_3^-$; (**c**) The ratio of $Ca^{2+}$ to $SO_4^{2-}$; (**d**) The ratio of $NO_3^-/Na^+$ to $Cl^-/Na^+$.

Figure 4c illustrates the relationship between $Ca^{2+}$ and $SO_4{}^{2-}$ in groundwater. The data points predominantly fall below the $Ca^{2+}/SO_4{}^{2-}$ = 1 line, with some points near the line, suggesting gypsum dissolution in the groundwater of the NBSB area. However, gypsum dissolution is not the primary source of $Ca^{2+}$ and $SO_4{}^{2-}$. In the end-member diagram of the $Cl^-/Na^+$ to $NO_3{}^-/Na^+$ ratio, the upper right corner indicates areas influenced by agricultural activities, the lower left corner denotes areas controlled by the weathering of carbonates and silicates, and the lower right corner represents areas affected by urban development and evaporite dissolution, as shown in Figure 4d. The end-member diagram of the $Cl^-/Na^+$ to $NO_3{}^-/Na^+$ ratio is commonly used to identify the sources of $Cl^-$ and $NO_3{}^-$ in groundwater [40–42]. Projecting the groundwater quality data from the NBSB area onto Figure 4d reveals that the water samples are primarily located in the agricultural, urban, and evaporite-controlled areas. This indicates that $NO_3{}^-$ in the groundwater is predominantly influenced by human activities such as agriculture and urban development, while some $Cl^-$ originates from evaporite dissolution.

The saturation index (SI) is defined as the logarithm (base 10) of the ratio between the ion activity product (IAP) and the solubility product constant (K), serving as an indicator of the equilibrium state between minerals and water [43]. An SI value of 0 signifies that the mineral is in a state of dissolution equilibrium with the aqueous solution. When SI < 0, the mineral is undersaturated and has a tendency to dissolve further; conversely, when SI > 0, the mineral is supersaturated and likely to precipitate. The SI values for groundwater samples from the NBSB were calculated, as depicted in Figure 5. The results indicate that the SI values for calcite, dolomite, and gypsum are all greater than 0, suggesting that these minerals are in a supersaturated state and are prone to precipitation. In contrast, the SI values for halite are all less than 0, indicating an undersaturated state and a tendency for continued dissolution.

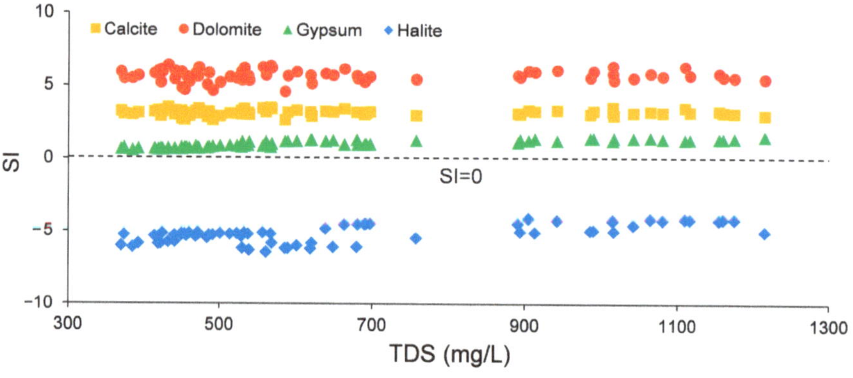


**Figure 5.** Relationship between the SI and TDS for groundwater samples.

### *4.2. Variation Characteristics of Major Ion Content in Groundwater*

Regular testing of groundwater indicators enables the timely detection of water quality trends and provides a comprehensive understanding of groundwater quality, offering a scientific basis for water resource management and protection [7,20,21]. In this study, five monitoring points (BQ01, BQ03, BQ04, BQ06, and BQ07) with the most complete groundwater data in the NBSB area were selected to analyze the temporal dynamics of their major ion contents, as shown in Figures 6 and 7.

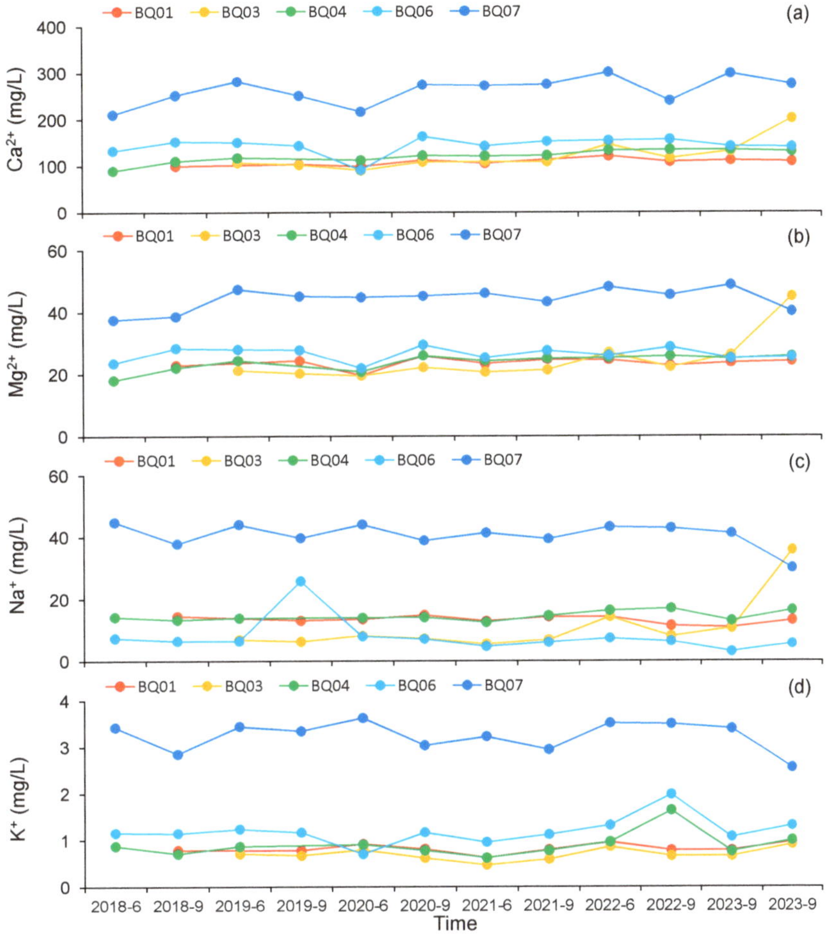


**Figure 6.** Temporal variation of the major cation content in groundwater. (**a**) Time variation of $Ca^{2+}$ content; (**b**) Time variation of $Mg^{2+}$ content; (**c**) Time variation of $Na^{2+}$ content; (**d**) Time variation of $K^+$ content.

Figure 6 presents the temporal variations of major cations in NBSB groundwater. Overall, the cation concentrations at each monitoring point exhibit minimal long-term trends and show some fluctuation. Compared to $Ca^{2+}$ and $Mg^{2+}$, $K^+$ and $Na^+$ display greater fluctuations, indicating higher instability, which is reflected in their larger CV values (Table 1). The cation concentrations at BQ01 are relatively stable compared to other points, whereas BQ03 shows a significant upward trend in $Ca^{2+}$, $Mg^{2+}$, and $Na^+$ concentrations after September 2022. This change in water quality at BQ03 warrants further investigation. Additionally, the cation concentrations at monitoring point BQ07 are notably higher than those at other points, followed by BQ06 (Figure 6).

Figure 7 illustrates the temporal variations in the concentrations of major anions in NBSB groundwater. Compared to cations, anions exhibit more pronounced fluctuations over time. For example, there was a significant decrease in $HCO_3^-$ content at BQ07 in September 2022 and in $Cl^-$ content in September 2023. The anion concentrations at BQ03 began to show an increasing trend starting in June 2022, particularly for $NO_3^-$ (Figure 7d). At BQ06, the $SO_4^{2-}$ concentration decreased significantly in June 2020, then increased, and has since exhibited a fluctuating downward trend. Conversely, the $SO_4^{2-}$ concentration at

BQ07 has shown a consistent upward trend (Figure 7b). Additionally, $NO_3^-$ concentrations at BQ07 have demonstrated significant fluctuations since 2021, suggesting potential impacts from human activities on groundwater at this location.

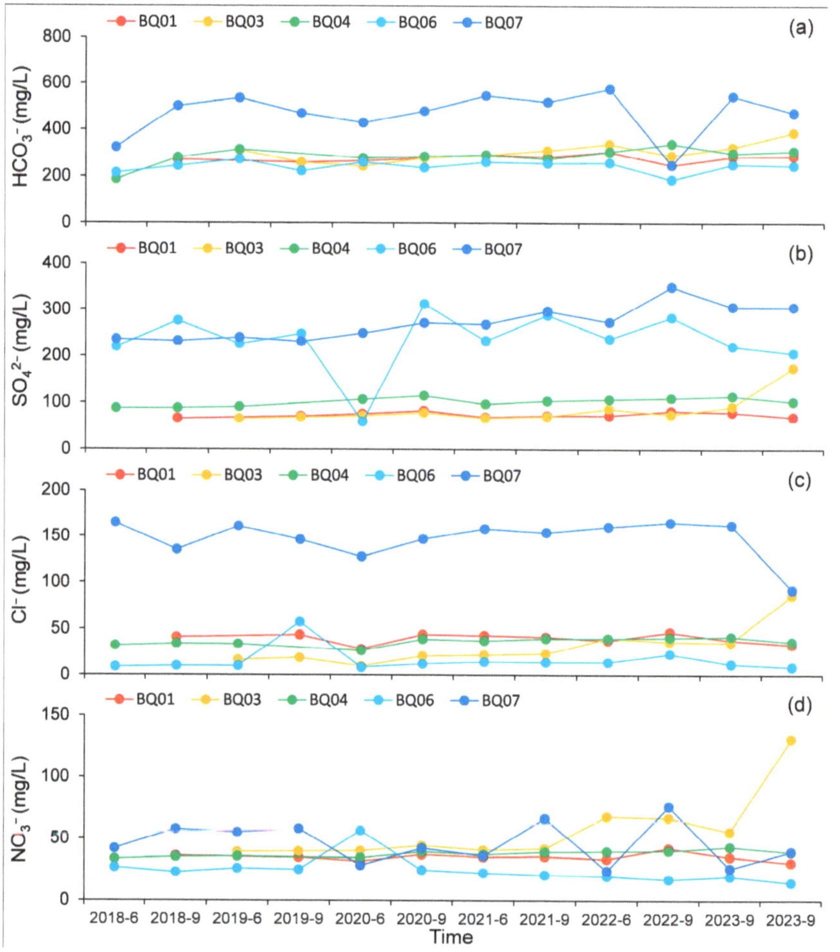


**Figure 7.** Temporal variation of the major anion content in groundwater. (**a**) Time variation of $HCO_3^-$ content; (**b**) Time variation of $SO_4^{2-}$ content; (**c**) Time variation of $Cl^-$ content; (**d**) Time variation of $NO_3^-$ content.

### *4.3. Variation Characteristics of Groundwater Quality*

Figure 8 illustrates the temporal variation in EWQI values for each monitoring point. Most points exhibited an increasing trend, except for BQ06, which showed a slight decrease, with BQ03 exhibiting the most significant increase. In terms of mean values, BQ01 demonstrated superior water quality compared to the other points and showed relatively stable changes. The EWQI at BQ03 generally increased annually, aligning with the trends observed in major ions at this location (Figures 6 and 7). The EWQI at BQ04 also showed an upward trend, but with minimal fluctuation. Conversely, EWQI values at BQ06 and BQ07 exhibited more pronounced variability. In summary, while groundwater quality in the NBSB remains high, it is important to note the gradual increase in ion content at certain monitoring points.

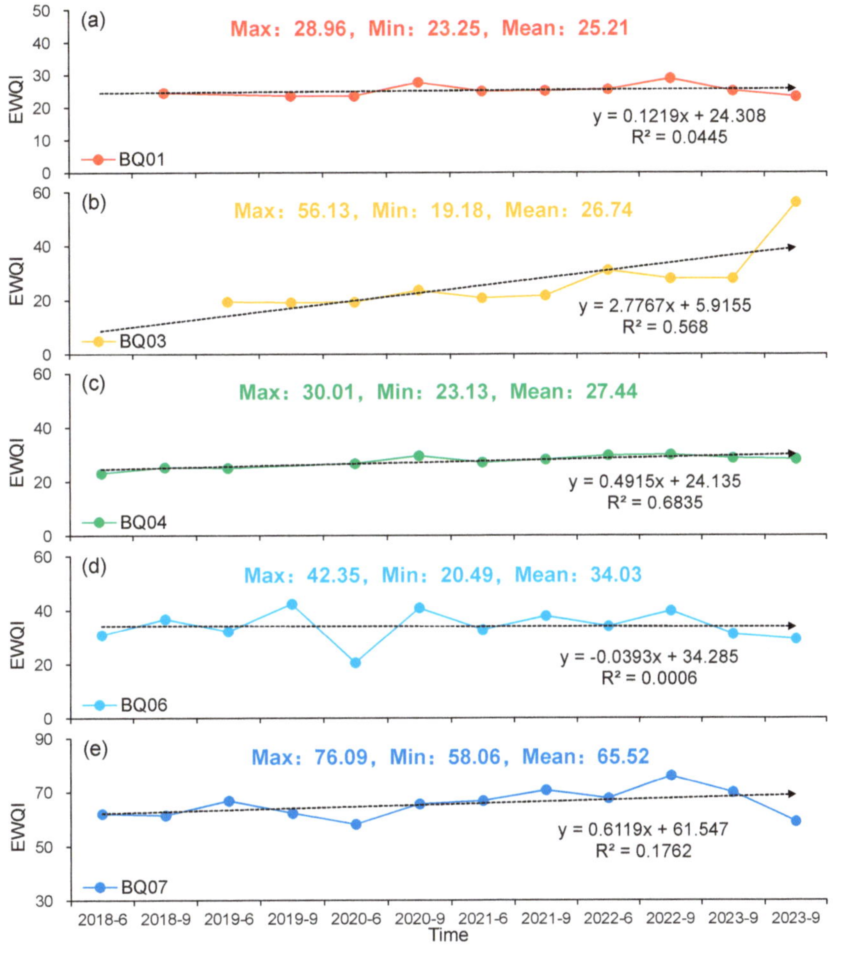


**Figure 8.** Temporal variation of the EWQI values in groundwater. (**a**) Time variation of EWQI value in BQ01; (**b**) Time variation of EWQI value in BQ03; (**c**) Time variation of EWQI value in BQ04; (**d**) Time variation of EWQI value in BQ06; (**e**) Time variation of EWQI value in BQ07.

### *4.4. Suggestions for the Management, Utilization, and Protection of Groundwater Resources in NBSB*

Groundwater is a vital water supply resource for the NBSB, significantly contributing to daily life, industrial production, and agricultural development. To ensure the effective management, utilization, and protection of groundwater and to achieve sustainable use of these resources, the following measures are recommended:

(1) Establish Comprehensive Legislation and Policies: Develop robust laws, regulations, and policies to define groundwater management responsibilities and rights clearly. Regulate groundwater development and use, and enhance supervision over groundwater resources.

(2) Enhance Groundwater Monitoring: Strengthen groundwater monitoring by establishing a comprehensive network to track groundwater levels, water quality, and quantity. This will enable the assessment of dynamic changes in groundwater resources and provide a scientific basis for their rational use.

(3) Control Groundwater Pollution Sources: Implement stringent measures to prevent the discharge of harmful substances into groundwater. Increase efforts to manage and mitigate groundwater pollution.
(4) Improve Groundwater Resource Management: Advance the integrated management of groundwater resources by refining management institutions, coordinating resources across stakeholders, and ensuring the rational allocation and use of groundwater.
(5) Designate Groundwater Protection Zones: Identify and protect groundwater protection zones by restricting development and utilization in sensitive areas. Prevent pollutant infiltration to safeguard groundwater quality and ensure its sustainable use.
(6) Promote Public Awareness and Education: Enhance public education and awareness about groundwater resources. Encourage water conservation practices and foster a greater understanding of groundwater protection among the public.

In conclusion, groundwater is a crucial resource for human activities and well-being. The rational development and management of groundwater are essential for the sustainable advancement of society. It is imperative for everyone to contribute to the protection of groundwater resources.

## 5. Conclusions

This study employed long-term dynamic monitoring data of groundwater quality from the NBSB region to ascertain the hydrochemical characteristics and ion sources of groundwater were determined, the temporal changes in the concentration of major ions in groundwater were elucidated, and the quality and evolution trend of groundwater were evaluated. The main conclusions are as follows:

(1) From 2018 to 2023, the average pH of groundwater in the NBSB was 7.59, indicating a generally weak alkalinity. The average TDS and TH were 665.63 mg/L and 527.80 mg/L, respectively, classifying the water predominantly as hard-fresh. The predominant anions and cations were $HCO_3^-$ and $Ca^{2+}$, with mean concentrations of 160.62 mg/L and 329.97 mg/L, respectively, which classifies the groundwater hydrochemical type in the NBSB as $HCO_3$-Ca.
(2) The primary ions in the NBSB's groundwater are derived from mineral dissolution, primarily from carbonate and silicate minerals. There is also dissolution of halite and gypsum. Additionally, anthropogenic activities have influenced the chemical composition of the groundwater, with $NO_3^-$ primarily originating from agricultural practices.
(3) The fluctuation in anion concentrations over time was more pronounced compared to cations. Overall, while the main ions at the BQ03 monitoring point exhibited a significant increasing trend after 2021, other monitoring points did not show notable increases or decreases over time.
(4) The overall groundwater quality in the NBSB is satisfactory and suitable for use as a water supply source. However, with the exception of the BQ06 monitoring point, the EWQI values at all other points have shown an upward trend over time, particularly at BQ03, which warrants attention. It is recommended to enhance the management, utilization, and protection of groundwater resources in the NBSB to ensure their sustainable use.

**Author Contributions:** Conceptualization, D.W.; methodology, B.L. and Y.L.; software, Q.L.; visualization, C.S.; writing—review and editing, J.L.; validation, M.W., Y.M. and J.F.; formal analysis, Y.Z.; supervision, J.Y. All authors have read and agreed to the published version of the manuscript.

**Funding:** This study was financially supported by the National Natural Science Foundation of China (No. 42272288) and the Shandong Provincial Geological Exploration Project ((Lu Kan Zi (2023) No. 4).

**Data Availability Statement:** The datasets used or analyzed during the current study are available from the corresponding author on reasonable request.

**Acknowledgments:** We would like to express our sincere thanks to the editors and reviewers for their very helpful comments for the paper.

**Conflicts of Interest:** The authors declare no competing interests.

## References


1. Gao, Z.; Li, Q.; Liu, J.; Su, Q.; Tan, M.; Wang, Y. Assessment of Groundwater Hydrogeochemistry, Controlling Factors, Water Quality, and Nitrate-Related Health Risks in the Longkou Bay, North China. *Water Air Soil Pollut.* **2024**, *235*, 392. [CrossRef]
2. Liu, J.; Gao, Z.; Feng, J.; Wang, M. Identification of the hydrochemical features, genesis, water quality and potential health hazards of groundwater in Dawen River Basin, North China. *Ecol. Indic.* **2023**, *149*, 110175. [CrossRef]
3. Makhlouf, A.; Sharaan, M.; El-Rawy, M.; Kanae, S.; Ibrahim, M.G. Investigating the effects of surface water recharge on groundwater quality using hydrochemistry and ANFIS model: A case study Minia Governorate, Egypt. *J. Environ. Manag.* **2024**, *362*, 121269. [CrossRef] [PubMed]
4. Chaudhari, M.P.; George, A.; Sanyal, M.; Shrivastav, P.S. Hydrochemistry and groundwater quality assessment of Gujarat, India: A compendious review. *Phys. Chem. Earth Parts A/B/C* **2024**, *135*, 103635. [CrossRef]
5. Khan, M.H.; Xiao, Y.; Yang, H.; Wang, L.; Zhang, Y.; Hu, W.; Wang, J.; Liu, G.; Liu, W. Identification of hydrochemical fingerprints, quality and formation dynamics of groundwater in western high Himalayas. *Environ. Monit. Assess.* **2024**, *196*, 305. [CrossRef] [PubMed]
6. Kouser, B.; Bala, A.; Verma, O.; Prashanth, M.; Khosla, A.; Pir, R.A. Hydrochemistry for the assessment of groundwater quality in the Kathua region, Jammu and Kashmir, India. *Appl. Water Sci.* **2022**, *12*, 143. [CrossRef]
7. Liu, J.; Gao, Z.; Wang, M.; Li, Y.; Ma, Y.; Shi, M.; Zhang, H. Study on the dynamic characteristics of groundwater in the valley plain of Lhasa City. *Environ. Earth Sci.* **2018**, *77*, 646. [CrossRef]
8. Xie, C.; Liu, H.; Li, X.; Zhao, H.; Dong, X.; Ma, K.; Wang, N.; Zhao, L. Spatial characteristics of hydrochemistry and stable isotopes in river and groundwater, and runoff components in the Shule River Basin, Northeastern of Tibet Plateau. *J. Environ. Manag.* **2024**, *349*, 119512. [CrossRef]
9. Bayou, W.T.; Mohammed, M.; Ayenew, T.; Mebrahtu, T.K.; Wohnlich, S. Hydrochemistry and isotope hydrology of groundwater and surface water in the Sor and Gebba watershed, southwestern Ethiopia. *Environ. Earth Sci.* **2024**, *83*, 316. [CrossRef]
10. Liu, J.; Peng, Y.; Li, C.; Gao, Z.; Chen, S. Characterization of the hydrochemistry of water resources of the Weibei Plain, Northern China, as well as an assessment of the risk of high groundwater nitrate levels to human health. *Environ. Pollut.* **2021**, *268*, 115947. [CrossRef]
11. Shang, Y.; Fu, C.; Zhang, W.; Li, X.; Li, X. Groundwater hydrochemistry, source identification and health assessment based on self-organizing map in an intensive mining area in Shanxi, China. *Environ. Res.* **2024**, *252*, 118934. [CrossRef]
12. Liu, J.; Lou, K.; Gao, Z.; Wang, Y.; Li, Q.; Tan, M. Comprehending hydrochemical fingerprint, spatial patterns, and driving forces of groundwater in a topical coastal plain of Northern China based on hydrochemical and isotopic evaluations. *J. Clean. Prod.* **2024**, *461*, 142640. [CrossRef]
13. Meena, P.L. Appraisal of hydrogeochemical and quality of groundwater in the Banas river basin in North West India. *Arab. J. Geosci.* **2024**, *17*, 128. [CrossRef]
14. Wang, S.; Chen, J.; Zhang, S.; Bai, Y.; Zhang, X.; Chen, D.; Hu, J. Groundwater hydrochemical signatures, nitrate sources, and potential health risks in a typical karst catchment of North China using hydrochemistry and multiple stable isotopes. *Environ. Geochem. Health* **2024**, *46*, 173. [CrossRef] [PubMed]
15. Chen, Z.; Zhou, P.; Wang, G.; Mao, H.; Feng, X.; Huang, H. Groundwater chemistry and isotope for interpreting the hydrogeological conditions and hydrochemical evolution of multilayer aquifer system of Donghai island, China. *Appl. Geochem.* **2023**, *159*, 105833. [CrossRef]
16. Rind, I.K.; Khuhawar, M.Y.; Lanjwani, M.F.; Khuhawar, T.M.J.; Samtio, M.S.; Soomro, W.A.; Baloch, A.R. Spatial variability of hydrochemistry and health risk assessment of groundwater of taluka Hala, district Matiari, Sindh, Pakistan. *Arab. J. Geosci.* **2023**, *16*, 99. [CrossRef]
17. Ismail, A.H.; Shareef, M.A.; Hassan, G.; Alatar, F.M. Hydrochemistry and water quality of shallow groundwater in the Tikrit area of Salah Al Din Province, Iraq. *Appl. Water Sci.* **2023**, *13*, 197. [CrossRef]
18. Obiri-Nyarko, F.; Duah, A.A.; Karikari, A.Y.; Tagoe, R. Characterization of leachate, groundwater quality analysis, and evaluation of hydrogeochemical processes at the Kpone engineered landfill site, Ghana. *Sustain. Water Resour. Manag.* **2022**, *9*, 15. [CrossRef]
19. Poetra, R.P.; Adji, T.N.; Haryono, E.; Cahyadi, A.; Agniy, R.F.; Riyanto, I.A.; Fatchurohman, H.; Kholis, A.N.; Tivianton, T.A.; Febriarta, E. Discovering water relations using hydrochemistry, isotopic data, and mass-balance calculation approaches to reduce the risk of impacts from limestone mining. *Groundw. Sustain. Dev.* **2023**, *21*, 100949. [CrossRef]
20. Mishra, S.; Chauhan, M.S.; Sundaramurthy, S. Assessing groundwater quality dynamics in Madhya Pradesh: Chemical contaminants and their temporal patterns. *Environ. Res.* **2024**, *252*, 118887. [CrossRef]
21. Ismail, S.; Ahmed, M.F.; Abu Bakar, M.Z. Assessing the impact of urbanization on groundwater quality of lahore region, Pakistan. *Environ. Sci. Pollut. Res.* **2023**, *30*, 83929–83949. [CrossRef] [PubMed]
22. Pacheco Castro, R.; Pacheco Ávila, J.; Ye, M.; Cabrera Sansores, A. Groundwater Quality: Analysis of Its Temporal and Spatial Variability in a Karst Aquifer. *Groundwater* **2018**, *56*, 62–72. [CrossRef] [PubMed]
23. Zhang, L.; Xu, Y.; Zhang, H. Study on Formation Mechanism of Baiquan Spring Groups in Jinan City. *Shandong Land Resour.* **2017**, *33*, 58–62.

24. Chen, H.; Cao, H. Evaluation of Groundwater Resources and Withdrawal Potential in Bai Spring Area, Jinan City, China. *Front. Earth Sci.* **2021**, *11*, 1250–1258.
25. Shu, L.; Wang, X.; Li, H.; Ni, H.; Li, G.; Yu, Y.; Wang, X.; Zhang, M. Lmpact of Subway Construction on Flow of Baiquan Spring Group in Jinan. *J. Jilin Univ.* **2021**, *51*, 192–200.
26. Qi, X.; Li, W.; Yang, L.; Shang, H.; Yi, F. The Lag Analysis of Groundwater Level Anomaliesto Precipitation Anomaly of Jinan Baiquan Springs Watershed. *Earth Environ.* **2015**, *43*, 619–627.
27. Zhang, Y.; He, Z.; Tian, H.; Huang, X.; Zhang, Z.; Liu, Y.; Xiao, Y.; Li, R. Hydrochemistry appraisal, quality assessment and health risk evaluation of shallow groundwater in the Mianyang area of Sichuan Basin, southwestern China. *Environ. Earth Sci.* **2021**, *80*, 576. [CrossRef]
28. Apollaro, C.; Buccianti, A.; Vespasiano, G.; Vardè, M.; Fuoco, I.; Barca, D.; Bloise, A.; Miriello, D.; Cofone, F.; Servidio, A.; et al. Comparative geochemical study between the tap waters and the bottled mineral waters in Calabria (Southern Italy) by compositional data analysis (CoDA) developments. *Appl. Geochem.* **2019**, *107*, 19–33. [CrossRef]
29. Aghazadeh, N.; Chitsazan, M.; Golestan, Y. Hydrochemistry and quality assessment of groundwater in the Ardabil area, Iran. *Appl. Water Sci.* **2017**, *7*, 3599–3616. [CrossRef]
30. Zhang, M.; Han, S.; Wang, Y.; Wang, Z.; Li, H.; Wang, X.; Liu, J.; Li, C.; Gao, Z. Characteristics and Controlling Factors of Groundwater Hydrochemistry in Dongzhi Tableland Area of the Loess Plateau of Eastern Gansu—A Case Study of Ning County Area, North China. *Water* **2022**, *14*, 3601. [CrossRef]
31. Ramaroson, V.; Randriantsivery, J.R.; Rajaobelison, J.; Fareze, L.P.; Rakotomalala, C.U.; Razafitsalama, F.A.; Rasolofonirina, M. Nitrate contamination of groundwater in Ambohidrapeto–Antananarivo-Madagascar using hydrochemistry and multivariate analysis. *Appl. Water Sci.* **2020**, *10*, 178. [CrossRef]
32. Yang, F.; Jia, C.; Yang, X.; Yang, H.; Chang, W. Probabilistic potential health risk quantification, hydrochemistry, driving forces of groundwater nitrate and fluoride in a typical irrigation district, Northern China. *J. Environ. Manag.* **2022**, *323*, 116171. [CrossRef] [PubMed]
33. Durov, S.A. Classification of natural waters and graphical representation of their composition. *Dokl. Akad. Nauk. SSSR* **1948**, *59*, 87–90.
34. Jalees, M.I.; Farooq, M.U.; Anis, M.; Hussain, G.; Iqbal, A.; Saleem, S. Hydrochemistry modelling: Evaluation of groundwater quality deterioration due to anthropogenic activities in Lahore, Pakistan. *Environ. Dev. Sustain.* **2021**, *23*, 3062–3076. [CrossRef]
35. Zhao, C.; Li, X.; Wang, Z.; Hou, X.; Ma, J. Integration of self-organizing maps and entropy-weighted quality index methods to study hydrochemical processes and quality assessment in Qilian coal base of China. *Environ. Earth Sci.* **2024**, *83*, 92. [CrossRef]
36. Gao, Z.; Liu, J.; Feng, J.; Wang, M.; Wu, G. Hydrogeochemical Characteristics and the Suitability of Groundwater in the Alluvial-Diluvial Plain of Southwest Shandong Province, China. *Water* **2019**, *11*, 1577. [CrossRef]
37. Fu, T.; Fu, Y.; Li, C.; Dong, M.; Qi, C.; Wang, Z.; Chen, G.; Su, Q.; Xu, X.; Yu, H. Seawater intrusion-triggered high fluoride groundwater development on the eastern coast of China. *Environ. Sci. Pollut. Res.* **2024**, *31*, 11307–11320. [CrossRef]
38. Roy, S.K.; Zahid, A. Assessment of river water–groundwater–seawater interactions in the coastal delta of Bangladesh based on hydrochemistry and salinity distribution. *SN Appl. Sci.* **2021**, *3*, 411. [CrossRef]
39. Karunanidhi, D.; Subramani, T.; Roy, P.D.; Li, H. Impact of groundwater contamination on human health. *Environ. Geochem. Health* **2021**, *43*, 643–647. [CrossRef]
40. Liu, J.; Peng, Y.; Li, C.; Gao, Z.; Chen, S. An investigation into the hydrochemistry, quality and risk to human health of groundwater in the central region of Shandong Province, North China. *J. Clean. Prod.* **2021**, *282*, 125416. [CrossRef]
41. Xiao, J.; Wang, L.; Chai, N.; Liu, T.; Jin, Z.; Rinklebe, J. Groundwater hydrochemistry, source identification and pollution assessment in intensive industrial areas, eastern Chinese loess plateau. *Environ. Pollut.* **2021**, *278*, 116930. [CrossRef] [PubMed]
42. Hua, K.; Xiao, J.; Li, S.; Li, Z. Analysis of hydrochemical characteristics and their controlling factors in the Fen River of China. *Sust. Cities Soc.* **2020**, *52*, 101827. [CrossRef]
43. Apollaro, C.; Caracausi, A.; Paternoster, M.; Randazzo, P.; Aiuppa, A.; De Rosa, R.; Fuoco, I.; Mongelli, G.; Muto, F.; Vanni, E.; et al. Fluid geochemistry in a low-enthalpy geothermal field along a sector of southern Apennines chain (Italy). *J. Geochem. Explor.* **2020**, *219*, 106618. [CrossRef]

water

MDPI

Article

# Optimization of Advance Drainage Borehole Layout Based on Visual Modflow

Yue Li [1,*], Yunpeng Zhang [1], Yajie Ma [1] and Fangang Meng [2]

1 College of Mining Engineering, North China University of Science and Technology, 21 Bohai Avenue, Caofeidian New Town, Tangshan 063210, China; mengf147852@126.com (Y.Z.); myj147852@126.com (Y.M.)
2 Donghuantuo Mining Branch of Kailuan Group, Tangshan 063210, China; mengfangang@kailuan.com.cn
* Correspondence: 147@stu.ncst.edu.cn

**Abstract:** It is an effective measure to prevent water damage in coal mines in order to construct drainage boreholes in water-filled aquifers of a working face roof. The hydrogeological parameters of the roof water-filled aquifer and the parameters of the drainage borehole are closely related to the underground drainage effect. Taking the 3085 working face of the Donghuantuo Mine in Kailuan as an example, the influence degree of hydrogeological parameters and hydrophobic borehole parameters on the amount of drainage water was analyzed using the generalized gray correlation degree. Based on Visual Modflow, the 3D groundwater visualization model was established and the dredging borehole was generalized into the pumping borehole. By changing the main influencing factors, the design optimization of the advanced hydrophobic borehole was discussed. The results showed that the aquifer thickness had a great influence on the amount of water discharged, and the influence degree of the sharp angle between the formation and the direction of drilling, the depth of the final hole, the azimuth angle of drilling, the dip angle of drilling, the elevation of the final hole and the elevation of the borehole on the amount of water discharged decreased successively. Based on the simulation calculation, it could be observed that the hydrophobic borehole should be placed in a position with a larger accumulated thickness of the aquifer to have a better effect of hydrophobic depressurization.

**Keywords:** hydrophobic drilling; hydrogeological parameters; drilling parameters; Visual Modflow

**Citation:** Li, Y.; Zhang, Y.; Ma, Y.; Meng, F. Optimization of Advance Drainage Borehole Layout Based on Visual Modflow. *Water* **2024**, *16*, 2613. https://doi.org/10.3390/w16182613

Academic Editors: Zongjun Gao and Jiutan Liu

Received: 26 July 2024
Revised: 1 September 2024
Accepted: 6 September 2024
Published: 14 September 2024



## 1. Introduction

Coal mine water disasters seriously threaten the safe production of a mining area. At present, China's mine water prevention and control work adopts comprehensive control measures of prevention, plugging, dredging, drainage and interception [1]. According to the Detailed Rules for Coal Mine Water Prevention and Control, in order to realize the safe discharge of water inflow after mining, it is necessary to adopt advanced drainage measures for the fractured aquifer of the coal mine roof [2]. The commonly used method of hydrophobic depressurization is to construct drainage boreholes underground [3]. Drainage borehole water inflow, aquifer water pressure and water storage can directly evaluate the effect of borehole drainage. At present, the point source and line sink theory is mostly adopted for the calculation of drainage water from coal seam roof boreholes, but the flow pattern characteristics in the drainage holes are more complex, and the water amount calculation is not accurate enough [4–7]. Many scholars have analyzed the influence of borehole design parameters and roof aquifer hydrogeological parameters on the advance drainage effect using correlation analysis, gray correlation degree analysis, the single index model, the well flow calculation method, the numerical simulation method and other methods [8–13]. Taking the water drainage project of the Donghuantuo 3085 working face as an example, the generalized gray relational degree analysis method is used to study the correlation between the aquifer hydrogeological parameters, drainage borehole parameters and drainage volume and determine the main influencing factors. Based on Visual Modflow, the groundwater flow field model of the pre-drainage of the working face was established, the main influencing factors were changed, the trend of the final water level of the aquifer under different drainage borehole distribution conditions was analyzed, the drainage capacity of the working face under different drainage schemes was summarized and the pre-drainage schemes were optimized.

## 2. Hydrogeological Survey of the Study Area

### *2.1. Water-Filled Aquifer*

The coal measure strata and the Ordovician limestone strata at the bottom of Donghuantuo Mine are hidden under thick alluvium. In the mine field, the three-dimensional water-filling geological structure of the multi-layer aquifer with internal and external boundary communication is formed [14]. The 3085 working face is located in the central mining area. As the main mining coal seam of the working face, the water inflow of coal seam 8 mainly travels from the weak aquifer section (IVc) of coal seam 8 to coal seam 5 and the strong water-bearing section (Va+b) of coal seam 5.

Roof caving and fractures in coal seam mining destroy the roof aquiclude and lead to the upper aquifer. The range of the water-conducting fracture zone determines the range of the direct water-filling aquifer. The aquifer above the fracture zone also participates in the water inflow of the working face under the condition of natural vertical fracture communication [15]. The mining height of the coal seam in the 3085 working face is 2.9 m. The ratio of measured height to mining thickness in Donghuantuo Mine is 15.4, and the height of the roof water-conducting fracture zone is 46.2 m. Considering the upward conduction of natural vertical fracture development, the aquifer within the cumulative thickness of 100 m above coal seam 8 is used as the water source for roof filling. The hydrophobic borehole in the 3085 working face mainly drains the water above the roof of coal seam 5.

### *2.2. Arrangement of Drill Sockets and Hydrophobic Boreholes*

The mining of coal seam 8 in Donghuantuo Mine of Kailuan has included a large number of drainage boreholes in the roof fractured aquifer in its construction, and the drainage boreholes have revealed the aquifer layer and its water-rich change [16]. A total of 12 drilling holes were laid on the 3085 working face, 7 drilling holes were laid on the air duct, and 5 drilling holes were laid on the haulage road. The specific distribution of the drilling holes is shown in Figure 1.

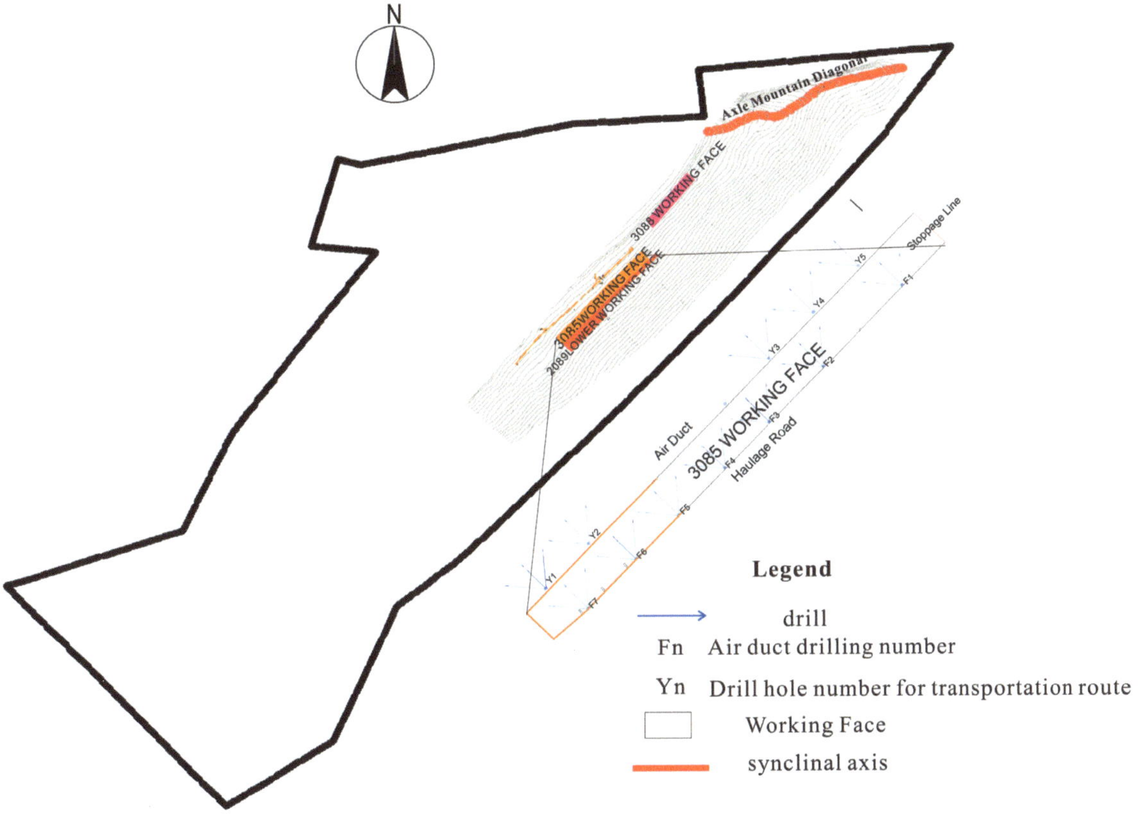


**Figure 1.** 3085 working face drainage borehole location diagram.

## 3. The Variation Law and Influencing Factors of Underground Borehole Drainage Water Quantity

### 3.1. Dynamic Variation Law of Borehole Water Inflow with Time

There is a matching and corresponding relationship between the dynamic water inflow of the advance drainage borehole and the hydrogeological conditions. By observing the dynamic water inflow of the borehole and analyzing the hydrogeological conditions of the working face, it can not only predict the degree of water damage but also determine whether it is suitable for mining. It is suitable for mining when the drilling water volume decreases rapidly or when the drilling water volume enters a stable period in a small case. If the amount of drilling water is large and does not drop continuously, additional drilling is needed to increase the intensity of drainage.

The dynamic curve of the water yield of the advance drainage borehole in the 3085 working face is illustrated in Figure 2. Taking Ss3085-1, Ss3085-2, Ss3085-3, Ss3085-19, Ss3085-20 and Ss3085-21 as examples, it can be seen from the figure that the initial water yield of boreholes Ss3085-1, Ss3085-2 and Ss3085-3 is large, but the decline rate is fast and the water yield of the boreholes is then kept at 0.1 $m^3$/min and enters a stable period. The water inflow of boreholes Ss3085-19, Ss3085-20 and Ss3085-21 remain at 0.05 $m^3$/min after three rapid declines. It can be observed that the borehole water inflow of the 3085 working face is a typical step-down type, and the initial water pressure of the working face is still high, with a stable supply.

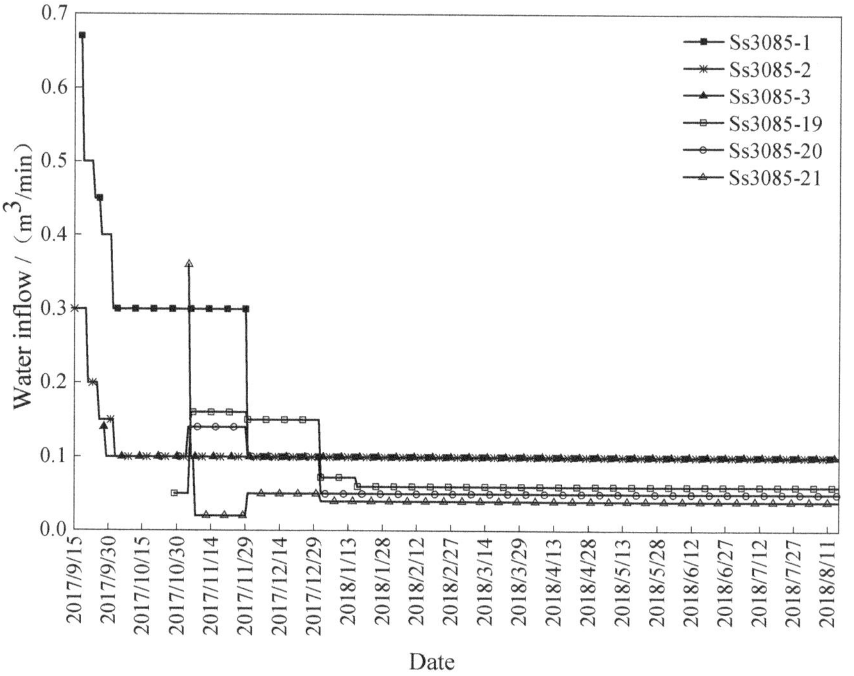


**Figure 2.** 3085 typical dynamic curve of drainage drilling in the working face.

### *3.2. Generalized Gray Correlation Analysis*

Gray system theory is based on the similarity of the geometric shape of each factor to determine the degree of correlation between various factors. The generalized gray correlation degree combines the absolute gray correlation degree and the relative gray correlation degree and uses the area value between the two sequence curves to measure the similarity of the two sequence curves as a whole, which can analyze the correlation factors more comprehensively [17].

Absolute degree of gray incidence.

Let $X_i = (x_i(1),\ x_i(2),\ \cdots,\ x_i(n))$, be the sequence of system behavior, $D_1$ be the sequence of related factors.

$$X_iD_1 = (x_i(1)d_1,\ x_i(2)d_1,\ \cdots,\ x_i(n)d_1) \tag{1}$$

where $x_i(k)d_1 = x_i(k) - x_i(1)\ (k = 1, 2, \ldots, n)$ then $D_1$ is called the starting point annihilator, $X_iD_1$ is the starting point annihilator of $X_i$, denoted by:

$$X_iD_1 = X_i^0 = \left(x_i^0(1), x_i^0(2),\ \cdots,\ x_i^0(n)\right) \tag{2}$$

Assuming that the system behavior sequences $X_i$ and $X_j$ have the same length, they are both time-distance sequences, and their starting points are respectively:

$$X_i^0 = \left(x_i^0(1), x_i^0(2),\ \cdots,\ x_i^0(n)\right) \tag{3}$$

$$X_j^0 = \left(x_j^0(1), x_j^0(2),\ \cdots,\ x_j^0(n)\right) \tag{4}$$

$$|s_i| = \left|\int_1^n X_i^0 dt\right| = \left|\sum_{k=2}^{n-1} x_{i_i}^0(k) + \frac{1}{2}x_i^0(\mathrm{n})\right| \tag{5}$$

$$|s_j| = \left|\int_1^n X_j^0 dt\right| = \left|\sum_{k=2}^{n-1} x_j^0(k) + \frac{1}{2}x_j^0(\mathrm{n})\right| \tag{6}$$

$$|s_i - s_j| = \left|\sum_{k=2}^{n-1}(x_{i_i}^0(k) - x_j^0) + (x_i^0(\mathrm{n}) - x_j^0(\mathrm{n}))\right| \tag{7}$$

The formula for calculating the gray absolute correlation degree is as follows:

$$\varepsilon_{ij} = \frac{1 + |s_i| + |s_j|}{1 + |s_i| + |s_j| + |s_i - s_j|} \tag{8}$$

Gray relative degree

Let $X_i = (x_i(1), x_i(2), \cdots, x_i(n))$ be the sequence of system behaviors, $D_2$ be the sequence of related factors.

$$X_i D_2 = (x_i(1)d_2, x_i(2)d_2, \cdots, x_i(n)d_2) \tag{9}$$

where, $x_i(k)d_2 = x_i(k)/x_i(1),\ x_i(1) \neq 0(k = 1, 2, \ldots, n)$, then $D_2$ is called the initializer, $X_i D_2$ is the initial image of $X_i$, denoted by:

$$X_i D_2 = X_i' = \left(x_i'(1), x_i'(2), \cdots, x_i'(n)\right) \tag{10}$$

The calculation formula of the gray relative correlation degree is as follows:

$$\gamma_{ij} = \frac{1 + |s_i'| + |s_j'|}{1 + |s_i'| + |s_j'| + |s_i' - s_j'|} \tag{11}$$

The gray comprehensive correlation formula is as follows:

$$\rho_{ij} = \theta\varepsilon_{ij} + (1 - \theta)\gamma_{ij} \tag{12}$$

$\rho_{ij}$ is the gray comprehensive correlation degree between $X_i$ and $X_j$, where $\theta \in [0, 1]$. Generally, $\theta = 0.5$.

### *3.3. Factors Affecting the Water Yield of the Hydrophobic Borehole*

Taking the water yield of advanced drainage boreholes as the system characteristic sequence, the borehole orifice elevation, borehole orientation, borehole inclination, the final hole elevation, sharp angle between stratum and borehole direction, final hole depth and aquifer thickness are related factors. The seven influencing factors of the water discharge of the advanced drainage borehole are analyzed and calculated, and the results are shown in Table 1. (Please refer to the Supplementary Materials for drilling related data.)

**Table 1.** Generalized correlation analysis of influencing factors of drainage borehole water yield.

| Influencing Factor | Gray Absolute Correlation Degree | Gray Relative Correlation Degree | Comprehensive Correlation Degree |
|---|---|---|---|
| The elevation of the borehole $X_1$ (m) | 0.50005 | 0.50381 | 0.50193 |
| The azimuth angle of drilling $X_2$ (°) | 0.50015 | 0.53886 | 0.51951 |
| The dip angle of drilling $X_3$ (°) | 0.50041 | 0.50567 | 0.50304 |
| The elevation of the final hole $X_4$ (m) | 0.50006 | 0.50381 | 0.50194 |
| The sharp angle between the formation and the direction of drilling $X_5$ (°) | 0.50062 | 0.56743 | 0.53402 |
| The depth of the final hole $X_6$ (m) | 0.51283 | 0.52856 | 0.52069 |
| The aquifer thickness $X_7$ (m) | 0.50532 | 0.70263 | 0.60397 |

The gray absolute correlation degree is only applicable to the case where the dimensions of each variable are consistent or the magnitude changes little. For this study, the data of each main control factor are quite different, and the results may have errors; the gray relative correlation degree focuses on the degree of change of the representation relative to the starting point, which is also one-sided. The gray comprehensive correlation degree combines the characteristics of the two methods, and the results are more accurate. From the table, it can be seen that the gray comprehensive correlation degree of the influencing factors of the water discharge of the advanced drainage borehole is $X_7 > X_5 > X_6 > X_2 > X_3 > X_4 > X_1$, indicating that the aquifer thickness has a great influence on the amount of water discharged, and the influence degree of the sharp angle between the formation and the direction of drilling, the depth of the final hole, the azimuth angle of drilling, the dip angle of drilling, the elevation of the final hole and the elevation of the borehole on the amount of water discharged decreases successively.

## 4. Establish a Numerical Model of Groundwater Flow

Collect geological and hydrogeological data, establish a mathematical model of the groundwater system based on Modflow and simulate groundwater flow. In this model, the default groundwater density is a fixed value, and the differential equation for the three-dimensional unsteady flow of groundwater can be expressed as follows:

$$\frac{\partial H}{\partial t} = \frac{\partial}{\partial x}\left(K_{xx}\frac{\partial H}{\partial x}\right) + \frac{\partial}{\partial y}\left(K_{yy}\frac{\partial H}{\partial y}\right) + \frac{\partial}{\partial z}\left(K_{zz}\frac{\partial H}{\partial z}\right) + W \tag{13}$$

where $K_{xx}$, $K_{yy}$, $K_{zz}$ are the permeability coefficients in the $(x, y, z)$ directions ($LT^{-1}$), $\mu_s$ is the water storage rate ($L^{-1}$), H is the water head (L), t is time (T), W is the source sink term, with positive inflow and negative outflow ($T^{-1}$).

Based on the above generalization of aquifer and boundary conditions, a mathematical model for groundwater flow can be established as follows:

$$\begin{cases} \mu_s \frac{\partial H}{\partial t} = \frac{\partial}{\partial x}\left(K_{xx}\frac{\partial H}{\partial x}\right) + \frac{\partial}{\partial y}\left(K_{yy}\frac{\partial H}{\partial y}\right) + \frac{\partial}{\partial z}\left(K_{zz}\frac{\partial H}{\partial z}\right) + W & (x,y,z) \in \Omega \\ h(x,y,z) = h_0(x,y,z) & (x,y,z) \in \Omega \\ \left.\frac{\partial H}{\partial n}\right|_{\Gamma_i} = 0 \quad (i = 1,2,3,4) & (x,y,z) \in \Gamma \end{cases} \tag{14}$$

where $h_0(x,y,z)$ is the initial water head (L), $\Omega$ is the research area, $\Gamma_i$ is the second type of boundary (waterproof boundary), $i = 1, 2, 3, 4$.

After investigating and analyzing the regional geological structure and hydrogeological conditions in the study area, the conceptual model and boundary conditions of groundwater movement in the study area are generalized, and the numerical model of groundwater flow is established.

### 4.1. Model Boundary Condition

The three-dimensional model of groundwater seepage based on Modflow requires a reasonable generalized boundary condition to describe the flow exchange between the study area and the surrounding environment [18]. There is water exchange around the simulation area, so it is considered the flow boundary, which is usually called the second boundary condition [19]. The Donghuantuo Mine is regarded as a typical mine with a developed inner boundary of planar fractures. Under the condition of vertical fracture communication in planar distribution, there is a certain hydraulic connection between the upper and lower water-bearing sections.

(1) Taking the 2089 lower working face as the shallow boundary, the 2089 lower face was completed in 2012. The study area of the coal seam 8 roof aquifer flows freely to the goaf of the haulage road, which is the constant head outflow boundary. The water

head value adopts the elevation value of the coal seam 8 floor and decreases from NE to SW.

(2) The −950 contour line of the coal seam 8 floor and the fault F18 are the deep boundaries. The coal seam 8 roof aquifer runs from NW to SE and is the mining area confluence. The boundary is the inflow boundary, which is far away from the mining area, and the groundwater level drops more evenly, which is the constant flow inflow boundary. The strike of the F18 fault is perpendicular to the direction of groundwater runoff. Depending on the understanding of 3D seismic exploration and production practice, most of the faults in the Donghuantuo Mine are transverse water-resisting faults, so the section of the F18 fault is the 0 flow boundary of water-resisting.

(3) The cutting hole along the formation trend line of the 3085 working face is used as the WS boundary. The boundary is basically consistent with the flow direction of groundwater, which is the inflow boundary with a small flow. However, there are few mining projects in the southern part of the boundary, the water level is higher than that in the study area, and the flow rate increases at the southern end.

(4) Taking the stop line of the 3085 working face along the stratigraphic dip line as the NE boundary, the northern area is a syncline axis and a steeply inclined wing, which is strongly affected by the 3088 working face and the north second mining area. The runoff direction is consistent with the boundary, which is a small flow boundary.

### 4.2. Generalization of Aquifer and Aquiclude

When the numerical method is used to establish the model, the aquifer group of coal measure strata is divided according to the stable development of the impermeable layer in the whole mining area [20]. The aquifer group contains multiple water-bearing layers and water-resisting layers, resulting in the insufficient accuracy of groundwater numerical model simulation. The lithology change of sedimentary strata can reflect the sedimentary environment. By comparing the stratification of borehole lithology, the aquifer and aquiclude can be divided in detail [21,22]. The lithology of the strata within 100 m above the roof of coal seam 8 in the 3085 working face mainly includes sandstone, siltstone, clay rock, shale and coal. Through a comparison of lithology, the roof of coal seam 8 in the study area is subdivided into 27 lithologic layers from bottom to top, as shown in Table 2. According to the comprehensive lithology, the water content and impermeability of each layer are determined. The strong aquifer represents the lithology of medium coarse sandstone, mainly sandstone; the representative lithology of the aquifer is fine sandstone, fine sandstone with clay or siltstone; the weak permeable layer represents the lithology of siltstone and siltstone with clay rock; the representative lithology of the aquiclude is mainly clay rock and shale.

The main strong aquifers in the 3085 working face are seven layers, which are the 3rd, the 7th, the 14th, the 16th, the 20th, the 22nd and the 24th layers. The comprehensive thickness of the strong aquifers is 10.44~55.65 m, and the average thickness is 31.16 m. The 3rd layer and the 7th layer are near coal seam 8 and within the scope of the caving zone. The thicknesses of the 3rd layer and the 7th layer are thicker, and the local thickness is extra thick. The local thickness of the 3rd layer reaches 25.50 m, which is the maximum thickness of the strong aquifers. The average thickness of the 14th and 16th strong aquifers in the middle section are thicker. There are four layers of aquifers in the 3085 working face, which are the 5th, 10th, 13th and 26th layers. The comprehensive thickness of the aquifers is 2.90~19.80 m, and the average thickness is 10.18 m. The average thickness of the 5th aquifer near coal seam 8 is thicker and uniform. There are 10 layers of weak permeables in the 3085 working face, which are the 2nd, 4th, 8th, 12th, 15th, 17th, 19th, 23rd, 25th and 27th layers. The comprehensive thickness is 19.72~52.95 m, and the average thickness is 40.39 m. The average thickness of the 2nd weak permeable layer near coal seam 8 is 2.02 m, and the thickness is quite different. There are six layers of aquicludes in the 3085 working face, which are the 1st layer, 6th layer, 9th layer, 11th layer, 18th layer and 21st layer. The comprehensive thickness is 2.80~35.50 m, and the average thickness is 19.32 m. The average thickness of the 6th aquiclude near coal seam 8 is thicker. The average thickness of the 6th aquiclude is 5.33 m, and the local thickness can reach 22.53 m. The 3rd strong aquifer, the 5th strong aquifer and the 7th strong aquifer are the main water filling layers of coal seam 8, and the water source above the roof of coal seam 8 is abundant.

**Table 2.** Characteristics of aquifer thickness in model area.

| Level Number | Comprehensive Lithology | Average Thickness/m | Maximum Thickness/m |
|---|---|---|---|
| The 24th strong aquifer layer | Medium sandstone and fine sandstone | 2.86 | 8.30 |
| The 22nd strong aquifer layer | Sandstone with a small amount of siltstone | 2.00 | 10.20 |
| The 20th strong aquifer layer | Sandstone with a small amount of siltstone | 3.30 | 7.35 |
| The 16th strong aquifer layer | Medium sandstone and fine sandstone | 6.33 | 20.47 |
| The 14th strong aquifer layer | Medium sandstone and fine sandstone | 6.26 | 27.21 |
| The 7th strong aquifer layer | Medium sandstone and fine sandstone | 5.52 | 13.30 |
| The 3rd strong aquifer layer | Medium sandstone and fine sandstone | 4.88 | 25.50 |
| The 26th aquifer layer | Fine sandstone | 1.41 | 5.28 |
| The 13th aquifer layer | Fine sandstone with a small amount of clay or siltstone | 3.08 | 8.70 |
| The 10th aquifer layer | Fine sandstone | 0.46 | 2.80 |
| The 5th aquifer layer | Fine sandstone | 5.23 | 10.83 |
| The 27th weak permeable layer | Siltstone with mudstone | 4.37 | 10.30 |
| The 25th weak permeable layer | Siltstone with mudstone | 7.51 | 22.59 |
| The 23rd weak permeable layer | Siltstone with mudstone | 1.35 | 6.40 |
| The 19th weak permeable layer | Siltstone with coal or clay | 3.13 | 10.77 |
| The 17th weak permeable layer | Siltstone | 1.57 | 8.39 |
| The 15th weak permeable layer | Siltstone with coal | 1.59 | 10.98 |
| The 12th weak permeable layer | Siltstone | 3.62 | 11.30 |
| The 8th weak permeable layer | Siltstone | 11.65 | 22.80 |
| The 4th weak permeable layer | Siltstone | 3.57 | 11.60 |
| The 2nd weak permeable layer | Siltstone | 2.02 | 5.50 |
| The 21st aquiclude layer | Siltstone with clay | 3.50 | 25.80 |
| The 18th aquiclude layer | Coal and clay | 1.11 | 6.98 |
| The 11th aquiclude layer | Coal and clay | 1.55 | 8.25 |
| The 9th aquiclude layer | Coal and clay | 5.50 | 20.75 |
| The 6th aquiclude layer | Coal and clay | 5.32 | 22.53 |
| The 1st aquiclude layer | Coal and clay | 2.33 | 11.36 |

### 4.3. Establishment of Geometric Model

Using the finite difference method, the simulation area is divided into several cells of 40 m × 40 m by rectangular subdivision, which is divided into 60 grids in the X direction and 58 grids in the Y direction.

The fault structure in the study area is relatively developed, and the existence of faults will affect the safety of coal mining. Affected by the fault structure, the coal seam floor strata are broken and moved, and the water-conducting fissures in the aquiclude increase, which reduces the water-resisting ability of the aquiclude and increases the risk of water inrush from the coal seam floor. The fault intensity index is the total of the product of the drop of all faults per unit area and its strike length, which can reflect the complexity of faults in the study area. The greater the fault intensity index, the stronger the risk of water inrush from the coal seam floor. The distribution of the fault intensity index is shown in Figure 3. Import the elevation data into the model to assign the elevation of each layer to generate a three-dimensional model, as shown in Figure 3. Different aquifers and impermeable layers are set with different colors, and each layer is generalized into a complete stratum in the simulation area without stratum loss. In the 3085 working face, 12 drilling holes and 32 drainage holes were constructed. The hydrophobic borehole of the same drilling socket is generalized into the same pumping well. The position of the well is determined by the final hole position of the borehole, and the position of the casing is penetrated from the elevation of the final hole to the model floor. All the casings penetrate the strata below the 10th aquifer; some penetrate the 14th aquifer, the 16th aquifer and the 20th aquifer; and a few penetrate the 22nd aquifer. The distribution of aquifers, aquicludes and wells in the model is shown in Figure 4.

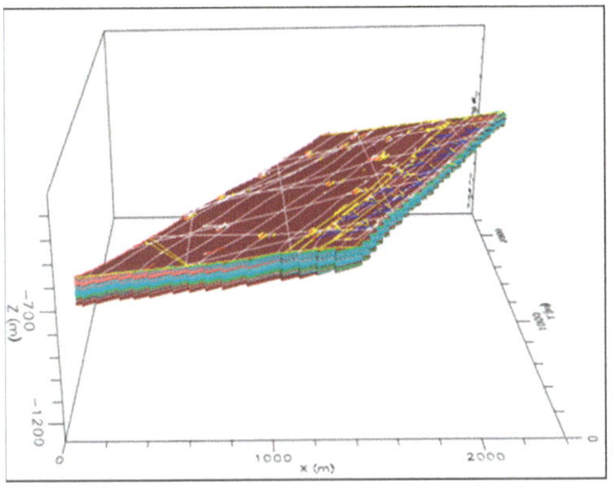


**Figure 3.** Schematic diagram of geometric model with a 1:1 horizontal longitudinal ratio.

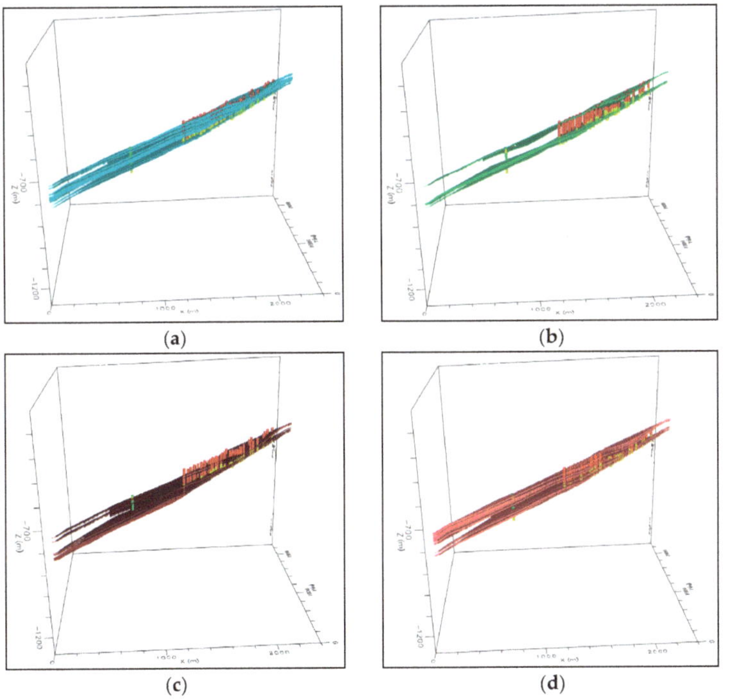

(a) (b) (c) (d)

**Figure 4.** Three-dimensional model aquifer and well display diagram. (**a**) Strong aquifer layers. (**b**) Aquifer layers. (**c**) Weak permeable layers. (**d**) Aquiclude layers.

The 3D model of the coal seam 8 floor contour line is fitted with the actual coal seam 8 floor contour map. The fitting results are shown in Figure 5. The lowest elevation of the simulation area is −950 m, the highest is −500 m and the contour interval is 50 m. It can be seen from the figure that the fitting effect is good, and the created 3D model is more realistic.

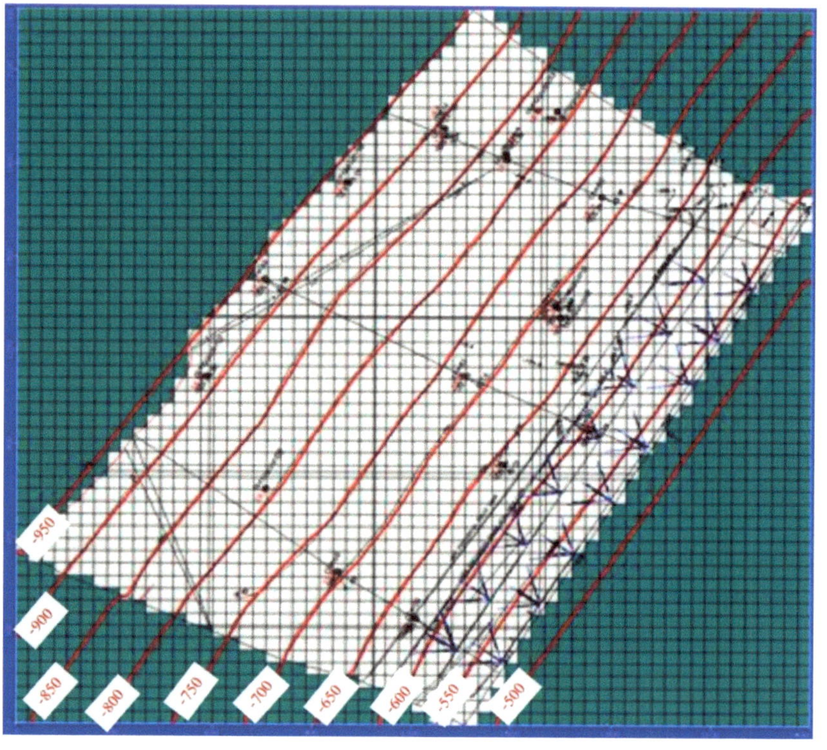


**Figure 5.** Elevation fitting result diagram of the coal seam 8 floor.

### *4.4. Model Identification and Verification*

The identification of the model is also called model inversion. By continuously adjusting the hydrogeological parameters and equilibrium terms, the measured values of the groundwater level in the same period are fitted [23]. Ensure that the constructed model can more truly reflect the hydrogeological conditions of the simulation area and more reasonably describe the groundwater flow state. The model is identified by fitting the long-term hole water level data. Model validation was carried out after the model fitting. Keeping the parameters unchanged, the correctness of the model is judged by using the long-term observation data after the fitting period. There is a long view hole in Dongguan 55 in the 3085 working face. The curve of fitting calculation water level and actual change water level of the Dongguan 55 hole is shown in Figure 6. The water level change of the model is in good agreement with the actual situation; a suitable fitting recognition effect is achieved.

The calculated water level and observed water level values of the Dongguan 55 hole are shown in Table 3. The error between the simulated water level and the actual water level is mostly controlled within 5 m, with a maximum error of only 7.81 m. Using formula 13 to calculate the simulated water level error rate, the maximum error rate is only 1.63%, indicating a good simulation effect.

$$\text{Error rate} = \frac{|\text{Calculated water level} - \text{Observing water level}|}{\text{Observing water level}} * 100\% \quad (15)$$

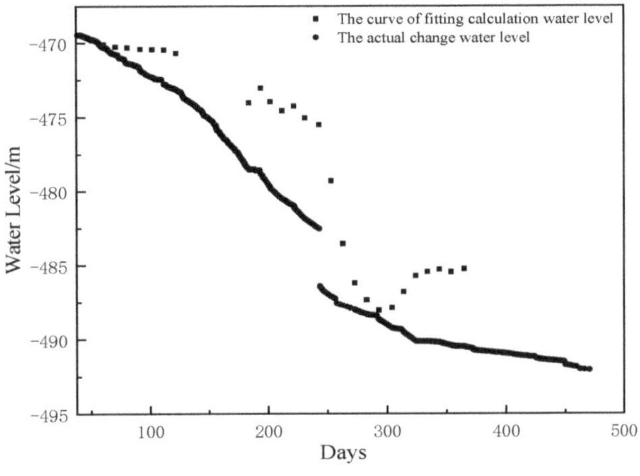


**Figure 6.** Dongguan 55 hole water level change fitting curve diagram.

**Table 3.** Table of simulated and measured water levels in the Dongguan 55 hole.

| Observed Time | Observing Water Level/m | Calculated Water Level/m | Error Value/m | Error Rate |
|---|---|---|---|---|
| 41 | −469.47 | −469.47 | 0.00 | 0.00 |
| 51 | −469.82 | −469.79 | 0.03 | 0.01 |
| 61 | −470.1 | −470.30 | −0.20 | 0.04 |
| 71 | −470.25 | −470.77 | −0.52 | 0.11 |
| 81 | −470.3 | −471.37 | −1.07 | 0.23 |
| 92 | −470.4 | −471.81 | −1.41 | 0.30 |
| 102 | −470.43 | −472.30 | −1.87 | 0.40 |
| 112 | −470.46 | −472.77 | −2.31 | 0.49 |
| 122 | −470.68 | −473.11 | −2.43 | 0.52 |
| 184 | −474.03 | −478.54 | −4.51 | 0.95 |
| 194 | −473.04 | −478.84 | −5.80 | 1.23 |
| 202 | −473.96 | −479.72 | −5.76 | 1.22 |
| 212 | −474.56 | −480.51 | −5.95 | 1.25 |
| 222 | −474.25 | −481.00 | −6.75 | 1.42 |
| 231 | −475.05 | −481.85 | −6.80 | 1.43 |
| 243 | −475.5 | −482.52 | −7.02 | 1.48 |
| 253 | −479.31 | −487.12 | −7.81 | 1.63 |
| 263 | −483.54 | −487.72 | −4.18 | 0.86 |
| 273 | −486.19 | −488.00 | −1.81 | 0.37 |
| 283 | −487.32 | −488.31 | −0.99 | 0.20 |
| 293 | −488.05 | −488.65 | −0.60 | 0.12 |
| 304 | −487.86 | −489.23 | −1.37 | 0.28 |
| 314 | −486.78 | −489.51 | −2.73 | 0.56 |
| 324 | −485.71 | −490.11 | −4.40 | 0.91 |
| 334 | −485.43 | −490.13 | −4.70 | 0.97 |
| 344 | −485.26 | −490.18 | −4.92 | 1.01 |
| 354 | −485.44 | −490.41 | −4.97 | 1.02 |
| 365 | −485.24 | −490.48 | −5.24 | 1.08 |

## 5. Simulation and Optimization of Advance Drainage Borehole Layout

The layout of advanced drainage boreholes in the roof of the 3085 working face was optimized to reduce the number of drainage boreholes and reduce the drainage time. Through gray correlation analysis, it can be seen that the thickness of the aquifer has a great influence on the water output of the advanced drainage borehole. Combined with the data of drainage boreholes, the total thicknesses of all aquifers and strong aquifers from the 3rd layer to the 16th layer of the 3085 working face are counted. The total thickness contour map of the roof aquifer of coal seam 8 is shown in Figure 7. The total thickness of the aquifer is more than 7 m, and the local thickness can reach 21 m.

A total of 12 drainage points were arranged in the 3085 working face, of which seven drainage points were arranged on the air duct, namely F1, F2, F3, F4, F5, F6 and F7. Five hydrophobic points were arranged on the haulage road, namely Y1, Y2, Y3, Y4 and Y5. It can be seen from Figure 7 that the F2 drainage point and the F4 drainage point are in the weak area of the roof aquifer thickness. Considering the thickness of the aquifer and the principle of uniform pore distribution, the F2 drainage point and the F4 drainage point are moved to the B1 drainage point and the B2 drainage point, respectively. The drainage project of the original drainage point position is cancelled, and the same drainage work is set at the moved drainage point position as the original drainage point, and the influence of the two groups of drainage projects on the aquifer water level is compared.

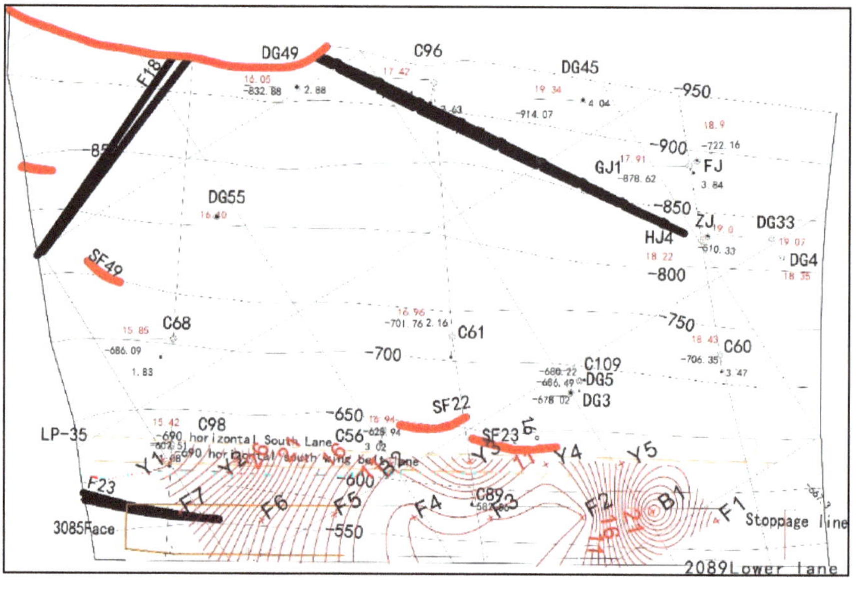


**Figure 7.** Contour map of the total thickness of the roof aquifer.

The day before the mining work, the water level line of the aquifer before and after the drilling optimization is illustrated in Figure 8. The water level at F1 is −460 m before optimization, and the water level at F1 is reduced to −465 m after optimization. After moving the hydrophobic point, the aquifer water level drops faster. Through numerical simulation, it can be observed that the drainage boreholes set at the larger cumulative thickness of the aquifer in the drainage layer can indeed carry out the roof advance drainage work more quickly.

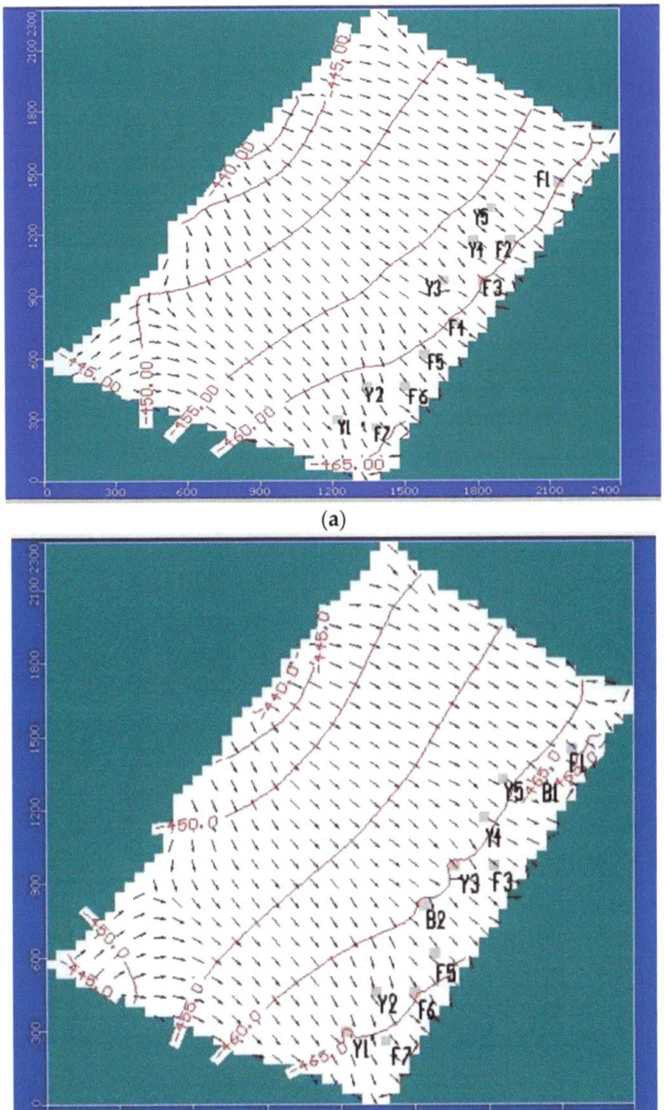


(a)

(b)

**Figure 8.** Distribution position and isoline diagram of aquifer drain hole. (**a**) Before optimization. (**b**) After optimization.

## 6. Conclusions

The gray comprehensive correlation method was used to analyze the factors affecting the water output of advanced drainage boreholes. The gray comprehensive correlation ranking of each influencing factor suggests that the main influencing factor on the water output of advanced drainage boreholes is the thickness of the aquifer.

The water source for filling the roof of the 3085 mining face is the aquifer within a cumulative thickness range of 100 m above the 8th coal seam. According to the lithology, it can be divided into 27 layers from bottom to top, with a total of 7 strong aquifers mainly composed of coarse to medium sandstone and 4 aquifers mainly composed of fine sand

and siltstone. The roof water source of this working face is abundant, and pre-drainage is required before mining.

Comparing the calculated water level and observed water level data of the Dongguan 55 hole, the two sets of values are relatively close with a small error rate. It is believed that the three-dimensional model of groundwater seepage established based on Visual Modflow numerical simulation software is in line with the actual water level dynamics and can optimize the advanced drainage drilling layout plan. Comparing the models before and after optimization, the layout of advanced drainage boreholes should fully consider the development range of the roof conductivity and the hydrogeological conditions of the roof aquifer. When the drainage boreholes are set at locations with thicker roof aquifers, the effect of advanced drainage work is better.

**Supplementary Materials:** The following supporting information can be downloaded at: https://www.mdpi.com/article/10.3390/w16182613/s1, Table S1: Drilling data.

**Author Contributions:** Conceptualization, Y.L. and Y.M.; methodology, Y.L.; software, Y.L.; validation, Y.L., Y.M. and Y.Z.; formal analysis, Y.L.; investigation, F.M.; resources, F.M.; data curation, Y.M.; writing—original draft preparation, Y.L.; writing—review and editing, Y.Z.; visualization, Y.L.; supervision, Y.M.; project administration, F.M.; funding acquisition, Y.L. All authors have read and agreed to the published version of the manuscript.

**Funding:** This paper did not receive funding support.

**Data Availability Statement:** All data, models, or code generated or used during the study are available from the corresponding author by request.

**Acknowledgments:** The authors would like to sincerely thank the Donghuantuo Coal Mine of Kailuan Group for providing the geological data.

**Conflicts of Interest:** Author Fangang Meng was employed by the company Donghuantuo Mining Branch of Kailuan Group. The remaining authors declare that the research was conducted in the absence of any commercial or financial relationships that could be construed as a potential conflict of interest.

## References


1. State Administration of Coal Mine Safety. *Detailed Rules for Coal Mine Water Prevention and Control*; China Coal Industry Publishing House: Beijing, China, 2018.
2. State Administration of Coal Mine Safety. *Regulations on Water Control in Coal Mines*; China University of Mining and Technology Publishing House: Xuzhou, China, 2009; p. 123.
3. Wu, Q. Progress, problems and prospects of prevention and control technology of mine water and reutilization in China. *J. China Coal Soc.* **2014**, *39*, 795–805. [CrossRef]
4. Chen, S.; Dong, S.; Li, J.; Wang, H.; Yang, J. Analytical solution for slanted well in the roof of coal mine working face. *J. China Coal Soc.* **2016**, *41*, 1517–1523. [CrossRef]
5. Wang, G.; Liu, B.; Li, Y. Analysis on Prediction Method of Water Inflow of Drilling Hole in Mailing Underground Drilling. *Coal Technol.* **2019**, *38*, 96–98. [CrossRef]
6. Zhan, H.; Cao, J. Analytical and semi-analytical solutions of horizontal well capture times under no-flow and constant-head boundaries. *Adv. Water Resour.* **2000**, *23*, 835–848. [CrossRef]
7. Zhan, H.; Wang, L.V.; Park, E. On the horizontal-well pumping tests in anisotropic confined aquifer. *J. Hydrol.* **2001**, *252*, 37–50. [CrossRef]
8. Zheng, Y.; Ma, Y.; Zhang, C. Water drainage law in No.8 coal seam roof of Donghuantuo Coal Mine. *Coal Eng.* **2017**, *49*, 67–70.
9. Zhao, B. Application of gray correlation in the effect analysis of underground drilling water drainage. *J. Liaoning Tech. Univ. (Nat. Sci.)* **2013**, *32*, 289–292.
10. Zhao, C.; Dong, S.; Wang, H.; Jin, D.; Li, C.; Wang, S.; Zhou, Z.; Liu, Y. Analysis of water inrush from boreholes for drainage of confined aquifer by upward boreholes in underground coal mining face. *J. China Coal Soc.* **2020**, *45*, 405–414. [CrossRef]
11. Zhou, Z. Application of the monoexponential decay function on analyzing water-inflow in underground boreholes. *J. Min. Saf. Eng.* **2018**, *35*, 642–648. [CrossRef]
12. Yang, Z.; Dou, L.; Liu, C.; Xu, M.; Lei, Z.; Yao, Y. Application of high-pressure water jet technology and the theory of rock burst control in roadway. *Int. J. Min. Sci. Technol.* **2016**, *26*, 929–935. [CrossRef]
13. Ma, Y.; Zuo, W.; Liu, B.; Gao, L.; Chang, J. Analysis on law of geological structures controlling groundwater in buried syncline nose:a case in the Donghuantuo Coal Mine, Kailuan. *J. China Coal Soc.* **2012**, *37*, 157–160. [CrossRef]

14. Guo, W.; Lou, G.; Zhao, B. Study on the height of water-conductive fracture zone in alternate overburden of soft and hard with top coal caving mining in Lugou coal mine. *J. Min. Saf. Eng.* **2019**, *36*, 519–526. [CrossRef]
15. Wang, F.; Wu, S.; Li, Y.; Meng, X.; Wang, Z. Tracing Design Method of Pectinate Directional Borehole of Tunnels in Underground Mine. *Saf. Coal Mines* **2017**, *48*, 153–156. [CrossRef]
16. Deng, J. *Grey System Theory Tutorial*; Wuhan University of Technology Publishing House: Wuhan, China, 1992.
17. Liu, S. *The Grey System Theory and Its Application*, 5th ed.; Science Publishing House: Beijing, China, 2010.
18. Ding, Y.; Ma, R.; Li, N. A simulation model for three-dimensional coupled wave-current flumes. *Eng. Mech.* **2015**, *32*, 68–74+88.
19. Li, X.; Li, Y.; Zhou, S. Study and Application of Forecasting System for Water Inrush Under High Pressure in Xiamen Submarine Tunnel Construction Based on GIS. *Procedia Environ. Sci.* **2011**, *10*, 999–1005. [CrossRef]
20. Wu, Q.; Dong, S.; Zhang, Z. *Water Hazard Prevention of Mine*; China University of Mining and Technology Publishing House: Xuzhou, China, 2007.
21. Shao, L.; Xu, X.; Wang, S.; Wang, D.; Gao, D.; Wang, X.; Lu, J. Research progress of palaeogeography and palaeoenvironmental evolution of coal-bearing series in China. *J. Palaeogeog Raphy (Chin. Ed.)* **2021**, *23*, 19–38. [CrossRef]
22. Lu, B.; Xiao, Y. Study on Lithological Division of Coal Bearing Fractured Aquifer in Heilongjiang Province. *J. Shandong Univ. Sci. Technol.* **2017**, *36*, 96–102. [CrossRef]
23. Thompson, K.; Cadle, S.; Roemer, G. Integrating Less Common Data Sources to Improve Groundwater Model Calibration. *Mine Water Environ.* **2015**, *34*, 153–161. [CrossRef]

sustainability

Article

# Spatial-Temporal Evaluation and Prediction of Water Resources Carrying Capacity in the Xiangjiang River Basin Using County Units and Entropy Weight TOPSIS-BP Neural Network

Jiacheng Wang [1,†], Zhixiang Wang [1,†], Zeding Fu [2,†], Yingchun Fang [3], Xuhong Zhao [1], Xiang Ding [1], Jing Huang [1], Zhiming Liu [4], Xiaohua Fu [1,*] and Junwu Liu [3,*]

1 Center for Ecological Environment Management and Assessment, Central South Forestry University of Science and Technology, Changsha 410004, China; 20221100485@csuft.edu.cn (J.W.); 20211200665@csuft.edu.cn (Z.W.); knoccc2023@163.com (X.Z.); 20231200771@csuft.edu.cn (X.D.); 20231200751@csuft.edu.cn (J.H.)
2 Yueyang City Water Resources Bureau, Yueyang 414000, China; 13975885159@163.com
3 Hunan Kaidi Engineering Technology Co., Ltd., Yueyang 414000, China; 5450599@163.com
4 Henan Academy of Sciences, Zhengzhou 450046, China; zhiming.liu@enmu.edu
* Correspondence: fxh4431491@csuft.edu.cn (X.F.); custfzd@163.com (J.L.)
† These authors contributed equally to this work.

**Abstract:** To improve the water resources carrying capacity of the Xiangjiang River Basin and achieve sustainable development, this article evaluates and predicts the Xiangjiang River Basin's water resources carrying capacity level based on county-level units. This article takes 44 county-level units in the Xiangjiang River Basin as the evaluation target, selects TOPSIS and the entropy weight method to determine weights, calculates the water resources carrying capacity level of the evaluation sample, uses a BP neural network model to calculate the predicted water resources carrying capacity level for the next 5 years, and adds the GIS method for spatiotemporal analysis.(1) The water resources carrying capacity of the Xiangjiang River Basin has remained relatively stable for a long period, with overloaded areas being the majority. (2) There are relatively significant spatial differences in the carrying capacity of water resources: Zixing City, located upstream of the tributary, is far ahead due to its possession of the Dongjiang Reservoir; the water resources carrying capacity in the middle and lower reaches (northern region) is generally higher than that in the upper reaches (southern region). (3) According to the BP neural network model prediction, the water resources carrying capacity of the Xiangjiang River Basin will maintain a stable development trend in 2022, while areas such as Changsha and Zixing City will be in a critical state, and other counties and cities will be in an overloaded state.This study has important references value for the evaluation and early warning work of the Xiangjiang River Basin and related research, providing a scientific and systematic evaluation method and providing strong support for water resource management and planning in Hunan Province and other regions.

**Keywords:** Xiangjiang River; water resources carrying capacity; TOPSIS model; entropy weighting method; BP neural network

**Citation:** Wang, J.; Wang, Z.; Fu, Z.; Fang, Y.; Zhao, X.; Ding, X.; Huang, J.; Liu, Z.; Fu, X.; Liu, J. Spatial-Temporal Evaluation and Prediction of Water Resources Carrying Capacity in the Xiangjiang River Basin Using County Units and Entropy Weight TOPSIS-BP Neural Network. *Sustainability* **2024**, *16*, 8184. https://doi.org/10.3390/su16188184

Academic Editors: Zongjun Gao and Jiutan Liu

Received: 28 July 2024
Revised: 10 September 2024
Accepted: 12 September 2024
Published: 19 September 2024



## 1. Introduction

The carrying capacity of water resources has always been a vital issue affecting a region's economic, social, and ecological environment. The "Outline of the National Water Network Construction Plan" proposes to build a national water network from 2021 to 2035 to solve the problems of water resources, water ecology, water environment, and water disasters [1]. Strengthening the rigid constraints of water resources carrying capacity is a crucial measure to ensure the construction of the national water network. At present, research on water resources carrying capacity is mostly focused on provincial, municipal, or administrative regions, with relatively few studies involving the water resources carrying capacity of complete river basins. For example, Xia Weijing et al. [2] used nine basins in Shaanxi as evaluation units and constructed a comprehensive evaluation model for water resources carrying capacity based on macro and comprehensive indicators; Ait Aoudia [3] assessed the population that can be supported in terms of water resources and household consumption patterns in Algeria, a country with scarce water resources.

At the level of research methodology, recent studies have shown a trend towards systematization, with a diverse array of tools being employed. For example, Zou Zonghua [4] evaluated the water resources carrying capacity of nine provinces and regions along the Yellow River through the construction of an indicator system. Yu Po [5] used the coupled water footprint theory and principal component analysis method to explore the water resources carrying capacity in the Xinjiang region based on water footprint. Liu Yan [6] used the system dynamics model to analyze the carrying capacity of water resources in Suibin County and solved the contradiction between supply and demand of water resources in Sanjiang Plain. Zuo et al. [7] evaluated the water resources carrying capacity of 15 regions in the Xinjiang Uygur Autonomous Region from 2004 to 2017 using the fuzzy comprehensive evaluation method. Although each of the above methods has its characteristics, for the Xiangjiang River Basin, the conventional trend method does not fully consider the interaction between factors affecting water resources carrying capacity, making it difficult to reflect the current situation. Principal component analysis can objectively determine the weights of indicators, but subjective factors also influence it in the grading criteria of evaluation parameters. The fuzzy comprehensive evaluation method can comprehensively process the discrete processes generated, but the information utilization rate is relatively low. The combination of the TOPSIS model and the entropy weight method can fully utilize indicator information to objectively and accurately evaluate the superiority and inferiority of different evaluation objects. Currently, it has been fully applied in research on water resources carrying capacity. Zhang Fan [8] evaluated the water resources carrying capacity of the Gansu section of the Yellow River Basin based on entropy weight TOPSIS, promoting economic development to match the water resources carrying capacity; Jiang Dejuan [9] quantitatively evaluated the water resources carrying capacity of 16 cities in Shandong Province from 2011 to 2020 based on comprehensive weighting and the TOPSIS model and proposed improvement suggestions. At the same time, choosing the BP neural network model as the prediction method is due to its strong self-learning ability, which can automatically extract features and patterns from data. In order to meet the needs of building a smart water network, the BP neural network model can achieve refined management and prediction of water resources, providing a scientific basis for the rational allocation and utilization of water resources. Xue Qing [10] conducted research on predicting and regulating the water resources carrying capacity in Jiangsu Province based on a BP neural network, providing a constructive basis for optimizing water resources regulation.

The Xiangjiang River is the largest in Hunan Province and is known as the "mother river" of Hunan. The basin is rich in mineral resources, and there are developed industries in the middle and lower reaches. Due to seasonal changes, the rainfall in the Xiangjiang River Basin fluctuates significantly throughout the year, resulting in varying degrees of droughts and floods throughout the basin. It is of great significance to evaluate and predict its water resources carrying capacity, which can better serve the basin's economy and provide data to support the construction of Hunan's smart water network.

## 2. Status of the Study Area

### *2.1. Research Modules*

#### 2.1.1. Selection of Research Units

This article evaluates the water resources carrying capacity of the Xiangjiang River Basin (Figure 1) at the county level, which can provide more detailed water resource data and facilitate the identification of specific water resource distribution and demand differences (Table 1). This detailed evaluation helps local governments formulate more effective water resource management policies, achieve rational allocation, and optimize resource utilization and thus improve the pertinence and operability of management.

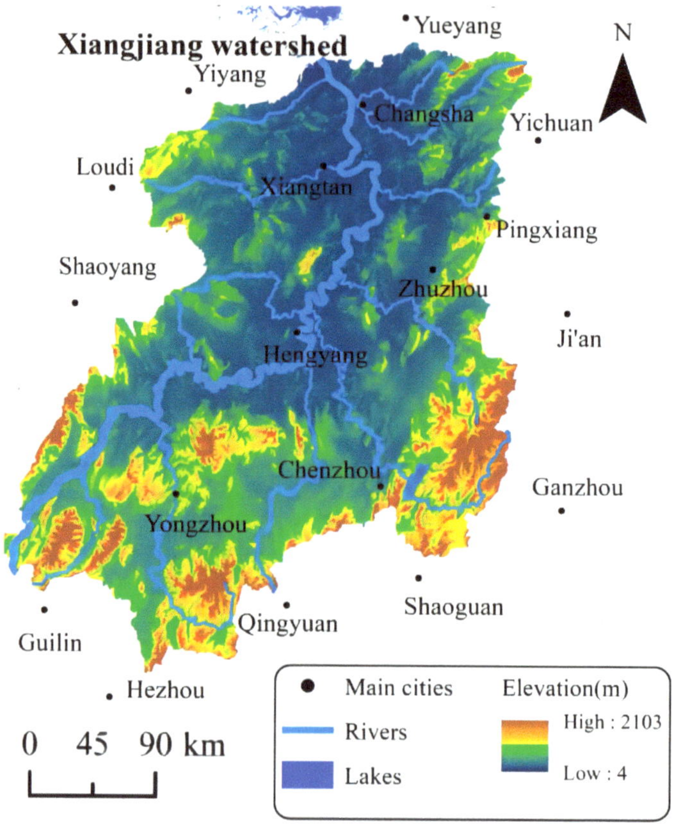


**Figure 1.** The Xiangjiang basin.

**Table 1.** The scales and their advantages.

| Scale | Advantage |
|---|---|
| County scale | Data precision: Data at the county scale are more specific, reflecting local water conditions, water demand, and water management issues.<br>Local features: able to identify local water resource issues in detail, such as overexploitation or pollution of water sources in a particular county.<br>Management measures: The county scale is conducive to formulating targeted management measures and policies that adapt to the local situation. |
| Basin scale | Comprehensive: The basin scale can comprehensively consider the water resources situation in the whole basin, including the flow and utilization of water resources upstream and downstream.<br>Regional collaboration: Considering upstream and downstream water use, the basin scale facilitates the development of transregional water management measures to address transboundary water issues. |
| Regional scale | Macro perspective: Regional scales often involve multiple basins, providing a more macro perspective to analyze water resources' spatial distribution and utilization.<br>Trend analysis: suitable for identifying long-term trends in water resource changes and potential water scarcity issues within a region. |
| National scale | Overall trend: National-scale analysis can reveal the supply–demand imbalance and overall trend of water resources nationwide, such as the uneven distribution of water resources in different regions.<br>Policymaking: It helps to formulate national-level water resource management policies and strategies, but may overlook specific local issues. |

2.1.2. Overview of the Research Unit

The vast majority of the Xiang River is within Hunan Province, accounting for more than 95% of the basin area year-round, with upstream (southern) tributaries extending into

Guangxi Province. The source to the mainstream is in Hunan, with a basin covering nine prefecture-level cities (Changsha, Xiangtan, Zhuzhou, Hengyang, Chenzhou, Yongzhou, Loudi, Shaoyang, and Yueyang). According to the Xiangjiang River Flood Control Planning Report (HND/A1g-1-001) of the Hunan Provincial Institute of Water Resources, under the Ministry of Water Resources, in 1999, the Xiangjiang River involves 58 counties (cities and districts). However, the area of the counties covered by the basin varies considerably (Table 2).

**Table 2.** Distribution and drainage area ratio of counties (cities) in the Xiangjiang River Basin.

| Name of Place or City | County and City Name and Percentage (%) | | | | | | | | | | |
|---|---|---|---|---|---|---|---|---|---|---|---|
| | 1 | 2 | 3 | 4 | 5 | 6 | 7 | 8 | 9 | 10 | 11 |
| Changsha | 100.0 Changsha | 100.0 Urban, suburban | 100.0 Wangcheng | 91.5 Ningxiang | 100.0 Liuyang | | | | | | |
| Zhuzhou | 100.0 Zhuzhou | 100.0 Liling | 100.0 Youxian | 100.0 Chaling | 100.0 Yanling | 100.0 Urban, suburban | | | | | |
| Xiangtan | 100.0 Xiangtan | 100.0 Urban, suburban | 100.0 Xiangxiang | 100.0 Shaoshan | | | | | | | |
| Hengyang | 100.0 Hennan | 100.0 Hengyang | 100.0 Hengshan | 100.0 Hengdong | 100.0 Leiyang | 100.0 Changning | 100.0 Qidong | 100.0 Urban, suburban | | | |
| Shaoyang | 6.3 Xinshao | 47.5 Shaodong | 2.0 Shaoyang | | | | | | | | |
| Yueyang | 3.5 Xiangyin | 14.0 Miluo | | | | | | | | | |
| Loudi | 100.0 Shuangfeng | 100.0 Lianyuan | 26.8 Lengshuijiang | | | | | | | | |
| Chenzhou | 100.0 Anren | 100.0 Yongxing | 100.0 Zixing | 91.0 Guidong | 79.0 Rucheng | 100.0 Jiahe | 98.5 Guiyang | 100.0 urban area | 25.0 Linwu | 5.5 Yizhang | |
| Yongzhou | 100.0 Urban area | 100.0 Shuangpai | 57.0 Jiang yong | 95.5 Jianghua | 100.0 Daoxian | 100.0 Lanshan | 100.0 Qiyang | 100.0 Xintian | 100.0 Ningyuan | 100.0 Dongan | 100.0 Jindong Forest Farm |

In order to assess the water resources carrying capacity of the Xiangjiang River in a more in-depth and precise manner, this study is the first to refine its analysis to the level of county units. However, the land area belonging to the Xiangjiang River varies among the county units. Combining the previous studies [11–14] with the actual situation of the Xiangjiang River, more than 50% of the area where the Xiangjiang River flows through is considered the "county unit" in this study. The urban area of a prefecture-level city within the assessment scope may be less than the land area of a county or township in relation to the basin area; therefore, it is more accurate and objective to combine prefecture-level cities' urban areas into a "county unit", which was verified in the earlier pre-study; Changsha's Tianxin and Yuhua Districts and Chenzhou's Suxian and Beihu Districts were taken as a separate "county unit". In this regard, the earlier pre-study has verified that the evaluation of Changsha's Tianxin District and Yuhua District and Chenzhou's Suxian District and Beihu District as separate "county units" basically makes no difference. Accordingly, this study involves seven prefecture-level cities' urban areas and 37 counties (cities), or 44 "county units" (Table 3).

**Table 3.** County-level units in the research area.

| Prefecture-Level City | County Level Unit |
|---|---|
| Changsha City | Changsha City District, Wangcheng District, Ningxiang County, Liuyang City, Changsha County |
| Xiangtan City | Xiangtan City District, Xiangxiang City, Shaoshan City, Xiangtan County |
| Zhuzhou City | Zhuzhou City District, Zhuzhou County, Liling City, You County, Chaling County, Yiling County |
| Hengyang City | Hengyang City District, Leiyang City, Changning City, Qidong County, Hengyang County, Hengdong County, Hengshan County, Hengnan County |
| Chenzhou City | Chenzhou City District, Zixing City, Anren County, Guiyang County, Rucheng County, Jiahe County, Guidong County, Yongxing County |
| Yongzhou City | Yongzhou City District, Qiyang City, Lanshan County, Xintian County, Jiangyong County, Shuangpai County, Ningyuan County, Dong'an County, Daocheng County, Jianghua Yao Autonomous County. |
| Loudi City | Loudi City District, Shuangfeng County, Lianyuan County |

## 2.2. Regional Water Resources

### 2.2.1. Hydrological Water Resources

The Xiangjiang River is the largest in the Dongting Lake system of the Yangtze River Basin. It originates from Ziliang Yao Autonomous Township in Lanshan County, Hunan

Province, and its mainstream runs from south to north through cities such as Yongzhou, Hengyang, Zhuzhou, Xiangtan, and Changsha before flowing into Dongting Lake at Lulintan in Xiangyin County (Figure 2). The main stream is 670 km long, with a drainage area of 85,225 km$^2$ and an annual runoff of 106 billion m$^3$, making it the largest river in Hunan. The Xiangjiang River has a developed water system and numerous reservoirs, with 124 primary tributaries and 16 main tributaries and a drainage area greater than 1000 km$^2$. Among them, Xiaoshui has a drainage area of 12,000 km$^2$, followed by Leishui and Yishui. Eleven controlled water conservancy hub projects have been built on the Xiangjiang River, including large-scale reservoirs with a storage capacity of over 100 million cubic meters, such as Centian River, Shuangpai, Ouyang Sea, Taoshui, Dongjiang, Qingshanlong, Jiubu River, Shuifu Temple, Zhushu Bridge, Guanzhuang, and Huangcai. The largest reservoir, Dongjiang Lake, is located in Zixing City, with a total storage capacity of 9.15 billion cubic meters and a catchment area of 26,000 square kilometers.

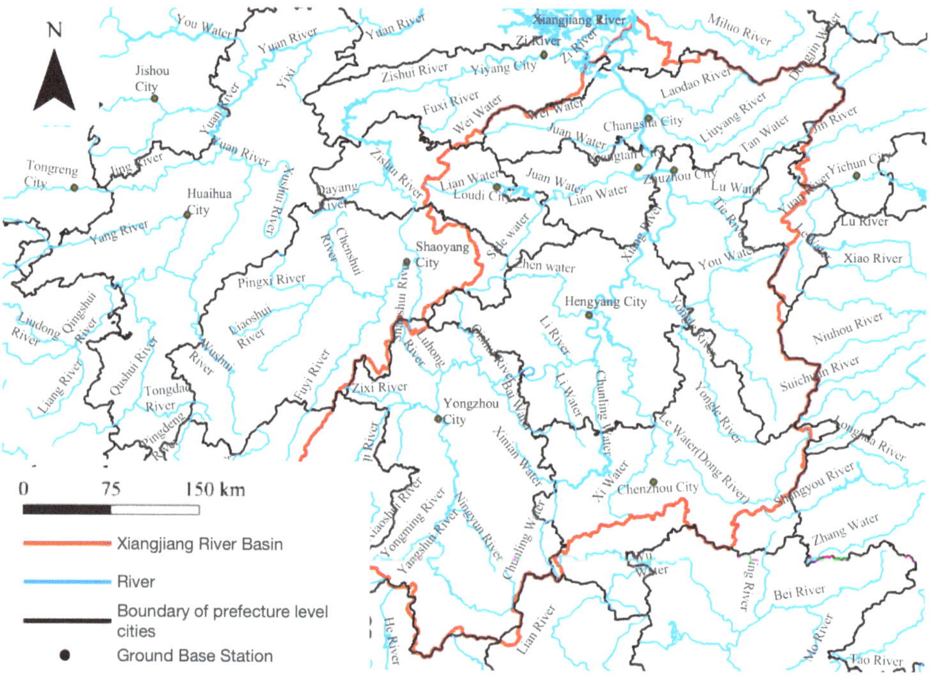


**Figure 2.** Branches of the Xiangjiang River Basin.

### 2.2.2. Water Quality and Environment

The Xiangjiang River has abundant mineral resources, with agriculture as the main upstream activity and industry as the main midstream and downstream activity; this poses hidden dangers to the water quality of the Xiangjiang River. The rainstorms in spring and summer wash the ground, pollutants are carried into the river through surface runoff, and total nitrogen and phosphorus gradually increase from upstream to downstream. Since 2012, efforts to improve the ecological environment have continuously increased, with over 90% of the Xiangjiang River's sections under national and provincial control having water quality above Class III (Figure 3).

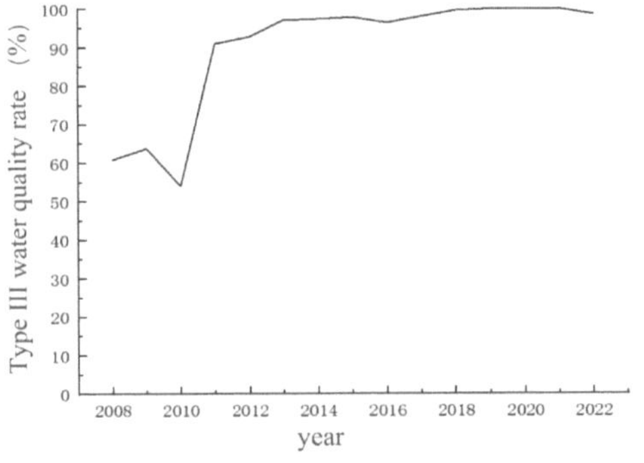


**Figure 3.** Proportion of water quality above Class III in the Xiangjiang River.

2.2.3. Economic and Social Conditions

The population of Xiangjiang increased from 32.6559 million in 2008 to 36.3745 million in 2022, accounting for approximately 55% of the province's population. The urbanization process is accelerating, and the urban population share has increased from 41.94% in 2008 to 57.05% in 2022. GDP has shown remarkable growth, increasing from CNY 714.087 billion in 2008 to CNY 3061.655 billion in 2022, and per capita GDP has increased from CNY 18,637.26 in 2008 to CNY 70,886.43 in 2022, demonstrating a sustained upward trend.

## 3. Evaluation and Forecasting Methods

### *3.1. Evaluation Models and Methods*

3.1.1. Evaluation Modelling Framework

The model architecture is defined as having three layers: the target layer, the criterion layer, and the indicator layer (Figure 4). The target layer is the water resources carrying capacity of the Xiangjiang River, which includes five interrelated and constrained criterion layers: resource support, economic support, social support, environmental support, and management support (Figure 5). Resource support is the most direct water resources endowment system, economic and social conditions are the most direct results of the basin water resources carrying capacity, the environmental status reflects the sustainable development level of "green water and green mountains are golden silver mountains", and the management capacity reflects the potential and soft power of the water resources carrying capacity, which provides support for the intelligent water network. The criterion layer is evaluated by approximating the ideal solution ranking (TOPSIS). The method has a clear and explicit model structure, is widely used, has mature calculation methods, and is highly adaptable.

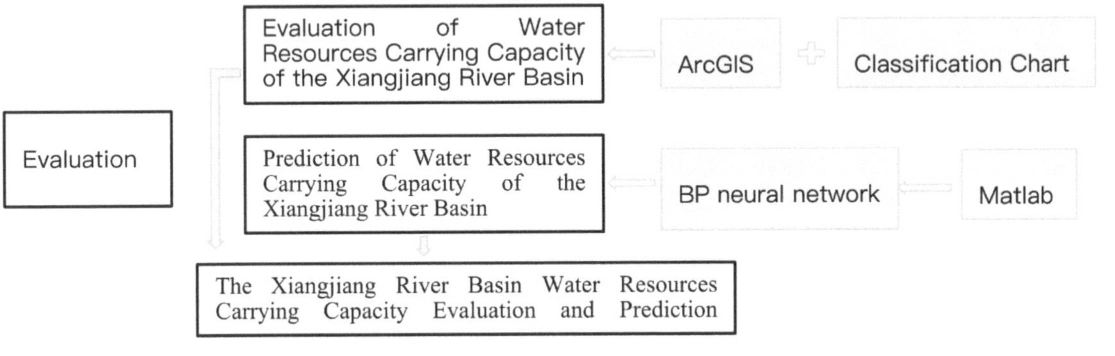


**Figure 4.** Figure technological route.

**Figure 5.** Guideline layer framework.

3.1.2. TOPSIS Method

TOPSIS (technique for order preference by similarity to an ideal solution), also known as the "ideal solution ranking method", is a method that ranks multiple evaluation objects by comparing their similarity to the ideal solution and determines their relative superiority or inferiority. This method has the advantages of a flexible and convenient calculation process, accurate and reasonable evaluation results, etc.

To improve the application of the TOPSIS model in water resources carrying capacity evaluation, many scholars have conducted in-depth research. For example, Du Xuefang et al. [15] combined subjective and objective weighting methods with the TOPSIS model to evaluate the water resources carrying capacity of Zhengzhou City from 2010 to 2019; Ma Jimin et al. [16] combined the entropy weight method and the CRITIC (criteria importance through intercriteria correlation) method and applied them to the improved grey relational approximation solution ranking (GRA-TOPSIS) model. Deng Quancheng et al. [17] used a TOPSIS model combining the entropy value method and the analytic hierarchy process to comprehensively evaluate the water resources carrying capacity of Xinxiang City from 2006 to 2018 and conducted obstacle analysis on the evaluation results. However, Xu Zhenghua [18] cited an improved entropy weight TOPSIS model evaluation method, using information entropy to calculate the weights of each evaluation indicator and using the optimal solution of Euclidean distance to evaluate the water resources carrying capacity.

In these studies, analysis was conducted based on the proximity of the evaluation values $C_i$. The closer $C_i$ is to 1, the closer the evaluation object is to the optimal solution, indicating that the object is relatively better. Finally, the water resources carrying capacity levels are classified based on the proximity of each object in descending order. Through the research on the improvement and application of the TOPSIS method in the evaluation of water resources carrying capacity, it can be seen that in the actual evaluation process, combining other weight determination methods can improve the accuracy and reliability of the evaluation results. The introduction of these methods makes the evaluation process more comprehensive and scientific, providing decision-makers with more in-depth and actionable information.

Based on TOPSIS and the entropy weight method to calculate the evaluation value of the water resources carrying capacity of the Xiangjiang River, the steps of the TOPSIS calculation are as follows:

1. A set of data with a sample size of m and a number of indicators of $n$ is processed with the same trend, and then the data are dimensionless according to the following formula, i.e., a dimensionless decision matrix:

$$Z = (z_{ij})_{m\times n}, \ z_{ij} = \frac{x_{ij}}{\sqrt{\sum\limits_{i=1}^{m} x_{ij}}} \tag{1}$$

2. Determine the optimal solution $z_j^+$ and the worst solution $z_j^-$ for each indicator:

$$z_j^+ = \max\{z_{1j}, z_{2j}, \cdots, z_{mj}\} \tag{2}$$

$$z_j^- = \min\{z_{1j}, z_{2j}, \cdots, z_{mj}\} \tag{3}$$

3. Determine the weighted Euclidean distance between each evaluation object and the optimal and worst solutions $D_i^+$ and $D_i^-$:

$$D_i^+ = \sqrt{\sum\limits_{j=1}^{n} [w_j(z_{ij} - z_j^+)]^2} \tag{4}$$

$$D_i^- = \sqrt{\sum\limits_{j=1}^{n} [w_j(z_{ij} - z_j^-)]^2} \tag{5}$$

where $w_j$ is the combination weight of indicator $j$.

4. Determine the proximity $C_i$:

$$C_i = \frac{D_i^-}{D_i^+ + D_i^-} \tag{6}$$

Finally, according to the degree of $C_i$'s closeness to 1, the closer $C_i$ is to 1, the closer the object is to the best solution, i.e., the object is relatively better, and ultimately, the objects are ranked according to the degree of proximity of each object from the largest to the smallest and divided into water resources carrying capacity classes.

### 3.2. BP Neural Network Prediction Method

The accurate prediction of the evolution trend of the water resources carrying capacity in the Xiangjiang River Basin is of great significance for promoting the construction of a smart water network in the region and achieving sustainable development strategies. Therefore, based on the established evaluation index system for water resources carrying capacity, an effective prediction model is adopted to predict the water resources carrying capacity of the Xiangjiang River Basin. In terms of selecting prediction models, both grey prediction models and neural network models can be used to predict water resources carrying capacity. Among them, the BP neural network is a widely used artificial neural network model which has the advantages of simple structure and strong operability. By adjusting the network structure and parameters, a BP neural network can solve most of the problems faced by neural networks, providing an innovative approach and way to enhance the carrying capacity of regional water resources. Therefore, selecting a BP neural network as the model for predicting the water resources carrying capacity of the Xiangjiang River Basin and applying a BP neural network to predict the water resources carrying capacity of the Xiangjiang River Basin can provide a scientific basis for the construction of a smart water network in the region and provide practical and effective analysis methods for the planning and decision-making of relevant government departments.

In this study, with a 15-year dataset of 26 indicators of 44 county units in the Xiangjiang River, 16,160 pieces of data were collected, which is a large amount of data, and the BP neural model was selected for prediction. The specific steps were performed as follows:

First, the network structure was determined. A three-layer BP neural network containing one hidden layer (i.e., input layer, hidden layer, and output layer) was used to predict the risk of the water carrying capacity of the Xiangjiang River.

Second, the number of neurons in each layer was determined. The water resources carrying capacity system of the 44 county units of the Xiangjiang River and the 5 subsystems of resources, economy, society, water environment, and water management carrying capacity assessment value served as the input part, and the number of nodes in the input layer was 5; the prediction part of the model was trained many times. We determined the training set for the assessment value of the data samples of 2008–2015, the prediction set for the assessment value of the data samples of 2016–2022, and the predicted water resources carrying capacity assessment value for 2023–2027. In this study, the output is the carrying capacity value of each system in the 44 county units of the Xiangjiang River, and the number of nodes in the output layer was determined to be 1. Through substitution and repeated tests, the number of nodes in the implied layer stratification of 6 was finally selected.

Third, the allowable error and learning rate were selected. Usually, the value of the learning rate ($\eta$) ranges from 0.01 to 0.80. In this study, we used a variable adaptive learning rate so that the network training in different phases of the network automatically set the sizes of the different learning rates; the initial learning rate was 0.01. The maximum number of training iterations was 1000 times.

Fourth was data processing and Matlab implementation design. There are some differences in the magnitudes of the sample data, and the sample data are mapped to $[-1, 1]$ for normalization, allowing us to input data in Matlab to repeatedly train to obtain a suitable model before substituting the values to obtain the predicted water resources carrying capacity of 44 county units of the Xiangjiang River in the next 5 years.

### 3.3. Screening the Indicator System

The selection of the indicator system adopts a combination of "universal" and "special" methods. This study is based on previous research and relevant literature to determine an evaluation index system for the actual situation of the Xiangjiang River. Firstly, 15 literature indicators closely related to the water resources carrying capacity of the basin were statistically analyzed [19–33]. The classification and frequency of the indicators were compared and analyzed, and 19 indicators with a frequency greater than 4, such as precipitation, per capita water resources, water production modulus, and population density, were directly selected into the indicator system (universal indicators). Based on field research and in-depth interviews with experts, a total of 26 evaluation indicators for the carrying capacity of water resources in the Xiangjiang River were selected, including seven specific indicators such as agricultural potential indicators, first-tier water prices, pesticide and fertilizer usage, and water conservancy practitioners (Table 4). It should be noted that the natural population growth rate is a negative indicator in other literature, following the idea of "family planning" to suppress population growth. Based on the current fertility situation and policies to encourage fertility, this study has determined it as a positive indicator.

**Table 4.** The Xiangjiang River Water Resources Carrying Capacity Index System.

| Target Level | System Level | Indicator Layer | Unit (of Measure) | Characteristic |
|---|---|---|---|---|
| Water Resources Carrying Capacity of the Xiangjiang River | Resource supportability | Precipitation | mm | Positive |
| | | Average per capita water resources | $m^3$/person | Positive |
| | | Coefficient of Storage | $10^4$ $m^3/km^2$ | Positive |
| | | Wodule of water supply | $10^4$ $m^3/km^2$ | Positive |
| | | Eco-environment water consumption | $10^4$ $m^3$ | Negative |
| | Economic support | GDP per capita | $10^4$ CNY | Positive |
| | | Effective irrigation area | % | Positive |
| | | Agricultural potential index | ton | Positive |
| | | Water consumption per 10,000 CNY GDP | $m^3/10^4$ CNY | Negative |
| | | The output value of agriculture, forestry, animal husbandry, and fishery | $10^4$ CNY | Positive |
| | | Sewage discharge per 10,000 CNY of GDP | $m^3/10^4$ CNY | Negative |
| | Social support | Density of population | Persons/$km^{-2}$ | Negative |
| | | Urbanization rate | % | Positive |
| | | Natural population growth rate | % | Positive |
| | | First-tier water price | CNY | Negative |
| | | Per capita water demand | $m^3$ | Negative |
| | Environmental support | Per capita domestic water consumption | $m^3$/person | Negative |
| | | consumption of pesticide and fertilize | kg | Negative |
| | | Proportion of Class III or above water | % | Positive |
| | | Wastewater treatment rate | % | Positive |
| | | Forest cover | % | greater than zero |
| | Management support | Water Utilization | % | Negative |
| | | Dam length | m | Positive |
| | | impoundage | $m^3$ | Positive |
| | | water conservation practitioner | person | Positive |
| | | Rural garbage management rate | % | Positive |

### *3.4. Data Sources and Pre-Processing*

#### 3.4.1. Data Sources

The evaluation period for the carrying capacity of water resources in the Xiangjiang River is from 2008 to 2022. The data sources for this period include the China Statistical Yearbook, Hunan Provincial Statistical Yearbook, Hunan Provincial Water Resources Bulletin, Yongzhou City Water Resources Bulletin, Chenzhou City Water Resources Bulletin, Hengyang City Water Resources Bulletin, Zhuzhou City Water Resources Bulletin, Xiangtan City Water Resources Bulletin, Changsha City Water Resources Bulletin, and Loudi City Water Resources Bulletin, as well as various county-level units' National Economic Statistics Bulletins and the EPS data platform. Due to the low level of openness of county-level unit data, some data sources were obtained through on-site consultation and telephone interviews. Very few missing data were imputed.

#### 3.4.2. Data Processing

1. Standardized treatment

The polar deviation method was chosen to standardize the data. After standardization, the values may have 0 values to ensure that the assigned weights are meaningful. The standardized values were summed with 0.001, thus obtaining $Z_{ij}$ as a positive indicator:

Positive indicators:

$$z_{ij} = \frac{x_{ij}(\max) - x_{ij}}{x_{ij}(\min) - x_{ij}(\min)} + 0.001 \quad (7)$$

Negative indicators:

$$z_{ij} = \frac{x_{ij} - x_{ij}(\min)}{x_{ij}(\max) - x_{ij}(\min)} + 0.001 \quad (8)$$

2. Weight calculation method

In this study, the entropy weighting method is used to determine the weights, which is classical and objective in dealing with multiple indicators.

$$S_{ij} = \frac{Z_{ij}}{\sum\limits_{i=1}^{n} Z_{ij}} \tag{9}$$

Calculate the entropy value of the $j$th indicator $H_j$:

$$H_j = -\frac{1}{\ln n}\sum_{i=1}^{n} S_{ij} \ln S_{ij} \tag{10}$$

Calculate the value of the coefficient of variation $E_j$ from the entropy value $H_j$:

$$E_j = 1 - H_j \tag{11}$$

Calculation of indicator weights $W_j$:

$$W_j = \frac{E_j}{\sum\limits_{j=1}^{m} E_j} \tag{12}$$

## 4. Measurement and Analysis of Water Carrying Capacity

### 4.1. Measurement of Water Carrying Capacity

#### 4.1.1. Weighting Values

This article calculates the weights of indicators for the Xiangjiang River from 2008 to 2022 using the entropy weight method. Due to the complexity of analyzing data spanning 15 years, in order to simplify the process and facilitate analysis, a 5-year interval is set to calculate the weight ratio of each criterion layer indicator (Table 5). The results indicate that the weight of management support has the highest proportion, while the rest of the system is relatively balanced. Each indicator showed fluctuating changes during the research period.

**Table 5.** Weight ratio of the criterion layer (%).

| Normative Layer | 2008 | 2013 | 2018 | 2022 | On Average |
|---|---|---|---|---|---|
| Resource support | 13.39 | 15.13 | 19.83 | 20.23 | 17.14 |
| Economic support | 15.00 | 17.16 | 16.72 | 18.19 | 16.77 |
| Social support | 8.27 | 9.87 | 10.18 | 7.92 | 9.06 |
| Environmental support | 10.73 | 9.53 | 11.98 | 9.36 | 10.40 |
| Management support | 52.61 | 48.39 | 41.30 | 44.31 | 46.65 |

#### 4.1.2. Evaluation of Water Resources Carrying Capacity

Due to the long time span of this study, centralized analysis may overlook the details of year-by-year changes, resulting in scattered information. Therefore, three "five-year stages" are set up, from the first five-year stage to the third five-year stage, to analyze the comprehensive evaluation results regarding the water resources carrying capacity of the Xiangjiang River with a moderate time span. The comprehensive evaluation values of the water resources carrying capacity for each "five-year stage" during the research period are shown in (Table 6).

**Table 6.** Evaluation values and rankings of water resources carrying capacity of each county unit in the Xiangjiang River.

| County Unit | 2008–2012 | 2013–2017 | 2018–2022 | On Average | Rankings |
|---|---|---|---|---|---|
| Changsha City District | 0.407 | 0.382 | 0.378 | 0.389 | 5 |
| Wangcheng District | 0.371 | 0.362 | 0.394 | 0.376 | 7 |
| Changsha County | 0.393 | 0.395 | 0.400 | 0.396 | 4 |
| Liuyang City | 0.409 | 0.441 | 0.457 | 0.436 | 2 |
| Ningxiang County | 0.394 | 0.452 | 0.452 | 0.433 | 3 |
| Zhuzhou City District | 0.364 | 0.360 | 0.358 | 0.360 | 8 |
| Zhuzhou County | 0.300 | 0.285 | 0.264 | 0.283 | 41 |
| Liling City | 0.303 | 0.374 | 0.364 | 0.347 | 12 |
| Youxian County | 0.306 | 0.329 | 0.309 | 0.314 | 25 |
| Chaling County | 0.287 | 0.317 | 0.306 | 0.303 | 30 |
| Yanling County | 0.314 | 0.337 | 0.338 | 0.330 | 17 |
| Xiangtan City District | 0.363 | 0.398 | 0.378 | 0.380 | 6 |
| Xiangxiang City | 0.303 | 0.356 | 0.359 | 0.339 | 14 |
| Shaoshan City | 0.267 | 0.288 | 0.301 | 0.285 | 40 |
| Xiangtan County | 0.343 | 0.311 | 0.311 | 0.321 | 20 |
| Hengyang City District | 0.368 | 0.358 | 0.337 | 0.354 | 10 |
| Leiyang City | 0.320 | 0.328 | 0.283 | 0.310 | 27 |
| Changning City | 0.289 | 0.342 | 0.333 | 0.321 | 21 |
| Qidong County | 0.298 | 0.386 | 0.359 | 0.348 | 11 |
| Hengyang County | 0.322 | 0.356 | 0.320 | 0.333 | 15 |
| Hengdong County | 0.295 | 0.300 | 0.277 | 0.291 | 36 |
| Hengshan County | 0.280 | 0.333 | 0.310 | 0.308 | 28 |
| Hengnan County | 0.321 | 0.330 | 0.302 | 0.318 | 23 |
| Loudi City District | 0.307 | 0.337 | 0.339 | 0.328 | 18 |
| Shuangfeng County | 0.283 | 0.295 | 0.289 | 0.289 | 39 |
| Lianyuan County | 0.274 | 0.299 | 0.295 | 0.289 | 38 |
| Chenzhou City | 0.327 | 0.364 | 0.374 | 0.355 | 9 |
| Zixing City | 0.516 | 0.520 | 0.510 | 0.515 | 1 |
| Anreng County | 0.271 | 0.281 | 0.290 | 0.281 | 43 |
| Guiyang County | 0.279 | 0.306 | 0.309 | 0.298 | 33 |
| Rucheng County | 0.281 | 0.325 | 0.355 | 0.320 | 22 |
| Jiahe County | 0.276 | 0.299 | 0.295 | 0.290 | 37 |
| Guidong County | 0.293 | 0.312 | 0.343 | 0.316 | 24 |
| Yongxing County | 0.282 | 0.305 | 0.300 | 0.296 | 35 |
| Yongzhou County | 0.348 | 0.345 | 0.344 | 0.346 | 13 |
| Qiyang City | 0.294 | 0.302 | 0.313 | 0.303 | 31 |
| Lanshan County | 0.278 | 0.309 | 0.337 | 0.308 | 29 |
| Xintian County | 0.247 | 0.279 | 0.288 | 0.271 | 44 |
| Jiangyong County | 0.284 | 0.310 | 0.348 | 0.314 | 26 |
| Shuangpai County | 0.304 | 0.321 | 0.342 | 0.322 | 19 |
| Ningyuan County | 0.248 | 0.289 | 0.306 | 0.281 | 42 |
| Dongan County | 0.266 | 0.310 | 0.314 | 0.297 | 34 |
| Daoxian County | 0.274 | 0.304 | 0.318 | 0.299 | 32 |
| Jianghua County | 0.311 | 0.335 | 0.350 | 0.332 | 16 |

### *4.2. Water Carrying Capacity Analysis*

The water resources carrying capacity of the Xiangjiang River Basin has shown a relatively stable growth trend from 2008 to 2022, with an overall carrying capacity level of level II. The carrying capacity of water resources is at a poor level, especially in the county units of Changsha City. The water resources carrying capacity has shown a continuous upward trend. In 2022, four county-level units reached level III, with a medium level of water resources carrying capacity. However, after the water resources carrying capacities of Changsha City, Xiangtan City, and Qidong County briefly remained at a medium level in 2008, 2009, and 2016, the water resources carrying capacities returned to a relatively poor level. The water resources carrying capacities of county-level units in Chenzhou and Yongzhou City have shown an upward trend, while Xiangtan City, Hengyang City, and Loudi City have lesser water resources carrying capacities. The upward trend of the carrying capacity is not obvious, and it fluctuates at the level of poor water resources carrying capacity.

#### 4.2.1. Time Trends

1. Changsha

The water resources carrying capacity of five county-level units in Changsha City is higher than the average level of the Xiangjiang River. Among them, Liuyang City and Ningxiang County have had a medium level of water resources carrying capacity for many years, and there is a fluctuating upward trend; Wangcheng District and Changsha County are at a moderate level and maintain relative stability. The carrying capacity of urban water resources has been fluctuating and declining since reaching its highest level in the city in 2008 (Figure 6), possibly due to the continuous influx of a large population and limited water supply capacity.

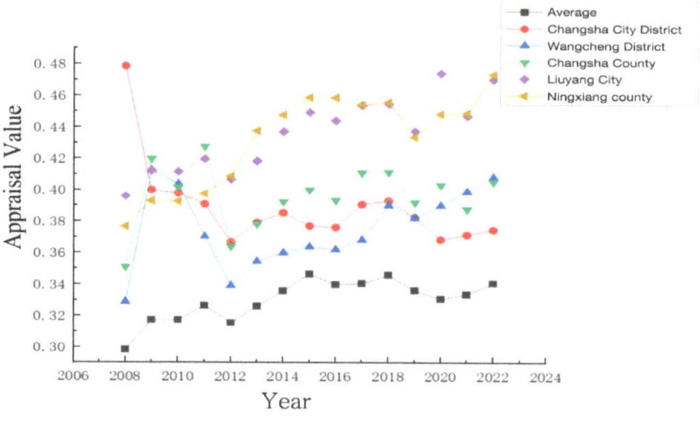


**Figure 6.** Trend Chart of Water Resources Carrying Capacity in Various County Units of Changsha City.

2. Zhuzhou

Among the six county units in Zhuzhou City, the water resources carrying capacity of Zhuzhou City District and Liling City is higher than the average level of the basin. Yiling and Youxian Counties fluctuate above and below the average level, while Zhuzhou and Chaling Counties have a lower-than-average water resources carrying capacity and a poorer level of water resources carrying capacity, respectively (Figure 7). As far as the current situation is concerned, the fluctuation in water resources carrying capacity of the six county units in Zhuzhou City is very significant. The government needs to increase investment and strengthen management to gradually improve the water resources carrying capacity level.

3. Xiangtan

The water resources carrying capacity of Xiangtan City is at a moderate level. Among the four county-level units in Xiangtan City, the water resources carrying capacity values of the urban area and Xiangxiang City are higher than the average level. Xiangtan County has only been above the average level for four years, while Shaoshan City has been consistently below the average level (Figure 8). The fluctuation trend of the water resources carrying capacity in Xiangxiang City and Shaoshan City is similar.

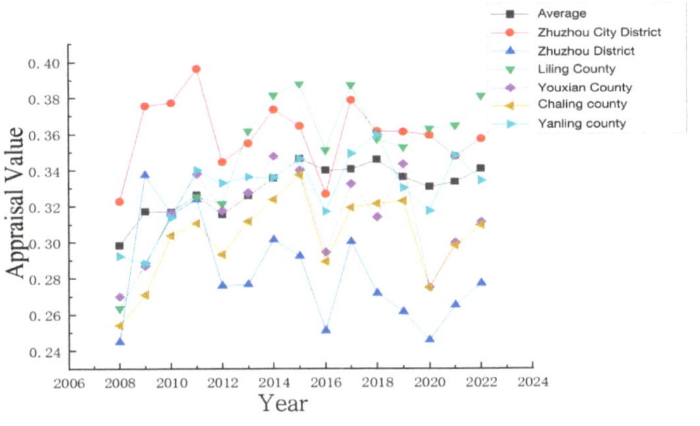


**Figure 7.** Trend Chart of Water Resources Carrying Capacity in Various County Units of Zhuzhou City.

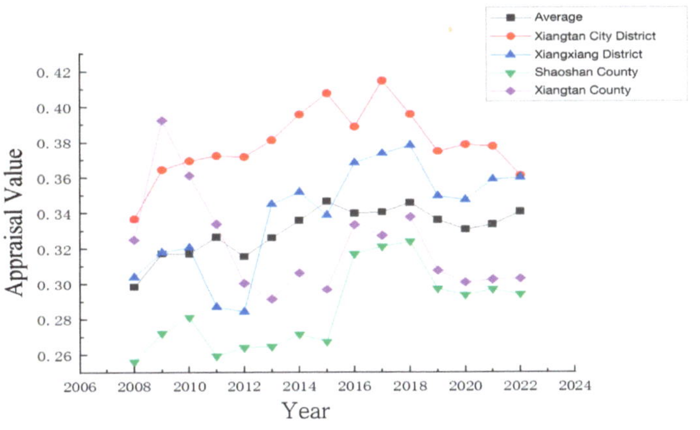


**Figure 8.** Trend Chart of Water Resources Carrying Capacity in Various County Units of Xiangtan City.

4. Hengyang

The overall performance of water resources carrying capacity in various county-level units of Hengyang City is relatively stable, but it still needs to be improved. Among the eight county-level units in Hengyang City, the water resources carrying capacity of the urban area and Qidong County is higher than the average level. Changning City has been above average for several years, while Leiyang City, Hengdong County, Hengshan County, and Hengnan County have been consistently below average (Figure 9). Qidong County has steadily improved since 2011 and reached a moderate level in 2016. The overall water resources carrying capacities of other county-level units have not fluctuated significantly.

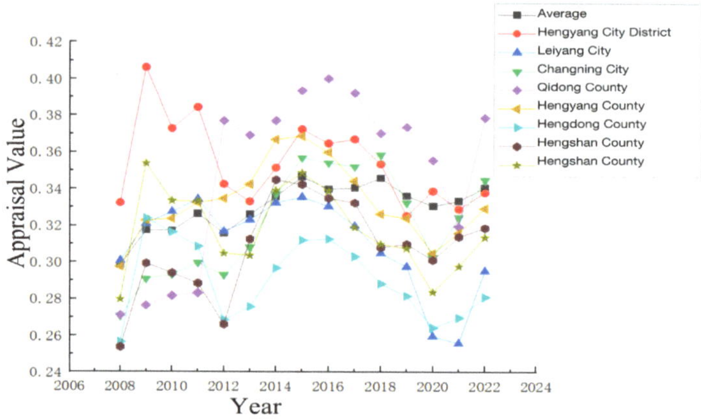


**Figure 9.** Trend Chart of Water Resources Carrying Capacity in Various County Units of Hengyang City.

5. Loudi

The overall water resources carrying capacity of Loudi City is at a poor level and fluctuates little (Figure 10). The water resources carrying capacity of the Loudi City area hovers at the average level all year round (Figure 10). Shuangfeng County and Lianyuan County, on the other hand, have water resources carrying capacities that are always below the average level, which may be mainly due to insufficient water resources supply.

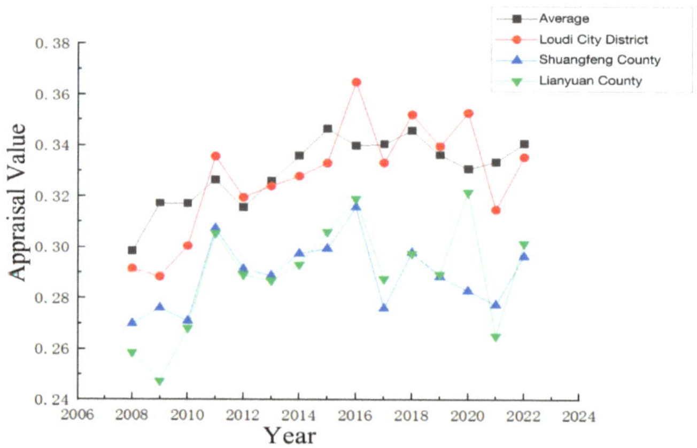


**Figure 10.** Trend Chart of Water Resources Carrying Capacity in Various County Units of Loudi City.

6. Chenzhou

Chenzhou City has a medium level of water resources carrying capacity. Zixing City has the highest water resources carrying capacity among the county units of the Xiangjiang River, mainly due to its Dongjiang Reservoir, the largest reservoir in Hunan. The Dongjiang Reservoir has a perennial water storage capacity dozens of times larger than other reservoirs, equivalent to 1/2 of the water demand of Dongting Lake. The water resources carrying capacity of the Chenzhou City area is always above average, while the rest of the county units are always below average. The carrying capacities of water resources in the seven county units of Chenzhou City reflect an overall stable state (Figure 11).

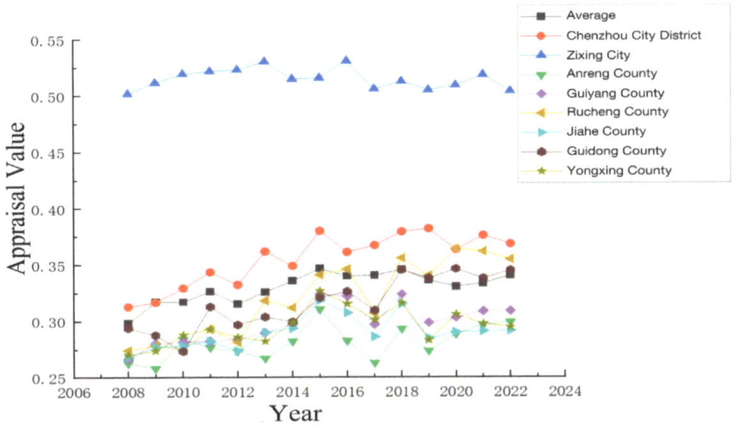


**Figure 11.** Trend Chart of Water Resources Carrying Capacity in Various County Units of Chenzhou City.

7. Yongzhou

Yongzhou City is the source of the Xiangjiang River, but its water resources carrying capacity is generally low. The water resources carrying capacity of the urban area of Yongzhou City is always higher than the average level, while Jianghua County fluctuates around the average level. The water resources carrying capacities of other county units are lower than the average level, but the water resources carrying capacities of most county units show a steady upward trend (Figure 12).

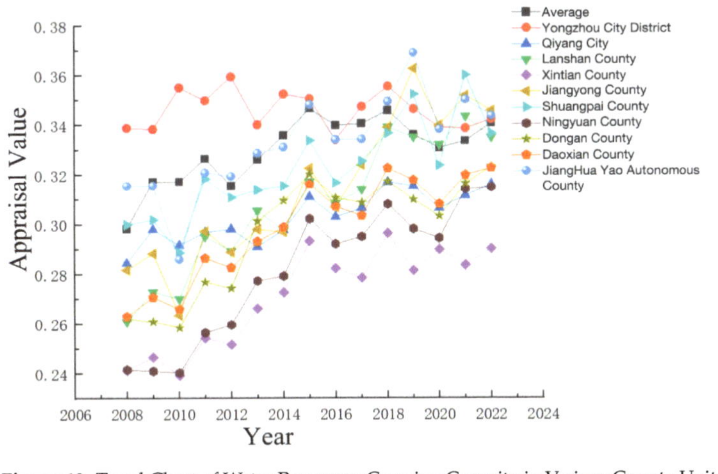


**Figure 12.** Trend Chart of Water Resources Carrying Capacity in Various County Units of Yongzhou City.

### 4.2.2. Spatial Evolution

The study imports the aforementioned evaluation values of the Xiangjiang River water resources carrying capacity into the ArcGIS PRO 10.6 software and uses the interval grading method to divide the water resources carrying capacity scores into five levels: poor (red), relatively poor (orange), moderate (yellow), relatively strong (blue), and strong (green) (Table 7), which are more convenient for internal comparison and trend analysis [34–38]. A software drawing was produced of a spatial evolution distribution map of the Xiangjiang River water resources carrying capacity based on long-term evaluation values (Figure 13).

**Table 7.** Evaluation Grading Table.

| Carrying Capacity Class | Appraise Value | Rating Color | Water Carrying Capacity |
|---|---|---|---|
| I | [0, 0.2] | red | Poor water carrying capacity |
| II | (0.2, 0.4] | orange | Relatively poor water carrying capacity |
| III | (0.4, 0.6] | yellow | Medium water carrying capacity |
| IV | (0.6, 0.8] | blue | Relatively strong water carrying capacity |
| V | (0.8, 1] | green | Strong water carrying capacity |

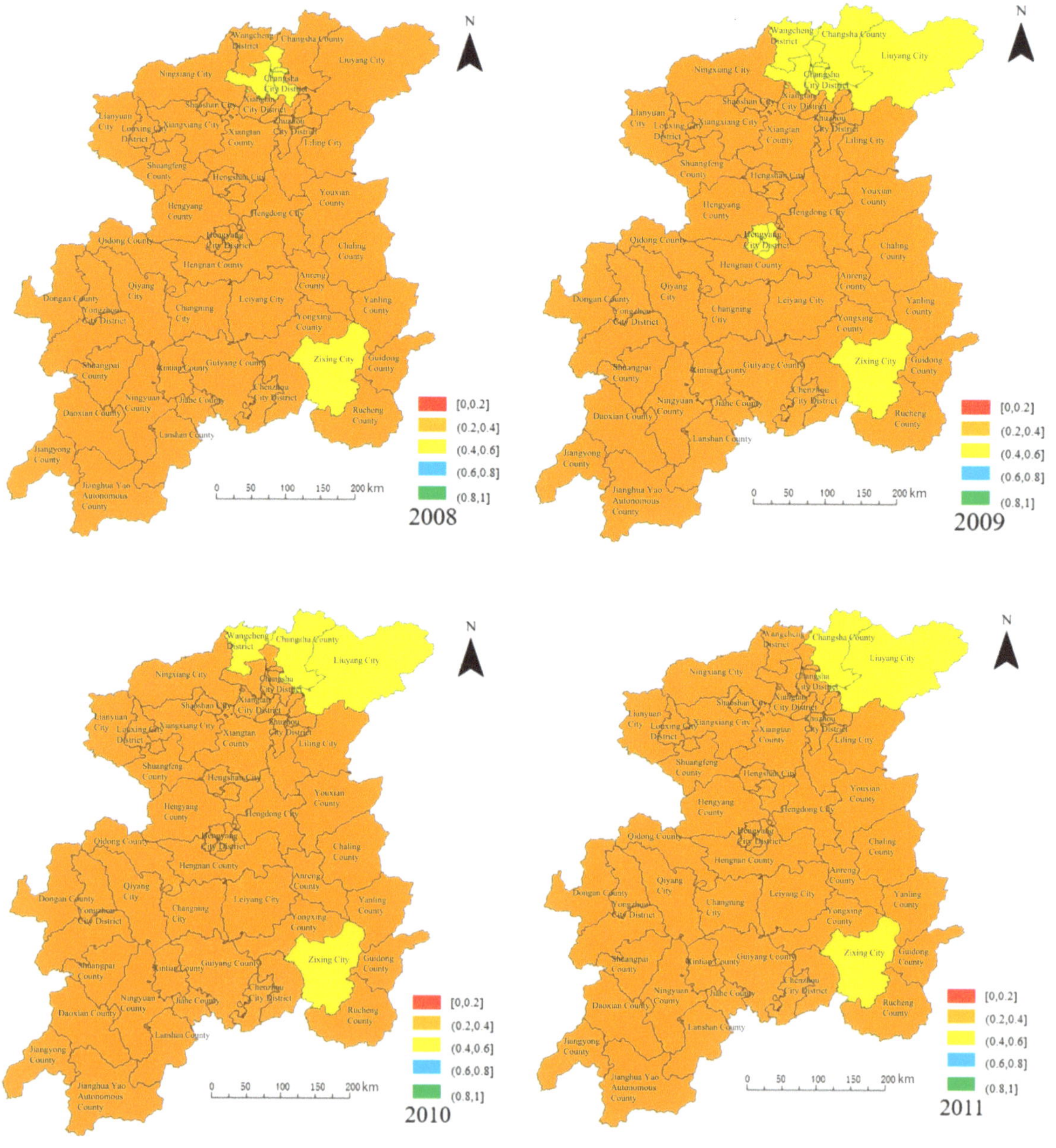


**Figure 13.** *Cont.*

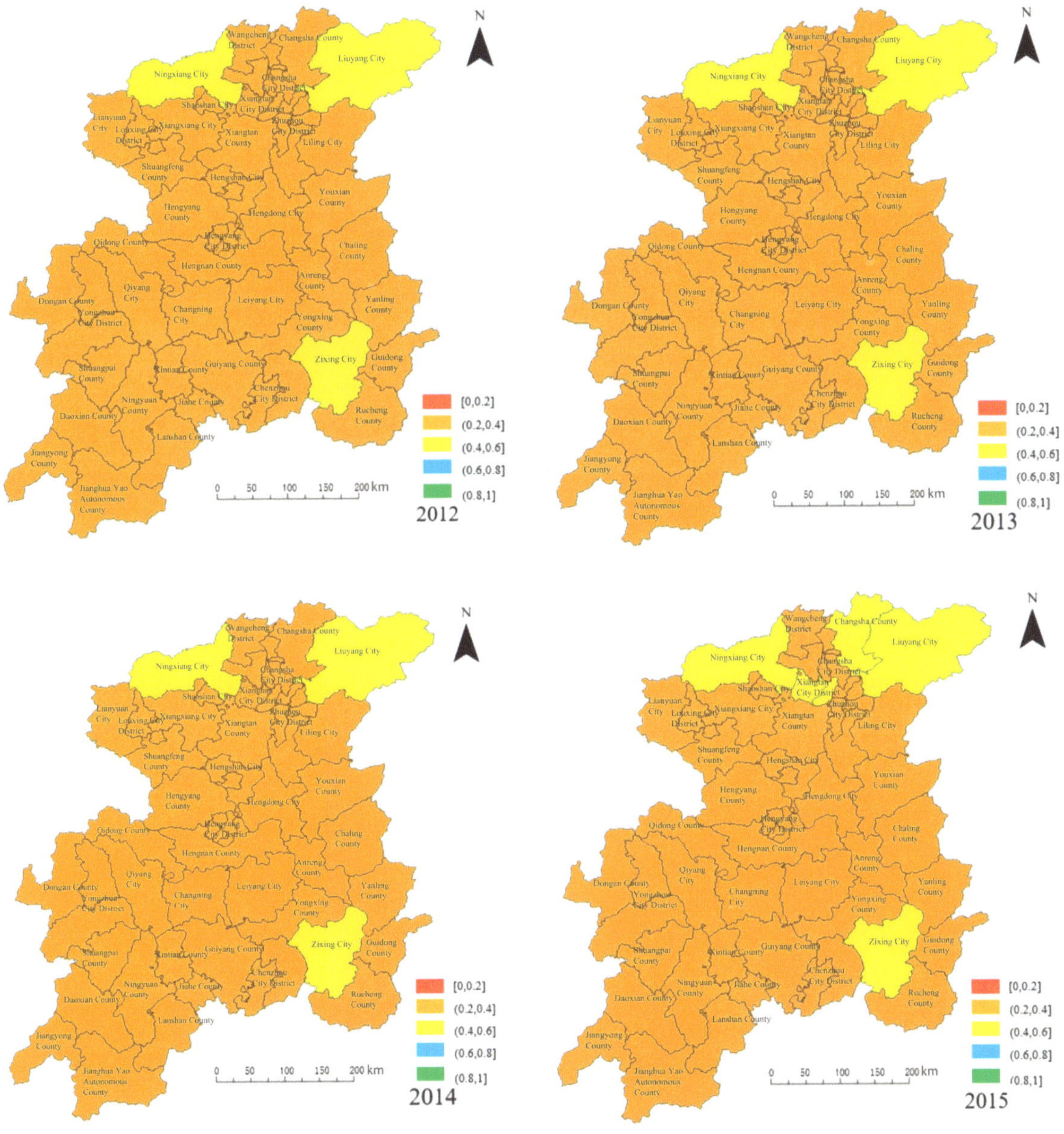


**Figure 13.** *Cont.*

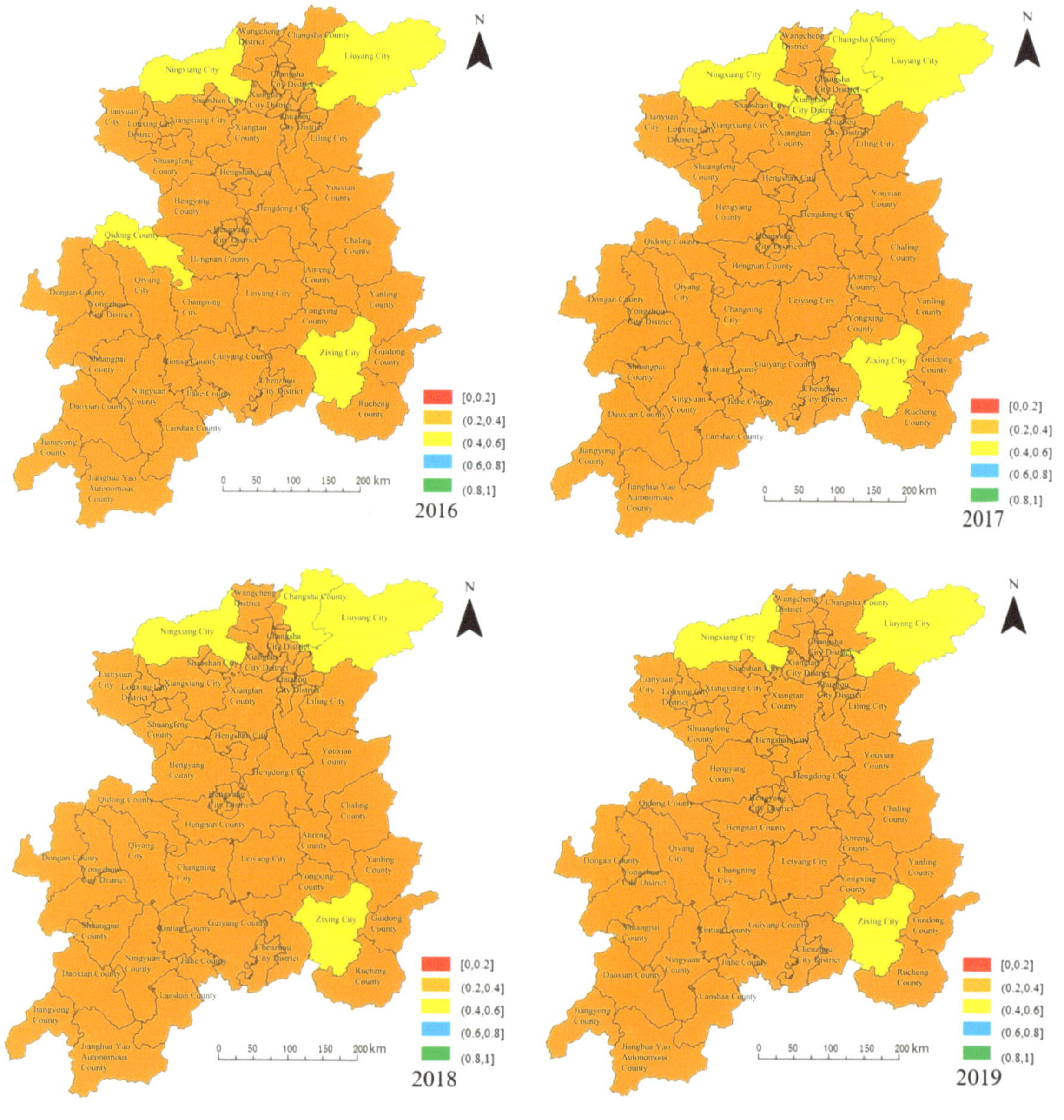


**Figure 13.** *Cont.*

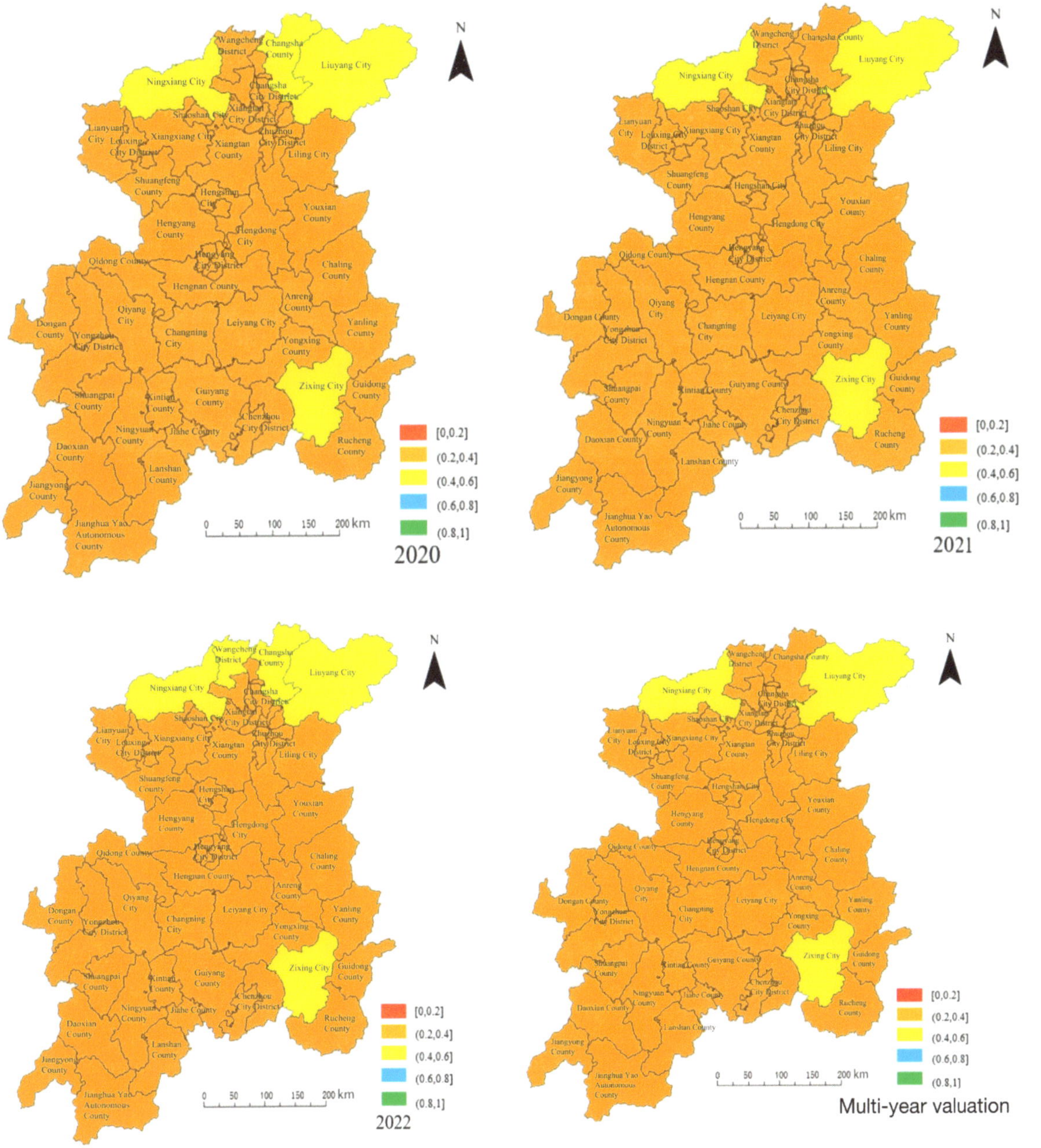


**Figure 13.** Spatial distribution map of 15-year average water resources carrying capacity of the Xiangjiang River.

From the above figure, it can be seen that there is a significant correlation between the spatial distribution of the water resources carrying capacity and the regional advantages in water resources during the evaluation period. The water resources carrying capacities of the five county-level units in Changsha City, the northernmost part of the Xiangjiang River, have fluctuated at a moderate level for many years. The water resources carrying capacities of Ningxiang County and Liuyang City have been at a moderate level all year round, while Zixing City has been above the moderate level due to the Dongjiang Reservoir. The areas with good water resources carrying capacities are mainly distributed in the lower reaches of the Xiang River, presenting an overall pattern of "high in the north and low in the south, low in the upstream and high in the downstream". The water resources carrying capacity of each urban area is basically higher than those of the counties and districts.

#### 4.2.3. Guideline Layer Analysis

The contribution of the water resources carrying capacity criterion layer to water resources carrying capacity varies year by year. In the indicator radar chart, the contribution of each subsystem to the carrying capacity of water resources in the Xiangjiang River and the differences between regions can be obtained from the five criterion layer dimensions. We use the radar chart comparison method to analyze the advantages and disadvantages of the criterion layer system. Based on the weight values of the five criterion layers, due to space limitations, only the radar chart of the average weight of the water resources carrying capacity criterion layers for seven prefecture-level city units from 2008 to 2022 was exported (Figure 14).

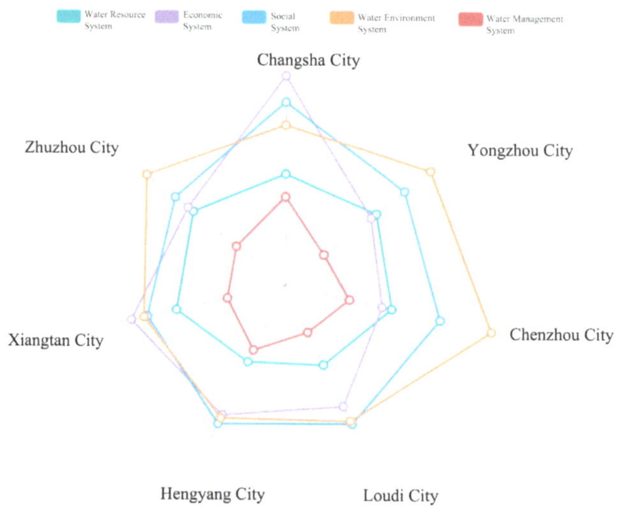


**Figure 14.** Radar chart of average evaluation value at the criterion level.

The radar chart clearly shows that the management support evaluation values of the seven prefecture-level cities in Xiangjiang have been lagging behind other systems for many years, indicating that the water management capacity of Xiangjiang is generally weak and needs to be further strengthened. In contrast, the resource support is only stronger than the management support, indicating that the endowment of water resources to the Xiangjiang River does not have many advantages. In terms of environmental support, the evaluation values of Yongzhou City, Chenzhou City, and Zhuzhou City are consistently better than those of other cities, possibly due to relatively fewer industries and the construction of water conservation forests. This indicates that good contributions have been made to protecting water sources, controlling water pollution, and protecting water resources for the Xiangjiang River. The economic system of Changsha has made the greatest contribution, but the water environment is not cause for optimism.

## 5. Water Carrying Capacity Projections

### 5.1. Forecasting Criteria

This article refers to the relevant literature on the risk warning of the water resources carrying capacity and the classification standard [39–44]. The risk warning level of the water resources environmental carrying capacity is divided into five levels (Table 8), which reflect the risk status of the water resources carrying capacity of the Xiangjiang River.

**Table 8.** Classification criteria for early warning results of water resources carrying capacity in the Xiangjiang River.

| Carrying Capacity Class | Appraise Value | State of Affairs | Early Warning Signal Signs |
|---|---|---|---|
| I | [0, 0.2] | Grossly overloaded | red |
| II | (0.2, 0.4] | Overloading | orange |
| III | (0.4, 0.6] | Boundary | yellow |
| IV | (0.6, 0.8] | Weak bearing | blue |
| V | (0.8, 1] | Can bear | green |

### 5.2. Projected Results

Overall, the water resources carrying capacity of the Xiangjiang River will increase slightly to a stable level, and the risk level will still be "overloaded" (Figure 15). From 2023 to 2027, the risk of the water resources carrying capacity of the Xiangjiang River will remain unchanged, and there will be a slight downward trend in the prediction interval, i.e., it is predicted that the carrying capacity of the water resources in the study area will decrease. Therefore, we should remain vigilant.

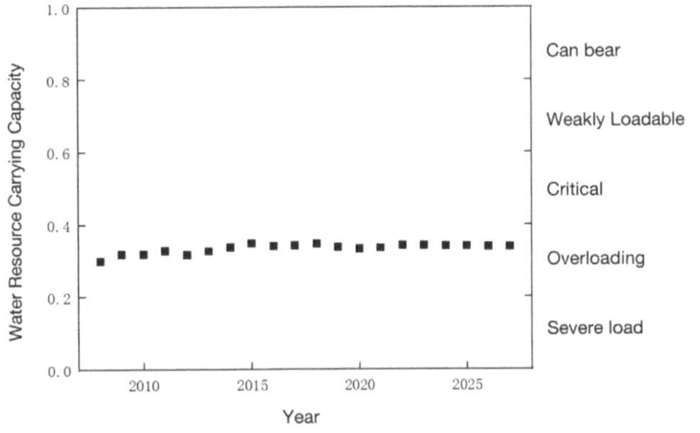


**Figure 15.** Prediction of the water resources carrying capacity of the Xiangjiang River by BP neural network.

Based on the predicted results, in-depth analysis shows that the sustained "overload" risk level in the Xiangjiang River Basin may be related to population migration and rapid economic development.

### 5.3. Spatial Distribution of Forecast Results

In 2022, the water resources carrying capacities of Changsha County, Wangcheng District, Ningxiang City, Liuyang City, and Zixing City were in the critical state, which is better than the other county units, which were classified under the overloaded state. Based on the aforementioned BP neural network prediction, in the next five years, the abovementioned five county units will continue to maintain their critical state, and the rest of the county units will remain in the overloaded state. In the predicted trend, the water resources carrying capacities of the Changsha urban area and Liling City will reach more than 0.39 in 2027, which is expected to be in the critical state later, and a specific spatial distribution is shown in Figure 16.

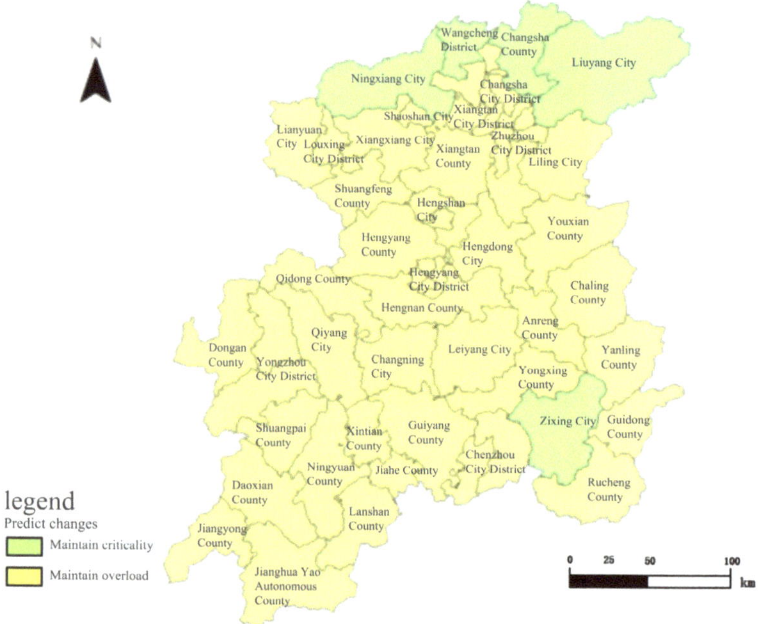


**Figure 16.** Forecast spatial distribution of water resources carrying capacity in the next 5 years.

## 6. Discussion

This article draws on Xiaohua Fu's [45] discussion of various algorithms' applications in water resource management, protection, and ecology, proposing future research directions and prospects. Xiaohua Fu [46] demonstrated that the particle-swarm-optimized support vector regression (PSO-SVR) model achieves high accuracy and broad applicability for predicting water quality; Honglei Chen [47] developed a model incorporating long short-term memory (LSTM) and attention mechanisms, enhancing the accuracy of water quality prediction for Australia's Burnett River. These papers have significantly aided our research regarding methodology and evaluation metric selection. Additionally, this article's evaluation and prediction of the Xiangjiang River Basin's water resources carrying capacity also incorporates Shengxin Lan's [48] early warning assessment and trend analysis of the resource and environmental carrying capacity in Altay, Xinjiang. These achievements strongly support our research conclusions, corroborating each other.

Based on the warning results, in-depth analysis shows that although the total amount of water resources in the Xiangjiang River Basin is relatively sufficient, factors such as population growth, economic development, and industrialization still exert severe pressure on water resources, especially in the middle and lower reaches. Changsha County and Wangcheng District in Changsha City are facing the risk of approaching the critical point of water resources carrying capacity, and the warning range is approaching the 0.4 warning line. This means that these two county-level units may enter an overloaded state at any time, which may be caused by the decline in the carrying capacity of the social system and water resources system due to future population inflows. Meanwhile, the water resources carrying capacities of Liuyang City, Ningxiang County, and Zixing City have also shown downward trends, which may be closely related to population migration and changes in economic activities. In contrast, other county-level units may achieve a steady increase in water resources carrying capacity in the future due to population outflow and optimization of water resource management.

For areas such as Changsha City, Wangcheng District, Ningxiang City, Liuyang City, and Zixing City that may continue to have critical water resources carrying capacities in the next five years, it is necessary to strengthen water resource management, such as by building or upgrading water supply systems, sewage treatment facilities, and water resource monitoring systems, to improve water resource utilization efficiency, reduce waste, and prevent further deterioration into an overloaded state. In addition, as these regions are in a critical state, local governments need to formulate and adjust policies based on the predicted results, with a focus on the sustainable use of water resources. Especially in Changsha, due to the increasing demand for water resources caused by population inflow and economic development, it is urgent to take measures to optimize water resource allocation to ensure sustainable development in the future. For areas like Zixing City, although that area is temporarily ahead of other regions due to the existence of Dongjiang Reservoir, it is still necessary to pay attention to ecological environment protection to avoid irreversible impacts on the environment caused by excessive resource development.

This article uses the TOPSIS model and entropy weight method to determine the water resources carrying capacities of 44 county-level units in the Xiangjiang River Basin and predicts the water resources carrying capacities for the next five years through a BP neural network model. This method mainly considers the specific situation of the Xiangjiang River Basin, involving details such as water resource distribution, economic development status, and environmental management, and provides scientific tools and methods for evaluating and predicting the water resources carrying capacity of the basin. However, for practical applications, adjustments and optimizations need to be made according to local conditions. Different basins may face different problems and challenges, such as water resource abundance and scarcity, differences in management capabilities, various types of pollution sources, and so on. Therefore, when applied to other watersheds, adjustments may need to be made based on local hydrological, economic, and social conditions, as well as environmental factors.

## 7. Conclusions

1. Based on 15 years of observation and research in the Xiangjiang River Basin, it was found that the water resources carrying capacity in the region maintained a relatively slow growth and then tended to flatten out during this period. However, the water resources carrying capacities of each county unit are not completely consistent, reflecting the differences and inconsistencies in local development trends, mainly due to various factors such as economic development, technological progress, and the implementation of water resource management policies.
2. In terms of spatial distribution, there are relatively significant differences in the carrying capacities of water resources in the area of the Xiangjiang River. The southern (upstream) counties of Chenzhou and Yongzhou have abundant water resources and a good ecological environment, but their water resources carrying capacities are generally lower than those of the northern (middle and downstream) counties, which may be closely related to the economic development levels of Changsha, Zhuzhou, and Xiangtan. The level of water resources carrying capacity is a comprehensive issue that goes beyond the endowment of water resources.
3. According to the BP neural network model, it is predicted that the water resources carrying capacity of the Xiangjiang River Basin in the coming years will continue the steady development trend of 2022. Changsha County, Wangcheng District, Ningxiang City, Liuyang City, Zixing City, and other areas will be in a critical state that is better than other counties and cities. Most county-level units are overloaded, indicating that the carrying capacity of water resources in the Xiangjiang River needs to be monitored vigilantly.

**Author Contributions:** Conceptualization, X.F.; Methodology, X.F.; Software, Y.F.; Validation, X.F.; Formal analysis, X.D. and Z.L.; Investigation, X.Z.; Resources, J.H.; Data curation, J.W.; Writing—original draft, Z.W. and Z.F.; Writing—review & editing, J.W.; Project administration, J.L. All authors have read and agreed to the published version of the manuscript.

**Funding:** This project was supported by the Hunan Key Open Laboratory of Lake Ecological Restoration and Sediment Resource Utilization Fund (KDKJ2023KFJJA03) and the Hunan Provincial Philosophy and Social Science Planning Fund (22YBA120).

**Institutional Review Board Statement:** Not applicable.

**Informed Consent Statement:** Not applicable.

**Data Availability Statement:** The data presented in this study are available on request from the corresponding author.

**Conflicts of Interest:** The authors Liu Junwu and Fang Yingchun are technicians of Hunan Kaidi Engineering Technology Co., LTD., and part-time tutors of Ecological and Environmental Management and Evaluation Center of Central South University of Forestry and Technology. The authors declare that this study received funding from the Hunan Key Open Laboratory of Lake Ecological Restoration and Sediment Resource Utilization Fund. The funder was not involved in the study design, collection, analysis, interpretation of data, the writing of this article or the decision to submit it for publication. The study was conducted without any business or financial relationship.

## References

1. Guo, X.; He, J.; Zhang, H.; Wang, P.; Li, Y.; Li, J. Some considerations on building a national water network system. *China Water Resour.* **2019**, *15*, 1–4.
2. Xia, W.; Wang, N.; Shen, Y. Study on water resources carrying capacity of Shaanxi Province based on watershed. *Arid Zone Geogr.* **2020**, *43*, 602–611.
3. Ait-Aoudia, M.N.; Berezowska-Azzag, E. Water resources carrying capacity assessment: The case of Algeria's capital city. *Habitat Int.* **2016**, *58*, 51–58. [CrossRef]
4. Zou, Z.; Zhao, B.; Chen, J.; Qian, H.; Gao, Y.; Li, C. Evaluation and spatiotemporal variation characteristics of water resources carrying capacity in nine provinces and regions along the Yellow River. *Earth Sci. Environ. J.* **2024**, *46*, 351363. [CrossRef]
5. Yu, P.; Shang, Y.T.; Yao, M.; Liu, P. Evaluation of Water Resources Carrying Capacity in Xinjiang by Coupling Water Footprint and Principal Component Analysis. *Hydrology* **2021**, *41*, 49–54.
6. Liu, Y.; Dong, H.; Tian, X. Dynamic simulation of water resources carrying capacity in Suibin County based on system dynamics model. *People's Yellow River* **2024**, *46* (Suppl. S1), 47–48+50.
7. Zuo, Q.; Guo, J.; Ma, J.; Cui, G.; Yang, R.; Yu, L. Assessment of regional-scale water resources carrying capacity based on fuzzy multiple attribute decision-making and scenario simulation. *Ecol. Indic.* **2021**, *130*, 108034. [CrossRef]
8. Zhang, F.; Yin, M.; Zhang, J. Evaluation of water resources carrying capacity in the Gansu section of the Yellow River Basin based on entropy weight TOPSIS. *People's Yellow River* **2024**, *46*, 79–85.
9. Jiang, D.; Yu, H.; Li, L. Dynamic evaluation of water resources carrying capacity in Shandong Province based on comprehensive weighting and TOPSIS model. *Resour. Sci.* **2024**, *46*, 538–548.
10. Xue, Q.; Yang, K. Research on the prediction and regulation of water resources carrying capacity in Jiangsu Province based on BP neural network system dynamics coupling model. *Water Resour. Hydropower Technol. (Chin. Eng.)* **2022**, *53*, 86–99. [CrossRef]
11. Fu, Z.; Hu, X.; Huang, C.; Chen, B.; Li, M.; You, Y. Study on the assessment of water resources carrying capacity in Hunan Province based on SPR model. *J. Chang. Univ. Technol. (Nat. Sci. Ed.)* **2023**, *20*, 92–103.
12. Chen, M.; Xiong, G.; Zhang, J.; Li, L.; Wang, J. Comprehensive evaluation of water quality in the Xiangjiang River Basin and its spatio-temporal evolution analysis. *Environ. Eng.* **2019**, *37*, 83–90+104.
13. Fu, X. Research on the Ecological Compensation Mechanism of Water Rights Handover in the Xiangjiang River. Master's Thesis, Hunan Agricultural University, Changsha, China, 2014.
14. Zhang, N.; Su, X.; Zhou, Y.; Niu, J. Evaluation of water resources carrying capacity in the Yellow River Basin. *J. Nat. Resour.* **2019**, *34*, 1759–1770.
15. Du, X.; Li, Y.; Zhang, X. Research on Water Resources Carrying Capacity of Zhengzhou City Based on TOPSIS Model. *People's Yellow River* **2022**, *44*, 84–88.
16. Ma, J.; De, Y.; Wang, Q.; Wang, F.; Zheng, Y.; Du, W. Evaluation of Water Resources Carrying Capacity and Diagnosis of Obstacle Factors in Yunnan Province Based on GRA-TOPSIS Method. *J. Water Resour. Water Eng.* **2022**, *33*, 11–17+26.
17. Deng, Q.; Ping, J.; Mei, X.; Lv, P.; Miao, C. Evaluation of Water Resources Carrying Capacity in Xinxiang City Based on Combination Weight TOPSIS Model. *J. Water Resour. Water Eng.* **2022**, *33*, 81–87.
18. Xu, Z.; Cao, Y. Evaluation of Water Resources Carrying Capacity in Changchun City Based on Entropy Weight TOPSIS Model. *J. Saf. Environ.* **2022**, *22*, 2900–2907. [CrossRef]

19. Chen, L.; Zhou, H. Evaluation of Water Resources Carrying Capacity in Karst Watersheds Based on Fuzzy Comprehensive Evaluation and Principal Component Analysis. *Saf. Environ. Eng.* **2021**, *28*, 159–173.
20. Zhou, X.; Luo, H. Evaluation of water resources carrying capacity and risk identification in the the Pearl River basin. *People's Pearl River* **2023**, *44*, 28–36+69.
21. Zhang, G. Analysis of Water Resources Carrying Capacity in Baiyang River Basin, Xinjiang. *Water Resour. Technol. Econ.* **2022**, *28*, 57–62.
22. Bian, J.; Xia, Y.; Mao, X.; Zhang, H. Adaptability Evaluation of Water Resources Carrying Capacity and Economic and Social Development in the Taihu Lake Lake Basin. *Water Resour. Hydropower News* **2022**, *43*, 31–37.
23. Han, L.; Men, B. Evaluation of Water Resources Carrying Capacity in Haihe River Basin Based on Combination Game Theory. *Hydroelectr. Energy* **2021**, *39*, 61–64.
24. Liu, X.; Zhang, M.; Xu, J.; Guo, Y.; Duan, W.; Shen, Y. Study on water resources carrying capacity of Tarim River Basin based on system dynamics model. *Arid Zone Geogr.* **2021**, *44*, 1407–1416.
25. Wang, L.; Huang, X.; Li, H. Evaluation of water resources carrying capacity of nine provinces in the Yellow River Basin based on CW-FSPA. *China Rural Water Conserv. Hydropower* **2021**, *63*, 67–75.
26. Shi, Z.; Feng, W.; Qian, H. Evaluation of water resources carrying capacity in Gansu Province based on basin scale. *Ecol. Sci.* **2021**, *40*, 51–57.
27. Yu, G. Intelligent monitoring method of water resources ecological carrying capacity based on neural network algorithm. *Constr. Manag.* **2022**, *30*, 97–100+125.
28. Men, H.-Q. Evaluation of water resources carrying capacity in Ningxia based on artificial neural network method. *Ningxia Agric. For. Sci. Technol.* **2014**, *55*, 65–68.
29. Zhang, X.; Cao, Z.; Zhang, R.; Han, R.; Li, X. Evaluation of water resources carrying capacity and trend prediction in the Yellow River Delta ecological and economic zone. *World Geogr. Res.* **2022**, *31*, 549–560.
30. Wang, W.; Yang, L.; Zheng, Y.; Xu, D. Evaluation of water resources carrying capacity of Shiyang River Basin based on game theory and cloud model. *Water Resour. Hydropower Technol.* **2021**, *52*, 35–45.
31. Wang, J.; Geng, Y.; Zheng, X. Evaluation of water resources carrying capacity based on PSR-MDS model—Taking Hebei Province as an example. *People's Yangtze River* **2021**, *52*, 81–86.
32. Lu, Y.; Xu, H.W.; Wang, Y.X.; Yang, Y. Evaluation of water environmental carrying capacity of city in Huaihe River Basin based on the AHP method: A case in Huai'an City. *Water Resour. Ind.* **2017**, *18*, 71–77.
33. Wang, C.H.; Hou, Y.L.; Xue, Y.J. Water resources carrying capacity of wetlands in Beijing: Analysis of policy optimization for urban wetland water resources management. *J. Clean. Prod.* **2017**, *161*, 1180–1191. [CrossRef]
34. Yang, J.F.; Lei, K.; Khu, S.T.; Qiao, F.; Liu, Q. Assessment of water ecological carrying capacity using an indicator-based method applied to Tieling City, China. *Water Sci. Technol. Water Supply* **2015**, *15*, 940–947.
35. Zhang, S.H.; Xiang, M.S.; Yang, J.S.; Fan, W.; Yi, Y. Distributed hierarchical evaluation and carrying capacity models for water resources based on optimal water cycle theory. *Ecol. Ind.* **2019**, *101*, 432–443.
36. Zhou, Y.; Xie, Y. Analysis of sustainable use of water resources in Liaoning Province based on the ecological footprint model. *J. Ecol.* **2014**, *33*, 3157–3163.
37. Zuo, Q.; Zhang, Z.; Wu, B. Evaluation of water resources carrying capacity of nine provinces and districts in the Yellow River Basin based on the combined weights TOPSIS model. *Water Resour. Prot.* **2020**, *36*, 1–7.
38. Fu, X.; Yang, X.; Song, X. Evaluation of resource and environmental carrying capacity of Yangtze River economic belt based on DPSIR-TOPSIS model. *J. Cent. South For. Univ. Sci. Technol. Soc. Sci. Ed.* **2022**, *16*, 38–47.
39. Li, Z.; Hu, Y.; Liu, Q.; Liu, S. Early warning evaluation and simulation analysis of resource and environmental carrying capacity in inland arid zones-an empirical analysis from Ningxia. *Ecol. Econ.* **2021**, *37*, 209–215+229.
40. Lei, X.; Qiu, G. Empirical study on regional resource and environmental carrying capacity evaluation based on entropy weight TOPSIS model. *J. Environ. Sci.* **2016**, *36*, 314–323.
41. Yang, Q.; Zhang, S. Evaluation of water environment carrying capacity based on BP neural network. *Res. Land Nat. Resour.* **2009**, *43*, 70–72.
42. Josefina, L.; Alfredo, A.; Roberto, A. Simple estimation of available water capacity in soils of semiarid and subhumid environments. *Arid Land Res. Manag.* **2018**, *32*, 133–148.
43. Ganguli, P.; Merz, B. Extreme Coastal Water Levels Exacerbate Fluvial Flood Hazards in Northwestern Europe. *Sci. Rep.* **2019**, *9*, 13165. [CrossRef] [PubMed]
44. Gao, W.; Liu, Y.; He, S. Early warning study on the carrying capacity of water resources in watersheds based on the SD model with sub-quality. *J. Peking Univ.* **2018**, *54*, 673–679.
45. Fu, X.; Jiang, J.; Wu, X.; Huang, L.; Han, R.; Li, K.; Liu, C.; Roy, K.; Chen, J.; Mahmoud, N.T.; et al. Deep learning in water protection of resources, environment, and ecology: Achievement and challenges. *Environ. Sci. Pollut. Res.* **2024**, *31*, 14503–14536. [CrossRef] [PubMed]
46. Fu, X.; Zheng, Q.; Jiang, G.; Roy, K.; Huang, L.; Liu, C.; Li, K.; Chen, H.; Song, X.; Chen, J.; et al. Water quality prediction of copper-molybdenum mining-beneficiation wastewater based on the PSO-SVR model. *Front. Environ. Sci. Eng.* **2023**, *17*, 84–97. [CrossRef]

47. Chen, H.; Yang, J.; Fu, X.; Zheng, Q.; Song, X.; Fu, Z.; Wang, J.; Liang, Y.; Yin, H.; Liu, Z.; et al. Water Quality Prediction Based on LSTM and Attention Mechanism: A Case Study of the Burnett River, Australia. *Sustainability* **2022**, *14*, 13231. [CrossRef]
48. Lan, S.; Wang, X.; Li, M.; Fu, X.; Xu, M.; Zhu, J.; Wang, P.; Mao, Y.; Dong, Z.; Li, J.; et al. Early Warning Evaluation and Warning Trend Analysis of the Resource and Environment Carrying Capacity in Altay Prefecture, Xinjiang. *Sustainability* **2023**, *15*, 9825. [CrossRef]

water

MDPI

Article

# Longitudinal Investigation of Groundwater and Surface Water Interaction of Two Gravel Pit Lakes in Central Texas: Chemical and Flow Implications

**William A. Brewer [1,*], Claudia R. Dawson [2] and Joe C. Yelderman, Jr. [1,2]**

1 The Institute of Ecological, Earth, and Environmental Sciences, Baylor University, Waco, TX 76798, USA
2 Department of Geosciences, Baylor University, Waco, TX 76798, USA
* Correspondence: will_brewer@baylor.edu

**Abstract:** Two gravel pit lakes in central Texas were examined over the course of two years with upgradient and downgradient piezometer installations. Groundwater and lake water were sampled bimonthly for nutrients, water levels, and groundwater chemistry, and in addition, rain and lake gauges and mini-piezometers were installed, depth surveys were conducted, and a simple 2D flow model was constructed. The project goal was evaluating and examining flow dynamics and chemical effects as groundwater flows to surface water and back to groundwater with the intent to understand the effects that gravel pit lake systems have on connected shallow groundwater. Both lake systems were shown to be flow-through systems that influence the water quality by decreasing the dissolved nutrients in the groundwater in their vicinity while oxygenating the water and altering the pH. However, the lakes are also prone to high levels of evaporation, meaning that minor improvements to water quality come at the cost of decreasing the quantity of water in storage within the aquifer. Similar groundwater and mine lake systems may show comparable tendencies, providing new information for water managers, regulators, and stake-holders about the potential roles that non-remediated gravel pit lakes may play in local ecosystems and aquifer dynamics.

**Keywords:** groundwater; pit mine; alluvial aquifer; remediation; groundwater surface water interaction; groundwater chemistry; flow system

**Citation:** Brewer, W.A.; Dawson, C.R.; Yelderman, J.C., Jr. Longitudinal Investigation of Groundwater and Surface Water Interaction of Two Gravel Pit Lakes in Central Texas: Chemical and Flow Implications. *Water* **2024**, *16*, 3068. https://doi.org/10.3390/w16213068

Academic Editors: Zongjun Gao and Jiutan Liu

Received: 7 October 2024
Revised: 22 October 2024
Accepted: 23 October 2024
Published: 26 October 2024



## 1. Introduction

As the Texas population grows over the next few decades, demand for natural aggregate construction materials (crushed stone, gravel, sand, etc.) and water will increase simultaneously. The Texas population is projected to nearly double by 2050 (from 29 million to 50 million people) which will necessitate the large-scale construction of infrastructure and exacerbate problems already being experienced in the provision of water [1,2]. In central Texas, much of the sand and gravel used in construction material is mined using open pits in shallow alluvial aquifers. This study focuses on two former pit mines in one such aquifer, the Brazos River Alluvium Aquifer (BRAA), specifically the section within McLennan County near the City of Waco.

In the BRAA and other alluvial aquifers in Texas, aggregates in pit mines are frequently quite shallow corresponding to the thickness of the aquifer. The BRAA's maximum thickness never exceeds 175 feet at its absolute thickest, and is much thinner in McLennan County, averaging around 35 feet thick [3]. The BRAA is generally characterized as being composed of alluvial sediments that exhibit a fining upward sequence. The alluvial sediments that form the aquifer range in size from extremely coarse limestone and chert gravel to medium sands to clays. The large gravels and coarse sands are typically the goal in mining operations. When observed at the local scale, the geometry of sediments can be extremely complex after years of reworking by the river. In the study area, there are terraced deposits of various particle sizes with sediments showing partial fining upward sequences in succession and non-laterally extensive but with frequent lenses of clay and silt. Despite this complexity, in many cases most of the materials that would be viable for mining are in the lower layers of the aquifer, below the water table [3,4]. This leads to pit mines that intersect the water table where the coarse aquifer material is removed. In Texas, it is common for these types of pits to be left unremediated and allowed to fill with water from the aquifer, equilibrating to the water table surface elevation, and forming a "gravel pit lake" (GPL). These lakes are frequently used as water features for recreation or to provide a reliable water source for livestock. Owners often stock the lakes with fish which provide incentive for birds and other animals to nest nearby. Given enough time, a GPL eventually can grow to resemble a naturally occurring groundwater-fed lake, complete with a lacustrine ecosystem [5].

The effects of GPLs on groundwater systems vary widely; sometimes they can serve as point sources of aquifer contamination through surface runoff and atmospheric deposition while at other times they might provide a site for the bioremediation for nutrients [6,7]. GPLs also have varying effects on the evaporation regimes within the aquifer which can alter the flow dynamics of the system locally [8]. Previous studies have shown that the banks, flora, and fauna, as well as the size of the lake and the mineralogy of the aquifer system, all play a role in their effects on the groundwater [7,9–12]. However, few studies have examined small-scale lakes in thinner aquifers in dry regions. The primary goal of this study is to characterize the local aquifer and groundwater effects of small GPLs on the BRAA to help guide the future management, regulation, and monitoring of such lakes as their numbers continue to multiply with increasing demand for mined aggregate materials. This study also seeks to begin to fill the literature gap surrounding GPLs in shallow unconfined aquifers, particularly in Texas, where their effects remain largely unexplored.

This study focuses on two GPLs in the northern segment of the BRAA in McLennan County Texas (Figure 1).

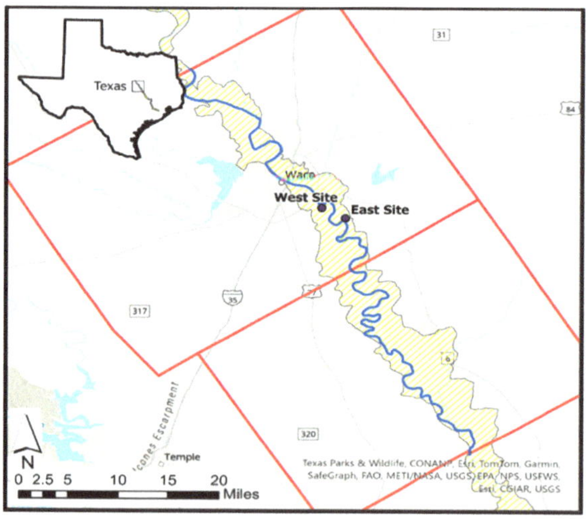


**Figure 1.** A map showing the extent of the northern segment of the Brazos River Alluvium Aquifer as well as the general lake locations. Inset: full extent of the Brazos River Alluvium Aquifer in the state of Texas.

The BRAA is a Quaternary emplacement of riverine sediments, a mix of clay, silt, sand, and gravel. The BRAA is typically characterized as a large fining upward sequence; however, the aquifer sediments are terraced in some locations leading to repetition of some sediment types where the aquifer is thicker. Beyond this, the local stratigraphic architecture in a given area may be quite complex, with lenses of gravel and sand creating perched water tables and clays that can be saturated and yet produce little to no water [13]. In the Pleistocene, the Brazos River was much larger, with discharges equaling between five to eight times current amounts. This larger volume of water during interglacial periods led to broad floodplains which were continually dissected and reworked as glacial fluctuations occurred, leading to a complex mélange of stratified alluvial sediments across the entire aquifer [14]. This complexity complicates modeling the aquifer. Some parts of the aquifer are more permeable and exhibit quicker flow paths than do others [15]; this has the potential to cause GPLs to function differently depending on their location within the aquifer, making the lake location within the BRAA an important consideration for future work in monitoring and for developing a regulatory framework.

The northern segment of the BRAA is characterized by thinner aquifer sediments and a shallower water table than that which occurs in the lower sections of the aquifer. The northern segment also features compartmentalization to a higher degree than the lower sections of the aquifer. In the BRAA, compartmentalization is due to the Brazos River intersecting low-permeability bedrock in some stream reaches. When the river channel intersects the bedrock, lateral aquifer flow is essentially cut off from one side of the aquifer to the other as the river channel serves as a hydrologic divide [16]. The bedrock is cretaceous aged chalk and marls and is not known to contribute appreciable volumes of water to the BRAA [13]. This compartmentalization is important to document because the two selected lakes in this study are on opposite sides of the Brazos in a stream reach that is bedrock dominated and can be considered to be impacted by independent flow systems [16]. This could have implications for the development of our understanding of gravel pit lakes. For the duration of this study, conditions in both lakes were similar. Such similar conditions are not guaranteed, particularly considering future climatic shifts towards dryness in the region [17] and differences in local deposition between the two compartments.

The two lakes are similar in terms of the observations made during this study but occur in two separate compartments of the BRAA which are divided from one another by the Brazos River acting as a hydrologic boundary [16]. At the time studied, both lakes were similar in terms of lateral dimensions with Lake 1 ranging from 420,000–450,000 sq feet (or 10 acres) (depending on water level), while Lake 2 had dimensions ranging from 380,000 to 390,000 sq feet (or nine acres). Both lakes showed similarly shaped depth profiles in the aquifer flow direction (Figure 2).

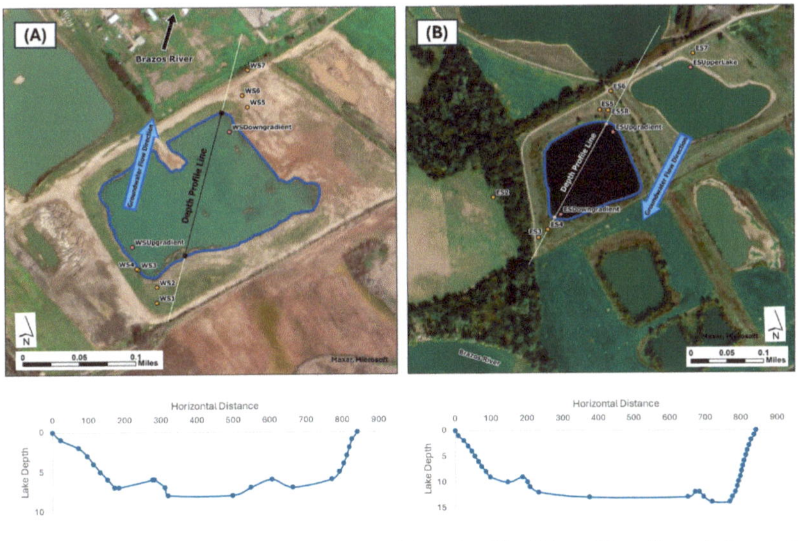


**Figure 2.** Aerial maps and depth profiles of the West Site (**A**) and East Site (**B**) Gravel Pit Lakes. Depth profiles were created in the groundwater flow direction for each site. For the West Site, this

means that the profile is from the southwest to the northeast; and for the East Site, this means that the profile is from the northeast to the southwest. The white line in each map represents the horizontal transects used to plot chemistry data in the Results and Discussion section.

The West Site lake had a maximum depth of about eight feet while the East Site lake had a maximum depth of about 14 feet. These depths are snapshots determined by lake depth measurements on a single trip to each site; the depths of the lakes, like their horizontal dimensions, are somewhat dependent on precipitation patterns, water table level, and drought conditions.

## 2. Materials and Methods

This study used a variety of hydrogeology field and lab techniques to characterize and explore the local aquifer effects that the GPLs have in the BRAA. Subsections describe chemical sampling methodologies, field methodologies, and modeling methodology, respectively.

### *2.1. Sampling Methods*

Prior to long-term monitoring, two-inch diameter piezometers were installed on both the up- and downgradient sides of each lake (Figure 3D–F) using completion methods described in hydrogeology field guides [18,19].

**Figure 3.** A selection of field photos: (**A**) an installed mini-piezometer at the West Site; (**B**) the rain gauge installed at the West Site; (**C**) a lake staff gauge installed at the West Site; (**D–F**) examples of the heterogenous aquifer material encountered during piezometer installation at both sites. Particle sizes ranging from clay and fine sand (**D**) to coarse sand (**E**) to gravel of one- to three-inch diameter mixed with very coarse sand (**F**).

Piezometers were arranged in a line in the interpolated flow direction of the BRAA aquifer (using groundwater contours and considering the Brazos River as the low point and interpreting a perpendicular line as the direction of groundwater flow) (Figure 2). Piezometers were installed using a Geoprobe 6620 DT drill rig with 3.25 inch diameter hollow stem auger flight (Geoprobe Systems Inc., Salina, KS, USA). Boreholes were augured to between 10 feet and 15 feet below the saturated zone in the aquifer, depending on the elevation difference between the top of the of saturated zone and the top of the bedrock on a per-hole basis. Piezometers were fully assembled before placement in boreholes. To reduce hole collapse during assembly, the auger flight was temporarily left in the boreholes until the well casing and screen were assembled. Piezometer length and screen length were dependent on the depth to the saturated zone. Piezometers were constructed from two-inch diameter interlocking PVC pipe with a pointed end cap. Each section of pipe measured 10 feet. The top section of each piezometer was sawed off after installation to minimize its surface visibility. Well screens were factory-machined PVC that used 0.02 inch slots. Screen lengths for all piezometers were dependent on the estimated elevation of the saturated zone but averaged 10 feet. After the piezometer was constructed and inserted into the borehole, the Geoprobe was used to push the screen into the saturated zone as completely as possible. Filter sand was used as a sand pack around the screen where necessary and when full insertion of the screen into the aquifer material was impossible. Bentonite pellets were poured into each borehole above the screen to serve as a plug after the filter sand. The remaining space in each borehole was then backfilled with construction sand. Piezometers were bailed between 30 to 50 times to help develop the material around the screen and reduce suspended sediment in future sampling. Auger cuttings were logged during drilling along with depths to provide insight about the nature of the aquifer material and saturated zone in each borehole. These cutting logs were used to construct piezometer diagrams (Figure 4) which were used for flow modeling in the aquifer.

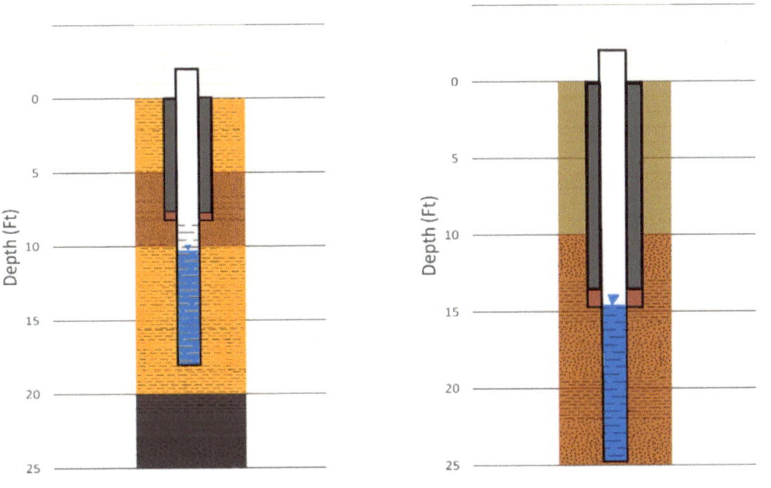


**Figure 4.** Example of piezometer completion diagrams created using recorded augur cuttings logged during piezometer installation. Different colors represent different sediment textures encountered while drilling. Diagrams show the water table at the time of piezometer installation as well as the bentonite (solid brown boxes) and sand (solid gray boxes) fill locations. Diagrams were used to inform the creation of the model domain used in TopoDrive and ParticleFlow 1.0. Only 2 of the 14 diagrams are included in this figure for brevity. Diagrams were arranged according to horizontal distance, and the cross sections were created by hand and then imported into the TopoDrive program.

Multiple types of water samples were taken from both the downgradient and upgradient sides of the lakes and from piezometers at different intervals over the course of the study. Standard ion samples and Oxygen-18 and Deuterium samples were collected in a 50 mL bottle at each piezometer and gravel pit lake. When collecting samples, there was no head space in the vials and the lids were taped closed. Isotope samples were analyzed using the Thermo-Electron Delta V Advantage Isotope Ratio Mass Spectrometer (IRMS) (Thermo Fisher Scientific Inc., Waltham, MA, USA) in the stable isotope lab at Baylor University while ion samples were analyzed by Biochem Inc., West, TX, USA. Dissolved nitrate, ammonia, and phosphate samples were analyzed in the Baylor Center for Reservoir and Aquatic Systems Research. Nutrient water samples were extracted from wells and lakes using a bailer. Wells were bailed a minimum of three times per sample to ensure fresh groundwater was in the well. Lake water was sampled by tossing the bailer 50 feet towards the center of the lake. Water was then pushed through 0.45 micron polypropylene syringe filters into 50 mL centrifuge tubes. Tubes were rinsed with filtered sample water 3 times before being filled with the samples and put on ice. Samples were transported back to the lab in an ice chest and frozen until being run on a Lachat Instruments Quikchem 8500 (Lachat Instruments, Milwaukee, WI, USA). For isotope samples, the water was unfiltered and the first bail from the wells was used to obtain unagitated water. Isotope samples were drawn from the bottom of the bailer in 50 mL centrifuge tubes that were sealed with no head and taped to prevent atmospheric exchange. Ion samples were collected in 250 mL bottles. Anion samples were acidified with 5 ml of nitric acid for transport. Titrations were performed immediately at each well site using the same waters captured for ion samples to reduce the degassing effects that caused changes in alkalinity [18,19].

### 2.2. Field Methods

Mini-piezometers (Figure 3A) were installed around the perimeters of both gravel pit lakes to examine the influx of groundwater from the banks. Mini-piezometers were installed according to the methodologies outlined in hydrogeology field manuals and field exercise publications [18–21]. Five-foot sections of one-inch internal diameter PVC were slotted using a pipe saw. The PVC sections were inserted into galvanized pipes and a stainless-steel bolt was secured in the PVC to prevent the warping of the piezometers and the galvanized pipe from hammering. The galvanized pipes were hammered into bank sediments using a rubber mallet until at least two feet of the PVC was fully enclosed by bank sediments. The galvanized pipes were removed, leaving the PVC behind. After construction, the mini-piezometers were bailed until empty and allowed to equilibrate with bank sediment pore water conditions. The mini-piezometers were measured biweekly through the beginning stages of the study using a small gauge Solinst electrical line in order to examine the presence or absence of groundwater flow into the lakes via their banks.

Slug tests were performed [22] on one upgradient (WS2 and ES5R) and one downgradient (WS5 and ES4) full-size piezometer at each lake to evaluate local hydraulic conductivity values around each of the lakes and compare them to Freeze and Cherry to verify the validity of the parameters used in the TopoDrive and ParticleFlow 1.0 model. Slugs were fitted into piezometers and water levels were allowed to equilibrate over 24 h. Data loggers were fitted into the piezometers and set to record water levels every 10 s. After equilibration, slugs were removed, and water levels were measured by hand using a Solinst water level indicator (e-line) every 10 s until static conditions were observed.

Depth surveys were performed from a kayak using a lead-weighted steel tape. Three equidistant lines of marked and buoyant nautical ropes were placed across each lake. Depth measurements were taken at pre-marked 30-foot intervals along each line and boundary conditions were measured at 0-foot depths (immediately where the water touched the shore), and the depths of each lake were measured at 1 foot from the shore around the rim of each lake. Data were input into ArcGIS pro 3.1.3 and interpolated using the IDW method to produce depth estimates for each lake. Depth profiles were constructed using the interpolation surfaces in the direction of groundwater flow for each lake using a 1-foot interval (Figure 2).

Lake gauges were installed at both sites to measure the lake water level elevation in relation to that of the piezometers (Figure 3C). Lake gauges, piezometers, and other measurement points were surveyed using a Leica kinematic GPS to establish their elevation in relation to sea level. All elevations are reported referenced to sea level. An auto-recording tipping bucket rain gauge was installed near the West Lake site (Figure 3B) to record rainfall events and examine their correlation to water levels in both the aquifer and lake. A piezometer on the upgradient and downgradient sides of each lake was also fitted with a Solinst pressure transducer data logger that recorded the groundwater level in each piezometer on an hourly basis to enable closer inspection of groundwater levels as they responded to rainfall.

### 2.3. Modeling Methods

TopoDrive and ParticleFlow 1.0, a two-dimensional, steady-state, finite element model [23], were used to estimate the time it takes for groundwater to move through the aquifer and the gravel pit lakes at each site. The two-dimensional model domains were aligned with the interpolated groundwater flow direction and extended from the farthest upgradient piezometer to the Brazos River. At the West Site, the model domain length was set to 705 m with a vertical exaggeration of 30 times. At the East Site, the model domain length was set to 875 m with a vertical exaggeration of 30 times. The water table points were made based on groundwater measurements from piezometers and surface-water measurements were taken from the lake gauges at both sites; the river elevations used for this model were obtained from Google Earth Pro. It should be noted that the groundwater elevation between the river and the gravel pit lake was only measured from two piezometers that are near the gravel pit lake and is interpreted for the rest of the model based on the river elevation. The shape of the lake was input into the model based on lake depth surveys that were conducted at each site. At the West and East Sites, the mesh was made from 100 columns and 50 rows. The sediment layer depths were added to the model domains based on auger cuttings that were recorded when the piezometers were installed (Figure 4). Figure 5 shows the TopoDrive color and the corresponding sediment description, hydraulic conductivities, and porosities that were used at the West Site and East Site.

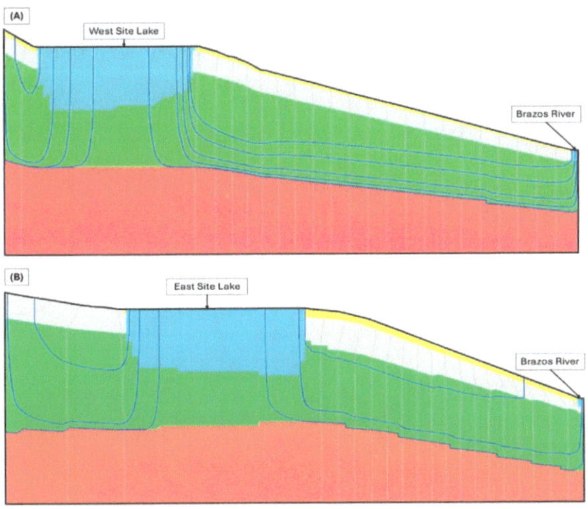


| TopoDrive Color | Sediment Description | Horizontal Conductivity (m/S) | Vertical Conductivity (m/S) | Porosity (%) |
|---|---|---|---|---|
| Blue | Water | 1 | 1 | 100 |
| Green | Gravel and Sand | $1\times10^{-3}$ | $2\times10^{-4}$ | 30 |
| Grey | Sandy Clay | $1\times10^{-5}$ | $1\times10^{-6}$ | 35 |
| Yellow | Clay Loam | $1\times10^{-6}$ | $1\times10^{-7}$ | 40 |
| Red | Bedrock (shale) | $1\times10^{-10}$ | $1\times10^{-10}$ | 5 |

**Figure 5.** TopoDrive and ParticleFlow 1.0 model domains for the West Site (**A**) and the East Site (**B**) showing flow lines for the general movement of groundwater through the model domain. The table describes the associated hydraulic conductivity of each color in the domain as calculated with slug tests and referenced to Freeze and Cherry (1979) [24].

The hydraulic conductivity of each sediment layer is based on slug tests conducted on the upgradient and downgradient side of both gravel pit lakes using methods and calculations from Bouwer and Rice [22]; the hydraulic conductivity and porosity of the sediments were then referenced to hydraulic conductivity values for sediment types as calculated by Freeze and Cherry [24]. The vertical hydraulic conductivity values are based on the data from the Final Numerical Model Report for the Brazos River Alluvium Aquifer Groundwater Availability Model [25,26].

## 3. Results and Discussion

Water sampling took place from May 2021 to September 2022 at the West Site and from July 2021 to September 2022 at the East Site to capture a full warm and cool seasonal cycle. This area of central Texas typically has hot dry summers and cooler and wetter winters [27]. Mini-piezometers (MPs) in both lakes were monitored for two months during the dry season in order to investigate groundwater flow into the lakes from bank filtration. All MPs in both lakes showed a consistent loss of water to the bank from the lake (mini-piezometer water levels were lower than the lake level) [18,21]; however, the gradient steepened in the predicted direction of groundwater flow in both lakes. This indicates that although the lakes are losing water to banks, the gradient of the lakes follows that which was found in the full-size piezometers, supporting the notion that the lakes are flow-through systems, at least partially, and likely are more connected to the BRAA system at depth than in the near surface portion of their banks.

Hand-measured data from both lakes and all piezometers compared to the tipping bucket rain gauge are displayed in Figure 6 for the first 6 months of the study.

West Site lake levels and groundwater levels have a weak positive correlation with rainfall events. However, the East Site does not share this relationship. This may be due to the monitoring period available for the East Site, given that the summer and fall of 2021 were so dry in this area that the intermittent rainfall slowed the decline of groundwater levels, flattening the slopes of the hydrographs rather than prompting recharge events and raising groundwater levels. The slowdown in water level decline could also be partially influenced by a reduction in regional pumping in the BRAA during the rain events or some combination thereof. Lake levels appear to respond to rainfall no more or less dramatically than do the piezometers. The lakes' water elevation consistently plots precisely where one might expect a piezometer's water elevation to fall along the gradient rather than being influenced by rainfall events. The responses of lake levels and piezometer levels to rainfall events implies a relationship between groundwater and lake levels; however, with the data gathered as part of this project, there could be further confounding factors that make it difficult to identify trends at the biweekly scale. However, when examined hourly, trends are more evident (Figure 7).

With the hourly groundwater data plotted against rainfall, it is possible to see the rapid increases in water level in the aquifer in response to rain events; however, these rises are frequently attenuated. Water levels return to previous trend levels quickly after the cessation of rain. This could indicate that rebound effects from pumping decrease during rain events (as mentioned above) rather than true recharge. Even with hourly data, the trend for the groundwater levels is decreasing over the course of the monitoring period as shown in Figure 6.

$^{18}O$ and $^{2}H$ isotopes samples are plotted for both lakes and their respective piezometers in Figure 8. The isotopes in most piezometers and between the lakes support the notion that the lakes are flow-through systems with the lakes showing the greatest amount of evaporation-driven change in their isotope ratios between all measurements; plotting farthest from the local meteoric water line from data gathered in Waco, Texas [28], the upgradient measurements show the least evaporation effects and plot most similarly to the precipitation values. The downgradient samples plot intermediately, signifying the influence of the lake/atmospheric exposure on the isotopic signature of the water flowing along the lake flow path in the aquifer. For $^{18}O$ and $^{2}H$ isotopes, evaporation of the sampled water is exhibited in a proliferation of heavier isotopes as lighter ones selectively evaporate first. In comparison to the local meteoric water line, this means waters that have undergone evaporation after rainfall plot either above or to the right of the line (depending on the primary enriched heavier isotope).

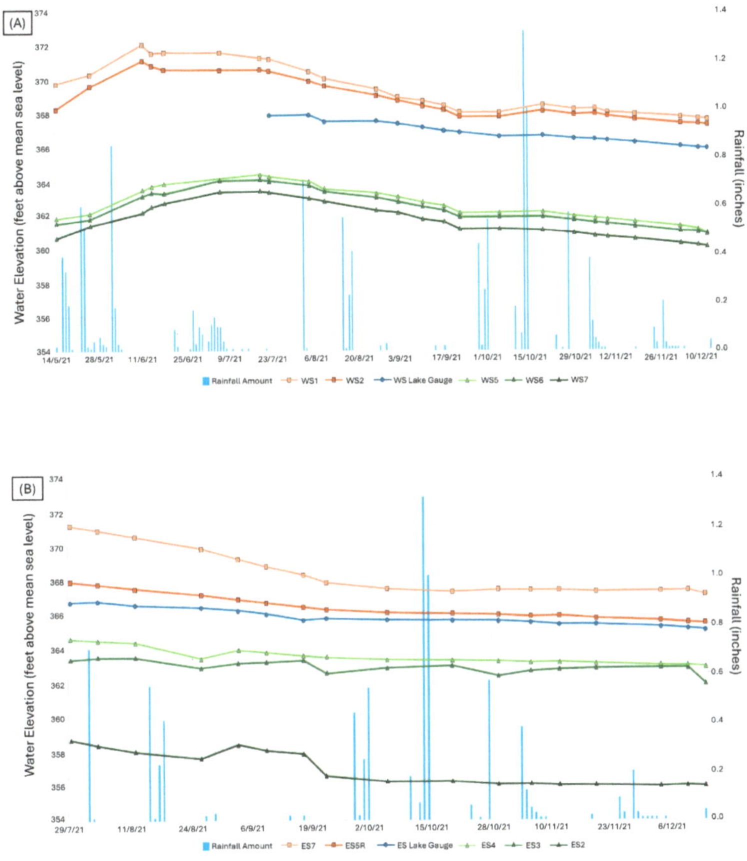


**Figure 6.** Piezometer and lake measurements taken by hand for the West Site (**A**) and East Site (**B**) as compared to the rainfall measurements recorded by the tipping bucket rain gauge installed near the West Site lake. Note that the East Site record covers the same period as the West Site lake gauge line (July to December). Orange shades represent upgradient measurements, blue represents lake gauge measurements, and green represents downgradient piezometer measurements.

In this area, the upgradient groundwater samples plot primarily as rainwater as the water that enters the aquifer system does not have time to evaporate under atmospheric conditions. In Figure 8, it is possible to see some locations where the data do not follow the trend. The downgradient measurements that plot with the upgradient measurements are sourced from ES2 and ES3 at the East Site. These piezometers were constructed with maximum depths deeper than the bottom of the gravel pit lake. It is likely that these isotope samples are indicative of a deeper flow path in the aquifer that does not intersect the flow path that provides water to the lake, like the deeper flow paths that McVea [15] found in his modeling efforts in the BRAA. The lake samples that plot closer to the downgradient measurements may have been a result of those samples being taken in the month of May, before the high temperatures and dryness of the following months began. This would lead to less evaporation than may have been evident had the samples been taken in the following months of the summer.

Cations and anions stay relatively consistent as the water flows from upgradient piezometers through the lake to downgradient piezometers (Figure 9).

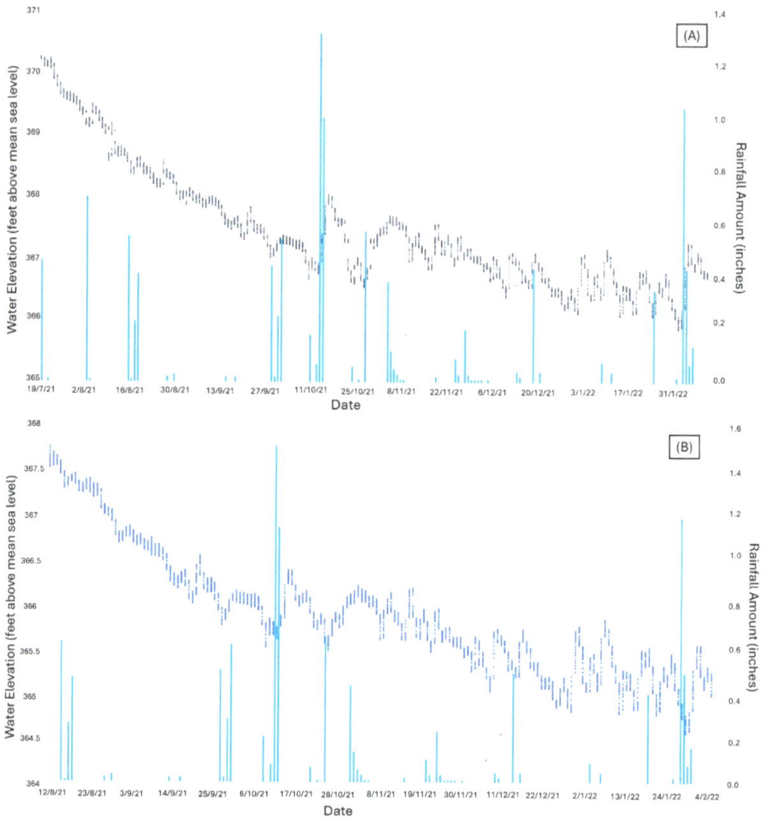


**Figure 7.** Data logger graphs from an upgradient piezometer at each site (West Site piezometer 2 (**A**) and East Site piezometer 5R (**B**)) compared to rainfall from the tipping bucket rain gauge at the East Site. On the map included in Figure 1, the piezometers are WS2 and ES5R, respectively. Overall, both sites show similar trends in groundwater response to rainfall events. Dark blue dots represent hourly groundwater levels plotted for each day in feet (daily water level fluctuations are visible for each day). Light blue bars represent total rainfall for each day in inches.

The notable exception to this is $HCO_3$, which is lowest in the surface water samples from both GPLs due to the degassing of carbon dioxide as the water leaves groundwater confinement. Calcium follows $HCO_3$ but with less dramatic differences between groundwater and surface water; this is also due to the changes in pH from groundwater to surface water. With the increase in pH, dissolved calcite in the groundwater is reprecipitating in the lake reducing the overall amounts of $HCO_3$ and Ca in the water in the lakes. More conservative elements [29,30], sodium and potassium, stay consistent as water moves from groundwater to the lake and back into groundwater; this indicates the connectivity between ground and surface waters in this system.

Nutrient sampling results across both lake systems also follow a distinctive pattern with some variation due to piezometer depth and the occurrence of seasonal algal blooms. Figures 10 and 11 show chemistry sampling results from the piezometers and lakes for the entire sampling period.

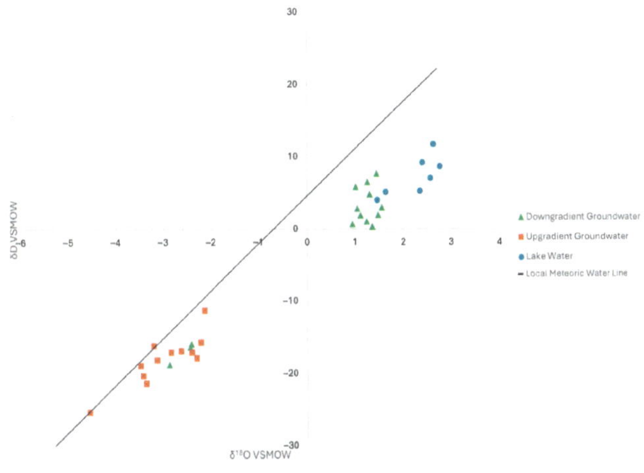


**Figure 8.** $\delta^{18}O$ and $\delta D$ samples from lakes and piezometers plotted against the local meteoric water line established in Waco from 1961 to 1980 [27]. All data collected are plotted; most sites were sampled multiple times over the course of observation in the study.

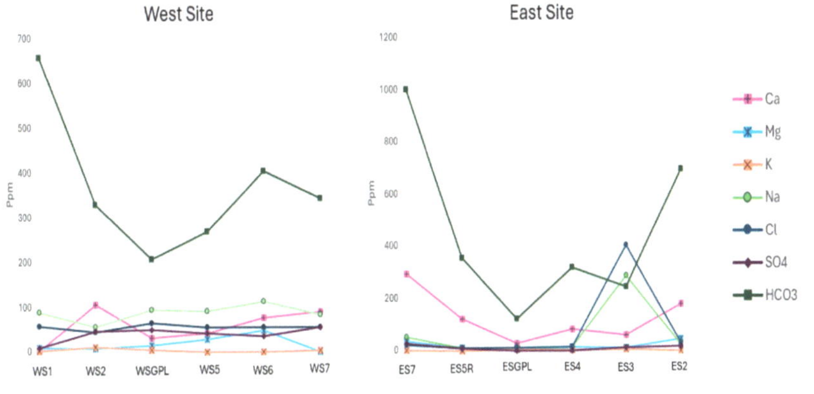


**Figure 9.** Cation and anion measurements for both gravel pit lake sites. Elevated sodium and chlorine levels at ES3 are due to the presence of a salt lick near the piezometer head to provide nutrients for cows.

The observed pH values show degassing effects from the water entering the surface driving the carbonate equilibria forward as $CO_2$. The exsolvation of $CO_2$ pushes the pH upwards by 1–2 units depending on the time of year (the warmer seasons appear to have a more dramatic degassing effect perhaps driven by temperature and solar irradiance). Specific conductance is consistent across the spatial dimension and the temporal dimension. Minor variation occurs between sample times and there is a relatively similar clustering of all samples in each area with the exception of the West Site's downgradient samples which are consistent per well but differ when viewed on the whole. This is likely due to differences between the aquifer flow paths accessed by each piezometer downgradient of the West Site lake. Temperature follows the expected pattern across the sampling period. Groundwater temperatures are consistent throughout the year with minor decreases following the winter months and increases when the air temperature rises in the summer; surface water temperatures have much greater spread (a spread of +/−10 C for surface water and +/−5 C for groundwater). Dissolved oxygen patterns are also as expected with the surface water showing a much higher average DO concentration than the surrounding groundwater. Anomalously high concentrations of DO were found in multiple piezometers over the course of the study, specifically, WS3 and ES5. These piezometers were both likely cut off from the surrounding groundwater supply due to a kink and subsequent break in the casing of the well in the case of ES5 and a too shallow installation in the case of WS3. In both cases, anomalous samples were collected during the warm and dry months when the water table likely dropped below the bottom of WS3 and below the kink in ES5. These values were included and are shown in Figure 10 in a black box.

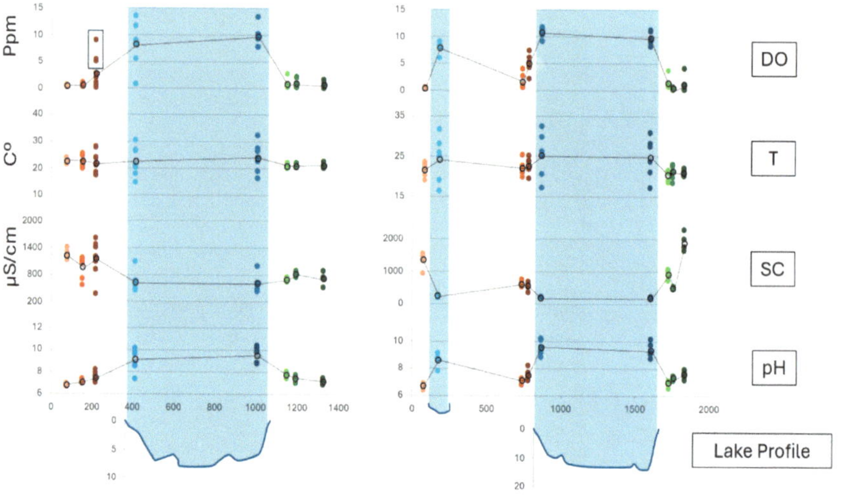


**Figure 10.** Water chemistry parameters for the West Site (**left**) and East Site (**right**) showing the evolution of the water chemistry from upgradient to downgradient along the flow path of the aquifer (left to right for both graphs) with the X-axis for all graphs representing feet along the transect line shown in Figure 1. Blue boxes show the samples taken from open water. Lake depth profiles from Figure 1 are reproduced to show the location of surface water samples in the lake. Average values for each metric in each piezometer are shown by black circles and connected by black lines to illustrate the trends from upgradient (shades of orange) to open water (shades of blue) to downgradient (shades of green). Color shades darken with the progression from more upgradient to more downgradient within each grouping.

For the nutrients analyzed (dissolved phosphorus, ammonia, and nitrate), if a sample was found to be below the MDL for a given analyte, the value for that sample was reported as half the MDL for graphical and data reporting purposes. Values were considered outliers for analysis purposes if they were higher than three times the average value for all other samples. In this case, the outliers were removed from the dataset and not included. This occurred once over the course of the study and an outlier value of 6710 µg/L of dissolved nitrate was thrown out. The piezometer WS4 at the West Site showed the outlier value in the October 2021 sample run. It is possible that this piezometer was uncapped and disturbed, leading to the anomalous value. Nutrient levels decline, on average, when water enters the surface waters from groundwater. The magnitude of decline depends on surface conditions in the lake (Figure 11).

Large algal blooms occurred in both lakes over the summer of 2021 which died off in the winter and regrew partially in the spring of 2022. These blooms are an influence on nutrient consumption and production in the lakes as has been shown in other lakes [31,32]. Many of the samples with the lowest observed nutrient levels are during these time periods. Clustered samples from downgradient, lakes, and upgradient piezometers suggest the remediation of nutrient contamination like that shown in similar work [7,11]. However, it must be acknowledged that the systems examined in those projects were quite different from the lakes described in this paper. Those lakes were deeper, more distinctly connected to the groundwater system, older, and larger. It is likely that lakes in this more shallow, alluvial system function differently depending on pre-existing conditions and are more responsive to changes in general conditions (rainfall) and wildlife disturbance. However, the prevalence of lakes like the ones examined in this project within the alluvial aquifers of Texas (and the continued expansion in their numbers), and the lack of regulation regarding their construction and use, means that their effects could be compounding quickly.

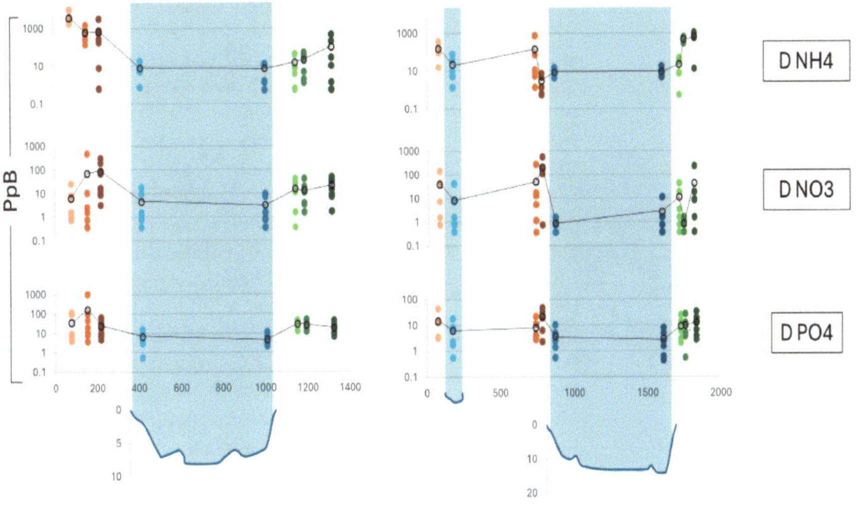


**Figure 11.** Dissolved ammonia, nitrate, and phosphate plotted along a transect for the West Site (**left**) and East Site (**right**). Surface water samples are shown in blue boxes. Upgradient samples are shown in shades of orange, surface water samples in shades of blue, and downgradient samples in shades of green. Color shades darken with the progression from more upgradient to more downgradient within each grouping. Th *x*-axis represents feet along the transect shown in Figure 1. All values are shown in µg/L.

According to the chemical analyses, the two GPLs studied appeared to be flow-through systems throughout the studied period. However, both lakes were especially close to the Brazos River. In this area, groundwater discharges into the river indicating a steeper groundwater gradient when compared to other sections of the aquifer that are farther from the river. Lakes farther from the river would likely have different dynamics and would require further investigation as to their effects on the aquifer. It is possible that they may behave more as evaporative sinks in late summer or other extended dry periods (which will increase in potency and length moving into the next 25 years of Texas climate [17]). Less flow through the lakes could also affect their potential to remediate nutrients or reverse their effects through "evapo-concentration", causing them to behave as sources of increased nutrients rather than as remediation sites.

To the authors' knowledge, the evaporative groundwater loss from GPLs or similar groundwater-fed anthropogenic water features is not considered in any groundwater availability models [25] or official water budgets for the BRAA. The exploratory modeled flow paths from the TopoDrive model using specific local lithology from the piezometers at each site (Figure 5) suggest again that water flows into each lake from the upgradient side of the lake and exits it from the downgradient side of the lake and that the deeper parts of the lake see a higher volume of groundwater inflow than do the banks [26]. At both sites, the downgradient flow paths saw, on average, an order of magnitude increase in the speed that a particle of water would travel in a day (0.87 ft/day upgradient vs. 12.6 ft/day downgradient at the West Site and 2.9 ft/day upgradient vs. 9.2 ft/day downgradient at the East Site) [26]. This is due to the steepening flow gradient in the direction of the river. This also means that although water is flowing into and out of the lakes at the edges, the water in the body of the lake is not experiencing significant movement. The lack of movement in the body of the lakes makes them more susceptible to surface conditions and evaporation effects.

The average total yearly evapotranspiration of the field including the East Site lake (for the period of record, 2019–present) is just over 38 inches per year [33,34]; this number is mediated by the grassy shoreline being included in the estimate. Although an open water calculation was used, the actual pure evaporation for just the lake surface may be a larger value. The average rainfall for the greater Waco area is around 32 inches per year (for the same period of record) [35]. The difference of six inches per year over the 10-acre lake is equivalent to an average loss of 5 acre-feet per year or 1.63 million gallons per year. This means that this lake is the equivalent of a well pumping at three gallons per minute 24 h per day, 365 days per year. In other terms, the evaporative loss from this lake is equivalent to the daily household water use of 20 single family homes for a year (an average use of 220 gallons per day) [36]. It is conceivable that lakes of similar sizes in the Waco area would produce similar evaporative patterns and result in an overall and continual decrease in water levels within the aquifer particularly during drier years. As of 2014, 4100 acres of the BRAA in McLennan County had been subject to gravel pit mining with no remediation [37]. Using the six-inch difference in rainfall and lake evaporation in the previous paragraph, the evaporative loss of all the 2014 open GPLs would be 2050 acre-feet of water per year, or the equivalent water use of 8300 single family households in Texas. Although some of these lakes have likely been partially backfilled or altered to enhance them aesthetically and some lakes have likely been remediated since 2014, the overall number of lakes has increased [1]. Increasing the lake surface area will yield increasing ET losses to the aquifer water supply and therefore could also affect the baseflow of the Brazos River itself. The presence or absence of such effects should be examined and quantified before any future policy adjustments are considered to account for GPLs or further water well drilling. While parallels can be drawn between the evaporative loss of water from GPLs and the losses from the aquifer due to irrigation, irrigation provides some return flow to the aquifer and is more seasonal. Evaporative loss from the surface of the GPLs might slow in the winter but, due to the Texas climate, it is continuous throughout the year and can be a considered a complete loss of water from the system as the lakes are not large enough to influence cloud formation or general humidity in the area. The estimated water use for agricultural irrigation in McLennan County is around 3500 acre-feet per year as of 2013 [38]. Water used for irrigation is not totally lost due to evapotranspiration, as there is return flow to the aquifer; however, this can be difficult to quantify, and a large fraction of water pumped for irrigation will be lost to evapotranspiration. The total evaporative loss of water from the lakes is 58% of all the water used for irrigation in the county, making it a significant and underreported source of water loss in the BRAA.

## 4. Conclusions

Further monitoring and expansion of our understanding of how these anthropogenic ecosystems affect groundwater quality is necessary; however, it is clear that they can function as flow-through systems under the right conditions and serve as potential areas of remediation for nutrients commonly used in the BRAA, improving overall water quality at the expense of serving as sites of high evaporative loss, essentially being wells with surface areas of multiple thousands of feet squared. Using satellite-based evapotranspiration data led to suspected losses of hundreds of millions of gallons of water over the course of the observation period in this study [33,34]. Trading water quality for water quantity in this aquifer may be detrimental for the long-term health of the water supply from the deeper flow paths, and the continued construction of GPLs may exponentially increase the evaporative losses to the point where any improvement in water quality is rendered moot.

Overall, this study does not manage to address the aggregate effects of all gravel pit lakes on the BRAA due to disparities between individual lakes, including proximity to the river and thus the groundwater gradient, the land use surrounding the lake, and the legacy and history of the lake and its age. Future work in similar GPL–groundwater systems should attempt to account for these differences by increasing the number of studied lakes, and studying lakes with more geographic variability, with a specific emphasis on lakes further from the river and in thicker aquifer sections with less compartmentalization.

The effects of GPLs will also continue to shift as the move towards urbanization continues; this makes it difficult to predict how the lakes will affect the aquifer moving forward, especially considering the lack of characterization of the lakes previously. This study highlights that the lakes are experiencing groundwater and surface water interaction as evidenced by shifts in water chemistry as water flows from groundwater to the lakes and back to the aquifer. The study attempts to examine the shape that such interactions take, which is important when considering the future impacts that the continuous construction of more such lakes will have.

**Author Contributions:** Individual contributions to this work are as follows: conceptualization, W.A.B. and J.C.Y.J.; methodology, W.A.B.; software, C.R.D. and W.A.B.; validation, W.A.B., C.R.D., and J.C.Y.J.; formal analysis, W.A.B. and C.R.D.; investigation, W.A.B. and C.R.D.; data curation, W.A.B. and C.R.D.; writing—original draft preparation, W.A.B.; writing—review and editing, W.A.B., J.C.Y.J., and C.R.D.; visualization, W.A.B. and C.R.D.; supervision, W.A.B. and J.C.Y.J.; project administration, J.C.Y.J. and W.A.B. All authors have read and agreed to the published version of the manuscript.

**Funding:** This project received no external funding.

**Data Availability Statement:** To preserve the privacy of the private landowners who allowed us to conduct this study on their properties, all data produced in this study are available by request from the corresponding author.

**Acknowledgments:** The authors wish to thank the Baylor Geosciences department for the use of departmental vehicles as well as the Baylor Center for Reservoir and Aquatic Systems Research and the Baylor Isotopes Laboratory for sample analyses. The authors also wish to thank the landowners for their support and access to their properties.

**Conflicts of Interest:** The authors declare no conflicts of interest.

## References


1. Kyle, J.R.; Elliott, B.A. Past, Present, and Future of Texas Industrial Minerals. *Min. Metall. Explor.* **2019**, *36*, 475–486. [CrossRef]
2. Paup, B.T.; Jackson, K. *Water for Texas: 2022 State Water Plan*; Texas Water Development Board: Austin, TX, USA, 2022; p. 11.
3. George, P.G.; Mace, R.E.; Petrossian, R. *Aquifers of Texas*; Texas Water Development Board: Austin, TX, USA, 2011; Volume 380.
4. Shah, S.D.; Houston, N.A.; Braun, C.L. *Hydrogeologic characterization of the Brazos River Alluvium Aquifer, Bosque County to Fort Bend County, Texas*; Texas Water Science Center: Austin, TX, USA, 2007.
5. Søndergaard, M.; Lauridsen, T.L.; Johansson, L.S.; Jeppesen, E. Gravel pit lakes in Denmark: Chemical and biological state. *Sci. Total Environ.* **2018**, *612*, 9–17. [CrossRef] [PubMed]
6. Kattner, E.; Schwarz, D.; Maier, G. Eutrophication of gravel pit lakes which are situated in close vicinity to the river Donau: Water and nutrient transport. *Limnologica* **2000**, *30*, 261–270. [CrossRef]
7. Muellegger, C.; Weilhartner, A.; Battin, T.J.; Hofmann, T. Positive and negative impacts of five Austrian gravel pit lakes on groundwater quality. *Sci. Total Environ.* **2013**, *443*, 14–23. [CrossRef] [PubMed]
8. Smerdon, B.D.; Mendoza, C.A.; Devito, K.J. The impact of gravel extraction on groundwater dependent wetlands and lakes in the Boreal Plains, Canada. *Environ. Earth Sci.* **2012**, *67*, 1249–1259. [CrossRef]
9. Weilhartner, A.; Muellegger, C.; Kainz, M.; Mathieu, F.; Hofmann, T.; Battin, T.J. Gravel pit lake ecosystems reduce nitrate and phosphate concentrations in the outflowing groundwater. *Sci. Total Environ.* **2012**, *420*, 222–228. [CrossRef] [PubMed]
10. Mollema, P.; Antonellini, M.; Hubeek, A.; Van Diepenbeek, P. The Effect of Artificial Recharge on Hydrochemistry: A Comparison of Two Fluvial Gravel Pit Lakes with Different Post-Excavation Uses in The Netherlands. *Water* **2016**, *8*, 409. [CrossRef]
11. Mollema, P.N.; Antonellini, M. Water and (bio)chemical cycling in gravel pit lakes: A review and outlook. *Earth-Sci. Rev.* **2016**, *159*, 247–270. [CrossRef]
12. Gottschall, N.; Boutin, C.; Crolla, A.; Kinsley, C.; Champagne, P. The role of plants in the removal of nutrients at a constructed wetland treating agricultural (dairy) wastewater, Ontario, Canada. *Ecol. Eng.* **2007**, *29*, 154–163. [CrossRef]
13. Cronin, J.G.; Follett, C.R.; Shafer, G.H.; Rettman, P.L. *Reconnaissance Investigation of the Groundwater Resources of the Brazos River Basin, Texas*; Texas Water Development Board: Austin, TX, USA, 1973.
14. Epps, L.W. *A Geologic History of the Brazos River*; Baylor University: Waco, TX, USA, 1973.
15. McVea, J. Unstressed Groundwater Flow in the Brazos River Alluvium Aquifer with Implications for Temporal Ranges in Groundwater to Surface Water Interactions. Master's Thesis, Baylor University, Waco, TX, USA, 2023. Unpublished.

16. Jarvis, J. Compartmentalization in the Northern Segment of the Brazos River Alluvium Aquifer. Master's Thesis, Baylor University, Waco, TX, USA, 2019. Unpublished.
17. Banner, J.L.; Jackson, C.S.; Yang, Z.-L.; Hayhoe, K.; Woodhouse, C.; Gulden, L.; Jacobs, K.; North, G.; Leung, R.; Washington, W.; et al. Climate Change Impacts on Texas Water: A White Paper Assessment of the Past, Present and Future and Recommendations for Action. *Tex. Water J.* **2010**, *1*, 1–19. [CrossRef]
18. Sanders, L.L. *A Manual of Field Hydrogeology*; Prentice Hall: Upper Saddle River, NJ, USA, 1998.
19. Weight, W.D. *Hydrogeology Field Manual*, 2nd ed.; McGraw-Hill: New York, NY, USA, 2008.
20. Lee, D.R.; Cherry, J.A. A Field Exercise on Groundwater Flow Using Seepage Meters and Mini-piezometers. *J. Geol. Educ.* **1979**, *27*, 6–10. [CrossRef]
21. Martinez, C.J. *Mini-Piezometers for Measuring Groundwater to Surface Water Exchange*; EDIS: Gainesville, FL, USA, 2010.
22. Bouwer, H.; Rice, R.C. A slug test for determining hydraulic conductivity of unconfined aquifers with completely or partially penetrating wells. *Water Resour. Res.* **1976**, *12*, 423–428. [CrossRef]
23. Hsieh, P. *TopoDrive and ParticleFlow—Two Computer Models for Simulation and Visualization of Ground-Water Flow and Transport of Fluid Particles in Two Dimensions*; US Geological Survey: Reston, VI, USA, 2001.
24. Freeze, R.A.; Cherry, J.A. *Groundwater, Illustrated ed.*; Prentice-Hall: Upper Saddle River, NJ, USA, 1979.
25. Ewing, J.E.; Jigmond, M. *Final Numerical Model Report for the Brazos River Alluvium Aquifer Groundwater Availability Model*; Texas Water Development Board: Austin, TX, USA, 2016.
26. Dawson, C. Groundwater/Surface Water Interactions: Gravel Pit Lakes in the Brazos River Alluvium Aquifer. Master's Thesis, Baylor University, Waco, TX, USA, 2022. Unpublished.
27. Bomar, G.W. *Weather in Texas: The Essential Handbook*, 3rd ed.; University of Texas Press: Austin, TX, USA, 2017.
28. IAEA. *Global Network of Isotopes in Precipitation*; IAEA: Vienna, Austria, 2001.
29. Meyer, M.M. Fate and Transport of Deicing Materials in an Unconfined Roadside Aquifer. Ph.D. Thesis, University of Massachusetts Amherst, Amherst, MA, USA, 1999.
30. Stinson, C.L.; Schwartz, B.F.; Gerard, B.R.; Schwinning, S.; Ramirez, P.; Timmins, G. Preliminary results from a multi-tracer epikarst recharge experiment; McCarty Cave, Texas, USA. *Abstr. Programs Geol. Soc. Am.* **2012**, *44*, 435.
31. Zou, R.; Wu, Z.; Zhao, L.; Elser, J.J.; Yu, Y.; Chen, Y.; Liu, Y. Seasonal algal blooms support sediment release of phosphorus via positive feedback in a eutrophic lake: Insights from a nutrient flux tracking modeling. *Ecol. Model.* **2020**, *416*, 108881. [CrossRef]
32. Qin, B.; Zhu, G. The nutrient forms, cycling and exchange flux in the sediment and overlying water system in lakes from the middle and lower reaches of Yangtze River. *Sci. China Ser. D* **2006**, *49*, 1–13. [CrossRef]
33. Melton, F.S.; Huntington, J.; Grimm, R.; Herring, J.; Hall, M.; Rollison, D.; Erickson, T.; Allen, R.; Anderson, M.; Fisher, J.B.; et al. OpenET: Filling a Critical Data Gap in Water Management for the Western United States. *JAWRA J. Am. Water Resour. Assoc.* **2021**, *58*, 971–994. [CrossRef]
34. Volk, J.M.; Huntington, J.L.; Melton, F.S.; Allen, R.; Anderson, M.; Fisher, J.B.; Kilic, A.; Ruhoff, A.; Senay, G.B.; Minor, B.; et al. Assessing the accuracy of OpenET satellite-based evapotranspiration data to support water resource and land management applications. *Nat. Water* **2024**, *2*, 193–205. [CrossRef]
35. NOAA. *Waco—Monthly and Annual Precipitation*; NOAA: Silver Spring, MD, USA, 2024.
36. Lake, P.; Jackson, K.; Paup, B.; Walker, J. *Water Use of Texas Water Utilities: 2020 Biennial Report*; Texas Water Development Board: Austin, Texas, USA, 2021.
37. Ju, D. *Aquifer Framework Restoration (AFR): A Protocol to Reverse Mining Effects on an Alluvial Aquifer, Central Texas*; Baylor University: Waco, TX, USA, 2014.
38. Blair, A.W. *Estimate of Agricultural Irrigation Water Use in McLennan County, Texas during 2013 Using LANDSAT 8 NDVI Images*; Southern Trinity Groundwater Conservation District: Austin, TX, USA, 2014.

*hydrology*

MDPI

*Article*

# Assessing the Origin and Mapping the Extension of Salinity Around Shrimp Culture Ponds in Rio Grande Do Norte (Brazil)

**José A. Beltrão-Sabadía [1], Albert Casas-Ponsatí [2], Evanimek Bernardo Sabino da Silva [3], Alex Sendrós [2,4,*], Josefina C. Tapias [4,5] and Francisco Pinheiro Lima-Filho [3]**

1 Geology Department, Universidade Federal do Ceará, Fortaleza-Ceará 60020-181, Brazil; cynthia.duarte@ufc.br
2 Mineralogy, Petrology and Applied Geology, Universitat de Barcelona, 08028 Barcelona, Spain; albert.casas@ub.edu
3 Geology Department, Universidade Federal do Rio Grande do Norte, Lagoa Nova 59078-970, Brazil; evanimek.sabinodasilva@usach.cl (E.B.S.d.S.); pinheiro.lima@ufrn.br (F.P.L.-F.)
4 Water Research Institute, Universitat de Barcelona, 08001 Barcelona, Spain; jtapias@ub.edu
5 Department of Biology, Health and Environment, Universitat de Barcelona, 08028 Barcelona, Spain
* Correspondence: alex.sendros@ub.edu

**Abstract:** The increasing installation of shrimp farms in vulnerable coastal areas around the world generates an environmental impact and makes it urgent to develop methodologies and studies for assessing and scaling the potential risks and sustainability of these activities. One of the main hazards of these activities is that the prolonged inundation of excavated ponds for shrimp farming allows the percolation of saltwater in the surroundings, resulting in increasing groundwater salinity. Saltwater intrusion in coastal aquifers, accompanied by salinization of soils, causes a decrease in available freshwater resources, a decline in crop productivity and the deterioration of the natural ecosystem. The coastal aquifer of Rio Grande do Norte State (Brazil) where, for years, several shrimp farm factories have been operating, reported some issues related to aquifer and soil salinization. The present study aims to assess the origin of and delineate groundwater salinization in a sector of this coastal aquifer using a low-budget procedure. The integration of hydrogeological and hydrogeochemical characterization by drilling shallow piezometers, measuring the hydrostatic level and analyzing the major ion concentrations of the groundwater has made it possible to establish that the origin of groundwater pollution in the studied area is caused by saltwater percolation from shrimp farms. The joint use of both characterization techniques has been shown to have an efficient cost–benefit ratio and less-intrusive methodology, which can be applied in other areas with similar environmental concerns.

**Keywords:** aquaculture; coastal aquifer; groundwater salinization; hydrochemical analysis

**Citation:** Beltrão-Sabadía, J.A.; Casas-Ponsatí, A.; Sabino da Silva, E.B.; Sendrós, A.; Tapias, J.C.; Pinheiro Lima-Filho, F. Assessing the Origin and Mapping the Extension of Salinity Around Shrimp Culture Ponds in Rio Grande Do Norte (Brazil). *Hydrology* **2024**, *11*, 188. https://doi.org/10.3390/hydrology11110188

Academic Editors: Zongjun Gao and Jiutan Liu

Received: 17 October 2024
Revised: 30 October 2024
Accepted: 4 November 2024
Published: 6 November 2024



## 1. Introduction

Increasing water salinity in coastal areas is a significant worldwide concern. Among the different activities threatening the natural environment, shrimp farming stands out as one of the most menacing anthropogenic changes for the ecological sustainability of sensitive coastal areas [1]. Shrimp can be farmed in a wide range of salinities, from freshwater to hypersaline waters [2,3]. Freshwater cultivation leads to a decrease in drinking water reserves [4] and, at the same time, data are accumulating that shrimp cultivation in saline waters also indirectly negatively affects freshwater resources [5,6]. In coastal areas, operation practises mostly consist of taking water directly from the sea by pumping through excavated channels on land near the sea and then discharging into ponds where the prawn farming takes place. As ponds normally do not have adequate waterproofing, stored water and effluents generated during shrimp farming percolate in the subsurface and reach groundwater. Consequently, despite the positive economic outcomes achieved by the entrepreneurs, shrimp farming activity still is frequently associated with severe negative social and environmental impacts such as affecting mangroves and wetlands ecosystems as well as water salinization due to seawater intrusion or poorly-developed pond drainage systems [7–11].

Shrimp farming has undergone extraordinary expansion in the last 25 years, from less than 9000 million tons (MT) to more than 1 million MT (with a first sale value of US$ 6 billion), which is equivalent to about one-third of the total world shrimp supply. About 80% of farmed shrimp is produced in Asia (in particular, in Thailand, followed by China and Indonesia), but the other 20% is produced mainly in Latin America, where Brazil, Ecuador, and Mexico are the largest producers. However, although the prawn aquaculture sector has demonstrated production and financial success, its rapid growth has resulted in several technical, environmental, economic, and social issues [12]. In many locations, the land will be incapable of supporting any agricultural or natural activity when prawn production ends, which frequently happens in a matter of years. The problem of abandoned shrimp ponds is becoming increasingly relevant. For example, it is estimated that in 1994, the top four producing countries abandoned 15,000 hectares of ponds [13], there were 250,000 ha of abandoned ponds in Indonesia [14], and 90% of Sri Lanka ponds [15] and 70% of Thailand ponds [16] have been abandoned recently after a few years of operation [17].

Due to its vast coastline acreage, year-round warm sea temperatures, robust agro-industrial foundation, and appealing domestic market, Brazil has long been acknowledged as having the perfect environment for the growth of prawn farming [18,19]. In 2021, Brazilian shrimp production reached 120,000 MT, mainly concentrated in the coastal areas of the northeast, where the area committed to shrimp farming is approximately 18,500 ha [20].

The first experiments with shrimp cultivated in Brazil date back to the 1970s when the Government of Rio Grande do Norte created the Shrimp Project to study the viability of the cultivation of this crustacean in place of salt extraction, a traditional activity of the State. Salt extraction at the time faced serious price and market crises with consequent widespread unemployment in the salting areas. At the same time, the Brazilian Institute of Colonization and Agrarian Reform (INCRA) has been developing a policy of land reform throughout the national territory intending to ensure social rescue to new settlers. However, often, their opportunities are threatened, as is the case of the serious salinization process in the settlement of Baixa Quixaba (State of Rio Grande do Norte, Brazil). There has been accelerated environmental degradation (including the death of native trees and natural flora), groundwater salinization, and decreased fertility of the soils for the livelihood of the settled families after several ponds for shrimp cultivation were installed.

The assessment of the extent of salinization and pollution of soils and groundwater by seepage, discharge, and flooding associated with shrimp farming requires a clear understanding of the morphodynamic behaviour of these complex environments with extensive, methodical, careful, and expensive surveillance (e.g., [21,22]). The development of cost-effective methodologies that help to establish environmental and sustainable policies in such environmentally sensitive coastal areas could be a suitable objective. For this purpose, the main goal of this research is to evaluate the usefulness of the joint use of widespread hydrogeological and hydrochemical methods to delineate the area affected by the impact of groundwater salinization and unveil the origin of salinization. The hypothesis was to determine whether salinization is the result of a natural process of marine intrusion or the effect of infiltration and recycling of hypersaline waters because of shrimp farming practises carried out in onshore ponds.

## 2. Geographical and Climate Setting

This study area is on the north coast of Rio Grande do Norte State (Brazil), next to a settlement administered by the INCRA called Baixa da Quixaba (Figure 1).

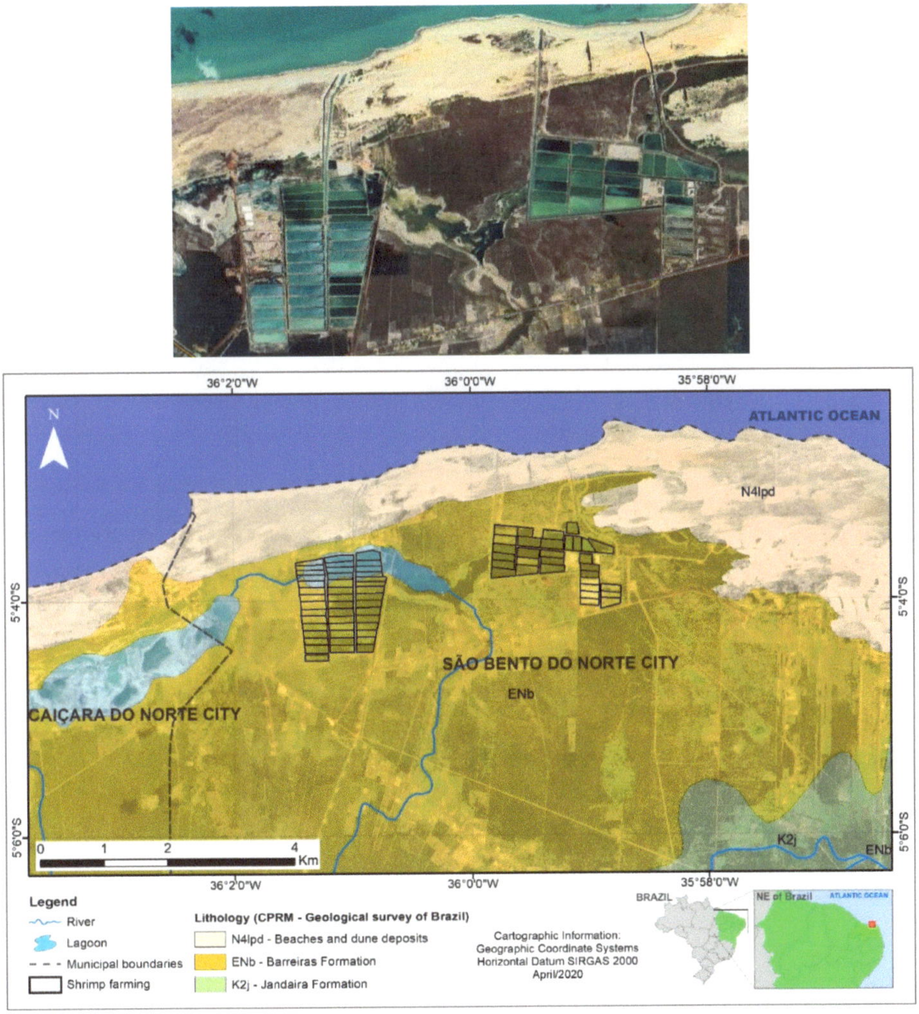


**Figure 1.** Aerial photography of the study area showing the location of the shrimp ponds and the respective drainage channels to the sea and the coastal environment with dunes and inter-dune lagoons (**upper part**) and a geological sketch of the study area (**lower part**).

The motivation of the study is due to the settlers' report that a large part of the natural vegetation had died, that the soils had become unproductive and that groundwater from the shallow wells used for the water supply was showing a progressive and persistent salinization process.

The salinization process became apparent after the installation close to the boundaries of the settlement of two shrimp culture companies that built several inland ponds. As the surface where the shrimp culture ponds are located at a level slightly higher than sea level, the water inlet is conducted by pumping through pipes from the sea, while the discharge is carried out by gravity directly to the sea through canals.

The climate of the region is tropical, with low variability between the minimum and maximum temperatures throughout the year that remain around 25 ± 1 °C and 30 ± 1 °C, with many hours of solar insolation resulting in high evaporation rates. The main factor causing rainfall is the Intertropical Convergence Zone (ITCZ) in the period from February to May. According to the Brazilian Instituto Nacional de Meteorologia (INMET), the annual rainfall is around 640 mm. The main climatic variation consists of the distribution of rainfall, with an arid period from August to December, while 70% of the annual precipitation is concentrated between February and May. Alongside the long periods of dryness, the region during the last decades has experienced a drastic reduction in annual precipitation, causing the expansion of desertification. The dominant wind direction is NE–SW, with an average annual speed of 6.2 m/s, which can reach speeds of 9 m/s in August. With these climatic characteristics of high temperature, sunstroke, and strong wind, the annual evaporation rate is greater than 1500 mm/year [23].

In addition to its impact on the high evaporation rate, winds are the main responsible for coastal dynamics. Its role is not limited to the generation of waves and, consequently, of sea currents. Subsequently, the sand deposited on the beaches is exposed to air, dried, and moved inland by the winds, resulting in its accumulation in the form of dune fields.

Geomorphologically, the study area is formed by a flattened coastal plain with a topography gently inclined towards the Atlantic Ocean, where beaches and fields of mobile dunes develop. The hydrographic network consists of small intermittent rivers that normally lead to interconnected lagoons, which do not flow directly into the sea. The presence of these lagoons is conditioned by the almost constant existence of dune fields along the coast, which act as natural reservoirs of rain freshwater [24]. Relatively low-relief and clastic, primarily siliciclastic, shorelines are characteristic of the littoral zone. Sand dunes from the Pleistocene to the Holocene, both vegetated and non-vegetated, typically enclose beaches. Tidal flats and lagoons are connected to the sea via tidal inlets, and the region is drained by minor streams and landwards [25].

## 3. Geological and Hydrogeological Setting

Geologically, the study area is located close to the shore of the Potiguar Basin. The Mesozoic sediments of the basin comprise siliciclastic rocks of the Açu Formation and carbonate rocks of the Jandaíra Formation [26,27], with a thin cover of Paleogene and Neogene age (Barreiras Group) and Quaternary sediments (Figure 1). Carbonate rocks of the Jandaira Formation are found in nearby locations deposited in tidal plains, shallow lagoons, shallow platforms, and deep-sea environments [28]. The Barreiras Formation is a lithostratigraphic unit from Oligocene (Chattian) to Miocene (Tortonian) age [29], distributed discontinuously along the east, northeast and north of the Brazilian coast (>4000 km), formed in a variety of environments, ranging from alluvial fans, fan delta, and meandering river to estuarine setting [30–33].

From the hydrogeological point of view, the Barreiras Aquifer is considered the most important groundwater resource in the coastal regions of NE Brazil. Its lower limit is marked by the top of the non-outcropping Mesozoic carbonate sequence, while its upper limit is occasionally represented by unconsolidated aeolian sediments formed by fine-to-medium sandy sediments, which play an important role in transmitting rainfall that infiltrates and recharges the aquifer.

The recent sandy deposits outcropping in this area are considered an essential aquifer (Dunas Aquifer). The Dunas Aquifer is composed of unconsolidated fine- and medium-grained quartz sand, and has a small thickness, a shallow water level, very low hydrogeological potential and a generally high hydraulic conductivity. The groundwater has electrical conductivity values below 350 µs/cm, reaching a minimum value of 200 µs/cm, and is contemplated as the unique hydrogeological unit with quality groundwater for direct human supply in the municipality of São Bento do Norte/RN [34]. The Barreiras Aquifer is directly connected with the Dunas Aquifer, lithologically is rather heterogeneous, ranging from slightly clayey sandstones and conglomerates to clays [35], has an unconfined to semi-confined behaviour, and is responsible for supplying most of the cities on the state's eastern coastline, including the capital Natal [24]. Accordingly, the Dunas/Barreiras aquifer can be considered as a single integrated system, although each unit has its specific hydrogeological characteristics.

## 4. Methods

The characterized area is located between the two factories that are involved in the cultivation of shrimp (located at the east and west boundaries of the studied site). Six shallow piezometers (PZ-09, PZ-16, PZ-17, PZ-18, PZ-21, and PZ-22) and the water of the channel that supplies ponds from seawater (C-10) were sampled. Moreover, eight additional piezometers were drilled and sampled to improve the hydrogeological characterization (PZ-10, PZ-11, PZ-12, PZ-13, PZ-14, PZ-15, PZ-19, and PZ-20). The depth of the new and previous piezometers ranges between 2 and 9 m. The drilling diameter was 8 inches and was quickly coated with a 4-inch PVC pipe. The filters were slotted into the PVC structure itself during drilling works varying in length from 1.0 to 2.5 metres. The annular space was filled with gravel and, in the shallowest 0.2 m, fine-grained material from the borehole drilling was used to isolate it. The location of all the piezometers used for the hydrogeological survey is shown in Figure 2.

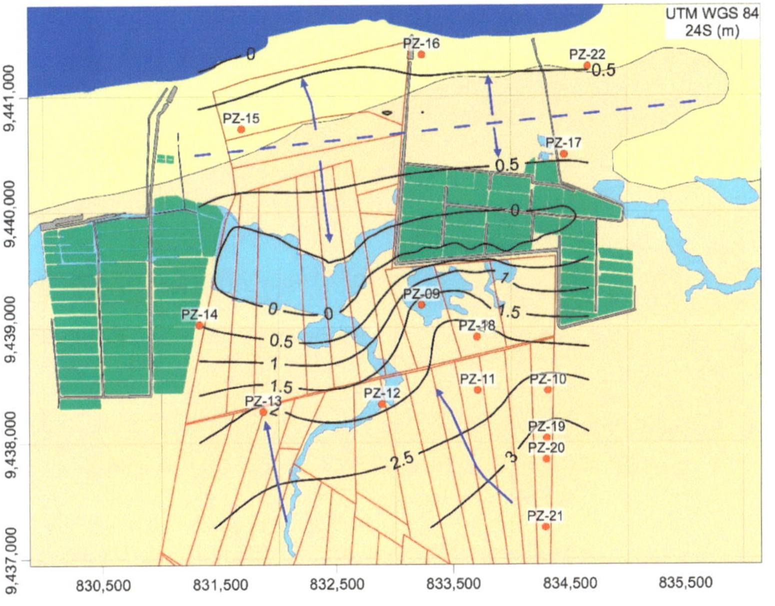


**Figure 2.** Piezometric map and main groundwater flow (blue arrows) in the studied area. The dashed blue line represents the inferred groundwater watershed in the coastal dune area. Hydraulic heads are presented in metres above mean sea level.

For the positioning of the piezometers and groundwater sample points, high-precision GPS (Geodesic GPS) was used to gather the coordinates and reference elevation as accurately as possible (Table 1). The water table's depth was measured with a standard analogic metre (resolution: ±1 cm).

Polythene bottles with a 1 L capacity were used to collect the samples. Before the collection, bottles were carefully cleaned in the lab using distilled water and diluted $HNO_3$ acid before being filled with samples. To prevent any potential contamination during bottling, each bottle was washed, stored and transported below 5 °C, and every other precautionary measure was taken [36]. Physico-chemical parameters including pH, temperature (T °C), and electrical conductivity (EC) were measured on-site.

Chemical analyses of major elements were conducted using standard methods [37] at the laboratories of Brazilian Agricultural Research Company EMBRAPA using ICP-AES for $Ca^{2+}$, $Mg^{2+}$, $Na^{+}$, $K^{+}$, and $Fe^{2+}$, ion liquid chromatography for $Cl^{-}$ and $SO_4^{2-}$, and automatic volumetric titration for $HCO_3^{-}$.

**Table 1.** Topographic level and water table depth of sampled piezometers.

| Well | Elevation (m.a.s.l.) | Water Table Depth (m) |
|---|---|---|
| PZ-09 | 4.5 | 2.70 |
| PZ-10 | 8.4 | 5.56 |
| PZ-11 | 6.8 | 4.70 |
| PZ-12 | 4.7 | 3.53 |
| PZ-13 | 8.4 | 6.20 |
| PZ-14 | 6.6 | 2.95 |
| PZ-15 | 5.0 | 2.85 |
| PZ-16 | 0.3 | 0.2 |
| PZ-17 | 5.2 | 4.59 |
| PZ-18 | 5.2 | 2.83 |
| PZ-19 | 10.1 | 6.93 |
| PZ-20 | 10.5 | 7.35 |
| PZ-21 | 11.5 | 8.13 |
| PZ-22 | 6.5 | 6.20 |

Careful quality controls were undertaken to obtain a reliable analytical dataset by checking blank samples, calibration standards, and duplicate sub-samples to assess that the analytical error was less than 5%. In addition, analytical precision for measurements of cations and anions, indicated by the ionic balance error (IBE), was computed based on ions expressed in meq/L for each sample as follows:

$$\text{IBE} = [(\Sigma c - \Sigma a)/(\Sigma c + \Sigma a)] \times 100$$

where $\Sigma c$ and $\Sigma a$ are the sums of cations and anions both expressed in (meq/L).

The value of IBE was observed to be below ±4% for all samples analyzed which, given the environmental circumstances at this study site, we see as an acknowledged value.

The software EASYQUIM v5.0 [38] was used to organize and plot the hydrogeochemical data. The software generates the main chemical diagrams such as Piper, to understand the geochemical evolution of groundwater [39] and Schöeller-Berkaloff, to assess the hydrochemical facies of the samples [40].

A third graphical diagram used to interpret the hydrogeochemical data is the Hydrochemical Facies Evolution diagram [41]. The Hydrochemical Facies Evolution (HFE) diagram assesses the salinization of coastal aquifers where the main modifying process is ion exchange. The HFE diagram is useful for highlighting two distinctive facies: the CaCl facies representative of saltwater intrusion and the $NaHCO_3$ facies representative of freshening.

The line of conservative mixing between seawater and freshwater is shown in the EFE diagram. The composition provided by the authors of [42] is used as a default for seawater, while the software chooses the freshwater composition once the set of waters has been stated. The fields of intrusion and freshening phases are determined by the mixing line's position, which is determined by these two fields.

To assess the distribution of hydraulic head charge or piezometric water levels, groundwater electrical conductivity, and major ion data in the studied site, we have used the software Surfer v11 (Golden Software, Golden, CO, USA). The kriging algorithm [43] was selected to interpolate the data, as it has been demonstrated to be well adapted to the interpolation of groundwater water table levels and water quality data [44,45].

## 5. Results and Discussion

### 5.1. Groundwater Flow

The assessment of groundwater piezometric levels is one of the most important parameters in the evaluation of groundwater flow of salt intrusion plumes, as they can indicate the direction of the predominant flow lines and their gradient.

Contour plots of the freshwater head, at constant depth and in a horizontal plane, are useful because the flow in isotropic aquifers is normal to these contour lines. The results of the piezometric map and the flow lines of the study area are shown in Figure 2. The distribution of piezometric levels shows two trends. A regional trend from south to north towards the sea is interrupted when reaching the area of the inter-dunes plain. The inter-dunes system is flooded in the rainy season, giving rise to lagoons, and is where the shrimp farms are installed. Near the shoreline, the coastal dune cordon, which can locally reach up to 10 m in height, represents a watershed that drains in both ways, towards the sea and inland towards the inter-dune lagoons. This distribution, where dunes act as a rainwater storage and drainage system, is an effective natural barrier to maintaining the fresh/saltwater balance.

Nevertheless, in the contour map of Figure 2, no variation in the piezometric lines justify clearly the groundwater flow from the ponds. This is because there is a strong gradient produced by the recharge areas located on both sides of the ponds, one in the coastal dunes and the other in the hills. The small gradient existing between the water table and the ponds should be only detected by a high-precision net of piezometers and or/combined with hydrochemical results.

### 5.2. Hydrogeochemical Data

The groundwater samples were collected on a single campaign carried out during the 2018 dry season. The hydrogeochemical characteristics of the samples have great variability, as applies to a coastal area where freshwater, saltwater, and mixing water converge. The results of the analyses are shown in Table 2.

**Table 2.** Hydrochemical results of analysis from groundwater samples from piezometers.

| Well | $HCO_3^-$ | $SO_4^{2-}$ | $Cl^-$ | $Ca^+$ | $Mg^{2+}$ | $Na^+$ | $Fe^{2+}$ | $K^+$ | EC |
|---|---|---|---|---|---|---|---|---|---|
| | meq/L | meq/L | meq/L | meq/L | meq/L | meq/L | meq/L | meq/L | µS/cm |
| PZ-09 | 3.8 | 10.6 | 281.8 | 34.4 | 33.8 | 136.7 | 0.55 | 1.8 | 17,090 |
| PZ-10 | 3.5 | 15.1 | 273.4 | 38.2 | 34.3 | 151.0 | 0.02 | 1.5 | 18,068 |
| PZ-11 | 4.5 | 15.0 | 245.8 | 30.5 | 31.3 | 141.9 | 0.01 | 1.3 | 16,677 |
| PZ-12 | 1.8 | 14.9 | 1,121.0 | 156.5 | 162.0 | 465.8 | 3.89 | 5.6 | 79,855 |
| PZ-13 | 3.7 | 3.1 | 135.7 | 22.0 | 17.8 | 62.1 | 0.11 | 2.2 | 10,421 |
| PZ-14 | 5.5 | 23.8 | 658.9 | 20.8 | 63.7 | 528.0 | 0.20 | 12.1 | 52,330 |
| PZ-15 | 1.7 | 0.0 | 0.8 | 0.5 | 0.4 | 0.4 | 0.02 | 0.0 | 222 |
| PZ-16 | 4.4 | 3.7 | 61.9 | 3.7 | 6.0 | 50.3 | 0.04 | 1.2 | 5927 |
| PZ-17 | 6.9 | 0.2 | 4.2 | 2.0 | 1.3 | 2.3 | 0.84 | 0.9 | 944 |
| PZ-18 | 2.8 | 3.0 | 66.3 | 7.1 | 8.3 | 40.4 | 0.01 | 1.3 | 5530 |
| PZ-19 | 7.6 | 9.4 | 149.5 | 21.2 | 21.5 | 83.9 | 0.03 | 1.5 | 11,825 |
| PZ-20 | 3.0 | 6.0 | 55.1 | 8.0 | 7.5 | 34.8 | 0.75 | 1.1 | 5232 |
| PZ-21 | 15.0 | 11.1 | 181.9 | 27.4 | 25.1 | 103.5 | 0.04 | 1.5 | 13,622 |
| PZ-22 | 2.2 | 0.0 | 0.6 | 0.8 | 0.4 | 0.4 | 0.00 | 0.0 | 261 |
| C-10 | 3.0 | 31.3 | 628.6 | 32.4 | 32.8 | 496.9 | 0.01 | 9.2 | 48,370 |

The Piper diagram gives significant information that allows us to discriminate between separate clusters (groups) of samples. The diagram shows that most of the groundwater samples analyzed fall in the field of the Calcium-Sodium Chloride type, except for samples from piezometers PZ-15 and PZ-17 that visibly stand out of this composition and can be classified as of the Magnesium Bicarbonate type (Figure 3). This method is very well suited for discriminating between hydrochemical facies because ion content is given as a percentage of milliequivalents, but Piper's diagram is not useful for describing water salinity concentrations, because it does not consider the water concentrations.

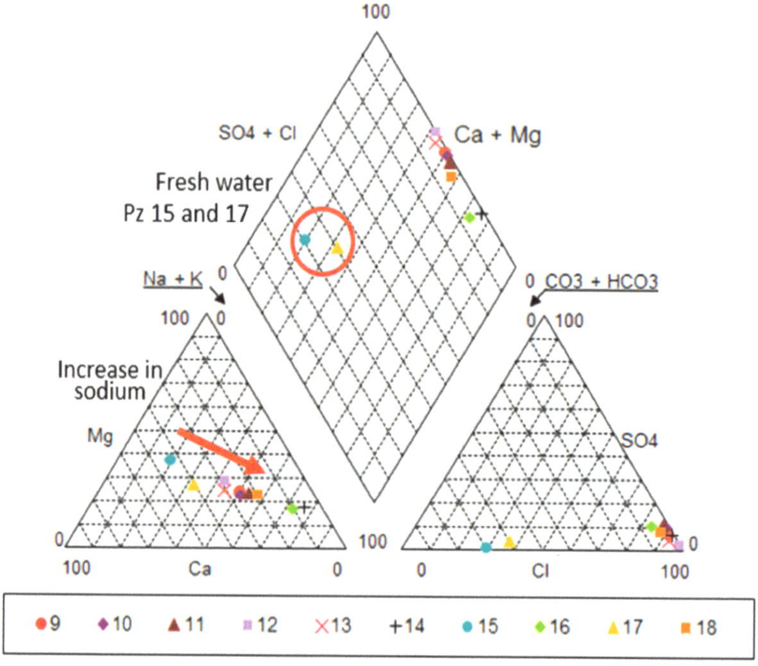


**Figure 3.** Piper trilinear diagram showing the different hydrogeochemical facies, where the freshwaters of the PZ-15 and PZ-17 piezometers located next to the coastline and further away from the aquaculture ponds are well isolated (within the red circle area).

In the Schoeller diagram, we have differentiated two different water facies. Water samples from coastal dunes (PZ-15 and PZ-17) could be classified as bicarbonate calcium facies and show values lower than 10 meq/L of the major ions plotted and values below 1 meq/L of $SO_4^{2-}$. The other water samples could be classified as chloride sodium facies and have concentrations of [$Cl^-$] and [$Na^+$] above 10 meq/L. There is a progressive increase in these values from the coastline (PZ-16) towards inland control points (PZ-12) (Figure 4).

The HFE diagram clusters the water of piezometers PZ-22, PZ-15 and PZ-17 (located close to the seashore) on the HCO3 zone indicating the effect of refreshing. Conversely, samples from piezometers located inland show values in the MixNa–Na range, indicating saltwater intrusion (Figure 5).

The main goal is to characterize the origin and distribution of a salinization process, and the first and key indicator of interest is the electrical conductivity of groundwater [46,47]. In the study area, as expected, the electrical conductivity values of groundwater have a high range of variation. The electrical conductivity of groundwater samples measured in the piezometers distributed in the aquifer show values ranging from a minimum of 222 µS/cm to a maximum of 79,855 µS/cm. The isolines resulting from the interpolation of sparse data points are shown in Figure 6. Unexpectedly, the lowest values of electrical conductivity and salinity are found on the PZ-15 piezometer located very close to the coastline over the dunes. In opposition, the highest values of electrical conductivity and salinity are found on the PZ-12 piezometer located in the central area of the study area between the two shrimp factories and almost 3 kilometres away from the sea. Therefore, the distribution of salinity does not appear to be directly related to a natural marine intrusion phenomenon, as freshwater piezometers are located near the coast, while the highest salinity is in an area between the shrimp culture ponds.

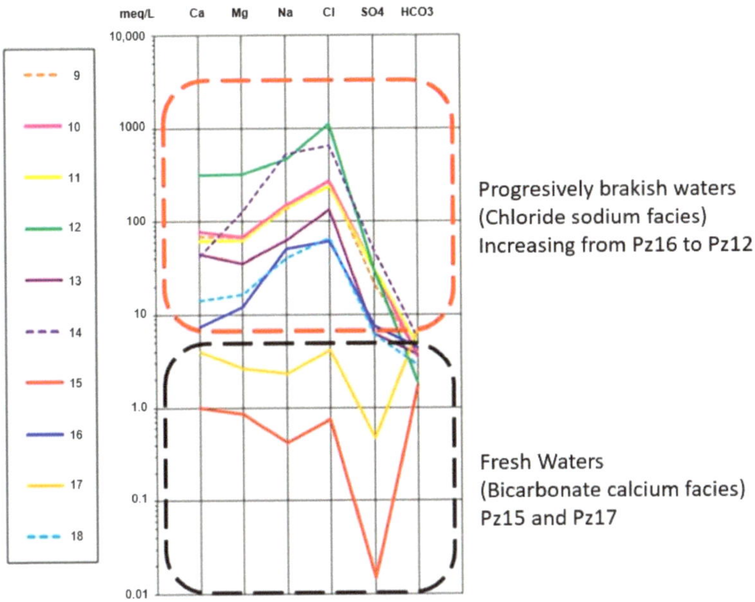


**Figure 4.** Schoeller diagram showing the different behaviour of the hydrochemical facies, including an increase in chloride and sodium of the salinized samples, and, conversely, the low sulphate content of the natural freshwater samples from the dunes aquifer.

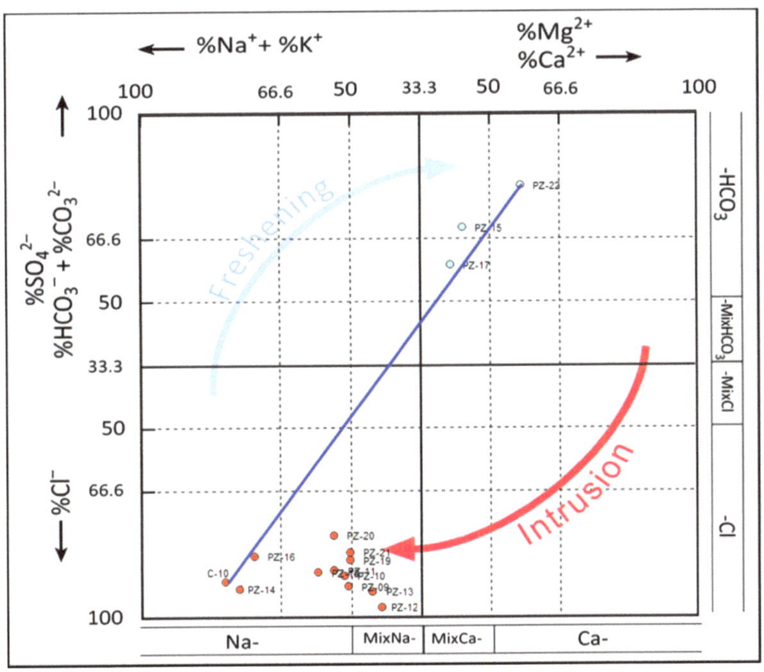


**Figure 5.** Representation of groundwater samples in the HFE diagram showing the effect of refreshing on piezometers PZ-22, PZ-15 and PZ-17 located close to the seashore. Conversely, samples from piezometers located inland show saltwater intrusion.

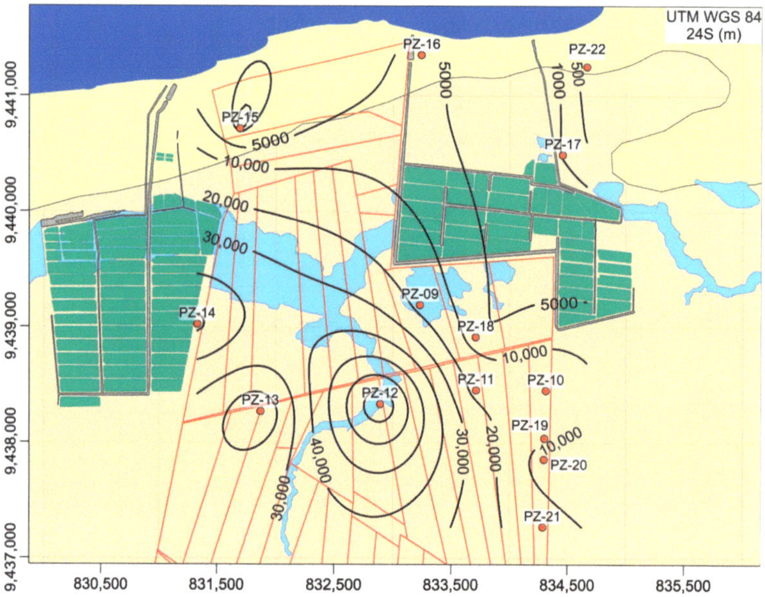


**Figure 6.** Map showing the distribution of the electrical conductivity of groundwater from water samples taken at the control piezometers. The area of the highest salinity is located between the ponds of the two aquiculture factories and decreases towards the coast. Isoline values are in μS/cm.

The hydrochemical concentration of groundwater also shows significant variability. Total dissolved solids (TDS) reasonably fluctuate similarly to electrical conductivity, from a minimum of 725 mg/L on the PZ-15 to a maximum of 62,078 mg/L on the PZ-14 piezometer, with a similar spatial distribution.

In Figure 4, Schoeller's diagram clearly shows the low sulphate content of groundwater samples from the dune aquifer. A better explanation of the origin and evolution of concentrations in sulphate can be seen in the form of a map depicting their distribution in the form of isolines. The distribution of the sulphate content of groundwater shows that the highest concentrations are in the central sector between the two shrimp factories and that the increase comes laterally from each of them (Figure 7).

The life and growth of shrimp may be adversely affected by large ion imbalances, even though natural marine water contains a salinity that is suitable for shrimp culture [48]. Farmers apply two fertilizers, potassium chloride and potassium magnesium sulphate, to add potassium ions to the pumped ocean water and salty water (Figure 7).

Shrimp production and survival were significantly increased by a single application of these fertilizers before or shortly after stocking ponds, with potassium accounting for the majority of the benefits [49].

Moreover, the joint interpretation of groundwater flow and hydrochemical results shows that the diffusion of the salty water of the ponds to the Dunas Aquifer freshwater is the contamination mechanism as the ponds are not properly waterproofed. The water level of the ponds is higher than the aquifer hydraulic head level and the coastal dunes act as a hydraulic barrier preventing seawater intrusion.

The study demonstrated that the salinization of groundwater around the shrimp factories of the Baixa de Queixaba sector in the state of Rio Grande do Norte State (Brazil) has its origin in the direct infiltration of seawater from the ponds and is not the result of a natural process of marine intrusion.

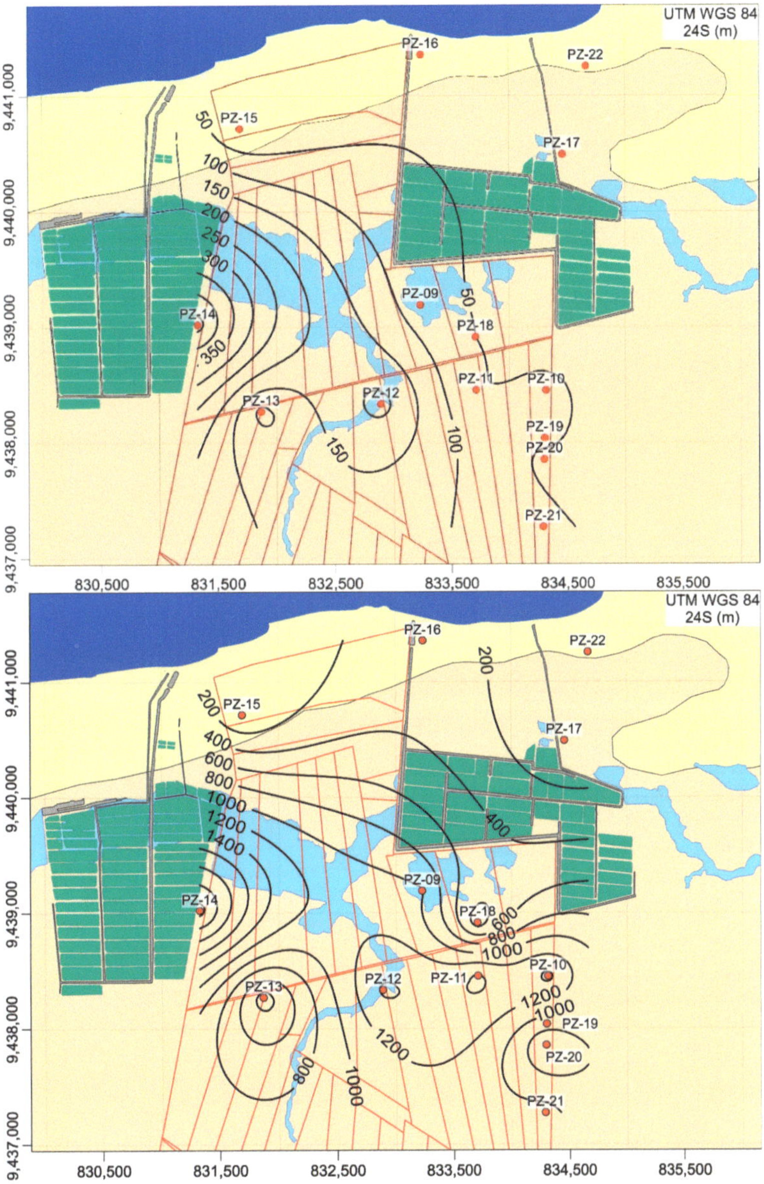


**Figure 7.** Map showing the distribution of potassium content of groundwater (**upper part**) and sulphate content (**lower part**) from water samples taken at the control piezometers. The highest concentration of potassium and sulphate is in piezometer PZ-14 located close to the western ponds and there is also an increasing trend from the south to the easter ponds. Isoline values are in mg/L.

## 6. Conclusions

The integration of hydrogeological and hydrogeochemical survey results has made it possible to establish that the origin of groundwater pollution in the studied area is caused by saltwater percolation from shrimp farms. Information derived from the piezometric map shows that rainwater recharge in the frontal dunes acts as a dynamic barrier to seawater intrusion, while saltwater flow must come from the ponds. Hydrochemical data also corroborate this interpretation, since the piezometers closest to the coast are those with the lowest salinization in terms of both electrical conductivity and concentration of chloride and sodium in groundwater. Another factor that reinforces this interpretation is that the plumes of sulphates and potassium escape from the edges of ponds and are added to natural seawater to increase the production and survival of shrimp larvae.

The joint use of both techniques has shown to be a cost-effective and low-intrusive methodology, which can be applied in other areas with similar environmental concerns; the results of this research have also highlighted the vulnerability of coastal aquifers affected by poorly planned and poorly managed human activities. In particular, the implementation of shrimp cultivation factories, which in principle could mean a socio-economic development for the native population, can cause a serious environmental impact, which is sometimes irreversible or comes with a very difficult and costly recovery process.

Sustainable shrimp farming requires the development of methodologies that allow the efficient design of production ponds and legal measures that require impact control through characterization procedures such as those carried out in this study.

**Author Contributions:** Conceptualization, A.C.-P. and F.P.L.-F.; methodology, A.C.-P. and J.A.B.-S. resources, F.P.L.-F. and J.A.B.-S.; data curation, A.C.-P., E.B.S.d.S. and A.S.; writing—original draft preparation, A.C.-P. and A.S.; writing—review and editing, A.C.-P., E.B.S.d.S., F.P.L.-F., A.S. and J.C.T. All authors have read and agreed to the published version of the manuscript.

**Funding:** This research received no external funding.

**Data Availability Statement:** The data presented in this study are available on request from the corresponding author.

**Acknowledgments:** We thank Nia Schamuells for designing and drawing the paper's graphical abstract.

**Conflicts of Interest:** The authors declare no conflicts of interest.

## References

1. Páez-Osuna, F. The environmental impact of shrimp aquaculture: A global perspective. *Environ. Pollut.* **2001**, *112*, 229–231. [CrossRef] [PubMed]
2. Sanz, J.; Suescun, J.; Molist, J.; Rubio, F.; Mujeriego, R.; Salgado, B. Reclaimed water for the Tarragona petrochemical park. *Water Supply* **2014**, *15*, 308–316. [CrossRef]
3. Yakovenko, V.; Shadrin, N.; Anufriieva, E. The Prawn Palaemon adspersus in the Hypersaline Lake Moynaki (Crimea): Ecology, Long-Term Changes, and Prospects for Aquaculture. *Water* **2022**, *14*, 2786. [CrossRef]
4. Anufriieva, E. V How can saline and hypersaline lakes contribute to aquaculture development? A review. *J. Oceanol. Limnol.* **2018**, *36*, 2002–2009. [CrossRef]
5. Hossain, M.S.; Uddin, M.J.; Fakhruddin, A.N.M. Impacts of shrimp farming on the coastal environment of Bangladesh and approach for management. *Rev. Environ. Sci. Bio/Technol.* **2013**, *12*, 313–332. [CrossRef]
6. HEIN, L. Toward Improved Environmental and Social Management of Indian Shrimp Farming*. *Environ. Manag.* **2002**, *29*, 349–359. [CrossRef]
7. Primavera, J.H. Overcoming the impacts of aquaculture on the coastal zone. *Ocean Coast. Manag.* **2006**, *49*, 531–545. [CrossRef]
8. Paul, B.G.; Vogl, C.R. Impacts of shrimp farming in Bangladesh: Challenges and alternatives. *Ocean Coast. Manag.* **2011**, *54*, 201–211. [CrossRef]
9. Hidayati, N.V.; Prudent, P.; Asia, L.; Vassalo, L.; Torre, F.; Widowati, I.; Sabdono, A.; Syakti, A.D.; Doumenq, P. Assessment of the ecological and human health risks from metals in shrimp aquaculture environments in Central Java, Indonesia. *Environ. Sci. Pollut. Res.* **2020**, *27*, 41668–41687. [CrossRef]
10. Ahmed, Z.; Ambinakudige, S. Does land use change, waterlogging, and salinity impact on sustainability of agriculture and food security? Evidence from southwestern coastal region of Bangladesh. *Environ. Monit. Assess.* **2022**, *195*, 74. [CrossRef]
11. Bijith, P.; Ramith, M.; Megha, T.; Shimod, K.P.; Pradeep, M.N. Shrimp farms as a threat to mangrove forests in Kannur district of Kerala, India. *Wetl. Ecol. Manag.* **2022**, *30*, 1281–1289. [CrossRef]
12. Neiland, A.E.; Soley, N.; Varley, J.B.; Whitmarsh, D.J. Shrimp aquaculture: Economic perspectives for policy development. *Mar. Policy* **2001**, *25*, 265–279. [CrossRef]
13. Lassen, T.J. Environmental extremes versus sustainable policies in aquaculture. *World Aquac.* **1997**, *28*, 49–51.
14. Gusmawati, N.; Soulard, B.; Selmaoui-Folcher, N.; Proisy, C.; Mustafa, A.; Le Gendre, R.; Laugier, T.; Lemonnier, H. Surveying shrimp aquaculture pond activity using multitemporal VHSR satellite images—Case study from the Perancak estuary, Bali, Indonesia. *Mar. Pollut. Bull.* **2018**, *131*, 49–60. [CrossRef]
15. Harkes, I.H.T.; Drengstig, A.; Kumara, M.P.; Jayasinghe, J.M.P.K.; Huxham, M. Shrimp aquaculture as a vehicle for Climate Compatible Development in Sri Lanka. The case of Puttalam Lagoon. *Mar. Policy* **2015**, *61*, 273–283. [CrossRef]

16. Elwin, A.; Jintana, V.; Feola, G. Characterizing shrimp-farm production intensity in Thailand: Beyond technical indices. *Ocean Coast. Manag.* **2020**, *185*, 105019. [CrossRef]
17. Dorber, M.; Verones, F.; Nakaoka, M.; Sudo, K. Can we locate shrimp aquaculture areas from space?—A case study for Thailand. *Remote Sens. Appl. Soc. Environ.* **2020**, *20*, 100416. [CrossRef]
18. Jablonski, S.; Filet, M. Coastal management in Brazil—A political riddle. *Ocean Coast. Manag.* **2008**, *51*, 536–543. [CrossRef]
19. Valenti, W.C.; Barros, H.P.; Moraes-Valenti, P.; Bueno, G.W.; Cavalli, R.O. Aquaculture in Brazil: Past, present and future. *Aquac. Reports* **2021**, *19*, 100611. [CrossRef]
20. Evans, J. Brazilian Shrimp Farmers Eye New Horizons 2023, 4. Available online: https://www.globalseafood.org/advocate/brazilian-shrimp-farmers-eye-new-horizons/ (accessed on 16 June 2024).
21. Callaghan, D.P.; Bouma, T.J.; Klaassen, P.; van der Wal, D.; Stive, M.J.F.; Herman, P.M.J. Hydrodynamic forcing on salt-marsh development: Distinguishing the relative importance of waves and tidal flows. *Estuar. Coast. Shelf Sci.* **2010**, *89*, 73–88. [CrossRef]
22. Roversi, F.; van Maanen, B.; Colonna Rosman, P.C.; Neves, C.F.; Scudelari, A.C. Numerical Modeling Evaluation of the Impacts of Shrimp Farming Operations on Long-term Coastal Lagoon Morphodynamics. *Estuaries Coasts* **2020**, *43*, 1853–1872. [CrossRef]
23. da Silva, V. de P.R. On climate variability in Northeast of Brazil. *J. Arid Environ.* **2004**, *58*, 575–596. [CrossRef]
24. Tabosa, W.F. Dinâmica Costeira da Região de São Bento do Norte e Caiçara do Norte—RN. MSc Thesis, Universidade Federal do Rio Grande do Norte, Natal, Brasil, 2000.
25. Bezerra, F.H.R.; Barreto, A.M.F.; Suguio, K. Holocene sea-level history on the Rio Grande do Norte State coast, Brazil. *Mar. Geol.* **2003**, *196*, 73–89. [CrossRef]
26. Neto, P.; Soares, U.; Silva, J.G.; Roesner, E.H.H.; Florencio, C.P.P.; Souza, C.A.V.A.V.; Neto, O.C. Bacia Potiguar. *Bol. Geociências Petrobras* **2007**, *15*, 357–369.
27. Melo, A.H.; Andrade, P.R.O.; Magalhães, A.J.C.; Fragoso, D.G.C.; Lima-Filho, F.P. Stratigraphic evolution from the early Albian to late Campanian of the Potiguar Basin, Northeast Brazil: An approach in seismic scale. *Basin Res.* **2020**, *32*, 1054–1080. [CrossRef]
28. Santos Filho, M.A.B.; Piovesan, E.K.; Fauth, G.; Srivastava, N.K. Paleoenvironmental interpretation through the analysis of ostracodes and carbonate microfacies: Study of the Jandaíra Formation, Upper Cretaceous, Potiguar Basin. *Brazilian J. Geol.* **2015**, *45*, 23–34. [CrossRef]
29. Nogueira, A.C.R.A.A.E.; Amorim, K.B.; Góes, A.M.; Truckenbrodt, W.; Petri, S.; Nogueira, A.C.R.A.A.E.; Bandeira, J.; Soares, J.L.; Baía, L.B.; Imbiriba Júnior, M.; et al. Upper Oligocene-Miocene deposits of Eastern Amazonia: Implications for the collapse of Neogene carbonate platforms along the coast of northern Brazil. *Palaeogeogr. Palaeoclimatol. Palaeoecol.* **2021**, *563*, 110178. [CrossRef]
30. Rossetti, D.F. Late Cenozoic sedimentary evolution in northeastern Pará, Brazil, within the context of sea level changes. *J. South Am. Earth Sci.* **2001**, *14*, 77–89. [CrossRef]
31. De Araújo, V.D.; Reyes-Peres, Y.A.; Lima, R.d.O.; Pelosi, A.P.d.M.R.; Menezes, L.; Córdoba, V.C.; Lima-Filho, F.P. Fácies e sistema deposicional da formação barreiras na região da Barreira do Inferno, Litoral Oriental do Rio Grande do Norte. *Geol. USP. Série Científica* **2006**, *6*, 43–49. [CrossRef]
32. Rossetti, D.F.; Bezerra, F.H.R.; Dominguez, J.M.L. Late Oligocene–Miocene transgressions along the equatorial and eastern margins of Brazil. *Earth-Sci. Rev.* **2013**, *123*, 87–112. [CrossRef]
33. Rodríguez-Escales, P.; Canelles, A.; Sanchez-Vila, X.; Folch, A.; Kurtzman, D.; Rossetto, R.; Fernández-Escalante, E.; Lobo-Ferreira, J.P.; Sapiano, M.; San-Sebastián, J.; et al. A risk assessment methodology to evaluate the risk failure of managed aquifer recharge in the Mediterranean Basin. *Hydrol. Earth Syst. Sci.* **2018**, *22*, 3213–3227. [CrossRef]
34. Vasconcelos, S.M.S.; Teixeira, Z.A.; Alvez, J.N. Caracterização do Aquífero Jandaíra, porção situada no estado do Ceará, Brasil. *Rev. Geol.* **2010**, *23*, 50–60.
35. Lucena, L.R.F.; Cabral, J.B.P.; Becegato, V.A. Aspectos hidroquímicos do aqüífero Barreiras na área da Bacia do rio Pirangi-RN/Brasil e adequações para uso humano e irrigação. *Geoambiente On-Line* **2004**, *1*, 1–14. [CrossRef]
36. Mondal, N.C.; Singh, V.P.; Singh, V.S.; Saxena, V.K. Determining the interaction between groundwater and saline water through groundwater major ions chemistry. *J. Hydrol.* **2010**, *388*, 100–111. [CrossRef]
37. Clesceri, L.S.; Greenberg, A.E.; Eaton, A.D. *Standard Methods for the Examination of Water and Wastewater*, 20th ed.; APHA American Public Health Association: Washington, DC, USA, 1998.
38. Vazquéz-Suñe, E.; Serrano-Juan, A. *EasyQuim*; Institute of Environmental Assessment and Water Research: Barcelona, Spain, 2012.
39. Piper, A.M. A graphic procedure in the geochemical interpretation of water-analyses. *Eos, Trans. Am. Geophys. Union* **1944**, *25*, 914–928. [CrossRef]
40. Schoeller, H. Qualitative Evaluation of Ground Water Resources. In *Water Resource Series No. 33*; UNESCO: Paris, France, 1967; pp. 44–52.
41. Giménez-Forcada, E. Dynamic of sea water interface using hydrochemical facies evolution diagram. *Ground Water* **2010**, *48*, 212–216. [CrossRef]
42. Turekian, K.K. *Oceans (Foundations of Earth Science Series)*, 2nd ed.; Pearson College: Upper Saddle River, NJ, USA, 1976; ISBN 978-0136304180.
43. Matheron, G. Traite de Geostatistique Apliquee. In *Tome I. Memoires Bureau de Recherches Geologiques et Minieres*; Editions Bureau de Recherche et Minieres: Paris, France, 1962.

44. Belkhiri, L.; Tiri, A.; Mouni, L. Spatial distribution of the groundwater quality using kriging and Co-kriging interpolations. *Groundw. Sustain. Dev.* **2020**, *11*, 100473. [CrossRef]
45. Nistor, M.M.; Rahardjo, H.; Satyanaga, A.; Hao, K.Z.; Xiaosheng, Q.; Sham, A.W.L. Investigation of groundwater table distribution using borehole piezometer data interpolation: Case study of Singapore. *Eng. Geol.* **2020**, *271*, 105590. [CrossRef]
46. Custodio, E.; Bruggerman, G.A. *Groundwater Problems in Coastal Areas*; UNESCO: Paris, France, 1987; ISBN 978-92-3-102415-3.
47. Kovalevsky, V.; Kruseman, G.; Rushton, K. *Groundwater Studies: An International Guide for Hydrogeological Investigations*; UNESCO: Paris, France, 2004; ISBN 92-9220-005-4.
48. Boyd, C.A.C.E.; Boyd, C.A.C.E.; Rouse, D.B. Potassium budget for inland, saline water shrimp ponds in Alabama. *Aquac. Eng.* **2007**, *36*, 45–50. [CrossRef]
49. McNevin, A.A.; Boyd, C.E.; Silapajarn, O.; Silapajarn, K. Ionic Supplementation of Pond Waters for Inland Culture of Marine Shrimp. *J. World Aquac. Soc.* **2004**, *35*, 460–467. [CrossRef]

earth

MDPI

Article

# Comparing Recovery Volumes of Steady and Unsteady Injections into an Aquifer Storage and Recovery Well

**Saeid Masoudiashtiani [1,*] and Richard C. Peralta [2]**

[1] Larry Walker Associates (LWA), 1480 Drew Avenue, Davis, CA 95618, USA
[2] Civil and Environmental Engineering Department, Utah State University, 4110 Old Main Hill, Logan, UT 84322-4110, USA; richard.peralta@usu.edu
* Correspondence: saeid.masoudiashtiani@usu.edu

**Abstract:** Aquifer Storage and Recovery (ASR) can involve injecting available surface water into an unconfined aquifer and then extracting it to provide secondary water for irrigation. This study demonstrates a method for evaluating the appropriateness of steady injection versus unsteady injection for an assumed situation. In design, it can be important to affect the transient: the proportion of the injected water that would be subsequently extracted (versus that remaining in the aquifer) and the proportion within the extracted water that would be an injectate (versus ambient groundwater). These proportions can be predicted from the predicted value of an ASR well's Recovery Effectiveness (REN)—the time-varying proportion of injectate that is extracted subsequently from the same fully penetrating well. Applying the demonstrated procedure with appropriately detailed data and simulation models can predict the REN values resulting from steady versus unsteady injection, followed by steady extraction. For convenience in displaying and computing REN, the injectate was assumed to have a 100 ppm conservative solute concentration. For this demonstration, a homogenous isotropic unconfined one-layer aquifer was assumed. The scenarios involved steady or unsteady injection for 61 days via a fully penetrating ASR well. Then, 91 days of steady pumping led to the extraction of a total volume equal to that injected. For the assumed hydrogeologic data—31 years of Salt Lake City, Utah, rainfall data and estimated captured runoff—the results show that steady injection is more likely to cause a predictable REN but might not cause a higher REN than daily varying injection of the same total volume. Assuming different runoff or hydrogeologic flows would lead to different REN values. Steady injection causes a predictable groundwater mound and can assure a sufficient vadose zone thickness for overlying plants. Augmentation and storage of captured rainwater can help to provide a steady injection rate. For a situation that requires REN management, appropriate simulations can help water managers design ASR systems that will achieve REN goals and increase sustainable groundwater availability.

**Keywords:** harvested rainwater; aquifer storage and recovery; unconfined aquifer; injectate recovery; WinSLAMM; MODFLOW; MT3DMS

**Citation:** Masoudiashtiani, S.; Peralta, R.C. Comparing Recovery Volumes of Steady and Unsteady Injections into an Aquifer Storage and Recovery Well. *Earth* **2024**, *16*, 990–1004. https://doi.org/10.3390/earth5040051

Academic Editors: Zongjun Gao and Jiutan Liu

Received: 7 October 2024
Revised: 28 November 2024
Accepted: 28 November 2024
Published: 9 December 2024



## 1. Introduction

Escalating withdrawals of ambient surface water and groundwater for various uses in agriculture, industry, etc., can harm water sustainability. Improving water sustainability involves increasing beneficial use and preventing harm to life and the environment. The management method for recharging aquifers (i.e., the MAR) is a key strategy in sustaining water for a long time by utilizing various water resources (stormwater, treated wastewater, etc.) to recharge aquifers [1]. The MAR tools for injecting water into an aquifer (i.e., a saturated zone) are aquifer recharge (AR) wells and aquifer storage and recovery (ASR) wells. Also, the tools allowing the infiltration of water into an unsaturated zone above a water table are bank filtration via dry wells (drywells) and infiltration galleries [1,2].

Drywells are infiltration wells that penetrate the vadose (unsaturated) zone and are screened above the water table [3,4]. Injecting water into the vadose zone will cause a greater rise in a well's water level than ASR at the same diameter [4,5]. Datry et al. [6] recommended minimizing contact between inflowing water and the organic sediments common in infiltration basins or galleries because organic sediments reduce the dissolved oxygen (DO) in the water that would otherwise re-oxygenate groundwater. Because of avoiding the sediments, ASR injection (i.e., the direct injection method) can increase groundwater DO content to a greater extent than infiltration basins or galleries.

ASR is a technological solution for treating and injecting clean water into an aquifer for temporary storage used by Minnesotans because most depend on groundwater [7]. In northern Cape May County, New Jersey, the USA, Wildwood Water Utility and Atlantic Electric Company have applied ASR to ensure sufficient water supply during the summer tourist season, with about 85% recovery, and prevent saltwater intrusion since 1967 [8]. The use of ASR in the United States, Europe, the Middle East, and Australia has been increased to secure and sustain the supply of freshwater seasonally [9,10].

ASR injectate could affect water quality within an aquifer, and subsequently, discharged water. The water quality of naturally discharged or ASR-extracted water can affect the well-being of the water users and the ecosystem. ASR wells are suitable for areas without sufficient space, with dense populations, and with growing use of groundwater extraction [11–13]. ASR can increase water storage in regions affected by both drought and intense precipitation [14]. When extra surface water is available, the considered ASR process involves injecting water into an aquifer via a vertical tube well. When surface water is unavailable, the well extracts a mixture of injected water (injectate) and native groundwater to meet water needs. Conservative contaminants such as chloride or dissolved nitrate [15–17] might exist within the injectate and/or the native groundwater. ASR well evaluation can involve predicting solute concentrations in the extracted water as well as estimating solute distribution and the remaining mass within the aquifer. Both estimation types are interesting to water users and environmental protection specialists. Ideally, the extracted water's solute concentration will satisfy the intended use, and enough injectate will have been removed to protect the aquifer. Modeling such solute concentrations or remaining masses involves groundwater flow and solute transport simulations. These simulations help obtain values of recovery effectiveness—REN (i.e., the time-varying proportion of injectate that has been extracted)—for an ASR well. These simulations numerically model groundwater flow and solute transport for ASR under both steady and unsteady injection conditions. The evaluation of injectate recovery for ASR involves quantifying (a) the remaining amount of injected solute within the aquifer (i.e., 1.00—REN) and (b) the recovered amount (REN) within ASR-extracted water [18,19]. REN differs from recovery efficiency (RE), which has been applied to estimate the performance of ASR wells in brackish, saline, and coastal aquifers [20,21]. The RE value can be greater than one because it is the product of dividing the extracted volume without any required treatment by the injectate volume [22–28]. The REN is crucial for understanding how much of the injected water can be extracted later or how much remains in the aquifer. Modest mixing of (un)treated surface water or stormwater with groundwater in an aquifer is sometimes legally and environmentally acceptable. The REN can be used to predict the extent of blending that will occur. In other situations, legal water rights are involved. For instance, the Jordan Valley Water Conservancy District (JVWCD) in Utah injects extra surface water into a confined aquifer during wet months to secure additional groundwater extraction rights during dry months. If the JVWCD does not extract the injectate volume within one year after injection, the volume corresponding to the new water right decreases due to assumed losses within the aquifer.

Rather than assessing how much injectate escapes subsequent capture by extracting an ASR well, Forghani and Peralta [18] and Masoudiashtiani and Peralta [19] employed a counting molecule method to predict the REN of an ASR well. The relationship between the REN and RE is expressed as follows:

$$\mathrm{RE} = \mathrm{REN}(\mathrm{V_E}) + \frac{\mathrm{V(E, amb)}}{\mathrm{V_{inj}}} \tag{1}$$

where V(E, amb) is the volume of ambient groundwater in the extracted water, $V_E$ is the total volume of extracted water, and $V_{inj}$ is the total volume of injectate.

Applying an MAR method can lead to groundwater improvement or deterioration based upon (a) the water itself; (b) the applied MAR technique; and (c) the interaction between the injectate and the aquifer materials [29]. USEPA Underground Injection Control (UIC) programs address Class V ASR wells, AR wells, and other injection wells in many states, including California [30]. In California, any treated wastewater ASR injectate must meet drinking water standards at the well. The USEPA requires that the regional water quality control boards and the department of health services establish ASR discharge requirements and approve applications in this state.

In this study, we simulate the daily harvested rainwater (i.e., daily varying stormwater runoff) volume of a Salt Lake City block located in the northeast of the confluence of the Red Butte Creek and the Jordan River in Utah from April to May of 1986–2016 (31 years) by using WinSLAMM 10.4.1 software [31]. The simulated volumes provide modeled ASR injection water at unsteady and steady rates. Then, the REN (i.e., injectate recovery) evaluations of the modeled ASR show the impacts of the injection types on the recovery in the following sections. The objectives of this study are to present a contrast of the simulated volumes and injectate recovery of aquifer storage and recovery (ASR) for secondary use after the unsteady and steady injections of harvested rainwater into an unconfined aquifer.

## 2. Materials and Methods

### 2.1. Overview

This section explains assumptions, parameters, and procedures relating to obtaining REN values of ASR via flow [32] and solute transport [33] simulations. Here, the goal is to improve available groundwater during the dry period (June–August) by injecting excess surface water (i.e., harvested rainwater) into an unconfined aquifer during a preceding wet period (April–May) in a city block. Based upon work by Fetter [34], Bedient et al. [35], Pavelic et al. [23], Ward et al. [25,26], Bakker [21], Brown et al. [28], Smith et al. [14], Forghani and Peralta [18], and Masoudiashtiani and Peralta [19], in the following section, we use values of the hydrogeologic factors that impact the REN (i.e., injectate recovery) of ASR.

After the selection of approximately half-month durations for the flow and solute transport simulations, standard methods were used in estimating the advective plume length after two months (61 days) of injection; the longitudinal dispersivity; courant number; maximum time step size; total number of simulation time steps; and time steps per stress period. These estimates enabled preparing a groundwater aquifer model domain that was sufficiently large and assumed injection and extraction rates would not appreciably affect any of the employed boundary conditions. Inputting the hydrogeologic parameters for the selected site preceded the execution of the simulations. This study assumes that the aquifer is both isotropic and homogeneous, which might be an oversimplification given that real-world conditions often exhibit significant heterogeneity and anisotropy that can impact flow and solute transport. A single-site location might limit the generalizability of the findings to other geographical or hydrogeological contexts.

### 2.2. Selected-Site Information

Figure 1 shows parts of Salt Lake City and Salt Lake County in UT, USA. The assumed ASR well location is within a 15.271-acre residential city block (Site 5). The site is located to the northeast of the confluence of the westward-flowing Red Butte Creek and the northward-flowing Jordan River. It is (a) bounded on the north and south by Fremont and Lucy Avenues, respectively; (b) almost bisected from north to south by Jeremy Street; and (c) bounded to the west and east by South 900 W and South 800 W, respectively.

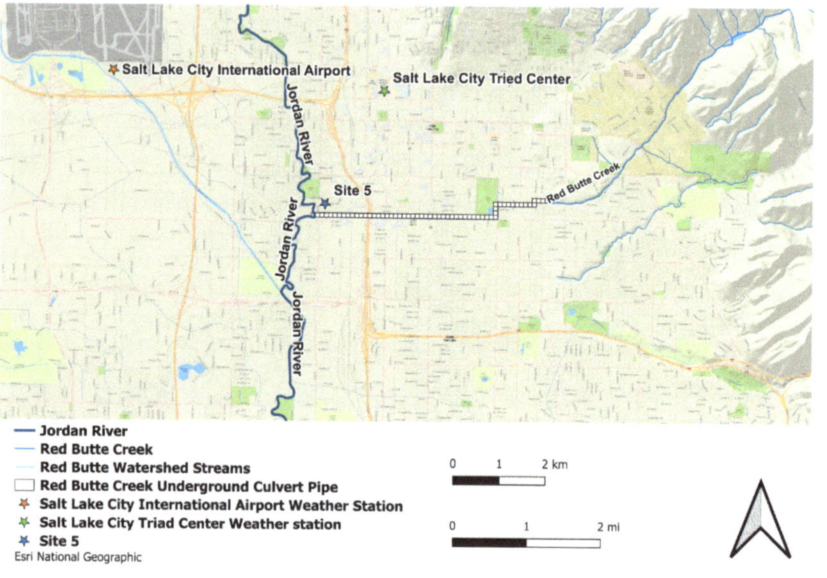


**Figure 1.** Residential area (Site 5) and weather stations at the international airport and Triad Center in Salt Lake County, UT, USA.

The hydrogeologic information on the properties Site 5 regarding the underlying shallow unconfined aquifer [36] includes (a) horizontal hydraulic conductivity of 18.17 m/d (59.61 ft/d); (b) initial aquifer (background) hydraulic gradient of 0.0029; (c) porosity of 0.3; (d) specific yield of 0.15; and (e) initial (original) saturated thickness of 10.23 m (33.56 ft). The basin-fill pre-historical geologic data were extracted from the Salt Lake Valley (SLV) groundwater model [36]. The observed groundwater level based on USGS observation well 404444111550501 data at Site 5 (Figure A1) varied between 1.1 (3.6) and 1.4 m (4.6 ft) below ground surface in April and May (Figure A2). The following was assumed: (a) there was no groundwater flow between the layer and lower strata because of the short term (61 days) of injection and subsequent extraction, (b) there were no seasonal changes in storage capacity, and (c) the ASR well diameter at the site was 15.24 cm (6 inches). The potential water resources for injection were harvested rainwater (i.e., stormwater runoff) at Site 5 and/or diversion from Red Butte Creek (RBC). In this study, WinSLAMM 10.4.1 software [31] helped simulate daily varying stormwater runoff (i.e., assumed-to-be unsteady injection water) of Site 5 from April through May. This software is not free or open-source code. WinSLAMM is a tool used to estimate the water availability/injection rate for ASR [31]. The U.S. EPA has utilized it, but it is not recommended. Included properties for the simulation were (a) silty soil and ground cover consisting of nearly flat roofs (2.238 acres, 14.6%); (b) sidewalks (2.421 acres, 15.8%); (c) streets of intermediate texture (2.741 acres, 17.9%); (d) small landscaping features (3.935 acres, 25.7%); and (e) undeveloped area (3.936 acres, 25.7%). The daily April–May precipitation data for 1986–2016 (31 years) from the Salt Lake City (SLC) International Airport weather station, which is 5.91 km (3.67 miles) away, were used (Figure 1). Rain was distinguished from snow using the Static Temperature method and a 1 °C (33.8 °F) threshold temperature [37] to simulate the runoff. To substitute for any missing precipitation data, the SLC Triad Center station, which is 3.38 km (2.1 miles) away from Site 5, was used (Figure 1).

To simulate a stormwater injection situation, the runoff volume during a particular day was injected into the aquifer at a steady rate during that day. To simulate a surface water situation, water was diverted from RBC at an assumed steady rate and then injected into the aquifer at the same rate.

For Site 5, the ASR-extracted water could irrigate turf based upon a defined irrigation schedule [38] from June through August.

*2.3. Modeled System and Simulation Tools*

For accurate REN prediction, the MODFLOW2005 [32] model, its Multi-Node Well (MNW2) package, and MT3DMS [33] were used. To distinguish the injected water from the native groundwater and facilitate solute transport simulation, a hypothetical concentration of 100 ppm of imaginary non-reactive solute was assigned to the injectate. In this study, a fully penetrating ASR well in a homogenous, isotropic, freshwater, one-layer, one-porosity, one-specific-yield, no-seasonal-change-in-storage-capacity, unconfined aquifer was modeled, and extraction volumes equaling injection volumes were used. Constant-head boundaries were defined along the eastern and western edges of a square model area, with no-flow boundaries along the northern and southern edges. The numbering of rows and columns starts from the top left in the MODFLOW grid. The numbers of rows and columns in the model grid are 129 and 129, respectively. Rows and columns from 1 to 55 and 75 to 120 include cell sizes of 10 m by 10 m. Rows and columns 56 to 74 have cell sizes of 7.5 by 7.5, 6.75 by 6.75, 5 by 5, 4 by 4, 3 by 3, 2 by 2, 1.25 by 1.25, 1 by 1, 0.75 by 0.75, 0.5 by 0.5, 0.75 by 0.75, 1 by 1, 1.25 by 1.25, 2 by 2, 3 by 3, 4 by 4, 5 by 5, 6.75 by 6.75, and 7.5 m by 7.5 m, respectively. An ASR well located at the center, as a block-center node in row 65 and column 65 (cell size of 0.5 m by 0.5 m), (Figures 2 and 3) was also included in the assumptions. Preliminary simulations helped determine the horizontal domain size required to avoid an appreciable boundary condition impact from groundwater pumping.

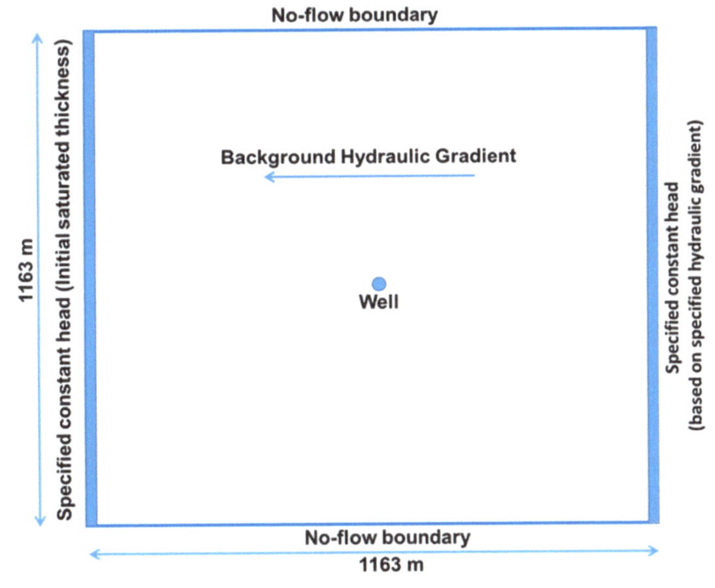


**Figure 2.** Modeled area for simulation of ASR well (top view) (not to scale).

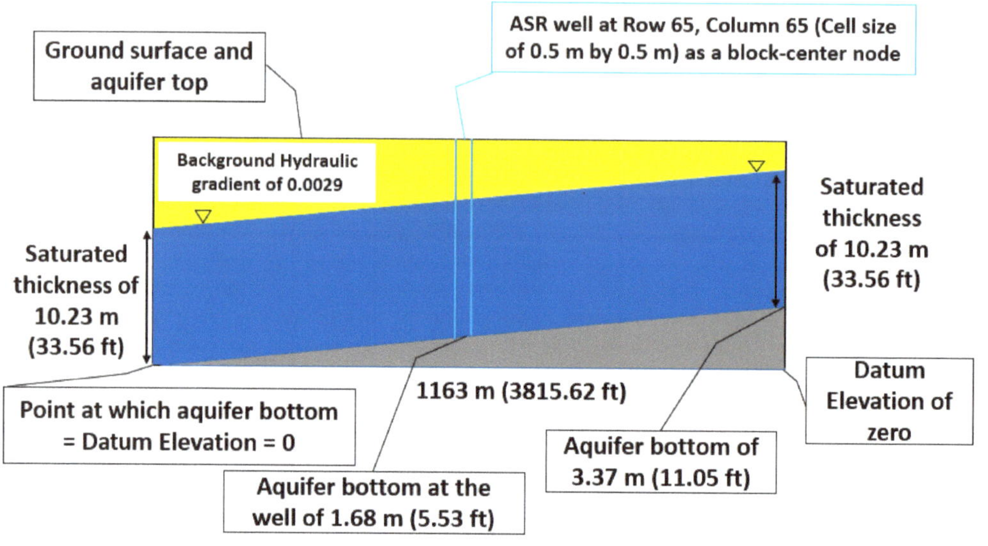


**Figure 3.** Modeled area for simulation of ASR well (side view) (not to scale).

For solute transport simulation, MT3DMS estimated longitudinal dispersivity via analytical equations found in a U.S. EPA online tool [39] that incorporates work by Gelhar et al. [40], Wilson et al. [41], and Xu and Eckstein [42]:

$$\text{If } L_p \leq 1\text{ m}: \ \alpha_L = 0.1 \times L_p \quad (2)$$

$$\text{If } L_p > 1\text{ m}: \ \alpha_L = 0.83 \times \left[\log_{10} L_p\right]^{2.414} \quad (3)$$

where $L_p$ denotes advective plume length (m), equal to (v × injection duration); v denotes linear pore velocity (m/d), equal to (K × i/n); K denotes horizontal hydraulic conductivity (m/d); i denotes aquifer initial (background) hydraulic gradient; n denotes porosity; and $\alpha_L$ denotes longitudinal dispersivity (m).

Also, the courant number (C) is an input parameter of the MT3DMS advective process used to reduce oscillations, enhance accuracy, and minimize numerical dispersion in cases where advection outweighs dispersion. The number is determined as follows: $(v \times \Delta t)/\Delta x$, where v represents linear pore velocity, $\Delta x$ is the grid cell size at 0.5 m (1.64 ft) inside the well, and $\Delta t$ is the maximum desired time step size [43].

To determine the appropriate simulation time step size in days for the preferred spatial discretization, the grid Peclet number, P, was estimated to be 2C [43]. By assuming that P is equal to $(\Delta x/\alpha_L)$, the maximum desirable time step size for use during injection can be computed.

For simulation of ASR injection (i.e., injection of available rainfall, streamflow, and plant water) into aquifer and extraction, MODFLOW2005 and MT3DMS include (a) period 1, in which steady-state background heads were simulated; (b) transient periods 2–5, in which injection was employed; (c) periods 6–11, in which extraction was simulated; (d) identical numbers of time steps per period; and (e) MODFLOW2005 PCG solver with a 0.01 m head change criterion and 0.01 m residual convergence criterion. For advection and dispersion simulation, MT3DMS utilizes the total variation diminishing (TVD) package and the generalized conjugate gradient (GCG) solver. To minimize processing time, the extraction period was assigned a single time step, as using even 300 time steps for extraction increases REN by less than 0.005.

As mentioned above, we assigned 100 ppm of an imaginary non-reactive solute as an injectate to distinguish it from native groundwater. MT3DMS simulations provide both the mass of solute injected into the ASR well and the mass of solute recovered from the well. The REN is equal to extracted solute mass divided by injected solute mass. Each successful simulation provided the total mass of solute injected through the ASR well and the mass of solute recovered from the well. $REN_t$ values were computed after 15, 30, 45, 61, 76, and 91 days of simulated extraction ($REN_{15}$, $REN_{30}$, $REN_{45}$, $REN_{61}$, $REN_{76}$, and $REN_{91}$, respectively) from simulation results. This study's intended use and purpose entail predicting the proportion of ASR recovery but not actual water quality. One hundred is a practical and convenient value for predicting the proportion.

As shown in Table 1, steady injected water could be diverted from RBC during April and May for ASR at Site 5. As the proposed RBC diversion for ASR steady injection, Demos 1 and 2 evaluate whether steady or unsteady injection results in extracting a predictable portion of the injectate.

**Table 1.** Characteristics of demonstrations (Demos) of ASR unsteady and steady injections and steady extraction.

| | I | II | III | IV | V | VI | VII |
|---|---|---|---|---|---|---|---|
| Demo | Number of MODFLOW-MT3DMS simulations | Injection volume, in Julian Days, of 61-day inj. period | Total 61-day injection vol. | Extraction vol. in each of 91 days | Total 91-day extraction vol. | Type(s) of water quality | Goal(s) |
| 1.1–1.31 | 31 | Unsteady. For 61-day simulations for each of 31 years, the injection volume of each day in the year is as follows: injection volume = Site 5 runoff of day in the year. | Total injection volume = sum of Column II daily injection volumes in the year | Steady. Daily extraction volume for year = (Total injection vol. for year/91 days) | Total extraction vol. = Total injection vol. | Assumed 100 ppm conservative injectate | Present recovery of unsteady stormwater injection via steady extraction |
| 2.1–2.31 | 31 | Steady injection volume in year = total injection volume/61 days | " | " | " | " | Present recovery of steady injection of diverted surface water via steady extraction |

## 3. Results and Discussion

Evaluating the responses of any aquifer storage and recovery (ASR) plan is required for water use and the environment (i.e., the aquifer). In this case, the ASR well was within a quasi-infinite, one-layer, one-porosity, one-specific yield, no-seasonal-change-in-storage-capacity, unconfined aquifer. The provided demonstrations (Demos) illustrate the responses of the flow (MODFLOW2005) and solute transport (MT3DMS) simulations for injectate recovery resulting from unsteady and steady ASR injections.

Here, for the selected site (Site 5), water-volume use of stormwater runoff (i.e., harvested rainwater) during the wet months (April and May) could provide ASR injection water. To quantify the volume, the April–May runoff volume of every year from 1986 to 2016 (31 years) at Site 5 was simulated using WinSLAMM 10.4.1 software. Figure 4 shows the annual volumes that were used for ASR injection from April to May of each year.

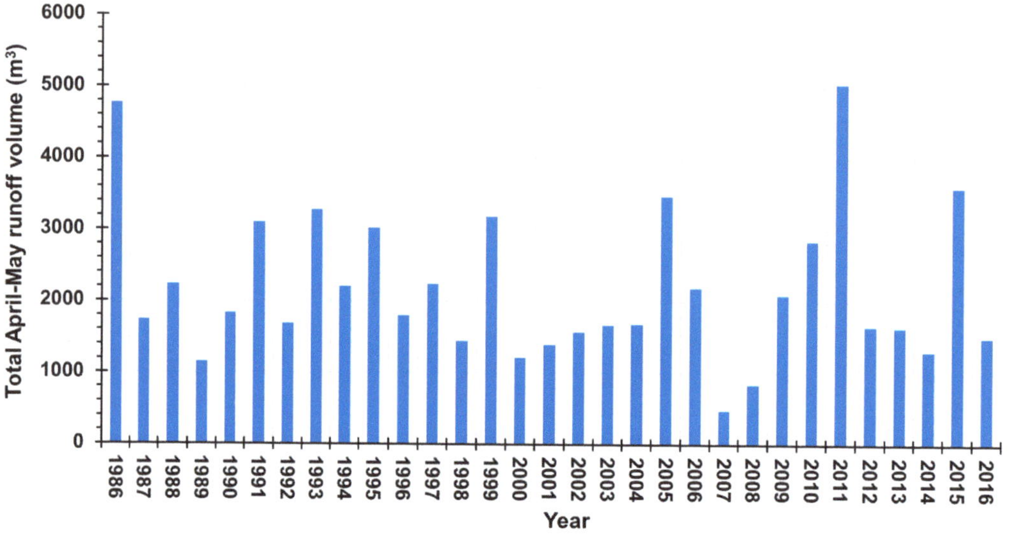


**Figure 4.** Total April–May runoff volumes of the 31 years in Site 5 simulated via WinSLAMM 10.4.1 software.

The steady injection of water into the aquifer could be diverted by surface water (Red Butte Creek, RBC) nearby. Signifying the beneficial use of RBC water steadily for ASR injection equaling the unsteady runoff volume for each of the 31 years is one of the goals of Demos 1 and 2 (Table 2). We acknowledged that a tributary can divert daily varying stormwater runoff (unsteady inflow) from Site 5 into RBC in exchange for steady RBC outflow into the ASR during the wet months.

**Table 2.** Selected results of demonstrations (Demos).

| Demo | Number of MODFLOW2005-MT3DMS Simulations | Description | Total Inj. Vol. / Avg. Flowrate | (Ext. Vol.)/(Inj. Vol.) (-) |
|---|---|---|---|---|
| 1.1–1.31 | 31 | Injecting values of daily varying runoff for 31 seasons to compare REN values | 478.23 to 5048.26 m$^3$<br>0.42 l/s or 6.58 gpm | 1.0 |
| 2.1–2.31 | 31 | Steady injection of the same annual volumes as Demo 2 to compare REN values | ″<br>″ | ″ |

Demos 1 and 2 explore which type of ASR injection can result in predictive recovery (REN) during extraction. For these Demos, the 100 ppm imaginary conservative contaminant for the ASR injectate was applied to evaluate the impacts of unsteady and steady injections on injectate recovery during extraction.

In Demo 1, the daily varying stormwater runoff (unsteady injection) during April and May (61 days) in each of the 31 years was injected, and the injection volume was steadily extracted during the dry months (91 days). For instance, Figure A4 presents unpredictable daily groundwater mounds that could be harmful to crops in 2001. Figure 5 shows that even though one year might have a greater total injection volume than another year, the REN resulting from the first year might not be greater than that of the other year.

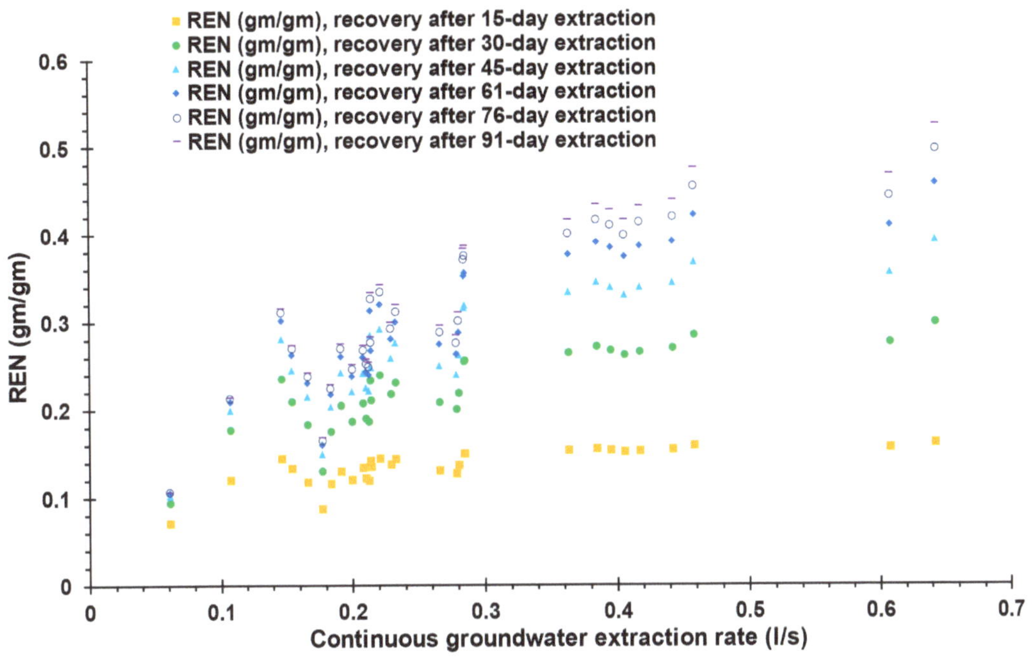


**Figure 5.** REN (gm/gm) values of every half month during extraction for Demo 1; continuous (steady) extraction rate equals the injection volume divided by 91 days for every 31 years. Extraction volume equals the injection volume.

For Demo 2, the runoff (i.e., harvested rainwater) volume in each of the 31 years was divided by the number of injection days (61 days) and calculated to provide the daily steady ASR injection rate. As a steady injection, the increasing rate of the continuous groundwater (steady) extraction results in increasing REN values (Figure 6). Thus, steady injection can provide greater confidence in including and extracting a predictable groundwater mound (Figure A4) to obtain a sufficient vadose zone for crops and a portion of the injectate (Figure 6), respectively.

When comparing the REN values resulting from the steady versus unsteady injection of the same total volume, it becomes clear that unsteady injection could achieve a higher or lower predicted REN than the steady one. For example, the simulated injection volumes for 1987 and 2001 (e.g., Figure A3 shows simulated flow rates of 61-day unsteady and steady injections and 91-day steady extraction for 2001) are 1392.81 m$^3$ and 1738.17 m$^3$, respectively. For 1987, the predicted REN values of the steady and unsteady injections are 0.288 (gm/gm) and 0.342 (gm/gm), respectively. For 2001, the respective REN values are 0.248 (gm/gm) for steady injection and 0.168 (gm/gm) for unsteady injection.

As mentioned, the water volume for ASR steady injection could be the 90%-probability runoff volume during the wet months (April and May) obtained after studying the historical data of any site. The volume can be diverted steadily from surface water nearby to apply the beneficial use of steady injection for ASR predictable injectate recovery. For Site 5, the 90%-probability runoff for the 31-year period (Figure 4) equals 1047.46 m$^3$ (230,409 gallons). The assumed pre-existing groundwater 91-day extraction volume is the same as the 61-day injection for ASR. The ASR-extracted water volume can irrigate 1896.71 m$^2$ (0.47 acre) of turf based on the irrigation schedule [38,44] from June through August (dry months, 91 days) at Site 5.

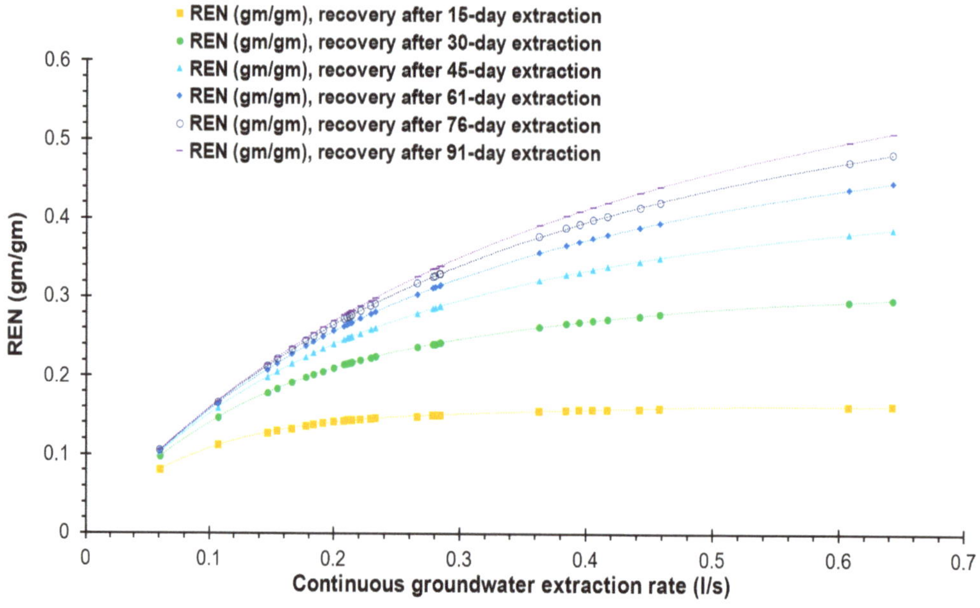


**Figure 6.** REN (gm/gm) values of every half month during extraction for Demo 2; continuous (steady) extraction rate equals the injection volume divided by 91 days for every 31 years. Extraction volume equals the injection volume.

The evaluations can help water managers predict whether an ASR system increases water supply in aquifers and for uses (e.g., secondary irrigation). The model results from ASR steady injection and the 90% probability of the 31-year harvested rainwater show that water managers could increase groundwater availability for use during three dry months (the 91-day period of high demand) via the ASR system at Site 5.

## 4. Conclusions

Presented are the simulated results of injecting specific total water volumes into an aquifer using steady and unsteady rates over time and then steadily extracting the same volumes. The intent was to compare the recovery effectiveness achieved after extraction was completed. The comparisons are intended to aid water managers in ASR planning. The assumptions and simulation processes are as follows.

For current land use in a specific Salt Lake City (SLC) block, precipitation data from 1986 to 2016 were used to create a 61-day time series of springtime daily runoff for each year. All the runoff was assumed to be fully available for injection via one fully penetrating well into an unconfined homogenous isotropic one-layer aquifer (the storage coefficient was assumed to be stationary, unchanging with the seasons). For these short-term simulations, possible vertical recharge from rainfall, seepage with adjacent strata, and lateral flows with extensions of the aquifer were ignored.

Two types of scenarios of groundwater flow and transport simulations were performed for each year from 1986 to 2016. In one type, the above time series was used to simulate 61 days of injection of estimated daily runoff (i.e., harvested rainwater), followed by 91 days of steady extraction that removed a total volume of groundwater equivalent to the injectate volume. The other scenario type involved the simulation of steady injection at a daily rate equaling 1/61 of the total April–May runoff of the year examined. Again, 91 days of steady extraction removed a volume equal to the total injection. Conclusions concerning the simulated final achieved REN are as follows.

(a) When comparing unsteady injection simulations, the unsteady injection time series has a greater impact on REN than the total injection volume; a year with a greater total injection volume might not yield a higher REN.
(b) When comparing steady injection simulations, the greater the total injection volume, the greater the resulting REN.
(c) When comparing unsteady versus steady injection simulations, the following was found:
    i. Scenarios in which water was injected at a steady rate usually yielded higher REN values than scenarios having random daily varying injection rates. For example, for the 1738 $m^3$ injection volume of 2001, the final REN values are 0.25 (gm/gm) for steady injection and 0.17 (gm/gm) for unsteady injection.
    ii. Unsteady injection sometimes yielded a higher REN than steady injection. For the 1392.81 $m^3$ injection volume in 1987, the REN values of the steady and unsteady injections are 0.288 (gm/gm) and 0.342 (gm/gm), respectively.

For one-well ASR situations where achieving a particular REN is desirable, steady injection through the well is preferable to injecting at random rates that depend upon daily runoff capture. Planned injection rates can be designed such that they do not exceed the likely amount of water available nor cause excessive groundwater mounding (preventing saturation in the root zone of the overlying vegetation). Steady injection and subsequent extraction rates can be designed to achieve RENs that prevent undesirable changes in ambient groundwater quality or to prevent undesirable salinity of the extracted water. In summary, utilizing REN values in an ASR system design can increase beneficial groundwater availability and use.

**Author Contributions:** Conceptualization, S.M. and R.C.P.; methodology, S.M. and R.C.P.; validation, S.M. and R.C.P.; formal analysis, S.M.; investigation, S.M. and R.C.P.; resources, S.M.; data curation, S.M.; writing—original draft preparation, S.M.; writing—review and editing, R.C.P.; visualization, S.M.; supervision, R.C.P.; project administration, R.C.P. All authors have read and agreed to the published version of the manuscript.

**Funding:** This research was funded by the U.S. EPA-STAR project under grant number [83582401], the Civil and Environmental Engineering (CEE) Department, the Utah Water Research Laboratory (UWRL), and the Utah Agricultural Experiment Station at Utah State University.

**Data Availability Statement:** All data generated and utilized during the study are included in the published article. Additional information can be obtained from the corresponding author upon request.

**Acknowledgments:** The authors express their gratitude for the support provided by the Center for High-Performance Computing (CHPC) at the University of Utah. This work was also supported by the Utah Agricultural Experiment Station at Utah State University. The authors extend their appreciation to R. Ryan Dupont for his leadership on the EPA-STAR grant project, as well as to Marv Halling and David G. Tarboton, the heads of the Civil and Environmental Department, and the Utah Water Research Laboratory at Utah State University, respectively.

**Conflicts of Interest:** Author Saeid Masoudiashtiani was employed by the company Larry Walker Associates (LWA). The remaining authors declare that the research was conducted in the absence of any commercial or financial relationships that could be construed as a potential conflict of interest.

## Appendix A

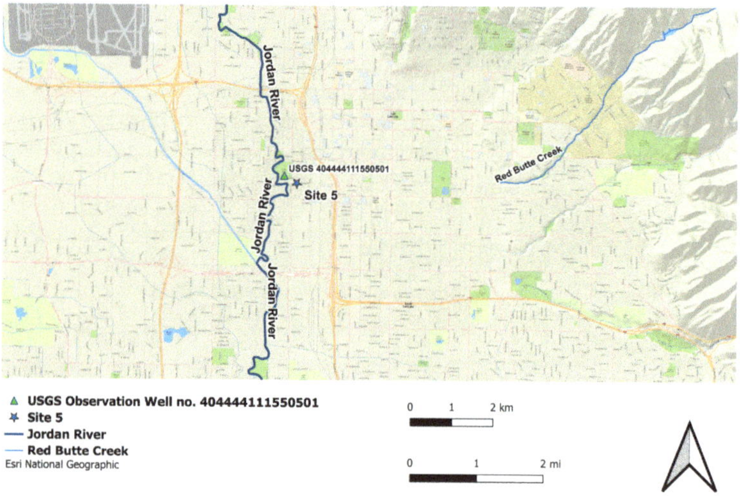


**Figure A1.** USGS Observation Well no. 4044441115505501 (Hydrologic Unit no. 16020204), with a latitude of 40°44′45″, a longitude of 111°55′06″ NAD27, and a land surface altitude of 1283.2 m (4210 ft.) above NGVD29. The well is 15.0 ft. (4.57 m) below the land surface and located about 350 m (1148.29 ft.) away from the residential area (Site 5) in Salt Lake County, Utah, USA (URL: http://waterdata.usgs.gov/nwis/inventory?agency_code=USGS&site_no=404444111550501, accessed on 1 October 2007).

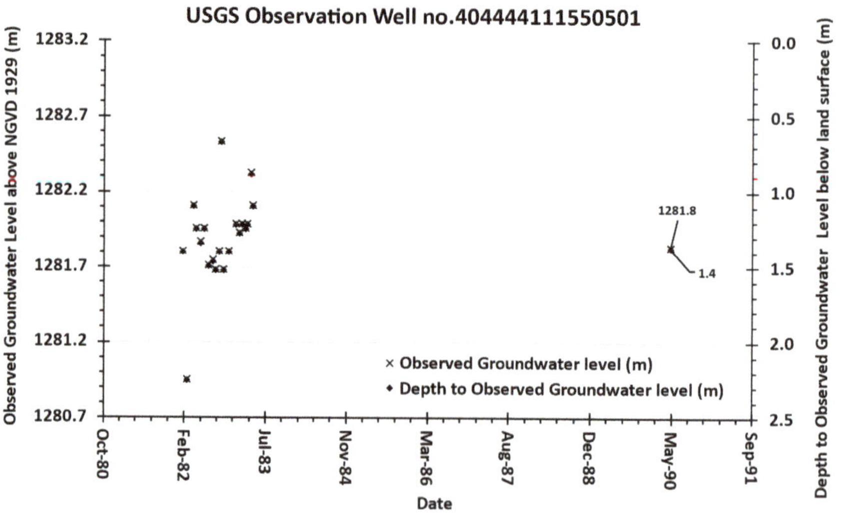


**Figure A2.** Available observed groundwater levels from 1982 to 1990 reported by USGS Observation Well no. 404444111550501 located about 350 m (1148.29 ft.) away from Site 5. The depth to groundwater level in April and May for the years is between 1.1 (3.6) and 1.4 m (4.6 ft.).

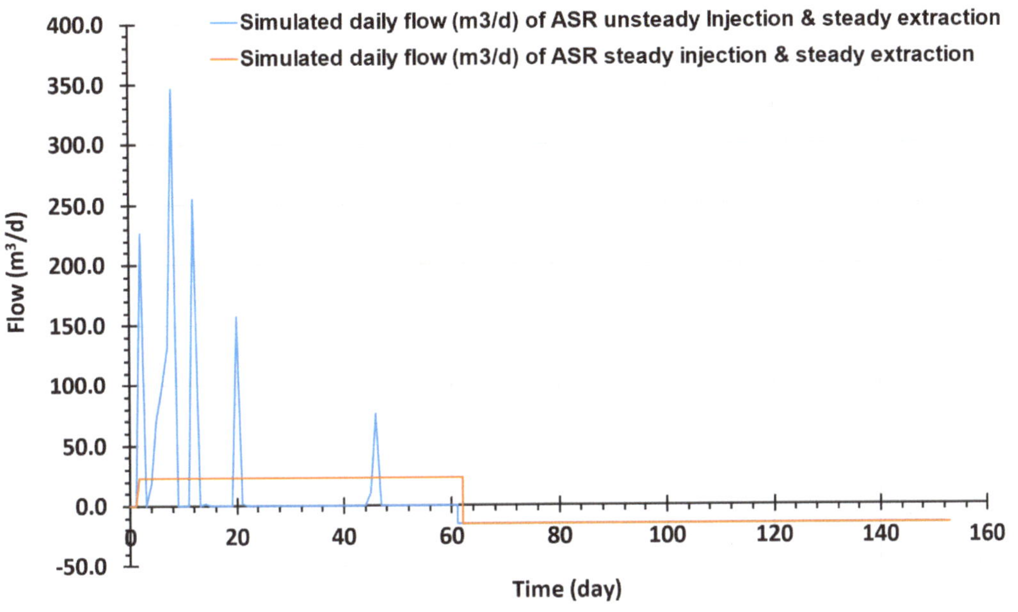


**Figure A3.** Simulated daily flows at the well resulting from 61 days of unsteady and steady injections (positive magnitudes), followed by 91 days of steady extraction (negative magnitudes), for an ASR well at Site 5 in 2001.

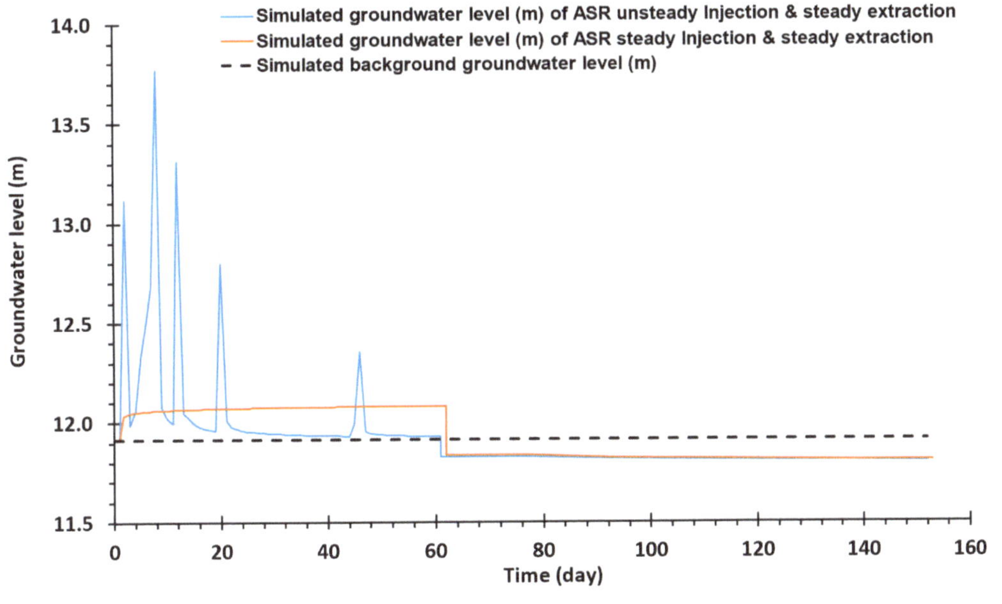


**Figure A4.** Simulated groundwater levels at the well resulting from 61 days of unsteady and steady injections, followed by 91 days of steady extraction, for an ASR well at Site 5 in 2001 (Note: the elevation datum is equal to zero).

## References


1. Alam, S.; Borthakur, A.; Ravi, S.; Gebremichael, M.; Mohanty, S.K. Managed aquifer recharge implementation criteria to achieve water sustainability. *Sci. Total Environ.* **2021**, *768*, 144992. [CrossRef]
2. Dillon, P. Future management of aquifer recharge. *Hydrogeol. J.* **2005**, *13*, 313–316. [CrossRef]
3. Edwards, E.C.; Harter, T.; Fogg, G.E.; Washburn, B.; Hamad, H. Assessing the effectiveness of drywells as tools for stormwater management and aquifer recharge and their groundwater contamination potential. *J. Hydrol.* **2016**, *539*, 539–553. [CrossRef]
4. Russo, D.; Kurtzman, D.; Nachshon, U. Hydraulic Issues Concerning Injection of Harvested Rainwater to the Subsurface Through Drywells: Insight from Numerical Simulations of Flow in a Realistic Combined Vadose Zone-Groundwater Flow System. *Water Resour. Res.* **2022**, *58*, e2021WR031881. [CrossRef]
5. Bouwer, H. Artificial recharge of groundwater: Hydrogeology and engineering. *Hydrogeol. J.* **2002**, *10*, 121–142. [CrossRef]
6. Datry, T.; Malard, F.; Gibert, J. Dynamics of solutes and dissolved oxygen in shallow urban groundwater below a stormwater infiltration basin. *Sci. Total Environ.* **2004**, *329*, 215–229. [CrossRef]
7. Jennings, C.; Bilotta, J.; Arnold, W.; Kang, P.; Yoon, S.; Shandilya, R.N.; Bresciani, E.; Lee, S.; Kirk, J.; Levers, L.; et al. Banking Groundwater: Managed Aquifer Recharge. A Study Examining Aquifer Storage and Recovery for Groundwater Sustainability in Minnesota. 2019. Available online: https://www.wrc.umn.edu/banking-groundwater-managed-aquifer-recharge (accessed on 1 July 2019).
8. Lacombe, P.J. *Artificial Recharge of Ground Water by Well Injection for Storage and Recovery, Cape May County, New Jersey, 1958–92*; Technical Report 96-313; U. S. Geological Survey: Reston, VA, USA, 1996. [CrossRef]
9. Dillon, P.; Stuyfzand, P.; Grischek, T.; Lluria, M.; Pyne, R.D.G.; Jain, R.C.; Bear, J.; Schwarz, J.; Wang, W.; Fernandez, E.; et al. Sixty years of global progress in managed aquifer recharge. *Hydrogeol. J.* **2019**, *27*, 1–30. [CrossRef]
10. Simbo, C.W. Hydrogeochemical Evaluation of Aquifer Storage and Recovery in Edwards Aquifer, New Braunfels, Texas. *Groundwater* **2023**, *62*, 560–577. [CrossRef]
11. Jakeman, A.J.; Barreteau, O.; Hunt, R.J.; Rinaudo, J.; Ross, A.; Arshad, M.; Hamilton, S. *An Overview of Issues and Options. Integrated Groundwater Management: Concepts, Approaches and Challenges*; Springer Open, National Centre for Groundwater Research and Training: Bedford Park, Australia, 2016; Part 16; pp. 413–434. Available online: https://link.springer.com/book/10.1007/978-3-319-23576-9 (accessed on 5 August 2016).
12. Daus, A.; GSI Environmental Inc. *Aquifer Storage and Recovery. Improving Water Supply Security in the Caribbean Opportunities and Challenges*; Discussion paper No. IDB-DP-00712; Inter-American Development Bank (IDB) Publication, Water and Sanitation Division: Washington, DC, USA, 2019. Available online: https://publications.iadb.org/en/aquifer-storage-and-recovery-improving-water-supply-security-caribbean-opportunities-and-challenges (accessed on 26 October 2022).
13. U.S. EPA. Underground Injection Control, Aquifer Recharge, and Aquifer Storage and Recovery. 2021. Available online: https://www.epa.gov/uic/aquifer-recharge-and-aquifer-storage-and-recovery (accessed on 23 October 2024).
14. Smith, W.B.; Miller, G.R.; Sheng, Z. Assessing aquifer storage and recovery feasibility in the Gulf Coastal Plains of Texas. *Hydrol. J.* **2017**, *14*, 92–108. [CrossRef]
15. Macpherson, G.L.; Townsend, M.A. Perspectives on Sustainable Development of Water Resources in Kansas, Chapter 5: Water Chemistry and Sustainable Yield; Kansas Geological Survey Bulletin 239. 1998. Available online: www.kgs.ku.edu/Publications/Bulletins/239/Macpherson/index.html (accessed on 27 May 2013).
16. AL-Hashimi, O.; Hashim, K.; Loffill, E.; Marolt Cebasek, T.; Nakouti, I.; Faisal, A.A.H.; Al-Ansari, N. A Comprehensive Review for Groundwater Contamination and Remediation: Occurrence, Migration and Adsorption Modelling. *Molecules* **2021**, *26*, 5913. [CrossRef]
17. Brindha, K.; Schneider, M. Chapter 13—Impact of Urbanization on Groundwater Quality. *GIS Geostat. Tech. Groundw. Sci.* **2019**, 179–196. [CrossRef]
18. Forghani, A.; Peralta, R.C. Intelligent performance evaluation of aquifer storage and recovery systems in freshwater aquifers. *J. Hydrol.* **2018**, *563*, 599–608. [CrossRef]
19. Masoudiashtiani, S.; Peralta, R.C. ANN-Based Predictors of ASR Well Recovery Effectiveness in Unconfined Aquifers. *Hydrology* **2023**, *10*, 151. [CrossRef]
20. Pyne, R.D.G. *Groundwater Recharge and Wells: A Guide to Aquifer Storage Recovery*; CRC Press: Boca Raton, FL, USA, 1995.
21. Bakker, M. Radial Dupuit interface flow to assess the aquifer storage and recovery potential of saltwater aquifers. *Hydrogeol. J.* **2010**, *18*, 107–115. [CrossRef]
22. Kimbler, O.K.; Kazmann, R.G.; Whitehead, W.R. *Cyclic Storage of Freshwater in Saline Aquifers*; Louisiana Water Resources Research Institute Bulletin: Baton Rouge, LA, USA, 1975; pp. 75–78.
23. Pavelic, P.; Dillon, P.J.; Simmons, C.T. Multiscale Characterization of a Heterogeneous Aquifer Using an ASR Operation. *Groundw. J.* **2005**, *44*, 155–164. [CrossRef]
24. Lowry, C.S.; Anderson, M.P. An Assessment of Aquifer Storage Recovery Using Ground Water Flow Models. *Ground Water J.* **2006**, *44*, 661–667. [CrossRef]
25. Ward, J.D.; Simmons, C.T.; Dillon, P.J. Variable-density modeling of multiple-cycle aquifer storage and recovery (ASR): Importance of anisotropy and layered heterogeneity in brackish aquifers. *Hydrol. J.* **2008**, *356*, 93–105. [CrossRef]
26. Ward, J.D.; Simmons, C.T.; Dillon, P.J.; Pavelic, P. Integrated assessment of lateral flow, density effects, and dispersion in aquifer storage and recovery. *Hydrol. J.* **2009**, *370*, 83–99. [CrossRef]

27. Lu, C.; Du, P.; Chen, Y.; Luo, J. Recovery efficiency of aquifer storage and recovery (ASR) with mass transfer limitation. *Water Resour. Res. J.* **2011**, *47*, 1–12. [CrossRef]
28. Brown, C.J.; Ward, J.; Mirecki, J. A Revised Brackish Water Aquifer Storage and Recovery (ASR) Site Selection Index for Water Resources Management. *Water Resour. Manag. J.* **2016**, *30*, 2465–2481. [CrossRef]
29. Luxem, K. Managed Aquifer Recharge. A Tool to Replenish Aquifers and Increase Underground Water Storage. American Geosciences Institute (AGI) 2017, Factsheet 2017-006, This work Is Licensed under a Creative Commons BY-NC-ND 4.0 License. Available online: https://www.americangeosciences.org/geoscience-currents/managed-aquifer-recharge (accessed on 25 September 2017).
30. U.S. Environmental Protection Agency (U.S. EPA). *The Class V Underground Injection Control Study 1999, Volume 21, Aquifer Recharge and Aquifer Storage and Recovery Wells. Office of Ground Water and Drinking Water, 4601, EPA/816-R-99-014u.* Available online: https://www.epa.gov/uic/class-v-underground-injection-control-study (accessed on 1 April 1998).
31. PV and Associates, LLC. WinSLAMM Model Algorithms. 2013. Available online: https://www.winslamm.net/ (accessed on 1 January 1996).
32. Harbaugh, A.W.; Langevin, C.D.; Hughes, J.D.; Niswonger, R.N.; Konikow, L.F. MODFLOW-2005 Version 1.12.00, the U.S. Geological Survey modular groundwater model: U.S. Geological Survey Software Release, 3 February 2017. Available online: https://www.usgs.gov/software/modflow-2005-usgs-three-dimensional-finite-difference-ground-water-model (accessed on 4 March 2019).
33. Zheng, C.; Wang, P.P. *MT3DMS: A Modular Three-Dimensional Multispecies Transport Model for Simulation of Advection, Dispersion, and Chemical Reactions of Contaminants in Groundwater Systems, Documentation and User's Guide*; Final Report, Contract Report SERDP-99-1; U.S. Army Engineer Research and Development Center Cataloging-in-Publication Data: Vicksburg, MI, USA, 1999.
34. Fetter, C.W. *Contaminant Hydrogeology*, 2nd ed.; Prentice-Hall Inc.: Upper Saddle River, NJ, USA, 1999; pp. 73–74.
35. Bedient, P.B.; Rifai, H.S.; Newell, C.J. *Ground Water Contamination, Transport and Remediation*, 2nd ed.; Prentice-Hall Inc.: Upper Saddle River, NJ, USA, 1999; pp. 179–180.
36. Lambert, P.M. Numerical Simulation of Ground-Water Flow in Basin-Fill Material in Salt Lake Valley, Utah. United States Geological Survey, Technical Publication No. 110-B 1995. Available online: https://pubs.er.usgs.gov/publication/70179464 (accessed on 24 August 2007).
37. Kienzle, S.W. A new temperature-based method to separate rain and snow. In *Hydrological Processes*; John Wiley, and Sons. Ltd.: Hoboken, NJ, USA, 2008; Volume 22, pp. 5067–5085. [CrossRef]
38. Kopp, K.; Allen, N.; Wagner, K. Simple Sprinkler Performance Testing for Salt Lake County. *Utah St. Univ. Coop. Ext. Svc.* **2013**. Available online: https://digitalcommons.usu.edu/extension_curall/339 (accessed on 1 May 2013).
39. U.S. Environmental Protection Agency (U.S. EPA). EPA Online Tools for Site Assessment Calculation 2019. Available online: https://www3.epa.gov/ceampubl/learn2model/part-two/onsite/longdisp.html (accessed on 31 August 2021).
40. Gelhar, L.W.; Welty, C.; Rehfeldt, K.R. A Critical Review of Data on Field-Scale Dispersion in Aquifers. *Water Resour. Res.* **1992**, *28*, 1955–1974. [CrossRef]
41. Wilson, J.L.; Conrad, S.H.; Mason, W.R.; Peplinski, W.; Hagan, E. *Laboratory Investigation of Residual Liquid Organics*; United States Environmental Protection Agency: Washington, DC, USA, 1990; EPA.600/6-90/004.
42. Xu, M.; Eckstein, Y. Use of Weighted Least-Squares Method in Evaluation of the Relationship between Dispersivity and Field Scale. *Ground Water* **1995**, *33*, 905–908. [CrossRef]
43. Daus, A.D.; Frind, E.O.; Sudicky, E.A. Comparative error analysis in finite element formulations of the advection-dispersion equation. *Adv. Water Resour.* **1985**, *8*, 86–95. [CrossRef]
44. Savva, A.P.; Frenken, K. Irrigation Development: A Multifaceted Process: Social, Economic, Engineering, Agronomic, Health and Environmental Issues to be Considered in a Feasibility Study. In *Irrigation Manual: Planning, Development Monitoring, and Evaluation of Irrigated Agriculture with Farmer Participation*; Food and Agriculture Organization of the United Nations (FAO) and Sub-Regional Office for East and Southern Africa (SAFR): Harare, Zimbabwe, 2001; Volume I, Module 1; p. 25. Available online: https://openknowledge.fao.org/server/api/core/bitstreams/2ca52c8b-a841-43b3-8983-803832882380/content (accessed on 1 January 2002).

*sustainability*

MDPI

*Article*

# Evaluation of Agricultural Measures to Safeguard the Vulnerable Karst Groundwater Habitat of the Black Olm (*Proteus anguinus parkelj*) from Nitrate Pollution

**Matjaž Glavan * and Rozalija Cvejić**

Agronomy Department, Biotechnical Faculty, University of Ljubljana, Jamnikarjeva 101, 1000 Ljubljana, Slovenia; rozalija.cvejic@bf.uni-lj.si
* Correspondence: matjaz.glavan@bf.uni-lj.si; Tel.: +386-1-320-3299

**Abstract:** The black olm (*Proteus anguinus parkelj* Sket & Arntzen) is an endemic species found exclusively in the Dobličica River subterranean water systems of the Dinaric karst in southern Slovenia. These unique habitats are vulnerable to contamination due to rapid water flow, primarily from nitrates from agricultural fertilisers and untreated urban wastewater. The safe limit of nitrate concentration for olms is 9.2 mg $NO_3^-$/L, yet measurements in karst springs have shown levels ranging from 3 mg to over 20 mg $NO_3^-$/L. The SWAT modelling tool assessed agri-environmental and land use scenarios for their impact on nitrate leaching. Using the model, we identified hotspots with high nitrogen leaching potential that require immediate attention and implementation of better agricultural practices for fertiliser use. For these hotspots, the most effective approach combines scenarios of cover crops (R2), reduced fertilisation (R3), crop rotation (R4), and conversion of cropland to grassland (E2, E4, E5), potentially decreasing nitrate leaching by up to 60%. Implementing the best scenarios is expected to reduce nitrogen levels below the limit value of 9.2 mg $NO_3^-$/L, essential for maintaining the black olm habitat.

**Keywords:** black olm; karst; groundwater; habitat; agriculture; measures; nitrate; SWAT model

**Citation:** Glavan, M.; Cvejić, R. Evaluation of Agricultural Measures to Safeguard the Vulnerable Karst Groundwater Habitat of the Black Olm (*Proteus anguinus parkelj*) from Nitrate Pollution. *Sustainability* **2024**, *16*, 11309. https://doi.org/10.3390/su162411309

Academic Editors: Zongjun Gao and Jiutan Liu

Received: 17 November 2024
Revised: 12 December 2024
Accepted: 19 December 2024
Published: 23 December 2024



## 1. Introduction

Slovenia is home to a remarkable diversity of subterranean species known as troglobionts, making it a hotspot for subterranean biodiversity [1,2]. These subterranean species often have very restricted habitats, frequently existing in only small areas or even a single location [3]. One notable species is the olm or proteus (*Proteus anguinus* Laurenti), an endemic amphibian that inhabits the subterranean water environments of the Dinaric Karst. The white olm (*Proteus anguinus anguinus* Laurenti) is commonly found throughout this region of the Western Balkans. The black olm (*Proteus anguinus parkelj* Sket & Arntzen) was discovered in 1986 in the karst spring of the Dobličica River, nearly three hundred years after the white species was first documented. The habitat of the black olm is confined to a small area (4 $km^2$) of karst springs in the Bela Krajina region in southern Slovenia, making it particularly vulnerable to environmental changes and human activities [4–6].

The black olm is characterised by its darker pigmentation and thrives in stable, oxygen-rich, and unpolluted water environments with temperatures ranging from 8 °C to 11 °C [7]. According to national legislation on protected wildlife, white and black olms are legally recognised as protected and vulnerable species in Slovenia. Due to their rarity, they are classified as vulnerable on the Red List of the Republic of Slovenia. The primary threat to their populations comes from high concentrations of nitrates originating from agricultural runoff and urban areas with poorly managed wastewater disposal and illegal waste dumps [6,7]. Elevated nitrate levels can lead to eutrophication, which causes a reduction in dissolved oxygen in the water, risking suffocation for the olms [8–10]. Notably, the acceptable limit for nitrate concentration in water for olm ecosystems is 9.2 mg $NO_3^-$/L, five times lower than the limit for drinking water at 50 mg $NO_3^-$/L [11].

The selection of black olm as the focal species of this study was driven by several ecological and conservation reasons, including (i) biodiversity maintenance contributing to overall ecosystem health and stability, (ii) indicator species providing valuable information about the state of the ecosystem, and (iii) cultural and ethical values, helping to preserve cultural heritage and promote ethical stewardship of the environment.

Karst is a landscape formed by the dissolution of limestone and dolomite by slightly acidic water, leading to features like lakes, caves, and sinkholes [12,13]. About 43% of the land in Slovenia is karst, mainly in the southwest, with Alpine, Dinaric, and Transitional karst types. The soil in karst areas varies, featuring undeveloped types like rendzinas in forests and jerovica or terra rossa in vineyards. Lower karst areas have more developed eutric brown soils [13]. Karst soils are generally shallow, diverse, rich in organic matter, and slightly acidic, but they often experience water shortages during droughts [14].

Karst springs connect to surface waters through subterranean channels. These systems are crucial for drinking water, supporting 20–25% of the world's population. They consist of interconnected areas with different shapes and water flow [15]. Water enters karst aquifers slowly through cracks or quickly through sinkholes, making them vulnerable to pollution but vital for ecological balance [6,16]. Pollutants from septic tanks and agricultural land can contaminate karst groundwater directly or via surface water [17–20]. Bellow the soil profile, the gravitational water flow transports pollutants into the aquifer, which reappear at the karst spring [19]. Fertilisers can leach into the groundwater and decrease water quality due to poor waste management practices and agricultural methods [21–23].

To safeguard groundwater, which constitutes a vital source of drinking water in Slovenia (exceeding 95%), the entire country is designated as a vulnerable nitrate area [24]. Following the Nitrates Directive (91/676/EEC) [25], aimed at protecting water from nitrate contamination from agricultural activities, all farms must adhere to specific regulations. These regulations mandate that fertilisers be applied only during periods when crops require nutrients for growth, that fertilisation rates be adjusted to align with the nutritional needs of the plants, that appropriate machinery be employed, and that sufficient storage capacities for livestock manure be maintained during the non-growing season when fertilisation of agricultural land is prohibited [26].

Current research on terrestrial subterranean karst systems primarily focuses on several key areas: defining hydrogeological characteristics [6,12,16]; identifying drivers of species richness [1,2]; assessing groundwater quality by measuring nitrogen concentrations and other contaminants, as well as determining pollution sources [7,15,18,20–23]; examining the toxicity of nitrogen to common organisms [9]; and using SWAT modelling to identify critical source areas and address non-point source pollution [19]. One of the most comprehensive works on agriculture and karst systems is the book *Karst Management*, published in 2011 [17], which discusses various agricultural activities that may impact soil and water in karst regions. This study advances our understanding of how implementing appropriate agricultural practices can protect the vulnerable karst groundwater habitat of the black olm from nitrate pollution. We investigate the potential nitrate concentrations in karst springs that are vital habitats for the black olm by implementing a range of alternative scenarios. We can better protect this unique species and its environment by understanding these dynamics.

This research seeks to identify the hotspots of nitrate leaching within the vulnerable groundwater habitat of the black olm residing in the Dobličica River aquifer. Additionally, it evaluates nitrate leaching by modelling various scenarios that incorporate adaptations in agricultural crop rotations and alterations in land use. Ultimately, the study aims to establish agricultural measures that will effectively reduce nitrate leaching and enhance the protection and improvement of the groundwater habitat of the black olm.

## 2. Materials and Methods

### 2.1. Study Area

The study area is located in south-eastern Slovenia (74 km$^2$), west of Črnomelj, within the Dobličica River Basin, which is part of the larger Lahinja River Basin. The altitude ranges from 140 to 1051 m, featuring several karst springs fed by the deep karst of Poljanska Gora (Figure 1a). Key springs include, from north to south, Otovski breg (157 m.a.s.l.), Pački breg (159 m.a.s.l.), Jelševniščica (142.5 m.a.s.l.), Obršec (145.1 m.a.s.l.), and Dobličica (141 m.a.s.l.), which is the primary drinking water source for the Bela Krajina region.

The terrain is structured by typical Dinaric folds and faults that spread in the northwest–southeast direction, so the movement of water through the system is relatively fast and can transfer pollutants from the surface to the aquifer in a short time [6]. Geologically, the area is characterised by Mesozoic-age carbonate rocks, which comprise the shallow marine carbonate platform. The climate of Bela Krajina is moderate continental, with average annual temperatures around 10.5 °C (July 21.2 °C, January −0.2 °C) and average annual rainfall of 1285 mm, peaking in spring and autumn.

The soil in the region is often shallow and stony, restricting its suitability for intensive agriculture (Figure 1d). The soil in the region formed on limestone and dolomite is often shallow and stony, with leached soils (Luvisol) covering 60% of the area (forest), while brown carbonate soil (Cambisol) makes up 30%, primarily in mixed forest and agricultural settings. The remaining 14% of the area includes district brown soils on Pliocene deposits (7%), rendzina (5%), and alluvial soils (2%). Land use is dominated by forest (81%), followed by meadows (8%), fields (3%), urban areas (2.5%), extensive orchards (2%), vineyards (1%), and other (3.5%) (Figure 1c). In dolinas and sinkholes, various crops are cultivated. Forest dominates steeper terrain and higher altitudes. Typical crop rotations consist of maise, cereals, and clover–grass mixtures for livestock farming. Winter wheat, silage maise, and winter barley comprise about 80% of cropland, with clovers, vegetables, and potatoes covering the rest [27]. The livestock density is 0.48 livestock units (LU) per hectare, including cattle and sheep.

### 2.2. SWAT Modelling

The SWAT 2012 (Soil and Water Assessment Tool) program, used in this study, is a robust model designed to predict hydrological processes in natural and human-influenced basins [28]. It integrates data from climate prediction models, soil characteristics, and land use scenarios to assess their effects on hydrology [29]. Developed by the USDA Agricultural Research Service, SWAT is widely applied for soil erosion control, non-point source pollution, and regional watershed management [30]. Its relevance has increased in the European Union since the 2000 Water Framework Directive, given its adaptability to various regions and environmental conditions [31,32].

The model's basic unit is the hydrological response unit (HRU), defined by consistent soil properties, land use, and slope. The number of HRUs varies based on the combination of these factors. SWAT simulates precipitation, runoff, infiltration, soil water content, evapotranspiration, and crop growth. It also models the transport of sediments, nutrients, and pesticides from land to water through surface runoff and groundwater flow [33]. Its flexibility allows for application in various basin sizes and land management practices [19].

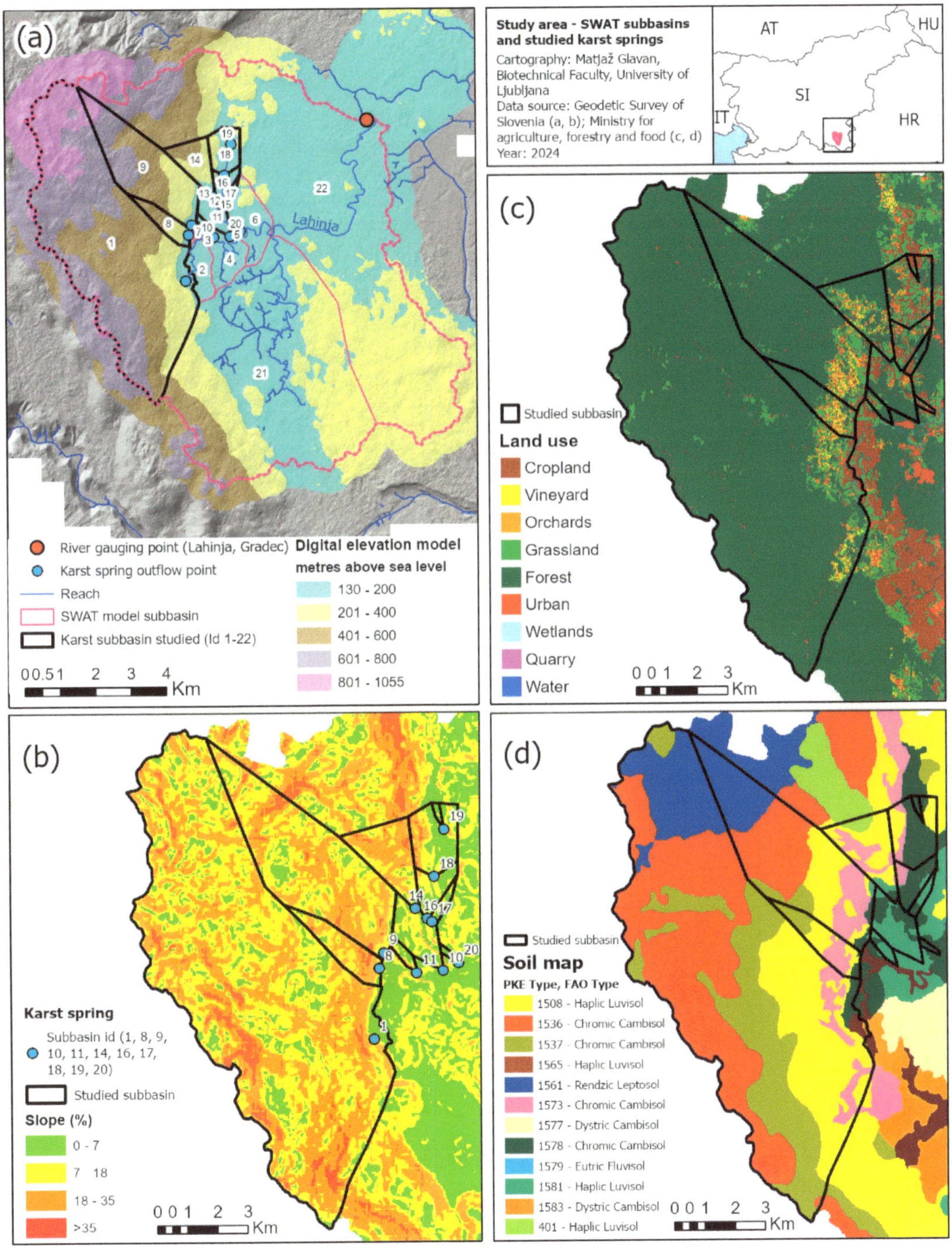


**Figure 1.** Geographic characteristics of study area presenting (**a**) topography, river network, SWAT model subbasin numbers (1–22), main river gauging point and karst springs outflow points, (**b**) slope and karst springs (with SWAT model subbasin id number), (**c**) land use, and (**d**) soil type.

Setting up the SWAT model entails several crucial steps to ensure its accuracy and reliability in simulating hydrological and nutrient cycling processes [28,30]. First, data collection is essential; this involves gathering observed data for the variables to be simulated, such as streamflow or nutrient concentrations, as this information is vital for calibration and validation. Next, the model configuration requires inputting the necessary data to reflect the studied river basin's physical characteristics accurately. The calibration phase adjusts model parameters to align simulated outputs with observed data, which can be accomplished manually or by utilising automated calibration tools like SWAT-CUP. This phase includes sensitivity analysis to pinpoint parameters that significantly affect model outputs, iterative adjustments of these sensitive parameters to enhance the fit between simulated and observed data, and the utilisation of performance metrics such as the coefficient of determination ($R^2$), Nash–Sutcliffe Efficiency (*NSE*), and Percent Bias (*PBIAS*) to assess the model's efficacy during calibration. Following calibration, validation is conducted to test the model with a distinct set of observed data not used in the calibration process, ensuring accurate simulations for conditions outside this period. This step comprises running the model using calibrated parameters for the validation period, comparing the simulated outputs against the observed data and employing statistical measures ($R^2$, *NSE*, *PBIAS*) to evaluate the model's performance and robustness, confirming consistent results across both calibration and validation periods. Adhering to these steps guarantees that the SWAT model is thoroughly calibrated and validated, yielding credible simulations for hydrological studies [28].

### 2.3. Model Input Database

The SWAT model requires a comprehensive dataset due to the complexities of simulating hydrological processes. The digital elevation model (DEM) serves as the basis for defining the basin area (Table 1). In this karst region, we utilised subbasin areas of karst springs developed by hydrogeologists from the Geological Institute of Slovenia and karstologists from the ZRC SAZU Institute for Karst Research (Figure 1).

We adjusted the parameters regulating water distribution among surface flow, subsurface lateral flow, and groundwater to accurately simulate water movement in karst systems. The official river network from the Water Directorate of the Republic of Slovenia was employed to map riverbeds and spring locations. Soil and land use data were obtained from the Ministry of Agriculture, Forestry, and Food. We also included local agricultural rotations and practices, such as ploughing and fertilisation, obtained from the regional Agricultural Advisory Service Novo mesto. River flow data and nitrate measurements were acquired from the Slovenian Environmental Agency and ZRC SAZU Karst Research Institute (Table 1). Data regarding nitrate concentration in water within the studied river basin were limited. The only reliable, accessible, and trusted measurements were conducted in 2021 and 2022.

Weather data were collected from the Črnomelj-Dobliče meteorological station, with solar radiation data sourced from the nearest station in Novo mesto.

Soil data in the SWAT model comprises physical and chemical properties. Physical properties influence water and air movement within the soil profile, affecting the hydrological cycle in hydrological response units (HRUs), while chemical properties establish initial concentrations of nitrogen (N), phosphorus (P), agronomic chemicals, and heavy metals. Although the national soil map provided initial input data, we utilised the pedotransfer function [34,35] to estimate additional missing physical properties of the soil, including bulk density, water retention, and hydraulic conductivity.

### 2.4. Model Setup, Calibration, and Validation

For modelling, we employed ArcSWAT version 2012.10.8.26, an ArcGIS extension and interface specifically designed for the SWAT model [28]. The reliability of the SWAT model was enhanced through a calibration process, as numerous model parameters could not be measured directly due to high equipment costs and a lack of available human resources.

The calibration was conducted utilising observed data and subsequently validated with an independent dataset. The model is considered appropriate if the observed and simulated data discrepancies are minimal.

**Table 1.** Input data for setting up a model in the SWAT program.

| Weather | Unit | Data Source |
|---|---|---|
| precipitation | mm $day^{-1}$ | Slovenian Environmental Agency (ARSO) Gauging stations Dobliče (precipitation, temperature, wind, relative humidity), Novo mesto (sun) |
| min. and max. temperature | °C $day^{-1}$ | |
| relative humidity | fraction | |
| global radiation energy (sun hours) | (h $day^{-1}$) | |
| average wind speed | km $day^{-1}$ | |
| **Soil** | | |
| digital soil map, soil horizons, colour | .shp | Ministry of Agriculture, Forestry and Food of the Republic of Slovenia (MKGP) |
| soil depth, roth depth | cm | |
| texture (clay, silt, sand), organic matter | mass % | |
| soil density | $cm^3$ $cm^{-3}$ | Calculation of pedotransfer functions based on digital soil map data (MKGP) |
| field capacity, wilting point | cm $cm^{-1}$ | |
| saturated hydraulic conductivity | cm $h^{-1}$ | |
| soil erosivity—MUSLE | | calculation |
| **Crop production techniques** | | |
| crop type and rotation, sowing/planting/harvesting time | | Farmers Agricultural advisory service (KGZ Novo mesto) |
| fertiliser type and method of application, equipment use scheduler, depth of cultivation | | |
| **Land** | | |
| actual land use | .shp | MKGP |
| digital elevation model | raster | Geodetic Survey RS (GURS) |
| **Water** | | |
| river flow—Lahinja gauging station Gradac (1992–2022) | $m^3/s$ | ARSO |
| karst springs (2021–2022) | | Institute for Karst Research (ZRC SAZU) |
| nitrate content—Lahinja gauging station Gradac (2021–2022) | (mg/L $NO_3^-$) | ARSO |
| monthly nitrate content (09/2021–11/2022) | | ZRC SAZU |
| karst springs subbasins | .shp | ZRC SAZU |

Despite its ecological significance, the availability of measured data for the study area is limited. Average daily flow data were obtained from a single gauging station on the Lahinja River (Gradac) managed by the Slovenian Environmental Agency, located downstream of the tributary Dobličica River, serving as a base model development reference. Nitrogen content data were available only for the year 2022. The model operated with a 30-year dataset from 1 January 1992 to 31 December 2022, divided into warm-up (1993–1997), calibration (1998–2010), and validation (2011–2022) phases. For the 11 karst springs observed (subbasins), average annual flow data were obtained from the Institute for Karst Research. They describe data as derived from water balance calculations for 2021–2022, and monthly nitrate data were collected from October 2021 to November

2022. Calibration was performed over the entire 2021–2022 dataset. Due to a lack of data, validation could not be performed.

The initial calibration was conducted manually to evaluate the effects of SWAT parameters on simulation results, followed by automatic calibration using SWAT-CUP Premium and then manual adjustments for accuracy. The SPE (SWAT Parameter Estimator) algorithm facilitated sensitivity analysis, calibration, and validation [36].

The base model for the Lahinja River Basin includes 22 subbasins and 1125 Hydrological Response Units (HRUs). Our analysis focused on 11 subbasins within the Dobličica River area (Figure 1). This focus was necessitated by the fact that only the hinterland of the selected subbasins (1, 8, 9, 10, 11, 14, 16, 17, 18, 19, 20) has a significant impact on the habitat of the black olm.

### 2.5. Scenarios

Based on a comprehensive literature review and analyses of data from prior studies concerning the black olm, supplemented by a field visit, scenarios were developed employing the SWAT model to evaluate the effects of crop rotations and changes in agricultural land use on nutrient leaching and groundwater quality at both the basin and subbasin levels [27]. A total of 11 alternative scenarios were developed (Table 2). The crop rotation change scenarios (R) focus on adapting agricultural technologies, expanding crop rotations, and managing fertilisation and tillage on arable land (scenarios 1–5). The agricultural land use change scenarios (E) encompass a broader range of measures that involve shifts in land use (such as converting cropland to grassland or forest), which subsequently affect cultivation intensity (scenarios 6–11). This scenario analysis thoroughly examines the impact of various potential measures on the amount of nitrogen leached from soils into groundwater and surface waters, thereby influencing their quality and the ecosystem of the black olm habitat.

When selecting the crop rotation change scenarios, the environmental effects of reducing nitrogen losses to the environment (effectiveness) were primarily followed, followed in second place by the preservation of the amount of biomass needed as feed for animal nutrition, and in third place, the ease of introduction into agricultural practice. Only environmental objectives were primarily followed when selecting the agricultural land use change scenarios. The exception is scenario E6, where the change sought to economically improve the situation of farms that were prepared to change extensive agricultural land (fields into meadows, meadows into forests) in hot spots that are shallow, steep, and at higher altitudes. Adverse economic effects on agricultural holdings (costs) that could arise from introducing measures would be mitigated by direct payments within the agricultural or environmental policy framework, providing a safety net against potential risks.

This research employed the SWAT model to simulate the nitrogen cycle in agricultural soil for each alternative scenario, thereby facilitating insights into the transport of nitrates within the soil and subbasins. The simulation results encompass hydrological processes, including surface runoff, lateral runoff, percolation, and nitrogen cycle processes, such as plant uptake, nitrification, denitrification, and nitrate leaching. The model effectively quantifies the movement of water and nitrogen below the root zone, where it becomes unavailable to plants, ultimately entering groundwater flow and, subsequently, surface waters. The data collection process for smaller, spatially defined hydrological response units (HRUs) has enabled the identification of critical nitrogen leaching points.

The results derived from the various scenarios were thoroughly analysed and presented in tabular and graphical formats. A map for the entire study area was generated, indicating regions where nutrient leaching is anticipated, given the constraints of the input data, along with management recommendations for critical areas.

**Table 2.** Scenarios of agricultural crop rotations and agricultural land use change.

| Scenarios | | | |
|---|---|---|---|
| **No.** | **Name** | **Description** | |
| | | **Arable Land** | **Grassland** |
| 0 | BASE (B) | 4-year rotation, no greening in 2nd year after winter wheat maize/wheat + no greening/maize/barley + clover–grass mix | 3 cuts |
| | | AGRICULTURAL CROP ROTATION CHANGE SCENARIOS | |
| 1 | ROTATION (R1) | 2-year rotation maise/winter barley + summer maize + clover–grass mix (CGM) | 3 cuts |
| 2 | ROTATION 1 (R2) | B + CGM in 2nd year (maize/wheat + CGM/maize/barley + CGM) | 3 cuts |
| 3 | ROTATION 2 (R3) | R2 + 20% decrease in fertilisation (maize/wheat + CGM/maize/barley + CGM) | 3 cuts |
| 4 | ROTATION 3 (R4) | 6-year rotation R3 + additional 2 years of CGM (maize/wheat + CGM/maize/barley + CGM/CGM/CGM | 3 cuts |
| 5 | ROTATION 4 (R5) | 6-year rotation R4 + winter fodder peas (WFP) replaced winter wheat (maize/WFP + CGM/maize/barley + CGM/CGM/CGM | 3 cuts |
| | | AGRICULTURAL LAND USE CHANGE SCENARIOS | |
| 6 | EXTENSIVE 1 (E1) | B + selected fields into grassland (3 cuts) (slope > 7%, Soil PKE type: 1508, 1536, 1537, 1561, 1573) | 3 cuts |
| 7 | EXTENSIVE 2 (E2) | B + all fields into grassland (3 cuts) | 3 cuts |
| 8 | EXTENSIVE 3 (E3) | B + selected fields into unfertilised grassland (1 cut) (slope > 7%, Soil PKE type: 1508, 1536, 1537, 1561, 1573) | 1 cut |
| 9 | EXTENSIVE 4 (E4) | B + all fields into unfertilised grassland (1 cut) | 1 cut |
| 10 | EXTENSIVE 5 (E5) | B + E4 + all grassland into forest | / |
| 11 | EXTENSIVE 6 (E6) | R5 + selected grasslands into fields (slope < 7, Soil PKE type: 1578, 1579, 1583) | 3 cuts |

The evaluation of the scenario results was based on the outputs of the SWAT model, employing the following indicators:

– Nitrate nitrogen transported into the main channel in the groundwater loading from the HRU (kg N/ha per year (NO3GW);
– Nitrate nitrogen leached from the soil profile (kg N/ha). Nitrate that leaches past the bottom of the soil profile during the time (NO3L);
– Average annual total biomass (dry matter) per HRU (metric tons/ha) (BIOM);
– Total nitrogen (TOT_N) transported with surface water flow from the subbasin (kg N/year);
– Nitrate nitrogen (NO3_OUT) transported by surface water flow from the subbasin (kg N/year).

### 2.6. Model Performance Objective Functions

To assess the model predictions for streamflow, we utilised three statistical methods: the coefficient of determination ($R^2$), the Nash–Sutcliffe efficiency (*NSE*), and percent bias (*PBIAS*). These methods are well-established for evaluating the performance of hydrological models [37].

The Coefficient of Determination ($R^2$) measures the proportion of the variance in the observed data that is predictable from the model. It ranges from 0 to 1, where a value closer to 1 indicates a better fit. An $R^2$ value greater than 0.60 is generally considered satisfactory for hydrological models, indicating that the model explains at least 60% of the variance in the observed data. The equation for $R^2$ is

$$R^2 = 1 - \left( \frac{\sum_{i=1}^{n} (O_i - P_i)^2}{\sum_{i=1}^{n} (O_i - \overline{O})^2} \right) \tag{1}$$

where $(Oi)$ is the observed value, $(Pi)$ is the predicted value, and $\overline{O}$ is the mean of the observed values.

Nash–Sutcliffe Efficiency (*NSE*) is a normalised statistic that determines the relative magnitude of the residual variance compared to the measured data variance. It ranges from $-\infty$ to 1, where 1 indicates a perfect match between the model and the observations. An *NSE* value greater than 0.50 is deemed satisfactory, suggesting that the model predictions are more accurate than the mean of the observed data. The equation for *NSE* is

$$NSE = 1 - \frac{\sum_{i=1}^{n} (O_i - O_i)^2}{\sum_{i=1}^{n} (O_i - \overline{O})^2} \tag{2}$$

Percent Bias (*PBIAS*) measures the average tendency of the simulated data to be larger or smaller than their observed counterparts. It is expressed as a percentage. Positive values indicate model overestimation bias, while negative values indicate underestimation bias. The equation for *PBIAS* is

$$PBIAS = 100 \times \frac{\sum_{i=1}^{n} (P_i - O_i)}{\sum_{i=1}^{n} O_i} \tag{3}$$

Moriasi et al. [37] established the following performance ratings for each statistical measure. Generally, a model's simulation is considered satisfactory if the following criteria are met: $R^2 > 0.60$, $NSE > 0.50$, *PBIAS* within ±15% for streamflow, and *PBIAS* within ±30% for nitrogen predictions [38].

## 3. Results and Discussion

### *3.1. Sensitivity Analysis, Calibration, and Validation*

In the sensitivity analysis process, greater emphasis was placed on parameters contributing to the results' variability during the calibration phase (Table 3). The performance indicators for the model calibration regarding the flow and nitrate nitrogen levels at the Lahinja River gauging station and karst springs are presented in Table 4. For the Lahinja River, daily measured data from the Gradac gauging station were used. The model calibration from 1998 to 2010 yielded the following results: *NSE* 0.59, $R^2$ 0.61, and *PBIAS* 0.09. During the validation period from 2011 to 2022, these values improved to *NSE* 0.69, $R^2$ 0.71, and *PBIAS* −2.28 (Table 4, Figure 2). A negative *PBIAS* during validation indicates that the model slightly overestimates the measured flow values. Overall, these results demonstrate a good agreement between the model and the measured data.

The annual average flows of karst springs from 2020 to 2022 exhibited differences between the measured and simulated values, allowing for an evaluation of flow excess or deficiency (Table 4). Except for three springs, all other values fell within the satisfactory range for *PBIAS* (±25%) [37]. It is important to note that the average annual flow values for karst springs were calculated based on the water balance between precipitation and evapotranspiration (ET), with occasional adjustments from flow measurements, which were then compared to the simulated flows. Regular flow measurements at karst springs should be conducted in the future to mitigate uncertainty. Additionally, the heterogeneity of subterranean water flows and the mixed overflows at varying water levels contribute to further uncertainty in flow assessments, a phenomenon that has also been observed by karstologists and hydrologists [39,40].

**Table 3.** SWAT parameters, range (min. and max. value), default and final values used in the model for flow calibration and validation and nitrogen calibration at Gradac gauging station.

| SWAT File | SWAT Parameters | | Range | Default Value | Final Value |
|---|---|---|---|---|---|
| | | River Flow | | | |
| .gw | GW_DELAY | Groundwater delay | 0–500 | 31 | 3 |
| | ALPHA_BF | Baseflow alpha factor | 0–1 | 0.048 | 0.70 |
| | GWQMN | Threshold depth of water in the shallow aquifer required for return flow to occur | 0–5000 | 1000 | 500 |
| | GW_REVAP | Groundwater "revap" coefficient | 0.02–0.2 | 0.02 | 0.02 |
| | RCHRG_DP | Deep aquifer percolation fraction | 0–1 | 0.05 | 0.01 |
| | REVAPMN | Threshold depth of water in the shallow aquifer for "revap" to occur | 0–1000 | 750 | 750 |
| .mgt | CN2 | SCS runoff curve number for moisture condition 2 | 0–100 | variable | −14.4% |
| .hoe | ESCO | Soil evaporation compensation factor | 0–1 | 0.95 | 1.00 |
| | SURLAG | Surface runoff lag time | 0.01–24 | 4 | 4.00 |
| .bsn | SFTMP | Snowfall temperature (°C) | –5–5 | 1 | 2.3 |
| | SMTMP | Snow melt base temperature (°C) | –5–5 | 0.5 | 3.515 |
| | SMFMX | Maximum melt rate for snow during the year (occurs on summer solstice) (°C) | 0–10 | 4.5 | 2.377 |
| | SMFMN | Minimum melt rate for snow during the year (occurs on winter solstice) (°C) | 0–10 | 4.5 | 3.457 |
| | TIMP | Snowpack temperature lag factor | 0–1 | 1 | 0.203 |
| | SNOCOVMX | Minimum snow water content that corresponds to 100% snow cover | 0–500 | 1 | 31.429 |
| | | Nitrogen load | | | |
| .bsn | CMN | Rate factor for humus mineralisation of active organic nitrogen | 0.001–0.003 | 0.0003 | 0.003 |
| | RCN | The concentration of nitrogen in rainfall | 0–15 | 0.9 | 1.150 |
| | CDN | Denitrification exponential rate coefficient | 0–3 | 0.0003 | 1.4 |
| .gw | HLIFE_NGW | Half-life of Nitrate in the shallow aquifer [days] | 1–365 | 0 | 32.5–365 |
| .sep | ISEP_TYP | The type of septic system | 1–100 | 1 | 1 |
| | SEP_DEN | Number of septic systems per square kilometre (only urban land use) | 0.001–500 | 1.5 | 280 |
| | ISP_OPT | Current condition of OWS (1 = active septic, 2 = failing septic, 0 = non-septic) | 0–2 | 0 | 1 |
| | SEP_CAP | Number of permanent residents in the house | 1–10,000 | 2.5 | 2.5 |
| Databases SepticWQ | SPTQ | Septic tank effluent (STE) flow rate ($m^3$/capita/day) | 0–1 | 0.227 | 0.227 |
| | IDSPTTYPE | Type of a septic system | 1–3 | 1 | 1 |

**Table 4.** Performance indicators of calibration for measured river flow at Gradac gauging station ($m^3/s$) and average annual flow of karst springs and comparison of nitrate nitrogen load (kg N/day) at karst springs.

| **Objective Functions of Model Performance Observed vs. Simulated** | | | |
|---|---|---|---|
| | *NSE* | $R^2$ | *PBIAS* |
| Gradac gauging station—River Lahinja—daily data | | | |
| Flow Calibration (1998–2010) | 0.59 | 0.61 | 0.09 |
| Flow Validation (2011–2022) | 0.69 | 0.71 | −2.28 |
| Nitrate nitrogen Calibration (2021–2022) | - | - | 12.74 |
| Karst springs—Calibration (2021–2022)—average annual values | | | |
| | | *PBIAS* | |
| **Subbasin** | **Name** | **River flow** | **Nitrate nitrogen** |
| 1 | Dobličica | 7.56 | 0.01 |
| 8 | Obršec | −5.60 | 0.04 |
| 9 | Jelševniščica | 3.14 | −0.15 |
| 11 | Janževe loke | 0.69 | −0.28 |
| 10 | Šprajcarjev zdenec | 27.86 | 5.66 |
| 14 | Talački breg | 1.02 | −0.04 |
| 16 | Pački breg | −14.25 | −0.08 |
| 18 | Otovski breg | −11.01 | −0.12 |
| 17 | Brežiček | −38.88 | 0.07 |
| 19 | Stobe | −80.54 | −0.20 |
| 20 | Planinc | −57.69 | −0.16 |

The calibration results for the daily transfer of nitrate nitrogen by water flow from 2021 to 2022 demonstrate a strong agreement, confirming the reliability of the model predictions (Table 4, Figure 3). Nitrate nitrogen levels in water are observed to increase during periods of high flow, a phenomenon associated with elevated precipitation that results in enhanced nitrogen leaching from agricultural soils into the aquifer (Figure 3a). In contrast, the nitrate nitrogen concentrations within the Jelševnik karst spring are characterised by low and stable values, indicating that the natural background is the primary source of nitrogen (Figure 3b). While the model does a good job of predicting nitrate nitrogen dynamics, some local variability may exist that it does not fully capture. The Percent Bias (*PBIAS*) values for the Gradac gauging station and karst springs are 12.74 and less than 6, respectively (Figure 2). This consistency between the measured and simulated values confirms the model's accuracy and usefulness in simulating nitrate nitrogen levels. Such a high level of accuracy is essential for effective water quality management in the studied karst area, as controlling nitrate concentrations is crucial for protecting both the environment and human health. It is crucial to acknowledge that simulated values were compared against observed data using monthly data from September 2021 to November 2022 (Table 1). The parameter HLIFE_NGW, which represents the half-life of nitrate nitrogen in a shallow aquifer, had the most significant impact on achieving this strong agreement (Table 3). This is in line with studies that provide insight into the response and transport of nitrate in karst groundwater to rainfall events and denitrification [41], as well as on karst hydrological and biogeochemical time lag impacts resulting in time-delayed impact of measures (rotation, land use change) on water quality [21].

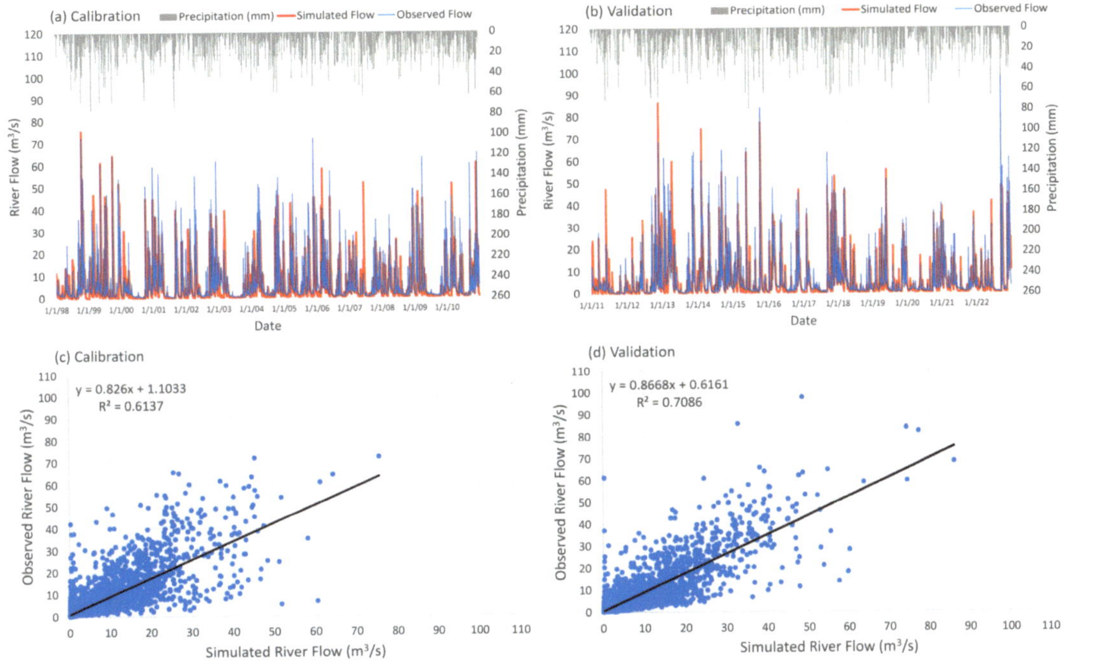


**Figure 2.** Display of measured and simulated flow data at the Gradac gauging station for the period of (**a**,**b**) calibration (1998–2010) and (**c**,**d**) validation (2011–2022) on a daily time step.

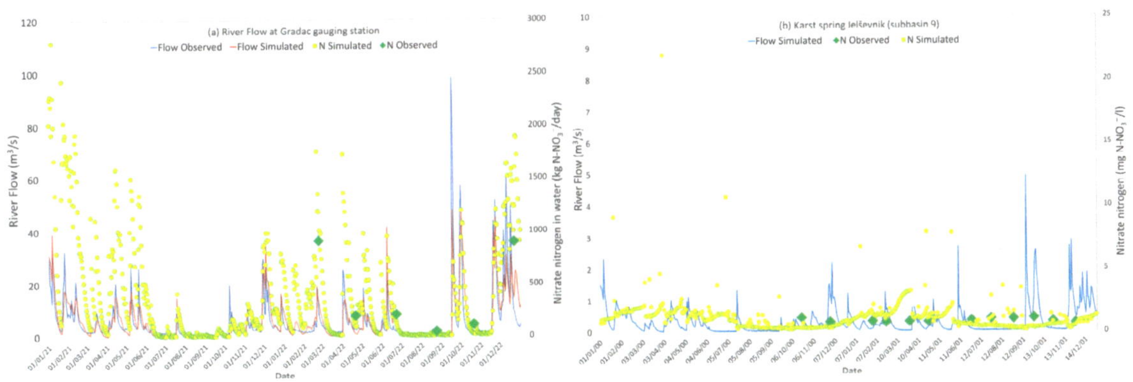


**Figure 3.** Comparison of observed and simulated data on (**a**) the nitrate nitrogen yield (kg N-$NO_3^-$/day) and water flow at the Gradac measuring station (subbasin 22) and (**b**) the nitrate nitrogen concentration (mg N-$NO_3^-$/day) and water flow at the Jelševnik karst spring (subbasin 9) on a daily time step.

### *3.2. Assessment of Agri-Environmental and Land Use Scenarios*

#### 3.2.1. Nitrate Leaching from the Hydrological Response Unit into Groundwater Flow

The results indicate that implementing alternative crop rotation scenarios (R2 to R5) can effectively reduce the leaching of nitrate nitrogen into groundwater. The most significant reductions are observed in scenarios R3 and R4. Detailed effects of the different scenarios are presented in Table 5.

**Table 5.** Comparison of results between baseline and alternative scenarios of agricultural crop rotation change for nitrate transported into the main channel in the groundwater loading from the HRU (NO3GW) (kg N/ha year), nitrate leached from the soil profile (NO3L) (kg N/ha year), and total plant biomass (BIOM) (metric tons/ha year).

| Simulated Annual Averages (1998–2022) | CROPLAND | | | | | | | | | | |
|---|---|---|---|---|---|---|---|---|---|---|---|
| | Subbasin Name and Number | | | | | | | | | | |
| | Dobličica | Obršec | Jelševnik | Janževe Loke | Šprajcarjev Zdenec | Talački Breg | Pački Breg | Brežiček | Otovski Breg | Stobe | Planinc |
| | 1 | 8 | 9 | 10 | 11 | 14 | 16 | 17 | 18 | 19 | 20 |
| | Nitrate transported into the main channel in the groundwater loading from the HRU (NO3GW) (kg N/ha year) | | | | | | | | | | |
| BASE model (kg N/ha year) | 23 | 21 | 15 | 8 | 32 | 25 | 7 | 18 | 16 | 8 | 22 |
| StDv | 17 | 16 | 12 | 7 | 21 | 18 | 7 | 14 | 13 | 7 | 15 |
| SCENARIOS | Change (%) in the amount of nitrate nitrogen (N-$NO_3^-$) | | | | | | | | | | |
| R1 | 67 | 64 | 63 | 70 | 68 | 69 | 70 | 66 | 66 | 69 | 62 |
| R2 | −16 | −15 | −12 | −18 | −16 | −17 | −17 | −18 | −18 | −17 | −16 |
| R3 | −38 | −38 | −35 | −39 | −37 | −39 | −40 | −39 | −40 | −36 | −36 |
| R4 | −68 | −68 | −66 | −67 | −66 | −68 | −68 | −67 | −68 | −65 | −66 |
| R5 | −52 | −52 | −50 | −50 | −51 | −52 | −52 | −51 | −52 | −48 | −49 |
| | Nitrate leached from the soil profile (NO3L) (kg N/ha year) | | | | | | | | | | |
| BASE model (kg N/ha year) | 61 | 65 | 73 | 66 | 64 | 60 | 66 | 68 | 61 | 72 | 75 |
| StDv | 46 | 48 | 58 | 45 | 44 | 44 | 47 | 46 | 45 | 46 | 47 |
| SCENARIOS | Change (%) in the amount of nitrate nitrogen (N-$NO_3^-$) | | | | | | | | | | |
| R1 | 70 | 68 | 64 | 69 | 70 | 72 | 70 | 70 | 71 | 66 | 64 |
| R2 | −13 | −12 | −8 | −13 | −14 | −15 | −14 | −13 | −14 | −12 | −11 |
| R3 | −35 | −35 | −31 | −34 | −35 | −37 | −36 | −34 | −36 | −32 | −31 |
| R4 | −66 | −65 | −62 | −64 | −64 | −66 | −66 | −64 | −65 | −62 | −62 |
| R5 | −50 | −49 | −47 | −48 | −48 | −50 | −50 | −48 | −49 | −46 | −46 |
| | Total plant biomass (BIOM) (metric tons/ha year) | | | | | | | | | | |
| BASE model (kg N/ha year) | 25 | 25 | 24 | 26 | 26 | 26 | 26 | 26 | 26 | 26 | 26 |
| StDv | 13 | 13 | 13 | 13 | 13 | 13 | 13 | 13 | 13 | 13 | 13 |
| SCENARIOS | Change (%) in the amount of biomass (dry matter) | | | | | | | | | | |
| R1 | 16 | 16 | 16 | 17 | 17 | 17 | 17 | 17 | 16 | 16 | 16 |
| R2 | 18 | 18 | 18 | 19 | 19 | 18 | 18 | 19 | 18 | 18 | 18 |
| R3 | 14 | 14 | 15 | 15 | 15 | 15 | 15 | 15 | 14 | 14 | 14 |
| R4 | −5 | −4 | −3 | −4 | −4 | −4 | −4 | −4 | −5 | −5 | −4 |
| R5 | −1 | 0 | 1 | −1 | −1 | −1 | 0 | −1 | −1 | −1 | −1 |

Scenario descriptions and abbreviations are explained in Table 2; red—increase in N leaching, blue—decrease in N leaching; green—increase in biomass, yellow—decrease in biomass.

The base scenario (B) represents the current situation without additional measures to reduce nitrate nitrogen leaching. The highest levels of nitrate nitrogen (kg N/ha per year) detected in groundwater transported from HRUs into surface waters (NO3GW) were observed in subbasins 1, 11, and 14 (Table 5).

Scenario R1 introduces a two-year corn crop rotation, which leads to more intensive fertiliser use and a significant increase in the amount of leached nitrate nitrogen across all subbasins. For instance, in subbasin 1, NO3GW rises by 67%, from approximately 23 kg/ha per year to about 40 kg/ha per year. The amount of nitrate leaching (NO3L) increases by 70%. Additionally, the average annual dry-weight biomass (BIOM) production increases by up to 17%.

Scenario R2 incorporates winter greening practices. This scenario reduces the amount of nitrate nitrogen in groundwater by 12% to 18%, while NO3L declines by 8% to 15%. With winter greening, the average annual biomass production increases by up to 19%.

Scenario R3 combines winter greening with a 20% reduction in fertiliser use. In this scenario, NO3GW decreases by 35% to 40%, and NO3L is reduced by 31% to 37%. The average annual biomass production increases by up to 15%.

Scenario R4 extends the CGM growing period by an additional two years. This scenario results in the highest overall reduction in nitrate nitrogen, with NO3GW decreasing by

65% to 68% and NO3L reduced by 62% to 66%. The average annual biomass production decreases by 5%, primarily due to less frequent sowing of maise.

Scenario R5 replaces winter wheat with winter fodder peas. In this case, NO3GW decreases by 48% to 52%, and NO3L is reduced by 46% to 50%. However, the average annual biomass production declines by up to 1%, as pea yield is lower than wheat.

Figure 4 illustrates the areas critical for nitrogen leaching from the soil. The green shades represent hydrological response units (HRUs) with lower leaching potential and, consequently, lower levels of nitrate nitrogen in groundwater. In contrast, the red shades highlight HRUs with high nitrogen leaching potential, requiring immediate attention and implementing better agricultural practices for fertiliser use. These critical areas can be identified as hotspots where more focused management and advice are necessary. Alternative scenarios, particularly R2, R3, and R4, show an increase in green areas, indicating lower nitrate nitrogen leaching compared to the baseline scenario characterised by intensive fertilisation. The measures proposed in scenarios R2 and R3 further demonstrate the effectiveness of these strategies in improving water quality and reducing the environmental impact of agriculture while also supporting enhanced biomass production.

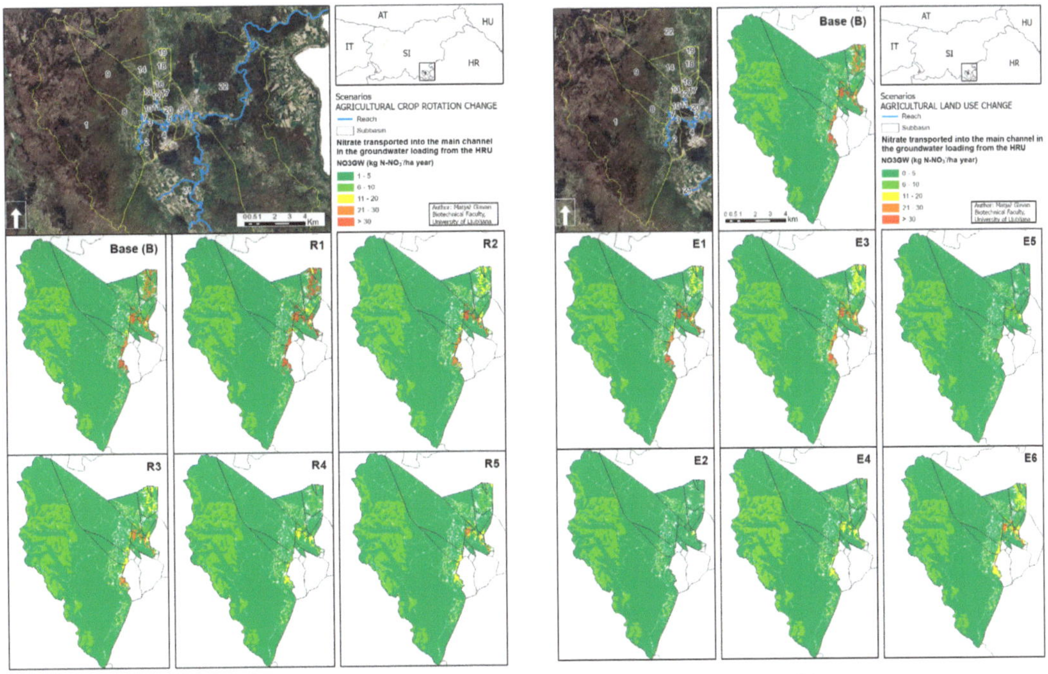


(a) (b)

**Figure 4.** Comparison of the basic scenario with alternative scenarios of (**a**) agricultural crop rotation change (R) and (**b**) agricultural land use change (E) in relation to the average nitrate nitrogen transported (kg N-$NO_3^-$/ha per year) from HRUs in the groundwater flow to the surface water flow in the study area of 11 subbasins.

Key factors influencing the rate of nitrate leaching and the soil's capacity for plant growth include soil depth, hydrological group, and soil texture (Table 6). Soils with greater water and nutrient retention capacity, such as deep loamy soils, yield higher biomass while simultaneously reducing nitrate nitrogen leaching [42,43]. This information is crucial for planning and implementing sustainable agricultural practices in a spatially considerate manner.

**Table 6.** The impact of soil properties in relation to land use (cropland grassland) on nitrogen leaching below the soil profile and the plant biomass yield for the base scenario (B).

| Soil Type | Soil Properties | | | | | | Nitrate Leached Below Soil Profile (NO3L) | | | Total Plant Biomass (BIOM) | |
|---|---|---|---|---|---|---|---|---|---|---|---|
| PKE ID | Depth (mm) | Hydrological Group | Texture (%) | | | | (kg N/ha Year) | | | (Metric Tons of Dry Matter/ha Year) | |
| | | | Description | Clay | Silt | Sand | Cropland | Grassland | Sum | Cropland | Grassland |
| 1508 | 1100 | B | silty loam | 21 | 66 | 13 | 56 | 6 | 27 | 26 | 9 |
| 1536 | 700 | B | silty loam | 17 | 66 | 17 | - | 11 | 11 | - | 8 |
| 1537 | 950 | C | silty loam | 24 | 69 | 7 | 57 | 6 | 21 | 25 | 8 |
| 1561 | 250 | D | silty clay loam | 37 | 57 | 6 | 118 | 16 | 37 | 17 | 7 |
| 1565 | 1000 | C | loam | 14 | 48 | 38 | 68 | 8 | 38 | 25 | 9 |
| 1573 | 1200 | D | silty clay loam | 27 | 64 | 9 | 59 | 7 | 29 | 26 | 9 |
| 1577 | 800 | C | silty loam | 20 | 73 | 7 | 70 | 8 | 36 | 24 | 8 |
| 1578 | 920 | B | silty loam | 15 | 74 | 10 | 71 | 8 | 38 | 26 | 9 |
| 1579 | 900 | C | sandy loam | 6 | 38 | 55 | 72 | 9 | 40 | 23 | 9 |
| 1581 | 1200 | B | silty loam | 18 | 62 | 20 | 59 | 6 | 29 | 27 | 9 |
| 1583 | 1200 | C | silty loam | 16 | 72 | 12 | 54 | 6 | 30 | 27 | 9 |
| 401 | 1300 | B | silty loam | 14 | 79 | 7 | 56 | 6 | 27 | 28 | 9 |
| | | | | | | Average | 64 | 8 | 32 | 25 | 9 |

PKE—pedocartographic unit of the soil map of Slovenia (Figure 1).

In contrast, soils with lower clay content (PKE 1579, 1578, 1565) and shallow soils (PKE 1561) exhibit the highest rates of nitrate leaching (ranging from 71 to 118 kg/ha per year) below the soil profile (NO3L). Consequently, stricter fertiliser application restrictions should be enforced on these soil types. Conversely, more lenient measures can be adopted for deeper loamy soils (PKE 1508 and PKE 401).

The crop rotation adaptation scenarios (R2 to R5) have been designed to reduce fertiliser use, introduce winter greening, and expand cover crop management, all of which help to minimise nitrate leaching. Table 6 allows a better understanding of which soils require more targeted adaptations. By implementing measures R1 to R5, we can tailor practices to meet the specific needs of different soil types, ultimately maximising both environmental and economic efficiency.

### 3.2.2. Nitrate Nitrogen Content in Surface Water at the Outlet of the Subbasin

Table 7 illustrates the variations in total nitrogen (TOT_N) and nitrate nitrogen (NO3_OUT) transported in surface water flow from the subbasins of karst springs. The base scenario (B) represents the average baseline values of total nitrogen from 1998 to 2022. The intensive scenario (R1) indicates an overall increase in nitrogen concentrations across all subbasins, which is attributed to the intensified application of fertilisers. In particular, subbasin 20 exhibits a 35% increase in NO3_OUT.

In contrast, alternative scenarios R1 to R4 demonstrate decreases in total nitrogen, with the most pronounced results occurring in scenario R3. This scenario achieves reductions of up to 51% in TOT_N and up to 32% in NO3_OUT within subbasin 19, which contains the highest proportion of arable land.

Scenarios E1 to E6 present variable effects, with scenario E2 identified as the most effective. This scenario entails a transition from entirely arable land to three-cut grasslands, reducing total nitrate nitrogen by as much as 44% in subbasin 19. Scenarios E4 and E5 exhibit comparable effectiveness.

**Table 7.** Comparison of results between baseline and alternative scenarios for total nitrogen (TOT_N) and nitrate nitrogen (NO3_OUT) transported by surface water flow from the karst springs subbasins (kg N/year, %).

| Simulated Annual Averages (1998–2022) | Subbasin Name and No. | | | | | | | | | | |
|---|---|---|---|---|---|---|---|---|---|---|---|
| | Dobličica | Obršec | Jelševnik | Janževe Loke | Šprajcarjev Zdenec | Talački Breg | Pački Breg | Brežiček | Otovski Breg | Stobe | Planinc |
| | 1 | 8 | 9 | 10 | 11 | 14 | 16 | 17 | 18 | 19 | 20 |
| Total nitrogen transported out of reach (TOT_N) at subbasin outflow (kg N/year) | | | | | | | | | | | |
| Base model (kg N/year) | 34,490 | 4210 | 10,511 | 360 | 6299 | 5705 | 9619 | 564 | 7342 | 267 | 265 |
| StDv | 10,791 | 1257 | 3206 | 106 | 1812 | 1837 | 2794 | 160 | 2164 | 75 | 82 |
| SCENARIO | Change (%) in the amount of nitrate nitrogen ($N\text{-}NO_3^-$) | | | | | | | | | | |
| R1 | 2 | 4 | 2 | 5 | 13 | 2 | 8 | 11 | 10 | 12 | 14 |
| R2 | −2 | −6 | −4 | −18 | −14 | −2 | −17 | −22 | −17 | −34 | −23 |
| R3 | −2 | −8 | −5 | −20 | −19 | −3 | −20 | −26 | −21 | −39 | −28 |
| R4 | −4 | −11 | −7 | −26 | −28 | −4 | −27 | −36 | −29 | −51 | −39 |
| R5 | −3 | −10 | −6 | −23 | −24 | −4 | −24 | −31 | −25 | −45 | −34 |
| E1 | −1 | −2 | −3 | 0 | 0 | −4 | −3 | 0 | −3 | 0 | 0 |
| E2 | −5 | −16 | −10 | −38 | −40 | −7 | −39 | −50 | −41 | −69 | −54 |
| E3 | 1 | 1 | −1 | 3 | 2 | 0 | 0 | 2 | 0 | 2 | 5 |
| E4 | −3 | −13 | −7 | −34 | −35 | −2 | −34 | −45 | −36 | −65 | −46 |
| E5 | −6 | −18 | −14 | −37 | −37 | −9 | −37 | −47 | −39 | −66 | −51 |
| E6 | −2 | −8 | −5 | −21 | −18 | −3 | −15 | −24 | −16 | −36 | −6 |
| Nitrate transported with water out of reach (NO3_OUT) at subbasin outflow (kg N/year) | | | | | | | | | | | |
| Base model (kg N/year) | 27,911 | 3041 | 7321 | 215 | 4787 | 4617 | 6416 | 355 | 4928 | 124 | 154 |
| StDv | 9620 | 1081 | 2715 | 86 | 1574 | 1663 | 2302 | 123 | 1788 | 45 | 64 |
| SCENARIO | Change (%) in the amount of nitrate nitrogen ($N\text{-}NO_3^-$) | | | | | | | | | | |
| R1 | 2 | 6 | 2 | 9 | 18 | 2 | 14 | 20 | 16 | 35 | 28 |
| R2 | 0 | −1 | −1 | −2 | −4 | −1 | −3 | −5 | −4 | −8 | −7 |
| R3 | −1 | −3 | −1 | −4 | −10 | −1 | −8 | −11 | −9 | −17 | −16 |
| R4 | −2 | −6 | −2 | −8 | −18 | −2 | −14 | −21 | −16 | −32 | −29 |
| R5 | −1 | −4 | −2 | −6 | −14 | −2 | −10 | −15 | −12 | −23 | −21 |
| E1 | 0 | 0 | −1 | 0 | 0 | −1 | −1 | 0 | −1 | 0 | 0 |
| E2 | −3 | −8 | −3 | −11 | −26 | −3 | −19 | −29 | −22 | −44 | −40 |
| E3 | 1 | 2 | 1 | 4 | 3 | 2 | 3 | 3 | 3 | 3 | 8 |
| E4 | −1 | −5 | −1 | −6 | −20 | 1 | −14 | −22 | −16 | −36 | −27 |
| E5 | −2 | −7 | −3 | −9 | −22 | −2 | −16 | −25 | −18 | −38 | −33 |
| E6 | −1 | −3 | −1 | −2 | −7 | −1 | 0 | −7 | 0 | −10 | 19 |

Scenario descriptions and abbreviations are explained in Table 2; red—increase in N, blue—decrease in N.

In general, the measures implemented in most scenarios demonstrate a high degree of efficacy in subbasins 10–11 and 16–20, which possess a significant proportion of agricultural land. Conversely, in subbasins 1, 8, and 9, where black olm populations are more prevalent and cropland is limited, the impact of the scenarios is minimal.

Scenario E3 assumes that there is only one grass cut per year, influencing the soil's formation and mineralisation of organic matter. This approach can be beneficial as it allows for the accumulation of organic matter, but it also presents specific challenges related to soil management and nutrient utilisation. A notable impact of this scenario is seen in soils currently used for arable land with better quality and fertility. These soils typically have a higher nutrient supply, are deeper, and retain more water, creating ideal conditions for greater biomass production. However, with reduced cutting frequency (only once a year), there may be less nutrient removal from the soil, which could increase the risk of excess nutrients and subsequent nitrogen leaching, particularly in more fertile areas. Over time, grassland species composition is expected to shift, leading to reduced nitrogen leaching. Alternatively, the optimal management of such soils would be to convert them into highly maintained grasslands (for example, those subjected to three cuts per year and additional grass sowing), which could mitigate the risk of excessive nutrient build-up in soil and nitrogen leaching. Adjusting agricultural practices in these regions could significantly enhance the balance between biomass production and the protection of water resources. The impacts of newly introduced measures may be time delayed due to specific soils, hydrology, and biochemical characteristics of the specific karst system [21,41].

Scenario E6 proposes converting permanent grasslands on flat terrain (slope < 7%) into cropland while simultaneously implementing an expanded crop rotation on the remaining cropland (R5). With measures outlined in scenario E6, we do not expect any increase in nitrogen levels in surface water, except for subbasin 19, where most agricultural land is on flat terrain. Scenario E5 demonstrates that converting all arable land to single-cut grassland and transforming all grassland into forests results in significantly lower nitrogen levels only in subbasins with a higher percentage of arable land (specifically, subbasins 16, 18, and 19). Additionally, it is important to highlight that the measures in scenario R4, which involve maintaining the current land use while reducing fertilisation rates by 20%, introducing winter greening, and extending the crop rotation to six years with an additional two years of cover crop management (CGM) cultivation, are similarly effective.

### 3.3. Evaluating the Potential for Improving Black Olm Habitat by Adapting Agricultural Practices

Various measures are needed when adapting agricultural practices to protect the black olm karst groundwater habitat, which must be implemented in a coordinated and spatially optimised manner to be effective and contribute to improving the habitat quality. It is essential to balance environmental protection with the maintenance of agricultural activity [17]. Particular attention should be given to adapting agricultural production technologies, such as reducing fertiliser use, expanding crop rotations, and implementing cover crops and sustainable grassland management practices. These approaches can significantly reduce nitrogen leaching into groundwater and surface water while maintaining agricultural productivity.

This study indicates that the soil and hydrological systems in karst areas, characterised by rapid infiltration and short water retention times during precipitation, possess unique features that must be considered when managing nitrogen [6,18]. Some measures aimed at protecting drinking water sources in Slovenia have already received financial support under the Strategic Plan of the Common Agricultural Policy (CAP) for 2023–2027 [25]. Through the current CAP agri-environment-climate payments, farmers can opt for interventions that preserve and improve the quality of water sources while enhancing soil properties and fertility. These measures help prevent nitrate leaching by adhering to specific requirements, such as mandatory five-year crop rotations, cultivating at least three different crops, greening arable land with year-round green cover using winter-freeze resistant crops, and restricting the use of phytopharmaceuticals in designated water protection areas.

The CAP also includes regulations related to the Nitrates Directive and water protection area legislation under Conditionality [26]. The "Regulation on the Protection of Waters against Pollution by Nitrates from Agricultural Sources" mandates landowners and land users in Slovenia to implement additional measures for balanced and targeted fertilisation. These include quantitative restrictions on annual nitrogen application, the creation of fertilisation plans, and bans on late autumn and winter fertilisation. These measures effectively reduce nitrate leaching and enhance the quality of water resources. Furthermore, other methods to mitigate nitrate leaching include using cover crops and intercrops in permanent crops, fertilising with composted organic fertilisers, applying smaller doses of fertiliser, utilising mulch or mechanical weed control, and practising conservation tillage.

The evaluation of the impact of various scenarios on the nitrate content in the subbasin outflow demonstrates that a more strategic application of these scenarios could substantially reduce nitrogen leaching and lower nitrate levels in water sources, particularly in areas with a higher share of cropland (Table 8). The benefits of reducing agricultural production intensity, as demonstrated in scenario R4, are also evident in modelling nitrate concentrations at subbasin 11, 17, 19, and 20 outlets, where cropland constitutes a larger portion. By implementing the R4 scenario, we can anticipate a reduction in nitrogen levels to below the threshold (target) value of 9.2 mg $NO_3^-$/L, which is suitable for the habitat of the black olm.

**Table 8.** Potential nitrate concentration in the reach of karst spring subbasin outflow upon implementation of the alternative scenarios (R, E).

| | Nitrate Concentration in the Reach at Karst Springs Subbasin Outflow (mg $NO_3^-$/L) | | | | | | | | | | |
|---|---|---|---|---|---|---|---|---|---|---|---|
| | Karst Springs Subbasin Name and No. | | | | | | | | | | |
| | **Dobličica** | **Obršec** | **Jelševnik** | **Janževe Loke** | **Šprajcarjev Zdenec** | **Talački Breg** | **Pački Breg** | **Brežiček** | **Otovski Breg** | **Stobe** | **Planinc** |
| | **1** | **8** | **9** | **10** | **11** | **14** | **16** | **17** | **18** | **19** | **20** |
| Average observed (ZRC SAZU) (9/2021–11/2022) | 3.3 | 10.1 | 3.6 | 4.2 | 10.3 | 10.3 | 13.6 | 10.2 | 14.3 | 12.5 | 11.1 |
| Target value for the black olm habitat | 9.2 mg $NO_3^-$/L | | | | | | | | | | |
| Required reduction | - | 1.1 | - | - | 1.3 | 1.3 | 4.6 | 1.2 | 5.3 | 3.5 | 2.1 |
| Scenario | Potential nitrate concentration upon implementation of the scenario | | | | | | | | | | |
| R1 | 3.4 | 10.7 | 3.7 | 4.6 | 12.2 | 10.5 | 15.5 | 12.2 | 16.6 | 16.9 | 14.2 |
| R2 | 3.3 | 10.0 | 3.6 | 4.1 | 9.9 | 10.2 | 13.2 | 9.7 | 13.7 | 11.5 | 10.3 |
| R3 | 3.3 | 9.8 | 3.6 | 4.0 | 9.3 | 10.2 | 12.5 | 9.1 | 13.0 | 10.4 | 9.3 |
| R4 | 3.2 | 9.5 | 3.5 | 3.9 | 8.4 | 10.1 | 11.7 | 8.1 | 12.0 | 8.5 | 7.9 |
| R5 | 3.3 | 9.7 | 3.5 | 3.9 | 8.9 | 10.1 | 12.2 | 8.7 | 12.6 | 9.6 | 8.8 |
| E1 | 3.3 | 10.1 | 3.6 | 4.2 | 10.3 | 10.2 | 13.5 | 10.2 | 14.2 | 12.5 | 11.1 |
| E2 | 3.2 | 9.3 | 3.5 | 3.7 | 7.6 | 10.0 | 11.0 | 7.2 | 11.2 | 7.0 | 6.7 |
| E3 | 3.3 | 10.3 | 3.6 | 4.4 | 10.6 | 10.5 | 14.0 | 10.5 | 14.7 | 12.9 | 12.0 |
| E4 | 3.3 | 9.6 | 3.6 | 3.9 | 8.2 | 10.4 | 11.7 | 8.0 | 12.0 | 8.0 | 8.1 |
| E5 | 3.2 | 9.4 | 3.5 | 3.8 | 8.0 | 10.1 | 11.4 | 7.7 | 11.7 | 7.8 | 7.4 |
| E6 | 3.3 | 9.8 | 3.6 | 4.1 | 9.6 | 10.2 | 13.6 | 9.5 | 14.3 | 11.3 | 13.2 |

white—the limit value of 9.2 mg $NO_3^-$/L; blue—below the target value, favourable for the habitat; red—above the target value, unfavourable for the habitat.

In subbasins 1, 9, and 10, where there is a significant variation in altitude, and forest land use predominates, the impact of agriculture on water nitrate levels is minimal. The primary nitrogen sources in these areas are the mineralisation of forest residues and atmospheric deposition. In Slovenia, the natural background concentration of nitrogen in water varies by region and specific water body. Generally, the nitrate ($NO_3^-$) concentration in unpolluted waters is estimated to be around 1 mg N/L (approximately 4.4 mg $NO_3^-$/L) [24].

Additionally, in subbasins 8, 11, 14, 16, and 18, nitrogen sources may be related to uncontrolled wastewater discharge from residential buildings or inadequately designed storage for organic fertilisers, resulting in unregulated leaching into the groundwater. This conclusion is supported by water quality measurements conducted in the study area within the HAČLORI project [44,45]. Microbial source tracking, utilising molecular and cultivable methods [46] at four selected karst springs (8, 9, 18, and 19), has shown that human activities represent the predominant nitrogen source in the sampled water of subbasin 18. At the same time, subbasins 8 and 9 indicate that human activities are the only additional nitrogen source. Conversely, in subbasin 19, animal sources are identified as the exclusive contributors to nitrogen levels.

To effectively adapt agricultural practices and land use, measures must be implemented in a coordinated manner alongside other initiatives tailored to the specific soil characteristics and karst hydrogeological conditions [16,18,43]. This strategy will facilitate improved nitrogen management and mitigate adverse environmental impacts, which are essential for preserving the black olm habitat and enhancing water quality within its ecosystem.

Karst springs are interconnected with the surface through a complex network of underground channels and tunnels, resulting in a hydraulic continuum that extends from the surface to the springs [16]. This connectivity renders them susceptible to pollution; however, they also play a vital role in sustaining the ecological balance of groundwater ecosystems [6]. Major precipitation events can intensify the geochemical processes inherent in karst geology. Consequently, developing agricultural measures for karst areas necessitates a specialised approach to manage nitrogen balance effectively [17].

The modelling results highlight specific agri-environmental measures that could help reduce nitrogen leaching from agriculture.

(a) These measures include supporting farms in acquiring the knowledge and financial resources needed to transition to sustainable farming practices [17,18,47]. Key recommendations are as follows: (i) Implement a diverse crop rotation that spans at least four years and includes a minimum of three different crops; (ii) Ensure that the soil is covered with cover crops or greening between the main summer or winter crops, preferably using winter resistant cover crops; (iii) Adopt shallow minimum tillage techniques, limiting tillage depth to 10 cm. It is important to note that bare soil without plant cover is more susceptible to nutrient leaching, and ploughing can accelerate the mineralisation of organic matter. Therefore, it is advisable to carry out tillage and sowing within a few days of each other. The proposed measures will involve short-term costs for purchasing seeds for cover crops or additional plants in the rotation. The shift from ploughing to minimum tillage is the most significant expense, which only the largest farmers can typically afford on their own. However, the long-term economic benefits of transitioning to sustainable tillage methods will provide much greater advantages for the environment and improve the productive capacity of agricultural soils.

(b) Including legumes, such as DTM and clover, in the existing crop rotation can help reduce the use of fertilisers. Research from the URAVIVO project has indicated that in some cases, nitrogen fertiliser doses can be reduced by 20% without adversely affecting yields [48,49]. This consideration was incorporated when developing the proposal for the extended R3 scenario rotation. It is recommended that professional services put more effort into promoting crops that require lower nitrogen inputs for growth, such as clover and clover–grass mixes. Raising awareness about the necessity of reducing both the frequency and amount of fertiliser applications while promoting controlled-release fertilisers will yield beneficial economic outcomes for farms. Legumes that require added fertiliser in their production technology, such as fodder peas and soybeans, should not be promoted in the study area to minimise fertiliser use further. Additionally, the mineralisation of underground plant residues can impact N leaching [50].

(c) A comparison of various farming scenarios indicates that cultivation practices involving higher nitrogen (N) inputs can adversely affect the nitrogen balance in the soil, leading to increased nitrogen leaching. It is crucial to note that a single cultivation technology may not be universally effective across different soil types. While a particular practice may have a minimal impact on the nitrogen balance in deeper soils or those with a clay-loam texture, it can result in significantly greater nitrogen leaching in shallower soils. The findings highlight the importance of managing agricultural practices according to specific soil characteristics, such as depth, texture, and skeletalness, to mitigate potential negative nitrogen balances. Developing a precise soil map, ideally at scales of 1:10,000 or 1:5000, grounded in field soil sampling is one of the essential tools to estimate groundwater vulnerability in karst and to implement targeted management [29,42,43].

(d) A comparative analysis of crop rotations following the fertiliser guidelines has revealed that, in certain instances, fertilisers are applied at doses that are technically appropriate but may not correspond to the specific soil types and geological conditions prevalent in the region. It is essential to strengthen oversight in implementing existing measures delineated in the nature protection and groundwater protection legislation and CAP, particularly concerning Conditionality [26,51–53]. To this end, it is recommended that one or more representative experimental plots showcasing diverse soil types be established in critical areas. These plots will enable the evaluation of agri-environmental measures on plant growth, yield, nitrogen leaching, pesticide usage and cost of production. Furthermore, they can function as effective educational tools for farmers, illustrating the practical benefits of these measures on environmental and economic performance. In addition, it is imperative to augment the number of individual consultations available to farmers, focusing on agricultural technologies tailored to each farm's distinctive characteristics.

(e) Organic farming can exhibit similar nitrogen (N) leaching levels as conventional practices when pursuing equivalent yields per hectare. However, in promoting organic farming, it is essential to avoid nitrogen surpluses from fresh organic fertilisers (livestock) and to promote composted manure and green manure. Weather conditions significantly influence nitrogen mineralisation from organic fertilisers and cannot be fully controlled [50,54]. Additionally, soil characteristics, including depth, texture, and water-holding capacity, must be considered when developing fertiliser strategies and agricultural practices.

(f) Intensive grasslands with more frequent cuts show a positive nitrogen (N) balance across various scenarios. Soil properties play a crucial role, especially with manure, slurry, or grazing on shallow soils typical of karst regions [55]. In areas with significant grassland, promoting extensive grazing by herbivores is beneficial. Incentives should support the breeding of suckler cows and small ruminants like sheep and goats, which have lower stocking densities (currently 0.48 livestock units per hectare). This helps improve the distribution of organic fertilisers and nutrient utilisation. Agricultural professionals should focus on educating farmers about grassland management, including aeration, seeding, liming, and proper fertilisation practices. A diverse and dense grassland enhances harvested yield and nutrient use, reduces nutrient losses and minimises environmental impact [56,57]. It is essential to define the type of livestock, stocking density, and appropriate fertilisers with the help of experts, ensuring sustainable environmental and economic outcomes [58].

(g) In the studied area, a significant portion of the land is dedicated to orchards and vineyards, which share similarities with grasslands due to their inter-row cultivation type. From a nitrogen balance perspective, fruit farming is a favourable option. Over recent decades, fertiliser application methods in fruit and grape cultivation have evolved, driven by new insights into the adverse effects of excessive fertilisation on crop quality and storage capacity [59,60]. However, viticulture and fruit growing face challenges related to irrigation and pesticide use. Ensuring a reliable water source is crucial for achieving optimal yield in quantity and quality. However, extracting water from subterranean aquifers could further strain the habitat of the black olm. Phytopharmaceuticals, including fungicides, insecticides, and herbicides, play a vital role in producing fruits and grapes. Therefore, it is essential to manage pesticide application effectively, utilising professional services that guide growers toward sustainable practices. These practices should emphasise preventive measures such as repellents, disruptors, and baits. To evaluate these measures, it is important to assess their impact on subterranean habitats and sensitive species, including the black olm.

Studies on urban areas in the study region [44,45] found that agricultural activities contribute between 50% and 90% of the nitrogen to groundwater within flatter, agriculturally active subbasins. In key karst springs (subbasins 8 and 9), which host the largest populations of the black olm, the contributions from agriculture and unregulated urban wastewater systems are each 50%. To protect the black olm habitat, it is essential to implement measures in both agriculture and residential areas. Specifically, residential buildings should be equipped with waterproof septic tanks, and a public wastewater discharge system should be established to direct effluent to treatment facilities. Additionally, maintaining records of wastewater devices, ensuring regular wastewater removal, and monitoring the integrity of septic tanks are critical for success. Studies on karst water quality from around the world propose that the upstream human populations must be advised of the risks resulting from their daily activities related to wastewater management [18,21,23,39,41–43].

Ensuring the safety and conservation of the endemic black olm (*Proteus anguinus parkelj*) is vital for maintaining the health of the subterranean karst aquatic ecosystem in the studied region. This species acts as an indicator of groundwater quality and overall ecosystem health. Protecting the black olm aids in preserving the unique biodiversity of karst environments, home to many specialised and often endemic species that rely on the same habitat, such as various species of stygobiont amphipods (i.e., *Niphargus* spp.) [10]. Additionally, the black olm supports ecosystem stability by maintaining food web balance. Its presence and health reflect the condition of the ecosystem, and its conservation can have cascading effects on other species. The black olm also provides valuable scientific insights into evolutionary biology, genetics, and ecology due to its unique adaptations to a subterranean lifestyle [10].

Beyond ecological importance, the black olm holds cultural and educational significance, promoting environmental awareness and stewardship. It symbolises the natural heritage of Slovenia and the Dinaric Karst region. By focusing on its conservation, we benefit numerous other species and the broader environment, fostering a sense of stewardship for natural resources [61].

When evaluating the impact of the scenarios, several uncertainties associated with the establishment of the model were duly considered. These uncertainties arise from various factors, including production technologies, crop rotations, harvest schedules, mowing dates, machine operation timelines, and fertiliser application practices. As a result, the outcomes produced by the model represent average estimates; individual farmers adapt their production schedules and technologies to accommodate specific crops, animal species, production intensities, and fluctuating weather conditions.

The final recommendations, formulated based on the model scenarios, should be interpreted as potential ecosystem responses to alterations in land use and agricultural technologies, as well as the anticipated effects of agri-environmental measures on the nitrogen balance within soil and water resources. The results and recommendations should be used for the management of karst areas in an interdisciplinary way, involving science as well as socio-economic and socio-political domains [62]

## 4. Conclusions

This research emphasises the significant impact of agricultural practices on soil and groundwater resources in karst areas, particularly within the unique habitat of the black olm. The water dynamics in this region are complex due to karst geology, making modelling challenging. Our scenarios demonstrated that adjusting crop rotation (R2, R4), fertilisation (R3), and land use (E2, E4, E5) can effectively reduce nitrate leaching into groundwater while considering environmental and socio-economic factors. This is vital for maintaining water quality, safeguarding endemic species like the black olm, and ensuring future sustainable coexistence with human populations. Furthermore, the results indicate that the target value of 9.2 mg $NO_3^-$/L cannot be achieved solely through agricultural measures; urban point sources must also be managed.

However, it is important to acknowledge some limitations and uncertainties in our research. One key area for improvement is the limited availability of accurate data on soil physical properties, which could impact the precision of our simulations. Additionally, the modelling is based on certain assumptions that may not fully capture the variability of natural processes in karst systems. To enhance prediction accuracy, it would be beneficial to incorporate a larger number of measurements, such as flow discharge and nitrogen content in the water of Karst spring, to extend the model calibration and validation period.

This research offers the scientific community a new approach to modelling subterranean karst aquifers and understanding how agricultural practices affect them. This knowledge is crucial for future research and managing environmentally sensitive subterranean groundwater habitats. For professionals, particularly decision-makers and agricultural policy planners, our findings provide concrete guidelines for directing agricultural practices that can mitigate adverse impacts on groundwater and surface water quality.

In the future, we could focus on refining models to predict the effects of climate change on karst aquifers, particularly concerning extreme weather events like droughts and floods, and evaluate their influence on groundwater resources and biodiversity.

Future efforts should also aim to predict the effects of climate change, analyse long-term agri-environmental measures, and assess the impact of urbanisation on karst ecosystems and water resources. An important aspect to explore is the socio-economic effects of proposed measures as part of environmental policies on local rural communities, where agriculture and land use are closely linked to natural resource conservation and settlement.

Finally, interdisciplinary studies examining the relationship between human activities, biodiversity, and groundwater quality could enhance our understanding of how biodiversity in karst regions contributes to natural water purification, thereby improving groundwater quality. Such research could inform more sustainable management of karst ecosystems and their water resources.

**Author Contributions:** Conceptualisation, M.G. and R.C.; methodology, M.G.; software, M.G.; validation, M.G.; formal analysis, M.G.; investigation, M.G. and R.C.; resources, M.G.; data curation, M.G.; writing—original draft preparation, M.G.; writing—review and editing, R.C.; visualisation, M.G.; supervision, R.C.; project administration, R.C.; funding acquisition, R.C. All authors have read and agreed to the published version of the manuscript.

**Funding:** This research was funded by the Slovenian Research and Innovation Agency, Ministry of Natural Resources and Spatial Planning and Ministry of Agriculture, Forestry and Food, grant number CRP 2021 V1-2139 (HAČLORI).

**Institutional Review Board Statement:** Not applicable.

**Informed Consent Statement:** Not applicable.

**Data Availability Statement:** The raw data supporting the conclusions of this article will be made available by the authors on request.

**Conflicts of Interest:** The authors declare no conflicts of interest. The funders had no role in the design of the study; in the collection, analyses, or interpretation of data; in the writing of the manuscript; or in the decision to publish the results.

## References

1. Culver, D.C.; Pipan, T. *The Biology of Caves and Other Subterranean Habitats*, 2nd ed.; Biology of habitats series; Oxford Academic: Oxford, UK, 2019. [CrossRef]
2. Zagmajster, M.; Malard, F.; Eme, D.; Culver, D.C. Subterranean Biodiversity Patterns from Global to Regional Scales. In *Cave Ecology*; Moldovan, O.T., Kováč, L., Halse, S., Eds.; Springer: Berlin/Heidelberg, Germany, 2018; pp. 195–227. [CrossRef]
3. Bregović, P.; Zagmajster, M. Understanding hotspots within a global hotspot—Identifying the drivers of regional species richness patterns in terrestrial subterranean habitats. *Insect Conserv. Divers.* **2016**, *9*, 268–281. [CrossRef]
4. Aljančič, G. History of research on *Proteus anguinus* Laurenti 1768 in Slovenia/Zgodovina raziskovanja človeške ribice (Proteus anguinus Laurenti 1768) v Sloveniji. *Folia Biol. Geol.* **2019**, *60*, 39–69. [CrossRef]
5. The Black Olm/Črna Človeška Ribica. Available online: https://repozitorij.uni-lj.si/IzpisGradiva.php?id=37125&lang=eng (accessed on 8 September 2024).
6. Koren, K.; Brajkovič, R.; Bajuk, M.; Vraničar, Š.; Fabjan, V. Hydrogeological characterisation of karst springs of the white (*Proteus anguinus* anguinus) and black olm (*Proteus anguinus* parkelj) habitat in Bela krajina (SE Slovenia). *Geologija* **2023**, *66*, 151–166. [CrossRef]
7. Mezga, K.; Janža, M.; Prestor, J.; Koren, K.; Šram, D. Groundwater dependent ecosystems—Groundwater status indicators. *Nat. Slov.* **2016**, *18*, 35–42. [CrossRef]
8. Jensen, K.M.; Jensen, M.H.; Kristensen, E. Nitrification and denitrification in Wadden Sea sediments (Königshafen, Island of Sylt, Germany) as measured by nitrogen isotope pairing and isotope dilution. *Aquat. Microb. Ecol.* **1996**, *11*, 181–191. [CrossRef]
9. Scott, G.; Crunkilton, R.L. Acute and chronic toxicity of nitrate to fathead minnows (*Pimephales promelas*), ceriodaphnia dubia, and Daphnia magna. *Environ. Toxicol. Chem.* **2000**, *19*, 2918–2922. [CrossRef]
10. Mali, L.B.; Sket, B. History and biology of the «black proteus« (*Proteus anguinus parkelj* Sket & Arntzen 1994; Amphibia: Proteidae): A review/Zgodovina in biologija črnega močerila (*Proteus anguinus parkelj* Sket & Arntzen 1994; Amphibia: Proteidae): Pregledni članek. *Folia Biol. Geol.* **2019**, *60*, 5–37. [CrossRef]

11. Kolar, B. The threshold concentration for nitrate in groundwater as a habitat of Proteus anguinus. *Nat. Slov.* **2018**, *20*, 39–42. [CrossRef]
12. Palmer, A.N. Geomorphic interpretation of karst features. In *Groundwater as a Geomorphic Agent*; Routledge: London, UK, 1984; pp. 173–209. [CrossRef]
13. Lisec, A.; Montanarella, L.; Vrščaj, B.; Kralj, T.; Lobnik, F.; Suhadolc, M.; Prus, T.; Mihelič, R.; Rupreht, J.; Zupan, M.; et al. Tla Slovenije s Pedološko Karto v Merilu 1:250000 = Soils of Slovenia with Soil Map 1:250000. European Commission Joint Reaearch Centre (JRC): Publications Office of the European Union, Luxembourg. 2005. Available online: https://op.europa.eu/sl/publication-detail/-/publication/538dee5b-dfbf-45d6-bc3c-d4b6fab3110d (accessed on 25 October 2024).
14. Urbančič, M.; Simončič, P.; Prus, T.; Kutnar, L. Atlas of Slovenian Forest Soils/Atlas Gozdnih tal Slovenije. Gozdarski Inštitut Slovenije, Ljubljana. 2005. Available online: https://plus.cobiss.net/cobiss/si/sl/bib/223504896 (accessed on 25 October 2024).
15. Tuyet, D. Characteristics of Karst Ecosystems of Vietnam and Their Vulnerability to Human Impact. *Acta Geol. Sin.—Engl. Ed.* **2001**, *75*, 325–329. [CrossRef]
16. Bakalowicz, M. Karst groundwater: A challenge for new resources. *Hydrogeol. J.* **2005**, *13*, 148–160. [CrossRef]
17. Coxon, C. Agriculture and Karst. In *Karst Management*; Beynen, P.E., Ed.; Springer: Berlin/Heidelberg, Germany, 2011; pp. 103–138. [CrossRef]
18. Guo, F.; Yuan, D.; Qin, Z. Groundwater Contamination in Karst Areas of Southwestern China and Recommended Countermeasures. *Acta Carsologica* **2010**, *39*, 389–399. [CrossRef]
19. Zeiger, S.J.; Owen, M.R.; Pavlowsky, R.T. Simulating non-point source pollutant loading in a karst basin: A SWAT modeling application. *Sci. Total Environ.* **2021**, *785*, 147295. [CrossRef]
20. Lukač Reberski, J.; Terzić, J.; Maurice, L.D.; Lapworth, D.J. Emerging organic contaminants in karst groundwater: A global level assessment. *J. Hydrol.* **2022**, *604*, 127242. [CrossRef]
21. Fenton, O.; Mellander, P.E.; Daly, K.; Wall, D.P.; Jahangir, M.M.R.; Jordan, P.; Hennessey, D.; Huebsch, M.; Blum, P.; Vero, S.; et al. Integrated assessment of agricultural nutrient pressures and legacies in karst landscapes. *Agric. Ecosyst. Environ.* **2017**, *239*, 246–256. [CrossRef]
22. Stevanović, Z.; Stevanović, A.M. Monitoring as the Key Factor for Sustainable Use and Protection of Groundwater in Karst Environments—An Overview. *Sustainability* **2021**, *13*, 5468. [CrossRef]
23. Brad, T.; Bizic, M.; Ionescu, D.; Chiriac, C.M.; Kenesz, M.; Roba, C.; Ionescu, A.; Fekete, A.; Mirea, I.C.; Moldovan, O.T. Potential for Natural Attenuation of Domestic and Agricultural Pollution in Karst Groundwater Environments. *Water* **2022**, *14*, 1597. [CrossRef]
24. Quality of Groundwater/Kakovost Podzemne Vode. Available online: https://kazalci.arso.gov.si/sl/content/kakovost-podzemne-vode-1?tid=16 (accessed on 20 October 2024).
25. Nitrates Directive (91/676/EEC). Available online: https://environment.ec.europa.eu/topics/water/nitrates_en (accessed on 20 October 2024).
26. Slovenian Strategic Plan for the Common Agricultural Policy 2023–2027. Available online: https://skp.si/skupna-kmetijska-politika-2023-2027 (accessed on 10 September 2024).
27. Glavan, M.; Cvejić, R.; Zupanc, V.; Vodnik, K.; Pintar, M. *Modeling the Environmental Impact of Agricultural Measures to Protect the Habitat of the Black Mantis (Proteus anguinus) from Nitrate Pollution/Modeliranje Okoljskega Vpliva Kmetijskih Ukrepov za Varovanje Habitata Črne Človeške (Proteus anguinus Parkelj) pred Onesnaženjem z Nitrati*; Biotechnical Faculty, University of Ljubljana: Ljubljana, Slovenia, 2024; Available online: https://plus.cobiss.net/cobiss/si/sl/bib/210724355 (accessed on 20 October 2024).
28. Arnold, J.G.; Moriasi, D.N.; Gassman, P.W.; Abbaspour, K.C.; White, M.J.; Srinivasan, R.; Santhi, C.; Harmel, R.D.; Griensven, A.; van Liew, M.W.; et al. SWAT: Model Use, Calibration, and Validation. *Trans. ASABE* **2012**, *55*, 1491–1508. [CrossRef]
29. Busico, G.; Colombani, N.; Fronzi, D.; Pellegrini, M.; Tazioli, A.; Mastrocicco, M. Evaluating SWAT model performance, considering different soils data input, to quantify actual and future runoff susceptibility in a highly urbanised basin. *J. Environ. Manag.* **2020**, *266*, 110625. [CrossRef]
30. Gassman, P.W.; Sadeghi, A.M.; Srinivasan, R. Applications of the SWAT Model Special Section: Overview and Insights. *J. Environ. Qual.* **2014**, *43*, 1–8. [CrossRef] [PubMed]
31. Čerkasova, N.; Mėžinė, J.; Idzelytė-, R.; Lesutienė, J.; Ertürk, A.; Umgiesser, G. Exploring variability in climate change projections on the Nemunas River and Curonian Lagoon: Coupled SWAT and SHYFEM modeling approach. *Ocean. Sci.* **2024**, *20*, 1123–1147. [CrossRef]
32. Plunge, S.; Schürz, C.; Čerkasova, N.; Strauch, M.; Piniewski, M. SWAT+ model setup verification tool: SWATdoctR. *Environ. Model. Softw.* **2024**, *171*, 105878. [CrossRef]
33. Zhang, Z.; Chen, S.; Wan, L.; Cao, J.; Zhang, Q.; Yang, C. The effects of landscape pattern evolution on runoff and sediment based on SWAT model. *Environ. Earth Sci.* **2021**, *80*, 2. [CrossRef]
34. Saxton, K.E.; Rawls, W.J. Soil Water Characteristic Estimates by Texture and Organic Matter for Hydrologic Solutions. *Soil Sci. Soc. Am. J.* **2006**, *70*, 1569–1578. [CrossRef]
35. Saxton, K.E.; Rawls, W.J.; Romberger, J.S.; Papendick, R.I. Estimating Generalized Soil-water Characteristics from Texture. *Soil Sci. Soc. Am. J.* **1986**, *50*, 1031–1036. [CrossRef]
36. Abbaspour, K.C.; Vaghefi, S.A.; Srinivasan, R. A Guideline for Successful Calibration and Uncertainty Analysis for Soil and Water Assessment: A Review of Papers from the 2016 International SWAT Conference. *Water* **2018**, *10*, 6. [CrossRef]

37. Moriasi, D.N.; Arnold, J.G.; Liew MW van Bingner, R.L.; Harmel, R.D.; Veith, T.L. Model Evaluation Guidelines for Systematic Quantification of Accuracy in Watershed Simulations. *Trans. ASABE* **2007**, *50*, 885–900. [CrossRef]
38. Moriasi, D.N.; Gitau, M.W.; Pai, N.; Daggupati, P. Hydrologic and Water Quality Models: Performance Measures and Evaluation Criteria. *Trans. ASABE* **2015**, *58*, 1763–1785. [CrossRef]
39. Binet, S.; Joigneaux, E.; Pauwels, H.; Albéric, P.; Fléhoc, C.; Bruand, A. Water exchange, mixing and transient storage between a saturated karstic conduit and the surrounding aquifer: Groundwater flow modeling and inputs from stable water isotopes. *J. Hydrol.* **2017**, *544*, 278–289. [CrossRef]
40. Rusjan, S.; Sapač, K.; Petrič, M.; Lojen, S.; Bezak, N. Identifying the hydrological behavior of a complex karst system using stable isotopes. *J. Hydrol.* **2019**, *577*, 123956. [CrossRef]
41. Wang, Z.J.; Yue, F.J.; Lu, J.; Wang, Y.C.; Qin, C.Q.; Ding, H.; Xue, L.L.; Li, S.L. New insight into the response and transport of nitrate in karst groundwater to rainfall events. *Sci. Total Environ.* **2022**, *818*, 151727. [CrossRef]
42. Moreno-Gómez, M.; Liedl, R.; Stefan, C.; Pacheco, J. Theoretical analysis and considerations of the main parameters used to evaluate intrinsic karst groundwater vulnerability to surface pollution. *Sci. Total Environ.* **2024**, *907*, 167947. [CrossRef] [PubMed]
43. Bakalowicz, M. Management of Karst Groundwater Resources. In *Karst Management*; Beynen, P.E., Ed.; Springer: Berlin/Heidelberg, Germany, 2011; pp. 263–282. [CrossRef]
44. Research to Identify and Prevent Pollution of the Catchment Hinterland of Jelševniščica and Otovski breg, with Special Attention to the Habitat of the Black Olm/Raziskave za opredelitev in preprečevanje obremenjevanja vodozbirnega zaledja Jelševniščice in Otovskega brega, s posebnim ozirom na habitat črne človeške ribice. ZRC SAZU, Ljubljana. 2024. Available online: https://izrkp.zrc-sazu.si/sl/programi-in-projekti/raziskave-za-opredelitev-in-preprecevanje-obremenjevanja-vodozbirnega-zaledja (accessed on 5 November 2024).
45. Prelovšek, M.; Curk, M.; Mali, N.; Mulec, J.; Petrič, M.; Pintar, M.; Urbanc, J.; Cvejić, R. Research to identify and prevent pollution of the catchment hinterland of Jelševniščica and Otovski breg, with special attention to the habitat of the black olm/Raziskave za opredelitev in preprečevanje obremenjevanja vodozbirnega zaledja Jelševniščice in Otovskega brega, s posebnim ozirom na habitat črne človeške ribice. In Proceedings of the 33. Mišičev Vodarski Dan 2022, Maribor, Slovenia, 5 October 2022; Available online: https://www.mvd20.com/LETO2022/R16.pdf (accessed on 20 October 2024).
46. Sresung, M.; Paisantham, P.; Ruksakul, P.; Kongprajug, A.; Chyerochana, N.; Gallage, T.P.; Srathongneam, T.; Rattanakul, S.; Maneein, S.; Surasen, C.; et al. Microbial source tracking using molecular and cultivable methods in a tropical mixed-use drinking water source to support water safety plans. *Sci. Total Environ.* **2023**, *876*, 162689. [CrossRef] [PubMed]
47. Guo, F.; Jiang, G. Karst Groundwater Management through Science and Education. *Open J. Geol.* **2011**, *1*, 45–50. [CrossRef]
48. Pintar, M.; Arh Marinčič, Š.; Cerar, S.; Cesar, A.; Curk, M.; Cvejić, R.; Glavan, M.; Kastelec, D.; Koroša, A.; Korpar, P.; et al. *Efficient Use of Water and Nutrients in Crop Production to Protect and Improve Drinking Water Resources: URaViVO: Final Report/Učinkovitejša Raba Vode in Hranil v Rastlinski Pridelavi za Varovanje in Izboljšanje Virov Pitne Vode: URaViVO: Končno Poročilo*; Biotechnical Faculty, University of Ljubljana: Ljubljana, Slovenia, 2021; Available online: https://plus.cobiss.net/cobiss/si/sl/bib/101641475 (accessed on 1 November 2024).
49. Curk, M.; Glavan, M. Assessing and Mapping the Environmental Impacts of Best Management Practices in Nitrate-Vulnerable Areas. *Water* **2023**, *15*, 2364. [CrossRef]
50. Benoit, M.; Garnier, J.; Billen, G.; Tournebize, J.; Gréhan, E.; Mary, B. Nitrous oxide emissions and nitrate leaching in an organic and a conventional cropping system (Seine basin, France). *Agric. Ecosyst. Environ.* **2015**, *213*, 131–141. [CrossRef]
51. Giordano, R.; D'Agostino, D.; Apollonio, C.; Scardigno, A.; Pagano, A.; Portoghese, I.; Lamaddalena, N.; Piccinni, A.F.; Vurro, M. Evaluating acceptability of groundwater protection measures under different agricultural policies. *Agric. Water Manag.* **2015**, *147*, 54–66. [CrossRef]
52. Goldscheider, N. A holistic approach to groundwater protection and ecosystem services in karst terrains. *Carbonates Evaporites* **2019**, *34*, 1241–1249. [CrossRef]
53. Ravbar, N.; Šebela, S. The effectiveness of protection policies and legislative framework with special regard to karst landscapes: Insights from Slovenia. *Environ. Sci. Policy* **2015**, *51*, 106–116. [CrossRef]
54. Pandey, A.; Li, F.; Askegaard, M.; Rasmussen, I.A.; Olesen, J.E. Nitrogen balances in organic and conventional arable crop rotations and their relations to nitrogen yield and nitrate leaching losses. *Agric. Ecosyst. Environ.* **2018**, *265*, 350–362. [CrossRef]
55. Valkama, E.; Rankinen, K.; Virkajärvi, P.; Salo, T.; Kapuinen, P.; Turtola, E. Nitrogen fertilisation of grass leys: Yield production and risk of N leaching. *Agric. Ecosyst. Environ.* **2016**, *230*, 341–352. [CrossRef]
56. Egan, G.; McKenzie, P.; Crawley, M.; Fornara, D.A. Effects of grassland management on plant nitrogen use efficiency (NUE): Evidence from a long-term experiment. *Basic Appl. Ecol.* **2019**, *41*, 33–43. [CrossRef]
57. Leimer, S.; Oelmann, Y.; Wirth, C.; Wilcke, W. Time matters for plant diversity effects on nitrate leaching from temperate grassland. *Agric. Ecosyst. Environ.* **2015**, *211*, 155–163. [CrossRef]
58. Klaus, V.H.; Friedritz, L.; Hamer, U.; Kleinebecker, T. Drought boosts risk of nitrate leaching from grassland fertilisation. *Sci. Total Environ.* **2020**, *726*, 137877. [CrossRef] [PubMed]
59. Brunetto, G.; Oliveira, B.S.; Ambrosini, V.G.; Couto, R.d.R.; Sete, P.B.; dos Santos Junior, E.; Loss, A.; da Silva, L.O.S.; Gatiboni, L.C. Nitrogen availability in an apple orchard with weed management. *Ciência Rural* **2018**, *48*, e20160895. [CrossRef]
60. He, Z.; Hu, Q.; Zhang, Y.; Cao, H.; Nan, X. Effects of irrigation and nitrogen management strategies on soil nitrogen and apple yields in loess plateau of China. *Agric. Water Manag.* **2023**, *280*, 108220. [CrossRef]

61. Sket, B. Discovering the black proteus *Proteus anguinus* parkelj (Amphibia: Caudata). *Nat. Slov.* **2017**, *19*, 27–28. [CrossRef]
62. Kosič Ficco, K.; Sasowsky, I.D. An interdisciplinary framework for the protection of karst aquifers. *Environ. Sci. Policy* **2018**, *89*, 41–48. [CrossRef]

MDPI AG
Grosspeteranlage 5
4052 Basel
Switzerland
Tel.: +41 61 683 77 34

MDPI Books Editorial Office
E-mail: books@mdpi.com
www.mdpi.com/books

www.ingramcontent.com/pod-product-compliance
Lightning Source LLC
Chambersburg PA
CBHW041429210926
51850CB00013B/3387
* 9 7 8 3 7 2 5 8 3 3 9 9 3 *